Multivariable Feedback Control

MULTIVARIABLE FEEDBACK CONTROL
Analysis and Design

Second Edition

Sigurd Skogestad
Norwegian University of Science and Technology

Ian Postlethwaite
University of Leicester

John Wiley & Sons, Ltd

Reprinted with corrections February 2007
Reprinted November 2007, September 2008, June 2009, June 2010

Other Wiley Editorial Offices

John Wiley & Sons Inc., 111 River Street, Hoboken, NJ 07030, USA

Jossey-Bass, 989 Market Street, San Francisco, CA 94103-1741, USA

Wiley-VCH Verlag GmbH, Boschstr. 12, D-69469 Weinheim, Germany

John Wiley & Sons Australia Ltd, 42 McDougall Street, Milton, Queensland 4064, Australia

John Wiley & Sons (Asia) Pte Ltd, 2 Clementi Loop #02-01, Jin Xing Distripark, Singapore 129809

John Wiley & Sons Canada Ltd, 22 Worcester Road, Etobicoke, Ontario, Canada M9W 1L1

Wiley also publishes its books in a variety of electronic formats. Some content that appears in print
may not be available in electronic books.

British Library Cataloguing in Publication Data

A catalogue record for this book is available from the British Library

ISBN-13 978-0-470-01167-6 (H/B)   978-0-470-01168-3 (P/B)

Produced from camera-ready copy supplied by the authors

To my parents, Ingulf and Patricia
S.S.

To Elizabeth, Stella and Rosa
I.P.

CONTENTS

PREFACE

This is a book on practical feedback control and not on system theory in general. Feedback is used in control systems to change the dynamics of the system (usually to make the response stable and sufficiently fast), and to reduce the sensitivity of the system to signal uncertainty (disturbances) and model uncertainty. Important topics covered in the book, include

- classical frequency domain methods
- analysis of directions in multivariable systems using the singular value decomposition
- input–output controllability (inherent control limitations in the plant)
- model uncertainty and robustness
- performance requirements
- methods for controller design and model reduction
- control structure selection and decentralized control
- linear matrix inequalities, LMIs

The treatment is for linear systems. The theory is then much simpler and more well developed, and a large amount of practical experience tells us that in many cases linear controllers designed using linear methods provide satisfactory performance when applied to real nonlinear plants.

We have attempted to keep the mathematics at a reasonably simple level, and we emphasize results that enhance *insight* and *intuition*. The design methods currently available for linear systems are well developed, and with associated software it is relatively straightforward to design controllers for most multivariable plants. However, without insight and intuition it is difficult to judge a solution, and to know how to proceed (e.g. how to change weights) in order to improve a design.

The book is appropriate for use as a text for an introductory graduate course in multivariable control or for an advanced undergraduate course. We also think it will be useful for engineers who want to understand multivariable control, its limitations, and how it can be applied in industrial practice. The analysis techniques and the material on control structure design should prove very useful in the new emerging area of systems biology. There are numerous worked examples, exercises and case studies which make frequent use of Matlab[TM][1].

The prerequisites for reading this book are an introductory course in classical single-input single-output (SISO) control and some elementary knowledge of matrices and linear algebra. Parts of the book can be studied alone, and provide an appropriate background for a number of linear control courses at both undergraduate and graduate levels: classical loop-shaping control, an introduction to multivariable control, advanced multivariable control,

---

[1] Matlab is a registered trademark of The MathWorks, Inc.

robust control, controller design, control structure design and controllability analysis. It may be desirable to teach the material in a different order from that given in the book. For example, in his course at ETH Zurich, Professor Manfred Morari has chosen to start with SISO systems (Chapters 1, 2, 5 and 7) and then system theory (Chapter 4), before moving on to MIMO systems (Chapters 3, 6, 8 and 9).

The book is partly based on a graduate multivariable control course given by the first author in the Cybernetics Department at the Norwegian University of Science and Technology in Trondheim. The course, attended by students from Electrical, Chemical and Mechanical Engineering, has usually consisted of 3 lectures a week for 12 weeks. In addition to regular assignments, the students have been required to complete a 50-hour design project using Matlab. In Appendix B, a project outline is given together with a sample exam.

Examples and Internet

All of the numerical examples have been solved using Matlab. Some sample files are included in the text to illustrate the steps involved. All these files use either the new Robust Control toolbox or the Control toolbox, but the problems could have been solved easily using other software packages.

The following are available over the Internet:

- Matlab files for examples and figures
- Solutions to selected exercises (those marked with a *)[2]
- Linear state-space models for plants used in the case studies
- Corrections, comments, extra exercises and exam sets
- Lecture notes for courses based on the book

This information can be accessed from the authors' home pages, which are easily found using a search engine like Google. The current addresses are:

- `http://www.nt.ntnu.no/users/skoge`
- `http://www.le.ac.uk/engineering/staff/Postlethwaite`

Comments and questions

Please send questions, information on any errors and any comments you may have to the authors. Their email addresses are:

- `skoge@chemeng.ntnu.no`
- `ixp@le.ac.uk`

Acknowledgements

The contents of the book are strongly influenced by the ideas and courses of Professors John Doyle and Manfred Morari from the first author's time as a graduate student at Caltech during the period 1983–1986, and by the formative years, 1975–1981, the second author spent at Cambridge University with Professor Alistair MacFarlane. We thank the organizers of the

---

[2] Solutions to the remaining exercises are available to course lecturers by contacting the authors.

1993 European Control Conference for inviting us to present a short course on applied $\mathcal{H}_\infty$ control, which was the starting point for our collaboration. The final manuscript for the first edition began to take shape in 1994–1995 during a stay the authors had at the University of California at Berkeley – thanks to Andy Packard, Kameshwar Poolla, Masayoshi Tomizuka and others at the BCCI-lab, and to the stimulating coffee at *Brewed Awakening*.

We are grateful for the numerous technical and editorial contributions of Yi Cao, Kjetil Havre, Ghassan Murad and Ying Zhao. The computations for Example 4.5 were performed by Roy S. Smith who shared an office with the authors at Berkeley. Helpful comments, contributions and corrections were provided by Richard Braatz, Jie Chen, Atle C. Christiansen, Wankyun Chung, Bjørn Glemmestad, John Morten Godhavn, Finn Are Michelsen and Per Johan Nicklasson. A number of people have assisted in editing and typing various versions of the manuscript, including Zi-Qin Wang, Yongjiang Yu, Greg Becker, Fen Wu, Regina Raag and Anneli Laur. We also acknowledge the contributions from our graduate students, notably Neale Foster, Morten Hovd, Elling W. Jacobsen, Petter Lundström, John Morud, Raza Samar and Erik A. Wolff.

For the *second edition*, we are indebted to Vinay Kariwala for many technical contributions and editorial changes. Other researchers at Trondheim have also been helpful and we are especially grateful to Vidar Alstad and Espen Storkaas. From Leicester, Matthew Turner and Guido Herrmann were extremely helpful with the preparation of the new chapter on LMIs. Finally, thanks to colleagues and former colleagues at Trondheim and Caltech from the first author, and at Leicester, Oxford and Cambridge from the second author.

The aero-engine model (Chapters 11 and 13) and the helicopter model (Chapter 13) are provided with the kind permission of Rolls-Royce Military Aero Engines Ltd and the UK Ministry of Defence, DRA (now QinetiQ) Bedford, respectively.

We have made use of material from several books. In particular, we recommend Zhou et al. (1996) as an excellent reference on system theory and $\mathcal{H}_\infty$ control and *The Control Handbook* (Levine, 1996) as a good general reference. Of the others we would like to acknowledge, and recommend for further reading, the following: Rosenbrock (1970), Rosenbrock (1974), Kwakernaak and Sivan (1972), Kailath (1980), Chen (1984), Francis (1987), Anderson and Moore (1989), Maciejowski (1989), Morari and Zafiriou (1989), Boyd and Barratt (1991), Doyle et al. (1992), Boyd et al. (1994), Green and Limebeer (1995), and the Matlab toolbox manuals of Grace et al. (1992), Balas et al. (1993), Chiang and Safonov (1992) and Balas et al. (2005).

Second edition

In this second edition, we have corrected a number of minor mistakes and made numerous changes and additions throughout the text, partly arising from the many questions and comments we have received from interested readers and partly to reflect developments in the field. The main additions and changes are:

Chapter 2: Material has been included on unstable plants, the feedback amplifier, the lower gain margin, simple IMC tuning rules for PID control, and the half rule for estimating the effective delay.

Chapter 3: Some material on the relative gain array has been moved in from Chapter 10.

Chapter 4: Changes have been made to the tests of state controllability and observability (of course, they are equivalent to the old ones).

Chapters 5 and 6: New results have been included on fundamental performance limitations introduced by RHP-poles and RHP-zeros.

Chapter 6: The section on limitations imposed by uncertainty has been rewritten

Chapter 7: The examples of parametric uncertainty have been introduced earlier and shortened.

Chapter 9: A clear strategy is given for incorporating integral action into LQG control.

Chapter 10: The chapter has been reorganized. New material has been included on the selection of controlled variables and self-optimizing control. The section on decentralized control has been rewritten and several examples have been added.

Chapter 12: A complete new chapter on LMIs.

Appendix: Minor changes to positive definite matrices and the all-pass factorization.

In reality, the book has been expanded by more than 100 pages, but this is not reflected in the number of pages in the second edition because the page size has also been increased.

All the Matlab programs have been updated for compatibility with the new Robust Control toolbox.

Sigurd Skogestad
Ian Postlethwaite
August 2005

December 2006: Minor corrections and changes (see book home page for details).

BORGHEIM, an engineer:

Herregud, en kan da ikke gjøre noe bedre enn leke i denne velsignede verden. Jeg synes hele livet er som en lek, jeg!

Good heavens, one can't do anything better than play in this blessed world. The whole of life seems like playing to me!

Act one, LITTLE EYOLF, Henrik Ibsen.

1

INTRODUCTION

In this chapter, we begin with a brief outline of the design process for control systems. We then discuss linear models and transfer functions which are the basic building blocks for the analysis and design techniques presented in this book. The scaling of variables is critical in applications and so we provide a simple procedure for this. An example is given to show how to derive a linear model in terms of deviation variables for a practical application. Finally, we summarize the most important notation used in the book.

1.1 The process of control system design

Control is the adjustment of the available degrees of freedom (manipulated variables) to assist in achieving acceptable operation of a system (process, plant). The process of designing (automatic) control systems usually makes many demands on the engineer or engineering team. These demands often emerge in a step-by-step design procedure as follows:

1. Study the system (process, plant) to be controlled and obtain initial information about the control objectives.
2. Model the system and simplify the model, if necessary.
3. Scale the variables and analyze the resulting model; determine its properties.
4. Decide which variables are to be controlled (controlled outputs).
5. Decide on the measurements and manipulated variables: what sensors and actuators will be used and where will they be placed?
6. Select the control configuration.
7. Decide on the type of controller to be used.
8. Decide on performance specifications, based on the overall control objectives.
9. Design a controller.
10. Analyze the resulting controlled system to see if the specifications are satisfied; and if they are not satisfied modify the specifications or the type of controller.
11. Simulate the resulting controlled system, on either a computer or a pilot plant.
12. Repeat from step 2, if necessary.
13. Choose hardware and software and implement the controller.
14. Test and validate the control system, and tune the controller on-line, if necessary.

Control courses and textbooks usually focus on steps 9 and 10 in the above procedure; that is, on methods for controller design and control system analysis. Interestingly, many real control systems are designed without any consideration of these two steps. For example, even for complex systems with many inputs and outputs, it may be possible to design workable

Multivariable Feedback Control: Analysis and Design. Second Edition
S. Skogestad and I. Postlethwaite   © 2005, 2006 John Wiley & Sons, Ltd

control systems, often based on a hierarchy of cascaded control loops, using only on-line tuning (involving steps 1, 4, 5, 6, 7, 13 and 14). However, even in such cases a suitable control structure may not be known at the outset, and there is a need for systematic tools and insights to assist the designer with steps 4, 5 and 6. A special feature of this book is the provision of tools for *input–output controllability analysis* (step 3) and for *control structure design* (steps 4, 5, 6 and 7).

Input–output controllability is the ability to achieve acceptable control performance. It is affected by the locations of the sensors and actuators, but otherwise it cannot be changed by the control engineer. Simply stated, "even the best control system cannot make a Ferrari out of a Volkswagen". Therefore, the process of control system design should in some cases also include a step 0, involving the design of the process equipment itself. The idea of looking at process equipment design and control system design as an integrated whole is not new, as is clear from the following quote taken from a paper by Ziegler and Nichols (1943):

> In the application of automatic controllers, it is important to realize that controller and process form a unit; credit or discredit for results obtained are attributable to one as much as the other. A poor controller is often able to perform acceptably on a process which is easily controlled. The finest controller made, when applied to a miserably designed process, may not deliver the desired performance. True, on badly designed processes, advanced controllers are able to eke out better results than older models, but on these processes, there is a definite end point which can be approached by instrumentation and it falls short of perfection.

Ziegler and Nichols then proceed to observe that there is a factor in equipment design that is neglected, and state that

> the missing characteristic can be called the "controllability", the ability of the process to achieve and maintain the desired equilibrium value.

To derive simple tools with which to quantify the inherent input–output controllability of a plant is the goal of Chapters 5 and 6.

1.2   The control problem

The objective of a control system is to make the output y behave in a desired way by manipulating the plant input u. The *regulator problem* is to manipulate u to counteract the effect of a disturbance d. The *servo problem* is to manipulate u to keep the output close to a given reference input r. Thus, in both cases we want the *control error* $e = y - r$ to be small. The algorithm for adjusting u based on the available information is the controller K. To arrive at a good design for K we need *a priori* information about the expected disturbances and reference inputs, and of the plant model (G) and disturbance model (G_d). In this book, we make use of linear models of the form

$$y = Gu + G_d d \tag{1.1}$$

A major source of difficulty is that the models (G, G_d) may be inaccurate or may change with time. In particular, inaccuracy in G may cause problems because the plant will be part of a feedback loop. To deal with such a problem we will make use of the concept of model uncertainty. For example, instead of a single model G we may study the behaviour of a class of models, $G_p = G + E$, where the model "uncertainty" or "perturbation" E is bounded, but otherwise unknown. In most cases weighting functions, $w(s)$, are used to express $E = w\Delta$ in terms of normalized perturbations, Δ, where the magnitude (norm) of Δ is less than or equal to 1. The following terms are useful:

Nominal stability (NS). The system is stable with no model uncertainty.

Nominal performance (NP). The system satisfies the performance specifications with no model uncertainty.

Robust stability (RS). The system is stable for all perturbed plants about the nominal model up to the worst-case model uncertainty.

Robust performance (RP). The system satisfies the performance specifications for all perturbed plants about the nominal model up to the worst-case model uncertainty.

1.3 Transfer functions

The book makes extensive use of transfer functions, $G(s)$, and of the frequency domain, which are very useful in applications for the following reasons:

- Invaluable insights are obtained from simple frequency-dependent plots.
- Important concepts for feedback such as bandwidth and peaks of closed-loop transfer functions may be defined.
- $G(j\omega)$ gives the response to a sinusoidal input of frequency ω.
- A series interconnection of systems corresponds in the frequency domain to the multiplication of the individual system transfer functions, whereas in the time domain, the evaluation of complicated convolution integrals is required.
- Poles and zeros appear explicitly in factorized scalar transfer functions.
- Uncertainty is more easily handled in the frequency domain. This is related to the fact that two systems can be described as close (i.e. have similar behaviour) if their frequency responses are similar. On the other hand, a small change in a parameter in a state-space description can result in an entirely different system response.

We consider linear, time-invariant systems whose input–output responses are governed by linear ordinary differential equations with constant coefficients. An example of such a system is

$$
\begin{aligned}
\dot{x}_1(t) &= -a_1 x_1(t) + x_2(t) + \beta_1 u(t) \\
\dot{x}_2(t) &= -a_0 x_1(t) + \beta_0 u(t) \\
y(t) &= x_1(t)
\end{aligned}
\tag{1.2}
$$

where $\dot{x}(t) \equiv dx/dt$. Here $u(t)$ represents the input signal, $x_1(t)$ and $x_2(t)$ the states, and $y(t)$ the output signal. The system is time-invariant since the coefficients a_1, a_0, β_1 and β_0 are independent of time. If we apply the Laplace transform to (1.2) we obtain

$$
\begin{aligned}
s\bar{x}_1(s) - x_1(t=0) &= -a_1\bar{x}_1(s) + \bar{x}_2(s) + \beta_1\bar{u}(s) \\
s\bar{x}_2(s) - x_2(t=0) &= -a_0\bar{x}_1(s) + \beta_0\bar{u}(s) \\
\bar{y}(s) &= \bar{x}_1(s)
\end{aligned}
\tag{1.3}
$$

where $\bar{y}(s)$ denotes the Laplace transform of $y(t)$, and so on. To simplify our presentation we will make the usual abuse of notation and replace $\bar{y}(s)$ by $y(s)$, etc. In addition, we will omit the independent variables s and t when the meaning is clear.

If $u(t), x_1(t), x_2(t)$ and $y(t)$ represent deviation variables away from a nominal operating point or trajectory, then we can assume $x_1(t = 0) = x_2(t = 0) = 0$. The elimination of $\bar{x}_1(s)$ and $\bar{x}_2(s)$ from (1.3) then yields the transfer function

$$\frac{y(s)}{u(s)} = G(s) = \frac{\beta_1 s + \beta_0}{s^2 + a_1 s + a_0} \tag{1.4}$$

Importantly, for linear systems, the transfer function is independent of the input signal (forcing function). Notice that the transfer function in (1.4) may also represent the following system:

$$\ddot{y}(t) + a_1 \dot{y}(t) + a_0 y(t) = \beta_1 \dot{u}(t) + \beta_0 u(t) \tag{1.5}$$

with input $u(t)$ and output $y(t)$.

Transfer functions, such as $G(s)$ in (1.4), will be used throughout the book to model systems and their components. More generally, we consider rational transfer functions of the form

$$G(s) = \frac{\beta_{n_z} s^{n_z} + \cdots + \beta_1 s + \beta_0}{s^n + a_{n-1} s^{n-1} + \cdots + a_1 s + a_0} \tag{1.6}$$

For multivariable systems, $G(s)$ is a matrix of transfer functions. In (1.6) n is the order of the denominator (or pole polynomial) and is also called the *order of the system*, and n_z is the order of the numerator (or zero polynomial). Then $n - n_z$ is referred to as the pole excess or *relative order*.

Definition 1.1

- *A system $G(s)$ is* **strictly proper** *if $G(j\omega) \to 0$ as $\omega \to \infty$.*
- *A system $G(s)$ is* **semi-proper** *or* **bi-proper** *if $G(j\omega) \to D \neq 0$ as $\omega \to \infty$.*
- *A system $G(s)$ which is strictly proper or semi-proper is* **proper**.
- *A system $G(s)$ is* **improper** *if $G(j\omega) \to \infty$ as $\omega \to \infty$.*

For a proper system, with $n \geq n_z$, we may realize (1.6) by a state-space description, $\dot{x} = Ax + Bu, \ y = Cx + Du$, similar to (1.2). The transfer function may then be written as

$$G(s) = C(sI - A)^{-1}B + D \tag{1.7}$$

Remark. All practical systems have zero gain at a sufficiently high frequency, and are therefore strictly proper. It is often convenient, however, to model high-frequency effects by a non-zero D-term, and hence semi-proper models are frequently used. Furthermore, certain derived transfer functions, such as $S = (I + GK)^{-1}$, are semi-proper.

Usually we let $G(s)$ represent the effect of the inputs u on the outputs y, whereas $G_d(s)$ represents the effect on y of the disturbances d ("process noise"). We then have the following linear process model in terms of deviation variables

$$y(s) = G(s)u(s) + G_d(s)d(s) \tag{1.8}$$

We have here made use of the superposition principle for linear systems, which implies that a change in a dependent variable (here y) can simply be found by adding together the separate effects resulting from changes in the independent variables (here u and d) considered one at a time.

All the signals $u(s)$, $d(s)$ and $y(s)$ are deviation variables. This is sometimes shown explicitly, for example, by use of the notation $\delta u(s)$, but since we always use deviation variables when we consider Laplace transforms, the δ is normally omitted.

1.4   Scaling

Scaling is very important in practical applications as it makes model analysis and controller design (weight selection) much simpler. It requires the engineer to make a judgement at the start of the design process about the required performance of the system. To do this, decisions are made on the expected magnitudes of disturbances and reference changes, on the allowed magnitude of each input signal, and on the allowed deviation of each output.

Let the unscaled (or originally scaled) linear model of the process in deviation variables be

$$\widehat{y} = \widehat{G}\widehat{u} + \widehat{G}_d\widehat{d}; \quad \widehat{e} = \widehat{y} - \widehat{r} \tag{1.9}$$

where a hat ($\widehat{}$) is used to show that the variables are in their unscaled units. A useful approach for scaling is to make the variables less than 1 in magnitude. This is done by *dividing each variable by its maximum expected or allowed change*. For disturbances and manipulated inputs, we use the scaled variables

$$d = \widehat{d}/\widehat{d}_{\max}, \quad u = \widehat{u}/\widehat{u}_{\max} \tag{1.10}$$

where:

- $\widehat{d}_{\max}$ – largest expected change in disturbance
- $\widehat{u}_{\max}$ – largest allowed input change

The maximum deviation from a nominal value should be chosen by thinking of the maximum value one can expect, or allow, as a function of time.

The variables $\widehat{y}$, $\widehat{e}$ and $\widehat{r}$ are in the same units, so the same scaling factor should be applied to each. Two alternatives are possible:

- $\widehat{e}_{\max}$ – largest allowed control error
- $\widehat{r}_{\max}$ – largest expected change in reference value

Since a major objective of control is to minimize the control error $\widehat{e}$, we here usually choose to scale with respect to the maximum control error:

$$y = \widehat{y}/\widehat{e}_{\max}, \quad r = \widehat{r}/\widehat{e}_{\max}, \quad e = \widehat{e}/\widehat{e}_{\max} \tag{1.11}$$

To formalize the scaling procedure, we introduce the scaling factors

$$D_e = \widehat{e}_{\max}, \ D_u = \widehat{u}_{\max}, \ D_d = \widehat{d}_{\max}, \ D_r = \widehat{r}_{\max} \tag{1.12}$$

For multi-input multi-output (MIMO) systems, each variable in the vectors $\widehat{d}$, $\widehat{r}$, $\widehat{u}$ and $\widehat{e}$ may have a different maximum value, in which case D_e, D_u, D_d and D_r become diagonal scaling *matrices.* This ensures, for example, that all errors (outputs) are of about equal importance in terms of their magnitude.

The corresponding scaled variables to use for control purposes are then

$$d = D_d^{-1}\widehat{d}, \; u = D_u^{-1}\widehat{u}, \; y = D_e^{-1}\widehat{y}, \; e = D_e^{-1}\widehat{e}, \; r = D_e^{-1}\widehat{r} \qquad (1.13)$$

On substituting (1.13) into (1.9) we get

$$D_e y = \widehat{G} D_u u + \widehat{G}_d D_d d; \quad D_e e = D_e y - D_e r$$

and introduction of the scaled transfer functions

$$\boxed{G = D_e^{-1}\widehat{G}D_u, \quad G_d = D_e^{-1}\widehat{G}_d D_d} \qquad (1.14)$$

yields the following model in terms of scaled variables:

$$y = Gu + G_d d; \quad e = y - r \qquad (1.15)$$

Here u and d should be less than 1 in magnitude, and it is useful in some cases to introduce a scaled reference $\widetilde{r}$, which is less than 1 in magnitude. This is done by dividing the reference by the maximum expected reference change

$$\widetilde{r} = \widehat{r}/\widehat{r}_{\max} = D_r^{-1}\widehat{r} \qquad (1.16)$$

We then have that

$$r = R\widetilde{r} \quad \text{where} \quad R \triangleq D_e^{-1}D_r = \widehat{r}_{\max}/\widehat{e}_{\max} \qquad (1.17)$$

Here R is the largest expected change in reference relative to the allowed control error (typically, $R \geq 1$). The block diagram for the system in terms of scaled variables may then be written as shown in Figure 1.1, for which the following control objective is relevant:

- In terms of scaled variables we have that $|d(t)| \leq 1$ and $|\widetilde{r}(t)| \leq 1$, and our control objective is to manipulate u with $|u(t)| \leq 1$ such that $|e(t)| = |y(t) - r(t)| \leq 1$ (at least most of the time).

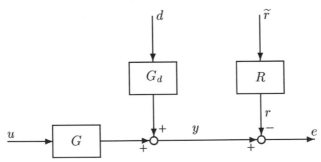

Figure 1.1: Model in terms of scaled variables

Remark 1 A number of the interpretations used in the book depend critically on a correct scaling. In particular, this applies to the input–output controllability analysis presented in Chapters 5 and 6. Furthermore, for a MIMO system one cannot correctly make use of the sensitivity function $S = (I + GK)^{-1}$ unless the output errors are of comparable magnitude.

Remark 2 With the above scalings, the worst-case behaviour of a system is analyzed by considering disturbances d of magnitude 1, and references $\tilde{r}$ of magnitude 1.

Remark 3 The control error is

$$e = y - r = Gu + G_d d - R\tilde{r} \tag{1.18}$$

and we see that a normalized reference change $\tilde{r}$ may be viewed as a special case of a disturbance with $G_d = -R$, where R is usually a constant diagonal matrix. We will sometimes use this observation to unify our treatment of disturbances and references.

Remark 4 The scaling of the outputs in (1.11) in terms of the control error is used when analyzing a given plant. However, if the issue is to *select* which outputs to control, see Section 10.3, then one may choose to scale the outputs with respect to their expected variation (which is usually similar to $\hat{r}_{\max}$).

Remark 5 If the expected or allowed variation of a variable about its nominal value is not symmetric, then to allow for the worst case, we should use the largest variation for the scaling $\hat{d}_{\max}$ and the smallest variations for the scalings $\hat{u}_{\max}$ and $\hat{e}_{\max}$.

Specifically, let $\tilde{\ }$ denote the original physical variable (before introducing any deviation or scaling), and let * denote the nominal value. Furthermore, assume that in terms of the physical variables we have that

$$\tilde{d}_{\min} \leq \tilde{d} \leq \tilde{d}_{\max}$$
$$\tilde{u}_{\min} \leq \tilde{u} \leq \tilde{u}_{\max}$$
$$-|\tilde{e}_-| \leq \tilde{e} \leq \tilde{e}_+$$

where $\tilde{e} = \tilde{y} - \tilde{r}$. Then we have the following scalings (or "ranges" or "spans"):

$$\hat{d}_{\max} = \max\left(|\tilde{d}_{\max} - \tilde{d}^*|, |\tilde{d}_{\min} - \tilde{d}^*|\right) \tag{1.19}$$

$$\hat{u}_{\max} = \min\left(|\tilde{u}_{\max} - \tilde{u}^*|, |\tilde{u}_{\min} - \tilde{u}^*|\right) \tag{1.20}$$

$$\hat{e}_{\max} = \min\left(|\tilde{e}_-|, |\tilde{e}_+|\right) \tag{1.21}$$

For example, if for the unscaled physical input we have $0 \leq \tilde{u} \leq 10$ with nominal value $\tilde{u}^* = 4$, then the input scaling is $\hat{u}_{\max} = \min(|10 - 4|, |0 - 4|) = \min(6, 4) = 4$.

Note that to get the worst case, we take the "max" for disturbances and "min" for inputs and outputs. For example, if the disturbance is $-5 \leq \tilde{d} \leq 10$ with zero nominal value ($\tilde{d}^* = 0$), then $\hat{d}_{\max} = 10$, whereas if the manipulated input is $-5 \leq \tilde{u} \leq 10$ with zero nominal value ($\tilde{u}^* = 0$), then $\hat{u}_{\max} = 5$. This approach may be conservative when the variations for *several* variables are not symmetric. The resulting scaled variables are

$$d = (\tilde{d} - \tilde{d}^*)/\hat{d}_{\max} \tag{1.22}$$

$$u = (\tilde{u} - \tilde{u}^*)/\hat{u}_{\max} \tag{1.23}$$

$$y = (\tilde{y} - \tilde{y}^*)/\hat{e}_{\max} \tag{1.24}$$

A further discussion on scaling and performance is given in Chapter 5 on page 165.

1.5   Deriving linear models

Linear models may be obtained from physical "first-principle" models, from analyzing input–output data, or from a combination of these two approaches. Although modelling and system identification are not covered in this book, it is always important for a control engineer to have a good understanding of a model's origin. The following steps are usually taken when deriving a linear model for controller design based on a first-principle approach:

1. Formulate a nonlinear state-space model based on physical knowledge.
2. Determine the steady-state operating point (or trajectory) about which to linearize.
3. Introduce deviation variables and linearize the model. There are essentially three parts to this step:
   (a) Linearize the equations using a Taylor expansion where second- and higher-order terms are omitted.
   (b) Introduce the deviation variables, e.g. $\delta x(t)$ defined by

   $$\delta x(t) = x(t) - x^*$$

   where the superscript * denotes the steady-state operating point or trajectory along which we are linearizing.
   (c) Subtract the steady-state (or trajectory) to eliminate the terms involving only steady-state quantities.

   These parts are usually accomplished together. For example, for a nonlinear state-space model of the form

   $$\frac{dx}{dt} = f(x, u) \tag{1.25}$$

   the linearized model in deviation variables $(\delta x, \delta u)$ is

   $$\frac{d\delta x(t)}{dt} = \underbrace{\left(\frac{\partial f}{\partial x}\right)^*}_{A} \delta x(t) + \underbrace{\left(\frac{\partial f}{\partial u}\right)^*}_{B} \delta u(t) \tag{1.26}$$

   Here x and u may be vectors, in which case the Jacobians A and B are matrices.
4. Scale the variables to obtain scaled models which are more suitable for control purposes.

In most cases steps 2 and 3 are performed numerically based on the model obtained in step 1. Also, since (1.26) is in terms of deviation variables, its Laplace transform becomes $s\delta x(s) = A\delta x(s) + B\delta u(s)$, or

$$\delta x(s) = (sI - A)^{-1} B\delta u(s) \tag{1.27}$$

Example 1.1 Physical model of a room heating process. *The above steps for deriving a linear model will be illustrated on the simple example depicted in Figure 1.2, where the control problem is to adjust the heat input Q to maintain constant room temperature T (within ± 1 K). The outdoor temperature T_o is the main disturbance. Units are shown in square brackets.*

1. Physical model. An energy balance for the room requires that the change in energy in the room must equal the net inflow of energy to the room (per unit of time). This yields the following state-space model:

$$\frac{d}{dt}(C_V T) = Q + \alpha(T_o - T) \tag{1.28}$$

where T [K] is the room temperature, C_V [J/K] is the heat capacity of the room, Q [W] is the heat input (from some heat source), and the term $\alpha(T_o - T)$ [W] represents the net heat loss due to exchange of air and heat conduction through the walls.

2. Operating point. Consider a case where the heat input Q^* is 2000 W and the difference between indoor and outdoor temperatures $T^* - T_o^*$ is 20 K. Then the steady-state energy balance yields $\alpha^* = 2000/20 = 100$ W/K. We assume the room heat capacity is constant, $C_V = 100$ kJ/K. (This value corresponds approximately to the heat capacity of air in a room of about 100 m^3; thus we neglect heat accumulation in the walls.)

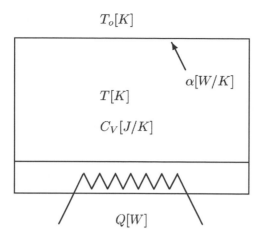

$T_o[K]$

$\alpha[W/K]$

$T[K]$

$C_V[J/K]$

$Q[W]$

Figure 1.2: Room heating process

3. Linear model in deviation variables. *If we assume α is constant, the model in (1.28) is already linear. Then introducing deviation variables*

$$\delta T(t) = T(t) - T^*(t), \ \delta Q(t) = Q(t) - Q^*(t), \ \delta T_o(t) = T_o(t) - T_o^*(t)$$

yields

$$C_V \frac{d}{dt}\delta T(t) = \delta Q(t) + \alpha(\delta T_o(t) - \delta T'(t)) \tag{1.29}$$

Remark. *If α may change then one would have to include an extra term $(T_o^* - T^*)\delta\alpha(t)$ on the right hand side of (1.29). Depending on the physics, $\alpha(t)$ could be an independent variable, or it could, e.g. for a temperature-dependent heat transfer coefficient, depend on the state variable (in which case we have $\delta alpha(t) = (\partial alpha/\partial T)\delta T(t)$).*

On taking Laplace transforms in (1.29), assuming $\delta T(t) = 0$ at $t = 0$, and rearranging we get

$$\delta T(s) = \frac{1}{\tau s + 1}\left(\frac{1}{\alpha}\delta Q(s) + \delta T_o(s)\right); \quad \tau = \frac{C_V}{\alpha} \tag{1.30}$$

The time constant for this example is $\tau = 100 \cdot 10^3/100 = 1000 \ s \approx 17$ min which is reasonable. It means that for a step increase in heat input it will take about 17 min for the temperature to reach 63% of its steady-state increase.

4. Linear model in scaled variables. *We introduce the following scaled variables:*

$$y(s) = \frac{\delta T(s)}{\delta T_{\max}}; \quad u(s) = \frac{\delta Q(s)}{\delta Q_{\max}}; \quad d(s) = \frac{\delta T_o(s)}{\delta T_{o,\max}} \tag{1.31}$$

In our case the acceptable variations in room temperature T are ± 1 K, i.e. $\delta T_{\max} = \delta e_{\max} = 1$ K. Furthermore, the heat input can vary between 0 W and 6000 W, and since its nominal value is 2000 W we have $\delta Q_{\max} = 2000$ W (see Remark 5 on page 7). Finally, the expected variations in outdoor temperature are ± 10 K, i.e. $\delta T_{o,\max} = 10$ K. The model in terms of scaled variables then becomes

$$\begin{aligned} G(s) &= \frac{1}{\tau s + 1}\frac{\delta Q_{\max}}{\delta T_{\max}}\frac{1}{\alpha} = \frac{20}{1000s + 1} \\ G_d(s) &= \frac{1}{\tau s + 1}\frac{\delta T_{o,\max}}{\delta T_{\max}} = \frac{10}{1000s + 1} \end{aligned} \tag{1.32}$$

Note that the static gain for the input is $k = 20$, whereas the static gain for the disturbance is $k_d = 10$. The fact that $|k_d| > 1$ means that we need some control (feedback or feedforward) to keep the output within its allowed bound ($|e| \leq 1$) when there is a disturbance of magnitude $|d| = 1$. The fact that $|k| > |k_d|$ means that we have enough "power" in the inputs to reject the disturbance at steady state; that is, we can, using an input of magnitude $|u| \leq 1$, have perfect disturbance rejection ($e = 0$) for the maximum disturbance ($|d| = 1$). We will return with a detailed discussion of this in Section 5.15.2 where we analyze the input–output controllability of the room heating process.

1.6   Notation

There is no standard notation to cover all of the topics covered in this book. We have tried to use the most familiar notation from the literature whenever possible, but an overriding concern has been to be consistent within the book, to ensure that the reader can follow the ideas and techniques through from one chapter to another.

The most important notation is summarized in Figure 1.3, which shows a one degree-of-freedom control configuration with negative feedback, a two degrees-of-freedom control configuration[1], and a general control configuration. The last configuration can be used to represent a wide class of controllers, including the one and two degrees-of-freedom configurations, as well as feedforward and estimation schemes and many others; and, as we will see, it can also be used to formulate optimization problems for controller design. The symbols used in Figure 1.3 are defined in Table 1.1. Apart from the use of v to represent the controller inputs for the general configuration, this notation is reasonably standard.

Lower-case letters are used for vectors and signals (e.g. u, y, n), and upper-case letters for matrices, transfer functions and systems (e.g. G, K). Matrix elements are usually denoted by lower-case letters, so g_{ij} is the ij'th element in the matrix G. However, sometimes we use upper-case letters G_{ij}, e.g. if G is partitioned so that G_{ij} is itself a matrix, or to avoid conflicts in notation. The Laplace variable s is often omitted for simplicity, so we often write G when we mean $G(s)$.

For state-space realizations we use the standard (A, B, C, D) notation. That is, a system G with a state-space realization (A, B, C, D) has a transfer function $G(s) = C(sI - A)^{-1}B + D$. We sometimes write

$$G(s) \overset{s}{=} \left[\begin{array}{c|c} A & B \\ \hline C & D \end{array} \right] \tag{1.33}$$

to mean that the transfer function $G(s)$ has a state-space realization given by the quadruple (A, B, C, D).

For closed-loop transfer functions we use S to denote sensitivity at the plant output, and $T = I - S$ to denote complementary sensitivity. With negative feedback, $S = (I + L)^{-1}$ and $T = L(I + L)^{-1}$, where L is the transfer function around the loop as seen from the output. In most cases $L = GK$, but if we also include measurement dynamics ($y_m = G_m y + n$) then $L = GKG_m$. The corresponding transfer functions as seen from the input of the plant are $L_I = KG$ (or $L_I = KG_m G$), $S_I = (I + L_I)^{-1}$ and $T_I = L_I(I + L_I)^{-1}$.

To represent uncertainty we use perturbations E (not normalized) or perturbations Δ (normalized such that their magnitude (norm) is less than or equal to 1). The nominal plant model is G, whereas the perturbed model with uncertainty is denoted G_p (usually for a set

---

[1] The one degree-of-freedom controller has only the control error $r - y_m$ as its input, whereas the two degrees-of-freedom controller has two inputs, namely r and y_m.

of possible perturbed plants) or G' (usually for a particular perturbed plant). For example, with additive uncertainty we may have $G_p = G + E_A = G + w_A \Delta_A$, where w_A is a weight representing the magnitude of the uncertainty.

By the right-half plane (RHP) we mean the closed right half of the complex plane, including the imaginary axis ($j\omega$-axis). The left-half plane (LHP) is the open left half of the complex plane, excluding the imaginary axis. A RHP-pole (unstable pole) is a pole located in the right-half plane, and thus includes poles on the imaginary axis. Similarly, a RHP-zero ("unstable" zero) is a zero located in the right-half plane.

We use A^T to denote the transpose of a matrix A, and A^H to represent its complex conjugate transpose.

Mathematical terminology

The symbol $\triangleq$ is used to denote *equal by definition*, $\overset{\text{def}}{\Leftrightarrow}$ is used to denote equivalent by definition, and $A \equiv B$ means that A is identically equal to B.

Let A and B be logic statements. Then the following expressions are equivalent:
$$A \Leftarrow B$$
A if B, or: If B then A
A is necessary for B
$$B \Rightarrow A, \text{ or: } B \text{ implies A}$$
B is sufficient for A
B only if A
$$\text{not A} \Rightarrow \text{not B}$$

The remaining notation, special terminology and abbreviations will be defined in the text.

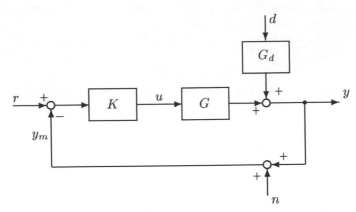

(a) One degree-of-freedom control configuration

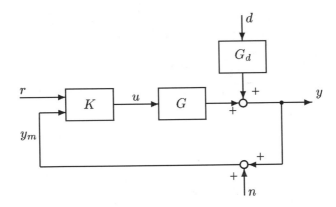

(b) Two degrees-of-freedom control configuration

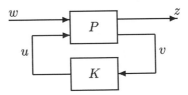

(c) General control configuration

Figure 1.3: Control configurations

Table 1.1: Nomenclature

K	controller, in whatever configuration. Sometimes the controller is broken down into its constituent parts. For example, in the two degrees-of-freedom controller in Figure 1.3(b), $K = [\, K_r \; K_y \,]$ where K_r is a prefilter and K_y is the feedback controller.

For the conventional control configurations (Figure 1.3(a) and (b)):

G	plant model
G_d	disturbance model
r	reference inputs (commands, setpoints)
d	disturbances (process noise, DV)
n	measurement noise
y	plant outputs (controlled variables, CV)
y_m	measured y
u	plant inputs (manipulated variables, MV, control signals)

For the general control configuration (Figure 1.3(c)):

P	generalized plant model. It will include G and G_d and the interconnection structure between the plant and the controller. In addition, if P is being used to formulate a design problem, then it will also include weighting functions.
w	exogenous inputs: commands, disturbances and noise
z	exogenous outputs; "error" signals to be minimized, e.g. $y - r$
v	controller inputs for the general configuration, e.g. commands, measured plant outputs, measured disturbances, etc. For the special case of a one degree-of-freedom controller with perfect measurements we have $v = r - y$.
u	control signals

2

CLASSICAL FEEDBACK CONTROL

In this chapter, we review the classical frequency response techniques for the analysis and design of single-loop (single-input single-output, SISO) feedback control systems. These loop-shaping techniques have been successfully used by industrial control engineers for decades, and have proved to be indispensable when it comes to providing insight into the benefits, limitations and problems of feedback control. During the 1980's the classical methods were extended to a more formal method based on shaping closed-loop transfer functions; for example, by considering the $\mathcal{H}_\infty$ norm of the weighted sensitivity function. We introduce this method at the end of the chapter.

The same underlying ideas and techniques will recur throughout the book as we present practical procedures for the analysis and design of multivariable (multi-input multi-output, MIMO) control systems.

2.1 Frequency response

On replacing s by $j\omega$ in a transfer function model $G(s)$ we get the so-called frequency response description. Frequency responses can be used to describe:

1. A system's response to sinusoids of varying frequency.
2. The frequency content of a deterministic signal via the Fourier transform.
3. The frequency distribution of a stochastic signal via the power spectral density function.

In this book, we use the first interpretation, namely that of frequency-by-frequency sinusoidal response. This interpretation has the advantage of being directly linked to the time domain, and at each frequency ω the complex number $G(j\omega)$ (or complex matrix for a MIMO system) has a clear physical interpretation. It gives the response to an input sinusoid of frequency ω. This will be explained in more detail below. For the other two interpretations we cannot assign a clear physical meaning to $G(j\omega)$ or $y(j\omega)$ at a particular frequency – it is the distribution relative to other frequencies which matters then.

One important advantage of a frequency response analysis of a system is that it provides insight into the benefits and trade-offs of feedback control. Although this insight may be obtained by viewing the frequency response in terms of its relationship between power spectral densities, as is evident from the excellent treatment by Kwakernaak and Sivan (1972), we believe that the frequency-by-frequency sinusoidal response interpretation is the most transparent and useful.

Multivariable Feedback Control: Analysis and Design. Second Edition
S. Skogestad and I. Postlethwaite © 2005, 2006 John Wiley & Sons, Ltd

Frequency-by-frequency sinusoids

We now want to give a physical picture of frequency response in terms of a system's response to persistent sinusoids. It is important that the reader has this picture in mind when reading the rest of the book. For example, it is needed to understand the response of a multivariable system in terms of its singular value decomposition. A physical interpretation of the frequency response for a stable linear system $y = G(s)u$ is as follows. Apply a sinusoidal input signal with frequency ω [rad/s] and magnitude u_0, such that

$$u(t) = u_0 \sin(\omega t + \alpha)$$

This input signal is persistent; that is, it has been applied since $t = -\infty$. Then the output signal is also a persistent sinusoid of the same frequency, namely

$$y(t) = y_0 \sin(\omega t + \beta)$$

Here u_0 and y_0 represent magnitudes and are therefore both non-negative. Note that the output sinusoid has a different amplitude y_0 and is also shifted in phase from the input by

$$\phi \triangleq \beta - \alpha$$

Importantly, it can be shown that y_0/u_0 and ϕ can be obtained directly from the Laplace transform $G(s)$ after inserting the imaginary number $s = j\omega$ and evaluating the magnitude and phase of the resulting complex number $G(j\omega)$. We have

$$y_0/u_0 = |G(j\omega)|; \quad \phi = \angle G(j\omega) \text{ [rad]} \tag{2.1}$$

For example, let $G(j\omega) = a + jb$, with real part $a = \text{Re } G(j\omega)$ and imaginary part $b = \text{Im } G(j\omega)$, then

$$|G(j\omega)| = \sqrt{a^2 + b^2}; \quad \angle G(j\omega) = \arctan(b/a) \tag{2.2}$$

In words, (2.1) says that *after sending a sinusoidal signal through a system $G(s)$, the signal's magnitude is amplified by a factor $|G(j\omega)|$ and its phase is shifted by $\angle G(j\omega)$*. In Figure 2.1, this statement is illustrated for the following first-order delay system (time in seconds):

$$G(s) = \frac{ke^{-\theta s}}{\tau s + 1}; \quad k = 5, \theta = 2, \tau = 10 \tag{2.3}$$

At frequency $\omega = 0.2$ rad/s, we see that the output y lags behind the input by about a quarter of a period and that the amplitude of the output is approximately twice that of the input. More accurately, the amplification is

$$|G(j\omega)| = k/\sqrt{(\tau\omega)^2 + 1} = 5/\sqrt{(10\omega)^2 + 1} = 2.24$$

and the phase shift is

$$\phi = \angle G(j\omega) = -\arctan(\tau\omega) - \theta\omega = -\arctan(10\omega) - 2\omega = -1.51 \text{ rad} = -86.5°$$

$G(j\omega)$ is called the *frequency response* of the system $G(s)$. It describes how the system responds to persistent sinusoidal inputs of frequency ω. The magnitude of the frequency

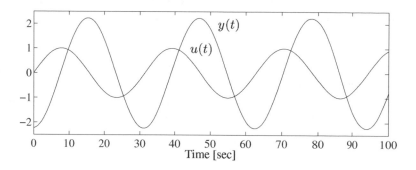

Figure 2.1: Sinusoidal response for system $G(s) = 5e^{-2s}/(10s + 1)$ at frequency $\omega = 0.2$ rad/s

response, $|G(j\omega)|$, being equal to $|y_0(\omega)|/|u_0(\omega)|$, is also referred to as the *system gain*. Sometimes the gain is given in units of dB (decibel) defined as

$$A \, [\text{dB}] = 20 \log_{10} A \tag{2.4}$$

For example, $A = 2$ corresponds to $A = 6.02$ dB, and $A = \sqrt{2}$ corresponds to $A = 3.01$ dB, and $A = 1$ corresponds to $A = 0$ dB.

Both $|G(j\omega)|$ and $\angle G(j\omega)$ depend on the frequency ω. This dependency may be plotted explicitly in Bode plots (with ω as independent variable) or somewhat implicitly in a Nyquist plot (phase plane plot). In Bode plots we usually employ a log-scale for frequency and gain, and a linear scale for the phase.

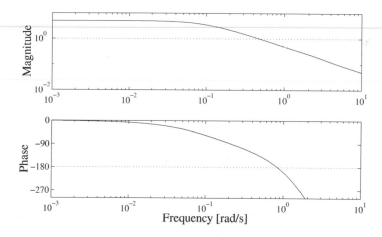

Figure 2.2: Frequency response (Bode plots) of $G(s) = 5e^{-2s}/(10s + 1)$

In Figure 2.2, the Bode plots are shown for the system in (2.3). We note that in this case both the gain and phase fall monotonically with frequency. This is quite common for process control applications. The delay θ only shifts the sinusoid in time, and thus affects the phase but not the gain. The system gain $|G(j\omega)|$ is equal to k at low frequencies; this is the steady-state gain and is obtained by setting $s = 0$ (or $\omega = 0$). The gain remains relatively constant up to the break frequency $1/\tau$ where it starts falling sharply. Physically, the system responds too slowly to let high-frequency ("fast") inputs have much effect on the outputs.

The frequency response is also useful for an *unstable plant* $G(s)$, which by itself has no steady-state response. Let $G(s)$ be stabilized by feedback control, and consider applying a sinusoidal forcing signal to the stabilized system. In this case all signals within the system are persistent sinusoids with the same frequency ω, and $G(j\omega)$ yields as before the sinusoidal response from the input to the output of $G(s)$.

Phasor notation. For any sinusoidal signal

$$u(t) = u_0 \sin(\omega t + \alpha)$$

we may introduce the phasor notation by defining the complex number

$$\boxed{u(\omega) \triangleq u_0 e^{j\alpha}} \qquad (2.5)$$

We then have that

$$u_0 = |u(\omega)|; \quad \alpha = \angle u(\omega) \qquad (2.6)$$

We use ω as an argument to show explicitly that this notation is used for sinusoidal signals, and also because u_0 and α generally depend on ω. Note that $u(\omega)$ is *not* equal to $u(s)$ evaluated at $s = \omega$ or $s = j\omega$, nor is it equal to $u(t)$ evaluated at $t = \omega$. From Euler's formula for complex numbers, we have that $e^{jz} = \cos z + j \sin z$. It then follows that $\sin(\omega t)$ is equal to the imaginary part of the complex function $e^{j\omega t}$, and we can write the time domain sinusoidal response in complex form as follows:

$$u(t) = u_0 \mathrm{Im}\, e^{j(\omega t + \alpha)} \text{ gives, as } t \to \infty: \quad y(t) = y_0 \mathrm{Im}\, e^{j(\omega t + \beta)} \qquad (2.7)$$

where

$$y_0 = |G(j\omega)| u_0, \quad \beta = \angle G(j\omega) + \alpha \qquad (2.8)$$

and $|G(j\omega)|$ and $\angle G(j\omega)$ are defined in (2.2). Since $G(j\omega) = |G(j\omega)|\, e^{j\angle G(j\omega)}$, the sinusoidal response in (2.7) and (2.8) can be compactly written in phasor notation as follows:

$$y(\omega)e^{j\omega t} = G(j\omega)u(\omega)e^{j\omega t} \qquad (2.9)$$

or because the term $e^{j\omega t}$ appears on both sides

$$\boxed{y(\omega) = G(j\omega)u(\omega)} \qquad (2.10)$$

At each frequency, $u(\omega)$, $y(\omega)$ and $G(j\omega)$ are complex numbers, and the usual rules for multiplying complex numbers apply. We will use this phasor notation throughout the book. Thus *whenever we use notation such as $u(\omega)$ (with ω and not $j\omega$ as an argument), the reader should interpret this as a (complex) sinusoidal signal, $u(\omega)e^{j\omega t}$*. The expression (2.10) also applies to MIMO systems where $u(\omega)$ and $y(\omega)$ are complex vectors representing the sinusoidal signals in the input and output channels, respectively, and $G(j\omega)$ is a complex matrix.

Minimum-phase systems. For stable systems which are minimum-phase (no time delays or right-half plane (RHP) zeros) there is a unique relationship between the gain and phase of the frequency response. This may be quantified by the Bode gain–phase relationship which gives the phase of G (normalized[1] such that $G(0) > 0$) at a given frequency ω_0 as a function of $|G(j\omega)|$ over the entire frequency range:

$$\angle G(j\omega_0) = \frac{1}{\pi} \int_{-\infty}^{\infty} \underbrace{\frac{d\ln|G(j\omega)|}{d\ln\omega}}_{N(\omega)} \ln\left|\frac{\omega + \omega_0}{\omega - \omega_0}\right| \cdot \frac{d\omega}{\omega} \qquad (2.11)$$

The name *minimum-phase* refers to the fact that such a system has the minimum possible phase lag for the given magnitude response $|G(j\omega)|$. The term $N(\omega)$ is the slope of the magnitude in log-variables at frequency ω. In particular, the local slope at frequency ω_0 is

$$N(\omega_0) = \left(\frac{d\ln|G(j\omega)|}{d\ln\omega}\right)_{\omega=\omega_0}$$

The term $\ln\left|\frac{\omega+\omega_0}{\omega-\omega_0}\right|$ in (2.11) is infinite at $\omega = \omega_0$, so it follows that $\angle G(j\omega_0)$ is primarily determined by the local slope $N(\omega_0)$. Also $\int_{-\infty}^{\infty} \ln\left|\frac{\omega+\omega_0}{\omega-\omega_0}\right| \cdot \frac{d\omega}{\omega} = \frac{\pi^2}{2}$ which justifies the commonly used approximation for stable minimum-phase systems

$$\angle G(j\omega_0) \approx \frac{\pi}{2} N(\omega_0) \text{ [rad]} = 90° \cdot N(\omega_0) \qquad (2.12)$$

The approximation is exact for the system $G(s) = 1/s^n$ (where $N(\omega) = -n$), and it is good for stable minimum-phase systems except at frequencies close to those of resonant (complex) poles or zeros.

RHP-zeros and time delays contribute additional phase lag to a system when compared to that of a minimum-phase system with the same gain (hence the term *non-minimum-phase* system). For example, the system $G(s) = \frac{-s+a}{s+a}$ with a RHP-zero at $s = a$ has a constant gain of 1, but its phase is $-2\arctan(\omega/a)$ [rad] (and not 0 [rad] as it would be for the minimum-phase system $G(s) = 1$ of the same gain). Similarly, the time delay system $e^{-\theta s}$ has a constant gain of 1, but its phase is $-\omega\theta$ [rad].

Straight-line approximations (asymptotes). For the design methods used in this book it is useful to be able to sketch Bode plots quickly, and in particular the magnitude (gain) diagram. The reader is therefore advised to become familiar with asymptotic Bode plots (straight-line approximations). For example, for a transfer function

$$G(s) = k\frac{(s+z_1)(s+z_2)\cdots}{(s+p_1)(s+p_2)\cdots} \qquad (2.13)$$

the asymptotic Bode plots of $G(j\omega)$ are obtained by using for each term $(s + a)$ the approximation $j\omega + a \approx a$ for $\omega < a$ and by $j\omega + a \approx j\omega$ for $\omega > a$. These approximations yield straight lines on a log–log plot which meet at the so-called *break point frequency* $\omega = a$. In (2.13) therefore, the frequencies $z_1, z_2, \ldots, p_1, p_2, \ldots$ are the break points where the asymptotes meet. For complex poles or zeros, the term $s^2 + 2\zeta s\omega_0 + \omega_0^2$ (where $|\zeta| < 1$) is approximated by ω_0^2 for $\omega < \omega_0$ and by $s^2 = (j\omega)^2 = -\omega^2$ for $\omega > \omega_0$. The magnitude of a transfer function is usually close to its asymptotic value, and the only case when there is

---

[1] The normalization of $G(s)$ is necessary to handle systems such as $\frac{1}{s+2}$ and $\frac{-1}{s+2}$, which have equal gain, are stable and minimum-phase, but their phases differ by $180°$. Systems with integrators may be treated by replacing $\frac{1}{s}$ by $\frac{1}{s+\epsilon}$ where ϵ is a small positive number.

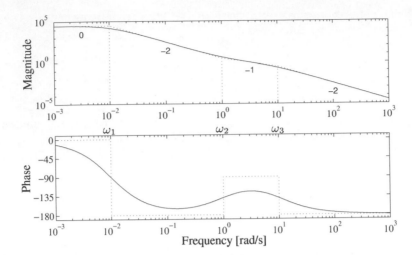

Figure 2.3: Bode plots of transfer function $L_1 = 30\frac{s+1}{(s+0.01)^2(s+10)}$. The asymptotes are given by dotted lines. The vertical dotted lines on the upper plot indicate the break frequencies ω_1, ω_2 and ω_3.

significant deviation is around the resonance frequency ω_0 for complex poles or zeros with a damping $|\zeta|$ of about 0.3 or less. In Figure 2.3, the Bode plots are shown for

$$L_1(s) = 30\frac{(s+1)}{(s+0.01)^2(s+10)} \tag{2.14}$$

The asymptotes (straight-line approximations) are shown by dotted lines. In this example the asymptotic slope of $|L_1|$ is 0 up to the first break frequency at $\omega_1 = 0.01$ rad/s where we have two poles and then the slope changes to $N = -2$. Then at $\omega_2 = 1$ rad/s there is a zero and the slope changes to $N = -1$. Finally, there is a break frequency corresponding to a pole at $\omega_3 = 10$ rad/s and so the slope is $N = -2$ at this and higher frequencies. We note that the magnitude follows the asymptotes closely, whereas the phase does not. The asymptotic phase jumps at the break frequency by -90° (LHP-pole or RHP-zero) or $+90^\circ$ (LHP-zero or RHP-pole),

Remark. The phase approximation can be significantly improved if, for each term $j\omega + a$, we let the phase contribution be zero for $\omega \leq 0.1a$ and $\pi/2$ (90°) for $\omega \geq 10a$, and then connect these two lines by a third line from $(0, \omega = 0.1a)$ to $(\pi/2, \omega = 10a)$, which of course passes through the correct phase $\pi/4$ at $\omega = a$. For the terms $s^2 + 2\zeta s\omega_0 + \omega_0^2$, $\zeta < 1$, we can better approximate the phase by letting it be zero for $\omega \leq 0.1\omega_0$ and π for $\omega \geq 10\omega_0$, with a third line connecting $(0, \omega = 0.1\omega_0)$ to $(\pi, \omega = 10\omega_0)$, which passes through the correct phase $\pi/2$ at $\omega = \omega_0$.

2.2 Feedback control

2.2.1 One degree-of-freedom controller

In most of this chapter, we examine the simple one degree-of-freedom negative feedback structure shown in Figure 2.4. The input to the controller $K(s)$ is $r - y_m$ where $y_m = y + n$

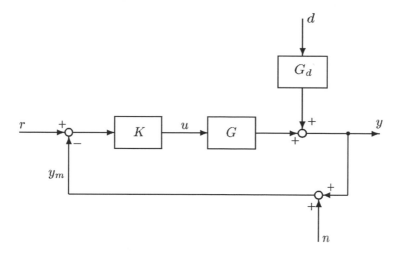

Figure 2.4: Block diagram of one degree-of-freedom feedback control system

is the measured output and n is the measurement noise. Thus, the input to the plant is

$$u = K(s)(r - y - n) \tag{2.15}$$

The objective of control is to manipulate u (design K) such that the control error e remains small in spite of disturbances d. The control error e is defined as

$$e = y - r \tag{2.16}$$

where r denotes the reference value (setpoint) for the output.

Remark. In the literature, the control error is frequently defined as $r - y_m$ which is often the controller input. However, this is not a good definition of an error variable. First, the error is normally defined as the actual value (here y) minus the desired value (here r). Second, the error should involve the actual value (y) and not the measured value (y_m). We therefore use the definition in (2.16).

2.2.2 Closed-loop transfer functions

The plant model is written as

$$y = G(s)u + G_d(s)d \tag{2.17}$$

and for a one degree-of-freedom controller the substitution of (2.15) into (2.17) yields

$$y = GK(r - y - n) + G_d d$$

or

$$(I + GK)y = GKr + G_d d - GKn \tag{2.18}$$

and hence the closed-loop response is

$$y = \underbrace{(I + GK)^{-1}GK}_{T} r + \underbrace{(I + GK)^{-1}}_{S} G_d d - \underbrace{(I + GK)^{-1}GK}_{T} n \tag{2.19}$$

The control error is

$$e = y - r = -Sr + SG_dd - Tn \tag{2.20}$$

where we have used the fact that $T - I = -S$. The corresponding plant input signal is

$$u = KSr - KSG_dd - KSn \tag{2.21}$$

The following notation and terminology are used:

$$L = GK \quad \text{loop transfer function}$$
$$S = (I + GK)^{-1} = (I + L)^{-1} \quad \text{sensitivity function}$$
$$T = (I + GK)^{-1}GK = (I + L)^{-1}L \quad \text{complementary sensitivity function}$$

We see that S is the closed-loop transfer function from the output disturbances to the outputs, while T is the closed-loop transfer function from the reference signals to the outputs. The term complementary sensitivity for T follows from the identity

$$S + T = I \tag{2.22}$$

To derive (2.22), we write $S + T = (I+L)^{-1} + (I+L)^{-1}L$ and factor out the term $(I+L)^{-1}$. The term sensitivity function is natural because S gives the sensitivity reduction afforded by feedback. To see this, consider the "open-loop" case, i.e. with no control ($K = 0$). Then the error is

$$e = y - r = -r + G_dd + 0 \cdot n \tag{2.23}$$

and a comparison with (2.20) shows that, with the exception of noise, the response with feedback is obtained by premultiplying the right hand side by S.

Remark 1 Actually, the above explanation is not the original reason for the name "sensitivity". Bode first called S sensitivity because it gives the relative sensitivity of the closed-loop transfer function T to the relative plant model error. In particular, at a given frequency ω we have for a SISO plant, by straightforward differentiation of T, that

$$\frac{dT/T}{dG/G} = S \tag{2.24}$$

Remark 2 Equations (2.15)–(2.23) are written in matrix form because they also apply to MIMO systems. Of course, for SISO systems we may write $S + T = 1$, $S = \frac{1}{1+L}$, $T = \frac{L}{1+L}$ and so on.

Remark 3 In general, closed-loop transfer functions for SISO systems with *negative* feedback may be obtained from the rule

$$\text{OUTPUT} = \frac{\text{"direct"}}{1 + \text{"loop"}} \cdot \text{INPUT} \tag{2.25}$$

where "direct" represents the transfer function for the direct effect of the input on the output (with the feedback path open) and "loop" is the transfer function around the loop (denoted $L(s)$). In the above case $L = GK$. If there is also a measurement device, $G_m(s)$, in the loop, then $L(s) = GKG_m$. The rule in (2.25) is easily derived by generalizing (2.18). In Section 3.2, we present a more general form of this rule which also applies to multivariable systems.

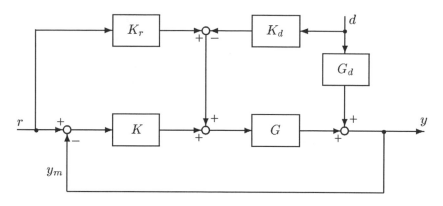

Figure 2.5: Two degrees-of-freedom feedback and feedforward control. Perfect measurements of y and d assumed.

2.2.3 Two degrees-of-freedom and feedforward control

The control structure in Figure 2.4 is called one degree-of-freedom because the controller K acts on a single signal, namely the difference $r - y_m$. In the two degrees-of-freedom structure of Figure 2.5, we treat the two signals y_m and r independently by introducing a "feedforward" controller K_r on the reference[2]. In Figure 2.5 we have also introduced a feedforward controller K_d for the measured disturbance d. The plant input in Figure 2.5 is the sum of the contributions from the feedback controller and the two feedforward controllers,

$$u = \underbrace{K(r - y)}_{\text{feedback}} + \underbrace{K_r r - K_d d}_{\text{feedforward}} \qquad (2.26)$$

where for simplicity we have assumed perfect measurements of y and d. After substituting (2.26) into (2.17) and solving with respect to y,

$$y = (I + GK)^{-1}[G(K + K_r)r + (G_d - GK_d)d] \qquad (2.27)$$

Using $SGK - I = T - I = -S$, the resulting control error is

$$e = y - r = S(-S_r r + S_d G_d d) \qquad (2.28)$$

where the three "sensitivity" functions, giving the effect of control, are defined by

$$S = (I + GK)^{-1}, \quad S_r = I - GK_r, \quad S_d = I - GK_d G_d^{-1} \qquad (2.29)$$

S is the classical feedback sensitivity function, whereas S_r and S_d are the "feedforward sensitivity functions" for reference and disturbance, respectively. Without feedback control ($K = 0$) we have $S = I$, and correspondingly without feedforward control ($K_d = 0$ and $K_r = 0$) we have $S_d = I$ and $S_r = I$. We want the sensitivities to be small to get a small error e. More precisely,

---

[2] There are many other ways of introducing two degrees-of-freedom control, see e.g. Figure 2.25 (page 52) for a "prefilter" structure. The form in Figure 2.5 is preferred here because it unifies the treatment of references and disturbances.

- For reference tracking we want the product SS_r to be small.
- For disturbance rejection we want the product SS_d to be small.

From this we get the important insight that that the primary objective of feedforward control is to improve performance (when required) at frequencies where feedback is not effective (i.e. where $|S| \geq 1$).

2.2.4   Why feedback?

"Perfect" control can be obtained, even without feedback ($K = 0$), by using the feedforward controllers

$$K_r(s) = G^{-1}(s); \quad K_d(s) = G^{-1}(s)G_d(s) \tag{2.30}$$

To confirm this, let $u = K_r r - K_d d$ and we get

$$y = G(G^{-1}r - G^{-1}G_d d) + G_d d = r$$

These controllers also give $S_r = 0$ and $S_d = 0$ in (2.28). However, note that in (2.30) we must assume that it is possible to realize physically the plant inverse G^{-1} and that both the plant G and the resulting controller containing the term G^{-1} are stable. These are serious considerations, but of more general concern is the loss of performance that inevitably arises because (1) the disturbances are never known (measured) exactly, and (2) G is never an exact model. *The fundamental reasons for using feedback control are therefore the presence of*

1. Signal uncertainty – unknown disturbance (d)
2. Model uncertainty (Δ)
3. An unstable plant

The third reason follows because unstable plants can only be stabilized by feedback (see Remark 2 on internal stability, page 145). In addition, for a nonlinear plant, feedback control provides a linearizing effect on the system's behaviour. This is discussed in the next section.

2.2.5   High-gain feedback

The benefits of feedback control require the use of "high" gains. As seen from (2.30), the perfect feedforward controller uses an *explicit* model of the plant inverse as part of the controller. With feedback, on the other hand, the use of high gains in GK *implicitly* generates an inverse. To see this, note that with $L = GK$ large, we get $S = (1 + GK)^{-1} \approx 0$ and $T = I - S \approx I$. From (2.21) the input signal generated by feedback is $u = KS(r - G_d d - n)$, and from the identity $KS = G^{-1}T$ it follows that with high-gain feedback the input signal is $u \approx G^{-1}(r - G_d d - n)$ and we get $y \approx r - n$. Thus, high-gain feedback generates the inverse without the need for an explicit model, and this also explains why feedback control is much less sensitive to uncertainty than feedforward control.

This is one of the beauties of feedback control; the problem is that high-gain feedback may induce instability. The solution is to use high feedback gains only over a limited frequency range (typically, at low frequencies), and to ensure that the gains "roll off" at higher frequencies where stability is a problem. The design is most critical around the "bandwidth" frequency where the loop gain $|L|$ drops below 1. The design of feedback controllers therefore depends primarily on a good model description G around the "bandwidth" frequency. Closed-loop stability is discussed briefly in the next section, and is a recurring issue throughout the book.

As mentioned earlier, an additional benefit of feedback control is its ability to "linearize" a system's behaviour. Actually, there are two different *linearizing effects*:

1. *A "local" linearizing effect in terms of the validity model*: By use of feedback we can control the output y about an operating point and prevent the system from drifting too far away from its desired state. In this way, the system remains in the "linear region" where the linear models $G(s)$ and $G_d(s)$ are valid. This local linearizing effect justifies the use of linear models in feedback controller design and analysis, as presented in this book and as used by most practising control engineers.
2. *A "global" linearizing effect in terms of the tracking response from the reference r to the output y*: As just discussed, the use of high-gain feedback yields $y \approx r - n$. This holds also for cases where nonlinear effects cause the linear model G to change significantly as we change r. Thus, even though the underlying system is strongly nonlinear (and uncertain) the input–output response from y to r is approximately linear (and certain) with a constant gain of 1.

Example 2.1 Feedback amplifier. *The "global" linearizing effect of negative feedback is the basis for feedback amplifiers, first developed by Harold Black in 1927 for telephone communication (Kline, 1993). In the feedback amplifier, we want to magnify the input signal r by a factor α by sending it through an amplifier G with a large gain. In an open loop (feedforward) arrangement $y = Gr$ and we must adjust the amplifier such that $G = \alpha$. Black's idea was to leave the high-gain amplifier unchanged, and instead modify the input signal r by subtracting $(1/\alpha)y$, where y is the measured output signal. This corresponds to inserting a controller $K_2 = 1/\alpha$ in the negative feedback path (e.g. see Figure 4.4(d) on page 147) to get $y = G(r - K_2 y)$. The closed-loop response becomes $y = \frac{G}{1+GK_2} r$ and for $|GK_2| \gg 1$ (which requires $|G| \gg \alpha$) we get $y \approx \frac{1}{K_2} r = \alpha \cdot r$, as desired. Note that the closed-loop gain α is set by the feedback network $(K_2 = 1/\alpha)$ and is independent of amplifier (G) parameter changes. Furthermore, within the system's closed-loop bandwidth, all signals (with any magnitude or frequency) are amplified by the same amount α, and this property is independent of the amplifier dynamics $G(s)$. Apparently, Black's claimed improvements, with simple negative feedback, over the then-standard feedforward approach, seemed so unlikely that his patent application was initially rejected.*

Remark. *In Black's design, the amplifier gain must be much larger than the desired closed-loop amplification (i.e. $|G| \gg \alpha$). This seems unnecessary, because with feedforward control, it is sufficient to require $|G| = \alpha$. Indeed, the requirement $|G| \gg \alpha$ can be avoided, if we add integral action to the loop. This may be done by use of a "two degrees-of-freedom" controller where we add a controller K_1 before the plant (amplifier) to get $y = GK_1(r - K_2 y)$ (see Figure 4.4 on page 147). The closed-loop response becomes $y = \frac{GK_1}{1+GK_1K_2} r$, and for $|GK_1K_2| \gg 1$ (which requires $|GK_1| \gg \alpha$) we get $y \approx \frac{1}{K_2} r = \alpha r$. The requirement $|GK_1K_2| \gg 1$ only needs to hold at those frequencies for which amplification is desired, and may be obtained by choosing K_1 as a simple PI (proportional–integral) controller with a proportional gain of 1; that is, $K_1 = 1 + \frac{1}{\tau_I s}$ where τ_I is the adjustable integral time.*

Of course, the "global" linearizing effect of negative feedback assumes that high-gain feedback is possible and does not result in closed-loop instability. The latter is well known with audio amplifiers as "singing", "ringing", "squalling" or "howling". In the next section, we consider conditions for closed-loop stability.

2.3 Closed-loop stability

A critical issue in designing feedback controllers is to achieve stability. As noted earlier, if the feedback gain is too large, then the controller may "overreact" and the closed-loop system becomes unstable. This is illustrated next by a simple example.

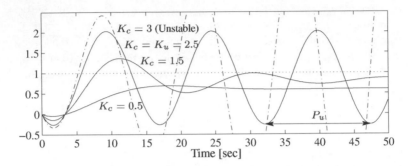

Figure 2.6: Effect of proportional gain K_c on the closed-loop response $y(t)$ for the inverse response process

Example 2.2 Inverse response process. *Consider the plant (time in seconds)*

$$G(s) = \frac{3(-2s + 1)}{(10s + 1)(5s + 1)} \qquad (2.31)$$

This is one of two main example processes used in this chapter to illustrate the techniques of classical control. The model has a right-half plane (RHP) zero at $s = 0.5$ rad/s. This imposes a fundamental limitation on control, and high controller gains will induce closed-loop instability.

This is illustrated for a proportional (P) controller $K(s) = K_c$ in Figure 2.6, where the response $y = Tr = GK_c(1 + GK_c)^{-1}r$ to a step change in the reference ($r(t) = 1$ for $t > 0$) is shown for four different values of K_c. The system is seen to be stable for $K_c < 2.5$, and unstable for $K_c > 2.5$. The controller gain at the limit of instability, $K_u = 2.5$, is sometimes called the ultimate gain and for this value the system is seen to cycle continuously with a period $P_u = 15.2$ s, corresponding to the frequency $\omega_u \triangleq 2\pi/P_u = 0.42$ rad/s.

Two methods are commonly used to determine closed-loop stability:

1. The poles of the closed-loop system are evaluated. That is, the roots of $1 + L(s) = 0$ are found, where L is the transfer function around the loop. The system is stable *if and only if* all the closed-loop poles are in the open left-half plane (LHP) (i.e. poles on the imaginary axis are considered "unstable"). The poles are also equal to the eigenvalues of the state-space A-matrix, and this is usually how the poles are computed numerically.
2. The frequency response (including negative frequencies) of $L(j\omega)$ is plotted in the complex plane and the number of encirclements it makes of the critical point -1 is counted. By *Nyquist's stability criterion* (for which a detailed statement is given in Theorem 4.9) closed-loop stability is inferred by equating the number of encirclements to the number of open-loop unstable poles (RHP-poles).

   For open-loop stable systems where $\angle L(j\omega)$ falls with frequency such that $\angle L(j\omega)$ crosses $-180°$ only once (from above at frequency ω_{180}), one may equivalently use *Bode's stability condition* which says that the closed-loop system is stable if and only if the loop gain $|L|$ is less than 1 at this frequency; that is

$$\text{Stability} \quad \Leftrightarrow \quad |L(j\omega_{180})| < 1 \qquad (2.32)$$

where ω_{180} is the phase crossover frequency defined by $\angle L(j\omega_{180}) = -180°$.

Method 1, which involves computing the poles, is best suited for numerical calculations. However, time delays must first be approximated as rational transfer functions, e.g. Padé approximations. Method 2, which is based on the frequency response, has a nice graphical interpretation, and may also be used for systems with time delays. Furthermore, it provides useful measures of relative stability and forms the basis for several of the robustness tests used later in this book.

Example 2.3 **Stability of inverse response process with proportional control.** *Let us determine the condition for closed-loop stability of the plant G in (2.31) with proportional control; that is, with $K(s) = K_c$ (a constant) and loop transfer function $L(s) = K_c G(s)$.*

1. The system is stable if and only if all the closed-loop poles are in the LHP. The poles are solutions to $1 + L(s) = 0$ or equivalently the roots of

$$(10s + 1)(5s + 1) + K_c 3(-2s + 1) = 0$$

$$\Leftrightarrow \quad 50s^2 + (15 - 6K_c)s + (1 + 3K_c) = 0 \qquad (2.33)$$

But since we are only interested in the half plane location of the poles, it is not necessary to solve (2.33). Rather, one may consider the coefficients a_i of the characteristic equation $a_n s^n + \cdots + a_1 s + a_0 = 0$ in (2.33), and use the Routh–Hurwitz test to check for stability. For second-order systems, this test says that we have stability if and only if all the coefficients have the same sign. This yields the following stability conditions:

$$(15 - 6K_c) > 0; \quad (1 + 3K_c) > 0$$

or equivalently $-1/3 < K_c < 2.5$. With negative feedback ($K_c \geq 0$) only the upper bound is of practical interest, and we find that the maximum allowed gain ("ultimate gain") is $K_u = 2.5$ which agrees with the simulation in Figure 2.6. The poles at the onset of instability may be found by substituting $K_c = K_u = 2.5$ into (2.33) to get $50s^2 + 8.5 = 0$, i.e. $s = \pm j\sqrt{8.5/50} = \pm j0.412$. Thus, at the onset of instability we have two poles on the imaginary axis, and the system will be continuously cycling with a frequency $\omega = 0.412$ rad/s corresponding to a period $P_u = 2\pi/\omega = 15.2$ s. This agrees with the simulation results in Figure 2.6.

2. Stability may also be evaluated from the frequency response of $L(s)$. A graphical evaluation is most enlightening. The Bode plots of the plant (i.e. $L(s)$ with $K_c = 1$) are shown in Figure 2.7. From these one finds the frequency ω_{180} where $\angle L$ is $-180°$ and then reads off the corresponding gain. This yields $|L(j\omega_{180})| = K_c|G(j\omega_{180})| = 0.4K_c$, and we get from (2.32) that the system is stable if and only if $|L(j\omega_{180})| < 1 \Leftrightarrow K_c < 2.5$ (as found above). Alternatively, the phase crossover frequency may be obtained analytically from

$$\angle L(j\omega_{180}) = -\arctan(2\omega_{180}) - \arctan(5\omega_{180}) - \arctan(10\omega_{180}) = -180°$$

which gives $\omega_{180} = 0.412$ rad/s as found in the pole calculation above. The loop gain at this frequency is

$$|L(j\omega_{180})| = K_c \frac{3 \cdot \sqrt{(2\omega_{180})^2 + 1}}{\sqrt{(5\omega_{180})^2 + 1} \cdot \sqrt{(10\omega_{180})^2 + 1}} = 0.4K_c$$

which is the same as found from the graph in Figure 2.7. The stability condition $|L(j\omega_{180})| < 1$ then yields $K_c < 2.5$ as expected.

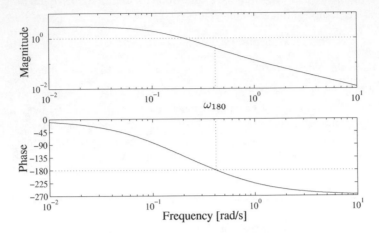

Figure 2.7: Bode plots for $L(s) = K_c \frac{3(-2s+1)}{(10s+1)(5s+1)}$ with $K_c = 1$

2.4 Evaluating closed-loop performance

Although closed-loop stability is an important issue, the real objective of control is to improve performance; that is, to make the output $y(t)$ behave in a more desirable manner. Actually, the possibility of inducing instability is one of the disadvantages of feedback control which has to be traded off against performance improvement. The objective of this section is to discuss ways of evaluating closed-loop performance.

2.4.1 Typical closed-loop responses

The following example, which considers proportional plus integral (PI) control of the stable inverse response process in (2.31), illustrates what type of closed-loop performance one might expect.

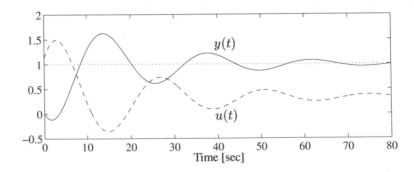

Figure 2.8: Closed-loop response to a unit step change in reference for the stable inverse response process (2.31) with PI control

Example 2.4 PI control of the inverse response process. *We have already studied the use of a proportional controller for the process in (2.31). We found that a controller gain of $K_c = 1.5$ gave a reasonably good response, except for a steady-state offset (see Figure 2.6). The reason for this offset is the non-zero steady-state sensitivity function, $S(0) = \frac{1}{1+K_c G(0)} = 0.18$ (where $G(0) = 3$ is the steady-state gain of the plant). From $e = -Sr$ in (2.20) it follows that for $r = 1$ the steady-state control error is -0.18 (as is confirmed by the simulation in Figure 2.6). To remove the steady-state offset we add integral action in the form of a PI controller*

$$K(s) = K_c \left(1 + \frac{1}{\tau_I s} \right) \tag{2.34}$$

The settings for K_c and τ_I can be determined from the classical tuning rules of Ziegler and Nichols (1942):

$$K_c = K_u/2.2, \quad \tau_I = P_u/1.2 \tag{2.35}$$

where K_u is the maximum (ultimate) P controller gain and P_u is the corresponding period of oscillations. In our case $K_u = 2.5$ and $P_u = 15.2$ s (as observed from the simulation in Figure 2.6), and we get $K_c = 1.14$ and $\tau_I = 12.7$ s. Alternatively, K_u and P_u can be obtained analytically from the model $G(s)$,

$$K_u = 1/|G(j\omega_u)|, \quad P_u = 2\pi/\omega_u \tag{2.36}$$

where ω_u is defined by $\angle G(j\omega_u) = -180°$.

The closed-loop response, with PI control, to a step change in reference is shown in Figure 2.8. The output $y(t)$ has an initial inverse response due to the RHP-zero, but it then rises quickly and $y(t) = 0.9$ at $t = 8.0$ s (the rise time). The response is quite oscillatory and it does not settle to within $\pm 5\%$ of the final value until after $t = 65$ s (the settling time). The overshoot (height of peak relative to the final value) is about 62% which is much larger than one would normally like for reference tracking. The overshoot is due to controller tuning, and could have been avoided by reducing the controller gain. The decay ratio, which is the ratio between subsequent peaks, is about 0.35 which is also a bit large.

Exercise 2.1 * *Use (2.36) to compute K_u and P_u for the process in (2.31).*

In summary, for this example, the Ziegler–Nichols PI tunings are somewhat "aggressive" and give a closed-loop system with smaller stability margins and a more oscillatory response than would normally be regarded as acceptable. For disturbance rejection the controller settings may be more reasonable, and one can add a prefilter to improve the response for reference tracking, resulting in a two degrees-of-freedom controller. However, this will not change the stability robustness of the system.

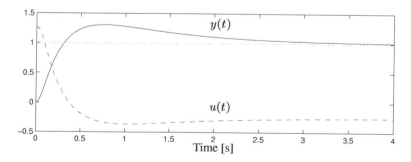

Figure 2.9: Closed-loop response to a unit step change in reference for the unstable process (2.37) with PI control

Example 2.5 PI control of unstable process. *Consider the <u>unstable</u> process,*

$$G(s) = \frac{4}{(s-1)(0.02s+1)^2} \tag{2.37}$$

Without control ($K = 0$), the output in response to any input change will eventually go out of bounds. To stabilize, we use a PI controller (2.34) with settings[3]

$$K_c = 1.25, \quad \tau_I = 1.5 \tag{2.38}$$

The resulting stable closed-loop response to a step change in the reference is shown in Figure 2.9. The response is not oscillatory and the selected tunings are robust with a large gain margin of 18.7 (see Section 2.4.3). The output $y(t)$ has some overshoot (about 30%), which is unavoidable for an unstable process. We note with interest that the input $u(t)$ starts out positive, but that the final steady-state value is negative. That is, the input has an inverse response. This is expected for an unstable process, since the transfer function KS (from the plant output to the plant input) must have a RHP-zero, see page 146.

2.4.2   Time domain performance

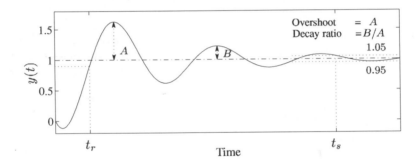

Figure 2.10: Characteristics of closed-loop response to step in reference

Step response analysis. The above examples illustrate the approach often taken by engineers when evaluating the performance of a control system. That is, one simulates the response to a step in the reference input, and considers the following characteristics (see Figure 2.10):

- *Rise time (t_r):* the time it takes for the output to first reach 90% of its final value, which is usually required to be small.
- *Settling time (t_s):* the time after which the output remains within $\pm\epsilon\%$ of its final value (typically $\epsilon = 5$), which is usually required to be small.
- *Overshoot:* the peak value divided by the final value, which should typically be 1.2 (20%) or less.
- *Decay ratio:* the ratio of the second and first peaks, which should typically be 0.3 or less.
- *Steady-state offset:* the difference between the final value and the desired final value, which is usually required to be small.

---

[3] The PI controller for this unstable process is almost identical to the H-infinity ($\mathcal{H}_\infty$) S/KS controller obtained using the weights $w_u = 1$ and $w_P = 1/M + \omega_B^*/s$ with $M = 1.5$ and $\omega_B^* = 10$ in (2.112) and (2.113); see Exercise 2.5 (page 65)

The rise time and settling time are measures of the *speed of the response*, whereas the overshoot, decay ratio and steady-state offset are related to the *quality of the response*. Another measure of the quality of the response is:

- *Total variation (TV):* the total up and down movement of the signal (input or output), which should be as small as possible. The computation of total variation is illustrated in Figure 2.11. In Matlab, TV = sum(abs(diff(y))).

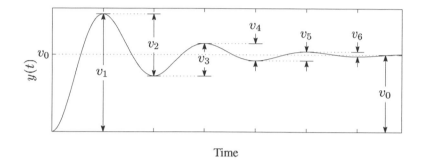

Figure 2.11: Total variation is TV $= \sum_{i=1}^{\infty} |v_i|$

The above measures address the output response, $y(t)$. In addition, one should consider the magnitude of the manipulated input (control signal, u), which usually should be as small and smooth as possible. One measure of "smoothness" is to have a small total variation. Note that attempting to reduce the total variation of the input signal is equivalent to adding a penalty on input movement, as is commonly done when using model predictive control (MPC). If there are important disturbances, then the response to these should also be considered. Finally, one may investigate in simulation how the controller works if the plant model parameters are different from their nominal values.

Remark 1 Another way of quantifying time domain performance is in terms of some norm of the error signal $e(t) = y(t) - r(t)$. For example, one might use the integral squared error (ISE), or its square root which is the 2-norm of the error signal, $\|e(t)\|_2 = \sqrt{\int_0^\infty |e(\tau)|^2 d\tau}$. In this way, the various objectives related to both the speed and quality of response are combined into one number. Actually, in most cases minimizing the 2-norm seems to give a reasonable trade-off between the various objectives listed above. Another advantage of the 2-norm is that the resulting optimization problems (such as minimizing ISE) are numerically easy to solve. One can also take input magnitudes into account by considering, for example, $J = \sqrt{\int_0^\infty (Q|e(t)|^2 + R|u(t)|^2) dt}$ where Q and R are positive constants. This is similar to linear quadratic (LQ) optimal control, but in LQ control one normally considers an impulse rather than a step change in $r(t)$.

Remark 2 The step response is equal to the integral of the corresponding impulse response, e.g. set $u(\tau) = 1$ in (4.11). Some thought then reveals that one can compute the total variation as the integrated absolute area (1-norm) of the corresponding impulse response (Boyd and Barratt, 1991, p. 98). That is, let $y = Tr$, then the total variation in y for a step change in r is

$$\text{TV} = \int_0^\infty |g_T(\tau)| d\tau \triangleq \|g_T(t)\|_1 \qquad (2.39)$$

where $g_T(t)$ is the impulse response of T, i.e. $y(t)$ resulting from an impulse change in $r(t)$.

2.4.3 Frequency domain performance

The frequency response of the loop transfer function, $L(j\omega)$, or of various closed-loop transfer functions, may also be used to characterize closed-loop performance. Typical Bode plots of L, T and S are shown in Figure 2.14. One advantage of the frequency domain compared to a step response analysis is that it considers a broader class of signals (sinusoids of any frequency). This makes it easier to characterize feedback properties, and in particular system behaviour in the crossover (bandwidth) region. We will now describe some of the important frequency domain measures used to assess performance, e.g. gain and phase margins, the maximum peaks of S and T, and the various definitions of crossover and bandwidth frequencies used to characterize speed of response.

Gain and phase margins

Let $L(s)$ denote the loop transfer function of a system which is closed-loop stable under negative feedback. A typical Bode plot and a typical Nyquist plot of $L(j\omega)$ illustrating the gain margin (GM) and phase margin (PM) are given in Figures 2.12 and 2.13, respectively. From Nyquist's stability condition, the closeness of the curve $L(j\omega)$ to the point -1 in the complex plane is a good measure of how close a stable closed-loop system is to instability. We see from Figure 2.13 that GM measures the closeness of $L(j\omega)$ to -1 along the real axis, whereas PM is a measure along the unit circle.

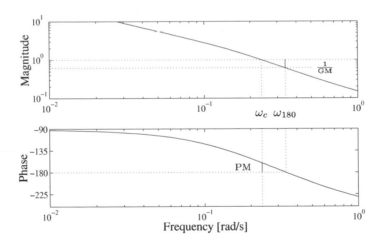

Figure 2.12: Typical Bode plot of $L(j\omega)$ with PM and GM indicated

More precisely, if the Nyquist plot of L crosses the negative real axis between -1 and 0, then the (upper) *gain margin* is defined as

$$\text{GM} = 1/|L(j\omega_{180})| \tag{2.40}$$

where the *phase crossover frequency* ω_{180} is where the Nyquist curve of $L(j\omega)$ crosses the negative real axis between -1 and 0, i.e.

$$\angle L(j\omega_{180}) = -180° \tag{2.41}$$

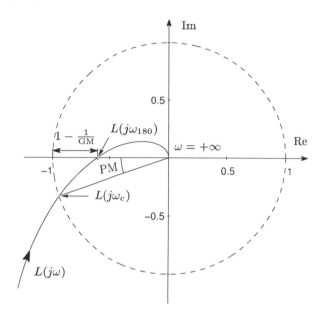

Figure 2.13: Typical Nyquist plot of $L(j\omega)$ for stable plant with PM and GM indicated. Closed-loop instability occurs if $L(j\omega)$ encircles the critical point -1.

If there is more than one such crossing between -1 and 0, then we take the closest crossing to -1, corresponding to the largest value of $|L(j\omega_{180})|$. For some systems, e.g. for low-order minimum-phase plants, there is no such crossing and GM $= \infty$. The GM is the factor by which the loop gain $|L(j\omega)|$ may be increased before the closed-loop system becomes unstable. The GM is thus a direct safeguard against steady-state gain uncertainty (error). Typically, we require GM > 2. On a Bode plot with a logarithmic axis for $|L|$, we have that GM is the vertical distance (in dB) from the unit magnitude line down to $|L(j\omega_{180})|$, see Figure 2.12. Note that $20\log_{10}$ GM is the GM in dB.

In some cases, e.g. for an unstable plant, the Nyquist plot of L crosses the negative real axis between $-\infty$ and -1, and a *lower gain margin* (or gain reduction margin) can be similarly defined,

$$\text{GM}_L = 1/|L(j\omega_{L180})| \tag{2.42}$$

where ω_{L180} is the frequency where the Nyquist curve of $L(j\omega)$ crosses the negative real axis between $-\infty$ and -1. If there is more than one such crossing, then we take the closest crossing to -1, corresponding to the smallest value of $|L(j\omega_{180})|$. For many systems, e.g. for most stable plants, there is no such crossing and $\text{GM}_L = 0$. The value of GM_L is the factor by which the loop gain $|L(j\omega)|$ may be *decreased* before the closed-loop system becomes unstable.

The *phase margin* is defined as

$$\text{PM} = \angle L(j\omega_c) + 180° \tag{2.43}$$

where the *gain crossover frequency* ω_c is the frequency where $|L(j\omega)|$ crosses 1, i.e.

$$|L(j\omega_c)| = 1 \tag{2.44}$$

If there is more than one such crossing, then the one giving the smallest value of PM is taken. The PM tells us how much negative phase (phase lag) we can add to $L(s)$ at frequency ω_c before the phase at this frequency becomes $-180°$ which corresponds to closed-loop instability (see Figure 2.13). Typically, we require PM larger than $30°$ or more. The PM is a direct safeguard against time delay uncertainty; the system becomes unstable if we add a time delay of

$$\theta_{\max} = \text{PM}/\omega_c \tag{2.45}$$

Note that the units must be consistent, and so if ω_c is in [rad/s] then PM must be in radians. It is also important to note that by decreasing the value of ω_c (lowering the closed-loop bandwidth, resulting in a slower response) the system can tolerate larger time delay errors.

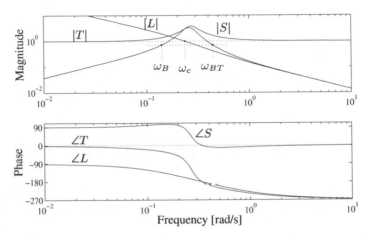

Figure 2.14: Bode magnitude and phase plots of $L = GK$, S and T when $G(s) = \frac{3(-2s+1)}{(10s+1)(5s+1)}$, and $K(s) = 1.136(1 + \frac{1}{12.7s})$ (a Ziegler–Nichols PI controller)

Example 2.4 (page 28) continued. *For the PI-controlled inverse response process example, the corresponding Bode plots for L, S and T are shown in Figure 2.14. From the plot of $L(j\omega)$, we find that the phase margin (PM) is $19.4°$ (0.34 rad), the gain margin (GM) is 1.63 and ω_c is 0.236 rad/s. The allowed time delay error is then $\theta_{\max} = 0.34$ rad/0.236 rad/s $= 1.44$ s. These margins are too small according to common rules of thumb. As defined later in the text, the peak value of $|S|$ is $M_S = 3.92$, and the peak value of $|T|$ is $M_T = 3.35$, which again are high according to normal design rules.*

Example 2.5 (page 30) continued. *The Bode plots of L, S and T for the PI-controlled unstable process are shown in Figure 2.15. The gain margin (GM), lower gain margin (GM_L), phase margin (PM) and peak values of S (M_S) and T (M_T) are*

$$GM = 18.7, GM_L = 0.21, PM = 59.5°, M_S = 1.19, M_T = 1.38$$

In this case, the phase of $L(j\omega)$ crosses $-180°$ twice. First, $\angle L$ crosses $-180°$ at a low frequency (ω about 0.9) where $|L|$ is about 4.8, and we have that the lower gain margin is $GM_L = 1/4.8 = 0.21$. Second, $\angle L$ crosses $-180°$ at a high frequency (ω about 40) where $|L|$ is about 0.054, and we have that the (upper) gain margin is GM= $1/0.054 = 18.7$. Thus, instability is induced by decreasing the loop gain by a factor 4.8 or increasing it by a factor 18.7.

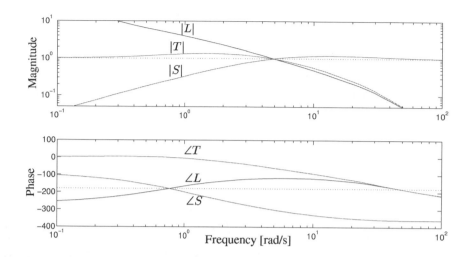

Figure 2.15: Bode magnitude and phase plots of $L = GK$, S and T for PI control of unstable process, $G(s) = \frac{4}{(s-1)(0.02s+1)^2}$, $K(s) = 1.25(1 + \frac{1}{1.5s})$

Exercise 2.2 *Prove that the maximum additional delay for which closed-loop stability is maintained is given by (2.45).*

Exercise 2.3 * *Derive the approximation for $K_u = 1/|G(j\omega_u)|$ given in (5.96) for a first-order delay system.*

Stability margins are measures of how close a stable closed-loop system is to instability. From the above arguments we see that the GM and PM provide stability margins for gain and delay uncertainty. More generally, to maintain closed-loop stability, the Nyquist stability condition tells us that the number of encirclements of the critical point -1 by $L(j\omega)$ must not change. Thus, the closeness of the frequency response $L(j\omega)$ to the critical point -1 is a good measure of closeness to instability. The GMs represent the closeness along the negative real axis, and the PM along the unit circle. As discussed next, the actual closest distance is equal to $1/M_S$, where M_S is the peak value of the sensitivity $|S(j\omega)|$. As expected, the GM and PM are closely related to M_S, and since $|S|$ is also a measure of performance, they are therefore also useful in terms of *performance*. In summary, specifications on the GM and PM (e.g. GM > 2 and PM > 30°) are used to provide the appropriate trade-off between performance and stability robustness.

Maximum peak criteria

The maximum peaks of the sensitivity and complementary sensitivity functions are defined as

$$M_S = \max_{\omega} |S(j\omega)|; \quad M_T = \max_{\omega} |T(j\omega)| \tag{2.46}$$

(Note that $M_S = \|S\|_\infty$ and $M_T = \|T\|_\infty$ in terms of the $\mathcal{H}_\infty$ norm introduced later.) Since $S + T = 1$, using (A.51), it follows that at any frequency

$$\big|\, |S| - |T| \,\big| \leq |S + T| = 1$$

so M_S and M_T differ at most by 1. A large value of M_S therefore occurs if and only if M_T is large. For a stable plant, the peak value for $|S|$ is usually higher than for $|T|$ ($M_S > M_T$) and occurs at a higher frequency (see Figure 2.14). For unstable plants, M_T is usually larger than M_S (see Figure 2.15). Note that these are not general rules.

Typically, it is required that M_S is less than about 2 (6 dB) and M_T is less than about 1.25 (2 dB). A large value of M_S or M_T (larger than about 4) indicates poor performance as well as poor robustness. An upper bound on M_T has been a common design specification in classical control and the reader may be familiar with the use of M-circles on a Nyquist plot or a Nichols chart used to determine M_T from $L(j\omega)$.

We now give some justification for why we may want to bound the value of M_S. Without control ($u = 0$), we have $e = y - r = G_d d - r$, and with feedback control $e = S(G_d d - r)$. Thus, feedback control improves performance in terms of reducing $|e|$ at all frequencies where $|S| < 1$. Usually, $|S|$ is small at low frequencies: for example, $|S(0)| = 0$ for systems with integral action. But because all real systems are strictly proper we must at high frequencies have that $L \to 0$ or equivalently $S \to 1$. At intermediate frequencies one cannot avoid in practice a peak value, M_S, larger than 1 (e.g. see the remark below). Thus, there is an intermediate frequency range where feedback control degrades performance, and the value of M_S is a measure of the worst-case performance degradation. One may also view M_S as a robustness measure. To maintain closed-loop stability, we want $L(j\omega)$ to stay away from the critical -1 point. The smallest distance between $L(j\omega)$ and -1 is M_S^{-1}, and therefore for robustness, the smaller M_S, the better. In summary, both for stability and performance we want M_S close to 1.

There is a close relationship between these maximum peaks and the GM and PM. Specifically, for a given M_S we are guaranteed

$$\text{GM} \geq \frac{M_S}{M_S - 1}; \qquad \text{PM} \geq 2\arcsin\left(\frac{1}{2M_S}\right) \geq \frac{1}{M_S} \text{ [rad]} \qquad (2.47)$$

For example, with $M_S = 2$ we are guaranteed $\text{GM} \geq 2$ and $\text{PM} \geq 29.0°$. Similarly, for a given value of M_T we are guaranteed

$$\text{GM} \geq 1 + \frac{1}{M_T}; \qquad \text{PM} \geq 2\arcsin\left(\frac{1}{2M_T}\right) \geq \frac{1}{M_T} \text{ [rad]} \qquad (2.48)$$

and specifically with $M_T = 2$ we have $\text{GM} \geq 1.5$ and $\text{PM} \geq 29.0°$.

Proof of (2.47) and (2.48): To derive the GM inequalities notice that $L(j\omega_{180}) = -1/\text{GM}$ (since $\text{GM} = 1/|L(j\omega_{180})|$ and L is real and negative at ω_{180}), from which we get

$$T(j\omega_{180}) = \frac{-1}{\text{GM} - 1}; \quad S(j\omega_{180}) = \frac{1}{1 - \frac{1}{\text{GM}}} \qquad (2.49)$$

and the GM results follow. To derive the PM inequalities in (2.47) and (2.48) consider Figure 2.16 where we have $|S(j\omega_c)| = 1/|1 + L(j\omega_c)| = 1/|-1 - L(j\omega_c)|$ and we obtain

$$|S(j\omega_c)| = |T(j\omega_c)| = \frac{1}{2\sin(\text{PM}/2)} \qquad (2.50)$$

and the inequalities follow. Alternative formulae, which are sometimes used, follow from the identity $2\sin(\text{PM}/2) = \sqrt{2(1 - \cos(\text{PM}))}$. □

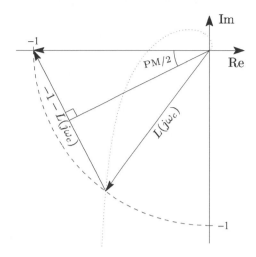

Figure 2.16: Nyquist plot of vector $L(j\omega)$. At frequency ω_c we see that $|1 + L(j\omega_c)| = 2\sin(\mathrm{PM}/2)$.

Remark. Since GM > 1, we note with interest that (2.49) requires $|S|$ to be larger than 1 at frequency ω_{180}. This means that provided ω_{180} exists, i.e. $L(j\omega)$ has more than $-180°$ phase lag at some frequency (which is the case for any real system), then the peak of $|S(j\omega)|$ must exceed 1.

In conclusion, we see that specifications on the peaks of $|S(j\omega)|$ or $|T(j\omega)|$ (M_S or M_T) can make specifications on the GM and PM unnecessary. For instance, requiring $M_S < 2$ implies the common rules of thumb GM > 2 and PM > 30°.

2.4.4 Relationship between time and frequency domain peaks

For a change in reference r, the output is $y(s) = T(s)r(s)$. Is there any relationship between the frequency domain peak of $T(j\omega)$, M_T, and any characteristic of the time domain step response, e.g. the overshoot or the total variation? To answer this consider a prototype second-order system with complementary sensitivity function

$$T(s) = \frac{1}{\tau^2 s^2 + 2\tau\zeta s + 1} \tag{2.51}$$

For underdamped systems with $\zeta < 1$ the poles are complex and yield oscillatory step responses. With $r(t) = 1$ (a unit step change) the values of the overshoot and total variation for $y(t)$ are given, together with M_T and M_S, as a function of ζ in Table 2.1. From Table 2.1, we see that the total variation TV correlates quite well with M_T. This is further confirmed by (A.137) and (2.39) which together yield the following general bounds:

$$M_T \leq \mathrm{TV} \leq (2n + 1)M_T \tag{2.52}$$

Here n is the order of $T(s)$, which is 2 for our prototype system in (2.51). Given that the response of many systems can be crudely approximated by fairly low-order systems, the bound in (2.52) suggests that M_T may provide a reasonable approximation to the total variation. This provides some justification for the use of M_T in classical control to evaluate the quality of the response.

Table 2.1: Step response characteristics and frequency peaks of prototype second-order system (2.51), see also Table 2.2

ζ	Time domain, $y(t)$		Frequency domain	
	Overshoot	Total variation	M_T	M_S
2.0	1	1	1	1.05
1.5	1	1	1	1.08
1.0	1	1	1	1.15
0.8	1.02	1.03	1	1.22
0.6	1.09	1.21	1.04	1.35
0.4	1.25	1.68	1.36	1.66
0.2	1.53	3.22	2.55	2.73
0.1	1.73	6.39	5.03	5.12
0.01	1.97	63.7	50.0	50.0

Table 2.2: Matlab program to generate Table 2.1

```
% Uses the Control toolbox
tau=1;zeta=0.1;t=0:0.01:100;
T = tf(1,[tau*tau 2*tau*zeta 1]); S = 1-T;
[A,B,C,D]=ssdata(T); y = step(A,B,C,D,1,t);
overshoot=max(y),tv=sum(abs(diff(y)))
Mt=norm(T,inf,1e-4),Ms=norm(S,inf,1e-4)
```

2.4.5   Bandwidth and crossover frequency

The concept of bandwidth is very important in understanding the benefits and trade-offs involved when applying feedback control. Above we considered peaks of closed-loop transfer functions, M_S and M_T, which are related to the quality of the response. However, for performance we must also consider the speed of the response, and this leads to considering the bandwidth frequency of the system. In general, a large bandwidth corresponds to a smaller rise time, since high-frequency signals are more easily passed on to the outputs. A high bandwidth also indicates a system which is sensitive to noise. Conversely, if the bandwidth is small, the time response will generally be slow, and the system will usually be more robust.

Loosely speaking, *bandwidth* may be defined as the frequency range $[\omega_1, \omega_2]$ over which control is effective. In most cases we require tight control at steady-state so $\omega_1 = 0$, and we then simply call $\omega_2 = \omega_B$ the bandwidth.

The word "effective" may be interpreted in different ways, and this may give rise to different definitions of bandwidth. The interpretation we use is that control is *effective* if we obtain some *benefit* in terms of performance. For tracking performance the error is $e = y - r = -Sr$ and we get that feedback is effective (in terms of improving performance) as long as the relative error $|e|/|r| = |S|$ is reasonably small, which we may define to be $|S| \leq 0.707.$[4] We then get the following definition:

Definition 2.1 *The (closed-loop) bandwidth, ω_B, is the frequency where $|S(j\omega)|$ first crosses $1/\sqrt{2} = 0.707 \; (\approx -3 \; dB)$ from below.*

Remark. Another interpretation is to say that control is *effective* if it significantly *changes* the output

---

[4] The reason for choosing the value 0.707 when defining the bandwidth ω_B is that, for the simple case of a first-order closed-loop response with $S = s/(s + a)$, the low-frequency asymptote s/a of S crosses magnitude 1 at frequency $\omega = a$, and at this frequency $|S(j\omega)| = 1/\sqrt{2} = 0.707$.

response. For tracking performance, the output is $y = Tr$ and since without control $T = 0$, we may say that control is effective as long as T is reasonably large, which we may define to be larger than 0.707. This leads to an alternative definition which has been traditionally used to define the bandwidth of a control system: *The bandwidth in terms of T, ω_{BT}, is the highest frequency at which $|T(j\omega)|$ crosses $1/\sqrt{2} = 0.707 \ (\approx -3 \ dB)$ from above.* However, we would argue that this alternative definition, although being closer to how the term is used in some other fields, is less useful for feedback control.

The *gain crossover frequency*, ω_c, defined as the frequency where $|L(j\omega_c)|$ first crosses 1 from above, is also sometimes used to define closed-loop bandwidth. It has the advantage of being simple to compute and usually gives a value between ω_B and ω_{BT}. Specifically, for systems with PM < 90° (most practical systems) we have

$$\omega_B < \omega_c < \omega_{BT} \tag{2.53}$$

Proof of (2.53): Note that $|L(j\omega_c)| = 1$ so $|S(j\omega_c)| = |T(j\omega_c)|$. Thus, when PM $= 90°$ we get $|S(j\omega_c)| = |T(j\omega_c)| = 0.707$ (see (2.50)), and we have $\omega_B = \omega_c = \omega_{BT}$. For PM < 90° we get $|S(j\omega_c)| = |T(j\omega_c)| > 0.707$, and since ω_B is the frequency where $|S(j\omega)|$ crosses 0.707 from below we must have $\omega_B < \omega_c$. Similarly, since ω_{BT} is the frequency where $|T(j\omega)|$ crosses 0.707 from above, we must have $\omega_{BT} > \omega_c$. □

From this we have that the situation is generally as follows: Up to the frequency ω_B, $|S|$ is less than 0.7 and control is effective in terms of improving performance. In the frequency range $[\omega_B, \omega_{BT}]$ control still affects the response, but does not improve performance – in most cases we find that in this frequency range $|S|$ is larger than 1 and control degrades performance. Finally, at frequencies higher than ω_{BT} we have $S \approx 1$ and control has no significant effect on the response. The situation just described is illustrated in Example 2.7 below (see Figure 2.18).

Example 2.4 (pages 28 and 34) continued. *The plant $G(s) = \frac{3(-2s+1)}{(10s+1)(5s+1)}$ has a RHP-zero and the Ziegler–Nichols PI tunings ($K_c = 1.14$, $\tau_I = 12.7$) are quite aggressive with GM = 1.63 and PM = 19.4°. The bandwidth and crossover frequencies are $\omega_B = 0.14$, $\omega_c = 0.24$ and $\omega_{BT} = 0.44$, which is in agreement with (2.53).*

Example 2.6 *Consider the simple case of a first-order closed-loop system,*

$$L(s) = \frac{k}{s}, \quad S(s) = \frac{s}{s+k}; \quad T(s) = \frac{k}{s+k}$$

In this ideal case, all bandwidth and crossover frequencies are identical: $\omega_c = \omega_B = \omega_{BT} = k$. Furthermore, the phase of L remains constant at $-90°$, so PM $= 90°$, $\omega_{180} = \infty$ (or really undefined) and GM $= \infty$.

Example 2.7 Comparison of ω_B and ω_{BT} as indicators of performance. *An example where ω_{BT} is a poor indicator of performance is the following (we are not suggesting this as a good controller design!):*

$$L = \frac{-s+z}{s(\tau s + \tau z + 2)}; \quad T = \frac{-s+z}{s+z}\frac{1}{\tau s + 1}; \quad z = 0.1, \ \tau = 1 \tag{2.54}$$

For this system, both L and T have a RHP-zero at $z = 0.1$, and we have GM $= 2.1$, $PM = 60.1°$, $M_S = 1.93$ and $M_T = 1$. We find that $\omega_B = 0.036$ and $\omega_c = 0.054$ are both less than $z = 0.1$ (as one should expect because speed of response is limited by the presence of RHP-zeros), whereas $\omega_{BT} = 1/\tau = 1.0$ is ten times larger than z. The closed-loop response to a unit step change in the reference is shown in Figure 2.17. The rise time is 31.0 s, which is close to $1/\omega_B = 28.0$ s, but very

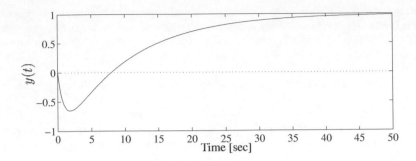

Figure 2.17: Step response for system $T = \frac{-s+0.1}{s+0.1}\frac{1}{s+1}$

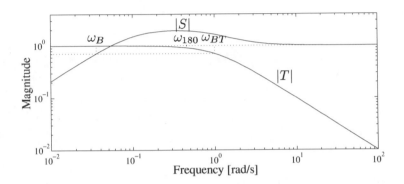

Figure 2.18: Plots of $|S|$ and $|T|$ for system $T = \frac{-s+0.1}{s+0.1}\frac{1}{s+1}$

different from $1/\omega_{BT} = 1.0$ s, illustrating that ω_B is a better indicator of closed-loop performance than ω_{BT}.

The magnitude Bode plots of S and T are shown in Figure 2.18. We see that $|T| \approx 1$ up to about ω_{BT}. However, in the frequency range from ω_B to ω_{BT} the phase of T (not shown) drops from about $-40°$ to about $-220°$, so in practice tracking is out of phase and control is poor in this frequency range.

In conclusion, ω_B (which is defined in terms of $|S|$) and ω_c (in terms of $|L|$) are good indicators of closed-loop performance, while ω_{BT} (in terms of $|T|$) may be misleading in some cases. The reason is that we want $T \approx 1$ in order to have good performance, and it is not sufficient that $|T| \approx 1$; we must also consider its phase. On the other hand, for good performance we want S close to 0, and this will be the case if $|S| \approx 0$ irrespective of the phase of S.

2.5   Controller design

Many methods exist for controller design and some of these will be discussed in Chapter 9. In addition to heuristic rules and on-line tuning we can distinguish between three main approaches to controller design:

1. **Shaping of transfer functions.** In this approach the designer specifies the *magnitude* of some transfer function(s) as a function of frequency, and then finds a controller which gives the desired shape(s).

   (a) **Loop shaping.** This is the classical approach in which the magnitude of the open-loop transfer function, $L(j\omega)$, is shaped. Usually no optimization is involved and the designer aims to obtain $|L(j\omega)|$ with desired bandwidth, slopes, etc. We will look at this approach in detail later in Section 2.6. However, classical loop shaping is difficult to apply for complicated systems, and one may then instead use the Glover–McFarlane $\mathcal{H}_\infty$ loop-shaping design presented in Chapter 9. The method consists of a second step where optimization is used to make an initial loop-shaping design more robust.

   (b) **Shaping of closed-loop transfer functions, such as S, T and KS.** One analytical approach is the internal model control (IMC) design procedure, where one aims to specify directly $T(s)$. This works well for simple systems and is discussed in Section 2.7. However, optimization is more generally used, resulting in various $\mathcal{H}_\infty$ optimal control problems such as mixed weighted sensitivity; see Section 2.8 and later chapters.

2. **The signal-based approach.** This involves time domain problem formulations resulting in the minimization of a norm of a transfer function. Here one considers a particular disturbance or reference change and then one tries to optimize the closed-loop response. The "modern" state-space methods from the 1960's, such as linear quadratic Gaussian (LQG) control, are based on this signal-oriented approach. In LQG the input signals are assumed to be stochastic (or alternatively impulses in a deterministic setting) and the expected value of the output variance (or the 2-norm) is minimized. These methods may be generalized to include frequency dependent weights on the signals leading to what is called the Wiener–Hopf (or $\mathcal{H}_2$ norm) design method.

   By considering sinusoidal signals, frequency by frequency, a signal-based $\mathcal{H}_\infty$ optimal control methodology can be derived in which the $\mathcal{H}_\infty$ norm of a combination of closed-loop transfer functions is minimized. This approach has attracted significant interest, and may be combined with model uncertainty representations to yield quite complex robust performance problems requiring μ-synthesis, an important topic which will be addressed in later chapters.

   In approaches 1 and 2, the overall design process is iterative between controller design and performance (or cost) evaluation. If performance is not satisfactory then one must either adjust the controller parameters directly (e.g. by reducing K_c from the value obtained by the Ziegler–Nichols rules) or adjust some weighting factor in the objective function used to synthesize the controller.

3. **Numerical optimization.** This often involves attempts to optimize directly the true objectives, such as minimizing the rise time subject to satisfying given values for the stability margins, etc. Computationally, such optimization problems may be difficult to solve, especially if one does not have convexity in the controller parameters. Also, by effectively including performance evaluation and controller design in a single-step procedure, the problem formulation is far more critical than in iterative two-step approaches.

The above *off-line* methods are used to precompute a feedback controller which is later implemented on the plant. This is the main focus of this book. In addition, there exist computational methods where the optimization problem is solved *on-line*. These methods are well suited for certain nonlinear problems where an explicit feedback controller does not exist or is difficult to obtain. For example, in the process industry, *model predictive control* is used to handle problems with constraints on the inputs and outputs. On-line optimization approaches are expected to become more popular in the future as faster computers and more efficient and reliable computational algorithms are developed.

2.6   Loop shaping

In the classical loop-shaping approach to feedback controller design, "loop shape" refers to the magnitude of the loop transfer function $L = GK$ as a function of frequency. An understanding of how K can be selected to shape this loop gain provides invaluable insight into the multivariable techniques and concepts which will be presented later in the book, and so we will discuss loop shaping in some detail in this section.

2.6.1   Trade-offs in terms of L

Recall (2.20), which yields the closed-loop response in terms of the control error $e = y - r$:

$$e = -\underbrace{(I+L)^{-1}}_{S}r + \underbrace{(I+L)^{-1}}_{S}G_d d - \underbrace{(I+L)^{-1}L}_{T}n \qquad (2.55)$$

For "perfect control" we want $e = y - r = 0$; that is, we would like

$$e \approx 0 \cdot d + 0 \cdot r + 0 \cdot n$$

The first two requirements in this equation, namely disturbance rejection and command tracking, are obtained with $S \approx 0$, or equivalently, $T \approx I$. Since $S = (I+L)^{-1}$, this implies that the loop transfer function L must be large in magnitude. On the other hand, the requirement for zero noise transmission implies that $T \approx 0$, or equivalently, $S \approx I$, which is obtained with $L \approx 0$. This illustrates the fundamental nature of feedback design which always involves a trade-off between conflicting objectives; in this case between large loop gains for disturbance rejection and tracking, and small loop gains to reduce the effect of noise.

It is also important to consider the magnitude of the control action u (which is the input to the plant). We want u small because this causes less wear and saves input energy, and also because u is often a disturbance to other parts of the system (e.g. consider opening a window in your office to adjust your comfort and the undesirable disturbance this will impose on the air conditioning system for the building). In particular, we usually want to avoid fast changes in u. The control action is given by $u = K(r - y_m)$ and we find as expected that a small u corresponds to small controller gains and a small $L = GK$.

The most important design objectives which necessitate trade-offs in feedback control are summarized below:

1. Performance, good disturbance rejection: needs large controller gains, i.e. L large.
2. Performance, good command following: L large.
3. Stabilization of unstable plant: L large.
4. Mitigation of measurement noise on plant outputs: L small.
5. Small magnitude of input signals: K small and L small.

6. Physical system must be strictly proper: L has to approach 0 at high frequencies.
7. Stability (stable plant): L small.

Fortunately, the conflicting design objectives mentioned above are generally in different frequency ranges, and we can meet most of the objectives by using a large loop gain ($|L| > 1$) at low frequencies below crossover, and a small gain ($|L| < 1$) at high frequencies above crossover.

2.6.2 Fundamentals of loop-shaping design

By *loop shaping* we mean a design procedure that involves explicitly shaping the magnitude of the loop transfer function, $|L(j\omega)|$. Here $L(s) = G(s)K(s)$ where $K(s)$ is the feedback controller to be designed and $G(s)$ is the product of all other transfer functions around the loop, including the plant, the actuator and the measurement device. Essentially, to get the benefits of feedback control we want the loop gain, $|L(j\omega)|$, to be as large as possible within the bandwidth region. However, due to time delays, RHP-zeros, unmodelled high-frequency dynamics and limitations on the allowed manipulated inputs, the loop gain has to drop below 1 at and above some frequency which we call the crossover frequency ω_c. Thus, disregarding stability for the moment, it is desirable that $|L(j\omega)|$ falls sharply with frequency. To measure how $|L|$ falls with frequency we consider the logarithmic slope $N = d\ln|L|/d\ln\omega$. For example, a slope $N = -1$ implies that $|L|$ drops by a factor of 10 when ω increases by a factor of 10. If the gain is measured in decibels (dB) then a slope of $N = -1$ corresponds to -20 dB/ decade. The value of $-N$ at high frequencies is often called the *roll-off* rate.

The design of $L(s)$ is most crucial and difficult in the crossover region between ω_c (where $|L| = 1$) and ω_{180} (where $\angle L = -180°$). For stability, we at least need the loop gain to be less than 1 at frequency ω_{180}, i.e. $|L(j\omega_{180})| < 1$. Thus, to get a high bandwidth (fast response) we want ω_c and therefore ω_{180} large; that is, we want the phase lag in L to be small. Unfortunately, this is not consistent with the desire that $|L(j\omega)|$ should fall sharply. For example, the loop transfer function $L = 1/s^n$ (which has a slope $N = -n$ on a log–log plot) has a phase $\angle L = -n \cdot 90°$. Thus, to have a PM of 45° we need $\angle L > -135°$, and the slope of $|L|$ cannot exceed $N = -1.5$.

In addition, if the slope is made steeper at lower or higher frequencies, then this will add unwanted phase lag at intermediate frequencies. As an example, consider $L_1(s)$ given in (2.14) with the Bode plot shown in Figure 2.3 on page 20. Here the slope of the asymptote of $|L|$ is -1 at the gain crossover frequency (where $|L_1(j\omega_c)| = 1$), which by itself gives $-90°$ phase lag. However, due to the influence of the steeper slopes of -2 at lower and higher frequencies, there is a "penalty" of about $-35°$ at crossover, so the actual phase of L_1 at ω_c is approximately $-125°$.

The situation becomes even worse for cases with delays or RHP-zeros in $L(s)$ which add undesirable phase lag to L without contributing to a desirable negative slope in L. At the gain crossover frequency ω_c, the additional phase lag from delays and RHP-zeros may in practice be $-30°$ or more.

In summary, a desired loop shape for $|L(j\omega)|$ typically has a slope of about -1 in the crossover region, and a slope of -2 or higher beyond this frequency; that is, the roll-off is 2 or larger. Also, with a proper controller, which is required for any real system, we must have that $L = GK$ rolls off at least as fast as G. At low frequencies, the desired shape of $|L|$ depends on what disturbances and references we are designing for. For example, if we are considering step changes in the references or disturbances which affect the outputs as steps, then a slope for $|L|$ of -1 at low frequencies is acceptable. If the references or disturbances require the outputs to change in a ramp-like fashion then a slope of -2 is required. In practice, integrators are included in the controller to get the desired low-frequency performance, and for offset-free reference tracking the rule is that

- $L(s)$ *must contain at least one integrator for each integrator in* $r(s)$.

Proof: Let $L(s) = \widehat{L}(s)/s^{n_I}$ where $\widehat{L}(0)$ is non-zero and finite and n_I is the number of integrators in $L(s)$ – sometimes n_I is called the *system type*. Consider a reference signal of the form $r(s) = 1/s^{n_r}$. For example, if $r(t)$ is a unit step, then $r(s) = 1/s$ ($n_r = 1$), and if $r(t)$ is a ramp then $r(s) = 1/s^2$ ($n_r = 2$). The final value theorem for Laplace transforms is

$$\lim_{t \to \infty} e(t) = \lim_{s \to 0} se(s) \tag{2.56}$$

In our case, the control error is

$$e(s) = -\frac{1}{1 + L(s)} r(s) = -\frac{s^{n_I - n_r}}{s^{n_I} + \widehat{L}(s)} \tag{2.57}$$

and to get zero offset (i.e. $e(t \to \infty) = 0$) we must from (2.56) require that $n_I \geq n_r$, and the rule follows.   □

In conclusion, one can define the desired loop transfer function in terms of the following specifications:

1. The gain crossover frequency, ω_c, where $|L(j\omega_c)| = 1$.
2. The shape of $L(j\omega)$, e.g. in terms of the slope of $|L(j\omega)|$ in certain frequency ranges. Typically, we desire a slope of about $N = -1$ around crossover, and a larger roll-off at higher frequencies. The desired slope at lower frequencies depends on the nature of the disturbance or reference signal.
3. The system type, defined as the number of pure integrators in $L(s)$.

In Section 2.6.4, we discuss how to specify the loop shape when disturbance rejection is the primary objective of control. Loop-shaping design is typically an iterative procedure where the designer shapes and reshapes $|L(j\omega)|$ after computing the PM and GM, the peaks of closed-loop frequency responses (M_T and M_S), selected closed-loop time responses, the magnitude of the input signal, etc. The procedure is illustrated next by an example.

Example 2.8 Loop-shaping design for the inverse response process. *We will now design a loop-shaping controller for the example process in (2.31) which has a RHP-zero at $s = 0.5$. The RHP-zero cannot be cancelled by the controller, because otherwise the system is internally unstable. Thus L must contain the RHP-zeros of G. In addition, the RHP-zero limits the achievable bandwidth and so the crossover region (defined as the frequencies between ω_c and ω_{180}) will be at about 0.5 rad/s. We require the system to have one integrator (type 1 system), and therefore a reasonable approach is to let the loop transfer function have a slope of -1 at low frequencies, and then to roll off with a higher slope at frequencies beyond 0.5 rad/s. The plant and our choice for the loop shape is*

$$G(s) = \frac{3(-2s + 1)}{(10s + 1)(5s + 1)}; \quad L(s) = 3K_c \frac{(-2s + 1)}{s(2s + 1)(0.33s + 1)} \tag{2.58}$$

The frequency response (Bode plots) of L is shown in Figure 2.19 for $K_c = 0.05$. The controller gain K_c was selected to get reasonable stability margins (PM and GM). The asymptotic slope of $|L|$ is -1 up to 3 rad/s where it changes to -2. The controller corresponding to the loop shape in (2.58) is

$$K(s) = K_c \frac{(10s + 1)(5s + 1)}{s(2s + 1)(0.33s + 1)}, \quad K_c = 0.05 \tag{2.59}$$

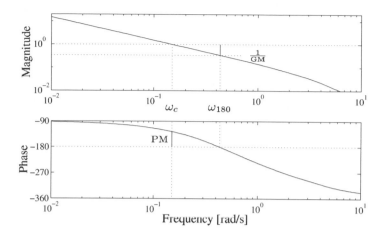

Figure 2.19: Frequency response of $L(s)$ in (2.58) for loop-shaping design with $K_c = 0.05$ ($GM = 2.92$, $PM = 54°$, $\omega_c = 0.15$, $\omega_{180} = 0.43$, $M_S = 1.75$, $M_T = 1.11$)

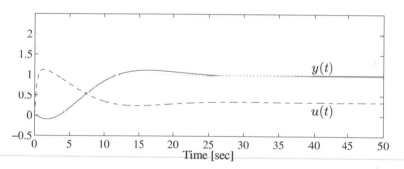

Figure 2.20: Response to step in reference for loop-shaping design (2.59)

The controller has zeros at the locations of the plant poles. This is desired in this case because we do not want the slope of the loop shape to drop at the break frequencies $1/10 = 0.1$ rad/s and $1/5 = 0.2$ rad/s just before crossover. The phase of L is $-90°$ at low frequency, and at $\omega = 0.5$ rad/s the additional contribution from the term $\frac{-2s+1}{2s+1}$ in (2.58) is $-90°$, so for stability we need $\omega_c < 0.5$ rad/s. The choice $K_c = 0.05$ yields $\omega_c = 0.15$ rad/s corresponding to $GM = 2.92$ and $PM = 54°$. The corresponding time response is shown in Figure 2.20. It is seen to be much better than the responses with either the simple PI controller in Figure 2.8 (page 28) or with the P controller in Figure 2.6 (page 26). Figure 2.20 also shows that the magnitude of the input signal remains less than 1 in magnitude most of the time. This means that the controller gain is not too large at high frequencies. The magnitude Bode plot for the controller (2.59) is shown in Figure 2.21. It is interesting to note that in the crossover region around $\omega = 0.5$ rad/s the controller gain is quite constant, around 1 in magnitude, which is similar to the "best" gain found using a P controller (see Figure 2.6).

Limitations imposed by RHP-zeros and time delays

Based on the above loop-shaping arguments we can now examine how the presence of delays and RHP-zeros limits the achievable control performance. We have already argued that if

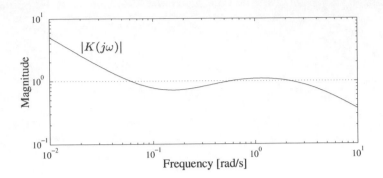

Figure 2.21: Magnitude Bode plot of controller (2.59) for loop-shaping design

we want the loop shape to have a slope of -1 around crossover (ω_c), with preferably a steeper slope before and after crossover, then the phase lag of L at ω_c will necessarily be at least $-90°$, even when there are no RHP-zeros or delays. Therefore, if we assume that for performance and robustness we want a PM of about $35°$ or more, then the additional phase contribution from any delays and RHP-zeros at frequency ω_c cannot exceed about $-55°$.

First consider a time delay θ. It yields an additional phase contribution of $-\theta\omega$, which at frequency $\omega = 1/\theta$ is -1 rad $= -57°$ (which is slightly more than $-55°$). Thus, for acceptable control performance we need $\omega_c < 1/\theta$, approximately.

Next consider a real RHP-zero at $s = z$. To avoid an increase in slope caused by this zero we place a pole at $s = -z$ such that the loop transfer function contains the term $\frac{-s+z}{s+z}$, the form of which is referred to as all-pass since its magnitude equals 1 at all frequencies. The phase contribution from the all-pass term at $\omega = z/2$ is $-2\arctan(0.5) = -53°$ (which is very close to $-55°$), so for acceptable control performance we need $\omega_c < z/2$, approximately.

2.6.3   Inverse-based controller design

In Example 2.8, we made sure that $L(s)$ contained the RHP-zero of $G(s)$, but otherwise the specified $L(s)$ was independent of $G(s)$. This suggests the following possible approach for a minimum-phase plant (i.e. one with no RHP-zeros or time delays). We select a loop shape which has a slope of -1 throughout the frequency range, namely

$$L(s) = \frac{\omega_c}{s} \tag{2.60}$$

where ω_c is the desired gain crossover frequency. This loop shape yields a PM of $90°$ and an infinite GM since the phase of $L(j\omega)$ never reaches $-180°$. The controller corresponding to (2.60) is

$$K(s) = \frac{\omega_c}{s}G^{-1}(s) \tag{2.61}$$

That is, the controller inverts the plant and adds an integrator ($1/s$). This is an old idea, and is also the essential part of the internal model control (IMC) design procedure (Morari and Zafiriou, 1989) (page 55), which has proved successful in many applications. However, there are at least three good reasons why this inverse-based controller may not be a good choice:

1. RHP-zeros or a time delay in $G(s)$ cannot be inverted.
2. The controller will not be realizable if $G(s)$ has a pole excess of two or larger, and may in any case yield large input signals. These problems may be partly fixed by adding high-frequency dynamics to the controller.
3. The loop shape resulting from (2.60) and (2.61) is *not* generally desirable, unless the references and disturbances affect the outputs as steps. This is illustrated by the following example.

Example 2.9 Disturbance process. *We now introduce our second SISO example control problem in which disturbance rejection is an important objective in addition to command tracking. We assume that the plant has been appropriately scaled as outlined in Section 1.4.*
   Problem formulation. *Consider the disturbance process described by*

$$G(s) = \frac{200}{10s+1}\frac{1}{(0.05s+1)^2}, \quad G_d(s) = \frac{100}{10s+1} \tag{2.62}$$

with time in seconds (a block diagram is shown in Figure 2.23 below). The control objectives are:

1. *Command tracking: The rise time (to reach 90% of the final value) should be less than 0.3 s and the overshoot should be less than 5%.*
2. *Disturbance rejection: The output in response to a unit step disturbance should remain within the range $[-1, 1]$ at all times, and it should return to 0 as quickly as possible ($|y(t)|$ should at least be less than 0.1 after 3 s).*
3. *Input constraints: $u(t)$ should remain within the range $[-1, 1]$ at all times to avoid input saturation (this is easily satisfied for most designs).*

   Analysis. *Since $G_d(0) = 100$ we have that without control the output response to a unit disturbance ($d = 1$) will be 100 times larger than what is deemed to be acceptable. The magnitude $|G_d(j\omega)|$ is lower at higher frequencies, but it remains larger than 1 up to $\omega_d \approx 10$ rad/s (where $|G_d(j\omega_d)| = 1$). Thus, feedback control is needed up to frequency ω_d, so we need ω_c to be approximately equal to 10 rad/s for disturbance rejection. On the other hand, we do not want ω_c to be larger than necessary because of sensitivity to noise and stability problems associated with high-gain feedback. We will thus aim at a design with $\omega_c \approx 10$ rad/s.*
   Inverse-based controller design. *We will consider the inverse-based design as given by (2.60) and (2.61) with $\omega_c = 10$. Since $G(s)$ has a pole excess of three this yields an unrealizable controller, and therefore we choose to approximate the plant term $(0.05s+1)^2$ by $(0.1s+1)$ and then in the controller we let this term be effective over one decade, i.e. we use $(0.1s+1)/(0.01s+1)$ to give the realizable design*

$$K_0(s) = \frac{\omega_c}{s}\frac{10s+1}{200}\frac{0.1s+1}{0.01s+1}, \quad L_0(s) = \frac{\omega_c}{s}\frac{0.1s+1}{(0.05s+1)^2(0.01s+1)}, \quad \omega_c = 10 \tag{2.63}$$

The response to a step reference is excellent as shown in Figure 2.22(a). The rise time is about 0.16 s and there is no overshoot so the specifications are more than satisfied. However, the response to a step disturbance (Figure 2.22(b)) is much too sluggish. Although the output stays within the range $[-1, 1]$, it is still 0.75 at $t = 3$ s (whereas it should be less than 0.1). Because of the integral action the output does eventually return to zero, but it does not drop below 0.1 until after 23 s.

The above example illustrates that the simple inverse-based design method, where L has a slope of about $N = -1$ at all frequencies, does not always yield satisfactory designs. In the example, reference tracking was excellent, but disturbance rejection was poor. The objective of the next section is to understand why the disturbance response was so poor, and to propose a more desirable loop shape for disturbance rejection.

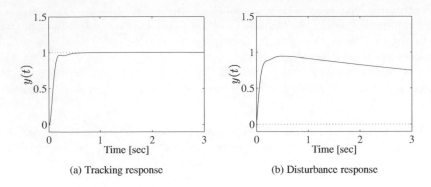

(a) Tracking response   (b) Disturbance response

Figure 2.22: Responses with "inverse-based" controller $K_0(s)$ for the disturbance process

2.6.4   Loop shaping for disturbance rejection

At the outset we assume that the disturbance has been scaled such that at each frequency $|d(\omega)| \leq 1$, and the main control objective is to achieve $|e(\omega)| < 1$. With feedback control we have $e = y = SG_d d$, so to achieve $|e(\omega)| < 1$ for $|d(\omega)| = 1$ (the worst-case disturbance) we require $|SG_d(j\omega)| < 1, \forall \omega$, or equivalently,

$$|1 + L| > |G_d| \quad \forall \omega \tag{2.64}$$

At frequencies where $|G_d| > 1$, this is approximately the same as requiring $|L| > |G_d|$. However, in order to minimize the input signals, thereby reducing the sensitivity to noise and avoiding stability problems, we do not want to use larger loop gains than necessary (at least at frequencies around crossover). A reasonable initial loop shape $L_{\min}(s)$ is then one that just satisfies the condition

$$|L_{\min}| \approx |G_d| \tag{2.65}$$

where the subscript min signifies that $L_{\min}$ is the smallest loop gain to satisfy $|e(\omega)| < 1$. Since $L = GK$, the controller must satisfy

$$|K| > |K_{\min}| \approx |G^{-1}G_d| \tag{2.66}$$

Note that this bound assumes that the models G and G_d are scaled such that the worst-case disturbance d is of unit magnitude and the desired control error e is less than 1. The implications of the bound are as follows:

- For disturbance rejection a good choice for the controller is one which contains the dynamics (G_d) of the disturbance and inverts the dynamics (G) of the inputs (at least at frequencies just before crossover).
- For disturbances entering directly at the plant output, $G_d = 1$, we get $|K_{\min}| = |G^{-1}|$, so an inverse-based design provides the best trade-off between performance (disturbance rejection) and minimum use of feedback.
- For disturbances entering directly at the plant input (which is a common situation in practice – often referred to as a load disturbance), we have $G_d = G$ and we get $|K_{\min}| = 1$, so a simple proportional controller with unit gain yields a good trade-off between output performance and input usage.

- Notice that a reference change may be viewed as a disturbance directly affecting the output. This follows from (1.18), from which we get that a maximum reference change $r = R$ may be viewed as a disturbance $d = 1$ with $G_d(s) = -R$ where R is usually a constant. This explains why selecting K to be like G^{-1} (an inverse-based controller) yields good responses to step changes in the reference.

In addition to satisfying $|L| \approx |G_d|$ (see (2.65)) at frequencies around crossover, the desired loop shape $L(s)$ may be modified as follows:

1. Increase the loop gain at low frequencies to improve the performance. For example, we could use

$$|K| = k \cdot \left| \frac{s + \omega_I}{s} \right| \cdot |G^{-1}G_d| \qquad (2.67)$$

   where $k > 1$ is used to speed up the response, and the integrator term is added to get zero steady-state offset to a step disturbance.
2. Around crossover make the slope N of $|L|$ to be about -1. This is to achieve good transient behaviour with acceptable GM and PM.
3. Let $L(s)$ roll off faster at higher frequencies (beyond the bandwidth) in order to reduce the use of manipulated inputs, to make the controller realizable and to reduce the effects of noise.

The above requirements are concerned with the magnitude, $|L(j\omega)|$. In addition, the dynamics (phase) of $L(s)$ must be selected such that the closed-loop system is stable. When selecting $L(s)$ to satisfy $|L| \approx |G_d|$ one should replace $G_d(s)$ by the corresponding minimum phase transfer function with the same magnitude; that is, time delays and RHP-zeros in $G_d(s)$ should not be included in $L(s)$ as this will impose undesirable limitations on feedback. On the other hand, any time delays or RHP-zeros in $G(s)$ must be included in $L = GK$ because RHP pole–zero cancellations between $G(s)$ and $K(s)$ yield internal instability; see Chapter 4. The final feedback controller has the form

$$K(s) = kw(s)G(s)^{-1}G_d(s) \qquad (2.68)$$

where $w(s)$ incorporates the various shaping and stabilizing ideas introduced above. Usually, $w(s)$ is also selected with the aim of getting a simple final controller $K(s)$.

Remark. The idea of including a disturbance model in the controller is well known and is more rigorously presented in, for example, research on the internal model principle (Wonham, 1974), or the internal model control design for disturbances (Morari and Zafiriou, 1989). However, our development is simple and sufficient for gaining the insight needed for later chapters.

Example 2.10 Loop-shaping design for the disturbance process. *Consider again the plant described by (2.62). The plant can be represented by the block diagram in Figure 2.23, and we see that the disturbance enters at the plant input in the sense that G and G_d share the same dominating dynamics as represented by the term $200/(10s + 1)$.*

   Step 1. Initial design. *From (2.65) we know that a good initial loop shape looks like $|L_{\min}| = |G_d| = \left| \frac{100}{10s+1} \right|$ at frequencies up to crossover. The corresponding controller is $K(s) = G^{-1}L_{\min} = 0.5(0.05s + 1)^2$. This controller is not proper (i.e. it has more zeros than poles), but since the term $(0.05s + 1)^2$ only comes into effect at $1/0.05 = 20$ rad/s, which is beyond the desired gain crossover frequency $\omega_c = 10$ rad/s, we may replace it by a constant gain of 1 resulting in a proportional controller*

$$K_1(s) = 0.5 \qquad (2.69)$$

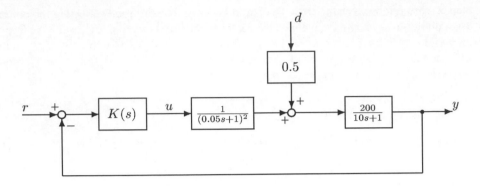

Figure 2.23: Block diagram representation of the disturbance process in (2.62)

The magnitude of the corresponding loop transfer function, $|L_1(j\omega)|$, and the response ($y_1(t)$) to a step change in the disturbance are shown in Figure 2.24. This simple controller works surprisingly well, and for $t < 3$ s the response to a step change in the disturbance is not much different from that with the more complicated inverse-based controller $K_0(s)$ of (2.63) as shown earlier in Figure 2.22. However, there is no integral action and $y_1(t) \rightarrow 1$ as $t \rightarrow \infty$.

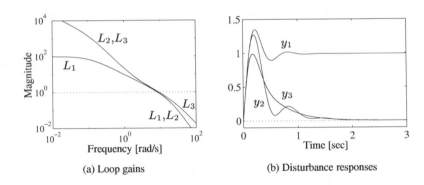

(a) Loop gains   (b) Disturbance responses

Figure 2.24: Loop shapes and disturbance responses for controllers K_1, K_2 and K_3 for the disturbance process

Step 2. More gain at low frequency. *To get integral action we multiply the controller by the term $\frac{s+\omega_I}{s}$, see (2.67), where ω_I is the frequency up to which the term is effective (the asymptotic value of the term is 1 for $\omega > \omega_I$). For performance we want large gains at low frequencies, so we want ω_I to be large, but in order to maintain an acceptable PM (which is 44.7° for controller K_1) the term should not add too much negative phase at frequency ω_c, so ω_I should not be too large. A reasonable value is $\omega_I = 0.2\omega_c$ for which the phase contribution from $\frac{s+\omega_I}{s}$ is $\arctan(1/0.2) - 90° = -11°$ at ω_c. In our case $\omega_c \approx 10$ rad/s, so we select the following controller:*

$$K_2(s) = 0.5\frac{s+2}{s} \tag{2.70}$$

The resulting disturbance response (y_2) shown in Figure 2.24(b) satisfies the requirement that $|y(t)| < 0.1$ at time $t = 3$ s, but $y(t)$ exceeds 1 for a short time. Also, the response is slightly oscillatory as might be expected since the PM is only 31° and the peak values for $|S|$ and $|T|$ are $M_S = 2.28$ and $M_T = 1.89$.

Step 3. High-frequency correction. *To increase the PM and improve the transient response we supplement the controller with "derivative action" by multiplying $K_2(s)$ by a lead–lag term which is effective over one decade starting at 20 rad/s:*

$$K_3(s) = 0.5 \frac{s+2}{s} \frac{0.05s+1}{0.005s+1} \qquad (2.71)$$

This gives a PM of 51°, and peak values $M_S = 1.43$ and $M_T = 1.23$. From Figure 2.24(b), it is seen that the controller $K_3(s)$ reacts quicker than $K_2(s)$ and the disturbance response $y_3(t)$ stays below 1.

Table 2.3: Alternative loop-shaping designs for the disturbance process

						Reference		Disturbance	
	GM	PM	ω_c	M_S	M_T	t_r	y_{max}	y_{max}	$y(t=3)$
Spec.→			≈ 10			≤ 0.3	≤ 1.05	≤ 1	≤ 0.1
K_0	9.95	72.9°	11.4	1.34	1	0.16	1.00	0.95	0.75
K_1	4.04	44.7°	8.48	1.83	1.33	0.21	1.24	1.35	0.99
K_2	3.24	30.9°	8.65	2.28	1.89	0.19	1.51	1.27	0.001
K_3	19.7	50.9°	9.27	1.43	1.23	0.16	1.24	0.99	0.001

Table 2.3 summarizes the results for the four loop-shaping designs; the inverse-based design K_0 for reference tracking and the three designs K_1, K_2 and K_3 for disturbance rejection. Although controller K_3 satisfies the requirements for disturbance rejection, it is not satisfactory for reference tracking; the overshoot is 24% which is significantly higher than the maximum value of 5%. On the other hand, the inverse-based controller K_0 inverts the term $1/(10s+1)$ which is also in the disturbance model, and therefore yields a very sluggish response to disturbances (the output is still 0.75 at $t=3$ s whereas it should be less than 0.1).

In summary, for this process none of the controller designs meet all the objectives for both reference tracking and disturbance rejection. The solution is to use a two degrees-of-freedom controller as discussed next.

2.6.5 Two degrees-of-freedom design

For reference tracking we typically want the controller to look like $\frac{1}{s}G^{-1}$, see (2.61), whereas for disturbance rejection we want the controller to look like $\frac{1}{s}G^{-1}G_d$, see (2.67). We cannot generally achieve both of these simultaneously with a single (feedback) controller.

The solution is to use a two degrees-of-freedom controller where the reference signal r and output measurement y_m are treated independently by the controller, rather than operating on their difference $r - y_m$ as in a one degree-of-freedom controller. There exist several alternative implementations of a two degrees-of-freedom controller. The most general form is shown in Figure 1.3(b) on page 12 where the controller has two inputs (r and y_m) and one output (u). However, the controller is often split into two separate blocks. One form was shown in Figure 2.5, but here we will use the form in Figure 2.25 where K_y denotes the feedback part of the controller and K_r the reference prefilter. The feedback controller K_y is used to reduce the effect of uncertainty (disturbances and model error) whereas the prefilter K_r shapes the commands r to improve tracking performance. In general, it is optimal to design the combined two degrees-of-freedom controller K in one step. However, in practice K_y is often designed first for disturbance rejection, and then K_r is designed to improve reference tracking. This is the approach taken here.

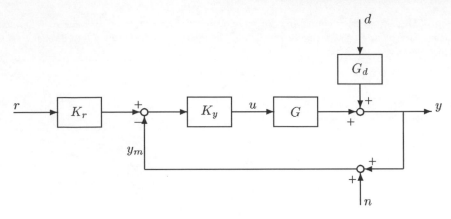

Figure 2.25: Two degrees-of-freedom controller

Let $T = L(1 + L)^{-1}$ (with $L = GK_y$) denote the complementary sensitivity function for the feedback system. Then for a one degree-of-freedom controller $y = Tr$, whereas for a two degrees-of-freedom controller $y = TK_r r$. If the desired transfer function for reference tracking (often denoted the reference model) is T_{ref}, then the corresponding ideal reference prefilter K_r satisfies $TK_r = T_{\text{ref}}$, or

$$K_r(s) = T^{-1}(s)T_{\text{ref}}(s) \tag{2.72}$$

Thus, in theory we may design $K_r(s)$ to get any desired tracking response $T_{\text{ref}}(s)$. However, in practice it is not so simple because the resulting $K_r(s)$ may be unstable (if $G(s)$ has RHP-zeros) or unrealizable, and also $TK_r \neq T_{\text{ref}}$ if $G(s)$ and thus $T(s)$ is not known exactly.

Remark. A convenient practical choice of prefilter is the lead–lag network

$$K_r(s) = \frac{\tau_{\text{lead}}s + 1}{\tau_{\text{lag}}s + 1} \tag{2.73}$$

Here we select $\tau_{\text{lead}} > \tau_{\text{lag}}$ if we want to speed up the response, and $\tau_{\text{lead}} < \tau_{\text{lag}}$ if we want to slow down the response. If one does not require fast reference tracking, which is the case in many process control applications, a simple lag is often used (with $\tau_{\text{lead}} = 0$).

Example 2.11 Two degrees-of-freedom design for the disturbance process. *In Example 2.10 we designed a loop-shaping controller $K_3(s)$ for the plant in (2.62) which gave good performance with respect to disturbances. However, the command tracking performance was not quite acceptable as is shown by y_3 in Figure 2.26. The rise time is 0.16 s which is better than the required value of 0.3 s, but the overshoot is 24% which is significantly higher than the maximum value of 5%. To improve upon this we can use a two degrees-of-freedom controller with $K_y = K_3$, and we design $K_r(s)$ based on (2.72) with reference model $T_{\text{ref}} = 1/(0.1s + 1)$ (a first-order response with no overshoot). To get a low-order $K_r(s)$, we either may use the actual $T(s)$ and then use a low-order approximation of $K_r(s)$, or we may start with a low-order approximation of $T(s)$. We will do the latter. From the step response y_3 in Figure 2.26 we approximate the response by two parts: a fast response with time constant 0.1 s and gain 1.5, and a slower response with time constant 0.5 s and gain -0.5 (the sum of the gains is 1). Thus we use $T(s) \approx \frac{1.5}{0.1s+1} - \frac{0.5}{0.5s+1} = \frac{(0.7s+1)}{(0.1s+1)(0.5s+1)}$, from which (2.72) yields $K_r(s) = \frac{0.5s+1}{0.7s+1}$. Following closed-loop simulations we modified this slightly to arrive at the design*

$$K_{r3}(s) = \frac{0.5s + 1}{0.65s + 1} \cdot \frac{1}{0.03s + 1} \tag{2.74}$$

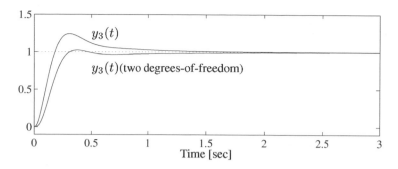

Figure 2.26: Tracking responses with the one degree-of-freedom controller (K_3) and the two degrees-of-freedom controller (K_3, K_{r3}) for the disturbance process

where the term $1/(0.03s + 1)$ was included to avoid the initial peaking of the input signal $u(t)$ above 1. The tracking response with this two degrees-of-freedom controller is shown in Figure 2.26. The rise time is 0.25 s which is better than the requirement of 0.3 s, and the overshoot is only 2.3% which is better than the requirement of 5%. The disturbance response is the same as curve y_3 in Figure 2.24. In conclusion, we are able to satisfy all specifications using a low-order two degrees-of-freedom controller.

Loop shaping applied to a flexible structure

The following example shows how the loop-shaping procedure for disturbance rejection can be used to design a one degree-of-freedom controller for a very different kind of plant.

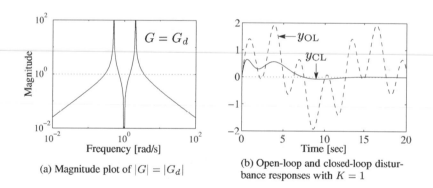

(a) Magnitude plot of $|G| = |G_d|$

(b) Open-loop and closed-loop disturbance responses with $K = 1$

Figure 2.27: Flexible structure in (2.75)

Example 2.12 Loop shaping for a flexible structure. *Consider the following model of a flexible structure with a disturbance occurring at the plant input:*

$$G(s) = G_d(s) = \frac{2.5s(s^2 + 1)}{(s^2 + 0.5^2)(s^2 + 2^2)} \qquad (2.75)$$

From the Bode magnitude plot in Figure 2.27(a) we see that $|G_d(j\omega)| \gg 1$ around the resonance frequencies of 0.5 and 2 rad/s, so control is needed at these frequencies. The dashed line in Figure 2.27(b) shows the open-loop response to a unit step disturbance. The output is seen to cycle

between −2 *and* 2 *(outside the allowed range* −1 *to* 1*), which confirms that control is needed.
From* (2.66) *a controller which meets the specification* $|y(\omega)| \leq 1$ *for* $|d(\omega)| = 1$ *is given by*
$|K_{\min}(j\omega)| = |G^{-1}G_d| = 1$. *Indeed the controller*

$$K(s) = 1 \qquad\qquad\qquad (2.76)$$

*turns out to be a good choice as is verified by the closed-loop disturbance response (solid line) in
Figure 2.27(b); the output goes up to about* 0.5 *and then returns to zero. The fact that the choice
$L(s) = G(s)$ gives closed-loop stability is not immediately obvious since* $|G|$ *has four gain crossover
frequencies. However, instability cannot occur because the plant is "passive" with* $\angle G > -180°$ *at all
frequencies.*

2.6.6   Conclusions on loop shaping

The loop-shaping procedure outlined and illustrated by the examples above is well suited for
relatively simple problems, as might arise for stable plants where $L(s)$ crosses the negative
real axis only once. Although the procedure may be extended to more complicated systems
the effort required by the engineer is considerably greater. In particular, it may be very
difficult to achieve stability.

Fortunately, there exist alternative methods where the burden on the engineer is much less.
One such approach is the Glover–McFarlane $\mathcal{H}_\infty$ loop-shaping procedure which is discussed
in detail in Chapter 9. It is essentially a two-step procedure, where in the first step the
engineer, as outlined in this section, decides on a loop shape, $|L|$ (denoted the "shaped plant"
G_s), and in the second step an optimization provides the necessary phase corrections to get a
stable and robust design. The method is applied to the disturbance process in Example 9.3 on
page 368.

An alternative to shaping the open-loop transfer function $L(s)$ is to shape closed-loop
transfer functions. This is discussed next in Sections 2.7 and 2.8.

2.7   IMC design procedure and PID control for stable plants

Specifications directly on the *open-loop transfer function $L = GK$*, as in the loop-shaping
design procedures of the previous section, make the design process transparent as it is clear
how changes in $L(s)$ affect the controller $K(s)$ and vice versa. An apparent problem with
this approach, however, is that it does not consider directly the *closed-loop transfer functions*,
such as S and T, which determine the final response. The following approximations apply:

$$|L(j\omega)| \gg 1 \quad \Rightarrow \quad S \approx L^{-1}; \quad T \approx 1$$
$$|L(j\omega)| \ll 1 \quad \Rightarrow \quad S \approx 1; \qquad T \approx L$$

but in the important crossover region where $|L(j\omega)|$ is close to 1, one cannot infer anything
about S and T from the magnitude of the loop shape, $|L(j\omega)|$.

An alternative design strategy is to shape directly the relevant closed-loop transfer
functions. In this section, we present such a strategy in the form of internal model control
(IMC), with a focus on the design of PID controllers. The more general approach of shaping
closed-loop transfer functions using $\mathcal{H}_\infty$ optimal control is discussed in the next section.

The IMC design method (e.g. Morari and Zafiriou, 1989) is simple and has proven to be successful in applications. The idea is to specify the desired closed-loop response and solve for the resulting controller. This simple idea, also known as "direct synthesis", results in an "inverse-based" controller design. The key step is to specify a good closed-loop response. To do so, one needs to understand what closed-loop responses are achievable and desirable.

The first step in the IMC procedure for a stable plant is to factorize the plant model into an invertible minimum-phase part (G_m) and a non-invertible all-pass part (G_a). A time delay θ and non-minimum-phase (RHP) zeros z_i cannot be inverted, because the inverse would be non-causal and unstable, respectively. We therefore have

$$G(s) = G_m G_a \tag{2.77}$$

$$G_a(s) = e^{-\theta s} \prod_i \frac{-s + z_i}{s + z_i}, \quad \text{Re}(z_i) > 0; \theta > 0 \tag{2.78}$$

The second step is to specify the desired closed-loop transfer function T from references to outputs, $y = Tr$. There is no way we can prevent T from including the non-minimum-phase elements of G_a, so we specify

$$T(s) = f(s)G_a(s) \tag{2.79}$$

where $f(s)$ is a low-pass filter selected by the designer, typically of the form $f(s) = 1/(\tau_c s + 1)^n$. The rest is algebra. We have from (2.19) that

$$T = GK(1 + GK)^{-1} \tag{2.80}$$

so combining (2.77), (2.79) and (2.80), and solving for the controller, yields

$$K = G^{-1} \frac{T}{1 - T} = G_m^{-1} \frac{1}{f^{-1} - G_a} \tag{2.81}$$

We note that the controller inverts the minimum-phase part G_m of the plant.

Example 2.13 *We apply the IMC design method to a stable second-order plus time delay process*

$$G(s) = k \frac{e^{-\theta s}}{\tau_0^2 s^2 + 2\tau_0 \zeta s + 1} \tag{2.82}$$

where ζ is the damping factor. $|\zeta| < 1$ gives an underdamped process with oscillations. We consider a stable process where τ_0 and ζ are non-negative. Factorization yields $G_a(s) = e^{-\theta s}$, $G_m(s) = \frac{k}{\tau_0^2 s^2 + 2\tau_0 \zeta s + 1}$. We select a first-order filter $f(s) = 1/(\tau_c s + 1)$. From (2.79) this specifies that we desire, following the unavoidable time delay, a simple first-order tracking response with time constant τ_c:

$$T(s) = \frac{1}{\tau_c s + 1} e^{-\theta s} \tag{2.83}$$

From (2.81), the resulting controller becomes

$$K(s) = G_m^{-1} \frac{1}{f^{-1} - G_a} = \frac{\tau_0^2 s^2 + 2\tau_0 \zeta s + 1}{k} \frac{1}{\tau_c s + 1 - e^{-\theta s}} \tag{2.84}$$

where τ_c is a tuning parameter. This controller is not rational and cannot be written in standard state-space form. However, it can be easily realized in discrete form as a "Smith predictor". Alternatively, we can approximate the time delay and derive a rational controller. For example, we may use a first-order Taylor approximation $e^{-\theta s} \approx 1 - \theta s$. This gives

$$K(s) = \frac{\tau_0^2 s^2 + 2\tau_0 \zeta s + 1}{k} \frac{1}{(\tau_c + \theta)s} \tag{2.85}$$

which can be implemented as a PID controller.

PID control. The PID controller, with three adjustable parameters, is the most widely used control algorithm in industry. There are many variations, but the most common is probably the "ideal" (or parallel) form

$$K_{\text{PID}}(s) = K_c \left(1 + \frac{1}{\tau_I s} + \tau_D s \right) \tag{2.86}$$

where the parameters are the gain K_c, integral time τ_I and derivative time τ_D. Another common implementation is the *cascade form*

$$K_{\text{PID,cascade}}(s) = \tilde{K}_c \frac{\tilde{\tau}_I s + 1}{\tilde{\tau}_I s} (\tilde{\tau}_D s + 1) \tag{2.87}$$

The cascade form is somewhat less general than (2.86) as it does not allow for complex zeros. To translate the cascade PID settings in (2.87) to the ideal form in (2.86), we introduce the factor $\alpha = 1 + \tilde{\tau}_D / \tilde{\tau}_I$, and we have

$$K_c = \tilde{K}_c \cdot \alpha, \ \tau_I = \tilde{\tau}_I \cdot \alpha, \ \tau_D = \tilde{\tau}_D / \alpha \tag{2.88}$$

As indicated, the reverse translation is not always possible.

With derivative action, the practical implementation is not as given in (2.86) and (2.87). First, with τ_D non-zero, the controllers in (2.86) and (2.87) are improper. In practice, one needs to add a filter to the controller itself or to the controller input (measurement). The filter is typically of the form $\frac{1}{\epsilon \tau_D s + 1}$ with ϵ about 0.1. In most cases, the addition of this filter will not noticeably change the closed-loop response. Second, to avoid "derivative kick" the reference signal is usually not differentiated, which in effect corresponds to a two degrees-of-freedom implementation. In summary, a typical practical implementation of the PID controller (2.86) is

$$u = K_c \left[\left(1 + \frac{1}{\tau_I s} \right) (r - y_m) - \frac{\tau_D s}{\epsilon \tau_D s + 1} y_m \right] \tag{2.89}$$

where u is the plant input, y_m the plant measurement and r the reference.

Example 2.13 continued. *The IMC controller (2.85) for the second-order process (2.82) can be realized as an ideal PID controller (2.86) with*

$$K_c = \frac{1}{k} \frac{2\tau_0 \zeta}{\tau_c + \theta}, \quad \tau_I = 2\tau_0 \zeta, \quad \tau_D = 0.5\tau_0 / \zeta \tag{2.90}$$

For $\zeta < 1$ we have complex zeros in the controller and it cannot be realized in the cascade PID form (2.87). However, for overdamped plants with $\zeta > 1$, we can write the model (2.82) in the form

$$G(s) = k \frac{e^{-\theta s}}{(\tau_1 s + 1)(\tau_2 s + 1)} \tag{2.91}$$

resulting from (2.85) in the controller $K(s) = \frac{(\tau_1 s + 1)(\tau_2 s + 1)}{k} \frac{1}{(\tau_c + \theta)s}$. Comparing with (2.87), the cascade PID settings become

$$\tilde{K}_c = \frac{1}{k} \frac{\tau_1}{\tau_c + \theta}, \quad \tilde{\tau}_I = \tau_1, \quad \tilde{\tau}_D = \tau_2 \tag{2.92}$$

Using (2.88), the corresponding ideal PID settings become

$$K_c = \frac{1}{k} \frac{(\tau_1 + \tau_2)}{\tau_c + \theta}, \quad \tau_I = \tau_1 + \tau_2, \quad \tau_D = \frac{\tau_2}{1 + \tau_2 / \tau_1} \tag{2.93}$$

We note that the PID settings are simpler if we use the cascade form.

SIMC (Skogestad/Simple IMC) PID design for first- or second-order plus delay process. Skogestad (2003) has derived simple rules for model reduction and PID tuning based on the above ideas. He claims these are "probably the best simple PID tuning rules in the world" ☺ . In process control, it is common to approximate the process with a first-order plus time delay model,

$$G(s) = \frac{k}{\tau s + 1} e^{-\theta s} \tag{2.94}$$

Specifying a first-order reference tracking response and using a first-order approximation of the time delay then gives $K_c = \frac{1}{k} \frac{\tau}{\tau_c + \theta}$ and $\tau_I = \tau$ (set $\tau_1 = \tau$ and $\tau_2 = 0$ in (2.93)). These settings are derived for step references and also work well for step disturbances entering directly at the plant output. However, for nearly integrating processes with large τ, say $\tau \geq 8\theta$, step disturbances entering at the plant input will affect the output in a ramp-like fashion. To counteract this, one may modify (increase) the integral action by decreasing τ_I. However, to avoid undesired closed-loop oscillations, τ_I cannot be decreased too much, and Skogestad (2003) recommends the following SIMC PI settings for the plant model (2.94):

$$\boxed{K_c = \frac{1}{k} \frac{\tau}{\tau_c + \theta}, \quad \tau_I = \min\left(\tau, 4(\tau_c + \theta)\right)} \tag{2.95}$$

For PI control there is no difference between the ideal (2.86) and cascade (2.87) forms. The use of *derivative action* (PID control) is uncommon in process control applications, where most plants are stable with simple overdamped responses. This is because the performance improvement is usually too small to justify the added complexity and the increased sensitivity to measurement noise. One exception is for a "dominant" second-order process,

$$G(s) = k \frac{e^{-\theta s}}{(\tau_1 s + 1)(\tau_2 s + 1)} \tag{2.96}$$

where $\tau_1 \geq \tau_2$ and "dominant" means roughly speaking that $\tau_2 > \theta$. We derived cascade PID settings for this model in (2.92). Again, to improve the performance for integrating disturbances, we need to modify the integral time for an integrating process. Thus, for the plant model (2.96) the recommended SIMC settings with the cascade PID controller (2.87) are

$$\boxed{\tilde{K}_c = \frac{1}{k} \frac{\tau_1}{\tau_c + \theta}, \quad \tilde{\tau}_I = \min\left(\tau_1, 4(\tau_c + \theta)\right), \quad \tilde{\tau}_D = \tau_2} \tag{2.97}$$

The corresponding settings for the ideal-form PID controller are obtained using (2.88), but are more complicated.

The settings in (2.95) and (2.97) follow directly from the model, except for the single tuning parameter τ_c that affects the controller gain (and the integral time for near-integrating processes). The choice of the tuning parameter τ_c is based on a trade-off between output performance (favoured by a small τ_c) and robustness and input usage (favoured by a large τ_c). To achieve "fast" control with acceptable robustness, Skogestad (2003) recommends $\boxed{\tau_c = \theta}$, which for the model (2.96) gives a sensitivity peak $M_S \approx 1.7$, gain margin GM ≈ 3 and crossover frequency $\omega_c = 0.5/\theta$.

Model reduction and effective delay. To derive a model in the form (2.94) or (2.96), where θ is the *effective delay*, Skogestad (2003) provides some simple analytic rules for model reduction. These are based on the approximations $e^{-\theta s} \approx 1 - \theta s$ (for approximating a RHP-zero as a delay) and $e^{-\theta s} \approx 1/(1 + \theta s)$ (for approximating a lag as a delay). The lag approximation is conservative, because in terms of control a delay is worse than a lag of equal magnitude. Thus, when approximating the largest lag, Skogestad (2003) recommends the use of the simple *half rule*:

- **Half rule.** *The largest neglected lag (denominator) time constant is distributed equally to the effective delay and the smallest retained time constant.*

To illustrate, let the original model be in the form

$$G_0(s) = \frac{\prod_j (-T_{j0}^{\mathrm{inv}} s + 1)}{\prod_i (\tau_{i0} s + 1)} e^{-\theta_0 s} \tag{2.98}$$

where the lags τ_{i0} are ordered according to their magnitude, and $T_{j0}^{\mathrm{inv}} = 1/z_{j0} > 0$ denote the inverse response (negative numerator) time constants corresponding to the RHP-zeros located at $s = z_{j0}$. Then, according to the half rule, to obtain a first-order model $G(s) = e^{-\theta s}/(\tau_1 s + 1)$ (for PI control), we use

$$\tau_1 = \tau_{10} + \frac{\tau_{20}}{2}; \quad \theta = \theta_0 + \frac{\tau_{20}}{2} + \sum_{i \geq 3} \tau_{i0} + \sum_j T_{j0}^{\mathrm{inv}} + \frac{h}{2} \tag{2.99}$$

and, to obtain a second-order model (2.96) (for PID control), we use

$$\tau_1 = \tau_{10}; \quad \tau_2 = \tau_{20} + \frac{\tau_{30}}{2}; \quad \theta = \theta_0 + \frac{\tau_{30}}{2} + \sum_{i \geq 4} \tau_{i0} + \sum_j T_{j0}^{\mathrm{inv}} + \frac{h}{2} \tag{2.100}$$

where h is the sampling period (for cases with digital implementation). The main objective of the empirical half rule is to maintain the robustness of the proposed PI and PID tuning rules, which with $\tau_c = \theta$ give M_S about 1.7. This is discussed by Skogestad (2003), who also provides additional rules for approximating positive numerator time constants (LHP-zeros).

Example 2.14 *The process*

$$G_0(s) = \frac{2}{(s + 1)(0.2s + 1)}$$

is approximated using the half rule as a first-order with time delay process, $G(s) = ke^{-\theta s + 1}/(\tau s + 1)$, with $k = 2, \theta = 0.2/2 = 0.1$ and $\tau = 1 + 0.2/2 = 1.1$. Choosing $\tau_c = \theta = 0.1$ the SIMC PI settings (2.95) become $K_c = \frac{1}{2} \frac{1.1}{2 \cdot 0.1} = 2.75$ and $\tau_I = \min(1.1, 4 \cdot 2 \cdot 0.1) = 0.8$.

In this case, we may also consider using a second-order model (2.96) with $k = 2, \tau_1 = 1, \tau_2 = 0.2, \theta = 0$ (no approximation). Since $\theta = 0$, we cannot choose $\tau_c = \theta$ as it would yield an infinite controller gain. However, the controller gain is limited by other factors, such as the allowed input magnitude, measurement noise and unmodelled dynamics. Because of such factors, let us assume that the largest allowed controller gain is $\widetilde{K}_c = 10$. From (2.97) this corresponds to $\tau_c = 0.05$, and we get $\widetilde{\tau}_I = \min(1, 4 \cdot 0.05) = 0.2$ and $\widetilde{\tau}_D = \tau_2 = 0.2$. Using (2.88), the corresponding ideal PID settings are $K_c = 20, \tau_I = 0.4$ and $\tau_D = 0.1$.

Example 2.15 *Consider the process*

$$G(s) = 3 \frac{(-0.8s + 1)}{(6s + 1)(2.5s + 1)^2 (0.4s + 1)} e^{-1.2s}$$

Using the half rule, the process is approximated as a first-order time delay model with

$$k = 3, \ \tau_1 = 6 + 2.5/2 = 7.25, \ \theta = 1.2 + 0.8 + 2.5/2 + 2.5 + 0.4 = 6.15$$

or as a second-order time delay model with

$$k = 3, \ \tau_1 = 6, \ \tau_2 = 2.5 + 2.5/2 = 3.75, \ \theta = 1.2 + 0.8 + 2.5/2 + 0.4 = 3.65$$

The PI settings based on the first-order model are (choosing $\tau_c = \theta = 6.15$)

$$K_c = \frac{1}{3} \frac{7.25}{2 \cdot 7.15} = 0.169, \quad \tau_I = \min{(7.25, 8 \cdot 6.15)} = 7.25$$

and the cascade PID settings based on the second-order model are (choosing $\tau_c = \theta = 3.65$)

$$\widetilde{K}_c = 0.274, \quad \widetilde{\tau}_I = 6, \quad \widetilde{\tau}_D = 3.75$$

We note that a PI controller results from a first-order model, and a PID controller from a second-order model. Since the effective delay θ is the main limiting factor in terms of control performance, its value gives invaluable insight into the inherent controllability of the process. With the effective delay computed using the half rule, it follows that PI control performance is limited by (half of) the magnitude of the second-largest time constant τ_2, whereas PID control performance is limited by (half of) the magnitude of the third-largest time constant, τ_3.

Example 2.16 *Let us finally consider the "disturbance process" in (2.62)*

$$G(s) = \frac{200}{10s + 1} \frac{1}{(0.05s + 1)^2}$$

Using the half rule, the process is approximated as a first-order time delay model with $k = 200, \tau_1 = 10.025$ and $\theta = 0.075$. The recommended choice for "fast" control is $\tau_c = \theta = 0.075$. However, on page 47 it was stated that we aim for a gain crossover frequency of about $w_c - 10$ [rad/s]. Since we desire a first-order closed-loop response, this corresponds to $\tau_c = 1/\omega_c = 0.1$. With $\tau_c = 0.1$ the corresponding SIMC PI settings are $K_c = \frac{1}{200} \frac{10.025}{(0.1 + 0.075)} = 0.286$ and $\tau_I = \min{(10.025, 4 \cdot (0.1 + 0.075))} = 0.7$. This is an almost-integrating process, and we note that we reduce the integral time from 10.025 (which would be good for tracking step references) to 0.7 in order to get acceptable performance for input disturbances.

To improve further the performance, we use the half rule to obtain a second-order model ($k = 200, \tau_1 = 10, \tau_2 = 0.075, \theta = 0.025$) and choose $\omega_c = 0.1$ to derive SIMC PID settings ($\widetilde{K}_c = 0.4, \widetilde{\tau}_I = 0.5, \widetilde{\tau}_D = 0.075$). Interestingly, the corresponding controller

$$K(s) = 0.4 \frac{s + 2}{s}(0.075s + 1)$$

is almost identical to the final controller K_3 in (2.71), designed previously using loop-shaping ideas.

2.8 Shaping closed-loop transfer functions

In Section 2.6, we discussed the shaping of the magnitude of the open-loop transfer function $L(s)$. In this section, we introduce the reader to the shaping of the magnitudes of *closed-loop* transfer functions, where we synthesize a controller by minimizing an $\mathcal{H}_\infty$ performance objective. The topic is discussed further in Section 3.5.7 and in more detail in Chapter 9. Such a design strategy automates the actual controller design, leaving the engineer with the task of selecting reasonable bounds ("weights") on the desired closed-loop transfer functions. Before explaining how this may be done in practice, we discuss the terms $\mathcal{H}_\infty$ and $\mathcal{H}_2$.

2.8.1   The terms $\mathcal{H}_\infty$ and $\mathcal{H}_2$

The $\mathcal{H}_\infty$ norm of a stable scalar transfer function $f(s)$ is simply the peak value of $|f(j\omega)|$ as a function of frequency, i.e.

$$\|f(s)\|_\infty \triangleq \max_\omega |f(j\omega)| \tag{2.101}$$

Remark. Strictly speaking, we should here replace "max" (the maximum value) by "sup" (the supremum, the least upper bound). This is because the maximum may only be approached as $w \to \infty$ and may therefore not actually be achieved. However, for engineering purposes there is no difference between "sup" and "max".

The terms $\mathcal{H}_\infty$ norm and $\mathcal{H}_\infty$ control are intimidating at first, and a name conveying the engineering significance of $\mathcal{H}_\infty$ would have been better. After all, we are simply talking about a design method which aims to press down the peak(s) of one or more selected transfer functions. However, the term $\mathcal{H}_\infty$, which is purely mathematical, has now established itself in the control community. To make the term less forbidding, an explanation of its background may help. First, the symbol ∞ comes from the fact that the maximum magnitude over frequency may be written as

$$\max_\omega |f(j\omega)| = \lim_{p \to \infty} \left(\int_{-\infty}^{\infty} |f(j\omega)|^p d\omega \right)^{1/p}$$

Essentially, by raising $|f|$ to an infinite power we pick out its peak value. Next, the symbol $\mathcal{H}$ stands for "Hardy space", and $\mathcal{H}_\infty$ in the context of this book is the set of transfer functions with bounded ∞-norm, which is simply the set of *stable and proper* transfer functions.

Similarly, the symbol $\mathcal{H}_2$ stands for the Hardy space of transfer functions with bounded 2-norm, which is the set of *stable and strictly proper* transfer functions. The $\mathcal{H}_2$ norm of a strictly proper stable scalar transfer function is defined as

$$\|f(s)\|_2 \triangleq \left(\frac{1}{2\pi} \int_{-\infty}^{\infty} |f(j\omega)|^2 d\omega \right)^{1/2} \tag{2.102}$$

The factor $1/\sqrt{2\pi}$ is introduced to get consistency with the 2-norm of the corresponding impulse response; see (4.120). Note that the $\mathcal{H}_2$ norm of a semi-proper (or bi-proper) transfer function (where $\lim_{s \to \infty} f(s)$ is a non-zero constant) is infinite, whereas its $\mathcal{H}_\infty$ norm is finite. An example of a semi-proper transfer function (with an infinite $\mathcal{H}_2$ norm) is the sensitivity function $S = (I + GK)^{-1}$.

2.8.2   Weighted sensitivity

As already discussed, the sensitivity function S is a very good indicator of closed-loop performance, both for SISO and MIMO systems. The main advantage of considering S is that because we ideally want S small, it is sufficient to consider just its magnitude $|S|$; that is, we need not worry about its phase. Typical specifications in terms of S include:

1. Minimum bandwidth frequency ω_B^* (defined as the frequency where $|S(j\omega)|$ crosses 0.707 from below).
2. Maximum tracking error at selected frequencies.
3. System type, or alternatively the maximum steady-state tracking error, A.

4. Shape of S over selected frequency ranges.
5. Maximum peak magnitude of S, $\|S(j\omega)\|_\infty \leq M$.

The peak specification prevents amplification of noise at high frequencies, and also introduces a margin of robustness; typically we select $M = 2$. Mathematically, these specifications may be captured by an upper bound, $1/|w_P(s)|$, on the magnitude of S, where $w_P(s)$ is a weight selected by the designer. The subscript P stands for *performance* since S is mainly used as a performance indicator, and the performance requirement becomes

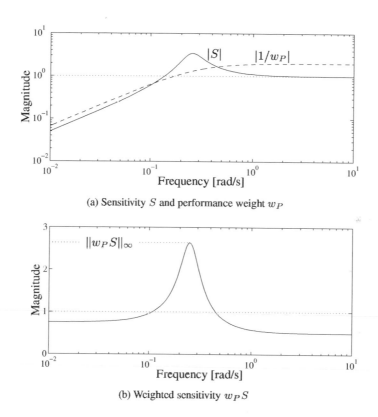

(a) Sensitivity S and performance weight w_P

(b) Weighted sensitivity $w_P S$

Figure 2.28: Case where $|S|$ exceeds its bound $1/|w_P|$, resulting in $\|w_P S\|_\infty > 1$

$$|S(j\omega)| < 1/|w_P(j\omega)|, \ \forall\omega \tag{2.103}$$

$$\Leftrightarrow \ |w_P S| < 1, \ \forall\omega \quad \Leftrightarrow \quad \boxed{\|w_P S\|_\infty < 1} \tag{2.104}$$

The last equivalence follows from the definition of the $\mathcal{H}_\infty$ norm, and in words the performance requirement is that the $\mathcal{H}_\infty$ norm of the weighted sensitivity, $w_P S$, must be less than 1. In Figure 2.28(a), an example is shown where the sensitivity, $|S|$, exceeds its upper bound, $1/|w_P|$, at some frequencies. The resulting weighted sensitivity, $|w_P S|$, therefore exceeds 1 at the same frequencies as is illustrated in Figure 2.28(b). Note that we usually do *not* use a log-scale for the magnitude when plotting *weighted* transfer functions, such as $|w_P S|$.

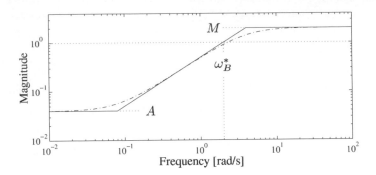

Figure 2.29: Inverse of performance weight: exact and asymptotic plot of $1/|w_P(j\omega)|$ in (2.105)

Weight selection. An asymptotic plot of a typical upper bound, $1/|w_P|$, is shown in Figure 2.29. The weight illustrated may be represented by

$$w_P(s) = \frac{s/M + \omega_B^*}{s + \omega_B^* A} \tag{2.105}$$

and we see that $1/|w_P(j\omega)|$ (the upper bound on $|S|$) is equal to A (typically $A \approx 0$) at low frequencies, is equal to $M \geq 1$ at high frequencies, and the asymptote crosses 1 at the frequency ω_B^*, which is approximately the bandwidth requirement.

Remark. For this weight the loop shape $L = \omega_B^*/s$ yields an S which exactly matches the bound (2.104) at frequencies below the bandwidth and easily satisfies (by a factor M) the bound at higher frequencies.

In some cases, in order to improve performance, we may require a steeper slope for L (and S) below the bandwidth, and then a higher-order weight may be selected. A weight which goes as $(w_B/s)^n$ at frequencies below crossover is

$$w_P(s) = \frac{(s/M^{1/n} + \omega_B^*)^n}{(s + \omega_B^* A^{1/n})^n} \tag{2.106}$$

Exercise 2.4 *For $n = 2$, make an asymptotic plot of $1/|w_P|$ in (2.106) and compare with the asymptotic plot of $1/|w_P|$ in (2.105).*

The insights gained in the previous section on loop-shaping design are very useful for selecting weights. For example, for disturbance rejection we must satisfy $|SG_d(j\omega)| < 1$ at all frequencies (assuming the variables have been scaled to be less than 1 in magnitude). It then follows that a good initial choice for the performance weight is to let $|w_P(j\omega)|$ look like $|G_d(j\omega)|$ at frequencies where $|G_d| > 1$. In other cases, one may first obtain an initial controller using another design procedure, such as LQG, and the resulting sensitivity $|S(j\omega)|$ may then be used to select a performance weight for a subsequent $\mathcal{H}_\infty$ design.

2.8.3   Stacked requirements: mixed sensitivity

The specification $\|w_P S\|_\infty < 1$ puts a lower bound on the bandwidth, but not an upper one, and nor does it allow us to specify the roll-off of $L(s)$ above the bandwidth. To do this one can make demands on another closed-loop transfer function, e.g. on the complementary sensitivity $T = I - S = GKS$. For instance, one might specify an upper bound $1/|w_T|$ on the magnitude of T to make sure that L rolls off sufficiently fast at high frequencies. Also, to achieve robustness or to restrict the magnitude of the input signals, $u = KS(r - G_d d)$, one may place an upper bound, $1/|w_u|$, on the magnitude of KS. To combine these "mixed sensitivity" specifications, a "stacking approach" is usually used, resulting in the following overall specification:

$$\|N\|_\infty = \max_\omega \bar{\sigma}(N(j\omega)) < 1; \quad N = \begin{bmatrix} w_P S \\ w_T T \\ w_u KS \end{bmatrix} \quad (2.107)$$

Here we use the maximum singular value, $\bar{\sigma}(N(j\omega))$, to measure the size of the matrix N at each frequency. For SISO systems, N is a vector and $\bar{\sigma}(N)$ is the usual Euclidean vector norm:

$$\bar{\sigma}(N) = \sqrt{|w_P S|^2 + |w_T T|^2 + |w_u KS|^2} \quad (2.108)$$

After selecting the form of N and the weights, the $\mathcal{H}_\infty$ optimal controller is obtained by solving the problem

$$\min_K \|N(K)\|_\infty \quad (2.109)$$

where K is a stabilizing controller. A good tutorial introduction to $\mathcal{H}_\infty$ control is given by Kwakernaak (1993).

Remark 1 The stacking procedure is selected for mathematical convenience as it does not allow us to specify exactly the bounds on the individual transfer functions as described above. For example, assume that $\phi_1(K)$ and $\phi_2(K)$ are two functions of K (which might represent $\phi_1(K) = w_P S$ and $\phi_2(K) = w_T T$) and that we want to achieve

$$|\phi_1| < 1 \quad \text{and} \quad |\phi_2| < 1 \quad (2.110)$$

This is similar to, but not quite the same as, the stacked requirement

$$\bar{\sigma} \begin{bmatrix} \phi_1 \\ \phi_2 \end{bmatrix} = \sqrt{|\phi_1|^2 + |\phi_2|^2} < 1 \quad (2.111)$$

Objectives (2.110) and (2.111) are very similar when either $|\phi_1|$ or $|\phi_2|$ is small, but in the "worst" case when $|\phi_1| = |\phi_2|$, we get from (2.111) that $|\phi_1| \leq 0.707$ and $|\phi_2| \leq 0.707$. That is, there is a possible "error" in each specification equal to at most a factor $\sqrt{2} \approx 3$ dB. In general, with n stacked requirements the resulting error is at most $\sqrt{n}$. This inaccuracy in the specifications is something we are probably willing to sacrifice in the interests of mathematical convenience. In any case, the specifications are in general rather rough, and are effectively knobs for the engineer to select and adjust until a satisfactory design is reached.

Remark 2 Let $\gamma_{\min} = \min_K \|N(K)\|_\infty$ denote the optimal $\mathcal{H}_\infty$ norm. An important property of $\mathcal{H}_\infty$ optimal controllers is that they yield a flat frequency response; that is, $\bar{\sigma}(N(j\omega)) = \gamma_{\min}$ at all frequencies. The practical implication is that, except for at most a factor $\sqrt{n}$, the transfer functions resulting from a solution to (2.109) will be close to $\gamma_{\min}$ times the bounds selected by the designer. This gives the designer a mechanism for directly shaping the magnitudes of $\bar{\sigma}(S)$, $\bar{\sigma}(T)$, $\bar{\sigma}(KS)$, and so on.

Example 2.17 $\mathcal{H}_\infty$ **mixed sensitivity design for the disturbance process.** *Consider again the plant in (2.62), and consider an $\mathcal{H}_\infty$ mixed sensitivity S/KS design in which*

$$N = \begin{bmatrix} w_P S \\ w_u K S \end{bmatrix} \tag{2.112}$$

Appropriate scaling of the plant has been performed so that the inputs should be about 1 or less in magnitude, and we therefore select a simple input weight $w_u = 1$. The performance weight is chosen, in the form of (2.105), as

$$w_{P1}(s) = \frac{s/M + \omega_B^*}{s + \omega_B^* A}; \quad M = 1.5, \ \omega_B^* = 10, \quad A = 10^{-4} \tag{2.113}$$

A value of $A = 0$ would ask for integral action in the controller, but to get a stable weight and to prevent numerical problems in the algorithm used to synthesize the controller, we have moved the integrator slightly by using a small non-zero value for A. This has no practical significance in terms of control performance. The value $\omega_B^* = 10$ has been selected to achieve approximately the desired crossover frequency ω_c of 10 rad/s. The $\mathcal{H}_\infty$ problem is solved with the Robust Control toolbox in Matlab using the commands in Table 2.4.

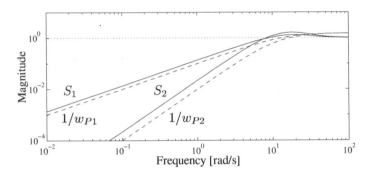

Figure 2.30: Inverse of performance weight (dashed line) and resulting sensitivity function (solid line) for two $\mathcal{H}_\infty$ designs (1 and 2) for the disturbance process

Table 2.4: Matlab program to synthesize $\mathcal{H}_\infty$ controller for Example 2.17

```
% Uses the Robust Control toolbox
G=tf(200,conv([10 1],conv([0.05 1],[0.05 1])));    % Plant is G.
M=1.5; wb=10; A=1.e-4;
Wp = tf([1/M wb], [1 wb*A]); Wu = 1;               % Weights.
% Find H-infinity optimal controller:
[khinf,ghinf,gopt] = mixsyn(G,Wp,Wu,[]);
Marg = allmargin(G*khinf)                          % Gain and phase margins
```

For this problem, we achieved an optimal $\mathcal{H}_\infty$ norm of 1.37, so the weighted sensitivity requirements are not quite satisfied (see design 1 in Figure 2.30 where the curve for $|S_1|$ is slightly above that for $1/|w_{P1}|$). Nevertheless, the design seems good with $\|S\|_\infty = M_S = 1.30$, $\|T\|_\infty = M_T = 1$, $GM = 8$, $PM = 71.19°$ and $\omega_c = 7.22$ rad/s, and the tracking response is very good as shown by curve y_1 in Figure 2.31(a). (The design is actually very similar to the loop-shaping design for references, K_0, which was an inverse-based controller.)

However, we see from curve y_1 in Figure 2.31(b) that the disturbance response is very sluggish. If disturbance rejection is the main concern, then from our earlier discussion in Section 2.6.4 this

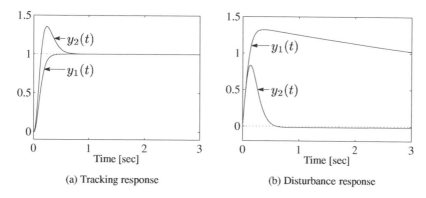

(a) Tracking response   (b) Disturbance response

Figure 2.31: Closed-loop step responses for two alternative $\mathcal{H}_\infty$ designs (1 and 2) for the disturbance process in Example 2.17

motivates the need for a performance weight that specifies higher gains at low frequencies. We therefore try

$$w_{P2}(s) = \frac{(s/M^{1/2} + \omega_B^*)^2}{(s + \omega_B^* A^{1/2})^2}, \quad M = 1.5, \omega_B^* = 10, A = 10^{-4} \tag{2.114}$$

The inverse of this weight is shown in Figure 2.30, and is seen from the dashed line to cross 1 in magnitude at about the same frequency as weight w_{P1}, but it specifies tighter control at lower frequencies. With the weight w_{P2}, we get a design with an optimal $\mathcal{H}_\infty$ norm of 2.19, yielding $M_S = 1.62$, $M_T = 1.42$, $GM = 4.77$, $PM = 43.8°$ and $\omega_c = 11.28$ rad/s. (The design is actually very similar to the loop-shaping design for disturbances, K_3.) The disturbance response is very good, whereas the tracking response has a somewhat high overshoot; see curve y_2 in Figure 2.31(a).

In conclusion, design 1 is best for reference tracking whereas design 2 is best for disturbance rejection. To get a design with both good tracking and good disturbance rejection we need a two degrees-of-freedom controller, as was discussed in Example 2.11 (page 52).

Exercise 2.5 $\mathcal{H}_\infty$ **design for unstable plant.** *Obtain S/KS $\mathcal{H}_\infty$ controllers for the unstable process (2.37) using $w_u = 1$ and the performance weights in (2.113) (design 1) and (2.114) (design 2). Plot the frequency response of the controller for design 1 together with the PI controller (2.38) to confirm that the two controllers are almost identical. You will find that the response with the design 2 (second-order weight) is faster, but on the other hand robustness margins are not quite as good:*

	$\gamma_{min} = \|N\|_\infty$	ω_c	M_S	M_T	GM	GM_L	PM
Design 1:	*3.24*	*4.96*	*1.17*	*1.35*	*18.48*	*0.20*	*61.7°*
Design 2:	*5.79*	*8.21*	*1.31*	*1.56*	*11.56*	*0.23*	*48.5°*

2.9 Conclusion

The main purpose of this chapter has been to present the classical ideas and techniques of feedback control. We have concentrated on SISO systems so that insights into the necessary design trade-offs, and the design approaches available, can be properly developed before MIMO systems are considered. We also introduced the $\mathcal{H}_\infty$ problem based on weighted sensitivity, for which typical performance weights are given in (2.105) and (2.106).

3

INTRODUCTION TO MULTIVARIABLE CONTROL

In this chapter, we introduce the reader to multi-input multi-output (MIMO) systems. It is almost "a book within the book" because a lot of topics are discussed in more detail in later chapters. Topics include transfer functions for MIMO systems, multivariable frequency response analysis and the singular value decomposition (SVD), relative gain array (RGA), multivariable control, and multivariable right-half plane (RHP) zeros. The need for a careful analysis of the effect of uncertainty in MIMO systems is motivated by two examples. Finally, we describe a general control configuration that can be used to formulate control problems. The chapter should be accessible to readers who have attended a classical SISO control course.

3.1 Introduction

We consider a MIMO plant with m inputs and l outputs. Thus, the basic transfer function model is $y(s) = G(s)u(s)$, where y is an $l \times 1$ vector, u is an $m \times 1$ vector and $G(s)$ is an $l \times m$ transfer function matrix.

If we make a change in the first input, u_1, then this will generally affect all the outputs, $y_1, y_2, \ldots, y_l$; that is, there is *interaction* between the inputs and outputs. A non-interacting plant would result if u_1 only affects y_1, u_2 only affects y_2, and so on.

The main difference between a scalar (SISO) system and a MIMO system is the presence of *directions* in the latter. Directions are relevant for vectors and matrices, but not for scalars. However, despite the complicating factor of directions, most of the ideas and techniques presented in the previous chapter on SISO systems may be extended to MIMO systems. The singular value decomposition (SVD) provides a useful way of quantifying multivariable directionality, and we will see that most SISO results involving the absolute value (magnitude) may be generalized to multivariable systems by considering the maximum singular value. An exception to this is Bode's stability condition which has no generalization in terms of singular values. This is related to the fact that it is difficult to find a good measure of phase for MIMO transfer functions.

The chapter is organized as follows. We start by presenting some rules for determining multivariable transfer functions from block diagrams. Although most of the formulae for scalar systems apply, we must exercise some care since matrix multiplication is not commutative: that is, in general $GK \neq KG$. Then we introduce the singular value decomposition and show how it may be used to study directions in multivariable systems.

Multivariable Feedback Control: Analysis and Design. Second Edition
S. Skogestad and I. Postlethwaite © 2005, 2006 John Wiley & Sons, Ltd

We also give a brief introduction to multivariable control and decoupling. We then consider a simple plant with a multivariable RHP-zero and show how the effect of this zero may be shifted from one output channel to another. After this we discuss robustness, and study two example plants, each 2×2, which demonstrate that the simple gain and phase margins used for SISO systems do not generalize easily to MIMO systems. Finally, we consider a general control problem formulation.

At this point, the reader may find it useful to browse through Appendix A where some important mathematical tools are described. Exercises to test understanding of this mathematics are given at the end of this chapter.

3.2   Transfer functions for MIMO systems

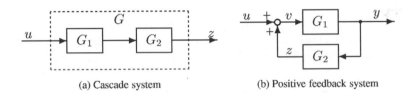

(a) Cascade system   (b) Positive feedback system

Figure 3.1: Block diagrams for the cascade rule and the feedback rule

The following three rules are useful when evaluating transfer functions for MIMO systems.

1. **Cascade rule.** *For the cascade (series) interconnection of G_1 and G_2 in Figure 3.1(a), the overall transfer function matrix is $G = G_2 G_1$.*

Remark. The order of the transfer function matrices in $G = G_2 G_1$ (first G_2 and then G_1) is the reverse of the order in which they appear in the block diagram of Figure 3.1(a) (first G_1 and then G_2). This has led some authors to use block diagrams in which the inputs enter at the right hand side. However, in this case the order of the transfer function blocks in a feedback path will be reversed compared with their order in the formula, so no fundamental benefit is obtained.

2. **Feedback rule.** *With reference to the positive feedback system in Figure 3.1(b), we have $v = (I - L)^{-1}u$ where $L = G_2 G_1$ is the transfer function around the loop.*
3. **Push-through rule.** *For matrices of appropriate dimensions*

$$G_1(I - G_2 G_1)^{-1} = (I - G_1 G_2)^{-1}G_1 \tag{3.1}$$

Proof: Equation (3.1) is verified by premultiplying both sides by $(I - G_1 G_2)$ and postmultiplying both sides by $(I - G_2 G_1)$.   □

Exercise 3.1 * *Derive the cascade and feedback rules.*

The cascade and feedback rules can be combined into the following MIMO rule for evaluating closed-loop transfer functions from block diagrams.

> **MIMO rule:** *Start from the output and write down the blocks as you meet them when moving backwards (against the signal flow) towards the input. If you exit*

from a feedback loop then include a term $(I - L)^{-1}$ for positive feedback (or $(I + L)^{-1}$ for negative feedback) where L is the transfer function around that loop (evaluated against the signal flow starting at the point of exit from the loop). Parallel branches should be treated independently and their contributions added together.

Care should be taken when applying this rule to systems with nested loops. For such systems it is probably safer to write down the signal equations and eliminate internal variables to get the transfer function of interest. The rule is best understood by considering an example.

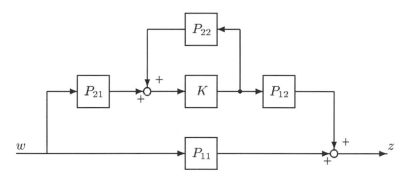

Figure 3.2: Block diagram corresponding to (3.2)

Example 3.1 *The transfer function for the block diagram in Figure 3.2 is given by*

$$z = (P_{11} + P_{12}K(I - P_{22}K)^{-1}P_{21})w \tag{3.2}$$

To derive this from the MIMO rule above we start at the output z and move backwards towards w. There are two branches, one of which gives the term P_{11} directly. In the other branch we move backwards and meet P_{12} and then K. We then exit from a feedback loop and get a term $(I - L)^{-1}$ (positive feedback) with $L = P_{22}K$, and finally we meet P_{21}.

Exercise 3.2 *Use the MIMO rule to derive the transfer functions from u to y and from u to z in Figure 3.1(b). Use the push-through rule to rewrite the two transfer functions.*

Exercise 3.3 * *Use the MIMO rule to show that (2.19) corresponds to the negative feedback system in Figure 2.4.*

Negative feedback control systems

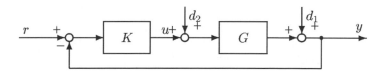

Figure 3.3: Conventional negative feedback control system

For the negative feedback system in Figure 3.3, we define L to be the loop transfer function as seen when breaking the loop at the *output* of the plant. Thus, for the case where the loop consists of a plant G and a feedback controller K we have

$$L = GK \tag{3.3}$$

The sensitivity and complementary sensitivity are then defined as

$$S \triangleq (I + L)^{-1}; \quad T \triangleq I - S = L(I + L)^{-1} \tag{3.4}$$

In Figure 3.3, T is the transfer function from r to y, and S is the transfer function from d_1 to y; also see (2.17) to (2.21) which apply to MIMO systems.

S and T are sometimes called the *output sensitivity* and *output complementary sensitivity*, respectively, and to make this explicit one may use the notation $L_O \equiv L$, $S_O \equiv S$ and $T_O \equiv T$. This is to distinguish them from the corresponding transfer functions evaluated at the *input* to the plant.

We define L_I to be the loop transfer function as seen when breaking the loop at the *input* to the plant with negative feedback assumed. In Figure 3.3

$$L_I = KG \tag{3.5}$$

The *input* sensitivity and *input* complementary sensitivity functions are then defined as

$$S_I \triangleq (I + L_I)^{-1}; \quad T_I \triangleq I - S_I = L_I(I + L_I)^{-1} \tag{3.6}$$

In Figure 3.3, $-T_I$ is the transfer function from d_2 to u. Of course, for SISO systems $L_I = L$, $S_I = S$ and $T_I = T$.

Exercise 3.4 *In Figure 3.3, what transfer function does S_I represent? Evaluate the transfer functions from d_1 and d_2 to $r - y$.*

The following relationships are useful:

$$(I + L)^{-1} + L(I + L)^{-1} = S + T = I \tag{3.7}$$

$$G(I + KG)^{-1} = (I + GK)^{-1}G \tag{3.8}$$

$$GK(I + GK)^{-1} = G(I + KG)^{-1}K = (I + GK)^{-1}GK \tag{3.9}$$

$$T = L(I + L)^{-1} = (I + (L)^{-1})^{-1} \tag{3.10}$$

Note that the matrices G and K in (3.7)–(3.10) need not be square whereas $L = GK$ is square: (3.7) follows trivially by factorizing out the term $(I + L)^{-1}$ from the right; (3.8) says that $GS_I = SG$ and follows from the push-through rule; (3.9) also follows from the push-through rule; (3.10) can be derived from the identity $M_1^{-1}M_2^{-1} = (M_2M_1)^{-1}$.

Similar relationships, but with G and K interchanged, apply for the transfer functions evaluated at the plant input. To assist in remembering (3.7)–(3.10) note that G comes first (because the transfer function is evaluated at the output) and then G and K alternate in sequence. A given transfer matrix never occurs twice in sequence. For example, the closed-loop transfer function $G(I+GK)^{-1}$ does *not* exist (unless G is repeated in the block diagram, but then these G's would actually represent two different physical entities).

Remark 1 The above identities are clearly useful when deriving transfer functions analytically, but they are also useful for numerical calculations involving state-space realizations, e.g. $L(s) = C(sI - A)^{-1}B + D$. For example, assume we have been given a state-space realization for $L = GK$ with n states (so A is an $n \times n$ matrix) and we want to find the state-space realization of T. Then we can first form $S = (I + L)^{-1}$ with n states, and then multiply it by L to obtain $T = SL$ with $2n$ states. However, a minimal realization of T has only n states. This may be obtained numerically using model reduction, but it is preferable to find it directly using $T = I - S$, see (3.7).

Remark 2 Note also that the right identity in (3.10) can only be used to compute the state-space realization of T if that of L^{-1} exists, so L must be semi-proper with $D \neq 0$ (which is rarely the case in practice). On the other hand, since L is square, we can always compute the frequency response of $L(j\omega)^{-1}$ (except at frequencies where $L(s)$ has $j\omega$-axis poles), and then obtain $T(j\omega)$ from (3.10).

Remark 3 In Appendix A.7 we present some factorizations of the sensitivity function which will be useful in later applications. For example, (A.147) relates the sensitivity of a perturbed plant, $S' = (I + G'K)^{-1}$, to that of the nominal plant, $S = (I + GK)^{-1}$. We have

$$S' = S(I + E_O T)^{-1}, \quad E_O \triangleq (G' - G)G^{-1} \tag{3.11}$$

where E_O is an output multiplicative perturbation representing the difference between G and G', and T is the nominal complementary sensitivity function.

3.3 Multivariable frequency response analysis

The transfer function $G(s)$ is a function of the Laplace variable s and can be used to represent a dynamic system. However, if we fix $s = s_0$ then we may view $G(s_0)$ simply as an $l \times m$ complex matrix (with m inputs and l outputs), which can be analyzed using standard tools in matrix algebra. In particular, the choice $s_0 = j\omega$ is of interest since $G(j\omega)$ represents the response to a sinusoidal signal of frequency ω.

3.3.1 Obtaining the frequency response from $G(s)$

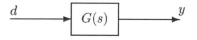

Figure 3.4: System $G(s)$ with input d and output y

The frequency domain is ideal for studying directions in multivariable systems at any given frequency. Consider the system $G(s)$ in Figure 3.4 with input $d(s)$ and output $y(s)$:

$$y(s) = G(s)d(s) \tag{3.12}$$

(We denote the input here by d rather than by u to avoid confusion with the matrix U used below in the singular value decomposition.) In Section 2.1 we considered the sinusoidal response of scalar systems. These results may be directly generalized to multivariable systems by considering the elements g_{ij} of the matrix G. We have

- $g_{ij}(j\omega)$ represents the sinusoidal response from input j to output i.

To be more specific, we apply to input channel j a scalar sinusoidal signal given by

$$d_j(t) = d_{j0}\sin(\omega t + \alpha_j) \tag{3.13}$$

This input signal is persistent: that is, it has been applied since $t = -\infty$. Then the corresponding persistent output signal in channel i is also a sinusoid with the same frequency

$$y_i(t) = y_{i0}\sin(\omega t + \beta_i) \tag{3.14}$$

where the amplification (gain) and phase shift may be obtained from the complex number $g_{ij}(j\omega)$ as follows:

$$\frac{y_{io}}{d_{jo}} = |g_{ij}(j\omega)|, \quad \beta_i - \alpha_j = \angle g_{ij}(j\omega) \tag{3.15}$$

In *phasor notation*, see (2.5) and (2.10), we may compactly represent the sinusoidal time response described in (3.13)–(3.15) by

$$y_i(\omega) = g_{ij}(j\omega)d_j(\omega) \tag{3.16}$$

where

$$d_j(\omega) = d_{jo}e^{j\alpha_j}, \quad y_i(\omega) = y_{io}e^{j\beta_i} \tag{3.17}$$

Here the use of ω (and not $j\omega$) as the argument of $d_j(\omega)$ and $y_i(\omega)$ implies that these are complex numbers, representing at each frequency ω the magnitude and phase of the sinusoidal signals in (3.13) and (3.14).

The overall response to simultaneous input signals of the same frequency in several input channels is, by the superposition principle for linear systems, equal to the sum of the individual responses, and we have from (3.16)

$$y_i(\omega) = g_{i1}(j\omega)d_1(\omega) + g_{i2}(j\omega)d_2(\omega) + \cdots = \sum_j g_{ij}(j\omega)d_j(\omega) \tag{3.18}$$

or in matrix form

$$\boxed{y(\omega) = G(j\omega)d(\omega)} \tag{3.19}$$

where

$$d(\omega) = \begin{bmatrix} d_1(\omega) \\ d_2(\omega) \\ \vdots \\ d_m(\omega) \end{bmatrix} \quad \text{and} \quad y(\omega) = \begin{bmatrix} y_1(\omega) \\ y_2(\omega) \\ \vdots \\ y_l(\omega) \end{bmatrix} \tag{3.20}$$

represent the vectors of sinusoidal input and output signals.

Example 3.2 *Consider a 2×2 multivariable system where we simultaneously apply sinusoidal signals of the same frequency ω to the two input channels:*

$$d(t) = \begin{bmatrix} d_1(t) \\ d_2(t) \end{bmatrix} = \begin{bmatrix} d_{10}\sin(\omega t + \alpha_1) \\ d_{20}\sin(\omega t + \alpha_2) \end{bmatrix} \quad \text{or} \quad d(\omega) = \begin{bmatrix} d_{10}e^{j\alpha_1} \\ d_{20}e^{j\alpha_2} \end{bmatrix} \tag{3.21}$$

The corresponding output signal is

$$y(t) = \begin{bmatrix} y_1(t) \\ y_2(t) \end{bmatrix} = \begin{bmatrix} y_{10}\sin(\omega t + \beta_1) \\ y_{20}\sin(\omega t + \beta_2) \end{bmatrix} \quad \text{or} \quad y(\omega) = \begin{bmatrix} y_{10}e^{j\beta_1} \\ y_{20}e^{j\beta_2} \end{bmatrix} \tag{3.22}$$

$y(\omega)$ is obtained by multiplying the complex matrix $G(j\omega)$ by the complex vector $d(\omega)$, as given in (3.19).

3.3.2 Directions in multivariable systems

For a SISO system, $y = Gd$, the gain at a given frequency is simply

$$\frac{|y(\omega)|}{|d(\omega)|} = \frac{|G(j\omega)d(\omega)|}{|d(\omega)|} = |G(j\omega)| \tag{3.23}$$

The gain depends on the frequency ω, but since the system is linear it is independent of the input magnitude $|d(\omega)|$.

Things are not quite as simple for MIMO systems where the input and output signals are both vectors, and we need to "sum up" the magnitudes of the elements in each vector by use of some norm, as discussed in Appendix A.5.1. If we select the vector 2-norm, the usual measure of length, then at a given frequency ω the magnitude of the vector input signal is

$$\|d(\omega)\|_2 = \sqrt{\sum_j |d_j(\omega)|^2} = \sqrt{d_{10}^2 + d_{20}^2 + \cdots} \tag{3.24}$$

and the magnitude of the vector output signal is

$$\|y(\omega)\|_2 = \sqrt{\sum_i |y_i(\omega)|^2} = \sqrt{y_{10}^2 + y_{20}^2 + \cdots} \tag{3.25}$$

The *gain* of the system $G(s)$ for a particular input signal $d(\omega)$ is then given by the ratio

$$\frac{\|y(\omega)\|_2}{\|d(\omega)\|_2} = \frac{\|G(j\omega)d(\omega)\|_2}{\|d(\omega)\|_2} = \frac{\sqrt{y_{10}^2 + y_{20}^2 + \cdots}}{\sqrt{d_{10}^2 + d_{20}^2 + \cdots}} \tag{3.26}$$

Again the gain depends on the frequency ω, and again it is independent of the input magnitude $\|d(\omega)\|_2$. However, for a MIMO system there are additional degrees of freedom and the gain depends also on the *direction* of the input d.[1] The maximum gain as the direction of the input is varied is the maximum singular value of G,

$$\max_{d \neq 0} \frac{\|Gd\|_2}{\|d\|_2} = \max_{\|d\|_2=1} \|Gd\|_2 = \bar{\sigma}(G) \tag{3.27}$$

whereas the minimum gain is the minimum singular value of G,

$$\min_{d \neq 0} \frac{\|Gd\|_2}{\|d\|_2} = \min_{\|d\|_2=1} \|Gd\|_2 = \underline{\sigma}(G) \tag{3.28}$$

The first identities in (3.27) and (3.28) follow because the gain is independent of the input magnitude for a linear system.

Example 3.3 *For a system with two inputs,* $d = \begin{bmatrix} d_{10} \\ d_{20} \end{bmatrix}$, *the gain is in general different for the following five inputs:*

$$d_1 = \begin{bmatrix} 1 \\ 0 \end{bmatrix}, \; d_2 = \begin{bmatrix} 0 \\ 1 \end{bmatrix}, \; d_3 = \begin{bmatrix} 0.707 \\ 0.707 \end{bmatrix}, \; d_4 = \begin{bmatrix} 0.707 \\ -0.707 \end{bmatrix}, \; d_5 = \begin{bmatrix} 0.6 \\ -0.8 \end{bmatrix}$$

---

[1] The term *direction* refers to a normalized vector of unit length.

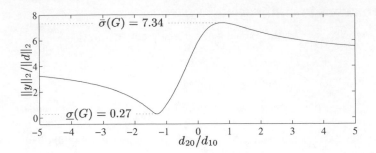

Figure 3.5: Gain $\|Gd\|_2/\|d\|_2$ as a function of d_{20}/d_{10} for G in (3.29)

*(which all have the same magnitude $\|d\|_2 = 1$ but are in different directions). For example, for the 2×2
system*

$$G = \begin{bmatrix} 5 & 4 \\ 3 & 2 \end{bmatrix} \tag{3.29}$$

(a constant matrix) we compute for the five inputs d_j the following output vectors:

$$y_1 = \begin{bmatrix} 5 \\ 3 \end{bmatrix}, \; y_2 = \begin{bmatrix} 4 \\ 2 \end{bmatrix}, \; y_3 = \begin{bmatrix} 6.36 \\ 3.54 \end{bmatrix}, \; y_4 = \begin{bmatrix} 0.707 \\ 0.707 \end{bmatrix}, \; y_5 = \begin{bmatrix} -0.2 \\ 0.2 \end{bmatrix}$$

and the 2-norms of these five outputs (i.e. the gains for the five inputs) are

$$\|y_1\|_2 = 5.83, \; \|y_2\|_2 = 4.47, \; \|y_3\|_2 = 7.30, \; \|y_4\|_2 = 1.00, \; \|y_5\|_2 = 0.28$$

*This dependency of the gain on the input direction is illustrated graphically in Figure 3.5 where we
have used the ratio d_{20}/d_{10} as an independent variable to represent the input direction. We see that,
depending on the ratio d_{20}/d_{10}, the gain varies between 0.27 and 7.34. These are the minimum and
maximum singular values of G, respectively.*

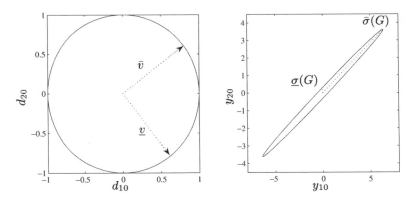

Figure 3.6: Outputs (right plot) resulting from use of $\|d\|_2 = 1$ (unit circle in left plot) for system G
in (3.29). The maximum ($\bar{\sigma}(G)$) and minimum ($\underline{\sigma}(G)$) gains are obtained for $d = (\bar{v})$ and $d = (\underline{v})$
respectively.

*An alternative plot, which shows the directions of the outputs more clearly, is shown in Figure 3.6.
From the shape of the output space (right plot), we see that it is easy to increase both y_{10} and
y_{20} simultaneously (gain $\bar{\sigma}(G) = 7.34$), but difficult to increase one and decrease the other (gain
$\underline{\sigma}(G) = 0.27$).*

3.3.3 Eigenvalues are a poor measure of gain

Before discussing in more detail the singular value decomposition, we want to demonstrate that the magnitudes of the eigenvalues of a transfer function matrix, e.g. $|\lambda_i(G(j\omega))|$, do *not* provide a useful means of generalizing the SISO gain, $|G(j\omega)|$. First of all, eigenvalues can only be computed for square systems, and even then they can be very misleading. To see this, consider the system $y = Gd$ with

$$G = \begin{bmatrix} 0 & 100 \\ 0 & 0 \end{bmatrix} \tag{3.30}$$

which has both eigenvalues λ_i equal to zero. However, to conclude from the eigenvalues that the system gain is zero is clearly misleading. For example, with an input vector $d = \begin{bmatrix} 0 & 1 \end{bmatrix}^T$ we get an output vector $y = \begin{bmatrix} 100 & 0 \end{bmatrix}^T$.

The "problem" is that the eigenvalues measure the gain for the special case when the inputs and the outputs are in the same direction, namely in the direction of the eigenvectors. To see this let t_i be an eigenvector of G and consider an input $d = t_i$. Then the output is $y = Gt_i = \lambda_i t_i$ where λ_i is the corresponding eigenvalue. We get

$$\|y\|/\|d\| = \|\lambda_i t_i\|/\|t_i\| = |\lambda_i|$$

so $|\lambda_i|$ measures the gain in the direction t_i. This may be useful for stability analysis, but not for performance.

To find useful generalizations of gain of G for the case when G is a matrix, we need the concept of a *matrix norm*, denoted $\|G\|$. Two important properties which must be satisfied for a matrix norm are the *triangle inequality*

$$\|G_1 + G_2\| \leq \|G_1\| + \|G_2\| \tag{3.31}$$

and the *multiplicative property*

$$\|G_1 G_2\| \leq \|G_1\| \cdot \|G_2\| \tag{3.32}$$

(see Appendix A.5 for more details). As we may expect, the magnitude of the largest eigenvalue, $\rho(G) \triangleq |\lambda_{\max}(G)|$ (the spectral radius), does *not* satisfy the properties of a matrix norm; also see (A.116).

In Appendix A.5.2 we introduce several matrix norms, such as the Frobenius norm $\|G\|_F$, the sum norm $\|G\|_{\text{sum}}$, the maximum column sum $\|G\|_{i1}$, the maximum row sum $\|G\|_{i\infty}$, and the maximum singular value $\|G\|_{i2} = \bar{\sigma}(G)$ (the latter three norms are induced by a vector norm, e.g. see (3.27); this is the reason for the subscript i). Actually, the choice of matrix norm among these is not critical because the various norms of an $l \times m$ matrix differ at most by a factor $\sqrt{ml}$, see (A.119)–(A.124). In this book, we will use all of the above norms, each depending on the situation. However, in this chapter we will mainly use the induced 2-norm, $\bar{\sigma}(G)$. Notice that $\bar{\sigma}(G) = 100$ for the matrix in (3.30).

Exercise 3.5 * *Compute the spectral radius and the five matrix norms mentioned above for the matrices in (3.29) and (3.30).*

3.3.4 Singular value decomposition

The singular value decomposition (SVD) is defined in Appendix A.3. Here we are interested in its physical interpretation when applied to the frequency response of a MIMO system $G(s)$ with m inputs and l outputs.

Consider a fixed frequency ω where $G(j\omega)$ is a constant $l \times m$ complex matrix, and denote $G(j\omega)$ by G for simplicity. Any matrix G may be decomposed into its singular value decomposition, and we write

$$G = U\Sigma V^H \tag{3.33}$$

where

Σ is an $l \times m$ matrix with $k = \min\{l, m\}$ non-negative singular values, σ_i, arranged in descending order along its *main diagonal*; the other entries are zero. The singular values are the positive square roots of the eigenvalues of $G^H G$, where G^H is the complex conjugate transpose of G,

$$\sigma_i(G) = \sqrt{\lambda_i(G^H G)} \tag{3.34}$$

U is an $l \times l$ unitary matrix of output singular vectors, u_i,

V is an $m \times m$ unitary matrix of input singular vectors, v_i.

In short, any matrix may be decomposed into an input rotation V, a scaling matrix Σ and an output rotation U. This is illustrated by the SVD of a real 2×2 matrix which can always be written in the form

$$G = \underbrace{\begin{bmatrix} \cos\theta_1 & -\sin\theta_1 \\ \sin\theta_1 & \cos\theta_1 \end{bmatrix}}_{U} \underbrace{\begin{bmatrix} \sigma_1 & 0 \\ 0 & \sigma_2 \end{bmatrix}}_{\Sigma} \underbrace{\begin{bmatrix} \cos\theta_2 & \pm\sin\theta_2 \\ -\sin\theta_2 & \pm\cos\theta_2 \end{bmatrix}^T}_{V^T} \tag{3.35}$$

where the angles θ_1 and θ_2 depend on the given matrix. From (3.35) we see that the matrices U and V involve rotations and that their columns are orthonormal.

The singular values are sometimes called the principal values or principal gains, and the associated directions are called principal directions. In general, the singular values must be computed numerically. For 2×2 matrices, however, analytic expressions for the singular values are given in (A.37).

Caution. It is standard notation to use the symbol U to denote the matrix of *output* singular vectors. This is unfortunate as it is also standard notation to use u (lower case) to represent the *input* signal. The reader should be careful not to confuse these two.

Input and output directions. The column vectors of U, denoted u_i, represent the *output directions* of the plant. They are orthogonal and of unit length (orthonormal), i.e.

$$\|u_i\|_2 = \sqrt{|u_{i1}|^2 + |u_{i2}|^2 + \cdots + |u_{il}|^2} = 1 \tag{3.36}$$

$$u_i^H u_i = 1, \quad u_i^H u_j = 0, \quad i \neq j \tag{3.37}$$

Likewise, the column vectors of V, denoted v_i, are orthogonal and of unit length, and represent the *input directions*. These input and output directions are related through the singular values. To see this, note that since V is unitary we have $V^H V = I$, so (3.33) may be written as $GV = U\Sigma$, which for column i becomes

$$Gv_i = \sigma_i u_i \tag{3.38}$$

where v_i and u_i are vectors, whereas σ_i is a scalar. That is, if we consider an *input* in the direction v_i, then the *output* is in the direction u_i. Furthermore, since $\|v_i\|_2 = 1$ and

$\|u_i\|_2 = 1$ we see that the i'th singular value σ_i directly gives the gain of the matrix G in this direction. In other words

$$\sigma_i(G) = \|Gv_i\|_2 = \frac{\|Gv_i\|_2}{\|v_i\|_2} \tag{3.39}$$

Some advantages of the SVD over the eigenvalue decomposition for analyzing gains and directionality of multivariable plants are:

1. The singular values give better information about the gains of the plant.
2. The plant directions obtained from the SVD are orthogonal.
3. The SVD also applies directly to non-square plants.

Maximum and minimum singular values. As already stated, it can be shown that the largest gain for *any* input direction is equal to the maximum singular value

$$\bar{\sigma}(G) \equiv \sigma_1(G) = \max_{d \neq 0} \frac{\|Gd\|_2}{\|d\|_2} = \frac{\|Gv_1\|_2}{\|v_1\|_2} \tag{3.40}$$

and that the smallest gain for any input direction (excluding the "wasted" inputs in the null space of G for cases with more inputs than outputs[2])
is equal to the minimum singular value

$$\underline{\sigma}(G) \equiv \sigma_k(G) = \min_{d \neq 0} \frac{\|Gd\|_2}{\|d\|_2} = \frac{\|Gv_k\|_2}{\|v_k\|_2} \tag{3.41}$$

where $k = \min\{l, m\}$. Thus, for any vector d, not in the null space of G, we have that

$$\underline{\sigma}(G) \leq \frac{\|Gd\|_2}{\|d\|_2} \leq \bar{\sigma}(G) \tag{3.42}$$

Defining $u_1 = \bar{u}, v_1 = \bar{v}, u_k = \underline{u}$ and $v_k = \underline{v}$, then it follows that

$$G\bar{v} = \bar{\sigma}\bar{u}, \qquad G\underline{v} = \underline{\sigma}\,\underline{u} \tag{3.43}$$

The vector $\bar{v}$ corresponds to the input direction with largest amplification, and $\bar{u}$ is the corresponding output direction in which the inputs are most effective. The directions involving $\bar{v}$ and $\bar{u}$ are sometimes referred to as the "strongest", "high-gain" or "most important" directions. The next most important directions are associated with v_2 and u_2, and so on (see Appendix A.3.5) until the "least important", "weak" or "low-gain" directions which are associated with $\underline{v}$ and $\underline{u}$.

Example 3.3 continued. *Consider again the system (3.29) with*

$$G = \begin{bmatrix} 5 & 4 \\ 3 & 2 \end{bmatrix} \tag{3.44}$$

The SVD of G_1 is

$$G = \underbrace{\begin{bmatrix} 0.872 & 0.490 \\ 0.490 & -0.872 \end{bmatrix}}_{U} \underbrace{\begin{bmatrix} 7.343 & 0 \\ 0 & 0.272 \end{bmatrix}}_{\Sigma} \underbrace{\begin{bmatrix} 0.794 & -0.608 \\ 0.608 & 0.794 \end{bmatrix}^H}_{V^H}$$

---

[2] For a "fat" matrix G with more inputs than outputs ($m > l$), we can always choose a non-zero input d in the null space of G such that $Gd = 0$.

The largest gain of 7.343 is for an input in the direction $\bar{v} = \begin{bmatrix} 0.794 \\ 0.608 \end{bmatrix}$. *The smallest gain of 0.272 is for*

an input in the direction $\underline{v} = \begin{bmatrix} -0.608 \\ 0.794 \end{bmatrix}$. *This confirms the findings on page 73 (see Figure 3.6).*

Note that the directions in terms of the singular vectors are not unique, in the sense that the elements in each pair of vectors (u_i, v_i) may be multiplied by a complex scalar c of magnitude 1 ($|c| = 1$). This is easily seen from (3.38). For example, we may change the sign of the vector $\bar{v}$ (multiply by $c = -1$) provided we also change the sign of the vector $\bar{u}$. Also, if you use Matlab to compute the SVD of the matrix in (3.44) (g=[5 4; 3 2]; [u,s,v]=svd(g)), then you will probably find that the signs of the elements in U and V are different from those given above.

Since in (3.44) both inputs affect both outputs, we say that the system is *interactive*. This follows from the relatively large off-diagonal elements in G in (3.44). Furthermore, the system is *ill-conditioned*: that is, some combinations of the inputs have a strong effect on the outputs, whereas other combinations have a weak effect on the outputs. This may be quantified by the *condition number*: the ratio between the gains in the strong and weak directions, which for the system in (3.44) is $\gamma = \bar{\sigma}/\underline{\sigma} = 7.343/0.272 = 27.0$.

Example 3.4 Shopping cart. *Consider a shopping cart (supermarket trolley) with fixed wheels which we may want to move in three directions: forwards, sideways and upwards. This is a simple illustrative example where we can easily figure out the principal directions from experience. The strongest direction, corresponding to the largest singular value, will clearly be in the forwards direction. The next direction, corresponding to the second singular value, will be sideways. Finally, the most "difficult" or "weak" direction, corresponding to the smallest singular value, will be upwards (lifting up the cart).*

For the shopping cart the gain depends strongly on the input direction, i.e. the plant is ill-conditioned. Control of ill-conditioned plants is sometimes difficult, and the control problem associated with the shopping cart can be described as follows. Assume we want to push the shopping cart sideways (maybe we are blocking someone). This is rather difficult (the plant has low gain in this direction) so a strong force is needed. However, if there is any uncertainty in our knowledge about the direction the cart is pointing, then some of our applied force will be directed forwards (where the plant gain is large) and the cart will suddenly move forward with an undesired large speed. We thus see that the control of an ill-conditioned plant may be especially difficult if there is input uncertainty which can cause the input signal to "spread" from one input direction to another. We will discuss this in more detail later.

Example 3.5 Distillation process. *Consider the following steady-state model of a distillation column:*

$$G = \begin{bmatrix} 87.8 & -86.4 \\ 108.2 & -109.6 \end{bmatrix} \tag{3.45}$$

The variables have been scaled as discussed in Section 1.4. Thus, since the elements are much larger than 1 in magnitude this suggests that there will be no problems with input constraints. However, this is somewhat misleading as the gain in the low-gain direction (corresponding to the smallest singular value) is actually only just above 1. To see this consider the SVD of G:

$$G = \underbrace{\begin{bmatrix} 0.625 & -0.781 \\ 0.781 & 0.625 \end{bmatrix}}_{U} \underbrace{\begin{bmatrix} 197.2 & 0 \\ 0 & 1.39 \end{bmatrix}}_{\Sigma} \underbrace{\begin{bmatrix} 0.707 & -0.708 \\ -0.708 & -0.707 \end{bmatrix}^{H}}_{V^H} \tag{3.46}$$

From the first input singular vector, $\bar{v} = [0.707 \quad -0.708]^T$, *we see that the gain is 197.2 when we increase one input and decrease the other input by a similar amount. On the other hand, from the second input singular vector,* $\underline{v} = [-0.708 \quad -0.707]^T$, *we see that if we change both inputs by the same amount then the gain is only 1.39. The reason for this is that the plant is such that the two inputs* counteract each other. Thus, the distillation process is ill-conditioned, at least at steady-state, and the condition number is $197.2/1.39 = 141.7$. The physics of this example is discussed in more detail below, and later in this chapter we will consider a simple controller design (see Motivating robustness example no. 2 in Section 3.7.2).

Example 3.6 Physics of the distillation process. *The model in (3.45) represents two-point (dual) composition control of a distillation column, where the top composition is to be controlled at $y_D = 0.99$ (output y_1) and the bottom composition at $x_B = 0.01$ (output y_2), using reflux L (input u_1) and boilup V (input u_2) as manipulated inputs (see Figure 10.6 on page 408). Note that we have here returned to the convention of using u_1 and u_2 to denote the manipulated inputs; the output singular vectors will be denoted by $\bar{u}$ and $\underline{u}$.*

The $1,1$-element of the gain matrix G is 87.8. Thus an increase in u_1 by 1 (with u_2 constant) yields a large steady-state change in y_1 of 87.8; that is, the outputs are very sensitive to changes in u_1. Similarly, an increase in u_2 by 1 (with u_1 constant) yields $y_1 = -86.4$. Again, this is a very large change, but in the opposite direction of that for the increase in u_1. We therefore see that changes in u_1 and u_2 counteract each other, and if we increase u_1 and u_2 simultaneously by 1, then the overall steady-state change in y_1 is only $87.8 - 86.4 = 1.4$.

Physically, the reason for this small change is that the compositions in the distillation column are only weakly dependent on changes in the internal *flows (i.e. simultaneous changes in the internal flows L and V). This can also be seen from the smallest singular value, $\underline{\sigma}(G) = 1.39$, which is obtained for inputs in the direction $\underline{v} = \begin{bmatrix} -0.708 \\ -0.707 \end{bmatrix}$. From the output singular vector $\underline{u} = \begin{bmatrix} -0.781 \\ 0.625 \end{bmatrix}$ we see that the effect is to move the outputs in different directions; that is, to change $y_1 - y_2$. Therefore, it takes a large control action to move the compositions in different directions; that is, to make both products purer simultaneously. This makes sense from a physical point of view.*

On the other hand, the distillation column is very sensitive to changes in external *flows (i.e. increase $u_1 - u_2 = L - V$). This can be seen from the input singular vector $\bar{v} = \begin{bmatrix} 0.707 \\ -0.708 \end{bmatrix}$ associated with the largest singular value, and is a general property of distillation columns where both products are of high purity. The reason for this is that the external distillate flow (which varies as $V - L$) has to be about equal to the amount of light component in the feed, and even a small imbalance leads to large changes in the product compositions.*

For dynamic systems the singular values and their associated directions vary with frequency, and for control purposes it is usually the frequency range corresponding to the closed-loop bandwidth which is of main interest. The singular values are usually plotted as a function of frequency in a Bode magnitude plot with a log-scale for frequency and magnitude. Typical plots are shown in Figure 3.7.

Non-square plant

The SVD is also useful for non-square plants. For example, consider a plant with two inputs and three outputs. In this case the third output singular vector, u_3, tells us in which output direction the plant cannot be controlled. Similarly, for a plant with more inputs than outputs, the additional input singular vectors tell us in which directions the input will have no effect.

Example 3.7 *Consider a non-square system with three inputs and two outputs,*

$$G_2 = \begin{bmatrix} 5 & 4 & 1 \\ 3 & 2 & -1 \end{bmatrix}$$

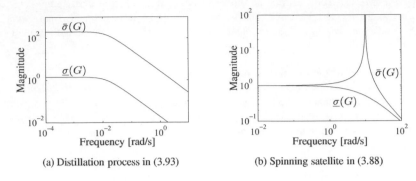

(a) Distillation process in (3.93)   (b) Spinning satellite in (3.88)

Figure 3.7: Typical plots of singular values

with SVD

$$G_2 = \underbrace{\begin{bmatrix} 0.877 & 0.481 \\ 0.481 & -0.877 \end{bmatrix}}_{U} \underbrace{\begin{bmatrix} 7.354 & 0 & 0 \\ 0 & 1.387 & 0 \end{bmatrix}}_{\Sigma} \underbrace{\begin{bmatrix} 0.792 & -0.161 & 0.588 \\ 0.608 & 0.124 & -0.785 \\ 0.054 & 0.979 & 0.196 \end{bmatrix}^{H}}_{V^{H}}$$

From our definition, the minimum singular value is $\underline{\sigma}(G_2) = 1.387$, but note that an input d in the direction $v_3 = \begin{bmatrix} 0.588 \\ -0.785 \\ 0.196 \end{bmatrix}$ is in the null space of G and yields a zero output, $y = Gd = 0$.

Exercise 3.6 *For a system with m inputs and one output, what is the interpretation of the singular values and the associated input directions (V)? What is U in this case?*

3.3.5   Singular values for performance

So far we have used the SVD primarily to gain insight into the directionality of MIMO systems. But the maximum singular value is also very useful in terms of frequency domain performance and robustness. We consider performance here.

For SISO systems we earlier found that $|S(j\omega)|$ evaluated as a function of frequency gives useful information about the effectiveness of feedback control. For example, it is the gain from a sinusoidal reference input (or output disturbance) $r(\omega)$[3] to the control error, $|e(\omega)| = |S(j\omega)| \cdots |r(\omega)|$.

For MIMO systems a useful generalization results if we consider the ratio $\|e(\omega)\|_2/\|r(\omega)\|_2$, where r is the vector of reference inputs, e is the vector of control errors, and $\|\cdot\|_2$ is the vector 2-norm. As explained above, this gain depends on the *direction* of $r(\omega)$ and we have from (3.42) that it is bounded by the maximum and minimum singular value of S,

$$\underline{\sigma}(S(j\omega)) \leq \frac{\|e(\omega)\|_2}{\|r(\omega)\|_2} \leq \bar{\sigma}(S(j\omega)) \tag{3.47}$$

In terms of *performance*, it is reasonable to require that the gain $\|e(\omega)\|_2/\|r(\omega)\|_2$ remains small for any direction of $r(\omega)$, including the "worst-case" direction which gives a gain of

---

[3] We use phasor notation here, see page 18, and $|r(\omega)|$ is the magnitude of the sinusoidal signal at frequency ω.

$\bar{\sigma}(S(j\omega))$. Let $1/|w_P(j\omega)|$ (the inverse of the performance weight) represent the maximum allowed magnitude of $\|e\|_2/\|r\|_2$ at each frequency. This results in the following performance requirement:

$$\bar{\sigma}(S(j\omega)) < 1/|w_P(j\omega)|, \; \forall\omega \quad \Leftrightarrow \quad \bar{\sigma}(w_P S) < 1, \; \forall\omega$$

$$\Leftrightarrow \quad \|w_P S\|_\infty < 1 \tag{3.48}$$

where the $\mathcal{H}_\infty$ norm (see also page 60) is defined as the peak of the maximum singular value of the frequency response

$$\|M(s)\|_\infty \triangleq \max_\omega \bar{\sigma}(M(j\omega)) \tag{3.49}$$

Typical performance weights $w_P(s)$ are given in Section 2.8.2, which should be studied carefully.

The singular values of $S(j\omega)$ may be plotted as functions of frequency, as illustrated later in Figure 3.12(a). Typically, they are small at low frequencies where feedback is effective, and they approach 1 at high frequencies because any real system is strictly proper:

$$\omega \to \infty: \quad L(j\omega) \to 0 \quad \Rightarrow \quad S(j\omega) \to I \tag{3.50}$$

The maximum singular value, $\bar{\sigma}(S(j\omega))$, usually has a peak larger than 1 around the crossover frequencies. This peak is undesirable, but it is unavoidable for real systems.

As for SISO systems we define the bandwidth as the frequency up to which feedback is effective. For MIMO systems the bandwidth will depend on directions, and we have a *bandwidth region* between a lower frequency where the maximum singular value, $\bar{\sigma}(S)$, reaches 0.7 (the "low-gain" or "worst-case" direction), and a higher frequency where the minimum singular value, $\underline{\sigma}(S)$, reaches 0.7 (the "high-gain" or "best-case")[4]. If we want to associate a single bandwidth frequency for a multivariable system, then we consider the worst-case (low-gain) direction, and define

- *Bandwidth*, ω_B: Frequency where $\bar{\sigma}(S)$ crosses $\frac{1}{\sqrt{2}} = 0.7$ from below.

It is then understood that the bandwidth is at least ω_B for any direction of the input (reference or disturbance) signal. Since $S = (I + L)^{-1}$, (A.54) yields

$$\underline{\sigma}(L) - 1 \leq \frac{1}{\bar{\sigma}(S)} \leq \underline{\sigma}(L) + 1 \tag{3.51}$$

Thus at frequencies where feedback is effective (namely where $\underline{\sigma}(L) \gg 1$) we have $\bar{\sigma}(S) \approx 1/\underline{\sigma}(L)$, and at the bandwidth frequency (where $1/\bar{\sigma}(S(j\omega_B)) = \sqrt{2} = 1.41$) we have that $\underline{\sigma}(L(j\omega_B))$ is between 0.41 and 2.41. Thus, the bandwidth is approximately where $\underline{\sigma}(L)$ crosses 1. Finally, at higher frequencies, where for any real system $\underline{\sigma}(L)$ (and $\bar{\sigma}(L)$) is small, we have that $\bar{\sigma}(S) \approx 1$.

3.3.6   Condition number

In Examples 3.4 and 3.5, we noted that the system's gain varied considerably with the input direction. Such systems are said to have strong directionality. Two measures which are used to

---

[4] The terms "low-gain" and "high-gain" refer to L, whereas the terms "worst-case" and "best-case" refer to the resulting speed of response for the closed-loop system.

quantify the degree of directionality and the level of (two-way) interactions in MIMO systems are the condition number and the relative gain array (RGA), respectively. We first consider the *condition number* of a matrix which is defined as the ratio between the maximum and minimum singular values,

$$\gamma(G) \triangleq \bar{\sigma}(G)/\underline{\sigma}(G) \tag{3.52}$$

A matrix with a large condition number is said to be *ill-conditioned*. For a non-singular (square) matrix $\underline{\sigma}(G) = 1/\bar{\sigma}(G^{-1})$, so $\gamma(G) = \bar{\sigma}(G)\bar{\sigma}(G^{-1})$. It then follows from (A.120) that the condition number is large if both G and G^{-1} have large elements.

The condition number depends strongly on the scaling of the inputs and outputs. To be more specific, if D_1 and D_2 are diagonal scaling matrices, then the condition numbers of the matrices G and $D_1 G D_2$ may be arbitrarily far apart. In general, the matrix G should be scaled on physical grounds, e.g. by dividing each input and output by its largest expected or desired value as discussed in Section 1.4.

One might also consider minimizing the condition number over all possible scalings. This results in the *minimized or optimal condition number* which is defined by

$$\gamma^*(G) = \min_{D_1, D_2} \gamma(D_1 G D_2) \tag{3.53}$$

and can be computed using (A.74).

The condition number has been used as an input–output controllability measure, and in particular it has been postulated that a large condition number indicates sensitivity to uncertainty. This is not true in general, but the reverse holds: if the condition number is small, then the multivariable effects of uncertainty are not likely to be serious (see (6.89)).

If the condition number is large (say, larger than 10), then this may *indicate* control problems:

1. A large condition number $\gamma(G) = \bar{\sigma}(G)/\underline{\sigma}(G)$ may be caused by a small value of $\underline{\sigma}(G)$, which is generally undesirable (on the other hand, a large value of $\bar{\sigma}(G)$ need not necessarily be a problem).
2. A large condition number may mean that the plant has a large minimized condition number, or equivalently, it has large RGA elements which indicate fundamental control problems; see below.
3. A large condition number *does* imply that the system is sensitive to "unstructured" (full-block) input uncertainty (e.g. with an inverse-based controller, see (8.136)), but this kind of uncertainty often does not occur in practice. We therefore *cannot* generally conclude that a plant with a large condition number is sensitive to uncertainty, e.g. see the diagonal plant in Example 3.12 (page 89).

3.4   Relative gain array (RGA)

The RGA (Bristol, 1966) of a non-singular square complex matrix G is a square complex matrix defined as

$$\text{RGA}(G) = \Lambda(G) \triangleq G \times (G^{-1})^T \tag{3.54}$$

where × denotes element-by-element multiplication (the Hadamard or Schur product). With Matlab, we write[5]

$$RGA = G.*pinv(G).'$$

The RGA of a transfer matrix is generally computed as a function of frequency (see Matlab program in Table 3.1). For a 2×2 matrix with elements g_{ij} the RGA is

$$\Lambda(G) = \begin{bmatrix} \lambda_{11} & \lambda_{12} \\ \lambda_{21} & \lambda_{22} \end{bmatrix} = \begin{bmatrix} \lambda_{11} & 1 - \lambda_{11} \\ 1 - \lambda_{11} & \lambda_{11} \end{bmatrix}; \quad \lambda_{11} = \frac{1}{1 - \frac{g_{12}g_{21}}{g_{11}g_{22}}} \quad (3.55)$$

The RGA is a very useful tool in practical applications. The RGA is treated in detail at three places in this book. First, we give a general introduction in this section (pages 82–91). The use of the RGA for decentralized control is discussed in more detail in Section 10.6 (pages 442–454). Finally, its algebraic properties and extension to non-square matrices are considered in Appendix A.4 (pages 526–529).

3.4.1 Original interpretation: RGA as an interaction measure

We follow Bristol (1966) here, and show that the RGA provides a measure of interactions. Let u_j and y_i denote a particular input–output pair for the multivariable plant $G(s)$, and assume that our task is to use u_j to control y_i. Bristol argued that there will be two extreme cases:

- All other loops open: $u_k = 0, \forall k \neq j$.
- All other loops closed with perfect control: $y_k - 0, \forall k \neq i$.

Perfect control is only possible at steady-state, but it is a good approximation at frequencies within the bandwidth of each loop. We now evaluate "our" gain $\partial y_i / \partial u_j$ for the two extreme cases:

$$\text{Other loops open:} \quad \left(\frac{\partial y_i}{\partial u_j} \right)_{u_k=0, k \neq j} = g_{ij} \quad (3.56)$$

$$\text{Other loops closed:} \quad \left(\frac{\partial y_i}{\partial u_j} \right)_{y_k=0, k \neq i} \triangleq \widehat{g}_{ij} \quad (3.57)$$

Here $g_{ij} = [G]_{ij}$ is the ij'th element of G, whereas $\widehat{g}_{ij}$ is the inverse of the ji'th element of G^{-1}

$$\widehat{g}_{ij} = 1/[G^{-1}]_{ji} \quad (3.58)$$

To derive (3.58) we note that

$$y = Gu \quad \Rightarrow \quad \left(\frac{\partial y_i}{\partial u_j} \right)_{u_k=0, k \neq j} = [G]_{ij} \quad (3.59)$$

and interchange the roles of G and G^{-1}, of u and y, and of i and j to get

$$u = G^{-1}y \quad \Rightarrow \quad \left(\frac{\partial u_j}{\partial y_i} \right)_{y_k=0, k \neq i} = [G^{-1}]_{ji} \quad (3.60)$$

---

[5] The symbol ' in Matlab gives the conjugate transpose (A^H), and we must use . ' to get the "regular" transpose (A^T).

and (3.58) follows. Bristol argued that the ratio between the gains in (3.56) and (3.57) is a useful measure of interactions, and defined the ij'th "relative gain" as

$$\lambda_{ij} \triangleq \frac{g_{ij}}{\widehat{g}_{ij}} = [G]_{ij}[G^{-1}]_{ji} \tag{3.61}$$

The RGA is the corresponding matrix of relative gains. From (3.61) we see that $\Lambda(G) = G \times (G^{-1})^T$ where $\times$ denotes element-by-element multiplication (the Schur product). This is identical to our definition of the RGA matrix in (3.54).

Remark. The assumption of $y_k = 0$ ("perfect control of y_k") in (3.57) is satisfied at steady-state ($\omega = 0$) provided we have integral action in the loop, but it will generally not hold exactly at other frequencies. Unfortunately, this has led many authors to dismiss the RGA as being "only useful at steady-state" or "only useful if we use integral action". On the contrary, in most cases it is the value of the RGA at frequencies close to crossover which is most important, and both the gain and the phase of the RGA elements are important. The derivation of the RGA in (3.56) to (3.61) was included to illustrate one useful interpretation of the RGA, but note that our definition of the RGA in (3.54) is purely algebraic and makes no assumption about "perfect control". The general usefulness of the RGA is further demonstrated by the additional general algebraic and control properties of the RGA listed on page 88.

Example 3.8 RGA for 2×2 **system.** *Consider a* 2×2 *system with the plant model*

$$y_1 = g_{11}(s)u_1 + g_{12}(s)u_2 \tag{3.62}$$
$$y_2 = g_{21}(s)u_1 + g_{22}(s)u_2 \tag{3.63}$$

Assume that "our" task is to use u_1 to control y_1. First consider the case when the other loop is open, i.e. u_2 is constant or equivalently $u_2 = 0$ in terms of deviation variables. We then have

$$u_2 = 0: \quad y_1 = g_{11}(s)u_1$$

Next consider the case when the other loop is closed with perfect control, i.e. $y_2 = 0$. In this case, u_2 will also change when we change u_1, due to interactions. More precisely, setting $y_2 = 0$ in (3.63) gives

$$u_2 = -\frac{g_{21}(s)}{g_{22}(s)}u_1$$

Substituting this into (3.62) gives

$$y_2 = 0: \quad y_1 = \underbrace{\left(g_{11} - \frac{g_{21}}{g_{22}}g_{12}\right)}_{\widehat{g}_{11}(s)} u_1$$

This means that "our gain" changes from $g_{11}(s)$ to $\widehat{g}_{11}(s)$ as we close the other loop, and the corresponding RGA element becomes

$$\lambda_{11}(s) = \frac{\text{"open-loop gain (with } u_2 = 0)\text{"}}{\text{"closed-loop gain (with } y_2 = 0)\text{"}} = \frac{g_{11}(s)}{\widehat{g}_{11}(s)} = \frac{1}{1 - \frac{g_{12}(s)g_{21}(s)}{g_{11}(s)g_{22}(s)}}$$

Intuitively, for decentralized control, we *prefer* to pair variables u_j and y_i so that λ_{ij} is close to 1 at all frequencies, because this means that the gain from u_j to y_i is unaffected by closing the other loops. More precisely, we have:

Pairing rule 1 (page 450): *Prefer pairings such that the rearranged system, with the selected pairings along the diagonal, has an RGA matrix close to identity at frequencies around the closed-loop bandwidth.*

However, one should avoid pairings where the sign of the steady-state gain from u_j to y_i may change depending on the control of the other outputs, because this will yield instability with integral action in the loop. Thus, $g_{ij}(0)$ and $\hat{g}_{11}(0)$ should have the same sign, and we have:

Pairing rule 2 (page 450): *Avoid (if possible) pairing on negative steady-state RGA elements.*

The reader is referred to Section 10.6.4 (page 438) for derivation and further discussion of these pairing rules.

3.4.2 Examples: RGA

Example 3.9 Blending process. *Consider a blending process where we mix sugar (u_1) and water (u_2) to make a given amount ($y_1 = F$) of a soft drink with a given sugar fraction ($y_2 = x$). The balances "mass in = mass out" for total mass and sugar mass are*

$$F_1 + F_2 = F$$

$$F_1 = xF$$

Note that the process itself has no dynamics. Linearization yields

$$dF_1 + dF_2 = dF$$

$$dF_1 = x^* dF + F^* dx$$

With $u_1 = dF_1, u_2 = dF_2, y_1 = dF$ and $y_2 = dx$ we then get the model

$$y_1 = u_1 + u_2$$

$$y_2 = \frac{1 - x^*}{F^*} u_1 - \frac{x^*}{F^*} u_2$$

where $x^* = 0.2$ is the nominal steady-state sugar fraction and $F^* = 2$ kg/s is the nominal amount. The transfer matrix then becomes

$$G(s) = \begin{bmatrix} 1 & 1 \\ \frac{1-x^*}{F^*} & -\frac{x^*}{F^*} \end{bmatrix} = \begin{bmatrix} 1 & 1 \\ 0.4 & -0.1 \end{bmatrix}$$

and the corresponding RGA matrix is (at all frequencies)

$$\Lambda = \begin{bmatrix} x^* & 1 - x^* \\ 1 - x^* & x^* \end{bmatrix} = \begin{bmatrix} 0.2 & 0.8 \\ 0.8 & 0.2 \end{bmatrix}$$

For decentralized control, it then follows from pairing rule 1 ("prefer pairing on RGA elements close to 1") that we should pair on the off-diagonal elements; that is, use u_1 to control y_2 and use u_2 to control y_1. This corresponds to using the largest stream (water, u_2) to control the amount ($y_1 = F$), which is reasonable from a physical point of view. Pairing rule 2 is also satisfied for this choice.

Example 3.10 Steady-state RGA. *Consider a* 3×3 *plant for which we have at steady-state*

$$G = \begin{bmatrix} 16.8 & 30.5 & 4.30 \\ -16.7 & 31.0 & -1.41 \\ 1.27 & 54.1 & 5.40 \end{bmatrix}, \ \Lambda(G) = \begin{bmatrix} 1.50 & 0.99 & -1.48 \\ -0.41 & 0.97 & 0.45 \\ -0.08 & -0.95 & 2.03 \end{bmatrix} \quad (3.64)$$

For decentralized control, we need to pair on one element in each column or row. It is then clear that the only choice that satisfies pairing rule 2 ("avoid pairing on negative RGA elements") is to pair on the diagonal elements; that is, use u_1 to control y_1, u_2 to control y_2 and u_3 to control y_3.

Remark. *The plant in (3.64) represents the steady-state model of a fluid catalytic cracking (FCC) process. A dynamic model of the FCC process in (3.64) is given in Exercise 6.17 (page 257).*

Some additional examples and exercises, that further illustrate the effectiveness of the steady-state RGA for selecting pairings, are given on page 443.

Example 3.11 Frequency-dependent RGA. *The following model describes a a large pressurized vessel (Skogestad and Wolff, 1991), for example, of the kind found in offshore oil-gas separations. The inputs are the valve positions for liquid (u_1) and vapour (u_2) flow, and the outputs are the liquid volume (y_1) and pressure (y_2).*

$$G(s) = \frac{0.01e^{-5s}}{(s + 1.72 \cdot 10^{-4})(4.32s + 1)} \begin{bmatrix} -34.54(s + 0.0572) & 1.913 \\ -30.22s & -9.188(s + 6.95 \cdot 10^{-4}) \end{bmatrix} \quad (3.65)$$

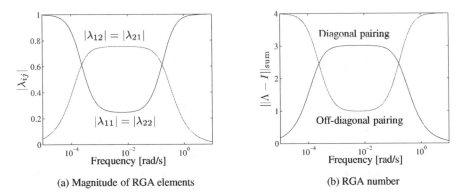

(a) Magnitude of RGA elements   (b) RGA number

Figure 3.8: Frequency-dependent RGA for $G(s)$ in (3.65)

The RGA matrix $\Lambda(s)$ depends on frequency. At steady-state ($s = 0$) the 2,1 element of $G(s)$ is zero, so $\Lambda(0) = I$. Similarly, at high frequencies the 1,2 element is small relative to the other elements, so $\Lambda(j\infty) = I$. This seems to suggest that the diagonal pairing should be used. However, at intermediate frequencies, the off-diagonal RGA elements are closest to 1, see Figure 3.8(a). For example, at frequency $\omega = 0.01$ rad/s the RGA matrix becomes (see Table 3.1)

$$\Lambda = \begin{bmatrix} 0.2469 + 0.0193i & 0.7531 - 0.0193i \\ 0.7531 - 0.0193i & 0.2469 + 0.0193i \end{bmatrix} \quad (3.66)$$

Thus, from pairing rule 1, the reverse pairings is probably best if we use decentralized control and the closed-loop bandwidth is around 0.01 rad/s. From a physical point of view the use of the reverse pairings is quite surprising, because it involves using the vapour flow (u_2) to control liquid level (y_1). and the liquid flow (u_1) to control pressure (y_2).

Table 3.1: **Matlab program to calculate frequency-dependent RGA**

```
% Plant model (3.65)
s = tf('s');
G = (0.01/(s+1.72e-4)/(4.32*s + 1))*[-34.54*(s+0.0572),....
omega = logspace(-5,2,61);
% RGA
for i = 1:length(omega)
    Gf = freqresp(G,omega(i));                    % G(jω)
    RGAw(:,:,i) = Gf.*inv(Gf).';                  % RGA at frequency omega
    RGAno(i) = sum(sum(abs(RGAw(:,:,i) - eye(2))));   % RGA number
end
RGA = frd(RGAw,omega);
```

Remark. *Although it is possible to use decentralized control for this interactive process, see the following exercise, one may achieve much better performance with multivariable control. If one insists on using decentralized control, then it is recommended to add a liquid flow measurement and use an "inner" (lower layer) flow controller. The resulting u_1 is then the liquid flow rate rather than the valve position. Then u_2 (vapour flow) has no effect on y_1 (liquid volume), and the plant is triangular with $g_{12} = 0$. In this case the diagonal pairing is clearly best.*

Exercise 3.7 * *Design decentralized single-loop controllers for the plant (3.65) using (a) the diagonal pairings and (b) the off-diagonal pairings. Use the delay θ (which is nominally 5 seconds) as a parameter. Use PI controllers independently tuned with the SIMC tuning rules (based on the paired elements).*

Outline of solution: For tuning purposes the elements in $G(s)$ are approximated using the half rule to get

$$G(s) \approx \begin{bmatrix} 0.0823\frac{e^{-\theta s}}{s} & 0.01913\frac{e^{-(\theta+2.16)s}}{s} \\ -0.3022\frac{e^{-\theta s}}{4.32s+1} & -0.09188\frac{e^{-\theta s}}{4.32s+1} \end{bmatrix}$$

For the diagonal pairings this gives the PI settings

$$K_{c1} = -12.1/(\tau_{c1} + \theta), \tau_{I1} = 4(\tau_{c1} + \theta); K_{c2} = -47.0/(\tau_{c2} + \theta), \tau_{I2} = 4.32$$

and for the off-diagonal pairings (the index refers to the output)

$$K_{c1} = 52.3/(\tau_{c1} + \theta + 2.16), \tau_{I1} = 4(\tau_{c1} + \theta + 2.16); K_{c2} = -14.3/(\tau_{c2} + \theta), \tau_{I2} = 4.32$$

For improved robustness, the level controller (y_1) is tuned about 3 times slower than the pressure controller (y_2), i.e. use $\tau_{c1} = 3\theta$ and $\tau_{c2} = \theta$. This gives a crossover frequency of about $0.5/\theta$ in the fastest loop. With a delay of about 5 s or larger you should find, as expected from the RGA at crossover frequencies (pairing rule 1), that the off-diagonal pairing is best. However, if the delay is decreased from 5 s to 1 s, then the diagonal pairing is best, as expected since the RGA for the diagonal pairing approaches 1 at frequencies above 1 rad/s.

3.4.3   RGA number and iterative RGA

Note that in Figure 3.8(a) we plot only the magnitudes of λ_{ij}, but this may be misleading when selecting pairings. For example, a magnitude of 1 (seemingly a desirable pairing) may correspond to an RGA element of -1 (an undesirable pairing). The phase of the RGA elements should therefore also be considered. An alternative is to compute the RGA number, as defined next.

RGA number. A simple measure for selecting pairings according to rule 1 is to prefer pairings with a small *RGA number*. For a diagonal pairing,

$$\text{RGA number} \triangleq \|\Lambda(G) - I\|_{\text{sum}} \tag{3.67}$$

where we have (somewhat arbitrarily) chosen the sum norm, $\|A\|_{\text{sum}} = \sum_{i,j} |a_{ij}|$. The RGA number for other pairings is obtained by subtracting 1 for the selected pairings; for example, $\Lambda(G) - \begin{bmatrix} 0 & 1 \\ 1 & 0 \end{bmatrix}$ for the off-diagonal pairing for a 2×2 plant. The disadvantage with the RGA number, at least for larger systems, is that it needs to be recomputed for each alternative pairing. On the other hand, the RGA elements need to be computed only once.

Example 3.11 continued. *The RGA number for the plant $G(s)$ in (3.65) is plotted for the two alternative pairings in Figure 3.8(b). As expected, we see that the off-diagonal pairing is preferred at intermediate frequencies.*

Exercise 3.8 *Compute the RGA number for the six alternate pairings for the plant in (3.64). Which pairing would you prefer?*

Remark. Diagonal dominance. A more precise statement of pairing rule 1 (page 85) would be to prefer pairings that have "diagonal dominance" (see definition on page 10.6.4). There is a close relationship between a small RGA number and diagonal dominance, but unfortunately there are exceptions for plants of size 4×4 or larger, so a small RGA number does not always guarantee diagonal dominance; see Example 10.18 on page 441.

Iterative RGA. An iterative evaluation of the RGA, $\Lambda^2(G) = \Lambda(\Lambda(G))$ etc., is very useful for choosing pairings with diagonal dominance for large systems. Wolff (1994) found numerically that

$$\Lambda^\infty \triangleq \lim_{k \to \infty} \Lambda^k(G) \tag{3.68}$$

is a permuted identity matrix (except for "borderline" cases). More importantly, Johnson and Shapiro (1986, Theorem 2) have proven that Λ^∞ *always converges to the identity matrix if G is a generalized diagonally dominant matrix* (see definition in Remark 10.6.4 on page 440) . Since permuting the matrix G causes similar permutations of $\Lambda(G)$, Λ^∞ may then be used as a candidate pairing choice. Typically, Λ^k approaches Λ^∞ for k between 4 and 8. For example, for $G = \begin{bmatrix} 1 & 2 \\ -1 & 1 \end{bmatrix}$ we get $\Lambda = \begin{bmatrix} 0.33 & 0.67 \\ 0.67 & 0.33 \end{bmatrix}$, $\Lambda^2 = \begin{bmatrix} -0.33 & 1.33 \\ 1.33 & -0.33 \end{bmatrix}$, $\Lambda^3 = \begin{bmatrix} -0.07 & 1.07 \\ 1.07 & -0.07 \end{bmatrix}$ and $\Lambda^4 = \begin{bmatrix} 0.00 & 1.00 \\ 1.00 & 0.00 \end{bmatrix}$, which indicates that the off-diagonal pairing is diagonally dominant. Note that Λ^∞ may sometimes "recommend" a pairing on negative RGA elements, even if a positive pairing is possible.

Exercise 3.9 *Test the iterative RGA method on the plant (3.64) and confirm that it gives the diagonally dominant pairing (as it should according to the theory).*

3.4.4   Summary of algebraic properties of the RGA

The (complex) RGA matrix has a number of interesting *algebraic properties*, of which the most important are (see Appendix A.4, page 526, for more details):

A1. It is independent of input and output scaling.
A2. Its rows and columns sum to 1.
A3. The RGA is the identity matrix if G is upper or lower triangular.
A4. A relative change in an element of G equal to the negative inverse of its corresponding RGA element, $g'_{ij} = g_{ij}(1 - 1/\lambda_{ij})$, yields singularity.

A5. From (A.80), plants with large RGA elements are always ill-conditioned (with a large value of $\gamma(G)$), but the reverse may not hold (i.e. a plant with a large $\gamma(G)$ may have small RGA elements).

From property A3, it follows that the RGA (or more precisely $\Lambda - I$) provides a measure of *two-way interaction*.

Example 3.12 *Consider a diagonal plant for which we have*

$$G = \begin{bmatrix} 100 & 0 \\ 0 & 1 \end{bmatrix}, \Lambda(G) = I, \gamma(G) = \frac{\bar{\sigma}(G)}{\underline{\sigma}(G)} = \frac{100}{1} = 100, \gamma^*(G) = 1 \qquad (3.69)$$

Here the condition number is 100 which means that the plant gain depends strongly on the input direction. However, since the plant is diagonal there are no interactions so $\Lambda(G) = I$ and the minimized condition number $\gamma^*(G) = 1$.

Example 3.13 *Consider a triangular plant G for which we get*

$$G = \begin{bmatrix} 1 & 2 \\ 0 & 1 \end{bmatrix}, G^{-1} = \begin{bmatrix} 1 & -2 \\ 0 & 1 \end{bmatrix}, \Lambda(G) = I, \gamma(G) = \frac{2.41}{0.41} = 5.83, \gamma^*(G) = 1 \qquad (3.70)$$

Note that for a triangular matrix, there is one-way interaction, but no two-way interaction, and the RGA is always the identity matrix.

Example 3.14 *Consider again the distillation process in (3.45) for which we have at steady-state*

$$G = \begin{bmatrix} 87.8 & -86.4 \\ 108.2 & -109.6 \end{bmatrix}, G^{-1} = \begin{bmatrix} 0.399 & -0.315 \\ 0.394 & -0.320 \end{bmatrix}, \Lambda(G) = \begin{bmatrix} 35.1 & -34.1 \\ -34.1 & 35.1 \end{bmatrix} \qquad (3.71)$$

In this case $\gamma(G) = 197.2/1.391 = 141.7$ is only slightly larger than $\gamma^*(G) = 138.268$. The magnitude sum of the elements in the RGA matrix is $\|\Lambda\|_{sum} = 138.275$. This confirms property A5 which states that, for 2×2 systems, $\|\Lambda(G)\|_{sum} \approx \gamma^*(G)$ when $\gamma^*(G)$ is large. The condition number is large, but since the minimum singular value $\underline{\sigma}(G) = 1.391$ is larger than 1 this does not by itself imply a control problem. However, the large RGA elements indicate problems, as discussed below (control property C1).

Example 3.15 *Consider again the FCC process in (3.64) with $\gamma = 69.6/1.63 = 42.6$ and $\gamma^* = 7.80$. The magnitude sum of the elements in the RGA is $\|\Lambda\|_{sum} = 8.86$ which is close to γ^* as expected from property A5. Note that the rows and the columns of Λ in (3.64) sums to 1. Since $\underline{\sigma}(G)$ is larger than 1 and the RGA elements are relatively small, this steady-state analysis does not indicate any particular control problems for the plant.*

3.4.5 Summary of control properties of the RGA

In addition to the algebraic properties listed above, the RGA has a surprising number of useful *control properties*:

C1. *Large RGA elements (typically, $5 - 10$ or larger) at frequencies important for control indicate that the plant is fundamentally difficult to control due to strong interactions and sensitivity to uncertainty.*

   (a) *Uncertainty in the input channels (diagonal input uncertainty).* Plants with large RGA elements (at crossover frequency) are fundamentally difficult to control because of sensitivity to input uncertainty, e.g. caused by uncertain or neglected actuator dynamics. In particular, decouplers or other inverse-based controllers should not be used for plants with large RGA elements (see page 251).

(b) *Element uncertainty.* As implied by algebraic property A4 above, large RGA elements imply sensitivity to element-by-element uncertainty. However, this kind of uncertainty may not occur in practice due to physical couplings between the transfer function elements. Therefore, diagonal input uncertainty (which is always present) is usually of more concern for plants with large RGA elements.

C2. *RGA and RHP-zeros.* If the sign of an RGA element changes as we go from $s = 0$ to $s = \infty$, then there is a RHP-zero in G or in some subsystem of G (see Theorem 10.7, page 446).

C3. *Non-square plants.* The definition of the RGA may be generalized to non-square matrices by using the pseudo-inverse; see Appendix A.4.2. Extra inputs: If the sum of the elements in a column of RGA is small ($\ll 1$), then one may consider deleting the corresponding input. Extra outputs: If all elements in a row of RGA are small ($\ll 1$), then the corresponding output cannot be controlled.

C4. *RGA and decentralized control.* The usefulness of the RGA is summarized by the two pairing rules on page 85.

Example 3.14 continued. *For the steady-state distillation model in (3.71), the large RGA element of 35.1 indicates a control problem. More precisely, fundamental control problems are expected if analysis shows that $G(j\omega)$ has large RGA elements also in the crossover frequency range. Indeed, with the idealized dynamic model (3.93) used below, the RGA elements are large at all frequencies, and we will confirm in simulations that there is a strong sensitivity to input channel uncertainty with an inverse-based controller, see page 100. For decentralized control, we should, according to rule 2, avoid pairing on the negative RGA elements. Thus, the diagonal pairing is preferred.*

Example 3.16 *Consider the plant*

$$G(s) = \frac{1}{5s + 1} \begin{pmatrix} s + 1 & s + 4 \\ 1 & 2 \end{pmatrix} \tag{3.72}$$

We find that $\lambda_{11}(\infty) = 2$ and $\lambda_{11}(0) = -1$ have different signs. Since none of the diagonal elements have RHP-zeros we conclude from property C2 that $G(s)$ must have a RHP-zero. This is indeed true and $G(s)$ has a zero at $s = 2$.

Let us elaborate a bit more on the use of RGA for decentralized control (control property C4). Assume we use decentralized control with integral action in each loop, and want to pair on one or more negative steady-state RGA elements. This may happen because this pairing is preferred for dynamic reasons or because there exists no pairing choice with only positive RGA elements, e.g. see the system in (10.81) on page 444. What will happen? Will the system be unstable? No, not necessarily. We may, for example, tune one loop at a time in a sequential manner (usually starting with the fastest loops), and we will end up with a stable overall system. However, due to the negative RGA element there will be some hidden problem, because the system is not *decentralized integral controllable* (DIC); see page 443. The stability of the overall system then depends on the individual loops being in service. This means that detuning one or more of the individual loops may result in instability for the overall system. Instability may also occur if an input saturates, because the corresponding loop is then effectively out of service. In summary, pairing on negative steady-state RGA elements should be avoided, and if it cannot be avoided then one should make sure that the loops remain in service.

For a detailed analysis of achievable performance of the plant (input–output controllability analysis), one must consider the singular values, as well as the RGA and condition number as functions of frequency. In particular, the crossover frequency range is important. In addition, disturbances and the presence of unstable (RHP) plant poles and zeros must be considered. All these issues are discussed in much more detail in Chapters 5 and 6 where we address achievable performance and input–output controllability analysis for SISO and MIMO plants, respectively.

3.5 Control of multivariable plants

3.5.1 Diagonal controller (decentralized control)

The simplest approach to multivariable controller design is to use a diagonal or block-diagonal controller $K(s)$. This is often referred to as decentralized control. Decentralized control works well if $G(s)$ is close to diagonal, because then the plant to be controlled is essentially a collection of independent sub-plants. However, if the off-diagonal elements in $G(s)$ are large, then the performance with decentralized diagonal control may be poor because no attempt is made to counteract the interactions. There are three basic approaches to the design of decentralized controllers:

- Fully coordinated design
- Independent design
- Sequential design

Decentralized control is discussed in more detail in Chapter 10 on page 429.

3.5.2 Two-step compensator design approach

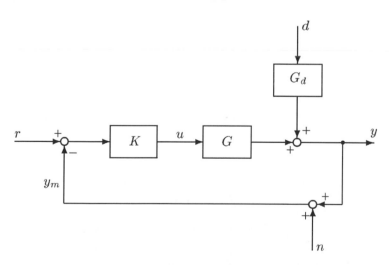

Figure 3.9: One degree-of-freedom feedback control configuration

Consider the simple feedback system in Figure 3.9. A conceptually simple approach to multivariable control is given by a two-step procedure in which we first design a "compensator" to deal with the interactions in G, and then design a *diagonal* controller using methods similar to those for SISO systems in Chapter 2. Several such approaches are discussed below.

The most common approach is to use a pre-compensator, $W_1(s)$, which counteracts the interactions in the plant and results in a "new" shaped plant:

$$G_s(s) = G(s)W_1(s) \tag{3.73}$$

which is more diagonal and easier to control than the original plant $G(s)$. After finding a suitable $W_1(s)$ we can design a *diagonal* controller $K_s(s)$ for the shaped plant $G_s(s)$. The overall controller is then

$$K(s) = W_1(s)K_s(s) \tag{3.74}$$

In many cases effective compensators may be derived on physical grounds and may include nonlinear elements such as ratios.

Remark 1 Some design approaches in this spirit are the Nyquist array technique of Rosenbrock (1974) and the characteristic loci technique of MacFarlane and Kouvaritakis (1977).

Remark 2 The $\mathcal{H}_\infty$ loop-shaping design procedure, described in detail in Section 9.4, is similar in that a pre-compensator is first chosen to yield a shaped plant, $G_s = GW_1$, with desirable properties, and then a controller $K_s(s)$ is designed. The main difference is that in $\mathcal{H}_\infty$ loop shaping, $K_s(s)$ is a full multivariable controller, designed and based on optimization (to optimize $\mathcal{H}_\infty$ robust stability).

3.5.3   Decoupling

Decoupling control results when the compensator W_1 is chosen such that $G_s = GW_1$ in (3.73) is diagonal at a selected frequency. The following different cases are possible:

1. **Dynamic decoupling:** $G_s(s)$ is diagonal at all frequencies. For example, with $G_s(s) = I$ and a square plant, we get $W_1 = G^{-1}(s)$ (disregarding the possible problems involved in realizing $G^{-1}(s)$). If we then select $K_s(s) = l(s)I$ (e.g. with $l(s) = k/s$), the overall controller is

$$K(s) = K_{\text{inv}}(s) \triangleq l(s)G^{-1}(s) \tag{3.75}$$

We will later refer to (3.75) as an *inverse-based* controller. It results in a decoupled nominal system with identical loops, i.e. $L(s) = l(s)I$, $S(s) = \frac{1}{1+l(s)}I$ and $T(s) = \frac{l(s)}{1+l(s)}I$.

Remark. In some cases we may want to keep the diagonal elements in the shaped plant unchanged by selecting $W_1 = G^{-1}G_{diag}$. In other cases we may want the diagonal elements in W_1 to be 1. This may be obtained by selecting $W_1 = G^{-1}((G^{-1})_{diag})^{-1}$, and the off-diagonal elements of W_1 are then called "decoupling elements".

2. **Steady-state decoupling:** $G_s(0)$ is diagonal. This may be obtained by selecting a constant pre-compensator $W_1 = G^{-1}(0)$ (and for a non-square plant we may use the pseudo-inverse provided $G(0)$ has full row (output) rank).

3. **Approximate decoupling at frequency** w_o: $G_s(j\omega_o)$ is as diagonal as possible. This is usually obtained by choosing a constant pre-compensator $W_1 = G_o^{-1}$ where G_o is a real approximation of $G(j\omega_o)$. G_o may be obtained, for example, using the align algorithm of Kouvaritakis (1974) (see file `align.m` available at the book's home page). The bandwidth frequency is a good selection for ω_o because the effect on performance of reducing interaction is normally greatest at this frequency.

The idea of decoupling control is appealing, but there are several difficulties:

1. As one might expect, decoupling may be very sensitive to modelling errors and uncertainties. This is illustrated below in Section 3.7.2 (page 100).
2. The requirement of decoupling and the use of an inverse-based controller may not be desirable for disturbance rejection. The reasons are similar to those given for SISO systems in Section 2.6.4, and are discussed further below; see (3.79).
3. If the plant has RHP-zeros then the requirement of decoupling generally introduces extra RHP-zeros into the closed-loop system (see Section 6.6.1, page 236).

Even though decoupling controllers may not always be desirable in practice, they are of interest from a theoretical point of view. They also yield insights into the limitations imposed by the multivariable interactions on achievable performance. One popular design method, which essentially yields a decoupling controller, is the internal model control (IMC) approach (Morari and Zafiriou, 1989).

Another common strategy, which avoids most of the problems just mentioned, is to use *partial (one-way) decoupling* where $G_s(s)$ in (3.73) is upper or lower triangular.

3.5.4 Pre- and post-compensators and the SVD controller

The above pre-compensator approach may be extended by introducing a post-compensator $W_2(s)$, as shown in Figure 3.10. One then designs a *diagonal* controller K_s for the shaped

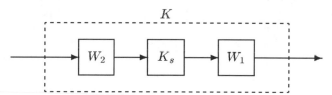

Figure 3.10: Pre- and post-compensators, W_1 and W_2. K_s is diagonal.

plant W_2GW_1. The overall controller is then

$$K(s) = W_1 K_s W_2 \tag{3.76}$$

The *SVD controller* is a special case of a pre- and post-compensator design. Here

$$W_1 = V_o \quad \text{and} \quad W_2 = U_o^T \tag{3.77}$$

where V_o and U_o are obtained from the SVD of $G_o = U_o\Sigma_o V_o^T$, where G_o is a real approximation of $G(j\omega_o)$ at a given frequency w_o (often around the bandwidth). SVD controllers are studied by Hung and MacFarlane (1982), and by Hovd et al. (1997) who found that the SVD-controller structure is optimal in some cases, e.g. for plants consisting of symmetrically interconnected subsystems.

In summary, the SVD controller provides a useful class of controllers. By selecting $K_s = l(s)\Sigma_o^{-1}$ a decoupling design is achieved, and selecting a diagonal K_s with a low condition number ($\gamma(K_s)$ small) generally results in a robust controller (see Section 6.10).

3.5.5   What is the shape of the "best" feedback controller?

Consider the problem of disturbance rejection. The closed-loop disturbance response is $y = SG_d d$. Suppose we have scaled the system (see Section 1.4) such that at each frequency the disturbances are of maximum magnitude 1, $\|d\|_2 \leq 1$, and our performance requirement is that $\|y\|_2 \leq 1$. This is equivalent to requiring $\bar{\sigma}(SG_d) \leq 1$. In many cases there is a trade-off between input usage and performance, such that the controller that minimizes the input magnitude is one that yields all singular values of SG_d equal to 1, i.e. $\sigma_i(SG_d) = 1, \forall \omega$. This corresponds to

$$S_{\min}G_d = U_1 \tag{3.78}$$

where $U_1(s)$ is some all-pass transfer function (which at each frequency has all its singular values equal to 1). The subscript min refers to the use of the smallest loop gain that satisfies the performance objective. For simplicity, we assume that G_d is square so $U_1(j\omega)$ is a unitary matrix. At frequencies where feedback is effective we have $S = (I + L)^{-1} \approx L^{-1}$, and (3.78) yields $L_{\min} = GK_{\min} \approx G_d U_1^{-1}$. In conclusion, the controller and loop shape with the minimum gain will often look like

$$K_{\min} \approx G^{-1}G_d U_2, \quad L_{\min} \approx G_d U_2 \tag{3.79}$$

where $U_2 = U_1^{-1}$ is some all-pass transfer function matrix. This provides a generalization of $|K_{\min}| \approx |G^{-1}G_d|$ which was derived in (2.66) for SISO systems, and the summary following (2.66) on page 48 therefore also applies to MIMO systems. For example, we see that for disturbances entering at the plant inputs, $G_d = G$, we get $K_{\min} = U_2$, so a simple constant unit gain controller yields a good trade-off between output performance and input usage. We also note with interest that it is generally not possible to select a unitary matrix U_2 such that $L_{\min} = G_d U_2$ is diagonal, so a decoupling design is generally not optimal for disturbance rejection. These insights can be used as a basis for a loop-shaping design; see more on $\mathcal{H}_\infty$ loop shaping in Chapter 9.

3.5.6   Multivariable controller synthesis

The above design methods are based on a two-step procedure in which we first design a pre-compensator (for decoupling control) or we make an input–output pairing selection (for decentralized control) and then we design a diagonal controller $K_s(s)$. Invariably this two-step procedure results in a suboptimal design.

The alternative is to synthesize directly a multivariable controller $K(s)$ based on minimizing some objective function (norm). Here we use the word *synthesize* rather than *design* to stress that this is a more formalized approach. Optimization in controller design became prominent in the 1960's with "optimal control theory" based on minimizing the expected value of the output variance in the face of stochastic disturbances. Later, other approaches and norms were introduced, such as $\mathcal{H}_\infty$ optimal control.

3.5.7   Summary of mixed-sensitivity $\mathcal{H}_\infty$ synthesis (S/KS)

We provide a brief summary here of one multivariable synthesis approach, namely the S/KS (mixed-sensitivity) $\mathcal{H}_\infty$ design method which is used in later examples in this chapter. In the

S/KS problem, the objective is to minimize the $\mathcal{H}_\infty$ norm of

$$N = \begin{bmatrix} W_P S \\ W_u K S \end{bmatrix} \tag{3.80}$$

This problem was discussed earlier for SISO systems, and another look at Section 2.8.3 would be useful now. A sample Matlab file is provided in Example 2.17, page 64.

The following issues and guidelines are relevant when selecting the weights W_P and W_u:

1. S is the transfer function from r to $-e = r - y$. A common choice for the performance weight is $W_P = \text{diag}\{w_{Pi}\}$ with

$$w_{Pi} = \frac{s/M_i + \omega_{Bi}^*}{s + \omega_{Bi}^* A_i}, \quad A_i \ll 1 \tag{3.81}$$

   (see also Figure 2.29 on page 62). Selecting $A_i \ll 1$ ensures approximate integral action with $S(0) \approx 0$. Often we select M_i about 2 for all outputs, whereas the desired closed-loop bandwidth ω_{Bi}^* may be different for each output. A large value of ω_{Bi}^* yields a faster response for output i.

2. KS is the transfer function from references r to inputs u in Figure 3.9, so for a system which has been scaled as in Section 1.4, a reasonable initial choice for the input weight is $W_u = I$. However, if we require tight control at low frequencies (i.e. A_i small in (3.81)), then input usage is unavoidable at low frequencies, and it may be better to use a weight of the form $W_u = s/(s + \omega_1)$, where the adjustable frequency ω_1 is approximately the closed-loop bandwidth. One could also include additional high-frequency penalty in W_u, but often this is not necessary due to the low gain of G at high frequencies. If one wants to bound KS at high frequencies, it is often better instead to put a bound on T (see below).

3. To find a reasonable initial choice for the weight W_P, one can first obtain a controller with some other design method, plot the magnitude of the resulting diagonal elements of S as a function of frequency, and select $w_{Pi}(s)$ as a rational approximation of $1/|S_{ii}|$.

4. For disturbance rejection, we may in some cases want a steeper slope for $w_{Pi}(s)$ at low frequencies than that given in (3.81), e.g. see the weight in (2.106). However, it may be better to consider the disturbances explicitly by considering the $\mathcal{H}_\infty$ norm of

$$N = \begin{bmatrix} W_P S & W_P S G_d \\ W_u K S & W_u K S G_d \end{bmatrix} \tag{3.82}$$

or equivalently

$$N = \begin{bmatrix} W_P S W_d \\ W_u K S W_d \end{bmatrix} \quad \text{with } W_d = \begin{bmatrix} I & G_d \end{bmatrix} \tag{3.83}$$

where N represents the transfer function from $\begin{bmatrix} r \\ d \end{bmatrix}$ to the weighted e and u. In some situations we may want to adjust W_P or G_d in order to satisfy better our original objectives. The helicopter case study in Section 13.2 illustrates this by introducing a scalar parameter α to adjust the magnitude of G_d.

5. T is the transfer function from $-n$ to y. To reduce sensitivity to noise and uncertainty, we want T small at high frequencies, and so we may want additional roll-off in L. This can be achieved in several ways. One approach is to add $W_T T$ to the stack for N in (3.80), where $W_T = \text{diag}\{w_{Ti}\}$ and $|w_{Ti}|$ is smaller than 1 at low frequencies and large

at high frequencies. A more direct approach is to add high-frequency dynamics, $W_1(s)$, to the plant model to ensure that the resulting shaped plant, $G_s = GW_1$, rolls off with the desired slope. We then obtain an $\mathcal{H}_\infty$ optimal controller K_s for this shaped plant, and finally include $W_1(s)$ in the controller, $K = W_1 K_s$.

Numerically, the problem $\min_K \|N\|_\infty$ is often solved by γ-iteration, where one solves for the controllers that achieve $\|N\|_\infty < \gamma$, and then reduces γ iteratively to obtain the smallest value γ_{min} for which a solution exists. More details about $\mathcal{H}_\infty$ design are given in Chapter 9.

3.6 Introduction to multivariable RHP-zeros

By means of an example, we now give the reader an appreciation of the fact that MIMO systems have zeros even though their presence may not be obvious from the elements of $G(s)$. As for SISO systems, we find that RHP-zeros impose fundamental limitations on control.

The zeros z of MIMO systems are defined as the values $s = z$ where $G(s)$ loses rank, and we can find the *direction* of a zero by looking at the direction in which the matrix $G(z)$ has zero gain. For square systems we essentially have that the poles and zeros of $G(s)$ are the poles and zeros of $\det G(s)$. However, this crude method may fail in some cases, as it may incorrectly cancel poles and zeros with the same location but different directions (see Sections 4.5 and 4.5.3 for more details).

Example 3.17 *Consider the following plant:*

$$G(s) = \frac{1}{(0.2s + 1)(s + 1)} \begin{bmatrix} 1 & 1 \\ 1 + 2s & 2 \end{bmatrix} \tag{3.84}$$

The responses to a step in each individual input are shown in Figure 3.11(a) and (b). We see that the

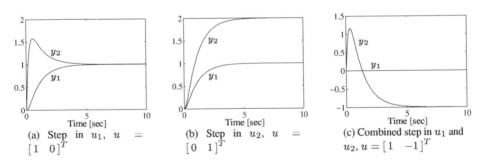

(a) Step in u_1, $u = \begin{bmatrix} 1 & 0 \end{bmatrix}^T$

(b) Step in u_2, $u = \begin{bmatrix} 0 & 1 \end{bmatrix}^T$

(c) Combined step in u_1 and u_2, $u = \begin{bmatrix} 1 & -1 \end{bmatrix}^T$

Figure 3.11: Open-loop response for $G(s)$ in (3.84)

plant is interactive, but for these two inputs there is no inverse response to indicate the presence of a RHP-zero. Nevertheless, the plant does have a multivariable RHP-zero at $z = 0.5$; that is, $G(s)$ loses rank at $s = 0.5$, and $\det G(0.5) = 0$. The SVD of $G(0.5)$ is

$$G(0.5) = \frac{1}{1.65} \begin{bmatrix} 1 & 1 \\ 2 & 2 \end{bmatrix} = \underbrace{\begin{bmatrix} 0.45 & 0.89 \\ 0.89 & -0.45 \end{bmatrix}}_{U} \underbrace{\begin{bmatrix} 1.92 & 0 \\ 0 & 0 \end{bmatrix}}_{\Sigma} \underbrace{\begin{bmatrix} 0.71 & 0.71 \\ 0.71 & -0.71 \end{bmatrix}^H}_{V^H} \tag{3.85}$$

and we have as expected $\underline{\sigma}(G(0.5)) = 0$. *The directions corresponding to the RHP-zero are* $\underline{v} = \begin{bmatrix} 0.71 \\ -0.71 \end{bmatrix}$ *(input direction) and* $\underline{u} = \begin{bmatrix} 0.89 \\ -0.45 \end{bmatrix}$ *(output direction). Thus, the RHP-zero is associated with both inputs and with both outputs. The presence of the multivariable RHP-zero is indeed observed from the time response in Figure 3.11(c), which is for a simultaneous input change in opposite directions,* $u = \begin{bmatrix} 1 \\ -1 \end{bmatrix}$. *We see that* y_2 *displays an inverse response whereas* y_1 *happens to remain at zero for this particular input change.*

To see how the RHP-zero affects the closed-loop response, we design a controller which minimizes the $\mathcal{H}_\infty$ *norm of the weighted* S/KS *matrix*

$$N = \begin{bmatrix} W_P S \\ W_u K S \end{bmatrix} \tag{3.86}$$

with weights

$$W_u = I, \ W_P = \begin{bmatrix} w_{P1} & 0 \\ 0 & w_{P2} \end{bmatrix}, w_{Pi} = \frac{s/M_i + \omega^*_{Bi}}{s + w^*_{Bi} A_i}, \quad A_i = 10^{-4} \tag{3.87}$$

The Matlab file for the design is the same as in Table 2.4 on page 64, except that we now have a 2×2 *system. Since there is a RHP-zero at* $z = 0.5$ *we expect that this will somehow limit the bandwidth of the closed-loop system.*

Design 1. *We weight the two outputs equally and select*

$$\text{Design 1}: \quad M_1 = M_2 = 1.5; \quad \omega^*_{B1} = \omega^*_{B2} = z/2 = 0.25$$

This yields an $\mathcal{H}_\infty$ *norm for* N *of 2.80 and the resulting singular values of* S *are shown by the solid lines in Figure 3.12(a). The closed-loop response to a reference change* $r = \begin{bmatrix} 1 & -1 \end{bmatrix}^T$ *is shown by the solid lines in Figure 3.12(b). We note that both outputs behave rather poorly and both display an inverse response.*

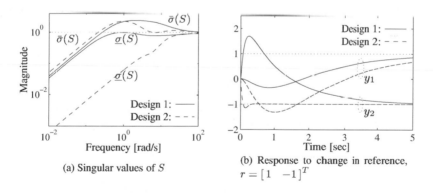

(a) Singular values of S

(b) Response to change in reference,
$r = \begin{bmatrix} 1 & -1 \end{bmatrix}^T$

Figure 3.12: Alternative designs for 2×2 plant (3.84) with RHP-zero

Design 2. *For MIMO plants, one can often move most of the deteriorating effect (e.g. inverse response) of a RHP-zero to a particular output channel. To illustrate this, we change the weight* w_{P2} *so that more emphasis is placed on output 2. We do this by increasing the bandwidth requirement in output channel 2 by a factor of 100:*

$$\text{Design 2}: \quad M_1 = M_2 = 1.5; \quad \omega^*_{B1} = 0.25, \ \omega^*_{B2} = 25$$

This yields an $\mathcal{H}_\infty$ norm for N of 2.92. In this case we see from the dashed line in Figure 3.12(b) that the response for output 2 (y_2) is excellent with no inverse response. However, this comes at the expense of output 1 (y_1) where the response is poorer than for Design 1.

Design 3. *We can also interchange the weights w_{P1} and w_{P2} to stress output 1 rather than output 2. In this case (not shown) we get an excellent response in output 1 with no inverse response, but output 2 responds very poorly (much poorer than output 1 for Design 2). Furthermore, the $\mathcal{H}_\infty$ norm for N is 6.73, whereas it was only 2.92 for Design 2.*

Thus, we see that it is easier, for this example, to get tight control of output 2 than of output 1. This may be expected from the output direction of the RHP-zero, $\underline{u} = \begin{bmatrix} 0.89 \\ -0.45 \end{bmatrix}$, which is mostly in the direction of output 1. We will discuss this in more detail in Section 6.6.1.

Remark 1 We find from this example that we can direct the effect of the RHP-zero to either of the two outputs. This is typical of multivariable RHP-zeros, but in other cases the RHP-zero is associated with a particular output channel and it is *not* possible to move its effect to another channel. The zero is then called a "pinned zero" (see Section 4.6).

Remark 2 It is observed from the plot of the singular values in Figure 3.12(a) that we were able to obtain by Design 2 a very large improvement in the "good" direction (corresponding to $\underline{\sigma}(S)$) at the expense of only a minor deterioration in the "bad" direction (corresponding to $\bar{\sigma}(S)$). Thus Design 1 demonstrates a shortcoming of the $\mathcal{H}_\infty$ norm: only the worst direction (maximum singular value) contributes to the $\mathcal{H}_\infty$ norm and it may not always be easy to get a good trade-off between the various directions.

3.7   Introduction to MIMO robustness

To motivate the need for a deeper understanding of robustness, we present two examples which illustrate that MIMO systems can display a sensitivity to uncertainty not found in SISO systems. We focus our attention on diagonal input uncertainty, which is present in any real system and often limits achievable performance because it enters between the controller and the plant.

3.7.1   Motivating robustness example no. 1: spinning satellite

Consider the following plant (Doyle, 1986; Packard et al., 1993) which can itself be motivated by considering the angular velocity control of a satellite spinning about one of its principal axes:

$$G(s) = \frac{1}{s^2 + a^2} \begin{bmatrix} s - a^2 & a(s+1) \\ -a(s+1) & s - a^2 \end{bmatrix}; \quad a = 10 \tag{3.88}$$

A minimal state-space realization, $G = C(sI - A)^{-1}B + D$, is

$$\left[\begin{array}{c|c} A & B \\ \hline C & D \end{array} \right] = \left[\begin{array}{cc|cc} 0 & a & 1 & 0 \\ -a & 0 & 0 & 1 \\ \hline 1 & a & 0 & 0 \\ -a & 1 & 0 & 0 \end{array} \right] \tag{3.89}$$

The plant has a pair of $j\omega$-axis poles at $s = \pm ja$ so it needs to be stabilized. Let us apply negative feedback and try the simple diagonal constant controller

$$K = I$$

The complementary sensitivity function is

$$T(s) = GK(I + GK)^{-1} = \frac{1}{s+1} \begin{bmatrix} 1 & a \\ -a & 1 \end{bmatrix} \qquad (3.90)$$

Nominal stability (NS). The closed-loop system has two poles at $s = -1$ and so it is stable. This can be verified by evaluating the closed-loop state matrix

$$A_{cl} = A - BKC = \begin{bmatrix} 0 & a \\ -a & 0 \end{bmatrix} - \begin{bmatrix} 1 & a \\ -a & 1 \end{bmatrix} = \begin{bmatrix} -1 & 0 \\ 0 & -1 \end{bmatrix}$$

(To derive A_{cl} use $\dot{x} = Ax + Bu$, $y = Cx$ and $u = -Ky$.)

Nominal performance (NP). The singular values of $L = GK = G$ are shown in Figure 3.7(a), page 80. We see that $\underline{\sigma}(L) = 1$ at low frequencies and starts dropping off at about $\omega = 10$. Since $\underline{\sigma}(L)$ never exceeds 1, we do not have tight control in the low-gain direction for this plant (recall the discussion following (3.51)), so we expect poor closed-loop performance. This is confirmed by considering S and T. For example, at steady-state $\bar{\sigma}(T) = 10.05$ and $\bar{\sigma}(S) = 10$. Furthermore, the large off-diagonal elements in $T(s)$ in (3.90) show that we have strong interactions in the closed-loop system. (For reference tracking, however, this may be counteracted by use of a two degrees-of-freedom controller.)

Robust stability (RS). Now let us consider stability robustness. In order to determine stability margins with respect to perturbations in each input channel, one may consider Figure 3.13 where we have broken the loop at the first input. The loop transfer function at this point (the transfer function from w_1 to z_1) is $L_1(s) = 1/s$ (which can be derived from $t_{11}(s) = \frac{1}{1+s} = \frac{L_1(s)}{1+L_1(s)}$). This corresponds to an infinite gain margin and a phase margin of $90°$. On breaking the loop at the second input we get the same result. This suggests good robustness properties irrespective of the value of a. However, the design is far from robust as a further analysis shows. Consider input gain uncertainty, and let ϵ_1 and ϵ_2 denote the relative

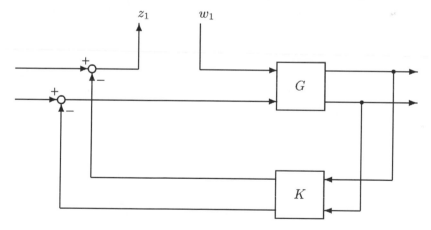

Figure 3.13: Checking stability margins "one-loop-at-a-time"

error in the gain in each input channel. Then

$$u'_1 = (1 + \epsilon_1)u_1, \quad u'_2 = (1 + \epsilon_2)u_2 \qquad (3.91)$$

where u_1' and u_2' are the actual changes in the manipulated inputs, while u_1 and u_2 are the desired changes as computed by the controller. It is important to stress that this diagonal input uncertainty, which stems from our inability to know the exact values of the manipulated inputs, is *always* present. In terms of a state-space description, (3.91) may be represented by replacing B by

$$B' = \begin{bmatrix} 1 + \epsilon_1 & 0 \\ 0 & 1 + \epsilon_2 \end{bmatrix}$$

The corresponding closed-loop state matrix is

$$A'_{cl} = A - B'KC = \begin{bmatrix} 0 & a \\ -a & 0 \end{bmatrix} - \begin{bmatrix} 1 + \epsilon_1 & 0 \\ 0 & 1 + \epsilon_2 \end{bmatrix} \begin{bmatrix} 1 & a \\ -a & 1 \end{bmatrix}$$

which has a characteristic polynomial given by

$$\det(sI - A'_{cl}) = s^2 + \underbrace{(2 + \epsilon_1 + \epsilon_2)}_{a_1} s + \underbrace{1 + \epsilon_1 + \epsilon_2 + (a^2 + 1)\epsilon_1\epsilon_2}_{a_0} \qquad (3.92)$$

The perturbed system is stable if and only if both the coefficients a_0 and a_1 are positive. We therefore see that *the system is always stable if we consider uncertainty in only one channel at a time* (at least as long as the channel gain is positive). More precisely, we have stability for $(-1 < \epsilon_1 < \infty, \epsilon_2 = 0)$ and $(\epsilon_1 = 0, -1 < \epsilon_2 < \infty)$. This confirms the infinite gain margin seen earlier. However, the system can only tolerate *small simultaneous changes* in the two channels. For example, let $\epsilon_1 = -\epsilon_2$, then the system is unstable ($a_0 < 0$) for

$$|\epsilon_1| > \frac{1}{\sqrt{a^2 + 1}} \approx 0.1$$

In summary, we have found that checking single-loop margins is inadequate for MIMO problems. We have also observed that large values of $\bar{\sigma}(T)$ or $\bar{\sigma}(S)$ indicate robustness problems. We will return to this in Chapter 8, where we show that with input uncertainty of magnitude $|\epsilon_i| < 1/\bar{\sigma}(T)$, we are guaranteed robust stability (even for "full-block complex perturbations").

In the next example we find that there can be sensitivity to diagonal input uncertainty even in cases where $\bar{\sigma}(T)$ and $\bar{\sigma}(S)$ have no large peaks. This cannot happen for a diagonal controller, see (6.92), but it will happen if we use an inverse-based controller for a plant with large RGA elements, see (6.93).

3.7.2 Motivating robustness example no. 2: distillation process

The following is an idealized dynamic model of a distillation column:

$$G(s) = \frac{1}{75s + 1} \begin{bmatrix} 87.8 & -86.4 \\ 108.2 & -109.6 \end{bmatrix} \qquad (3.93)$$

(time is in minutes). The physics of this example was discussed in Example 3.6. The plant is ill-conditioned with condition number $\gamma(G) = 141.7$ at all frequencies. The plant is also strongly two-way interactive and the RGA matrix at all frequencies is

$$\Lambda(G) = \begin{bmatrix} 35.1 & -34.1 \\ -34.1 & 35.1 \end{bmatrix} \qquad (3.94)$$

The large elements in this matrix indicate that this process is fundamentally difficult to control.

Remark. Equation (3.93) is admittedly a very crude model of a real distillation column; there should be a high-order lag in the transfer function from input 1 to output 2 to represent the liquid flow down to the column, and higher-order composition dynamics should also be included. Nevertheless, the model is simple and displays important features of distillation column behaviour. It should be noted that with a more detailed model, the RGA elements would approach 1 at frequencies around 1 rad/min, indicating less of a control problem.

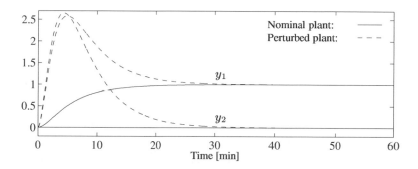

Figure 3.14: Response with decoupling controller to filtered reference input $r_1 = 1/(5s + 1)$. The perturbed plant has 20% gain uncertainty as given by (3.97).

We consider the following inverse-based controller, which may also be looked upon as a steady-state decoupler with a PI controller:

$$K_{\text{inv}}(s) = \frac{k_1}{s}G^{-1}(s) = \frac{k_1(1 + 75s)}{s}\begin{bmatrix} 0.3994 & -0.3149 \\ 0.3943 & -0.3200 \end{bmatrix}, \quad k_1 = 0.7 \qquad (3.95)$$

Nominal performance (NP). We have $GK_{\text{inv}} = K_{\text{inv}}G = \frac{0.7}{s}I$. With no model error this controller should counteract all the interactions in the plant and give rise to two decoupled first-order responses each with a time constant of $1/0.7 = 1.43$ min. This is confirmed by the solid line in Figure 3.14 which shows the simulated response to a reference change in y_1. The responses are clearly acceptable, and we conclude that *nominal performance (NP) is achieved with the decoupling controller*.

Robust stability (RS). The resulting sensitivity and complementary sensitivity functions with this controller are

$$S = S_I = \frac{s}{s + 0.7}I; \quad T = T_I = \frac{1}{1.43s + 1}I \qquad (3.96)$$

Thus, $\bar{\sigma}(S)$ and $\bar{\sigma}(T)$ are both less than 1 at all frequencies, so there are no peaks which would indicate robustness problems. We also find that this controller gives an infinite gain margin (GM) and a phase margin (PM) of 90° in each channel. Thus, use of the traditional margins and the peak values of S and T indicate no robustness problems. However, from the large RGA elements there is cause for concern, and this is confirmed in the following.

We consider again the input gain uncertainty (3.91) as in the previous example, and we select $\epsilon_1 = 0.2$ and $\epsilon_2 = -0.2$. We then have

$$u_1' = 1.2u_1, \quad u_2' = 0.8u_2 \qquad (3.97)$$

Note that we use deviation variables; for example, u_1 is actually the change Δu_1. This means that the uncertainty is on the *change* in the inputs (flow rates), and not on their absolute values. A 20% error is typical for process control applications (see Remark 2 on page 297). The uncertainty in (3.97) does not by itself yield instability. This is verified by computing the closed-loop poles, which, assuming no cancellations, are solutions to $\det(I + L(s)) = \det(I + L_I(s)) = 0$ (see (4.105) and (A.12)). In our case

$$L_I'(s) = K_{\mathrm{inv}} G' = K_{\mathrm{inv}} G \begin{bmatrix} 1 + \epsilon_1 & 0 \\ 0 & 1 + \epsilon_2 \end{bmatrix} = \frac{0.7}{s} \begin{bmatrix} 1 + \epsilon_1 & 0 \\ 0 & 1 + \epsilon_2 \end{bmatrix}$$

so the perturbed closed-loop poles are

$$s_1 = -0.7(1 + \epsilon_1), \quad s_2 = -0.7(1 + \epsilon_2) \tag{3.98}$$

and we have closed-loop stability as long as the input gains $1 + \epsilon_1$ and $1 + \epsilon_2$ remain positive, so we can have up to 100% error in each input channel. We thus conclude that *we have robust stability (RS) with respect to input gain errors for the decoupling controller.*

Robust performance (RP). For SISO systems we generally have that nominal performance (NP) and robust stability (RS) imply robust performance (RP), but this is not the case for MIMO systems. This is clearly seen from the dashed lines in Figure 3.14 which show the closed-loop response of the perturbed system. It differs drastically from the nominal response represented by the solid line, and even though it is stable, the response is clearly not acceptable; it is no longer decoupled, and $y_1(t)$ and $y_2(t)$ reach a value of about 2.5 before settling at their desired values of 1 and 0. *Thus RP is not achieved by the decoupling controller.*

Remark 1 There is a simple reason for the observed poor response to the reference change in y_1. To accomplish this change, which occurs mostly in the direction corresponding to the low plant gain, the inverse-based controller generates relatively *large* inputs u_1 and u_2, while trying to keep $u_1 - u_2$ very *small*. However, the input uncertainty makes this impossible – the result is an undesired *large* change in the actual value of $u_1' - u_2'$, which subsequently results in large changes in y_1 and y_2 because of the large plant gain ($\bar{\sigma}(G) = 197.2$) in this direction, as seen from (3.46).

Remark 2 The system remains stable for gain uncertainty up to 100% because the uncertainty occurs only at one side of the plant (at the input). If we also consider uncertainty at the output then we find that the decoupling controller yields instability for relatively small errors in the input and output gains. This is illustrated in Exercise 3.11 below.

Remark 3 It is also difficult to get a robust controller with other standard design techniques for this model. For example, an S/KS design as in (3.80) with $W_P = w_P I$ (using $M = 2$ and $\omega_B = 0.05$ in the performance weight (3.81)) and $W_u = I$ yields a good nominal response (although not decoupled), but the system is very sensitive to input uncertainty, and the outputs go up to about 3.4 and settle very slowly when there is 20% input gain error.

Remark 4 Attempts to make the inverse-based controller robust using the second step of the Glover–McFarlane $\mathcal{H}_\infty$ loop-shaping procedure are also unhelpful; see Exercise 3.12. This shows that robustness with respect to general coprime factor uncertainty does not necessarily imply robustness with respect to input uncertainty. In any case, the solution is to avoid inverse-based controllers for a plant with large RGA elements.

Exercise 3.10* *Design an SVD controller* $K = W_1 K_s W_2$ *for the distillation process in (3.93), i.e. select* $W_1 = V$ *and* $W_2 = U^T$ *where* U *and* V *are given in (3.46). Select* K_s *in the form*

$$K_s = \begin{bmatrix} c_1 \frac{75s+1}{s} & 0 \\ 0 & c_2 \frac{75s+1}{s} \end{bmatrix}$$

and try the following values:

(a) $c_1 = c_2 = 0.005$;

(b) $c_1 = 0.005, c_2 = 0.05$;

(c) $c_1 = 0.7/197 = 0.0036, c_2 = 0.7/1.39 = 0.504$.

Simulate the closed-loop reference response with and without uncertainty. Designs (a) and (b) should be robust. Which has the best performance? Design (c) should give the response in Figure 3.14. In the simulations, include high-order plant dynamics by replacing $G(s)$ *by* $\frac{1}{(0.02s+1)^5} G(s)$. *What is the condition number of the controller in the three cases? Discuss the results. (See also the conclusion on page 251.)*

Exercise 3.11 *Consider again the distillation process (3.93) with the decoupling controller, but also include output gain uncertainty* $\hat{\epsilon}_i$. *That is, let the perturbed loop transfer function be*

$$L'(s) = G' K_{\mathrm{inv}} = \frac{0.7}{s} \underbrace{\begin{bmatrix} 1+\hat{\epsilon}_1 & 0 \\ 0 & 1+\hat{\epsilon}_2 \end{bmatrix} G \begin{bmatrix} 1+\epsilon_1 & 0 \\ 0 & 1+\epsilon_2 \end{bmatrix} G^{-1}}_{L_0} \tag{3.99}$$

where L_0 *is a constant matrix for the distillation model (3.93), since all elements in* G *share the same dynamics,* $G(s) = g(s)G_0$. *The closed-loop poles of the perturbed system are solutions to* $\det(I + L'(s)) = \det(I + (k_1/s)L_0) = 0$, *or equivalently*

$$\det\left(\frac{s}{k_1}I + L_0\right) = (s/k_1)^2 + \mathrm{tr}(L_0)(s/k_1) + \det(L_0) = 0 \tag{3.100}$$

For $k_1 > 0$ *we have from the Routh–Hurwitz stability condition that instability occurs if and only if the trace and/or the determinant of* L_0 *are negative. Since* $\det(L_0) > 0$ *for any gain error less than 100%, instability can only occur if* $\mathrm{tr}(L_0) < 0$. *Evaluate* $\mathrm{tr}(L_0)$ *and show that with gain errors of equal magnitude the combination of errors which most easily yields instability is with* $\hat{\epsilon}_1 = -\hat{\epsilon}_2 = -\epsilon_1 = \epsilon_2 = \epsilon$. *Use this to show that the perturbed system is unstable if*

$$|\epsilon| > \sqrt{\frac{1}{2\lambda_{11} - 1}} \tag{3.101}$$

where $\lambda_{11} = g_{11}g_{22}/\det G$ *is the 1,1 element of the RGA of* G. *In our case* $\lambda_{11} = 35.1$ *and we get instability for* $|\epsilon| > 0.120$. *Check this numerically, e.g. using Matlab.*

Remark. The instability condition in (3.101) for simultaneous input and output gain uncertainty applies to the very special case of a 2×2 plant, in which all elements share the same dynamics, $G(s) = g(s)G_0$, and an inverse-based controller, $K(s) = (k_1/s)G^{-1}(s)$.

Exercise 3.12* *Consider again the distillation process* $G(s)$ *in (3.93). The response using the inverse-based controller* K_{inv} *in (3.95) was found to be sensitive to input gain errors. We want to see if the controller can be modified to yield a more robust system by using the Glover–McFarlane* $\mathcal{H}_\infty$ *loop-shaping procedure. To this effect, let the shaped plant be* $G_s = GK_{\mathrm{inv}}$, *i.e.* $W_1 = K_{\mathrm{inv}}$, *and design an* $\mathcal{H}_\infty$ *controller* K_s *for the shaped plant (see page 370 and Chapter 9), such that the overall controller becomes* $K = K_{\mathrm{inv}}K_s$. *(You will find that* $\gamma_{\min} = 1.414$ *which indicates good robustness with respect to coprime factor uncertainty, but the loop shape is almost unchanged and the system remains sensitive to input uncertainty.)*

3.7.3   Robustness conclusions

From the two motivating examples above we found that multivariable plants can display a
sensitivity to uncertainty (in this case input uncertainty) which is fundamentally different
from what is possible in SISO systems.

In the first example (spinning satellite), we had excellent stability margins (PM and GM)
when considering one loop at a time, but small simultaneous input gain errors gave instability.
This might have been expected from the peak values ($\mathcal{H}_\infty$ norms) of S and T, defined as

$$\|T\|_\infty = \max_\omega \bar{\sigma}(T(j\omega)), \quad \|S\|_\infty = \max_\omega \bar{\sigma}(S(j\omega)) \tag{3.102}$$

which were both large (about 10) for this example.

In the second example (distillation process), we again had excellent stability margins (PM
and GM), and the system was also robustly stable to errors (even simultaneous) of up to
100% in the input gains. However, in this case small input gain errors gave very poor output
performance, so robust performance was not satisfied, and adding simultaneous output gain
uncertainty resulted in instability (see Exercise 3.11). These problems with the decoupling
controller might have been expected because the plant has large RGA elements. For this
second example the $\mathcal{H}_\infty$ norms of S and T were both about 1, so the absence of peaks in S
and T does not guarantee robustness.

Although sensitivity peaks, RGA elements, etc., are useful indicators of robustness
problems, they provide no exact answer to whether a given source of uncertainty will yield
instability or poor performance. This motivates the need for better tools for analyzing the
effects of model uncertainty. We want to avoid a trial-and-error procedure based on checking
stability and performance for a large number of candidate plants. This is very time consuming,
and in the end one does not know whether those plants are the limiting ones. What is desired,
is a simple tool which is able to identify the worst-case plant. This will be the focus of
Chapters 7 and 8 where we show how to represent model uncertainty in the $\mathcal{H}_\infty$ framework,
and introduce the structured singular value μ as our tool. The two motivating examples are
studied in more detail in Example 8.10 and Section 8.11.3 where a μ-analysis predicts the
robustness problems found above.

3.8   General control problem formulation

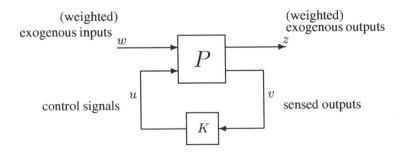

Figure 3.15: General control configuration for the case with no model uncertainty

In this section we consider a general method of formulating control problems introduced by Doyle (1983; 1984). The formulation makes use of the general control configuration in Figure 3.15, where P is the generalized plant and K is the generalized controller as explained in Table 1.1 on page 13. Note that positive feedback is used.

The overall control objective is to minimize some norm of the transfer function from w to z, e.g. the $\mathcal{H}_\infty$ norm. The controller design problem is then:

- Find a controller K, which, based on the information in v, generates a control signal u, which counteracts the influence of w on z, thereby minimizing the closed-loop norm from w to z.

The most important point of this section is to appreciate that almost any linear control problem can be formulated using the block diagram in Figure 3.15 (for the nominal case) or in Figure 3.23 (with model uncertainty).

Remark 1 The configuration in Figure 3.15 may at first glance seem restrictive. However, this is not the case, and we will demonstrate the generality of the setup with a few examples, including the design of observers (the estimation problem) and feedforward controllers.

Remark 2 We may generalize the control configuration still further by including diagnostics as additional outputs from the controller giving the *4-parameter controller* introduced by Nett (1986), but this is not considered in this book.

3.8.1 Obtaining the generalized plant P

The routines in Matlab for synthesizing $\mathcal{H}_\infty$ and $\mathcal{H}_2$ optimal controllers assume that the problem is in the general form of Figure 3.15; that is, they assume that P is given. To derive P (and K) for a specific case we must first find a block diagram representation and identify the signals w, z, u and v. To construct P one should note that it is an *open-loop* system and remember to break all "loops" entering and exiting the controller K. Some examples are given below and further examples are given in Section 9.3 (Figures 9.9, 9.10, 9.11 and 9.12).

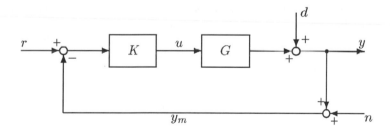

Figure 3.16: One degree-of-freedom control configuration

Example 3.18 One degree-of-freedom feedback control configuration. *We want to find P for the conventional one degree-of-freedom control configuration in Figure 3.16. The first step is to identify the signals for the generalized plant:*

$$w = \begin{bmatrix} w_1 \\ w_2 \\ w_3 \end{bmatrix} = \begin{bmatrix} d \\ r \\ n \end{bmatrix}; \quad z = e = y - r; \quad v = r - y_m = r - y - n \qquad (3.103)$$

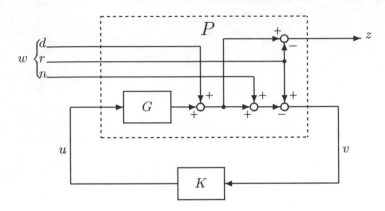

Figure 3.17: Equivalent representation of Figure 3.16 where the error signal to be minimized is $z = y - r$ and the input to the controller is $v = r - y_m$

With this choice of v, the controller only has information about the deviation $r - y_m$. Also note that $z = y - r$, which means that performance is specified in terms of the actual output y and not in terms of the measured output y_m. The block diagram in Figure 3.16 then yields

$$z = y - r = Gu + d - r = Iw_1 - Iw_2 + 0w_3 + Gu$$
$$v = r - y_m = r - Gu - d - n = -Iw_1 + Iw_2 - Iw_3 - Gu$$

and P which represents the transfer function matrix from $\begin{bmatrix} w & u \end{bmatrix}^T$ to $\begin{bmatrix} z & v \end{bmatrix}^T$ is

$$P = \begin{bmatrix} I & -I & 0 & G \\ -I & I & -I & -G \end{bmatrix} \tag{3.104}$$

Note that P does not contain the controller. Alternatively, P can be obtained by inspection from the representation in Figure 3.17.

Remark. Obtaining the generalized plant P may seem tedious. However, when performing numerical calculations P can be generated using software. For example, in Matlab we may use the `simulink` program, or we may use the `sysic` program in the Robust Control toolbox. The code in Table 3.2 generates the generalized plant P in (3.104) for Figure 3.16.

Table 3.2: Matlab program to generate P in (3.104)

```
% Uses the Robust Control toolbox
systemnames = 'G';                              % G is the SISO plant.
inputvar = '[d(1);r(1);n(1);u(1)]';            % Consists of vectors w and u.
input_to_G = '[u]';
outputvar = '[G+d-r; r-G-d-n]';                % Consists of vectors z and v.
sysoutname = 'P';
sysic;
```

3.8.2 Controller design: including weights in P

To get a meaningful controller synthesis problem, e.g. in terms of the $\mathcal{H}_\infty$ or $\mathcal{H}_2$ norms, we generally have to include weights W_z and W_w in the generalized plant P, see Figure 3.18.

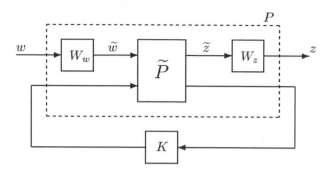

Figure 3.18: General control configuration for the case with no model uncertainty

That is, we consider the weighted or normalized exogenous inputs w (where $\tilde{w} = W_w w$ consists of the "physical" signals entering the system; disturbances, references and noise), and the weighted or normalized controlled outputs $z = W_z \tilde{z}$ (where $\tilde{z}$ often consists of the control error $y - r$ and the manipulated input u). The weighting matrices are usually frequency dependent and typically selected such that weighted signals w and z are of magnitude 1; that is, the norm from w to z should be less than 1. Thus, in most cases only the magnitude of the weights matter, and we may without loss of generality assume that $W_w(s)$ and $W_z(s)$ are stable and minimum-phase (they need not even be rational transfer functions but if not they will be unsuitable for controller synthesis using current software).

Example 3.19 **Stacked** $S/T/KS$ **problem.** *Consider an $\mathcal{H}_\infty$ problem where we want to bound $\bar{\sigma}(S)$ (for performance), $\bar{\sigma}(T)$ (for robustness and to avoid sensitivity to noise) and $\bar{\sigma}(KS)$ (to penalize large inputs). These requirements may be combined into a stacked $\mathcal{H}_\infty$ problem*

$$\min_K \|N(K)\|_\infty, \quad N = \begin{bmatrix} W_u KS \\ W_T T \\ W_P S \end{bmatrix} \tag{3.105}$$

where K is a stabilizing controller. In other words, we have $z = Nw$ and the objective is to minimize the $\mathcal{H}_\infty$ norm from w to z. Except for some negative signs which have no effect when evaluating $\|N\|_\infty$, the N in (3.105) may be represented by the block diagram in Figure 3.19 (convince yourself that this is true). Here w represents a reference command ($w = -r$, where the negative sign does not really matter) or a disturbance entering at the output ($w = d_y$), and z consists of the weighted input $z_1 = W_u u$, the weighted output $z_2 = W_T y$, and the weighted control error $z_3 = W_P(y - r)$. We get from Figure 3.19 the following set of equations:

$$\begin{aligned} z_1 &= W_u u \\ z_2 &= W_T G u \\ z_3 &- W_P w + W_P G u \\ v &= -w - Gu \end{aligned}$$

so the generalized plant P from $[\,w \quad u\,]^T$ to $[\,z \quad v\,]^T$ is

$$P = \begin{bmatrix} 0 & W_u I \\ 0 & W_T G \\ W_P I & W_P G \\ -I & -G \end{bmatrix} \tag{3.106}$$

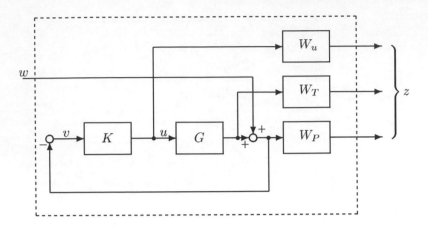

Figure 3.19: Block diagram corresponding to $z = Nw$ in (3.105)

3.8.3 Partitioning the generalized plant P

We often partition P as

$$P = \begin{bmatrix} P_{11} & P_{12} \\ P_{21} & P_{22} \end{bmatrix} \tag{3.107}$$

such that its parts are compatible with the signals w, z, u and v in the generalized control configuration,

$$z = P_{11}w + P_{12}u \tag{3.108}$$

$$v = P_{21}w + P_{22}u \tag{3.109}$$

The reader should become familiar with this notation. In Example 3.19 we get

$$P_{11} = \begin{bmatrix} 0 \\ 0 \\ W_P I \end{bmatrix}, \quad P_{12} = \begin{bmatrix} W_u I \\ W_T G \\ W_P G \end{bmatrix} \tag{3.110}$$

$$P_{21} = -I, \quad P_{22} = -G \tag{3.111}$$

Note that P_{22} has dimensions compatible with the controller, i.e. if K is an $n_u \times n_v$ matrix, then P_{22} is an $n_v \times n_u$ matrix. For cases with one degree-of-freedom negative feedback control we have $P_{22} = -G$.

3.8.4 Analysis: closing the loop to get N

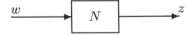

Figure 3.20: General block diagram for analysis with no uncertainty

The general feedback configurations in Figures 3.15 and 3.18 have the controller K as a separate block. This is useful when synthesizing the controller. However, for *analysis* of closed-loop performance the controller is given, and we may absorb K into the interconnection structure and obtain the system N as shown in Figure 3.20 where

$$z = Nw \tag{3.112}$$

where N is a function of K. To find N, we first partition the generalized plant P as given in (3.107)–(3.109), combine this with the controller equation

$$u = Kv \tag{3.113}$$

and eliminate u and v from (3.108), (3.109) and (3.113) to yield $z = Nw$ where N is given by

$$N = P_{11} + P_{12}K(I - P_{22}K)^{-1}P_{21} \triangleq F_l(P, K) \tag{3.114}$$

Here $F_l(P, K)$ denotes a lower *linear fractional transformation (LFT)* of P with K as the parameter. Some properties of LFTs are given in Appendix A.8. In words, N is obtained from Figure 3.15 by using K to close a lower feedback loop around P. Since positive feedback is used in the general configuration in Figure 3.15 the term $(I - P_{22}K)^{-1}$ has a negative sign.

Remark. To assist in remembering the sequence of P_{12} and P_{21} in (3.114), notice that the first (last) index in P_{11} is the same as the first (last) index in $P_{12}K(I - P_{22}K)^{-1}P_{21}$. The lower LFT in (3.114) is also represented by the block diagram in Figure 3.2.

The reader is advised to become comfortable with the above manipulations before progressing further.

Example 3.20 *We want to derive N for the partitioned P in (3.110) and (3.111) using the LFT formula in (3.114). We get*

$$N = \begin{bmatrix} 0 \\ 0 \\ W_P I \end{bmatrix} + \begin{bmatrix} W_u I \\ W_T G \\ W_P G \end{bmatrix} K(I + GK)^{-1}(-I) = \begin{bmatrix} -W_u KS \\ -W_T T \\ W_P S \end{bmatrix}$$

where we have made use of the identities $S = (I + GK)^{-1}$, $T = GKS$ and $I - T = S$. With the exception of the two negative signs, this is identical to N given in (3.105). Of course, the negative signs have no effect on the norm of N.

Again, it should be noted that deriving N from P is much simpler using available software. For example, in the Matlab Robust Control toolbox we can evaluate $N = F_l(P, K)$ using the command N=lft(P,K).

Exercise 3.13 *Consider the two degrees-of-freedom feedback configuration in Figure 1.3(b). (i) Find P when*

$$w = \begin{bmatrix} d \\ r \\ n \end{bmatrix}; \quad z = \begin{bmatrix} y - r \\ u \end{bmatrix}; \quad v = \begin{bmatrix} r \\ y_m \end{bmatrix} \tag{3.115}$$

(ii) Let $z = Nw$ and derive N in two different ways: directly from the block diagram and using $N = F_l(P, K)$.

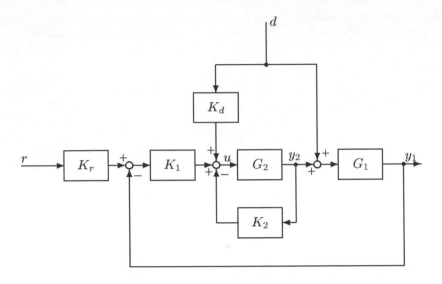

Figure 3.21: System with feedforward, local feedback and two degrees-of-freedom control

3.8.5   Generalized plant P: further examples

To illustrate the generality of the configuration in Figure 3.15, we now present two further examples: one in which we derive P for a problem involving feedforward control, and one for a problem involving estimation.

Example 3.21 *Consider the control system in Figure 3.21, where y_1 is the output we want to control, y_2 is a secondary output (extra measurement), and we also measure the disturbance d. By secondary we mean that y_2 is of secondary importance for control; that is, there is no control objective associated with it. The control configuration includes a two degrees-of-freedom controller, a feedforward controller and a local feedback controller based on the extra measurement y_2. To recast this into our standard configuration of Figure 3.15 we define*

$$w = \begin{bmatrix} d \\ r \end{bmatrix}; \quad z = y_1 - r; \quad v = \begin{bmatrix} r \\ y_1 \\ y_2 \\ d \end{bmatrix} \tag{3.116}$$

Note that d and r are both inputs and outputs to P and we have assumed a perfect measurement of the disturbance d. Since the controller has explicit information about r we have a two degrees-of-freedom controller. The generalized controller K may be written in terms of the individual controller blocks in Figure 3.21 as follows:

$$K = \begin{bmatrix} K_1 K_r & -K_1 & -K_2 & K_d \end{bmatrix} \tag{3.117}$$

By writing down the equations or by inspection from Figure 3.21 we get

$$P = \begin{bmatrix} G_1 & -I & G_1 G_2 \\ 0 & I & 0 \\ G_1 & 0 & G_1 G_2 \\ 0 & 0 & G_2 \\ I & 0 & 0 \end{bmatrix} \tag{3.118}$$

Then partitioning P as in (3.108) and (3.109) yields $P_{22} = \begin{bmatrix} 0^T & (G_1 G_2)^T & G_2^T & 0^T \end{bmatrix}^T$.

Exercise 3.14* Cascade implementation. *Consider Example 3.21 further. The local feedback based on y_2 is often implemented in a cascade manner; see also Figure 10.11. In this case the output from K_1 enters into K_2 and it may be viewed as a reference signal for y_2. Derive the generalized controller K and the generalized plant P in this case.*

Remark. From Example 3.21 and Exercise 3.14, we see that a cascade *implementation* does not usually limit the achievable performance since, unless the optimal K_2 or K_1 have RHP-zeros, we can obtain from the optimal overall K the subcontrollers K_2 and K_1 (although we may have to add a small D-term to K to make the controllers proper). However, if we impose restrictions on the *design* such that, for example, K_2 or K_1 are designed "locally" (without considering the whole problem), then this will limit the achievable performance. For example, for a *two degrees-of-freedom controller* a common approach is first to design the feedback controller K_y for disturbance rejection (without considering reference tracking) and then design K_r for reference tracking. This will generally give some performance loss compared to a simultaneous design of K_y and K_r.

Example 3.22 Output estimator. *Consider a situation where we have no measurement of the output y which we want to control. However, we do have a measurement of another output variable y_2. Let d denote the unknown external inputs (including noise and disturbances) and u_G the known plant inputs (a subscript G is used because in this case the output u from K is not the plant input). Let the model be*

$$y = Gu_G + G_d d; \quad y_2 = Fu_G + F_d d$$

The objective is to design an estimator, K_{est}, such that the estimated output $\widehat{y} = K_{\text{est}} \begin{bmatrix} y_2 \\ u_G \end{bmatrix}$ is as close as possible in some sense to the true output y; see Figure 3.22. This problem may be written in the general framework of Figure 3.15 with

$$w = \begin{bmatrix} d \\ u_G \end{bmatrix}, \ u = \widehat{y}, \ z = y - \widehat{y}, \ v = \begin{bmatrix} y_2 \\ u_G \end{bmatrix}$$

Note that $u = \widehat{y}$; that is, the output u from the generalized controller is the estimate of the plant output. Furthermore, $K = K_{\text{est}}$ and

$$P = \begin{bmatrix} G_d & G & -I \\ F_d & F & 0 \\ 0 & I & 0 \end{bmatrix} \tag{3.119}$$

We see that $P_{22} = \begin{bmatrix} 0 \\ 0 \end{bmatrix}$ since the estimator problem does not involve feedback.

Exercise 3.15 State estimator (observer). *In the Kalman filter problem studied in Section 9.2 the objective is to minimize $x - \widehat{x}$ (whereas in Example 3.22 the objective was to minimize $y - \widehat{y}$). Show how the Kalman filter problem can be represented in the general configuration of Figure 3.15 and find P.*

3.8.6 Deriving P from N

For cases where N is given and we wish to find a P such that

$$N = F_l(P, K) = P_{11} + P_{12}K(I - P_{22}K)^{-1}P_{21}$$

it is usually best to work from a block diagram representation. This was illustrated above for the stacked N in (3.105). Alternatively, the following procedure may be useful:

1. Set $K = 0$ in N to obtain P_{11}.

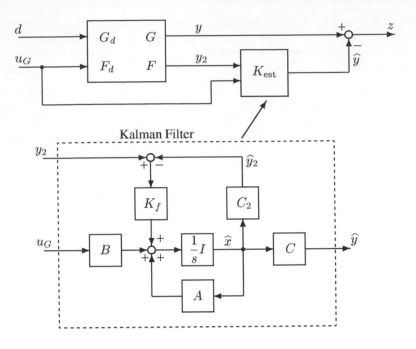

Figure 3.22: Output estimation problem. One particular estimator K_{est} is a Kalman filter

2. Define $Q = N - P_{11}$ and rewrite Q such that each term has a common factor $R = K(I - P_{22}K)^{-1}$ (this gives P_{22}).
3. Since $Q = P_{12}RP_{21}$, we can now usually obtain P_{12} and P_{21} by inspection.

Example 3.23 Weighted sensitivity. *We will use the above procedure to derive P when $N = w_P S = w_P(I + GK)^{-1}$, where w_P is a scalar weight.*

1. $P_{11} = N(K = 0) = w_P I$.
2. $Q = N - w_P I = w_P(S - I) = -w_P T = -w_P GK(I + GK)^{-1}$, and we have $R = K(I + GK)^{-1}$ so $P_{22} = -G$.
3. $Q = -w_P GR$ so we have $P_{12} = -w_P G$ and $P_{21} = I$, and we get

$$P = \begin{bmatrix} w_P I & -w_P G \\ I & -G \end{bmatrix} \tag{3.120}$$

Remark. When obtaining P from a given N, we have that P_{11} and P_{22} are unique, whereas from step 3 in the above procedure we see that P_{12} and P_{21} are not unique. For instance, let α be a real scalar, then we may instead choose $\tilde{P}_{12} = \alpha P_{12}$ and $\tilde{P}_{21} = (1/\alpha)P_{21}$. For P in (3.120) this means that we may move the negative sign of the scalar w_P from P_{12} to P_{21}.

Exercise 3.16 * **Mixed sensitivity.** *Use the above procedure to derive the generalized plant P for the stacked N in (3.105).*

3.8.7 Problems not covered by the general formulation

The above examples have demonstrated the generality of the control configuration in Figure 3.15. Nevertheless, there are some controller design problems which are not covered.

Let N be some closed-loop transfer function whose norm we want to minimize. To use the general form we must first obtain a P such that $N = F_l(P, K)$. However, this is not always possible, since there may not exist a block diagram representation for N. As a simple example, consider the stacked transfer function

$$N = \begin{bmatrix} (I + GK)^{-1} \\ (I + KG)^{-1} \end{bmatrix} \tag{3.121}$$

The transfer function $(I + GK)^{-1}$ may be represented on a block diagram with the input and output signals *after* the plant, whereas $(I + KG)^{-1}$ may be represented by another block diagram with input and output signals *before* the plant. However, in N there are no cross coupling terms between an input before the plant and an output after the plant (corresponding to $G(I + KG)^{-1}$), or between an input after the plant and an output before the plant (corresponding to $-K(I + GK)^{-1}$) so N cannot be represented in block diagram form. Equivalently, if we apply the procedure in Section 3.8.6 to N in (3.121), we are not able to find solutions to P_{12} and P_{21} in step 3.

Another stacked transfer function which *cannot* in general be represented in block diagram form is

$$N = \begin{bmatrix} W_P S \\ S G_d \end{bmatrix} \tag{3.122}$$

Remark. The case where N cannot be written as an LFT of K is a special case of the Hadamard-weighted $\mathcal{H}_\infty$ problem studied by van Diggelen and Glover (1994a). Although the solution to this $\mathcal{H}_\infty$ problem remains intractable, van Diggelen and Glover (1994b) present a solution for a similar problem where the Frobenius norm is used instead of the singular value to "sum up the channels".

Exercise 3.17 *Show that N in (3.122) can be represented in block diagram form if $W_P = w_P I$ where w_P is a scalar.*

3.8.8 A general control configuration including model uncertainty

The general control configuration in Figure 3.15 may be extended to include model uncertainty as shown by the block diagram in Figure 3.23. Here the matrix Δ is a *block-diagonal* matrix that includes all possible perturbations (representing uncertainty) to the system. It is usually normalized in such a way that $\|\Delta\|_\infty \leq 1$.

The block diagram in Figure 3.23 in terms of P (for synthesis) may be transformed into the block diagram in Figure 3.24 in terms of N (for analysis) by using K to close a lower loop around P. If we partition P to be compatible with the controller K, then the same *lower LFT* as found in (3.114) applies, and

$$N = F_l(P, K) = P_{11} + P_{12} K (I - P_{22} K)^{-1} P_{21} \tag{3.123}$$

To evaluate the perturbed (uncertain) transfer function from external inputs w to external outputs z, we use Δ to close the upper loop around N (see Figure 3.24), resulting in an *upper LFT* (see Appendix A.8):

$$z = F_u(N, \Delta) w; \quad F_u(N, \Delta) \triangleq N_{22} + N_{21} \Delta (I - N_{11} \Delta)^{-1} N_{12} \tag{3.124}$$

Remark 1 Controller synthesis based on Figure 3.23 is still an unsolved problem, although good practical approaches like DK-iteration to find the "μ-optimal" controller are in use (see Section 8.12).

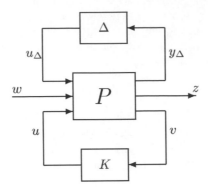

Figure 3.23: General control configuration for the case with model uncertainty

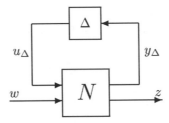

Figure 3.24: General block diagram for analysis with uncertainty included

For analysis (with a given controller), the situation is better and with the $\mathcal{H}_\infty$ norm an assessment of robust performance involves computing the structured singular value, μ. This is discussed in more detail in Chapter 8.

Remark 2 In (3.124) N has been partitioned to be compatible with Δ; that is, N_{11} has dimensions compatible with Δ. Usually, Δ is square, in which case N_{11} is a square matrix of the same dimension as Δ. For the nominal case with no uncertainty we have $F_u(N,\Delta) = F_u(N,0) = N_{22}$, so N_{22} is the nominal transfer function from w to z.

Remark 3 Note that P and N here also include information about how the uncertainty affects the system, so they are *not* the same P and N as used earlier, e.g. in (3.114). Actually, the parts P_{22} and N_{22} of P and N in (3.123) (with uncertainty) are equal to the P and N in (3.114) (without uncertainty). Strictly speaking, we should have used another symbol for N and P in (3.123), but for notational simplicity we did not.

Remark 4 The fact that almost any control problem with uncertainty can be represented by Figure 3.23 may seem surprising, so some explanation is in order. First, represent each source of uncertainty by a perturbation block, Δ_i, which is normalized such that $\|\Delta_i\| \leq 1$. These perturbations may result from parametric uncertainty, neglected dynamics, etc., as will be discussed in more detail in Chapters 7 and 8. Then "pull out" each of these blocks from the system so that an input and an output can be associated with each Δ_i as shown in Figure 3.25(a). Finally, collect these perturbation blocks into a large block-diagonal matrix having perturbation inputs and outputs as shown in Figure 3.25(b). In Chapter 8 we discuss in detail how to obtain N and Δ. Generally, it is difficult to perform these tasks manually, but this can be easily done using software; see examples in Chapters 7.

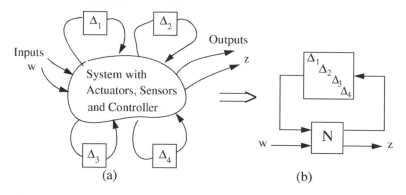

Figure 3.25: Rearranging a system with multiple perturbations into the $N\Delta$-structure

3.9   Additional exercises

Most of these exercises are based on material presented in Appendix A. The exercises illustrate material which the reader should know before reading the subsequent chapters.

Exercise 3.18 * *Consider the performance specification* $\|w_P S\|_\infty < 1$. *Suggest a rational transfer function weight* $w_P(s)$ *and sketch it as a function of frequency for the following two cases:*

1. *We desire no steady-state offset, a bandwidth better than 1 rad/s and a resonance peak (worst amplification caused by feedback) lower than 1.5.*
2. *We desire less than 1% steady-state offset, less than 10% error up to frequency 3 rad/s, a bandwidth better than 10 rad/s, and a resonance peak lower than 2. (Hint: See (2.105) and (2.106).)*

Exercise 3.19 *By* $\|M\|_\infty$ *one can mean either a spatial or temporal norm. Explain the difference between the two and illustrate by computing the appropriate infinity norm for*

$$M_1 = \begin{bmatrix} 3 & 4 \\ -2 & 6 \end{bmatrix}, \quad M_2(s) = \frac{s-1}{s+1}\frac{3}{s+2}$$

Exercise 3.20 * *What is the relationship between the RGA matrix and uncertainty in the individual elements? Illustrate this for perturbations in the* $1,1$ *element of the matrix*

$$A = \begin{bmatrix} 10 & 9 \\ 9 & 8 \end{bmatrix} \tag{3.125}$$

Exercise 3.21 *Assume that A is non-singular. (i) Formulate a condition in terms of the maximum singular value of E for the matrix $A + E$ to remain non-singular. Apply this to A in (3.125) and (ii) find an E of minimum magnitude which makes $A + E$ singular.*

Exercise 3.22 * *Compute* $\|A\|_{i1}$, $\bar{\sigma}(A) = \|A\|_{i2}$, $\|A\|_{i\infty}$, $\|A\|_F$, $\|A\|_{\max}$ *and* $\|A\|_{\text{sum}}$ *for the following matrices and tabulate your results:*

$$A_1 = I; \quad A_2 = \begin{bmatrix} 1 & 0 \\ 0 & 0 \end{bmatrix}; A_3 = \begin{bmatrix} 1 & 1 \\ 1 & 1 \end{bmatrix}; A_4 = \begin{bmatrix} 1 & 1 \\ 0 & 0 \end{bmatrix}; A_5 = \begin{bmatrix} 1 & 0 \\ 1 & 0 \end{bmatrix}$$

Show using the above matrices that the following bounds are tight (i.e. we may have equality) for 2×2 matrices ($m = 2$):

$$\bar{\sigma}(A) \le \|A\|_F \le \sqrt{m}\,\bar{\sigma}(A)$$
$$\|A\|_{\max} \le \bar{\sigma}(A) \le m\|A\|_{\max}$$

$$\|A\|_{i1}/\sqrt{m} \leq \bar{\sigma}(A) \leq \sqrt{m}\|A\|_{i1}$$
$$\|A\|_{i\infty}/\sqrt{m} \leq \bar{\sigma}(A) \leq \sqrt{m}\|A\|_{i\infty}$$
$$\|A\|_F \leq \|A\|_{\text{sum}}$$

Exercise 3.23 *Find example matrices to illustrate that the above bounds are also tight when A is a square m × m matrix with m > 2.*

Exercise 3.24 * *Do the extreme singular values bound the magnitudes of the elements of a matrix? That is, is $\bar{\sigma}(A)$ greater than the largest element (in magnitude), and is $\underline{\sigma}(A)$ smaller than the smallest element? For a non-singular matrix, how is $\underline{\sigma}(A)$ related to the largest element in A^{-1}?*

Exercise 3.25 *Consider a lower triangular m × m matrix A with $a_{ii} = -1$, $a_{ij} = 1$ for all $i > j$, and $a_{ij} = 0$ for all $i < j$.*

(a) *What is* det *A?*

(b) *What are the eigenvalues of A?*

(c) *What is the RGA of A?*

(d) *Let m = 4 and find an E with the smallest value of $\bar{\sigma}(E)$ such that A + E is singular.*

Exercise 3.26 * *Find two matrices A and B such that $\rho(A + B) > \rho(A) + \rho(B)$ which proves that the spectral radius does not satisfy the triangle inequality and is thus not a norm.*

Exercise 3.27 *Write $T = GK(I + GK)^{-1}$ as an LFT of K, i.e. find P such that $T = F_l(P, K)$.*

Exercise 3.28 * *Write K as an LFT of $T = GK(I + GK)^{-1}$, i.e. find J such that $K = F_l(J, T)$.*

Exercise 3.29 *State-space descriptions may be represented as LFTs. To demonstrate this find H for*

$$F_l(H, 1/s) = C(sI - A)^{-1}B + D$$

Exercise 3.30 * *Show that the set of all stabilizing controllers in (4.94) can be written as $K = F_l(J, Q)$ and find J.*

Exercise 3.31 *In (3.11) we stated that the sensitivity of a perturbed plant, $S' = (I + G'K)^{-1}$, is related to that of the nominal plant, $S = (I + GK)^{-1}$, by*

$$S' = S(I + E_O T)^{-1}$$

where $E_O = (G' - G)G^{-1}$. This exercise deals with how the above result may be derived in a systematic (though cumbersome) manner using LFTs (see also Skogestad and Morari, 1988a).

  (a) First find F such that $S' = (I + G'K)^{-1} = F_l(F, K)$, and find J such that $K = F_l(J, T)$ (see Exercise 3.28).

  (b) Combine these LFTs to find $S' = F_l(N, T)$. What is N in terms of G and G'? Note that since $J_{11} = 0$ we have from (A.164)

$$N = \begin{bmatrix} F_{11} & F_{12}J_{12} \\ J_{21}F_{21} & J_{22} + J_{21}F_{22}J_{12} \end{bmatrix}$$

  (c) Evaluate $S' = F_l(N, T)$ and show that

$$S' = I - G'G^{-1}T(I - (I - G'G^{-1})T)^{-1}$$

  (d) Finally, show that this may be rewritten as $S' = S(I + E_O T)^{-1}$.

3.10   Conclusion

The main purpose of this chapter has been to give an overview of methods for analysis and design of multivariable control systems.

In terms of analysis, we have shown how to evaluate MIMO transfer functions and how to use the singular value decomposition of the frequency-dependent plant transfer function matrix to provide insight into multivariable directionality. Other useful tools for analyzing directionality and interactions are the condition number and the RGA. Closed-loop performance may be analyzed in the frequency domain by evaluating the maximum singular value of the sensitivity function as a function of frequency. Multivariable RHP-zeros impose fundamental limitations on closed-loop performance, but for MIMO systems we can often direct the undesired effect of a RHP-zero to a subset of the outputs. MIMO systems are often more sensitive to uncertainty than SISO systems, and we demonstrated in two examples the possible sensitivity to input gain uncertainty.

In terms of controller design, we discussed some simple approaches such as decoupling and decentralized control. We also introduced a general control configuration in terms of the generalized plant P, which can be used as a basis for synthesizing multivariable controllers using a number of methods, including LQG, $\mathcal{H}_2$, $\mathcal{H}_\infty$ and μ-optimal control. These methods are discussed in much more detail in Chapters 8 and 9. In this chapter we have only discussed the $\mathcal{H}_\infty$ weighted sensitivity method.

4

ELEMENTS OF LINEAR SYSTEM THEORY

The main objective of this chapter is to summarize important results from linear system theory. The treatment is thorough, but readers are encouraged to consult other books, such as Kailath (1980) or Zhou et al. (1996), for more details and background information if these results are new to them.

4.1 System descriptions

The most important property of a linear system (operator) is that it satisfies the *superposition principle*. Let $f(u)$ be a linear operator, let u_1 and u_2 be two independent variables (e.g. input signals), and let α_1 and α_2 be two real scalars, then

$$f(\alpha_1 \cdot u_1 + \alpha_2 \cdot u_2) - \alpha_1 \cdot f(u_1) + \alpha_2 \cdot f(u_2) \tag{4.1}$$

We use in this book various representations of time-invariant linear systems, all of which are equivalent for systems that can be described by linear ordinary differential equations with constant coefficients and which do not require differentiation of the inputs (independent variables). The most important of these representations are discussed in this section.

4.1.1 State-space representation

Consider a system with m inputs (vector u) and l outputs (vector y) which has an internal description of n states (vector x). A natural way to represent many physical systems is by nonlinear state-space models of the form

$$\dot{x} = f(x, u); \quad y = g(x, u) \tag{4.2}$$

where $\dot{x} \equiv dx/dt$ and f and g are nonlinear functions. Linear state-space models may then be derived from the linearization of such models. In terms of deviation variables (where x represents a deviation from some nominal value or trajectory, etc.) we have

$$\dot{x}(t) = Ax(t) + Bu(t) \tag{4.3}$$

$$y(t) = Cx(t) + Du(t) \tag{4.4}$$

where A, B, C and D are real matrices. The dependence of x, u and y of time t is usually omitted to simplify notation. We consider *time-invariant linear systems* where these matrices

Multivariable Feedback Control: Analysis and Design. Second Edition
S. Skogestad and I. Postlethwaite © 2005, 2006 John Wiley & Sons, Ltd

are independent of time. If (4.3) is derived by linearizing (4.2) then $A = \partial f / \partial x$ and $B = \partial f / \partial u$ (see Section 1.5 for an example of such a derivation). A is sometimes called the state matrix. These equations provide a convenient means of describing the dynamic behaviour of proper, rational, linear systems. They may be rewritten as

$$\begin{bmatrix} \dot{x} \\ y \end{bmatrix} = \begin{bmatrix} A & B \\ C & D \end{bmatrix} \begin{bmatrix} x \\ u \end{bmatrix}$$

which gives rise to the shorthand notation

$$G \stackrel{s}{=} \left[\begin{array}{c|c} A & B \\ \hline C & D \end{array} \right] \tag{4.5}$$

which is frequently used to describe a state-space model of a system G. Note that the representation in (4.3)–(4.4) is *not* a unique description of the input–output behaviour of a linear system. First, there exist realizations with the same input–output behaviour, but with additional unobservable and/or uncontrollable states (modes). Second, even for a minimal realization (a realization with the fewest number of states and consequently no unobservable or uncontrollable modes) there are an infinite number of possibilities. To see this, let S be an invertible constant matrix, and introduce the new states $\tilde{x} = Sx$, i.e. $x = S^{-1}\tilde{x}$. Then an equivalent state-space realization (i.e. one with the same input–output behaviour) in terms of these new states is

$$\tilde{A} = SAS^{-1}, \quad \tilde{B} = SB, \quad \tilde{C} = CS^{-1}, \quad \tilde{D} = D$$

The most common realizations are given by a few canonical forms, such as the Jordan (diagonalized) canonical form, the observability canonical form, etc.; see page 126.

Given the linear dynamical system in (4.3) with an initial state condition $x(t_0)$ and an input $u(t)$, the dynamical system response $x(t)$ for $t \geq t_0$ can be determined from

$$x(t) = e^{A(t-t_0)} x(t_0) + \int_{t_0}^{t} e^{A(t-\tau)} B u(\tau) d\tau \tag{4.6}$$

where the matrix exponential is

$$e^{At} = I + \sum_{k=1}^{\infty} (At)^k / k! = \sum_{i=1}^{n} t_i e^{\lambda_i t} q_i^H \tag{4.7}$$

The latter dyadic expansion, involving the right (t_i) and left (q_i) eigenvectors of A, applies for cases with distinct eigenvalues λ_i of A, see (A.23). We will refer to the term $e^{\lambda_i t}$ as the *mode* associated with the eigenvalue $\lambda_i(A)$. For a diagonalized realization (where we select S such that $\tilde{A} = SAS^{-1} = \Lambda$ is a diagonal matrix) we have that $e^{\tilde{A}t} = \text{diag}\{e^{\lambda_i(A)t}\}$; see (A.22).

Remark 1 In the state-space model (4.3)–(4.4) u represents all independent variables. Usually, we consider three kinds of independent variables, namely the manipulated inputs (u), the disturbances (d) and the measurement noise n. The state-space model is then written as

$$\begin{aligned} \dot{x} &= Ax + Bu + B_d d \\ y &= Cx + Du + D_d d + n \end{aligned} \tag{4.8}$$

Note that the symbol n is used to represent both the noise signal and the number of states.

Remark 2 A more general state-space representation is the *descriptor* representation

$$E\dot{x} = Ax + Bu \qquad (4.9)$$

If E is non-singular, (4.9) is a special case of (4.3), because (4.9) may then be written as

$$\dot{x} = \bar{A}x + \bar{B}u$$

where $\bar{A} = E^{-1}A$ and $\bar{B} = E^{-1}B$. However, if the matrix E is singular then (4.9) allows for implicit algebraic relations between the states x. For example, if $E = [I\ 0]$ then (4.9) is equivalent to the following set of differential and algebraic equations:

$$
\begin{aligned}
\dot{x}_1 &= A_{11}x_1 + A_{12}x_2 + B_1 u \\
0 &= A_{21}x_1 + A_{22}x_2 + B_2 u
\end{aligned}
$$

It would be possible to eliminate the algebraic variables x_2 (by solving the algebraic equations) to get $x_2 = -A_{22}^{-1}(A_{21}x_1 + B_2 u)$ and thus derive a set of differential equations (4.3) in x_1 only. However, it is often more convenient to keep the system on the original descriptor form in (4.9).

4.1.2   Impulse response representation

The impulse response matrix is

$$g(t) = \begin{cases} 0 & t < 0 \\ Ce^{At}B + D\delta(t) & t \geq 0 \end{cases} \qquad (4.10)$$

where $\delta(t)$ is the unit impulse (delta) function which satisfies $\lim_{\epsilon \to 0} \int_0^\epsilon \delta(t)dt = 1$. The ij'th element of the impulse response matrix, $g_{ij}(t)$, represents the response $y_i(t)$ to an impulse $u_j(t) = \delta(t)$ for a system with a zero initial state.

With initial state $x(0) = 0$, the dynamic response to an arbitrary input $u(t)$ (which is zero for $t < 0$) may from (4.6) be written as

$$y(t) = g(t) * u(t) = \int_0^t g(t-\tau)u(\tau)d\tau \qquad (4.11)$$

where $*$ denotes the convolution operator.

4.1.3   Transfer function representation – Laplace transforms

The transfer function representation is unique and is very useful for directly obtaining insight into the properties of a system. It is defined as the Laplace transform of the impulse response matrix

$$G(s) = \int_0^\infty g(t)e^{-st}dt \qquad (4.12)$$

Alternatively, we may start from the state-space description. With the assumption of a zero initial state, $x(t = 0) = 0$, the Laplace transforms of (4.3) and (4.4) become[1]

$$sx(s) = Ax(s) + Bu(s) \quad \Rightarrow \quad x(s) = (sI - A)^{-1}Bu(s) \qquad (4.13)$$

---

[1] We make the usual abuse of notation and let $f(s)$ denote the Laplace transform of $f(t)$.

$$y(s) = Cx(s) + Du(s) \quad \Rightarrow \quad y(s) = \underbrace{\left(C(sI - A)^{-1}B + D\right)}_{G(s)} u(s) \tag{4.14}$$

where $G(s)$ is the transfer function matrix. Equivalently, from (A.1),

$$G(s) = \frac{1}{\det(sI - A)}[C \operatorname{adj}(sI - A)B + D \det(sI - A)] \tag{4.15}$$

where $\det(sI - A) = \prod_{i=1}^{n}(s - p_i)$ is the pole polynomial. The poles are equal to the eigenvalues of A, i.e. $p_i = \lambda_i(A)$. For cases where the eigenvalues of A are distinct, we may use the dyadic expansion of A given in (A.23), and derive

$$G(s) = \sum_{i=1}^{n} \frac{Ct_i q_i^H B}{s - p_i} + D \tag{4.16}$$

where q_i and t_i are the left and right eigenvectors of the state matrix A respectively. When disturbances are treated separately, see (4.8), the corresponding disturbance transfer function is

$$G_d(s) = C(sI - A)^{-1}B_d + D_d \tag{4.17}$$

Note that any system written in the state-space form of (4.3) and (4.4) has a transfer function, but the opposite is not true. For example, time delays and improper systems can be represented by Laplace transforms, but do not have a state-space representation. On the other hand, the state-space representation yields an internal description of the system which may be useful if the model is derived from physical principles. It is also more suitable for numerical calculations.

4.1.4 Frequency response

An important advantage of transfer functions is that the frequency response (Fourier transform) is directly obtained from the Laplace transform by setting $s = j\omega$ in $G(s)$. For more details on the frequency response, the reader is referred to Sections 2.1 and 3.3.

4.1.5 Coprime factorization

Another useful way of representing systems is the coprime factorization which may be used both in state-space and transfer function form. In the latter case a *right coprime factorization* of G is

$$G(s) = N_r(s)M_r^{-1}(s) \tag{4.18}$$

where $N_r(s)$ and $M_r(s)$ are stable coprime transfer functions. The stability implies that $N_r(s)$ should contain all the RHP-zeros of $G(s)$, and $M_r(s)$ should contain as RHP-zeros all the RHP-poles of $G(s)$. The coprimeness implies that there should be no common RHP-zeros (including the point at infinity) in N_r and M_r, which result in pole–zero cancellations when forming $N_r M_r^{-1}$. Mathematically, coprimeness means that there exist stable $U_r(s)$ and $V_r(s)$ such that the following Bezout identity is satisfied:

$$U_r N_r + V_r M_r = I \tag{4.19}$$

Similarly, a *left coprime factorization* of G is

$$G(s) = M_l^{-1}(s)N_l(s) \tag{4.20}$$

Here N_l and M_l are stable and coprime; that is, there exist stable $U_l(s)$ and $V_l(s)$ such that the following Bezout identity is satisfied:

$$N_l U_l + M_l V_l = I \tag{4.21}$$

For a scalar system, the left and right coprime factorizations are identical, $G = NM^{-1} = M^{-1}N$.

Remark. Two stable scalar transfer functions, $N(s)$ and $M(s)$, are coprime if and only if they have no common RHP-zeros including the point at $s = \infty$. In this case, we can always find stable U and V such that $NU + MV = 1$.

Example 4.1 *Consider the scalar system*

$$G(s) = \frac{(s-1)(s+2)}{(s-3)(s+4)} \tag{4.22}$$

To obtain a coprime factorization, we first make all the RHP-poles of G zeros of M, and all the RHP-zeros of G zeros of N. We then allocate the poles of N and M so that N and M are both proper and the identity $G = NM^{-1}$ holds. Thus

$$N(s) = \frac{s-1}{s+4}, \quad M(s) = \frac{s-3}{s+2}$$

is a coprime factorization. Usually, we select N and M to have the same poles as each other and the same order as $G(s)$. This gives the most degrees of freedom subject to having a realization of $[M(s) \quad N(s)]^T$ with the lowest order. We then have that

$$N(s) = k\frac{(s-1)(s+2)}{s^2 + k_1 s + k_2}, \quad M(s) = k\frac{(s-3)(s+4)}{s^2 + k_1 s + k_2} \tag{4.23}$$

is a coprime factorization of (4.22) for any k and for any $k_1, k_2 > 0$.

From the above example, we see that the coprime factorization is not unique. Now we introduce the operator M^* defined as $M^*(s) = M^T(-s)$ (which for $s = j\omega$ is the same as the complex conjugate transpose $M^H = \bar{M}^T$). Then $G(s) = N_r(s)M_r^{-1}(s)$ is called a *normalized* right coprime factorization if

$$M_r^* M_r + N_r^* N_r = I \tag{4.24}$$

In this case $X_r(s) = \begin{bmatrix} M_r \\ N_r \end{bmatrix}$ satisfies $X_r^* X_r = I$ and is called an *inner* transfer function. The normalized left coprime factorization $G(s) = M_l^{-1}(s)N_l(s)$ is defined similarly, requiring that

$$M_l M_l^* + N_l N_l^* = I \tag{4.25}$$

In this case $X_l(s) = [M_l \quad N_l]$ is *co-inner* which means that $X_l X_l^* = I$. The normalized coprime factorizations are unique to within a right (left) multiplication by a unitary matrix.

Exercise 4.1 * *We want to find the normalized coprime factorization for the scalar system in (4.22). Let N and M be as given in (4.23), and substitute them into (4.24). Show that after some algebra and comparing of terms one obtains: $k = \pm 0.71$, $k_1 = 5.67$ and $k_2 = 8.6$.*

To derive normalized coprime factorizations by hand, as in the above exercise, is in general difficult. Numerically, however, one can easily find a state-space realization. If G has a minimal state-space realization

$$G \stackrel{s}{=} \left[\begin{array}{c|c} A & B \\ \hline C & D \end{array} \right]$$

then a minimal state-space realization of a normalized left coprime factorization is given (Vidyasagar, 1988) by

$$\begin{bmatrix} N_l(s) & M_l(s) \end{bmatrix} \stackrel{s}{=} \left[\begin{array}{c|ccc} A + HC & B + HD & H \\ \hline R^{-1/2}C & R^{-1/2}D & R^{-1/2} \end{array} \right] \tag{4.26}$$

where

$$H \stackrel{\triangle}{=} -(BD^T + ZC^T)R^{-1}, \quad R \stackrel{\triangle}{=} I + DD^T$$

and the matrix Z is the unique positive definite solution to the algebraic Riccati equation

$$(A - BS^{-1}D^TC)Z + Z(A - BS^{-1}D^TC)^T - ZC^TR^{-1}CZ + BS^{-1}B^T = 0$$

where

$$S \stackrel{\triangle}{=} I + D^TD$$

Notice that the formulae simplify considerably for a strictly proper plant, i.e. when $D = 0$. The Matlab commands in Table 4.1 can be used to find the normalized coprime factorization for $G(s)$ using (4.26).

Table 4.1: Matlab commands to generate a normalized coprime factorization

```
% Uses the Robust Control toolbox
%
% Find Normalized Coprime factors of system [a,b,c,d] using (4.26)
%
S=eye(size(d'*d))+d'*d;
R=eye(size(d*d'))+d*d';
A1 = a-b*inv(S)*d'*c;
R1 = c'*inv(R)*c;
[R1s,R1err] = sqrtm(R1);
Q1 = b*inv(S)*b';
[Z,L,G]=care(A1',R1s,Q1); %Solve Riccati equation

H = -(b*d' + Z*c')*inv(R);
A = a + H*c;
Bn = b + H*d; Bm = H;
C = inv(sqrtm(R))*c;
Dn = inv(sqrtm(R))*d;
Dm = inv(sqrtm(R));
N = ss(A,Bn,C,Dn);
M = ss(A,Bm,C,Dm);
```

Exercise 4.2 *Verify numerically (e.g. using the Matlab file in Table 4.1 or the Robust Control toolbox command* ncfmr*) that the normalized coprime factors of $G(s)$ in (4.22) are as given in Exercise 4.1.*

4.1.6   More on state-space realizations

Inverse system. In some cases we may want to find a state-space description of the inverse of a system. For a square $G(s)$ we have

$$G^{-1} \stackrel{s}{=} \left[\begin{array}{c|c} A - BD^{-1}C & BD^{-1} \\ \hline -D^{-1}C & D^{-1} \end{array} \right] \tag{4.27}$$

where D is assumed to be non-singular. For a non-square $G(s)$ in which D has full row (or column) rank, a right (or left) inverse of $G(s)$ can be found by replacing D^{-1} by $D^{\dagger}$, the pseudo-inverse of D.

For a strictly proper system with $D = 0$, one may obtain an approximate inverse by including a small additional feed-through term D, preferably chosen on physical grounds. One should be careful, however, to select the signs of the terms in D such that one does not introduce RHP-zeros in $G(s)$ because this will make $G(s)^{-1}$ unstable.

Improper systems. Improper transfer functions, where the order of the s-polynomial in the numerator exceeds that of the denominator, cannot be represented in standard state-space form. To approximate improper systems by state-space models, we can include some high-frequency dynamics which we know from physical considerations will have little significance.

Realization of SISO transfer functions. Transfer functions are a good way of representing systems because they give more immediate insight into a system's behaviour. However, for numerical calculations a state-space realization is usually desired. One way of obtaining a state-space realization from a SISO transfer function is given next. Consider a strictly proper transfer function ($D = 0$) of the form

$$G(s) = \frac{\beta_{n-1}s^{n-1} + \cdots + \beta_1 s + \beta_0}{s^n + a_{n-1}s^{n-1} + \cdots + a_1 s + a_0} \tag{4.28}$$

Then, since multiplication by s corresponds to differentiation in the time domain, (4.28) and the relationship $y(s) = G(s)u(s)$ correspond to the following differential equation:

$$y^n(t) + a_{n-1}y^{n-1}(t) + \cdots + a_1 y'(t) + a_0 y(t) = \beta_{n-1}u^{n-1}(t) + \cdots + \beta_1 u'(t) + \beta_0 u(t)$$

where $y^{n-1}(t)$ and $u^{n-1}(t)$ represent $n - 1$'th order derivatives, etc. We can further write this as

$$y^n = \underbrace{(-a_{n-1}y^{n-1} + \beta_{n-1}u^{n-1}) + \cdots + \underbrace{(-a_1 y' + \beta_1 u') + \underbrace{(-a_0 y + \beta_0 u)}_{x'_n}}_{x^2_{n-1}}}_{x^n_1}$$

where we have introduced new variables $x_1, x_2, \ldots, x_n$ and we have $y = x_1$. Note that x^n_1 is the n'th derivative of $x_1(t)$. With the notation $\dot{x} \equiv x'(t) = dx/dt$, we have the following state-space equations:

$$\dot{x}_n = -a_0 x_1 + \beta_0 u$$
$$\dot{x}_{n-1} = -a_1 x_1 + x_n + \beta_1 u$$
$$\vdots$$
$$\dot{x}_1 = -a_{n-1}x_1 + x_2 + \beta_{n-1}u$$

corresponding to the realization

$$A = \begin{bmatrix} -a_{n-1} & 1 & 0 & \cdots & 0 & 0 \\ -a_{n-2} & 0 & 1 & & 0 & 0 \\ \vdots & \vdots & & \ddots & & \vdots \\ -a_2 & 0 & 0 & & 1 & 0 \\ -a_1 & 0 & 0 & \cdots & 0 & 1 \\ -a_0 & 0 & 0 & \cdots & 0 & 0 \end{bmatrix}, \quad B = \begin{bmatrix} \beta_{n-1} \\ \beta_{n-2} \\ \vdots \\ \beta_2 \\ \beta_1 \\ \beta_0 \end{bmatrix} \tag{4.29}$$

$$C = \begin{bmatrix} 1 & 0 & 0 & \cdots & 0 & 0 \end{bmatrix}$$

This is called the *observer canonical form*. Two advantages of this realization are that one can obtain the elements of the matrices directly from the transfer function, and that the output y is simply equal to the first state. Notice that if the transfer function is not strictly proper, then we must first bring out the constant term, i.e. write $G(s) = G_1(s) + D$, and then find the realization of $G_1(s)$ using (4.29).

Example 4.2 *To obtain the state-space realization, in observer canonical form, of the SISO transfer function $G(s) = \frac{s-a}{s+a}$, we first bring out a constant term by division to get*

$$G(s) = \frac{s-a}{s+a} = \frac{-2a}{s+a} + 1$$

Thus $D = 1$. For the term $\frac{-2a}{s+a}$ we get from (4.28) that $\beta_0 = -2a$ and $a_0 = a$, and therefore (4.29) yields $A = -a, B = -2a$ and $C = 1$.

Example 4.3 *Consider an ideal PID controller*

$$K(s) = K_c \left(1 + \frac{1}{\tau_I s} + \tau_D s \right) = K_c \frac{\tau_I \tau_D s^2 + \tau_I s + 1}{\tau_I s} \tag{4.30}$$

Since this involves differentiation of the input, it is an improper transfer function and cannot be written in state-space form. A proper PID controller may be obtained by letting the derivative action be effective over a limited frequency range. For example,

$$K(s) = K_c \left(1 + \frac{1}{\tau_I s} + \frac{\tau_D s}{1 + \epsilon \tau_D s} \right) \tag{4.31}$$

where ϵ is typically about 0.1 (see also page 56). This can now be realized in state-space form in an infinite number of ways. Four common forms are given below. In all cases, the D-matrix, which represents the controller gain at high frequencies ($s \to \infty$), is a scalar given by

$$D = K_c \frac{1+\epsilon}{\epsilon} \tag{4.32}$$

1. Diagonalized form (Jordan canonical form)

$$A = \begin{bmatrix} 0 & 0 \\ 0 & -\frac{1}{\epsilon \tau_D} \end{bmatrix}, \quad B = \begin{bmatrix} K_c/\tau_I \\ K_c/(\epsilon^2 \tau_D) \end{bmatrix}, \quad C = \begin{bmatrix} 1 & -1 \end{bmatrix} \tag{4.33}$$

2. Observability canonical form

$$A = \begin{bmatrix} 0 & 1 \\ 0 & -\frac{1}{\epsilon \tau_D} \end{bmatrix}, \quad B = \begin{bmatrix} \gamma_1 \\ \gamma_2 \end{bmatrix}, \quad C = \begin{bmatrix} 1 & 0 \end{bmatrix} \tag{4.34}$$

$$\text{where} \quad \gamma_1 = K_c \left(\frac{1}{\tau_I} - \frac{1}{\epsilon^2 \tau_D} \right), \quad \gamma_2 = \frac{K_c}{\epsilon^3 \tau_D^2}$$

3. Controllability canonical form

$$A = \begin{bmatrix} 0 & 0 \\ 1 & -\frac{1}{\epsilon \tau_D} \end{bmatrix}, \quad B = \begin{bmatrix} 1 \\ 0 \end{bmatrix}, \quad C = [\gamma_1 \quad \gamma_2] \tag{4.35}$$

where γ_1 and γ_2 are as given above.

4. Observer canonical form in (4.29)

$$A = \begin{bmatrix} -\frac{1}{\epsilon \tau_D} & 1 \\ 0 & 0 \end{bmatrix}, \quad B = \begin{bmatrix} \beta_1 \\ \beta_0 \end{bmatrix}, \quad C = [1 \quad 0] \tag{4.36}$$

$$\text{where} \quad \beta_0 = \frac{K_c}{\epsilon \tau_I \tau_D}, \quad \beta_1 = K_c \frac{\epsilon^2 \tau_D - \tau_I}{\epsilon^2 \tau_I \tau_D}$$

On comparing these four realizations with the transfer function model in (4.31), it is clear that the transfer function offers more immediate insight. One can at least see that it is a **PID** controller.

Time delay. A time delay (or dead time) is an infinite-dimensional system and not representable as a rational transfer function. For a state-space realization it must therefore be approximated. An n'th-order approximation of a time delay θ may be obtained by putting n first-order Padé approximations in series

$$e^{-\theta s} \approx \frac{(1 - \frac{\theta}{2n} s)^n}{(1 + \frac{\theta}{2n} s)^n} \tag{4.37}$$

Alternative (and possibly better) approximations are in use, but the above approximation is often preferred because of its simplicity.

4.2 State controllability and state observability

It is useful to introduce the concept of pole vectors. We define the i'th *input pole vector*

$$u_{p_i} \triangleq B^H q_i \tag{4.38}$$

and the i'th *output pole vector*

$$y_{p_i} \triangleq C t_i \tag{4.39}$$

(see Matlab commands in Table 4.2). For the case when A has distinct eigenvalues, we have from (4.16) the following dyadic expansion of the transfer function matrix from inputs to outputs:

$$G(s) = \sum_{i=1}^{n} \frac{C t_i q_i^H B}{s - p_i} + D = \sum_{i=1}^{n} \frac{y_{p_i} u_{p_i}^H}{s - p_i} + D \tag{4.40}$$

where we have scaled the eigenvectors such that $q_i^H t_i = 1$. From (4.40), $u_{p,i}$ is an indication of how much the i'th mode is excited (and thus may be "controlled") by the inputs, whereas $y_{p,i}$ indicates how much the i'th mode is observed in the outputs. Thus, the pole vectors may be used for checking the state controllability and observability of a system. This is explained in more detail below, but let us start by defining *state* controllability.

Definition 4.1 State controllability. *The dynamical system $\dot{x} = Ax + Bu$, or equivalently the pair (A, B), is said to be state controllable if, for any initial state $x(0) = x_0$, any time $t_1 > 0$ and any final state x_1, there exists an input $u(t)$ such that $x(t_1) = x_1$. Otherwise the system is said to be state uncontrollable.*

Table 4.2: Matlab commands to find pole vectors

```
% Find pole vectors of system [A,B,C,D]
%
[T,Po] = eig(A);
YP = C*T % output pole vectors (must normalize columns to obtain directions)
[Q,Pi] = eig(A');
UP = B'*Q % input pole vectors
Shouldbezero=Po-Pi % if not, the pole vectors refer to different poles
```

To test for state controllability it is instructive to consider the individual poles p_i and the associated input pole vectors $u_{p,i}$. Based on (4.40) we have (Zhou et al., 1996, p. 52):

Theorem 4.1 *Let p_i be an eigenvalue of A or, equivalently, a pole of the system.*

- *The pole p_i is state controllable if and only if*

$$u_{p,i} = B^H q_i \neq 0 \tag{4.41}$$

for all left eigenvectors q_i (including linear combinations) associated with p_i. Otherwise, the pole is uncontrollable.
- *A system is state controllable if and only if every pole p_i is controllable.*

Remark. The need to consider linear combinations of eigenvectors only applies when p_i is a repeated pole (with multiplicity greater than 1). In this case, we may collect the left eigenvectors associated with p_i in the matrix Q_i and collect the corresponding input pole vectors in the matrix $U_{p,i} = B^H Q_i$. The number of uncontrollable states corresponding to the pole p_i is then $\text{rank}(Q_i) - \text{rank}(U_{p,i})$.

In summary, a system is state controllable if and only if all its input pole vectors are non-zero.

There exist many other tests for state controllability. Two of these are:

1. The system (A, B) is state controllable if and only if the controllability matrix

$$\mathcal{C} \triangleq [\, B \quad AB \quad A^2 B \quad \cdots \quad A^{n-1} B \,] \tag{4.42}$$

has rank n (full row rank). Here n is the number of states.
2. From (4.6) one can verify that a particular input which achieves $x(t_1) = x_1$ is

$$u(t) = -B^T e^{A^T (t_1 - t)} W_c(t_1)^{-1} (e^{At_1} x_0 - x_1) \tag{4.43}$$

where $W_c(t)$ is the Gramian matrix at time t,

$$W_c(t) \triangleq \int_0^t e^{A\tau} BB^T e^{A^T \tau} d\tau$$

Therefore, the system (A, B) is state controllable if and only if the Gramian matrix $W_c(t)$ has full rank (and thus is positive definite) for any $t > 0$. For a stable system (A is stable) we only need to consider $P \triangleq W_c(\infty)$; that is, the pair (A, B) is state controllable if and only if the *controllability Gramian*

$$P \triangleq \int_0^\infty e^{A\tau} BB^T e^{A^T \tau} d\tau \tag{4.44}$$

is positive definite ($P > 0$) and thus has full rank n. P may also be obtained as the solution to the Lyapunov equation

$$AP + PA^T = -BB^T \tag{4.45}$$

Example 4.4 *Consider a scalar system with two states and the following state-space realization:*

$$A = \begin{bmatrix} -2 & -2 \\ 0 & -4 \end{bmatrix}, \ B = \begin{bmatrix} 1 \\ 1 \end{bmatrix}, \ C = [1 \ \ 0], \ D = 0$$

The transfer function (minimal realization) is

$$G(s) = C(sI - A)^{-1}B = \frac{1}{s+4}$$

which has only one state. In fact, the first state corresponding to the eigenvalue at -2 is not controllable. This is verified by considering state controllability.

1. *The eigenvalues of A, and thus the system poles, are $p_1 = -2$ and $p_2 = -4$. The corresponding left eigenvectors are $q_1 = [0.707 \ \ -0.707]^T$ and $q_2 = [0 \ \ 1]^T$. The two input pole vectors are*

$$u_{p_1} = B^H q_1 = 0, \quad u_{p_2} = B^H q_2 = 1$$

   and since u_{p_1} is zero we have that the first pole (eigenvalue) is not state controllable.
2. *The controllability matrix has rank 1 since it has two linearly dependent rows:*

$$\mathcal{C} = [B \ \ AB] = \begin{bmatrix} 1 & -4 \\ 1 & -4 \end{bmatrix}$$

3. *The controllability Gramian is also singular*

$$P = \begin{bmatrix} 0.125 & 0.125 \\ 0.125 & 0.125 \end{bmatrix}$$

Example 4.5 *Consider a scalar system $G(s) = 1/(\tau s + 1)^4$ with the following realization:*

$$A = \begin{bmatrix} -1/\tau & 0 & 0 & 0 \\ 1/\tau & -1/\tau & 0 & 0 \\ 0 & 1/\tau & -1/\tau & 0 \\ 0 & 0 & 1/\tau & -1/\tau \end{bmatrix}, \ B = \begin{bmatrix} 1/\tau \\ 0 \\ 0 \\ 0 \end{bmatrix}, C = [0 \ \ 0 \ \ 0 \ \ 1] \quad (4.46)$$

The system has four repeated eigenvalues at $-1/\tau$ (multiplicity 4), and the corresponding left eigenvectors of A are the columns of

$$Q = \begin{bmatrix} 1 & -1 & 1 & -1 \\ 0 & 0 & 0 & 0 \\ 0 & 0 & 0 & 0 \\ 0 & 0 & 0 & 0 \end{bmatrix}$$

Since the four eigenvectors q_i are linearly dependent, there is no need to consider linear combinations, and since all input pole vectors are non-zero ($u_{p,i} = B^H q_i = \pm 1/\tau, i = 1, \ldots, 4$), we conclude that the system is state controllable. The is confirmed by computing the controllability matrix $\mathcal{C}$ in (4.42) which has full rank.

In words, if a system is state controllable we can by use of its inputs u bring it from any initial state to any final state within any given finite time. State controllability would therefore seem to be an important property for practical control, but it rarely is for the following four reasons:

1. It says nothing about how the states behave at earlier and later times, e.g. it does not imply that one can hold (as $t \to \infty$) the states at a given value.
2. The required inputs may be very large with sudden changes.
3. Some of the states may be of no practical importance.

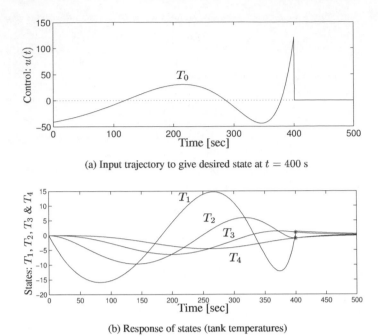

(a) Input trajectory to give desired state at $t = 400$ s

(b) Response of states (tank temperatures)

Figure 4.1: State controllability of four first-order systems in series

4. The definition is an existence result which provides no degree of controllability (see Hankel singular values for this).

The first two objections are illustrated by the following example.

Example 4.5 continued. State controllability of tanks in series. *Consider a system with one input and four states arising from four first-order systems in series:*

$$G(s) = 1/(\tau s + 1)^4$$

A state-space realization is given by (4.46). A physical example could be four identical tanks (e.g. bath tubs) in series where water flows from one tank to the next. Energy balances, assuming no heat loss, yield $T_4 = \frac{1}{\tau s + 1}T_3$, $T_3 = \frac{1}{\tau s + 1}T_2$, $T_2 = \frac{1}{\tau s + 1}T_1$, $T_1 = \frac{1}{\tau s + 1}T_0$ where the states $x = [T_1 \ \ T_2 \ \ T_3 \ \ T_4]^T$ are the four tank temperatures, the input $u = T_0$ is the inlet temperature, and $\tau = 100$ s is the residence time in each tank. In practice, we know that it is very difficult to control the four temperatures independently, since at steady-state all temperatures must be equal. However, we found above that the system is state controllable, so it must be possible to achieve at any given time any desired temperature in each of the four tanks simply by adjusting the inlet temperature. This sounds almost too good to be true, so let us consider a specific case.

Assume that the system is initially at steady-state (all temperatures are zero), and that we want to achieve at $t = 400$ s the following temperatures: $T_1(400) = 1$, $T_2(400) = -1$, $T_3(400) = 1$ and $T_4(400) = -1$. The change in inlet temperature, $T_0(t)$, to achieve this was computed from (4.43) and is shown as a function of time in Figure 4.1(a). The corresponding tank temperatures are shown in Figure 4.1(b). Two things are worth noting:

1. *The required change in inlet temperature T_0 is more than 100 times larger than the desired temperature changes in the tanks and it also varies widely with time.*

2. *Although the states (tank temperatures T_i) are indeed at their desired values of ± 1 at $t = 400$ s, it is not possible to hold them at these values, since at steady-state all the states must be equal (in our case, all states approach 0 as time goes to infinity, since $u = T_0$ is reset to 0 at $t = 400$ s).*

It is quite easy to explain the shape of the input $T_0(t)$. The fourth tank is furthest away and we want its temperature to decrease ($T_4(400) = -1$) and therefore the inlet temperature T_0 is initially decreased to about -40. Then, since $T_3(400) = 1$ is positive, T_0 is increased to about 30 at $t = 220$ s; it is subsequently decreased to about -40, since $T_2(400) = -1$, and finally increased to more than 100 to achieve $T_1(400) = 1$.

From the above example, we see clearly that the property of state controllability may not imply that the system is "controllable" in a practical sense[2]. This is because state controllability is concerned only with the value of the states at *discrete* values of time (target hitting), while in most cases we want the outputs to remain close to some desired value (or trajectory) for all values of time, and without using inappropriate control signals.

So now we know that state controllability does not imply that the system is controllable from a practical point of view. But what about the reverse: if we do *not* have state controllability, is this an indication that the system is not controllable in a practical sense? In other words, should we be concerned if a system is not state controllable? In many cases the answer is "no", since we may not be concerned with the behaviour of the uncontrollable states which may be outside our system boundary or of no practical importance. If we are indeed concerned about these states then they should be included in the output vector y. State uncontrollability will then appear as a rank deficiency in the transfer function matrix $G(s)$ (see functional controllability).

In conclusion, state controllability is neither a necessary nor sufficient condition for a system to be controllable in a practical sense (input–output controllability). So is the issue of state controllability of any value at all? Yes, because it tells us whether we have included some states in our model that we have no means of affecting. This is certainly a practical (and numerical) concern if the associated mode is unstable. It also tells us when we can save on computer time by deleting uncontrollable states which have no effect on the output for a zero initial state.

In summary, state controllability is a system theoretical concept which is important when it comes to computations and realizations. However, its name is somewhat misleading, and most of the above discussion might have been avoided if only Kalman, who originally defined (state) controllability, had used a different terminology. For example, better terms might have been "point-wise controllability" or "state affectability" from which it would have been understood that although all the states could be individually affected, we might not be able to control them independently over a period of time.

Definition 4.2 State observability. *The dynamical system $\dot{x} = Ax + Bu$, $y = Cx + Du$ (or the pair (A, C)) is said to be state observable if, for any time $t_1 > 0$, the initial state $x(0) = x_0$ can be determined from the time history of the input $u(t)$ and the output $y(t)$ in the interval $[0, t_1]$. Otherwise the system, or (A, C), is said to be state unobservable.*

To test for state observability it is instructive to consider the individual modes p_i and the associated output pole vectors $y_{p,i}$. Based on (4.40) we have (Zhou et al., 1996, p. 52):

Theorem 4.2 *Let p_i be an eigenvalue of A or, equivalently, a mode of the system.*

---

[2] In Chapter 5, we introduce a more practical concept of controllability which we call "input–output controllability".

- *The <u>mode</u> p_i is observable if and only if*

$$y_{p,i} = Ct_i \neq 0 \qquad\qquad (4.47)$$

for all right eigenvectors t_i (including linear combinations) associated with p_i. Otherwise, the mode is unobservable.
- *A <u>system</u> is observable if and only if every mode p_i is observable.*

Remark. The need to consider linear combinations of eigenvectors only applies when p_i is a repeated pole (with multiplicity greater than 1). In this case, we may collect the right eigenvectors associated with p_i in the matrix T_i and collect the corresponding input pole vectors in the matrix $Y_{p,i} = CT_i$. The number of unobservable states corresponding to the mode p_i is then $\mathrm{rank}(T_i) - \mathrm{rank}(Y_{p,i})$.

In summary, a system is observable if and only if all its output pole vectors are non-zero. The following example illustrates this, and what may happen if we have repeated poles.

Example 4.6 *Consider a system with two states, two inputs, one output and the following state-space realization:*

$$A = \begin{bmatrix} p_1 & 0 \\ 0 & p_2 \end{bmatrix}, \ B = \begin{bmatrix} 1 & 4 \\ 2 & 0 \end{bmatrix}, \ C = [0.5 \ \ 0.25], \ D = [0 \ \ 0]$$

The corresponding transfer function is

$$G(s) = C(sI - A)^{-1}B = \left[\frac{s - (p_1 + p_2)/2}{(s - p_1)(s - p_2)} \quad \frac{2}{s - p_1} \right]$$

The eigenvalues (poles) are p_1 and p_2 and the corresponding right and left eigenvector matrices are

$$T = \begin{bmatrix} 1 & 0 \\ 0 & 1 \end{bmatrix}, \quad Q = \begin{bmatrix} 1 & 0 \\ 0 & 1 \end{bmatrix}$$

(where the first column is associated with p_1 and the second with p_2). The associated output and input pole vectors may be collected in matrices,

$$Y_p = CT = [y_{p,1} \ \ y_{p,2}] = [0.5 \ \ 0.25], \quad U_p = B^H Q = [u_{p,1} \ \ u_{p,2}] = \begin{bmatrix} 1 & 2 \\ 4 & 0 \end{bmatrix}$$

Let us first consider the case with distinct poles, i.e. $p_1 \neq p_2$. We see that the two output pole "vectors" (columns in Y_p) are both non-zero, so both modes are observable. The two input pole vectors (columns in U_p) are also both non-zero, so both modes are state controllable. However, since the second element in $u_{p,2}$ is 0 it follows that mode p_2 is not state controllable from input 2 (which is also easily seen from the transfer function representation).

Next consider the case with two repeated poles, $p_1 = p_2$. In this case, both the columns of T and their linear combinations are the right eigenvectors of A. Since $\mathrm{rank}(T) - \mathrm{rank}(Y_p) = 2 - 1 = 1$, one of the two states is not observable (which is also easily seen from the transfer function representation as there is a pole–zero cancellation in the first element in $G(s)$). However, both states remain state controllable since $\mathrm{rank}(Q) - \mathrm{rank}(U_p) = 2 - 2 = 0$.

In the above example the poles are "in parallel" (as can been seen since the first element in $G(s)$ can be written $\frac{0.5}{s-p_1} + \frac{0.5}{s-p_2}$), and this may give problems with observability and controllability for repeated poles. However, if the repeated poles are "in series" there is no such problem, as illustrated in Example 4.5 and further in the following example.

Example 4.7 *Consider the scalar system*

$$G(s) = \frac{1}{(s-p)^2} \stackrel{s}{=} \left[\begin{array}{c|c} A & B \\ \hline C & D \end{array}\right] = \left[\begin{array}{cc|c} p & 1 & 0 \\ 0 & p & 1 \\ \hline 1 & 0 & 0 \end{array}\right]$$

There are two eigenvalues (poles) at p and the corresponding right and left eigenvector matrices are

$$T = \begin{bmatrix} 1 & -1 \\ 0 & 0 \end{bmatrix}, \quad Q = \begin{bmatrix} 0 & 0 \\ 1 & -1 \end{bmatrix}$$

Note that the two right (left) eigenvectors are linearly dependent. The associated output and input pole vectors are collected in matrices,

$$Y_p = CT = \begin{bmatrix} 1 & -1 \end{bmatrix}, \quad U_p = B^H Q = \begin{bmatrix} 1 & -1 \end{bmatrix}$$

Both states are observable since $\mathrm{rank}(T) - \mathrm{rank}(Y_p) = 1 - 1 = 0$, and both states are state controllable since $\mathrm{rank}(Q) - \mathrm{rank}(U_p) = 1 - 1 = 0$. This agrees with the transfer function representation.

Two other tests for state observability are:

1. The system (A, C) is state observable if and only if we have full column rank (rank n) of the observability matrix

$$\mathcal{O} \triangleq \begin{bmatrix} C \\ CA \\ \vdots \\ CA^{n-1} \end{bmatrix} \tag{4.48}$$

2. For a stable system we may consider the *observability Gramian*

$$Q \triangleq \int_0^\infty e^{A^T \tau} C^T C e^{A\tau} d\tau \tag{4.49}$$

which must have full rank n (and thus be positive definite) for the system to be state observable. Q can also be found as the solution to the following Lyapunov equation:

$$A^T Q + QA = -C^T C \tag{4.50}$$

A system is state observable if we can obtain the value of all individual states by measuring the output $y(t)$ over some time period. However, even if a system is state observable it may not be observable in a practical sense. For example, obtaining $x(0)$ may require taking high-order derivatives of $y(t)$ which may be numerically poor and sensitive to noise. This is illustrated in the following example.

Example 4.5 (tanks in series) continued. *We have $y = T_4$ (the temperature of the last tank), and, similar to Example 4.7, all states are observable from y. However, consider a case where the initial temperatures in the tanks, $T_i(0), i = 1, \ldots, 4$, are non-zero (and unknown), and the inlet temperature $T_0(t) = u(t)$ is zero for $t \geq 0$. Then, from a practical point of view, it is clear that it is numerically very difficult to back-calculate, for example, $T_1(0)$ based on measurements of $y(t) = T_4(t)$ over some interval $[0, t_1]$, although in theory all states are observable from the output.*

Definition 4.3 Minimal realization, McMillan degree and hidden mode. *A state-space realization (A, B, C, D) of $G(s)$ is said to be a **minimal realization** of $G(s)$ if A has the smallest possible dimension (i.e. the fewest number of states). The smallest dimension is called the **McMillan degree** of $G(s)$. A mode is **hidden** if it is not state controllable or observable and thus does not appear in the minimal realization.*

Since only controllable and observable states contribute to the input–output behaviour from u to y, it follows that a state-space realization is minimal if and only if (A, B) is state controllable and (A, C) is state observable.

Remark 1 Note that uncontrollable states will contribute to the output response $y(t)$ if the initial state is non-zero, $x(t = 0) \neq 0$, but this effect will die out if the uncontrollable states are stable.

Remark 2 Unobservable states have no effect on the outputs whatsoever, and may be viewed as outside the system boundary, and thus of no direct interest from a control point of view (unless the unobservable state is unstable, because we want to avoid the system "blowing up"). However, observability is important for measurement selection and when designing state estimators (observers).

4.3 Stability

There are a number of ways in which stability may be defined, e.g. see Willems (1970). Fortunately, for linear time-invariant systems these differences have no practical significance, and we use the following definition:

Definition 4.4 *A system is* (**internally**) **stable** *if none of its components contain hidden unstable modes and the injection of bounded external signals at any place in the system results in bounded output signals measured anywhere in the system.*

Here we define a signal $u(t)$ to be "bounded" if there exists a constant c such that $|u(t)| < c$ for all t. The word *internally* is included in the definition to stress that we do not only require the response from one particular input to another particular output to be stable, but require stability for signals injected or measured at any point of the system. This is discussed in more detail for feedback systems in Section 4.7. Similarly, the components must contain no hidden unstable modes; that is, any instability in the components must be contained in their input–output behaviour.

Definition 4.5 Stabilizable, detectable and hidden unstable modes. *A system is stabilizable if all unstable modes are state controllable. A system is detectable if all unstable modes are observable. A system with unstabilizable or undetectable modes is said to contain hidden unstable modes.*

A linear system is stabilizable (detectable) if and only if all input (output) pole vectors associated with the unstable modes are non-zero; see (4.41) and (4.47) for details. If a system is not detectable, then there is a state within the system which will eventually grow out of bounds, but we have no way of observing this from the outputs $y(t)$.

Remark 1 Any unstable linear system can be stabilized by feedback control (at least in theory) provided the system contains no hidden unstable mode(s). However, this may require an unstable controller, see also page 150.

Remark 2 Systems with hidden unstable modes must be avoided both in practice and in computations (since variables will eventually blow up on our computer if not on the factory floor). In the book we always assume, unless otherwise stated, that our systems contain no hidden unstable modes.

4.4   Poles

We have above used that the poles of a system are the eigenvalues of the state-space A-matrix, and this is the definition given below. More generally, the poles of $G(s)$ may be somewhat loosely defined as the finite values $s = p$ where $G(p)$ has a singularity ("is infinite"), see also Theorem 4.4 below.

Definition 4.6 Poles. *The poles p_i of a system with state-space description (4.3)–(4.4) are the eigenvalues $\lambda_i(A), i = 1, \ldots, n$, of the matrix A. The pole or characteristic polynomial $\phi(s)$ is defined as $\phi(s) \triangleq \det(sI - A) = \prod_{i=1}^{n}(s - p_i)$. Thus the poles are the roots of the characteristic equation*

$$\phi(s) \triangleq \det(sI - A) = 0 \tag{4.51}$$

To see that this definition is reasonable, recall (4.15) and see Appendix A.2.1. Note that if A does not correspond to a minimal realization then the poles by this definition will include the poles (eigenvalues) corresponding to uncontrollable and/or unobservable states.

4.4.1   Poles and stability

For linear systems, the poles determine stability:

Theorem 4.3 *A linear dynamic system $\dot{x} = Ax + Bu$ is stable if and only if all the poles are in the open left-half plane (LHP); that is, $\mathrm{Re}(p_i) = \mathrm{Re}\{\lambda_i(A)\} < 0, \forall i$. A matrix A with such a property is said to be "stable" or Hurwitz.*

Proof: From (4.7) we see that the time response (4.6) can be written as a sum of terms each containing a *mode* $e^{p_i t}$. Poles in the RHP with $\mathrm{Re}\{p_i\} > 0$ give rise to *unstable modes* since in this case $e^{p_i t}$ is unbounded as $t \to \infty$. Poles in the open LHP give rise to stable modes where $e^{p_i t} \to 0$ as $t \to \infty$. Systems with poles on the $j\omega$-axis, including integrators, are unstable from our Definition 4.4 of stability. For example, consider $y = Gu$ and assume $G(s)$ has imaginary poles $s = \pm j\omega_o$. Then with a bounded sinusoidal input, $u(t) = \sin \omega_o t$, the output $y(t)$ grows unbounded as $t \to \infty$. $\square$

4.4.2   Poles from state-space realizations

Poles are usually obtained numerically by computing the eigenvalues of the A-matrix. To get the fewest number of poles, without unstabilizable or uncontrollable modes, we should use a minimal realization of the system.

4.4.3   Poles from transfer functions

The following theorem from MacFarlane and Karcanias (1976) allows us to obtain the poles directly from the transfer function matrix $G(s)$ and is useful for hand calculations. It also has the advantage of yielding only the poles corresponding to a minimal realization of the system.

Theorem 4.4 *The pole polynomial $\phi(s)$ corresponding to a minimal realization of a system with transfer function $G(s)$ is the least common denominator of all non-identically zero minors of all orders of $G(s)$.*

A *minor* of a matrix is the determinant of the matrix obtained by deleting certain rows and/or columns of the matrix. We will use the notation M_c^r to denote the minor corresponding to the deletion of rows r and columns c in $G(s)$. In the procedure defined by the theorem we cancel common factors in the numerator and denominator of each minor. It then follows that only observable and controllable poles will appear in the pole polynomial.

Example 4.8 *Consider the plant* $G(s) = \frac{(3s+1)^2}{(s+1)}e^{-\theta s}$ *which has no state-space realization as it contains a delay and is also improper. Thus we cannot compute the poles from (4.51). However, from Theorem 4.4 we have that the denominator is* $(s+1)$ *and as expected* $G(s)$ *has a pole at* $s = -1$.

Example 4.9 *Consider the square transfer function matrix*

$$G(s) = \frac{1}{1.25(s+1)(s+2)}\begin{bmatrix} s-1 & s \\ -6 & s-2 \end{bmatrix} \tag{4.52}$$

The minors of order 1 are the four elements that all have $(s+1)(s+2)$ *in the denominator. The minor of order 2 is the determinant*

$$\det G(s) = \frac{(s-1)(s-2)+6s}{1.25^2(s+1)^2(s+2)^2} = \frac{1}{1.25^2(s+1)(s+2)} \tag{4.53}$$

Note the pole–zero cancellation when evaluating the determinant. The least common denominator of all the minors is then

$$\phi(s) = (s+1)(s+2) \tag{4.54}$$

so a minimal realization of the system has two poles: one at $s = -1$ *and one at* $s = -2$.

Example 4.10 *Consider the* 2×3 *system, with three inputs and two outputs,*

$$G(s) = \frac{1}{(s+1)(s+2)(s-1)}\begin{bmatrix} (s-1)(s+2) & 0 & (s-1)^2 \\ -(s+1)(s+2) & (s-1)(s+1) & (s-1)(s+1) \end{bmatrix} \tag{4.55}$$

The minors of order 1 are the five non-zero elements (e.g. $M_{2,3}^2 = g_{11}(s)$*):*

$$\frac{1}{s+1}, \ \frac{s-1}{(s+1)(s+2)}, \ \frac{-1}{s-1}, \ \frac{1}{s+2}, \ \frac{1}{s+2} \tag{4.56}$$

The minor of order 2 corresponding to the deletion of column 2 is

$$M_2 = \frac{(s-1)(s+2)(s-1)(s+1)+(s+1)(s+2)(s-1)^2}{((s+1)(s+2)(s-1))^2} = \frac{2}{(s+1)(s+2)} \tag{4.57}$$

The other two minors of order 2 are

$$M_1 = \frac{-(s-1)}{(s+1)(s+2)^2}, \quad M_3 = \frac{1}{(s+1)(s+2)} \tag{4.58}$$

By considering all minors we find their least common denominator to be

$$\phi(s) = (s+1)(s+2)^2(s-1) \tag{4.59}$$

The system therefore has four poles: one at $s = -1$, *one at* $s = 1$ *and two at* $s = -2$.

From the above examples we see that the MIMO poles are essentially the poles of the elements. However, by looking at only the elements it is not possible to determine the multiplicity of the poles. For instance, let $G_0(s)$ be a square $m \times m$ transfer function matrix with no pole at $s = -a$, and consider

$$G_1(s) = \frac{1}{s+a}G_0(s) \tag{4.60}$$

How many poles at $s = -a$ does a minimal realization of $G_1(s)$ have? From (A.10),

$$\det(G_1(s)) = \det\left(\frac{1}{s+a}G_0(s)\right) = \frac{1}{(s+a)^m}\det(G_0(s)) \tag{4.61}$$

so if G_0 has no zeros at $s = -a$, then $G_1(s)$ has m poles at $s = -a$. However, G_0 may have zeros at $s = -a$. As an example, consider a 2×2 plant in the form given by (4.60). It may have two poles at $s = -a$ (as in (3.93)), one pole at $s = -a$ (as in (4.52) where $\det G_0(s)$ has a zero at $s = -a$) or no pole at $s = -a$ (if all the elements of $G_0(s)$ have a zero at $s = -a$).

As noted above, the poles are obtained numerically by computing the eigenvalues of the A-matrix. Thus, to compute the poles of a transfer function $G(s)$, we must first obtain a state-space realization of the system. Preferably this should be a minimal realization. For example, if we make individual realizations of the five non-zero elements in Example 4.10 and then simply combine them to get an overall state-space realization, we will get a system with 15 states, where each of the three poles (in the common denominator) are repeated five times. A model reduction to obtain a minimal realization will subsequently yield a system with four poles as given in (4.59).

4.4.4   Pole vectors and directions

In multivariable systems poles have directions associated with them. To quantify them we use the input and output *pole vectors* defined in (4.38) and (4.39):

$$y_{p_i} = Ct_i, \quad u_{p_i} = B^H q_i \tag{4.62}$$

These give an indication of how much the i'th mode is excited in each output and input. *Pole directions* are defined as pole vectors normalized to have unit length, i.e.

$$y'_{p,i} = \frac{1}{\|y_{p_i}\|_2}y_{p_i}, \quad u'_{p,i} = \frac{1}{\|u_{p_i}\|_2}u_{p_i} \tag{4.63}$$

The pole directions may alternatively be obtained directly from the transfer function matrix by evaluating $G(s)$ at the pole p_i and considering the directions of the resulting complex matrix $G(p_i)$. The matrix is infinite in the direction of the pole, and we may somewhat crudely write

$$G(p_i)\,u'_{p_i} = \infty \cdot y'_{p_i} \tag{4.64}$$

where u'_{p_i} is the input pole direction, and y'_{p_i} is the output pole direction. The pole directions may then in principle be obtained from an SVD of $G(p_i) = U\Sigma V^H$. Then u'_{p_i} is the first column in V (corresponding to the infinite singular value), and y'_{p_i} the first column in U. For numerical calculations we may evaluate $G(s)$ at $s = p_i + \epsilon$ where ϵ is a small number.

Remark 1 As already mentioned, if $u_p = B^H q = 0$ then the corresponding pole is not state controllable, and if $y_p = Ct = 0$ the corresponding pole is not state observable (see also Zhou et al., 1996, p. 52).

Remark 2 For a multivariable plant the pole vectors defined in (4.62) provide a very useful tool for selecting inputs and outputs for stabilization; see Section 10.4.3 for details. For a single unstable mode, selecting the input corresponding to the largest element in u_p and the output corresponding to the largest element in y_p minimizes the input usage required for stabilization. More precisely, this choice minimizes the lower bound on both the $\mathcal{H}_2$ and $\mathcal{H}_\infty$ norms of the transfer function KS from measurement (output) noise to input (Havre and Skogestad, 2003).

Remark 3 Notice that there is difference between the non-normalized *pole vector* and the normalized *pole direction* (vector). Above we used a ' to show explicitly that the direction vector is normalized, but later in the book this is omitted. For zeros (see below) such problems do not arise because we are only interested in the normalized zero direction (vector).

4.5   Zeros

Zeros of a system arise when competing effects, internal to the system, are such that the output is zero even when the inputs (and the states) are not themselves identically zero. For a SISO system the zeros z_i are the solutions to $G(z_i) = 0$. In general, it can be argued that zeros are values of s at which $G(s)$ loses rank (from rank 1 to rank 0 for a SISO system). This is the basis for the following definition of zeros for a multivariable system (MacFarlane and Karcanias, 1976):

Definition 4.7 Zeros. z_i *is a zero of $G(s)$ if the rank of $G(z_i)$ is less than the normal rank of $G(s)$. The zero polynomial is defined as $z(s) = \prod_{i=1}^{n_z}(s - z_i)$ where n_z is the number of finite zeros of $G(s)$.*

In this book, we do not consider zeros at infinity; we require that z_i is finite. The normal rank of $G(s)$ is defined as the rank of $G(s)$ at all values of s except at a finite number of singularities (which are the zeros).

This definition of zeros is based on the transfer function matrix, corresponding to a minimal realization of a system. These zeros are sometimes called "transmission zeros", but we will simply call them "zeros". We may sometimes use the term "multivariable zeros" to distinguish them from the zeros of the elements of the transfer function matrix.

4.5.1   Zeros from state-space realizations

Zeros are usually computed from a state-space description of the system. First note that the state-space equations of a system may be written as

$$P(s)\begin{bmatrix} x \\ u \end{bmatrix} = \begin{bmatrix} 0 \\ y \end{bmatrix}, \quad P(s) = \begin{bmatrix} sI - A & -B \\ C & D \end{bmatrix} \tag{4.65}$$

The zeros are then the values $s = z$ for which the polynomial system matrix, $P(s)$, loses rank, resulting in zero output for some non-zero input. Numerically, the zeros are found as non-trivial solutions (with $u_z \neq 0$ and $x_z \neq 0$) to the following problem:

$$(zI_g - M)\begin{bmatrix} x_z \\ u_z \end{bmatrix} = 0 \tag{4.66}$$

$$M = \begin{bmatrix} A & B \\ C & D \end{bmatrix}; \quad I_g = \begin{bmatrix} I & 0 \\ 0 & 0 \end{bmatrix} \tag{4.67}$$

This is solved as a generalized eigenvalue problem – in the conventional eigenvalue problem we have $I_g = I$. Note that we usually get additional zeros if the realization is not minimal.

4.5.2   Zeros from transfer functions

The following theorem from MacFarlane and Karcanias (1976) is useful for hand calculating
the zeros of a transfer function matrix $G(s)$.

Theorem 4.5 *The zero polynomial $z(s)$, corresponding to a minimal realization of the
system, is the greatest common divisor of all the numerators of all order-r minors of $G(s)$,
where r is the normal rank of $G(s)$, provided that these minors have been adjusted in such a
way as to have the pole polynomial $\phi(s)$ as their denominator.*

Example 4.11 *Consider the 2×2 transfer function matrix*

$$G(s) = \frac{1}{s+2} \begin{bmatrix} s-1 & 4 \\ 4.5 & 2(s-1) \end{bmatrix} \tag{4.68}$$

*The normal rank of $G(s)$ is 2, and the minor of order 2 is the determinant, $\det G(s) = \frac{2(s-1)^2 - 18}{(s+2)^2} =
2\frac{s-4}{s+2}$. From Theorem 4.4, the pole polynomial is $\phi(s) = s + 2$ and therefore the zero polynomial is
$z(s) = s - 4$. Thus, $G(s)$ has a single RHP-zero at $s = 4$.*

This illustrates that in general multivariable zeros have no relationship with the zeros of the
transfer function elements. This is also shown by the following example where the system
has no zeros.

Example 4.9 continued. *Consider again the 2×2 system in (4.52) where $\det G(s)$ in (4.53) already
has $\phi(s)$ as its denominator. Thus the zero polynomial is given by the numerator of (4.53), which is 1,
and we find that the system has no multivariable zeros.*

The next two examples consider non-square systems.

Example 4.12 *Consider the 1×2 system*

$$G(s) = \begin{bmatrix} \dfrac{s-1}{s+1} & \dfrac{s-2}{s+2} \end{bmatrix} \tag{4.69}$$

*The normal rank of $G(s)$ is 1, and since there is no value of s for which both elements become zero,
$G(s)$ has no zeros.*

In general, non-square systems are less likely to have zeros than square systems. For instance,
for a square 2×2 system to have a zero, there must be a value of s for which the two columns
in $G(s)$ are linearly dependent. On the other hand, for a 2×3 system to have a zero, we need
all three columns in $G(s)$ to be linearly dependent.

The following is an example of a non-square system which does have a zero.

Example 4.10 continued. *Consider again the 2×3 system in (4.55), and adjust the minors of order
2 in (4.57) and (4.58) so that their denominators are $\phi(s) = (s + 1)(s + 2)^2(s - 1)$. We get*

$$M_1(s) = \frac{-(s-1)^2}{\phi(s)}, \quad M_2(s) = \frac{2(s-1)(s+2)}{\phi(s)}, \quad M_3(s) = \frac{(s-1)(s+2)}{\phi(s)} \tag{4.70}$$

*The common factor for these minors is the zero polynomial $z(s) = (s - 1)$. Thus, the system has a
single RHP-zero located at $s = 1$.*

We also see from the last example that a minimal realization of a MIMO system can have
poles and zeros at the same value of s, provided their directions are different.

4.5.3 Zero directions

In the following let s be a fixed complex scalar and consider $G(s)$ as a complex matrix. For example, given a state-space realization, we can evaluate $G(s) = C(sI - A)^{-1}B + D$. Let $G(s)$ have a zero at $s = z$. Then $G(s)$ loses rank at $s = z$, and there will exist non-zero vectors u_z and y_z such that

$$G(z)u_z = 0 \cdot y_z \qquad (4.71)$$

Here u_z is defined as the input zero direction, and y_z is defined as the output zero direction. We usually normalize the direction vectors to have unit length,

$$u_z^H u_z = 1; \quad y_z^H y_z = 1$$

From a practical point of view, the output zero direction, y_z, is usually of more interest than u_z, because y_z gives information about which output (or combination of outputs) may be difficult to control.

Remark 1 Taking the Hermitian (conjugate transpose) of (4.71) yields $u_z^H G^H(z) = 0 \cdot y_z^H$. Premultiplying by u_z and postmultiplying by y_z noting that $u_z^H u_z = 1$ and $y_z^H y_z = 1$ yields $G(z)^H y_z = 0 \cdot u_z$, or

$$y_z^H G(z) = 0 \cdot u_z^H \qquad (4.72)$$

Remark 2 In principle, we may obtain u_z and y_z from an SVD of $G(z) = U\Sigma V^H$, and we have that u_z is the last column in V (corresponding to the zero singular value of $G(z)$) and y_z is the last column of U. An example was given earlier in (3.85). A better approach numerically is to obtain u_z from a state-space description using the generalized eigenvalue problem in (4.66). Similarly, y_z may be obtained from the transposed state-space description, see (4.72), using M^T in (4.66).

Example 4.13 Zero and pole directions. *Consider the 2×2 plant in (4.68), which has a RHP-zero at $z = 4$ and a LHP-pole at $p = -2$. The pole and zero directions are usually found from a state-space realization using (4.38)–(4.39) and (4.65)–(4.67), respectively. However, we will here use an SVD of $G(z)$ and $G(p)$ to determine the zero and pole directions using the Matlab commands in Table 4.3, although we stress that this is generally not a reliable method numerically. An SVD of $G(z)$ gives*

$$G(z) = G(4) = \frac{1}{6}\begin{bmatrix} 3 & 4 \\ 4.5 & 6 \end{bmatrix} = \frac{1}{6}\begin{bmatrix} 0.55 & -0.83 \\ 0.83 & 0.55 \end{bmatrix}\begin{bmatrix} 9.01 & 0 \\ 0 & 0 \end{bmatrix}\begin{bmatrix} 0.6 & -0.8 \\ 0.8 & 0.6 \end{bmatrix}^H$$

The input and output zero directions are associated with the zero singular value of $G(z)$, see (4.71), and we get $u_z = \begin{bmatrix} -0.80 \\ 0.60 \end{bmatrix}$ and $y_z = \begin{bmatrix} -0.83 \\ 0.55 \end{bmatrix}$. We see from y_z that the zero has a slightly larger component in the first output. Next, to determine the pole directions consider

$$G(p + \epsilon) = G(-2 + \epsilon) = \frac{1}{\epsilon^2}\begin{bmatrix} -3 + \epsilon & 4 \\ 4.5 & 2(-3 + \epsilon) \end{bmatrix} \qquad (4.73)$$

The SVD as $\epsilon \to 0$ becomes

$$G(-2 + \epsilon) = \frac{1}{\epsilon^2}\begin{bmatrix} -0.55 & -0.83 \\ 0.83 & -0.55 \end{bmatrix}\begin{bmatrix} 9.01 & 0 \\ 0 & 0 \end{bmatrix}\begin{bmatrix} 0.6 & -0.8 \\ -0.8 & -0.6 \end{bmatrix}^H$$

The pole input and output directions are associated with the largest singular value, $\sigma_1 = 9.01/\epsilon^2$, and we get $u_p = \begin{bmatrix} 0.60 \\ -0.80 \end{bmatrix}$ and $y_p = \begin{bmatrix} -0.55 \\ 0.83 \end{bmatrix}$. We note from y_p that the pole has a slightly larger component in the second output.

It is important to note that although the locations of the poles and zeros are independent of input and output scalings, their directions are *not*. Thus, the inputs and outputs need to be scaled properly before making any interpretations based on pole and zero directions.

Table 4.3: **Matlab program to find pole and zero directions from transfer function**

```
%
s = tf('s'); G = [(s-1) 4; 4.5 2*(s-1)]/(s+2);p=-2;z=4;
% Crude method for computing pole and zero directions
Gz = evalfr(G,z); n = min(size(Gz));
[U,S,V] = svd(Gz); yz = U(:,n), uz = V(:,n)
Gp = evalfr(G,p+1.e-5);
[U,S,V] = svd(Gp); yp = U(:,1), up = V(:,1)
```

4.6   Some important remarks on poles and zeros

1. The zeros resulting from a minimal realization are sometimes called the *transmission zeros*. If one does *not* have a minimal realization, then numerical computations (e.g. using Matlab) may yield additional *invariant zeros*. These invariant zeros plus the transmission zeros are sometimes called the *system zeros*. The invariant zeros can be further subdivided into *input and output decoupling zeros*. These cancel poles associated with uncontrollable or unobservable states and hence have limited practical significance. To avoid all these complications, we recommend that a minimal realization is found before computing the zeros.

2. Rosenbrock (1966; 1970) first defined multivariable zeros using something similar to the Smith–McMillan form. Poles and zeros are defined in terms of the McMillan form in Zhou et al. (1996).

3. In the time domain, the presence of zeros implies blocking of certain input signals (MacFarlane and Karcanias, 1976). If z is a zero of $G(s)$, then there exists an input signal of the form $u_z e^{zt} 1_+(t)$, where u_z is a (complex) vector and $1_+(t)$ is a unit step, and a set of initial conditions (states) x_z, such that $y(t) = 0$ for $t > 0$.

4. For square systems we essentially have that the poles and zeros of $G(s)$ are the poles and zeros of $\det G(s)$. However, this crude definition may fail in a few cases, for instance when there is a zero and pole in different parts of the system which happen to cancel when forming $\det G(s)$. For example, the system

$$G(s) = \begin{bmatrix} (s+2)/(s+1) & 0 \\ 0 & (s+1)/(s+2) \end{bmatrix} \qquad (4.74)$$

   has $\det G(s) = 1$, although the system obviously has poles at -1 and -2 and (multivariable) zeros at -1 and -2.

5. $G(s)$ in (4.74) provides a good example for illustrating the importance of *directions* when discussing poles and zeros of multivariable systems. We note that although the system has poles and zeros at the same locations (at -1 and -2), their directions are different and so they do not cancel or otherwise interact with each other. In (4.74) the pole at -1 has directions $u_p = y_p = [1 \ 0]^T$, whereas the zero at -1 has directions $u_z = y_z = [0 \ 1]^T$.

6. For square systems with a non-singular D-matrix, the number of poles is the same as the number of zeros, and the zeros of $G(s)$ are equal to the poles $G^{-1}(s)$, and vice versa. Furthermore, if the inverse of $G(p)$ exists then it follows from the SVD that

$$G^{-1}(p)y_p = 0 \cdot u_p \qquad (4.75)$$

7. There are no zeros if the outputs y contain direct information about all the states; that is, if from y we can directly obtain x. For example, we have no zeros if $y = x$ or more generally if rank $C = n$ and $D = 0$ (see a proof in Example 4.15). This probably explains why zeros

were given very little attention in the optimal control theory of the 1960's which was based on state feedback.

8. Zeros usually appear when there are fewer inputs or outputs than states, or when $D \neq 0$. Consider a square $m \times m$ plant $G(s) = C(sI - A)^{-1}B + D$ with n states. We then have for the number of (finite) zeros of $G(s)$ (Maciejowski, 1989, p. 55)

$$
\begin{array}{lll}
D \neq 0: & \text{At most } n - m + \text{rank}(D) \text{ zeros} & \\
D = 0: & \text{At most } n - 2m + \text{rank}(CB) \text{ zeros} & (4.76) \\
D = 0 \text{ and } \text{rank}(CB) = m: & \text{Exactly } n - m \text{ zeros} &
\end{array}
$$

9. **Moving poles.** How are the poles affected by (a) feedback $(G(I + KG)^{-1})$, (b) series compensation (GK, feedforward control) and (c) parallel compensation ($G + K$)? The answer is that (a) feedback control moves the poles (e.g. $G = \frac{1}{s+a}, K = -2a$ moves the pole from $-a$ to $+a$), (b) series compensation cannot move the poles, but we may cancel poles in G by placing zeros in K (e.g. $G = \frac{1}{s+a}, K = \frac{s+a}{s+k}$), and (c) parallel compensation cannot move the poles, but we may cancel their effect by subtracting identical poles in K (e.g. $G = \frac{1}{s+a}, K = -\frac{1}{s+a}$).

10. For a strictly proper plant $G(s) = C(sI - A)^{-1}B$, the open-loop poles are determined by the characteristic polynomial $\phi_{ol}(s) = \det(sI - A)$. If we apply constant gain negative feedback $u = -K_0 y$, the poles are determined by the corresponding closed-loop characteristic polynomial $\phi_{cl}(s) = \det(sI - A + BK_0C)$. Thus, unstable plants may be stabilized by use of feedback control. See also Example 4.14.

11. **Moving zeros.** Consider next the effect of feedback, series and parallel compensation on the zeros.

(a) With feedback, the zeros of $G(I + KG)^{-1}$ are the zeros of G plus the poles of K. This means that the zeros in G, including their output directions y_z, are unaffected by feedback. However, even though y_z is fixed it is still possible with feedback control to move the deteriorating effect of a RHP-zero to a given output channel, provided y_z has a non-zero element for this output. This was illustrated by the example in Section 3.6, and is discussed in more detail in Section 6.6.1.

(b) Series compensation can counter the effect of zeros in G by placing poles in K to cancel them, but cancellations are not possible for RHP-zeros due to internal stability (see Section 4.7).

(c) The only way to move zeros is by parallel compensation, $y = (G + K)u$, which, if y is a physical output, can only be accomplished by adding an extra input (actuator).

12. **Pinned zeros.** A zero is pinned to a subset of the outputs if y_z has one or more elements equal to zero. In most cases, pinned zeros have a scalar origin. Pinned zeros are quite common in practice, and their effect cannot be moved freely to any output. For example, the effect of a measurement delay for output y_1 cannot be moved to output y_2. Similarly, a zero is pinned to certain inputs if u_z has one or more elements equal to zero. An example is $G(s)$ in (4.74), where the zero at -2 is pinned to input u_1 and to output y_1.

13. **Zeros of non-square systems.** The existence of zeros for non-square systems is common in practice in spite of what is sometimes claimed in the literature. In particular, they appear if we have a zero pinned to the side of the plant with the fewest number of channels. As an example consider a plant with three inputs and two outputs $G_1(s) = \begin{bmatrix} h_{11} & h_{12} & h_{13} \\ h_{21}(s - z) & h_{22}(s - z) & h_{23}(s - z) \end{bmatrix}$ which has a zero at $s = z$ which is pinned to output y_2, i.e. $y_z = [0 \quad 1]^T$. This follows because the second row of $G_1(z)$ is equal to zero, so

the rank of $G_1(z)$ is 1, which is less than the normal rank of $G_1(s)$, which is 2. On the other hand, $G_2(s) = \begin{bmatrix} h_{11}(s-z) & h_{12} & h_{13} \\ h_{21}(s-z) & h_{22} & h_{23} \end{bmatrix}$ does *not* have a zero at $s = z$ since $G_2(z)$ has rank 2 which is equal to the normal rank of $G_2(s)$ (assuming that the last two columns of $G_2(s)$ have rank 2).

14. The concept of functional controllability, see page 233, is related to zeros. Loosely speaking, one can say that a system which is functionally uncontrollable has in a certain output direction "a zero for all values of s".

The control implications of RHP-zeros and RHP-poles are discussed for SISO systems on pages 183–197 and for MIMO systems on pages 235–237.

Example 4.14 Effect of feedback on poles and zeros. *Consider a SISO negative feedback system with plant $G(s) = z(s)/\phi(s)$ and a constant gain controller, $K(s) = k$. The closed-loop response from reference r to output y is*

$$T(s) = \frac{L(s)}{1+L(s)} = \frac{kG(s)}{1+kG(s)} = \frac{kz(s)}{\phi(s)+kz(s)} = k\frac{z_{cl}(s)}{\phi_{cl}(s)} \qquad (4.77)$$

Note the following:

1. *The zero polynomial is $z_{cl}(s) = z(s)$, so the zero locations are unchanged by feedback.*
2. *The pole locations are changed by feedback. For example,*

$$k \to 0 \quad \Rightarrow \quad \phi_{cl}(s) \to \phi(s) \qquad (4.78)$$

$$k \to \infty \quad \Rightarrow \quad \phi_{cl}(s) \to kz(s) \qquad (4.79)$$

That is, as we increase the feedback gain, the closed-loop poles move from open-loop poles to the open-loop zeros. RHP-zeros therefore imply high gain instability. These results are well known from a classical root locus analysis.

Example 4.15 *We want to prove that $G(s) = C(sI - A)^{-1}B + D$ has no zeros if $D = 0$ and rank $(C) = n$, where n is the number of states. Solution: Consider the polynomial system matrix $P(s)$ in (4.65). The first n columns of P are independent because C has rank n. The last m columns are independent of s. Furthermore, the first n and last m columns are independent of each other, since $D = 0$ and C has full column rank and thus cannot have any columns equal to zero. In conclusion, $P(s)$ always has rank $n + m$ and there are no zeros. (We need $D = 0$ because if D is non-zero then the first n columns of P may depend on the last m columns for some value of s.)*

Exercise 4.3 * *(a) Consider a SISO system $G(s) = C(sI - A)^{-1}B + D$ with just one state, i.e. A is a scalar. Find the zeros. Does $G(s)$ have any zeros for $D = 0$? (b) Do GK and KG have the same poles and zeros for a SISO system? Ditto, for a MIMO system?*

Exercise 4.4 *Determine the poles and zeros of*

$$G(s) = \begin{bmatrix} \frac{11s^3 - 18s^2 - 70s - 50}{s(s+10)(s+1)(s-5)} & \frac{(s+2)}{(s+1)(s-5)} \\ \frac{5(s+2)}{(s+1)(s-5)} & \frac{5(s+2)}{(s+1)(s-5)} \end{bmatrix}$$

given that

$$\det G(s) = \frac{50(s^4 - s^3 - 15s^2 - 23s - 10)}{s(s+1)^2(s+10)(s-5)^2} = \frac{50(s+1)^2(s+2)(s-5)}{s(s+1)^2(s+10)(s-5)^2}$$

How many poles does $G(s)$ have?

Exercise 4.5 * Given $y(s) = G(s)u(s)$, with $G(s) = \frac{1-s}{1+s}$. Determine a state-space realization of $G(s)$ and then find the zeros of $G(s)$ using the generalized eigenvalue problem. What is the transfer function from $u(s)$ to $x(s)$, the single state of $G(s)$, and what are the zeros of this transfer function?

Exercise 4.6 Find the zeros for a 2×2 plant with

$$A = \begin{bmatrix} a_{11} & a_{12} \\ a_{21} & a_{22} \end{bmatrix}, \quad B = \begin{bmatrix} 1 & 1 \\ b_{21} & b_{22} \end{bmatrix}, \quad C = I, \quad D = 0$$

Exercise 4.7 * For what values of c_1 does the following plant have RHP-zeros?

$$A = \begin{bmatrix} 10 & 0 \\ 0 & -1 \end{bmatrix}, \quad B = I, \quad C = \begin{bmatrix} 10 & c_1 \\ 10 & 0 \end{bmatrix}, \quad D = \begin{bmatrix} 0 & 0 \\ 0 & 1 \end{bmatrix} \tag{4.80}$$

Exercise 4.8 Consider the plant in (4.80), but assume that both states are measured and used for feedback control, i.e. $y_m = x$ (but the controlled output is still $y = Cx + Du$). Can a RHP-zero in $G(s)$ give problems with stability in the feedback system? Can we achieve "perfect" control of y in this case? (Answers: No and no).

4.7 Internal stability of feedback systems

To test for closed-loop stability of a feedback system, it is usually enough to check just one closed-loop transfer function, e.g. $S = (I + GK)^{-1}$. However, this assumes that there are no internal RHP pole–zero cancellations between the controller and the plant. The point is best illustrated by an example.

Example 4.16 Consider the feedback system shown in Figure 4.2 where $G(s) = \frac{s-1}{s+1}$ and $K(s) =$

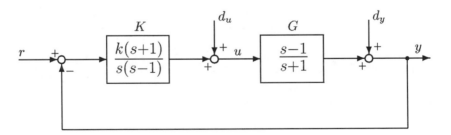

Figure 4.2: Internally unstable system

$\frac{k}{s}\frac{s+1}{s-1}$. In forming the loop transfer function $L = GK$ we then cancel the term $(s-1)$, a RHP pole–zero cancellation, to obtain

$$L = GK = \frac{k}{s}, \text{ and } S = (I + L)^{-1} = \frac{s}{s+k} \tag{4.81}$$

$S(s)$ is stable; that is, the transfer function from d_y to y is stable. However, the transfer function from d_y to u is unstable:

$$u = -K(I + GK)^{-1}d_y = -\frac{k(s+1)}{(s-1)(s+k)}d_y \tag{4.82}$$

Consequently, although the system appears to be stable when considering the output signal y, it is unstable when considering the "internal" signal u, so the system is (internally) unstable.

Remark 1 In practice, it is not possible to cancel exactly a plant zero or pole because of modelling errors. In the above example, therefore, L and S will in practice also be unstable. However, it is important to stress that even in the ideal case with a perfect RHP pole–zero cancellation, as in the above example, we would still get an *internally* unstable system. This is a subtle but important point. In this ideal case the state-space descriptions of L and S contain an unstable hidden mode corresponding to an unstabilizable or undetectable state.

Remark 2 By the same reasoning as in Example 4.16, we get an internally unstable system if we use *feedforward control* to cancel a RHP-zero or to stabilize an unstable plant. For example, consider Figure 4.2 with the feedback loop removed and K as the feedforward controller. For an unstable plant $G(s) = \frac{s+1}{s-1}$ we may use a feedforward controller $K(s) = \frac{s-1}{s+1}$ and get an (apparently) stable response $y = GKr = r$. First, this requires a perfect model with perfect cancellation of the unstable pole at $s = 1$. Second, even with a perfect model, we have $y = Gd_u$ where G is unstable, so any signal d_u entering between the controller and the plant will eventually drive the system out of bounds. Thus, the only way to stabilize an unstable plant is to *move* the unstable poles from the RHP to the LHP and this can only be accomplished by feedback control.

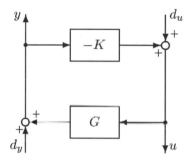

Figure 4.3. Block diagram used to check internal stability of feedback system

From the above example, it is clear that to be rigorous we must consider *internal* stability of the feedback system, see Definition 4.4. To this effect consider the system in Figure 4.3 where we inject and measure signals at both locations between the two components, G and K. We get

$$u = (I + KG)^{-1}d_u - K(I + GK)^{-1}d_y \qquad (4.83)$$

$$y = G(I + KG)^{-1}d_u + (I + GK)^{-1}d_y \qquad (4.84)$$

The theorem below follows immediately:

Theorem 4.6 *Assume that the components G and K contain no unstable hidden modes. Then the feedback system in Figure 4.3 is* **internally stable** *if and only if all four closed-loop transfer matrices in (4.83) and (4.84) are stable.*

The following can be proved using the above theorem (recall Example 4.16). *If there are RHP pole–zero cancellations between $G(s)$ and $K(s)$, i.e. if GK and KG do not both contain all the RHP-poles in G and K, then the system in Figure 4.3 is internally unstable.*

If we disallow RHP pole–zero cancellations between system components, such as G and K, then stability of *one* closed-loop transfer function implies stability of the others. This is stated in the following theorem.

Theorem 4.7 *Assume there are no RHP pole–zero cancellations between $G(s)$ and $K(s)$; that is, all RHP-poles in $G(s)$ and $K(s)$ are contained in the minimal realizations of GK and KG. Then the feedback system in Figure 4.3 is internally stable if and only if one of the four closed-loop transfer function matrices in (4.83) and (4.84) is stable.*

Proof: A proof is given by Zhou et al. (1996, p. 125).   □

Note how we define pole–zero cancellations in the above theorem. In this way, RHP pole–zero cancellations resulting from G or K not having full normal rank are also disallowed. For example, with $G(s) = 1/(s - a)$ and $K = 0$ we get $GK = 0$ so the RHP-pole at $s = a$ has disappeared and there is effectively a RHP pole–zero cancellation. In this case, we get $S(s) = 1$ which is stable, but internal stability is clearly not possible.

Exercise 4.9 * *Use (A.7) to show that the signal relationships (4.83) and (4.84) may also be written as*

$$\begin{bmatrix} u \\ y \end{bmatrix} = M(s) \begin{bmatrix} d_u \\ d_y \end{bmatrix}; \quad M(s) = \begin{bmatrix} I & K \\ -G & I \end{bmatrix}^{-1} \tag{4.85}$$

From this we get that the system in Figure 4.3 is internally stable if and only if $M(s)$ is stable.

4.7.1   Implications of the internal stability requirement

The requirement of internal stability in a feedback system leads to a number of interesting results, some of which are investigated below. Note in particular Exercise 4.12, where we discuss alternative ways of implementing a two degrees-of-freedom controller.

We first prove the following important statements which apply when the overall feedback system is internally stable (Youla et al., 1974):

1. *If $G(s)$ has a RHP-zero at z, then also $L = GK$, $T = GK(I + GK)^{-1}$, $SG = (I + GK)^{-1}G$, $L_I = KG$ and $T_I = KG(I + KG)^{-1}$ will each have a RHP-zero at z.*

2. *If $G(s)$ has a RHP-pole at p, then also $L = GK$ and $L_I = KG$ each have a RHP-pole at p, while $S = (I + GK)^{-1}$, $KS = K(I + GK)^{-1}$ and $S_I = (I + KG)^{-1}$ each have a RHP-zero at p.*

Proof of 1: To achieve internal stability, RHP pole–zero cancellations between system components, such as G and K, are not allowed. Thus $L = GK$ must have a RHP-zero when G has a RHP-zero. Now S is stable and thus has no RHP-pole which can cancel the RHP-zero in L, and so $T = LS$ must have a RHP-zero at z. Similarly, $SG = (I + GK)^{-1}G$ must have a RHP-zero, etc.   □

Proof of 2: Clearly, L has a RHP-pole at p. Since T is stable, it follows from $T = LS$ that S must have a RHP-zero which exactly cancels the RHP-pole in L, etc.   □

We notice from this that a RHP pole–zero cancellation between two transfer functions, such as between L and $S = (I + L)^{-1}$, does not necessarily imply internal instability. It is only between separate physical components (e.g. controller, plant) that RHP pole–zero cancellations are not allowed.

Exercise 4.10 **Interpolation constraints.** *Prove the following interpolation constraints which apply for SISO feedback systems when the plant $G(s)$ has a RHP-zero z or a RHP-pole p:*

$$G(z) = 0 \quad \Rightarrow \quad L(z) = 0 \quad \Leftrightarrow \quad T(z) = 0, S(z) = 1 \tag{4.86}$$

$$G^{-1}(p) = 0 \quad \Rightarrow \quad L(p) = \infty \quad \Leftrightarrow \quad T(p) = 1, S(p) = 0 \tag{4.87}$$

Exercise 4.11 * *Given the complementary sensitivity functions*

$$T_1(s) = \frac{2s+1}{s^2 + 0.8s + 1} \quad T_2(s) = \frac{-2s+1}{s^2 + 0.8s + 1}$$

what can you say about possible RHP-poles or RHP-zeros in the corresponding loop transfer functions, $L_1(s)$ *and* $L_2(s)$?

The following exercise demonstrates another application of the internal stability requirement.

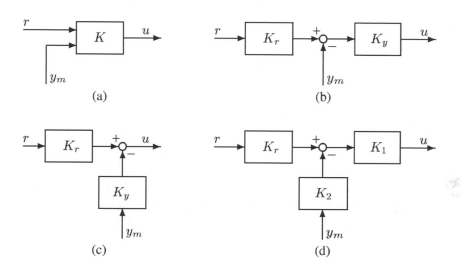

(a)   (b)

(c)   (d)

Figure 4.4: Different forms of two degrees-of-freedom controller:
> (a) General form
> (b) Suitable when $K_y(s)$ has no RHP-zeros
> (c) Suitable when $K_y(s)$ is stable (no RHP-poles)
> (d) $K_y(s) = K_1(s)K_2(s)$. Suitable when $K_1(s)$ contains
> no RHP-zeros and $K_2(s)$ no RHP poles
> (e) Yet another form is shown in Figure 2.5

Exercise 4.12 Internal stability of two degrees-of-freedom control configurations. *A two degrees-of-freedom controller allows one to improve performance by treating disturbance rejection and command tracking separately (at least to some degree). The general form shown in Figure 4.4(a) is usually preferred for both implementation and design. However, in some cases one may want first to design the pure feedback part of the controller, here denoted* $K_y(s)$, *for disturbance rejection, and then to add a simple pre-compensator,* $K_r(s)$, *for command tracking. This approach is in general not optimal, and may also yield problems when it comes to implementation, in particular if the feedback controller* $K_y(s)$ *contains RHP-poles or zeros, which can happen. This implementation issue is dealt with in this exercise by considering the three alternative forms in Figure 4.4(b)–4.4(d). In all these schemes* K_r *must clearly be stable.*

(a) The issue is to avoid "unnecessary" RHP-zeros in the transfer function from r to y. Show that (1) $K_y S$ *has a RHP-zero where* K_y *has a RHP-zero, and (2)* SG *has a RHP-zero where* K_y *has a RHP-pole.*

(b) Explain why the configuration in Figure 4.4(b) should not be used if K_y *contains RHP-zeros.*

(c) Explain why the configuration in Figure 4.4(c) should not be used if K_y contains RHP-poles. This implies that this configuration should not be used if we want integral action in K_y.

(d) Show that the configuration in Figure 4.4(d) may be used, provided the RHP-poles (including integrators) of K_y are contained in K_1 and the RHP-zeros in K_2. Discuss why one may often set $K_r = I$ in this case (to give a fourth possibility).

(e) A fifth form, where r goes to both K_r and K_y, is shown in Figure 2.5. When is this form suitable?

The requirement of internal stability also dictates that we must exercise care when we use a separate unstable disturbance model $G_d(s)$. To avoid this problem one should for state-space computations use a combined model for inputs and disturbances, i.e. write the model $y = Gu + G_d d$ in the form

$$y = [G \quad G_d]\begin{bmatrix} u \\ d \end{bmatrix}$$

where G and G_d share the same states, see (4.14) and (4.17).

4.8   Stabilizing controllers

In this section, we introduce a parameterization, known as the Q-parameterization or Youla-parameterization (Youla et al., 1976), of all stabilizing controllers for a plant. By all stabilizing controllers we mean all controllers that yield internal stability of the closed-loop system. We first consider stable plants, for which the parameterization is easily derived, and then unstable plants where we make use of the coprime factorization.

4.8.1   Stable plants

The following lemma forms the basis for parameterizing all stabilizing controllers for stable plants:

Lemma 4.8 *For a stable plant $G(s)$ the negative feedback system in Figure 4.3 is internally stable if and only if $Q = K(I + GK)^{-1}$ is stable.*

Proof: The four transfer functions in (4.83) and (4.84) are easily shown to be

$$K(I + GK)^{-1} = Q \tag{4.88}$$

$$(I + GK)^{-1} = I - GQ \tag{4.89}$$

$$(I + KG)^{-1} = I - QG \tag{4.90}$$

$$G(I + KG)^{-1} = G(I - QG) \tag{4.91}$$

which are clearly all stable if G and Q are stable. Thus, with G stable the system is internally stable if and only if Q is stable. □

As proposed by Zames (1981), by solving (4.88) with respect to the controller K, we find that a parameterization of *all stabilizing negative feedback controllers for the stable plant $G(s)$* is given by

$$K = (I - QG)^{-1}Q = Q(I - GQ)^{-1} \tag{4.92}$$

where the "parameter" Q is *any stable transfer function matrix.*

Remark 1 If only proper controllers are allowed then Q must be proper since the term $(I - QG)^{-1}$ is semi-proper.

Remark 2 We have shown that by varying Q freely (but stably) we will always have internal stability, and thus avoid internal RHP pole–zero cancellations between K and G. This means that although Q may generate unstable controllers K, there is no danger of getting a RHP-pole in K that cancels a RHP-zero in G.

The parameterization in (4.92) is identical to the internal model control (IMC) parameterization (Morari and Zafiriou, 1989) of stabilizing controllers. It may be derived directly from the IMC structure given in Figure 4.5. The idea behind the IMC structure is that the "controller" Q can be designed in an open-loop fashion since the feedback signal only contains information about the difference between the actual output and the output predicted from the model.

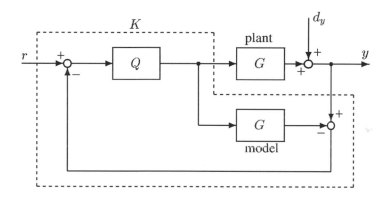

Figure 4.5: The internal model control (IMC) structure

Exercise 4.13 * *Show that the IMC structure in Figure 4.5 is internally unstable if either Q or G is unstable.*

Exercise 4.14 *Show that testing internal stability of the IMC structure is equivalent to testing for stability of the four closed-loop transfer functions in (4.88)–(4.91).*

Exercise 4.15 * *Given a stable controller K. What set of plants can be stabilized by this controller? (Hint: Interchange the roles of plant and controller.)*

4.8.2 Unstable plants

For an unstable plant $G(s)$, consider its left coprime factorization

$$G(s) = M_l^{-1} N_l \tag{4.93}$$

A parameterization of *all stabilizing negative feedback controllers for the plant* $G(s)$ is then (Vidyasagar, 1985)

$$K(s) = (V_r - QN_l)^{-1}(U_r + QM_l) \tag{4.94}$$

where V_r and U_r satisfy the Bezout identity (4.19) for the right coprime factorization, and $Q(s)$ is *any stable transfer function* satisfying the technical condition $\det(V_r(\infty) - Q(\infty)N_l(\infty)) \neq 0$. Similar to (4.94), the stabilizing negative feedback controllers can also be parameterized based on the right coprime factors, M_r, N_r (Vidyasagar, 1985).

Remark 1 With $Q = 0$ we have $K_0 = V_r^{-1}U_r$, so V_r and U_r can alternatively be obtained from a left coprime factorization of some initial stabilizing controller K_0.

Remark 2 For a stable plant, we may write $G(s) = N_l(s)$ corresponding to $M_l = I$. In this case $K_0 = 0$ is a stabilizing controller, so we may from (4.19) select $U_r = 0$ and $V_r = I$, and (4.94) yields $K = (I - QG)^{-1}Q$ as found before in (4.92).

Remark 3 We can also formulate the parameterization of all stabilizing controllers in state-space form, e.g. see page 312 in Zhou et al. (1996) for details.

The Q-parameterization may be very useful for controller synthesis. First, the search over all *stabilizing K's* (e.g. $S = (I + GK)^{-1}$ must be stable) is replaced by a search over *stable Q's*. Second, all *closed-loop* transfer functions (S, T, etc.) will be in the form $H_1 + H_2 Q H_3$, so they are affine[3] in Q. This further simplifies the optimization problem.

Strongly stabilizable. In theory, any linear plant may be stabilized irrespective of the location of its RHP-poles and RHP-zeros, provided the plant does not contain unstable hidden modes. However, this may require an unstable controller, and for practical purposes it is sometimes desirable that the controller is stable. If such a stable controller exists the plant is said to be *strongly stabilizable*. Youla et al. (1974) proved that *a strictly proper rational SISO plant is strongly stabilizable by a proper controller if and only if every real RHP-zero in $G(s)$ lies to the left of an even number (including zero) of real RHP-poles in $G(s)$*. Note that the presence of any complex RHP-poles or complex RHP-zeros does not affect this result. We then have:

- *A strictly proper rational plant with a single real RHP-zero z and a single real RHP-pole p, e.g.* $G(s) = \frac{s-z}{(s-p)(\tau s+1)}$, *can be stabilized by a stable proper controller if and only if* $z > p$.

Notice the requirement that $G(s)$ is strictly proper. For example, the plant $G(s) = \frac{s-1}{s-2}$ with $z = 1 < p = 2$ is stabilized with a stable constant gain controller $K(s) = K_c$ with $-2 < K_c < -1$. However, this plant is not strictly proper so the result by Youla et al. (1974) does not apply.

4.9 Stability analysis in the frequency domain

As noted above, the stability of a linear system is equivalent to the system having no poles in the closed RHP. This test may be used for any system, be it open-loop or closed-loop. In this section we will study the use of frequency domain techniques to derive information about *closed-loop* stability from the *open-loop* transfer matrix $L(j\omega)$. This provides a direct generalization of Nyquist's stability test for SISO systems.

Note that when we talk about eigenvalues in this section, we refer to the eigenvalues of a complex matrix, usually of $L(j\omega) = GK(j\omega)$, and not those of the state matrix A.

---

[3] A function $f(x)$ is affine in x if $f(x) = ax + b$, and is linear in x if $f(x) = ax$.

4.9.1 Open- and closed-loop characteristic polynomials

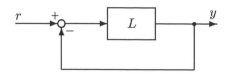

Figure 4.6: Negative feedback system

We first derive some preliminary results involving the determinant of the return difference operator $I + L$. Consider the feedback system shown in Figure 4.6, where $L(s)$ is the loop transfer function matrix. Stability of the open-loop system is determined by the poles of $L(s)$. If $L(s)$ has a state-space realization $\left[\begin{array}{c|c} A_{ol} & B_{ol} \\ \hline C_{ol} & D_{ol} \end{array}\right]$, i.e.

$$L(s) = C_{ol}(sI - A_{ol})^{-1}B_{ol} + D_{ol} \tag{4.95}$$

then the poles of $L(s)$ are the roots of the *open-loop* characteristic polynomial

$$\phi_{ol}(s) = \det(sI - A_{ol}) \tag{4.96}$$

Assume there are no RHP pole–zero cancellations between $G(s)$ and $K(s)$. Then from Theorem 4.7 internal stability of the *closed-loop* system is equivalent to the stability of $S(s) = (I + L(s))^{-1}$. The state matrix of $S(s)$ is given (assuming $L(s)$ is well-posed, i.e. $D_{ol} + I$ is invertible) by

$$A_{cl} = A_{ol} - B_{ol}(I + D_{ol})^{-1}C_{ol} \tag{4.97}$$

This equation may be derived by writing down the state-space equations for the transfer function from r to y in Figure 4.6

$$\dot{x} = A_{ol}x + B_{ol}(r - y) \tag{4.98}$$

$$y = C_{ol}x + D_{ol}(r - y) \tag{4.99}$$

and using (4.99) to eliminate y from (4.98). The closed-loop characteristic polynomial is thus given by

$$\phi_{cl}(s) \triangleq \det(sI - A_{cl}) = \det(sI - A_{ol} + B_{ol}(I + D_{ol})^{-1}C_{ol}) \tag{4.100}$$

Relationship between characteristic polynomials

The above identities may be used to express the determinant of the return difference operator, $I + L$, in terms of $\phi_{cl}(s)$ and $\phi_{ol}(s)$. From (4.95) we get

$$\det(I + L(s)) = \det(I + C_{ol}(sI - A_{ol})^{-1}B_{ol} + D_{ol}) \tag{4.101}$$

Schur's formula (A.14) then yields (with $A_{11} = I + D_{ol}, A_{12} = -C_{ol}, A_{22} = sI - A_{ol}, A_{21} = B_{ol}$)

$$\det(I + L(s)) = \frac{\phi_{cl}(s)}{\phi_{ol}(s)} \cdot c \tag{4.102}$$

where $c = \det(I + D_{ol})$ is a constant which is of no significance when evaluating the poles. Note that $\phi_{cl}(s)$ and $\phi_{ol}(s)$ are polynomials in s which have zeros only, whereas $\det(I + L(s))$ is a transfer function with both poles and zeros.

Example 4.17 *We will rederive expression (4.102) for SISO systems. Let $L(s) = k \frac{z(s)}{\phi_{ol}(s)}$. The sensitivity function is given by*

$$S(s) = \frac{1}{1 + L(s)} = \frac{\phi_{ol}(s)}{kz(s) + \phi_{ol}(s)} \tag{4.103}$$

and the denominator is

$$d(s) = kz(s) + \phi_{ol}(s) = \phi_{ol}(s)\left(1 + \frac{kz(s)}{\phi_{ol}(s)}\right) = \phi_{ol}(s)(1 + L(s)) \tag{4.104}$$

which is the same as $\phi_{cl}(s)$ in (4.102) (except for the constant c which is necessary to make the leading coefficient of $\phi_{cl}(s)$ equal to 1, as required by its definition).

Remark 1 One may be surprised to see from (4.103) that the zero polynomial of $S(s)$ is equal to the open-loop pole polynomial, $\phi_{ol}(s)$, but this is indeed correct. On the other hand, note from (4.77) that the zero polynomial of $T(s) = L(s)/(1 + L(s))$ is equal to $z(s)$, the open-loop zero polynomial.

Remark 2 From (4.102), for the case when there are no cancellations between $\phi_{ol}(s)$ and $\phi_{cl}(s)$, we have that the closed-loop poles are solutions to

$$\det(I + L(s)) = 0 \tag{4.105}$$

4.9.2 MIMO Nyquist stability criteria

We will consider the negative feedback system of Figure 4.6, and assume there are no internal RHP pole–zero cancellations in the loop transfer function matrix $L(s)$, i.e. $L(s)$ contains no unstable hidden modes. Expression (4.102) for $\det(I + L(s))$ then enables a straightforward generalization of Nyquist's stability condition to multivariable systems.

Theorem 4.9 Generalized (MIMO) Nyquist theorem. *Let P_{ol} denote the number of open-loop unstable poles in $L(s)$. The closed-loop system with loop transfer function $L(s)$ and negative feedback is stable if and only if the Nyquist plot of $\det(I + L(s))$*
   (i) makes P_{ol} anti-clockwise encirclements of the origin, and
   (ii) does not pass through the origin.

The theorem is proved below, but let us first make some important remarks.

Remark 1 By "Nyquist plot of $\det(I + L(s))$" we mean "the image of $\det(I + L(s))$ as s goes clockwise around the Nyquist D-contour". The Nyquist D-contour includes the entire $j\omega$-axis ($s = j\omega$) and an infinite semi-circle into the RHP as illustrated in Figure 4.7. The D-contour must also avoid locations where $L(s)$ has $j\omega$-axis poles by making small indentations (semi-circles) around these points.

Remark 2 In the following discussion, for practical reasons, we define *unstable poles* or *RHP-poles* as poles in the *open* RHP, excluding the $j\omega$-axis. In this case the Nyquist D-contour should make a small semi-circular indentation into the RHP at locations where $L(s)$ has $j\omega$-axis poles, thereby avoiding the extra count of encirclements due to $j\omega$-axis poles.

Remark 3 Another practical way of avoiding the indentation is to shift all $j\omega$-axis poles into the LHP, e.g. by replacing the integrator $1/s$ by $1/(s + \epsilon)$ where ϵ is a small positive number.

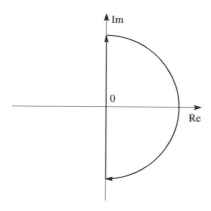

Figure 4.7: Nyquist D-contour for system with no open-loop $j\omega$-axis poles

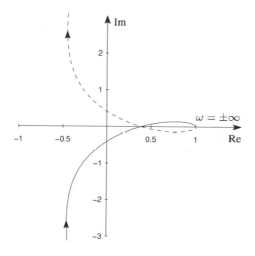

Figure 4.8: Typical Nyquist plot of $\det(I + L(j\omega))$

Remark 4 We see that for stability $\det(I + L(j\omega))$ should make no encirclements of the origin if $L(s)$ is open-loop stable, and should make P_{ol} anti-clockwise encirclements if $L(s)$ is unstable. If this condition is not satisfied then the number of closed-loop unstable poles of $(I + L(s))^{-1}$ is $P_{cl} = \mathcal{N} + P_{ol}$, where $\mathcal{N}$ is the number of clockwise encirclements of the origin by the Nyquist plot of $\det(I + L(j\omega))$.

Remark 5 For any real system, $L(s)$ is proper and so to plot $\det(I+L(s))$ as s traverses the D-contour we need to consider $s = j\omega$ only along the imaginary axis. This follows since $\lim_{s\to\infty} L(s) = D_{ol}$ is finite, and therefore for $s = \infty$ the Nyquist plot of $\det(I + L(s))$ converges to $\det(I + D_{ol})$ which is on the real axis.

Remark 6 In many cases $L(s)$ contains integrators so for $\omega = 0$ the plot of $\det(I + L(j\omega))$ may "start" from $\pm j\infty$. A typical plot for positive frequencies is shown in Figure 4.8 for the system

$$L = GK, \; G = \frac{3(-2s + 1)}{(10s + 1)(5s + 1)}, \; K = 1.14\frac{12.7s + 1}{12.7s} \qquad (4.106)$$

Note that the solid and dashed curves (positive and negative frequencies) need to be connected as ω approaches 0, so there is also a large (infinite) semi-circle (not shown) corresponding to the indentation of the D-contour into the RHP at $s = 0$ (the indentation is to avoid the integrator in $L(s)$). To find which way the large semi-circle goes, one can use the rule (based on conformal mapping arguments) that a right-angled turn in the D-contour will result in a right-angled turn in the Nyquist plot. It then follows for the example in (4.106) that there will be an infinite semi-circle into the RHP. There are therefore no encirclements of the origin. Since there are no open-loop unstable poles ($j\omega$-axis poles are excluded in the counting), $P_{ol} = 0$, and we conclude that the closed-loop system is stable.

Proof of Theorem 4.9: The proof makes use of the following result from complex variable theory (Churchill et al., 1974):

Lemma 4.10 Argument principle. *Consider a (transfer) function $f(s)$ and let C denote a closed contour in the complex plane. Assume that:*

1. $f(s)$ is "analytic" along C; that is, $f(s)$ has no poles on C.
2. $f(s)$ has Z zeros inside C.
3. $f(s)$ has P poles inside C.

Then the image $f(s)$ as the complex argument s traverses the contour C once in a clockwise direction will make $Z - P$ clockwise encirclements of the origin.

Let $\mathcal{N}(A, f(s), C)$ denote the number of clockwise encirclements of the point A by the image $f(s)$ as s traverses the contour C clockwise. Then a restatement of Lemma 4.10 is

$$\mathcal{N}(0, f(s), C) = Z - P \tag{4.107}$$

We now recall (4.102) and apply Lemma 4.10 to the function $f(s) = \det(I + L(s)) = \frac{\phi_{cl}(s)}{\phi_{ol}(s)} c$ selecting C to be the Nyquist D-contour. We assume $c = \det(I + D_{ol}) \neq 0$ since otherwise the feedback system would be ill-posed. The contour D goes along the $j\omega$-axis and around the entire RHP, but avoids open-loop poles of $L(s)$ on the $j\omega$-axis (where $\phi_{ol}(j\omega) = 0$) by making small semi-circles into the RHP. This is needed to make $f(s)$ analytic along D. We then have that $f(s)$ has $P = P_{ol}$ poles and $Z = P_{cl}$ zeros inside D. Here P_{cl} denotes the number of unstable closed-loop poles (in the open RHP). Equation (4.107) then gives

$$\mathcal{N}(0, \det(I + L(s)), D) = P_{cl} - P_{ol} \tag{4.108}$$

Since the system is stable if and only if $P_{cl} = 0$, condition (i) of Theorem 4.9 follows. However, we have not yet considered the possibility that $f(s) = \det(I + L(s))$, and hence $\phi_{cl}(s)$ has zeros on the D-contour itself, which will also correspond to a closed-loop unstable pole. To avoid this, $\det(I + L(j\omega))$ must not be zero for any value of ω and condition (ii) in Theorem 4.9 follows.   □

Example 4.18 SISO stability conditions. *Consider an open-loop stable SISO system. In this case, the Nyquist stability condition states that for closed-loop stability the Nyquist plot of $1 + L(s)$ should not encircle the origin. This is equivalent to the Nyquist plot of $L(j\omega)$ not encircling the point -1 in the complex plane*

4.9.3   Eigenvalue loci

The eigenvalue loci (sometimes called characteristic loci) are defined as the eigenvalues of the frequency response of the open-loop transfer function, $\lambda_i(L(j\omega))$. They partly provide a generalization of the Nyquist plot of $L(j\omega)$ from SISO to MIMO systems, and with them gain and phase margins can be defined as in the classical sense. However, these margins are not too useful as they only indicate stability with respect to a *simultaneous parameter change* in all of the loops. Therefore, although characteristic loci were well researched in the 1970's and greatly influenced the British developments in multivariable control, e.g. see Postlethwaite and MacFarlane (1979), they will not be considered further in this book.

4.9.4   Small-gain theorem

The small-gain theorem is a very general result which we will find useful in the book. We present first a generalized version of it in terms of the spectral radius, $\rho(L(j\omega))$, which at each frequency is defined as the maximum eigenvalue magnitude

$$\rho(L(j\omega)) \triangleq \max_i |\lambda_i(L(j\omega))| \tag{4.109}$$

Theorem 4.11 Spectral radius stability condition. *Consider a system with a stable loop transfer function $L(s)$. Then the closed-loop system is stable if*

$$\rho(L(j\omega)) < 1 \quad \forall \omega \tag{4.110}$$

Proof: The generalized Nyquist theorem (Theorem 4.9) says that if $L(s)$ is stable, then the closed-loop system is stable if and only if the Nyquist plot of $\det(I + L(s))$ does not encircle the origin. To prove condition (4.110) we will prove the "reverse"; that is, if the system is unstable and therefore $\det(I + L(s))$ does encircle the origin, then there is an eigenvalue, $\lambda_i(L(j\omega))$, which is larger than 1 at some frequency. If $\det(I + L(s))$ does encircle the origin, then there must exist a gain $\epsilon \in (0, 1]$ and a frequency ω' such that

$$\det(I + \epsilon L(j\omega')) = 0 \tag{4.111}$$

This is easily seen by geometric arguments since $\det(I + \epsilon L(j\omega')) = 1$ for $\epsilon = 0$. Expression (4.111) is equivalent to (see eigenvalue properties in Appendix A.2.1)

$$\prod_i \lambda_i(I + \epsilon L(j\omega')) = 0 \tag{4.112}$$

$$\Leftrightarrow \quad 1 + \epsilon\lambda_i(L(j\omega')) = 0 \quad \text{for some } i \tag{4.113}$$

$$\Leftrightarrow \quad \lambda_i(L(j\omega')) = -\frac{1}{\epsilon} \quad \text{for some } i \tag{4.114}$$

$$\Rightarrow \quad |\lambda_i(L(j\omega'))| \geq 1 \quad \text{for some } i \tag{4.115}$$

$$\Leftrightarrow \quad \rho(L(j\omega')) \geq 1 \tag{4.116}$$

$\square$

Theorem 4.11 is quite intuitive, as it simply says that if the system gain is less than 1 in all directions (all eigenvalues) and for all frequencies ($\forall \omega$), then all signal deviations will eventually die out, and the system is stable.

In general, the spectral radius theorem is conservative because phase information is not considered. For SISO systems $\rho(L(j\omega)) = |L(j\omega)|$, and consequently the above stability condition requires that $|L(j\omega)| < 1$ for all frequencies. This is clearly conservative, since from the Nyquist stability condition for a stable $L(s)$, we only require $|L(j\omega)| < 1$ at frequencies where the phase of $L(j\omega)$ is $-180° \pm n \cdot 360°$. As an example, let $L = k/(s+\epsilon)$. Since the phase never reaches $-180°$ the system is closed-loop stable for any value of $k > 0$. However, to satisfy (4.110) we need $k \leq \epsilon$, which for a small value of ϵ is very conservative indeed.

Remark. Later we will consider cases where the phase of L is allowed to vary freely, and in which case Theorem 4.11 is not conservative. Actually, a clever use of the above theorem is the main idea behind most of the conditions for robust stability and robust performance presented later in this book.

The small-gain theorem below follows directly from Theorem 4.11 if we consider a *matrix norm*, which by definition satisfies $\|AB\| \leq \|A\| \cdot \|B\|$. Then, at any frequency, we have $\rho(L) \leq \|L\|$ (see (A.117)).

Theorem 4.12 Small-gain theorem. *Consider a system with a stable loop transfer function $L(s)$. Then the closed-loop system is stable if*

$$\|L(j\omega)\| < 1 \quad \forall\omega \tag{4.117}$$

where $\|L\|$ denotes any matrix norm satisfying $\|AB\| \leq \|A\| \cdot \|B\|$.

Remark 1 This result is only a special case of a more general small-gain theorem which also applies to many nonlinear systems (Desoer and Vidyasagar, 1975).

Remark 2 The small-gain theorem does not consider phase information, and is therefore independent of the sign of the feedback.

Remark 3 Any induced norm can be used, e.g. the singular value, $\bar{\sigma}(L)$.

Remark 4 The small-gain theorem can be extended to include more than one block in the loop, e.g. $L = L_1 L_2$. In this case we get from (A.98) that the system is stable if $\|L_1\| \cdot \|L_2\| < 1$, $\forall\omega$.

Remark 5 The small-gain theorem is generally more conservative than the spectral radius condition in Theorem 4.11. Therefore, the arguments on conservatism made following Theorem 4.11 also apply to Theorem 4.12.

4.10   System norms

Figure 4.9: System G

Consider the system in Figure 4.9, with a stable transfer function matrix $G(s)$ and impulse response matrix $g(t)$. To evaluate the performance we ask the question: given information about the allowed input signals $w(t)$, how large can the outputs $z(t)$ become? To answer this, we must evaluate the relevant system norm.

We will here evaluate the output signal in terms of the usual 2-norm,

$$\|z(t)\|_2 = \sqrt{\sum_i \int_{-\infty}^{\infty} |z_i(\tau)|^2 d\tau} \tag{4.118}$$

and consider three different choices for the inputs:

1. $w(t)$ is a series of unit impulses.
2. $w(t)$ is any signal satisfying $\|w(t)\|_2 = 1$.

3. $w(t)$ is any signal satisfying $\|w(t)\|_2 = 1$, but $w(t) = 0$ for $t \geq 0$, and we only measure $z(t)$ for $t \geq 0$.

The relevant system norms in the three cases are the $\mathcal{H}_2$, $\mathcal{H}_\infty$ and Hankel norms, respectively. The $\mathcal{H}_2$ and $\mathcal{H}_\infty$ norms also have other interpretations as are discussed below. We introduced the $\mathcal{H}_2$ and $\mathcal{H}_\infty$ norms in Section 2.8, where we also discussed the terminology. In Appendix A.5.7 we present a more detailed interpretation and comparison of these and other norms.

4.10.1 $\mathcal{H}_2$ norm

Consider a strictly proper system $G(s)$, i.e. $D = 0$ in a state-space realization. For the $\mathcal{H}_2$ norm we use the Frobenius norm spatially (for the matrix) and integrate over frequency, i.e.

$$\|G(s)\|_2 \triangleq \sqrt{\frac{1}{2\pi} \int_{-\infty}^{\infty} \underbrace{\text{tr}(G(j\omega)^H G(j\omega))}_{\|G(j\omega)\|_F^2 = \sum_{ij} |G_{ij}(j\omega)|^2} \, d\omega} \tag{4.119}$$

We see that $G(s)$ must be strictly proper, otherwise the $\mathcal{H}_2$ norm is infinite. The $\mathcal{H}_2$ norm can also be given another interpretation. By Parseval's theorem, (4.119) is equal to the $\mathcal{H}_2$ norm of the impulse response

$$\|G(s)\|_2 = \|g(t)\|_2 \triangleq \sqrt{\int_0^{\infty} \underbrace{\text{tr}(g^T(\tau)g(\tau))}_{\|g(\tau)\|_F^2 = \sum_{ij} |g_{ij}(\tau)|^2} \, d\tau} \tag{4.120}$$

Remark 1 Note that $G(s)$ and $g(t)$ are dynamic systems while $G(j\omega)$ and $g(\tau)$ are constant matrices (for a given value of ω or τ).

Remark 2 We can change the order of integration and summation in (4.120) to get

$$\|G(s)\|_2 = \|g(t)\|_2 = \sqrt{\sum_{ij} \int_0^{\infty} |g_{ij}(\tau)|^2 d\tau} \tag{4.121}$$

where $g_{ij}(t)$ is the ij'th element of the impulse response matrix, $g(t)$. From this we see that the $\mathcal{H}_2$ norm can be interpreted as the 2-norm output resulting from applying unit impulses $\delta_j(t)$ to each input, one after another (allowing the output to settle to zero before applying an impulse to the next input). This is more clearly seen by writing $\|G(s)\|^2 = \sqrt{\sum_{i=1}^m \|z_i(t)\|_2^2}$ where $z_i(t)$ is the output vector resulting from applying a unit impulse $\delta_i(t)$ to the i'th input.

In summary, we have the following deterministic performance interpretation of the $\mathcal{H}_2$ norm:

$$\|G(s)\|_2 = \max_{w(t)= \text{ unit impulses}} \|z(t)\|_2 \tag{4.122}$$

The $\mathcal{H}_2$ norm can also be given a stochastic interpretation (see page 355) in terms of the quadratic criterion in optimal control (LQG) where we measure the expected root mean square (rms) value of the output in response to white noise excitation.

For numerical computations of the $\mathcal{H}_2$ norm, consider the state-space realization $G(s) = C(sI - A)^{-1}B$. By substituting (4.10) into (4.120) we find

$$\|G(s)\|_2 = \sqrt{\text{tr}(B^T Q B)} \quad \text{or} \quad \|G(s)\|_2 = \sqrt{\text{tr}(CPC^T)} \tag{4.123}$$

where P and Q are the controllability and observability Gramians, respectively, obtained as solutions to the Lyapunov equations (4.45) and (4.50).

4.10.2 $\mathcal{H}_\infty$ norm

Consider a proper linear stable system $G(s)$ (i.e. $D \neq 0$ is allowed). For the $\mathcal{H}_\infty$ norm we use the singular value (induced 2-norm) spatially (for the matrix) and pick out the peak value as a function of frequency

$$\|G(s)\|_\infty \triangleq \max_\omega \bar{\sigma}(G(j\omega)) \tag{4.124}$$

In terms of *performance* we see from (4.124) that the $\mathcal{H}_\infty$ norm is the peak of the transfer function "magnitude", and by introducing weights, the $\mathcal{H}_\infty$ norm can be interpreted as the magnitude of some closed-loop transfer function relative to a specified upper bound. This leads to specifying performance in terms of weighted sensitivity, mixed sensitivity, and so on.

However, the $\mathcal{H}_\infty$ norm also has several time domain performance interpretations. First, as discussed in Section 3.3.5, it is the worst-case gain for sinusoidal inputs at any frequency. As $t \to \infty$, let $z(\omega)$ denote the response of the system to a persistent sinusoidal input $w(\omega)$ (phasor notation). Then we have $z(\omega) = G(j\omega)w(\omega)$. At a given frequency ω, the amplification (gain) $\|z(\omega)\|_2/\|w(\omega)\|_2$ depends on the direction of $w(\omega)$, and the gain in the worst-case direction is given by the maximum singular value:

$$\bar{\sigma}(G(j\omega)) = \max_{w(\omega)\neq 0} \frac{\|z(\omega)\|_2}{\|w(\omega)\|_2}$$

The gain also depends on frequency, and the gain at the worst-case frequency is given by the $\mathcal{H}_\infty$ norm:

$$\|G(s)\|_\infty = \max_\omega \max_{w(\omega)\neq 0} \frac{\|z(\omega)\|_2}{\|w(\omega)\|_2} = \max_{\|w(\omega)\|_2=1} \|z(\omega)\|_2 \tag{4.125}$$

Second, from Tables A.1 and A.2 in the Appendix (page 540) we see that the $\mathcal{H}_\infty$ norm is equal to the induced (worst-case) 2-norm of any time domain signal:

$$\|G(s)\|_\infty = \max_{w(t)\neq 0} \frac{\|z(t)\|_2}{\|w(t)\|_2} = \max_{\|w(t)\|_2=1} \|z(t)\|_2 \tag{4.126}$$

The latter is a fortunate fact from functional analysis which is proved, for example, in Desoer and Vidyasagar (1975). In essence, (4.126) arises because the worst input signal $w(t)$ is a sinusoid with frequency ω^* and a direction which gives $\bar{\sigma}(G(j\omega^*))$ as the maximum gain.

Third, the $\mathcal{H}_\infty$ norm is equal to the induced power (rms) norm, and, fourth, has an interpretation as an induced norm in terms of the expected values of stochastic signals. All these various interpretations make the $\mathcal{H}_\infty$ norm useful in engineering applications.

The $\mathcal{H}_\infty$ norm is usually computed numerically from a state-space realization as the smallest value of γ such that the Hamiltonian matrix H has no eigenvalues on the imaginary axis, where

$$H = \begin{bmatrix} A + BR^{-1}D^T C & BR^{-1}B^T \\ -C^T(I + DR^{-1}D^T)C & -(A + BR^{-1}D^T C)^T \end{bmatrix} \tag{4.127}$$

and $R = \gamma^2 I - D^T D$, see Zhou et al. (1996, p. 115). This is an iterative procedure, where one may start with a large value of γ and reduce it until imaginary eigenvalues for H appear.

4.10.3 Difference between the $\mathcal{H}_2$ and $\mathcal{H}_\infty$ norms

To understand the difference between the $\mathcal{H}_2$ and $\mathcal{H}_\infty$ norms, note that from (A.127) we can write the Frobenius norm in terms of singular values. We then have

$$\|G(s)\|_2 = \sqrt{\frac{1}{2\pi} \int_{-\infty}^{\infty} \sum_i \sigma_i^2(G(j\omega)) d\omega} \tag{4.128}$$

From this we see that minimizing the $\mathcal{H}_\infty$ norm corresponds to minimizing the peak of the largest singular value ("worst direction, worst frequency"), whereas minimizing the $\mathcal{H}_2$ norm corresponds to minimizing the sum of the squares of all the singular values over all frequencies ("average direction, average frequency"). In summary, we have

- $\mathcal{H}_\infty$: "push down peak of largest singular value".
- $\mathcal{H}_2$: "push down whole thing" (all singular values over all frequencies).

Example 4.19 *We will compute the $\mathcal{H}_\infty$ and $\mathcal{H}_2$ norms for the following SISO plant:*

$$G(s) = \frac{1}{s+a} \tag{4.129}$$

The $\mathcal{H}_2$ norm is

$$\|G(s)\|_2 = \left(\frac{1}{2\pi} \int_{-\infty}^{\infty} \underbrace{|G(j\omega)|^2}_{\frac{1}{\omega^2+a^2}} d\omega \right)^{\frac{1}{2}} = \left(\frac{1}{2\pi a} \left[\tan^{-1}\left(\frac{\omega}{a}\right) \right]_{-\infty}^{\infty} \right)^{\frac{1}{2}} = \sqrt{\frac{1}{2a}} \tag{4.130}$$

To check Parseval's theorem we consider the impulse response

$$g(t) = \mathcal{L}^{-1}\left(\frac{1}{s+a}\right) = e^{-at}, t \ge 0 \tag{4.131}$$

and we get

$$\|g(t)\|_2 = \sqrt{\int_0^{\infty} (e^{-at})^2 dt} = \sqrt{\frac{1}{2a}} \tag{4.132}$$

as expected. The $\mathcal{H}_\infty$ norm is

$$\|G(s)\|_\infty = \max_\omega |G(j\omega)| = \max_\omega \frac{1}{(\omega^2+a^2)^{\frac{1}{2}}} = \frac{1}{a} \tag{4.133}$$

For interest, we also compute the 1-norm of the impulse response (which is equal to the induced ∞-norm in the time domain):

$$\|g(t)\|_1 = \int_0^{\infty} \underbrace{|g(t)|}_{e^{-at}} dt = \frac{1}{a} \tag{4.134}$$

In general, it can be shown that $\|G(s)\|_\infty \le \|g(t)\|_1$, and this example illustrates that we may have equality.

Example 4.20 *There exists no general relationship between the $\mathcal{H}_2$ and $\mathcal{H}_\infty$ norms. As an example consider the two systems*

$$f_1(s) = \frac{1}{\epsilon s + 1}, \quad f_2(s) = \frac{\epsilon s}{s^2 + \epsilon s + 1} \tag{4.135}$$

and let $\epsilon \to 0$. Then we have for f_1 that the $\mathcal{H}_\infty$ norm is 1 and the $\mathcal{H}_2$ norm is infinite. For f_2 the $\mathcal{H}_\infty$ norm is again 1 (at $\omega = 1$), but now the $\mathcal{H}_2$ norm is zero.

Why is the $\mathcal{H}_\infty$ norm so popular? In robust control we use the $\mathcal{H}_\infty$ norm mainly because it is convenient for representing unstructured model uncertainty, and because it satisfies the multiplicative property (A.98):

$$\|A(s)B(s)\|_\infty \le \|A(s)\|_\infty \cdot \|B(s)\|_\infty \tag{4.136}$$

This follows from (4.126) which shows that the $\mathcal{H}_\infty$ norm is an induced norm.

What is wrong with the $\mathcal{H}_2$ norm? The $\mathcal{H}_2$ norm has a number of good mathematical and numerical properties, and its minimization has important engineering implications. However, the $\mathcal{H}_2$ norm is *not* an induced norm and does *not* satisfy the multiplicative property. This implies that we cannot, by evaluating the $\mathcal{H}_2$ norm of individual components, say anything about how their series (cascade) interconnection will behave.

Example 4.21 *Consider again $G(s) = 1/(s+a)$ in (4.129), for which we found $\|G(s)\|_2 = \sqrt{1/2a}$. Now consider the $\mathcal{H}_2$ norm of $G(s)G(s)$:*

$$\|G(s)G(s)\|_2 = \sqrt{\int_0^\infty |\underbrace{\mathcal{L}^{-1}\left[\left(\frac{1}{s+a}\right)^2\right]}_{te^{-at}}|^2} = \sqrt{\frac{1}{a}\frac{1}{2a}} = \sqrt{\frac{1}{a}}\|G(s)\|_2^2$$

and we find, for $a < 1$, that

$$\|G(s)G(s)\|_2 > \|G(s)\|_2 \cdot \|G(s)\|_2 \tag{4.137}$$

which does not satisfy the multiplicative property (A.98). On the other hand, the $\mathcal{H}_\infty$ norm does satisfy the multiplicative property, and for the specific example we have equality with $\|G(s)G(s)\|_\infty = \frac{1}{a^2} = \|G(s)\|_\infty \cdot \|G(s)\|_\infty$.

4.10.4　Hankel norm

In the following discussion, we aim to develop an understanding of the Hankel norm. The Hankel norm of a stable system $G(s)$ is obtained when one applies an input $w(t)$ up to $t = 0$ and measures the output $z(t)$ for $t > 0$, and selects $w(t)$ to maximize the ratio of the 2-norms of these two signals:

$$\|G(s)\|_H \triangleq \max_{w(t)} \frac{\sqrt{\int_0^\infty \|z(\tau)\|_2^2 d\tau}}{\sqrt{\int_{-\infty}^0 \|w(\tau)\|_2^2 d\tau}} \tag{4.138}$$

The Hankel norm is a kind of induced norm from past inputs to future outputs. Its definition is analogous to trying to pump a swing with limited input energy such that the subsequent length of jump is maximized as illustrated (by the mythical creature) in Figure 4.10.

It may be shown that the *Hankel norm* is equal to

$$\|G(s)\|_H = \sqrt{\rho(PQ)} \tag{4.139}$$

where ρ is the spectral radius (absolute value of maximum eigenvalue), P is the controllability Gramian defined in (4.44) and Q the observability Gramian defined in (4.49). The name "Hankel" is used because the matrix PQ has the special structure of a Hankel matrix (which has identical elements along the "wrong-way" diagonals). The corresponding *Hankel singular values* are the positive square roots of the eigenvalues of PQ,

$$\sigma_i = \sqrt{\lambda_i(PQ)} \tag{4.140}$$

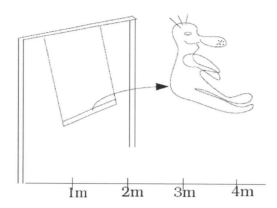

Figure 4.10: Pumping a swing: illustration of Hankel norm. The input is applied for $t \leq 0$ and the jump starts at $t = 0$.

The Hankel and $\mathcal{H}_\infty$ norms are closely related and we have (Zhou et al., 1996, p. 111)

$$\|G(s)\|_H \equiv \sigma_1 \leq \|G(s)\|_\infty \leq 2 \sum_{i=1}^{n} \sigma_i \qquad (4.141)$$

Thus, the Hankel norm is always smaller than (or equal to) the $\mathcal{H}_\infty$ norm, which is also reasonable by comparing the definitions in (4.126) and (4.138).

Model reduction. Consider the following problem: given a state-space description $G(s)$ of a system, find a model $G_a(s)$ with fewer states such that the input–output behaviour (from w to z) is changed as little as possible. Based on the discussion above it seems reasonable to make use of the Hankel norm, since the inputs only affect the outputs through the states at $t = 0$. For model reduction, we usually start with a realization of G which is internally balanced; that is, such that $Q = P = \Sigma$, where Σ is the matrix of Hankel singular values. We may then discard states (or rather combinations of states corresponding to certain subspaces) corresponding to the smallest Hankel singular values. The change in $\mathcal{H}_\infty$ norm caused by deleting states in $G(s)$ is less than twice the sum of the discarded Hankel singular values, i.e.

$$\|G(s) - G_a(s)\|_\infty \leq 2(\sigma_{k+1} + \sigma_{k+2} + \cdots) \qquad (4.142)$$

where $G_a(s)$ denotes a truncated or residualized balanced realization with k states; see Chapter 11. The method of Hankel norm minimization gives a somewhat improved error bound, where we are guaranteed that $\|G(s) - G_a(s)\|_\infty$ is less than the sum of the discarded Hankel singular values. This and other methods for model reduction are discussed in detail in Chapter 11 where a number of examples can be found.

Example 4.22 *We want to compute analytically the various system norms for $G(s) = 1/(s + a)$ using state-space methods. A state-space realization is $A = -a$, $B = 1$, $C = 1$ and $D = 0$. The controllability Gramian P is obtained from the Lyapunov equation $AP + PA^T = -BB^T$ $\Leftrightarrow$ $-aP - aP = -1$, so $P = 1/2a$. Similarly, the observability Gramian is $Q = 1/2a$. From (4.123) the $\mathcal{H}_2$ norm is then*

$$\|G(s)\|_2 = \sqrt{\operatorname{tr}(B^T Q B)} = \sqrt{1/2a}$$

The eigenvalues of the Hamiltonian matrix H in (4.127) are

$$\lambda(H) = \lambda \begin{bmatrix} -a & 1/\gamma^2 \\ -1 & a \end{bmatrix} = \pm\sqrt{a^2 - 1/\gamma^2}$$

We find that H has no imaginary eigenvalues for $\gamma > 1/a$, so

$$\|G(s)\|_\infty = 1/a$$

The Hankel matrix is $PQ = 1/4a^2$ and from (4.139) the Hankel norm is

$$\|G(s)\|_H = \sqrt{\rho(PQ)} = 1/2a$$

These results agree with the frequency domain calculations in Example 4.19.

Exercise 4.16 *Let $a = 0.5$ and $\epsilon = 0.0001$ and check numerically the results in Examples 4.19, 4.20, 4.21 and 4.22 using, for example, the Matlab Robust Control toolbox commands* `norm(sys,2)`, `norm(sys,inf)`, *and for the Hankel norm,* `max(hankelsv(sys))`.

4.11   Conclusion

This chapter has covered the following important elements of linear system theory: system descriptions, state controllability and observability, poles and zeros, stability and stabilization, and system norms. The topics are standard and the treatment is complete for the purposes of this book.

5

LIMITATIONS ON PERFORMANCE IN SISO SYSTEMS

In this chapter, we discuss the fundamental limitations on performance in SISO systems. We summarize these limitations in the form of a procedure for input–output controllability analysis, which is then applied to a series of examples. Input–output controllability of a plant is the ability to achieve acceptable control performance. Proper scaling of the input, output and disturbance variables prior to this analysis is critical.

5.1 Input–output controllability

In university courses on control, methods for controller design and stability analysis are usually emphasized. However, in practice the following three questions are often more important:

I. How well can the plant be controlled? Before starting any controller design one should first determine how easy the plant actually is to control. Is it a difficult control problem? Indeed, does there even exist a controller which meets the required performance objectives?

II. What control structure should be used? By this we mean what variables should we measure and control, which variables should we manipulate, and how are these variables best paired together? In other textbooks one can find qualitative rules for these problems. For example, in Seborg et al. (1989) in a chapter called "The art of process control", the following rules are given:

1. Control the outputs that are not self-regulating.
2. Control the outputs that have favourable dynamic and static characteristics, i.e. for each output, there should exist an input which has a significant, direct and rapid effect on it.
3. Select the inputs that have large effects on the outputs.
4. Select the inputs that rapidly affect the controlled variables

These rules are reasonable, but what are "self-regulating", "large", "rapid" and "direct"? A major objective of this chapter is to quantify these terms.

III. How might the process be changed to improve control? For example, to reduce the effects of a disturbance one might in process control consider changing the size of a buffer tank, or in automotive control one might decide to change the properties of a spring. In other

Multivariable Feedback Control: Analysis and Design. Second Edition
S. Skogestad and I. Postlethwaite   © 2005, 2006 John Wiley & Sons, Ltd

situations, the speed of response of a measurement device might be an important factor in achieving acceptable control.

The above three questions are each related to the inherent control characteristics of the process itself. We will introduce the term *input–output controllability* to capture these characteristics as described in the following definition.

Definition 5.1 (Input–output) controllability *is the ability to achieve acceptable control performance; that is, to keep the outputs (y) within specified bounds or displacements from their references (r), in spite of unknown but bounded variations, such as disturbances (d) and plant changes (including uncertainty), using available inputs (u) and available measurements* $(y_m$ *or* $d_m)$.

In summary, a plant is controllable if there *exists* a controller (connecting plant measurements and plant inputs) that yields acceptable performance for all expected plant variations. Thus, *controllability is independent of the controller, and is a property of the plant (or process) alone*. It can only be affected by changing the plant itself; that is, by (plant) design changes. These may include:

- changing the apparatus itself, e.g. type, size, etc.
- relocating sensors and actuators
- adding new equipment to dampen disturbances
- adding extra sensors
- adding extra actuators
- changing the control objectives
- changing the configuration of the lower layers of control already in place

Whether or not the last two actions are design modifications is arguable, but at least they address important issues which are relevant before the controller is designed.

Input–output controllability analysis is applied to a plant to find out what control performance can be expected. Another term for input–output controllability analysis is *performance targeting*. Early work on input–output controllability analysis includes that of Ziegler and Nichols (1943) and Rosenbrock (1970). Morari (1983) talked about "dynamic resilience" and made use of the concept of "perfect control". Important ideas on performance limitations are also found in Bode (1945), Horowitz (1963), Frank (1968a; 1968b), Kwakernaak and Sivan (1972), Horowitz and Shaked (1975), Zames (1981), Doyle and Stein (1981), Francis and Zames (1984), Boyd and Desoer (1985), Kwakernaak (1985), Freudenberg and Looze (1985; 1988), Engell (1988), Morari and Zafiriou (1989), Middleton (1991), Boyd and Barratt (1991), Chen (1995), Seron et al. (1997), Chen (2000) and Havre and Skogestad (2001). We also refer the reader to two IFAC workshops on *Interactions between process design and process control* (Perkins, 1992; Zafiriou, 1994) and the special issue of *IEEE Transactions on Automatic Control* on *Performance limitations* (Chen and Middleton, 2003).

5.1.1   Input–output controllability analysis

Surprisingly, given the plethora of mathematical methods available for control system design, the methods available for controllability analysis are largely qualitative. In most cases, the "simulation approach" is used, i.e. performance is assessed by exhaustive simulations.

However, this requires a specific controller design and specific values of disturbances and setpoint changes. Consequently, with this approach, one can never know if the result is a fundamental property of the plant, or if it depends on the specific controller designed, the disturbances or the setpoints.

A rigorous approach to controllability analysis would be to formulate mathematically the control objectives, the class of disturbances, the model uncertainty, etc., and then to synthesize controllers to see whether the objectives can be met. With model uncertainty this involves designing a μ-optimal controller (see Chapter 8). However, in practice such an approach is difficult and time consuming, especially if there are a large number of candidate measurements or actuators; see Chapter 10. Moreover, it provides little insight into the reasons for any controllability problems. More desirable is to have a few simple tools which can be used to get a rough idea of how easy the plant is to control, i.e. to determine whether or not a plant is controllable, without performing a detailed controller design. The main objective of this chapter is to derive such controllability tools based on appropriately scaled models of $G(s)$ and $G_d(s)$.

An apparent shortcoming of the controllability analysis presented in this book is that all the tools are linear. Recently, there has been some interest in analyzing the controller design trade-offs for nonlinear systems directly (see e.g. Middleton and Braslavsky, 2002), but we point out that usually the assumption of linearity is not restrictive. In fact, one of the most important nonlinearities, namely that associated with input constraints, can be handled quite well with a linear analysis. Also, to deal with slowly varying changes one may perform a controllability analysis at several selected operating points. Nonlinear simulations to validate the linear controllability analysis are of course still recommended. Experience from a large number of case studies confirms that the linear measures are often very good.

5.1.2 Scaling and performance

The above definition of controllability does not specify the allowed bounds for the displacements or the expected variations in the disturbance; that is, no definition of the desired performance is included. Throughout this chapter and the next, when we discuss controllability, we will assume that the variables and models have been scaled as outlined in Section 1.4, so that the requirement for acceptable performance is:

- For any reference $r(t)$ between $-R$ and R and any disturbance $d(t)$ between -1 and 1, keep the output $y(t)$ within the range $r(t) - 1$ to $r(t) + 1$ (at least most of the time), using an input $u(t)$ within the range -1 to 1.

We will interpret this definition from a frequency-by-frequency sinusoidal point of view, i.e. $d(t) = \sin \omega t$, and so on. With $e = y - r$ we then have:

> *For any disturbance* $|d(\omega)| \leq 1$ *and any reference* $|r(\omega)| \leq R(\omega)$, *the* **performance requirement** *is to keep at each frequency* ω *the control error* $|e(\omega)| \leq 1$, *using an input* $|u(\omega)| \leq 1$.

It is impossible to track very fast reference changes, so we will assume that $R(\omega)$ is frequency dependent; for simplicity, we assume that $R(\omega)$ is R (a constant) up to the frequency ω_r and is zero above that frequency.

It could also be argued that the magnitude of the sinusoidal disturbances should approach zero at high frequencies. While this may be true, we really only care about frequencies within the bandwidth of the system, and in most cases it is reasonable to assume that the plant experiences sinusoidal disturbances of constant magnitude up to this frequency. Similarly, it might also be argued that the allowed control error should be frequency dependent. For example, we may require no steady-state offset, i.e. e should be zero at low frequencies. However, including frequency variations is not recommended when doing a preliminary analysis (however, one may take such considerations into account when interpreting the results).

Recall that with $r = R\tilde{r}$ (see Section 1.4) the control error may be written as

$$e = y - r = Gu + G_d d - R\tilde{r} \tag{5.1}$$

where R is the magnitude of the reference and $|\tilde{r}(\omega)| \leq 1$ and $|d(\omega)| \leq 1$ are unknown signals. We will use (5.1) to unify our treatment of disturbances and references. Specifically, we will derive results for disturbances, which can then be applied directly to the references by replacing G_d by $-R$; see (5.1) .

5.1.3   Remarks on the term controllability

The definition of (input–output) controllability on page 164 is in tune with most engineers' intuitive feeling about what the term means, and was also how the term was used historically in the control literature. For example, Ziegler and Nichols (1943) defined *controllability* as *"the ability of the process to achieve and maintain the desired equilibrium value"*. Unfortunately, in the 1960's "controllability" became synonymous with the rather narrow concept of "state controllability" introduced by Kalman, and the term is still used in this restrictive manner by the systems theory community. *State controllability* is the ability to bring a system from a given initial state to any final state within a finite time. However, as shown in Example 4.5 this gives no regard to the quality of the response between these two states and later, and the required inputs may be excessive. The concept of *state controllability* is important for realizations and numerical calculations, but as long as we know that all the unstable modes are both controllable and observable, it usually has little practical significance. For example, Rosenbrock (1970, p. 177) notes that "most industrial plants are controlled quite satisfactorily though they are not [state] controllable". And conversely, there are many systems, like the tanks in series (Example 4.5), which are state controllable, but which are not input–output controllable. To avoid any confusion between practical controllability and Kalman's state controllability, Morari (1983) introduced the term *dynamic resilience*. However, this term does not capture the fact that it is related to control, so instead we prefer the term *input–output controllability*, or simply *controllability* when it is clear that we are not referring to state controllability.

Where are we heading? In this chapter we will discuss a number of results related to achievable performance. In Sections 5.2 and 5.3, we present some fundamental limitations imposed by RHP-poles and RHP-zeros. Readers who are more interested in the engineering implications of controllability may want to skip to Section 5.4. Many of the results can be formulated as upper and lower bounds on the bandwidth of the system. As noted in Section 2.4.5, there are several definitions of bandwidth (ω_B, ω_c and ω_{BT}) in terms of the transfer functions S, L and T, but since we are looking for approximate bounds we will not be too concerned with these differences. The main results are summarized at end of the chapter in terms of eight controllability rules.

5.2   Fundamental limitations on sensitivity

In this section, we present some fundamental algebraic and analytic constraints on the sensitivities S and T, including the waterbed effects. Bounds on the peak of $|S|$ and other closed-loop transfer functions are presented in Section 5.3.

5.2.1 S plus T is one

From the definitions $S = (I + L)^{-1}$ and $T = L(I + L)^{-1}$ we derive

$$S + T = I \tag{5.2}$$

(or $S + T = 1$ for a SISO system). Ideally, we want S small to obtain the benefits of feedback (small control error for commands and disturbances), and T small to avoid sensitivity to noise which is one of the disadvantages of feedback. Unfortunately, these requirements are not simultaneously possible at any frequency as is clear from (5.2). Specifically, (5.2) implies that at any frequency either $|S(j\omega)|$ or $|T(j\omega)|$ must be larger than or equal to 0.5, and also that $|S(j\omega)|$ and $|T(j\omega)|$ at any frequency can differ by at most 1.

5.2.2   Interpolation constraints

If p is a RHP-pole of the plant $G(s)$ then

$$\boxed{T(p) = 1, \quad S(p) = 0} \tag{5.3}$$

Similarly, if z is a RHP-zero of $G(s)$ then

$$\boxed{T(z) = 0, \quad S(z) = 1} \tag{5.4}$$

These *interpolation constraints* follow from the requirement of internal stability as shown in (4.86) and (4.87). The conditions clearly restrict the allowable S and T and prove very useful in Section 5.3.

   We can also formulate interpolation constraints resulting from the loop transfer function $L(s) = G(s)K(s)$. The fundamental constraints imposed by the RHP-poles and zeros of $G(s)$ will still be present, whereas the new constraints, identical to (5.3) and (5.4), arising from the RHP-poles and zeros of $K(s)$, are to some extent under our control and therefore not fundamental.

5.2.3   The waterbed effects (sensitivity integrals)

A typical sensitivity function is shown by the solid line in Figure 5.1. We note that $|S|$ has a peak value greater than 1; we will show that this peak is unavoidable in practice. Two formulae are given, in the form of theorems, which essentially say that if we push the sensitivity down at some frequencies then it will have to increase at others. The effect is similar to sitting on a waterbed: pushing it down at one point, which reduces the water level locally, will result in an increased level somewhere else on the bed. In general, a trade-off between sensitivity reduction and sensitivity increase must be performed whenever:

1. $L(s)$ has at least two more poles than zeros (first waterbed formula), or
2. $L(s)$ has a RHP-zero (second waterbed formula).

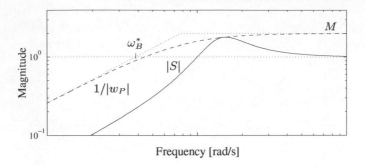

Figure 5.1: Plot of typical sensitivity, $|S|$, with upper bound $1/|w_P|$

Pole excess of two: first waterbed formula

To motivate the first waterbed formula consider the open-loop transfer function $L(s) = \frac{1}{s(s+1)}$. As shown in Figure 5.2, there exists a frequency range over which the Nyquist plot of $L(j\omega)$ is inside the unit circle centred on the point -1, such that $|1+L|$, which is the distance between L and -1, is less than 1, and thus $|S| = |1+L|^{-1}$ is greater than 1. In practice, $L(s)$ will have *at least* two more poles than zeros (at least at sufficiently high frequency, e.g. due to actuator and measurement dynamics), so there will always exist a frequency range over which $|S|$ is greater than 1. This behaviour may be quantified by the following theorem, of which the stable case is a classical result due to Bode.

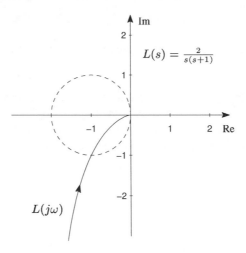

Figure 5.2: $|S| > 1$ whenever the Nyquist plot of L is inside the circle

Theorem 5.1 Bode sensitivity integral (first waterbed formula). *Suppose that the open-loop transfer function $L(s)$ is rational and has at least two more poles than zeros (relative degree of two or more). Suppose also that $L(s)$ has N_p RHP-poles at locations p_i. Then for*

closed-loop stability the sensitivity function must satisfy

$$\int_0^\infty \ln|S(j\omega)|dw = \pi \cdot \sum_{i=1}^{N_p} Re(p_i) \tag{5.5}$$

where $Re(p_i)$ denotes the real part of p_i.

Proof: See Doyle et al. (1992, p. 100) or Zhou et al. (1996). The generalization of Bode's criterion to unstable plants is due to Freudenberg and Looze (1985; 1988). □

For a graphical interpretation of (5.5) note that the magnitude scale is logarithmic whereas the frequency scale is *linear*.

Stable plant. For a stable plant (5.5) gives

$$\int_0^\infty \ln|S(j\omega)|dw = 0 \tag{5.6}$$

and the area of sensitivity reduction ($\ln|S|$ negative) must *equal* the area of sensitivity increase ($\ln|S|$ positive). In this respect, the benefits and costs of feedback are balanced exactly, as in the waterbed analogy. From this we expect that an increase in the bandwidth (S smaller than 1 over a larger frequency range) must come at the expense of a larger peak in $|S|$.

Remark. Although this is true in most practical cases, the effect may not be so striking in some cases, and it is not strictly implied by (5.5) anyway. This is because the increase in area may come over a large frequency range; imagine a very large waterbed. Consider $|S(j\omega)| = 1 + \delta$ for $\omega \in [\omega_1, \omega_2]$, where δ is arbitrarily small (small peak), then we can choose ω_1 arbitrary large (high bandwidth) simply by selecting the interval $[\omega_1, \omega_2]$ to be sufficiently large. However, in practice the frequency response of L has to roll off at frequencies above the bandwidth frequency ω_c and it is required that (Stein, 2003)

$$\int_0^{\omega_c} \ln|S(j\omega)|dw = 0 \tag{5.7}$$

Thus, (5.5) and (5.6) impose real design limitations. This is illustrated in Figure 5.5.

Unstable plant. The presence of unstable poles usually increases the peak of the sensitivity, as seen from the positive contribution $\pi \cdot \sum_{i=1}^{N_p} Re(p_i)$ in (5.5). Specifically, the area of sensitivity increase ($|S| > 1$) *exceeds* that of sensitivity reduction by an amount proportional to the sum of the distance from the unstable poles to the LHP. This is plausible since we might expect to have to pay a price for stabilizing the system.

RHP-zeros: second waterbed formula

For plants with RHP-zeros the sensitivity function must satisfy an additional integral relationship, which has stronger implications for the peak of S. Before stating the result, let us illustrate why the presence of a RHP-zero implies that the peak of S must exceed 1. First, consider the non-minimum-phase loop transfer function $L(s) = \frac{1}{1+s}\frac{1-s}{1+s}$ and its minimum-phase counterpart $L_m(s) = \frac{1}{1+s}$. From Figure 5.3 we see that the additional phase lag contributed by the RHP-zero and the extra pole causes the Nyquist plot to penetrate the unit circle and hence causes the sensitivity function to be larger than 1.

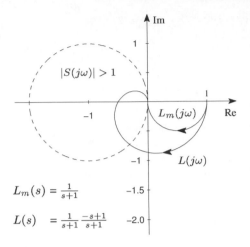

Figure 5.3: Additional phase lag contributed by RHP-zero causes $|S| > 1$

As a further example, consider Figure 5.4 which shows the magnitude of the sensitivity function for the following loop transfer function:

$$L(s) = \frac{k}{s}\frac{2-s}{2+s} \quad k = 0.1, 0.5, 1.0, 2.0 \tag{5.8}$$

The plant has a RHP-zero at $z = 2$, and we see that an increase in the controller gain k, corresponding to a higher bandwidth, results in a larger peak for S. For $k = 2$ the closed-loop system becomes unstable with a pair of complex conjugate poles on the imaginary axis, and the peak of S is infinite.

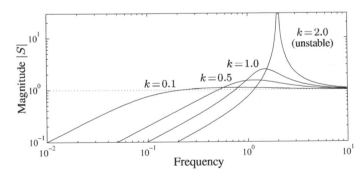

Figure 5.4: Effect of increased controller gain on $|S|$ for system with RHP-zero at $z = 2$, $L(s) = \frac{k}{s}\frac{2-s}{2+s}$

Theorem 5.2 Weighted sensitivity integral (second waterbed formula). *Suppose that $L(s)$ has a single real RHP-zero z or a complex conjugate pair of zeros $z = x \pm jy$, and has N_p RHP-poles, p_i. Let $\bar{p}_i$ denote the complex conjugate of p_i. Then for closed-loop stability* *the sensitivity function must satisfy*

$$\int_0^\infty \ln|S(j\omega)| \cdot w(z,\omega) d\omega = \pi \cdot \ln \prod_{i=1}^{N_p} \left| \frac{p_i + z}{\bar{p}_i - z} \right| \tag{5.9}$$

where if the zero is real

$$w(z,\omega) = \frac{2z}{z^2 + \omega^2} = \frac{2}{z} \frac{1}{1 + (\omega/z)^2} \tag{5.10}$$

and if the zero pair is complex $(z = x \pm jy)$

$$w(z,\omega) = \frac{x}{x^2 + (y - \omega)^2} + \frac{x}{x^2 + (y + \omega)^2} \tag{5.11}$$

Proof: See Freudenberg and Looze (1985; 1988).   □

Note that when there is a RHP-pole close to the RHP-zero $(p_i \to z)$ then $\frac{p_i+z}{p_i-z} \to \infty$. This is not surprising as such plants are in practice impossible to stabilize.

The weight $w(z,\omega)$ effectively "cuts off" the contribution from $\ln|S|$ to the sensitivity integral at frequencies $\omega > z$. Thus, for a stable plant where $|S|$ is reasonably close to 1 at high frequencies we have approximately

$$\int_0^z \ln|S(j\omega)| d\omega \approx 0 \tag{5.12}$$

This is similar to Bode's sensitivity integral relationship in (5.6), except that the trade-off between S less than 1 and S larger than 1 is done over a limited frequency range. Thus, in this case the waterbed is finite, and a large peak for $|S|$ is unavoidable if we try to push down $|S|$ at low frequencies. This is illustrated by the example in Figure 5.4 and further by the example in Figure 5.5. In Figure 5.5 we plot $\ln|S|$ as a function of ω (note the linear frequency scale) for two cases. In both cases, the areas of $\ln S$ below and above $10^0 = 1$ (dotted line) are equal, see (5.6), but for case 2 this must happen at frequencies below the RHP-zero at $z = 5$, see (5.12), and to achieve this the peak of $|S_2|$ must be higher.

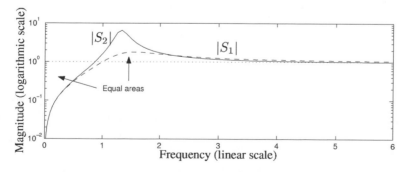

Figure 5.5: Sensitivity $S = \frac{1}{1+L}$ corresponding to $L_1 = \frac{2}{s(s+1)}$ (dashed line) and $L_2 = L_1 \frac{-s+5}{s+5}$ (with RHP-zero at $z = 5$) (solid line)

Exercise 5.1 * **Kalman inequality.** *The Kalman inequality for optimal state feedback, which also applies to unstable plants, says that $|S| \leq 1 \ \forall \omega$, see Example 9.2. Explain why this does not conflict with the above sensitivity integrals. (Solution: 1. Optimal control with state feedback yields a loop transfer function with a pole–zero excess of 1 so (5.5) does not apply. 2. There are no RHP-zeros when all states are measured so (5.9) does not apply.)*

5.3   Fundamental limitations: bounds on peaks

In Theorem 5.2, we found that a RHP-zero implies that a peak in $|S|$ is inevitable, and that the peak will increase if we reduce $|S|$ at other frequencies. Here we derive explicit bounds on the peaks of the important closed-loop transfer functions, which are more useful in applications than the integral relationships in Theorem 5.2. By the "peak" we mean the maximum value of the frequency response or $\mathcal{H}_\infty$ norm:

$$\|f(s)\|_\infty = \max_\omega |f(j\omega)|$$

We first consider bounds on the weighted sensitivity ($w_P S$) and the weighted complementary sensitivity ($w_T T$). The weights w_P and w_T are useful if we want to specify that $|S|$ and $|T|$ should be small in some selected frequency region.

5.3.1   Minimum peaks for S and T

Theorem 5.3 Sensitivity peak. *For closed-loop stability the sensitivity function must satisfy for each RHP-zero z of $G(s)$*

$$\|w_P S\|_\infty \geq |w_P(z)| \cdot \underbrace{\prod_{i=1}^{N_p} \frac{|z + p_i|}{|z - p_i|}}_{M_{zp_i}} \tag{5.13}$$

where p_i denote the N_p RHP-poles of $G(s)$. If $G(s)$ has no RHP-poles the bound simplifies to

$$\|w_P S\|_\infty \geq |w_P(z)| \tag{5.14}$$

Without a weight the bound (5.13) simplifies to

$$\|S\|_\infty = M_S \geq \underbrace{\prod_{i=1}^{N_p} \frac{|z + p_i|}{|z - p_i|}}_{M_{zp_i}} \tag{5.15}$$

The bounds (5.13), (5.14) and (5.15) are tight for the case with a single real RHP-zero z and no time delay. Here "tight" means that there exists a controller (possibly improper) that achieves the bound (with equality). For example, with a single RHP-zero and no time delay, $\min_K \|S\|_\infty = M_{S,\min} = M_{zp_i}$.

We note that the bound (5.15) approaches infinity, as the distance $|z - p_i|$ approaches zero. A time delay imposes additional problems for stabilization, but there does not exist a *tight* lower bound for S in terms of the time delay. However, similar bounds apply for the complementary sensitivity T and here the time delay also enters the tight bound.

Theorem 5.4 Complementary sensitivity peak. *The complementary sensitivity function must satisfy for each RHP-pole p of $G(s)$*

$$\|w_T T\|_\infty \geq |w_T(p)| \cdot \underbrace{\prod_{j=1}^{N_z} \frac{|z_j + p|}{|z_j - p|}}_{M_{pz_j}} \cdot |e^{p\theta}| \tag{5.16}$$

where z_j denote the N_z RHP-zeros of $G(s)$ and θ denotes the time delay of $G(s)$. If $G(s)$ has no RHP-zeros and time delay, the bound simplifies to

$$\|w_T T\|_\infty \geq |w_T(p)| \tag{5.17}$$

Without a weight the bound (5.16) simplifies to

$$\|T\|_\infty = M_T \geq \underbrace{\prod_{j=1}^{N_z} \frac{|z_j + p|}{|z_j - p|}}_{M_{pz_j}} \cdot |e^{p\theta}| \tag{5.18}$$

The bounds (5.16), (5.17) and (5.18) are tight for the case with a single real RHP-pole p. For example, with a single RHP-pole, $\min_K \|T\|_\infty = M_{T,\min} = M_{pz_j} \cdot |e^{p\theta}|$.

Note that (5.18) also imposes a bound on the peak of S for plants with a time delay. From (5.2), $|S|$ and $|T|$ differ by at most 1, so

$$\|S\|_\infty \geq \|T\|_\infty - 1 \tag{5.19}$$

and a peak in $|T|$ also implies a peak in $|S|$. Example 5.1 on page 175 further illustrates this point.

Proof of (5.13): The bounds for S were originally derived by Zames (1981). The results can be derived using the interpolation constraints $S(z) = 1$ and $T(p) = 1$ given above. In addition, we make use of the maximum modulus principle for complex analytic functions (e.g. see maximum principle in Churchill et al., 1974), which for our purposes can be stated as follows:

Maximum modulus principle. *Suppose $f(s)$ is stable (i.e. $f(s)$ is analytic in the complex RHP[1]). Then the maximum value of $|f(s)|$ for s in the RHP is attained on the region's boundary, i.e. somewhere along the $j\omega$-axis. Hence, we have for a stable $f(s)$*

$$\|f(j\omega)\|_\infty = \max_\omega |f(j\omega)| \geq |f(s_0)| \quad \forall s_0 \in \text{RHP} \tag{5.20}$$

Remark. Expression (5.20) can be understood by imagining a 3-D plot of $|f(s)|$ as a function of the complex variable s. In such a plot $|f(s)|$ has "peaks" at its poles and "valleys" at its zeros. Thus, if $f(s)$ has no poles (peaks) in the RHP, we find that $|f(s)|$ slopes downwards from the LHP and into the RHP.

---

[1] A function $f(s)$ of the complex variable s is *analytic* at a point s_0 if its derivative exists not only at s_0 but at each point s in some neighbourhood around s_0. If the derivative does not exist at s_0 but does so in some neighbourhood of s_0, then s_0 is called a *singular point*. We are considering a rational transfer function $f(s)$, which is analytic except at its poles ($s_0 = p$). The poles are singular points.

For a plant with a RHP-zero z, applying (5.20) to $f(s) = w_P(s)S(s)$ and using the interpolation constraint $S(z) = 1$ gives $\|w_P S\|_\infty \geq |w_P(z)S(z)| = |w_P(z)|$. To derive the additional penalty if the plant also has RHP-poles, we use a a "trick" where we first factor out the RHP-zeros in S into an all-pass part S_a (with magnitude 1 at all points on the $j\omega$-axis). Since G has RHP-poles at p_i, $S(s)$ has RHP-zeros at p_i (see (5.3)) and we may write

$$S = S_a S_m, \quad S_a(s) = \prod_i \frac{s - p_i}{s + \bar{p}_i} \tag{5.21}$$

Here S_m is the "minimum-phase version" of S with all RHP-zeros mirrored into the LHP. $S_a(s)$ is all-pass with $|S_a(j\omega)| = 1$ at all frequencies. (Remark: There is a technical problem here with $j\omega$-axis poles: these must first be moved slightly into the RHP.) The weight $w_P(s)$ is as usual assumed to be stable and minimum-phase. Consider a RHP-zero located at z, for which we get from the maximum modulus principle

$$\|w_P S\|_\infty = \max_\omega |w_P S(j\omega)| = \max_\omega |w_P S_m(j\omega)| \geq |w_P(z)S_m(z)|$$

where $S_m(z) = S(z)S_a(z)^{-1} = 1 \cdot S_a(z)^{-1}$. This proves (5.13). Chen (1995) and Chen (2000, p. 1107) provide an alternative proof of the bound, based on the integral relationship (5.9). The tightness of the bound was first proved by Havre and Skogestad (1998). An alternative proof is given by Chen (2000, p. 1109). $\quad\square$

Proof of (5.16): The proof of (5.16) is similar to the proof of (5.13). We write $T = T_a T_m$, where T_a contains the RHP-zeros z_j and the time delay; see also Theorem 5.5. $\quad\square$

From the bounds in Theorem 5.3 and 5.4, we note that

- S is primarily limited by RHP-zeros. The bound $|w_P S| \geq |w_P(z)|$ shows that we cannot freely specify the shape of $|S|$ for a plant with a RHP-zero z.
- T is primarily limited by RHP-poles. The bound $|w_T T| \geq |w_T(p)|$ shows that we cannot freely specify the shape of $|T|$ for a plant with a RHP-pole p.
- The terms M_{zp_i} and M_{pz_j} show that the limitations are more serious if we have both RHP-poles and RHP-zeros. Large peaks for S and T are unavoidable if we have a RHP-zero and RHP-pole located close to each other.

Remark 1 Let $M_{S,\min}$ and $M_{T,\min}$ denote the lowest achievable values for $\|S\|_\infty$ and $\|T\|_\infty$ respectively; that is, $\min_K \|S\|_\infty \triangleq M_{S,\min}$ and $\min_K \|T\|_\infty \triangleq M_{T,\min}$. Chen (2000) shows that the bound (5.15) is also tight for $\|T\|_\infty$ and the bound (5.18) (for the case with no time delay) is also tight for $\|S\|_\infty$. Then, for a plant with a single RHP-zero z (and no time delay) and any number of RHP-poles we have the following tight lower bound on $\|S\|_\infty$ and $\|T\|_\infty$:

$$M_{S,\min} = M_{T,\min} = \underbrace{\prod_{i=1}^{N_p} \frac{|z + p_i|}{|z - p_i|}}_{M_{zp_i}} \tag{5.22}$$

and for the case with a single RHP-pole p and any number of RHP-zeros (but no time delay) we have the following tight lower bound on $\|S\|_\infty$ and $\|T\|_\infty$:

$$M_{S,\min} = M_{T,\min} = \underbrace{\prod_{j=1}^{N_z} \frac{|z_j + p|}{|z_j - p|}}_{M_{pz_j}} \tag{5.23}$$

These tight bounds are further generalized in (6.8) (page 224) to any number of RHP-poles and RHP-zeros (including complex poles and zeros).

Remark 2 (5.22) and (5.23) provide non-tight lower bounds for plants with multiple RHP-zeros and multiple RHP-poles, respectively. For example, with multiple RHP-zeros, $\|S\|_\infty \geq \max_{z_j} M_{z_j p_i}$ and with multiple RHP-poles, $\|T\|_\infty \geq \max_{p_i} M_{p_i z_j} \cdot |e^{p_i \theta}|$. However, these bounds are generally not tight.

Remark 3 These bounds may be generalized to MIMO systems if the directions of poles and zeros are taken into account, see Chapter 6.

Example 5.1 **Unstable plant with time delay.** *The plant*

$$G(s) = \frac{e^{-0.5s}}{s - 3}$$

has $p = 3$ and $\theta = 0.5$. Since $p\theta = 1.5$ is larger than 1, the peak of $|T|$ will be large, and we will have difficulty in stabilizing the plant. Specifically, from (5.18) it follows that for any controller we must have

$$\|T\|_\infty \geq M_{T,\min} = e^{0.5 \cdot 3} = e^{1.5} = 4.48$$

This bound is tight in the sense that there exists a controller that achieves it. The peak of the sensitivity S must also be large since

$$\|S\|_\infty \geq M_{S,\min} \geq M_{T,\min} - 1 = 4.48 - 1 = 3.38$$

This bound is not tight, so the actual value of $M_{S,\min}$ may be higher than 3.38, but not higher than 5.38, since the peaks of $|S|$ and $|T|$ differ by at most 1. The unavoidable large values for $\|S\|_\infty$ and $\|T\|_\infty$ for this process imply poor performance and robustness problems.

Example 5.2 **Plant with complex RHP poles.** *The plant*

$$G(s) = 10 \cdot \frac{s - 2}{s^2 - 2s + 5} \tag{5.24}$$

has a RHP-zero at $z = 2$ and RHP-poles at $p = 1 \pm j2$. From (5.22), a tight lower bound on $\|S\|_\infty$ and $\|T\|_\infty$ is

$$M_{zp_i} = \frac{(2+1)^2 + 2^2}{(2-1)^2 + 2^2} = 2.6$$

We can also use (5.23), where $M_{pz_j} = \sqrt{2.6} = 1.61$, but this does not give a tight bound since we have two RHP poles.

The effect of combined RHP-poles and RHP-zeros is further illustrated by examples on page 179.

Stabilization. The results, e.g. (5.23), show that large peaks on S and T are unavoidable if we have a RHP-pole p located close to a RHP-zero z, such that $|z - p|$ is small. In practice, such a plant is impossible to stabilize. However, in theory, any linear plant may be stabilized irrespective of the location of its RHP-poles and RHP-zeros, provided the plant does not contain unstable hidden modes (e.g. corresponding to the situation $p = z$); see also page 150.

5.3.2 Minimum peaks for other closed-loop transfer functions

In this section, we provide bounds on peaks for some other closed-loop transfer functions. To motivate, recall from (2.19) and (2.20) that the closed-loop control error $e = y - r$ and the plant input u for the system in Figure 2.4 (page 21) are

$$e = -Sr - Tn + \sum_k SG_{d_k} d_k \tag{5.25}$$

$$u = KSr - KSn + \sum_k KSG_{d_k} d_k \tag{5.26}$$

Here we have considered the case with several disturbances d_k. One disturbance of particular interest is an input (or "load") disturbance $d_k = d_u$, for which $G_{d_k} = G$.

Ideally, we want both e and u small, so it is desirable that all the closed-loop transfer functions in (5.25) and (5.26) be small. This is generally not possible, because of algebraic constraints of the kind $S + T = I$. Nevertheless, we would like to avoid large peaks in any of these transfer functions.

In addition, we would like to bound these transfer functions for robustness reasons. In Figure 8.5 (page 293), we show six forms of uncertainty. To maintain robustness with respect to these uncertainties, we see from (8.53)–(8.58) that the following six transfer functions should be small:

$$KS, \ T_I = KSG, \ T = GKS, \ SG, \ S_I = (I + KG)^{-1}, \ S$$

Notice that for SISO systems, S_I is equal to S and $T = GKS$ is equal to $T_I = KSG$. We have already considered S and T, and will now derive bounds for the remaining transfer functions SG, SG_d, KS and KSG_d.

The results are summarized in Table 5.1, which also gives the performance and robustness reasons behind minimizing the peaks of each closed-loop transfer function. Note that the factors M_{zp_i} and M_{pz_j} express the additional penalty of *combined* RHP-zeros and RHP-poles.

Bounds on SG. The transfer function SG is required to be small to reduce the effect of the input disturbances on the control error signal (see (5.25)) and for robustness against pole uncertainty (see Figure 8.5(d) on page 293). Due to the interpolation constraints, any controller stabilizing S also stabilizes SG. Further, $\|SG\|_\infty = \|SG_{ms}\|_\infty$, where G_{ms}, the "minimum-phase, stable version" of G, is

$$G_{ms} \triangleq \underbrace{\prod_i \frac{s-p_i}{s+p_i} \cdot \overbrace{G(s) \cdot \prod_j \frac{s+z_j}{s-z_j}}^{\triangleq\, G_m(s)}}_{\triangleq\, G_s(s)} \tag{5.27}$$

Theorem 5.3 can be used to calculate the peak value for SG by treating G_{ms} as a weight. Specifically, for every RHP-zero of the system, $\|SG\|_\infty$ must satisfy

$$\|SG\|_\infty \geq |G_{ms}(z)| \cdot \underbrace{\prod_{i=1}^{N_p} \frac{|z+p_i|}{|z-p_i|}}_{M_{zp_i}} = |G_m(z)| \tag{5.28}$$

where G_m is the "minimum-phase version" of G; see (5.27). This bound is tight for plants with a single RHP-zero.

Table 5.1: Bounds on peak of important closed-loop transfer functions

	M	Want $\|M\|_\infty$ small for		Bound on $\|M\|_\infty$	
		Signals (see page 22)	Stability robus'ness (see page 3C3)	Special case (tight only for $N_z = 1$ and/or $N_p = 1$)	General case (including MIMO)
1.	S	Performance tracking ($e = -Sr$)	Relative inverse (pole) uncertainty (Δ_{io})	M_{zp_i} (5.15) or M_{pz_j} (5.23)	(6.8)
2.	T	Performance noise ($e = -Tn$)	Relative additive (zero) uncertainty (Δ_o)	M_{zr_i} (5.22) or $M_{pz_j} \cdot \lvert e^{p\theta} \rvert$ (5.18)	(6.8) and (6.16) for delay system
3.	KS	Input usage ($u = KS(r - n)$)	Additive (zero) uncertainty (Δ_A)	$\lvert G_s^{-1}(p) \rvert = \lvert G_{ms}^{-1}(p) \rvert \cdot M_{pz_j} \cdot \lvert e^{p\theta} \rvert$ (5.31)	$1/\underline{\sigma}_H (U(G)^*)$ (5.30)
4.*	SG_d	Performance disturbance ($e = SG_d d$)	$G_d = G$: Inverse (pole) uncertainty (Δ_{iA})	$\lvert G_{d,ms}(z) \rvert \cdot M_{zs_j}$ (5.29) $= \lvert G_m(z) \rvert$ for $G_d = G$ (5.28)	(6.12) with $W_1 = I$ and $W_2 = G_{d,ms}$
5.*	KSG_d	Input usage disturbance ($u = KSG_d d$)	$G_d = G$: Relative additive (zero) uncertainty (Δ_I)	$\lvert G_s(p)^{-1} G_{d,ms}(p) \rvert$ $= \lvert G_{ms}^{-1}(p) G_{d,ms}(p) \rvert \cdot M_{pz_j} \cdot \lvert e^{p\theta} \rvert$ (5.34)	$1/\underline{\sigma}_H (U(G_{d,ms}^{-1} G)^*)$ (5.33)

$$M_{zp_i} = \prod_{i=1}^{N_p} \frac{\lvert z + p_i \rvert}{\lvert z - p_i \rvert}, \quad M_{pz_j} = \prod_{j=1}^{N_z} \frac{\lvert z_j + p \rvert}{\lvert z_j - p \rvert}$$

* Special case: Input disturbance ($G_d = G$)

Bounds on SG_d. In the general disturbance case, $G_d \neq G$ and we want to keep $\|SG_d\|_\infty$ small to reduce the effect of disturbances on the outputs. This case can be handled similar to SG by replacing G_{ms} by $G_{d,ms}$ in (5.28) to get

$$\|SG_d\|_\infty \geq |G_{d,ms}(z)| \cdot \underbrace{\prod_{i=1}^{N_p} \frac{|z + p_i|}{|z - p_i|}}_{M_{zp_i}} \qquad (5.29)$$

Bounds on KS. The peak on the transfer function KS is required to be small to avoid large input signals in response to noise and disturbances; see (5.26). In particular, this is important for an unstable plant, where a large value of $\|KS\|_\infty$ is likely to cause saturation in u resulting in difficulties in stabilization.

Let $\underline{\sigma}_H$ denote the smallest Hankel singular value and $\mathcal{U}(G)^*$ be the mirror image of the anti-stable part of G. Glover (1986), who considered robustness against additive uncertainty, proved that

$$\|KS\|_\infty \geq 1/\underline{\sigma}_H(\mathcal{U}(G)^*) \qquad (5.30)$$

The bound (5.30) is tight, in the sense that there always exists a controller (possibly improper) that achieves the bound. For a stable plant there is no lower bound, as in this case, $\min_K \|KS\|_\infty = 0$, which can be achieved by $K = 0$.

A simpler bound is also available, since for any RHP-pole p, $\underline{\sigma}_H(\mathcal{U}(G)^*) \leq |G_s(p)|$, where $G_s(s)$ is the "stable version" of G with its RHP-poles mirrored into the LHP; see (5.27). Equality applies for a plant with a single real RHP-pole p. This gives the bound (Havre and Skogestad, 2001)

$$\|KS\|_\infty \geq |G_s^{-1}(p)| \qquad (5.31)$$

which is tight for plants with a single real RHP-pole p. This bound also applies to plants with time delay.

Proof of (5.31): We first prove the following generalized bound (Havre and Skogestad, 2001):

Theorem 5.5 *Let VT be a (weighted) closed-loop transfer function, where T is the complementary sensitivity function. Then for closed-loop stability we must require for each RHP-pole p in G,*

$$\|VT\|_\infty \geq |V_{ms}(p)| \cdot \prod_{j=1}^{N_z} \frac{|z_j + p|}{|z_j - p|} \cdot |e^{p\theta}| \qquad (5.32)$$

where V_{ms} is the "minimum-phase and stable version" of V (with its RHP-poles and RHP-zeros mirrored into the LHP), and z_j denote the N_z RHP-zeros of G. If G has no RHP-zeros the bound is simply $\|VT\|_\infty \geq |V_{ms}(p)|$. The bound (5.32) is tight (equality) for the case when G has only one RHP-pole.

Proof: G has RHP-zeros at z_j, and therefore T must have RHP-zeros at z_j, so we write $T = T_a T_m$ with $T_a(s) = \prod_j \frac{s - z_j}{s + \bar{z}_j}$. Next, note that $\|VT\|_\infty = \|V_{ms}T_{ms}\|_\infty = \|V_{ms}T_m\|_\infty$. Now, consider a RHP-pole located at p, and use the maximum modulus principle to show that $\|VT\|_\infty \geq |V_{ms}(p)T_m(p)| = |V_{ms}(p)T(p)T_a(p)^{-1}| = |V_{ms}(p) \cdot 1 \cdot \prod_j \frac{p + \bar{z}_j}{p - z_j}|$ which proves (5.32). To prove (5.31) we make use of the identity $KS = G^{-1}GKS = G^{-1}T$. Use of (5.32) with $V = G^{-1}$ then gives

$$\|KS\|_\infty \geq |G_{ms}(p)^{-1}| \cdot \underbrace{\prod_j \frac{|z_j + p|}{|z_j - p|}}_{M_{pz_j}} = |G_s(p)^{-1}|$$

which proves (5.31). □

Example 5.3 *For the unstable plant $G(s) = \frac{1}{s-3}$ we have $G_s(s) = \frac{1}{s+3}$ and from (5.31) $\|KS\|_\infty \geq |G_s(p)^{-1}| = 6$. That is, irrespective of the controller, the closed-loop transfer function KS, from plant output (e.g. measurement noise) to plant input, must exceed 6 in magnitude at some frequency.*

Exercise 5.2 *For a system with a single unstable pole p, show that the two bounds on $\|KS\|_\infty$, (5.30) and (5.31), are equivalent. (Hint: Use (4.140) to find the minimum (and only) Hankel singular value of $\mathcal{U}(G)^* = (G(s) \cdot (s - p))|_{s=p}/(s - p).$)*

Bounds on KSG_d. For arbitrary disturbances, the bound (5.30) can be generalized as (Kariwala, 2004)

$$\|KSG_d\|_\infty \geq 1/\underline{\sigma}_H(\mathcal{U}(G_{d,ms}^{-1}G)^*) \tag{5.33}$$

where $\mathcal{U}(G_{d,ms}^{-1}G)^*$ is the mirror image of the anti-stable part of $G_{d,ms}^{-1}G$. Note that any unstable modes in G_d must be contained in G such that they are stabilizable with feedback control. Under the same condition, the bound (5.31) may be generalized using (5.32) to get (Havre and Skogestad, 2001)

$$\|KSG_d\|_\infty \geq |G_{ms}(p)^{-1}G_{d,ms}(p)| \cdot M_{pz_j} \cdot |e^{p\theta}| = |G_s(p)^{-1}G_{d,ms}(p)| \tag{5.34}$$

Here $G_{d,ms}$ denotes the "stable and minimum-phase version" of G_d with both the RHP-poles and RHP-zeros mirrored into the LHP. The bound is tight for a single RHP-pole p. The bounds (5.30) and (5.33) can also be used for delay systems, since although the delay system itself is irrational, its anti-stable part is rational (Kariwala, 2004).

Example 5.4 *Consider a plant and disturbance model*

$$G(s) = \frac{5}{(s-3)(10s+1)}, \quad G_d = \frac{0.5}{(s-3)(0.2s+1)}e^{-1.5s}$$

We have $G_s(s) = \frac{5}{(s+3)(10s+1)}$ and $G_{d,ms} = \frac{0.5}{(s+3)(0.2s+1)}$. Notice that the time delay in G_d drops out in $G_{d,ms}$. With $p = 3$, (5.34) gives the following lower bound on the peak of the transfer function from a disturbance to plant input:

$$\|KSG_d\|_\infty \geq |G_{ms}(p)^{-1}G_{d,ms}(p)| = \frac{6 \cdot 31}{5} \cdot \frac{0.5}{6 \cdot 1.6} = 1.94$$

Example 5.5 *Consider an unstable plant ($p \geq 0$) with a RHP-zero ($z \geq 0$) and a time delay ($\theta \geq 0$), given by*

$$G(s) = \frac{k}{s-p}\frac{(s-z)}{(s+z)}e^{-\theta s} \tag{5.35}$$

We have $|G_s(p)| = |\frac{k}{s+p}\frac{(s-z)}{(s+z)}e^{-\theta s}|_{s=p} = \frac{k}{2p}\frac{|p-z|}{|p+z|}e^{-\theta p}$, and from (5.31) we must have for any stabilizing controller

$$\|KS\|_\infty \geq |G_s(p)^{-1}| = \left|\frac{2p}{k}\right| \cdot \frac{|p+z|}{|p-z|} \cdot |e^{\theta p}| \tag{5.36}$$

Since $u = -KS(G_d d + n)$, we see from the first term that the required input u is large if $|p|$ is large; that is, if the unstable mode is "fast". In addition, we note that the exponential term $e^{\theta p}$ grows sharply for $\theta > 1/p$.

For example, consider the following plant, which we will show is impossible to control in practice:

$$G(s) = \frac{1}{s-3}\frac{s-6}{s+6}e^{-0.5s}$$

First, from (5.36), we have that at some frequency $|KS(j\omega)| \geq \frac{2\cdot3}{1} \cdot \frac{|3+6|}{|3-6|} e^{0.5\cdot3} = 6\cdot3\cdot4.48 = 80.67.$
This gain is large, so the presence of noise or disturbances is likely to saturate the inputs, which again will most likely result in failure to stabilize the plant. In addition, we have from (5.18) that $M_{T,\min} = \frac{|p+z|}{|p-z|} \cdot e^{p\theta} = 3\cdot4.48 = 13.4$, *so* $|T|$ *must exceed 13.4 at some frequency and since* $S + T = 1$, $\|S\|_\infty$ *must exceed* $13.4 - 1 = 12.4$. *This is much larger than the typical maximum allowed value of about 2. The plant is therefore impossible to stabilize and control from a practical point of view.*

When considering controllability of a plant, all the closed-loop transfer functions in Table 5.1 should be considered. For a SISO plant, scaling of the signals y, u and d does not matter for S and T. However, for proper evaluation of KS, SG_d and KSG_d, it is recommended that the plant is scaled as outlined in Section 5.1.2. A peak value of KS, SG_d or KSG_d much larger than 1 will then imply possible control problems.

Exercise 5.3* *Consider again the plant (5.24) from Example 5.2. Compute the bounds on* $\|S\|_\infty$, $\|T\|_\infty$, $\|KS\|_\infty$ *and* $\|SG\|_\infty$ *using Table 5.1. Do you expect any difficulties in controlling this plant?*

Exercise 5.4 *For the plant* $G(s) = \frac{s^2-s+3}{s^2-3s+1}$, *compute the bounds on* $\|S\|_\infty$, $\|T\|_\infty$, $\|KS\|_\infty$ *and* $\|SG\|_\infty$ *using both the "special" and "general" cases in Table 5.1.*

This concludes the two sections on fundamental limitations. In the rest of this chapter, we will use these and other results in order to understand better what limits the (input–output) controllability of a plant. We start first with the simple idea of "perfect control".

5.4 Perfect control and plant inversion

A good way of obtaining insight into the inherent limitations on performance, which originate from the plant itself, is to consider the inputs needed to achieve *perfect control* (Morari, 1983). Let the plant model be

$$y = Gu + G_d d \tag{5.37}$$

"Perfect control" (which, of course, cannot be realized in practice) is achieved when the output is identically equal to the reference, i.e. $y = r$. To find the corresponding plant input, set $y = r$ and solve for u in (5.37):

$$u = G^{-1}r - G^{-1}G_d d \tag{5.38}$$

Equation (5.38) represents a perfect feedforward controller, assuming d is measurable. When feedback control $u = K(r - y)$ is used, we have from (2.21) that

$$u = KSr - KSG_d d$$

or since the complementary sensitivity function is $T = GKS$,

$$u = G^{-1}Tr - G^{-1}TG_d d \tag{5.39}$$

We see that at frequencies where feedback is effective and $T \approx I$ (these arguments also apply to MIMO systems and this is the reason why we here choose to use matrix notation), the input generated by feedback in (5.39) is the same as the perfect control input in (5.38).

That is, high-gain feedback generates an inverse of G even though the controller K may be very simple.

An important lesson therefore is that perfect control requires the controller somehow to generate an inverse of G. From this we get that perfect control *cannot* be achieved if

- G contains RHP-zeros (since then G^{-1} is unstable)
- G contains time delay (since then G^{-1} contains a non-causal prediction)
- G has more poles than zeros (since then G^{-1} is unrealizable)

In addition, for feedforward control we have that perfect control *cannot* be achieved because

- G is always uncertain (so G^{-1} cannot be obtained exactly)

The last restriction may be overcome by high-gain feedback, because then the model inverse is not generated from a model, but from output feedback. However, we know that we cannot have high-gain feedback at all frequencies.

The required input in (5.38) must not exceed the maximum physically allowed value. Therefore, perfect control *cannot* be achieved if

- $|G^{-1}G_d|$ is large
- $|G^{-1}R|$ is large

where "large" with our scaled models means larger than 1. There are also other situations which make control difficult such as

- G is unstable
- $|G_d|$ is large

If the plant is unstable, the outputs will "take off", and eventually hit physical constraints, unless feedback control is applied to stabilize the system. Similarly, if $|G_d|$ is large, then without control a disturbance will cause the outputs to move far away from their desired values. So in both cases control is required, and problems occur if this demand for control is somehow in conflict with the other factors mentioned above which also make control difficult. We have assumed perfect measurements in the discussion so far, but in practice, noise and uncertainty associated with the measurements of disturbances and outputs will present additional problems for feedforward and feedback control, respectively.

5.5   Ideal ISE optimal control

Another good way of obtaining insight into performance limitations is to consider an "ideal" controller which is integral square error (ISE) optimal. That is, for a given command $r(t)$ (which is zero for $t < 0$), the "ideal" controller is the one that generates the plant input $u(t)$ (zero for $t < 0$) which minimizes

$$\text{ISE} = \int_0^\infty |y(t) - r(t)|^2 dt \qquad (5.40)$$

This controller is "ideal" in the sense that it may not be realizable in practice because the cost function includes no penalty on the input $u(t)$. This particular problem is considered in detail by Frank (1968a; 1968b) and Morari and Zafiriou (1989), and also Qiu and Davison (1993) who study "cheap" linear quadratic regulator (LQR) control. Morari and Zafiriou show that for stable plants with RHP-zeros at z_j (real and/or complex) and a time delay θ, the "ideal" response $y = Tr$ when $r(t)$ is a *unit step* is given by

$$T(s) = \prod_j \frac{-s + z_j}{s + \bar{z}_j} e^{-\theta s} \tag{5.41}$$

and the corresponding optimal ISE value is (Goodwin et al., 2003)

$$\text{ISE}_{\min} = \min \int_0^\infty |y(t) - 1|^2 dt = \theta + 2 \sum_j \frac{1}{z_j} \tag{5.42}$$

The optimal ISE values for three simple stable plants are then:

1. with a delay θ : $\text{ISE}_{\min} = \theta$
2. with a RHP-zero z : $\text{ISE}_{\min} = 2/z$
3. with complex RHP-zeros $z = x \pm jy$: $\text{ISE}_{\min} = 4x/(x^2 + y^2)$

We see that the worst case is to have a RHP-zero at the origin ($z_j = 0$). This is reasonable because the steady-state gain is then zero, so it will not be possible to keep $y(t)$ at a steady-state value of 1 as $t \to \infty$ and ISE = ∞.

However, note that these ISE values are for step changes in the reference which emphasize the low-frequency behaviour. Alternatively, consider the tracking of a sinusoidal reference, $r(t) = \sin(\omega t)$. In this case, we get for a plant with RHP-zeros at z_j (Qiu and Davison, 1993)

$$\text{ISE}_{\min} = 2 \sum_j \left(\frac{1}{z_j - j\omega} + \frac{1}{z_j + j\omega} \right) \tag{5.43}$$

As expected, $\text{ISE}_{min} = \infty$ for a purely complex zero located at the frequency ω, $z_j = \pm jw$, because then $G(jw) = 0$. For a real RHP-zero z_j, the maximum (worst) value of $\text{ISE}_{\min}$ is achieved when $\omega = z_j$, and $\text{ISE}_{\min} = 0$ when $z_j = 0$ (zero located at the origin) or $z_j = \infty$ (zero located far out in the RHP). In summary, we find that a RHP-zero z_j mainly limits the performance around the frequency $|z_j|$, This interpretation is confirmed below when we consider the achievable bandwidth.

5.6   Limitations imposed by time delays

A time delay ($e^{-\theta s}$) imposes a serious limitation on achievable control performance. This is easy to understand, since no matter what controller we use, the effect of an input change on the output will be delayed by the time θ, see (5.42). In this section we consider the bandwidth implications.

The closed-loop bandwidth is limited to be less than $1/\theta$, approximately. To see this more clearly, consider the "ideal" $T(s) = e^{-\theta s}$ for the case of step changes in the reference, see (5.41). The corresponding "ideal" sensitivity function is

$$S = 1 - T = 1 - e^{-\theta s} \tag{5.44}$$

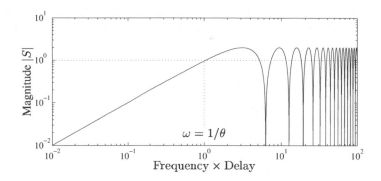

Figure 5.6: "Ideal" sensitivity function (5.44) for a plant with delay

The magnitude $|S|$ is plotted in Figure 5.6. At low frequencies, $\omega\theta < 1$, we have $1 - e^{-\theta s} \approx \theta s$ (by a Taylor series expansion of the exponential) and the low-frequency asymptote of $|S(j\omega)|$ crosses 1 at a frequency of about $1/\theta$ (the exact frequency where $|S(j\omega)|$ crosses 1 in Figure 5.6 is $\frac{\pi}{3}\frac{1}{\theta} = 1.05/\theta$). Since for $S = 1 - e^{-\theta s}$, we have $|S| = 1/|L|$, we also have that $1/\theta$ is equal to the gain crossover frequency for L. The "ideal" ISE optimal controller bounds the practically realizable controllers, so we expect this value to provide an approximate upper bound on w_c, namely (for a process with a time delay and performance requirements at low frequency)

$$\omega_c < 1/\theta \tag{5.45}$$

This approximate bound is the same as derived in Section 2.6.2 by considering the limitations imposed on a loop-shaping design by a time delay θ. In addition to this bandwidth limitation, we also have the limitations on the peak of the closed-loop transfer functions given in Table 5.1.

5.7 Limitations imposed by RHP-zeros

We will here consider plants with a zero z in the closed RHP (and no pure time delay). RHP-zeros typically appear when we have competing effects of slow and fast dynamics. For example, the plant

$$G(s) = \frac{1}{s+1} - \frac{2}{s+10} = \frac{-s+8}{(s+1)(s+10)}$$

has a real RHP-zero at $z = 8$. We may also have complex zeros, and since these always occur in complex conjugate pairs we have $z = x \pm jy$ where $x \geq 0$ for RHP-zeros.

The question here is: what control problems can be expected for a plant with a RHP-zero z? A good starting point for such a discussion is the fundamental constraint on the sensitivity function for internal instability:

$$S(z) = 1$$

We note immediately that this is not compatible with the desire to have $|S|$ small (compared to 1) in order to have tight control (good output performance). We therefore expect that the presence of a RHP-zero poses fundamental limitations in terms of the achievable output performance. In the following we attempt to build up insight into the performance limitations imposed by RHP-zeros using a number of different results in both the time and frequency domains.

5.7.1 Time response: inverse response and undershoot

RHP-zeros imply inverse response behaviour in the time domain. For a stable SISO plant with n_z real RHP-zeros, it may be proven (Leon de la Barra S., 1994) that the output in response to a step change in the input will cross zero (its original value) *at least* n_z times. Typical closed-loop responses for the case with one RHP-zero are shown in Figure 5.7(b). We see that the closed-loop output initially decreases before increasing to its positive steady-state value. With two real RHP-zeros the output will initially increase, then decrease below its original value, and finally increase to its positive steady-state value.

Similar to the overshoot defined in Section 2.4.2 on page 30, we may define the *undershoot* (y_{us}) as the "negative" peak value of the output signal y divided by the final value y_f. For a plant with a real RHP-zero z ($z > 0$), we have the following lower bound on the closed-loop undershoot (Middleton, 1991; Seron et al., 1997):

$$|y_{us}| \geq |y_f| \frac{1 - \epsilon}{e^{zt_s} - 1} \tag{5.46}$$

where t_s is the settling time and ϵ is the corresponding level for the settling time (typically $\epsilon = 0.05$); see Figure 2.10 on page 30. Relation (5.46) implies that the step response of a system with real RHP-zeros will display large undershoot as the settling time is reduced. This agrees with the simulation in Figure 5.7(b), which is further discussed in the following example.

Example 5.6 **Trade-off between undershoot and settling time.** *Consider the plant*

$$G(s) = \frac{-s + z}{s + z}, \quad z = 1$$

which is controlled by

$$K_1(s) = K_c \frac{s + 1}{s} \frac{1}{0.05s + 1}$$

The sensitivity function and the step response of the closed-loop system for $K_c = 0.2, 0.5, 0.8$ are shown in Figure 5.7. We note that as the controller becomes more aggressive (K_c increased), the settling time decreases, but this performance improvement comes at the cost of higher undershoot. This is expected from (5.46) and also from the fact that a higher value of K_c results in a higher bandwidth, but increased sensitivity peak; see Figure 5.7(a). However, (5.46) is conservative. With $\epsilon = 0.05$ and $K_c = 0.8$, the undershoot is approximately 1.8, whereas (5.46) gives a lower bound of only 0.106. The bound (5.46) is not tight, nevertheless it clearly illustrates the trade-off between undershoot and settling time for systems with real RHP-zeros.

5.7.2 High-gain instability

It is well known from classical root–locus analysis that as the feedback gain increases towards infinity, the closed-loop poles migrate to the positions of the open-loop zeros; also see (4.79). Thus, the presence of RHP-zeros implies high-gain instability. For example, the system in Example 5.6 is unstable for $K_c \geq 1$. Since high gain is required for performance, RHP-zeros limit the performance of a closed-loop system.

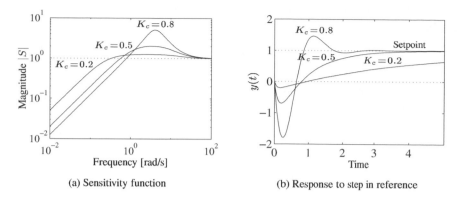

(a) Sensitivity function   (b) Response to step in reference

Figure 5.7: Control of plant with RHP-zero at $z = 1$ using negative feedback

5.7.3 Frequency response: bandwidth limitation

Consider the bound (5.14) on weighted sensitivity in Theorem 5.3. The idea is to select a form for the performance weight $w_P(s)$, and then to derive a bound for the "bandwidth parameter" in the weight.

The bandwidth is here defined as the frequency range where the asymptotic magnitude (straight-line approximation) of the sensitivity function is less than 1. To derive limitations on the achievable control bandwidth, we consider the interpolation constraints on the sensitivity function. As usual, we select $1/|w_P|$ as an upper bound on the sensitivity function (see Figure 5.1 on page 168); that is, we require

$$|S(j\omega)| < 1/|w_P(j\omega)| \quad \forall\omega \quad \Leftrightarrow \quad \|w_P S\|_\infty < 1 \qquad (5.47)$$

However, from the interpolation constraints $\boxed{S(z) = 1}$ and we have, as shown in (5.14), that $\|w_P S\|_\infty \geq |w_P(z)S(z)| = |w_P(z)|$, so to be able to satisfy (5.47) we must *at least* require that the weight satisfies

$$\boxed{|w_P(z)| < 1} \qquad (5.48)$$

(We say "at least" because condition (5.14) is not an equality.) We will now use (5.48) to gain insight into the limitations imposed by RHP-zeros: (A) by considering a weight that requires good performance at low frequencies, and (B) by considering a weight that requires good performance at high frequencies.

A. RHP-zero and performance at low frequencies

Consider the following performance weight:

$$w_P(s) = \frac{s/M + \omega_B^*}{s + \omega_B^* A} \qquad (5.49)$$

This weight emphasizes low-frequency performance. From (5.47) it specifies a minimum bandwidth ω_B^*, a maximum peak of $|S|$ less than M, a steady-state offset less than $A < 1$, and at frequencies lower than the bandwidth the sensitivity is required to improve by at least

20 dB/decade (i.e. $|S|$ has slope 1 or larger on a log–log plot); see Section 2.8.2 for further details. If the plant has a RHP-zero at $s = z$, then from (5.48) we must require

$$|w_P(z)| = \left| \frac{z/M + \omega_B^*}{z + \omega_B^* A} \right| < 1 \tag{5.50}$$

Real zero. Consider the case when z is real. Then all variables are real and positive and from (5.50) we derive the following bound on the achievable bandwidth:

$$\boxed{\omega_B^* < z \, \frac{1 - 1/M}{1 - A}} \tag{5.51}$$

For example, with $A = 0$ (no steady-state offset) and $M = 2$ ($\|S\|_\infty < 2$) we must at least require

$$\omega_B^* < 0.5z \tag{5.52}$$

Complex zeros. When the system has a pair of complex conjugate RHP-zeros $z = x \pm jy$, $x \geq 0$, a similar derivation with $A = 0$ yields

$$\boxed{\omega_B^* < -\frac{x}{M} + \sqrt{x^2 + y^2 \left(1 - \frac{1}{M^2}\right)}} \tag{5.53}$$

and with $M = 2$, we require that

$$\omega_B^* < -0.5x + \sqrt{x^2 + 0.75y^2} \tag{5.54}$$

The next two exercises show that the bound on ω_B^* does not depend much on the slope of the weight at low frequencies, or on how the weight behaves at high frequencies.

Exercise 5.5 *Consider the weight*

$$w_P(s) = \frac{s + M\omega_B^*}{s} \frac{s + fM\omega_B^*}{s + fM^2\omega_B^*} \tag{5.55}$$

with $f > 1$. This is the same weight as (5.49) with $A = 0$ except that it approaches 1 at high frequencies, and f gives the frequency range over which we allow a peak. Plot the weight for $f = 10$ and $M = 2$. Derive an upper bound on ω_B^* for the case with $f = 10$ and $M = 2$.

Exercise 5.6 * *Consider the weight $w_P(s) = \frac{1}{M} + (\frac{\omega_B^*}{s})^n$ which requires $|S|$ to have a slope of n at low frequencies and requires its low-frequency asymptote to cross 1 at a frequency ω_B^*. Note that $n = 1$ yields the weight (5.49) with $A = 0$. Derive an upper bound on ω_B^* when the plant has a RHP-zero at z. Show that the bound becomes $\omega_B^* \leq |z|$ as $n \to \infty$.*

Remark. The result for $n \to \infty$ in Exercise 5.6 is a bit surprising. It says that the bound $\omega_B^* < |z|$ is independent of the required slope (n) at low frequency and is also independent of M. This is surprising since from Bode's integral relationship (5.5) we expect to pay something for having the sensitivity smaller at low frequencies, so we would expect ω_B^* to be smaller for larger n. This illustrates that $|w_P(z)| < 1$ in (5.48) is a necessary condition on the weight (i.e. it must at least satisfy this condition), but since it is not sufficient it can be optimistic. For the simple weight (5.49), with $n = 1$, condition (5.48) is not very optimistic (as is confirmed by other results), but apparently it is optimistic for large n.

In summary, if we have a RHP-zero z and want tight control at low frequencies (frequency zero and upwards), then the upper bandwidth is limited to $|z|/2$, approximately. The reader is also referred to Exercise 5.11 for bandwidth limitations for plants having RHP-poles in addition to a RHP-zero.

B. RHP-zero and performance at high frequencies

We now consider the case where we want tight control at high frequencies, by use of the performance weight

$$w_P(s) = \frac{1}{M} + \frac{s}{\omega_B^*} \tag{5.56}$$

This requires tight control ($|S(j\omega)| < 1$) at frequencies *higher* than ω_B^*, whereas the only requirement at low frequencies is that the peak of $|S|$ is less than M. Admittedly, the weight in (5.56) is unrealistic in that it requires $S \to 0$ at high frequencies, but this does not affect the result as is confirmed in Exercise 5.9 where a more realistic weight is studied. In any case, to satisfy $\|w_P S\|_\infty < 1$ we must at least require that the weight satisfies $|w_P(z)| < 1$, and with a *real RHP-zero* we derive for the weight in (5.56)

$$\boxed{\omega_B^* > z\frac{1}{1 - 1/M}} \tag{5.57}$$

For example, with $M = 2$ the requirement is $\omega_B^* > 2z$, so we can only achieve tight control at frequencies *beyond* the frequency of the RHP-zero.

Exercise 5.7 *Draw an asymptotic magnitude Bode plot of $1/w_P(s)$ in (5.56).*

In summary, if we have a RHP-zero z and want tight control at high frequencies towards infinity, then the lower bandwidth is limited to $2|z|$, approximately.

RHP-zero: limitations around frequency $|z|$

Based on (5.51) and (5.57) we see that a RHP-zero will pose control limitations *either* at low *or* high frequencies. In most cases we desire tight control at low frequencies, and with a real RHP-zero this may be achieved at frequencies lower than about $|z|/2$. However, if we do not need tight control at low frequencies, then we may usually reverse the sign of the controller gain, and instead achieve tight control at frequencies higher than about $2|z|$.

Example 5.7 *To illustrate this, consider in Figures 5.7 and 5.8 the use of negative and positive feedback for the plant*

$$G(s) = \frac{-s + z}{s + z}, \quad z = 1 \tag{5.58}$$

Note that $G(s) \approx 1$ at low frequencies ($\omega \ll z$), whereas $G(s) \approx -1$ at high frequencies ($\omega \gg z$). The negative plant gain in the latter case explains why we then use positive feedback in order to achieve tight control at high frequencies.

More precisely, we show in the figures the sensitivity function and the time response to a step change in the reference using

1. PI control with negative feedback (Figure 5.7)
2. Derivative control with positive feedback (Figure 5.8).

Note that the time scales for the simulations are different. For positive feedback the step change in reference only has a duration of 0.1 s. This is because we cannot track references over longer times than this since the RHP-zero then causes the output to start drifting away (as can be seen in Figure 5.8(b)).

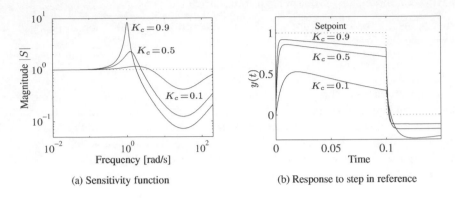

(a) Sensitivity function   (b) Response to step in reference

Figure 5.8: Control of plant with RHP-zero at $z = 1$ using *positive* feedback: $G(s) = \frac{-s+1}{s+1}$, $K_2(s) = -K_c \frac{s}{(0.05s+1)(0.02s+1)}$

Remark 1 The reversal of the sign in the controller is probably best understood by considering the inverse response behaviour of a plant with a RHP-zero. Normally, we want tight control at low frequencies, and the sign of the controller is based on the steady-state gain of the plant. However, if we instead want tight control at high frequencies (and have no requirements at low frequencies) then we base the controller design on the plant's initial response where the gain is reversed because of the inverse response.

Remark 2 An important case, where we can *only* achieve tight control at high frequencies, is characterized by plants with a zero at the origin, e.g. $G(s) = s/(5s + 1)$. In this case, good transient control is possible, but the control has no effect at steady-state. The only way to achieve tight control at low frequencies is to use an additional actuator (input) as is often done in practice.

Remark 3 Short-term control. In this book, we generally assume that the system behaviour as $t \to \infty$ is important. However, this is not true in some cases because the system may only be under closed-loop control for a finite time t_f. In this case, the presence of a "slow" RHP-zero (with $|z|$ small) may not be significant provided $t_f \ll 1/|z|$. For example, in Figure 5.8(b) if the total control time is $t_f = 0.01$ [s], then the RHP-zero at $z = 1$ [rad/s] is insignificant.

As an example of short-term control, consider treating a patient with some medication. Let u be the dosage of medication and y the condition of the patient. With most medications we find that in the short term the treatment has a positive effect, whereas in the long term the treatment has a negative effect (due to side effects which may eventually lead to death). However, this inverse response behaviour (characteristic of a plant with a RHP-zero) may be largely neglected during limited treatment, although one may find that the dosage has to be increased during the treatment to have the desired effect. Interestingly, the last point is illustrated by the upper left curve in Figure 5.9, which shows the input $u(t)$ using an internally unstable controller which over some finite time may eliminate the effect of the RHP-zero. In process control, similar conclusions are also applicable to the control of batch or semi-batch processes.

Exercise 5.8 *(a) Plot the plant input $u(t)$ corresponding to Figure 5.8 and discuss in the light of the above remark.*

(b) In the simulations in Figures 5.7 and 5.8, we use simple PI and derivative controllers. As an alternative, use the S/KS method in (3.80) to synthesize $\mathcal{H}_\infty$ controllers for both the negative and positive feedback cases. Use performance weights in the form given by (5.49) and (5.56), respectively. With $\omega_B^* = 1000$ and $M = 2$ in (5.56) and $w_u = 1$ (for the weight on KS) you will find that the

time response is quite similar to that in Figure 5.8 with $K_c = 0.5$. Try to improve the response, e.g. by letting the weight have a steeper slope at the crossover near the RHP-zero.

Exercise 5.9 * *Consider the case of a plant with a RHP-zero z where we want to limit the sensitivity function over some frequency range. To this effect let*

$$w_P(s) = \frac{(\frac{1000s}{\omega_B^*} + \frac{1}{M})(\frac{s}{M\omega_B^*} + 1)}{(\frac{10s}{\omega_B^*} + 1)(\frac{100s}{\omega_B^*} + 1)} \tag{5.59}$$

This weight is equal to $1/M$ at low and high frequencies, has a maximum value of about $10/M$ at intermediate frequencies, and the asymptote crosses 1 at frequencies $\omega_B^*/1000$ and ω_B^*. Thus we require "tight" control, $|S| < 1$, in the frequency range between $\omega_{BL}^* = \omega_B^*/1000$ and $\omega_{BH}^* = \omega_B^*$.

(a) Make a sketch of $1/|w_P|$ (which provides an upper bound on $|S|$).

(b) Show that the RHP-zero z cannot be in the frequency range where we require tight control, and that we can achieve tight control at frequencies either below about $z/2$ (the usual case) or above about $2z$. To see this, select $M = 2$ and evaluate $w_P(z)$ for various values of $\omega_B^* = kz$, e.g. $k = 0.1, 0.5, 1, 10, 100, 1000, 2000, 10000$. (You will find that $w_P(z) = 0.95$ (≈ 1) for $k = 0.5$ (corresponding to the requirement $\omega_{BH}^* < z/2$) and for $k = 2000$ (corresponding to the requirement $\omega_{BL}^* > 2z$).)

5.7.4 RHP-zeros and non-causal controllers

Perfect control can actually be achieved for a plant with a time delay or RHP-zero if we use a *non-causal controller*[2], i.e. a controller which uses information about the future. This is sometimes called "Preview Control" and may be relevant for certain servo problems, e.g. in robotics and for product changeovers in chemical plants. A brief discussion is given here, but non-causal controllers are not considered in the rest of the book since our focus is on feedback control.

Time delay. For a delay $e^{-\theta s}$ we may achieve perfect control with a non-causal feedforward controller $K_r = e^{\theta s}$ (a prediction). Such a controller may be used if we have knowledge about future changes in $r(t)$ or $d(t)$.

For example, if we know that we should be at work at 08:00, and we know that it takes 30 min to get to work, then we make a prediction and leave home at 07:30. We don't wait until 08:00 when we are suddenly told, by the appearance of a step change in our reference position, that we should be at work.

RHP-zero. Future knowledge can also be used to give perfect control in the presence of a RHP-zero. As an example, consider a plant with a real RHP-zero given by

$$G(s) = \frac{-s + z}{s + z}; \quad z > 0 \tag{5.60}$$

and a desired reference change

$$r(t) = \begin{cases} 0 & t < 0 \\ 1 & t \geq 0 \end{cases}$$

With a feedforward controller K_r the response from r to y is $y = G(s)K_r(s)r$. In theory, we may achieve perfect control $(y(t) = r(t))$ with the following two controllers (e.g. Eaton and Rawlings, 1992):

---

[2] A system is causal if its outputs depend only on past inputs, and non-causal if its outputs also depend on future inputs.

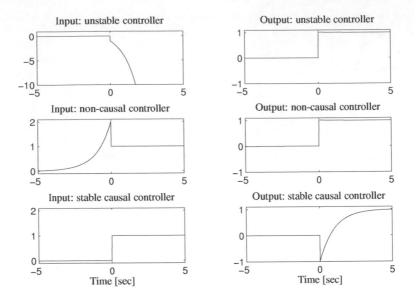

Figure 5.9: Control of plant with RHP-zero at $z = 1$

1. A causal *unstable feedback controller*

$$K_r(s) = \frac{s + z}{-s + z}$$

For a step in r from 0 to 1 at $t = 0$, this controller generates the following input signal:

$$u(t) = \begin{cases} 0 & t < 0 \\ 1 - 2e^{zt} & t \geq 0 \end{cases}$$

However, since the controller cancels the RHP-zero in the plant it yields an internally unstable system.

2. A stable *non-causal (feedforward) "preview" controller* that assumes that the future setpoint change is known. This controller cannot be represented in the usual transfer function form, but it will generate the following input:

$$u(t) = \begin{cases} 2e^{zt} & t < 0 \\ 1 & t \geq 0 \end{cases}$$

These input signals $u(t)$ and the corresponding outputs $y(t)$ are shown in Figure 5.9 for a plant with $z = 1$. Note that for perfect control the non-causal controller needs to start changing the input at $t = -\infty$, but for practical reasons we started the simulation at $t = -5$ where $u(t) = 2e^{-5} = 0.013$.

The first option, the unstable controller, is not acceptable as it yields an internally unstable system in which $u(t)$ goes to infinity as t increases (an exception may be if we want to control the system only over a limited time t_f; see page 188).

The second option, the non-causal controller, is usually not possible because future setpoint changes are unknown. However, if we have such information, it is certainly beneficial for plants with RHP-zeros. For example, for a system with single RHP-zero z (Middleton et al., 2004),

$$\|w_P S\|_\infty \geq |w_P(z)| e^{-z t_P} \tag{5.61}$$

where t_p is the preview time (reference change is known at time t_p before it occurs). Then, similar to (5.52),

$$\omega_B^* < 0.5 z e^{z t_P} \tag{5.62}$$

which shows that the non-causal controller can overcome the bandwidth limitation imposed by the RHP-zero (by having a large preview time).

3. In most cases we have to accept the poor performance resulting from the RHP-zero and use a *stable causal controller*. The ideal causal feedforward controller in terms of minimizing the ISE ($\mathcal{H}_2$ norm) of $y(t)$ for the plant in (5.60) is to use $K_r = 1$, and the corresponding plant input and output responses are shown in the lower plots in Figure 5.9.

5.7.5 LHP-zeros

Zeros in the LHP, usually corresponding to "overshoots" in the time response, do not present a *fundamental* limitation on control, but *in practice* a LHP-zero located close to the origin may cause problems. First, one may encounter problems with input constraints at low frequencies (because the steady-state gain is small). Second, a simple controller can probably not then be used. For example, a simple PID controller as in (2.93) contains no adjustable poles that can be used to counteract the effect of a LHP-zero.

For uncertain plants, zeros can cross from the LHP into the RHP, either through zero (which is worse if we want tight control at low frequencies) or through infinity. We discuss this in Section 7.4 (page 264).

5.8 Limitations imposed by phase lag

We know that the phase lag from RHP-zeros and time delays pose a fundamental problem, but are there any limitations imposed by the phase lag resulting from minimum-phase elements? The answer is both no and yes: *No*, there are no fundamental limitations, but *Yes*, there are often limitations on practical designs.

As an example, consider a minimum-phase plant of the form

$$G(s) = \frac{k}{(1 + \tau_1 s)(1 + \tau_2 s)(1 + \tau_3 s) \cdots} = \frac{k}{\prod_{i=1}^n (1 + \tau_i s)} \tag{5.63}$$

where n is 3 or larger. At high frequencies the gain drops sharply with frequency, $|G(j\omega)| \approx (k/\prod \tau_i)\omega^{-n}$. From condition (5.82) derived below, it is therefore likely (at least if k is small) that we encounter problems with *input saturation*. Otherwise, the presence of high-order lags *does not present any fundamental limitations*.

However, *in practice* a large *phase lag* at high frequencies, e.g. $\angle G(j\omega) \rightarrow -n \cdot 90°$ for the plant in (5.63), poses a problem (independent of K) even when input saturation is not an issue. This is because for stability we need a positive phase margin, i.e. the phase of $L = GK$ must be larger than $-180°$ at the gain crossover frequency ω_c. That is, for stability we need $\omega_c < \omega_{180}$; see (2.32).

In principle, ω_{180} (the frequency at which the phase lag around the feedback loop is $-180°$) is not directly related to phase lag in the plant, but in most practical cases there is a close relationship. Define ω_u as the frequency where the phase lag in the plant G is $-180°$, i.e.

$$\angle G(j\omega_u) \triangleq -180°$$

Note that ω_u depends only on the plant model. Then, with a proportional controller we have that $\omega_{180} = \omega_u$, and with a PI controller $\omega_{180} < \omega_u$. Thus with these two simple controllers a phase lag in the plant *does* pose a fundamental limitation:

Stability bound for P or PI control: $\boxed{\omega_c < \omega_u}$ (5.64)

Note that this is a strict bound to get stability, and for performance (phase and gain margin) we typically need ω_c less than about $0.5\omega_u$.

If we want to extend the gain crossover frequency ω_c beyond ω_u, we must place zeros in the controller (e.g. "derivative action") to provide phase lead which counteracts the negative phase in the plant. A commonly used controller is the PID controller which has a maximum phase lead of $90°$ at high frequencies. In practice, the maximum phase lead is smaller than $90°$. For example, an industrial *cascade PID controller* (2.87) typically has derivative action over only one decade, and the maximum phase lead is $55°$ (which is the maximum phase lead of the term $\frac{\tau_D s+1}{0.1\tau_D s+1}$). This is also a reasonable value for the phase margin, so for performance we approximately require

Practical performance bound (PID control): $\omega_c < \omega_u$ (5.65)

We stress again that plant phase lag does *not* pose a *fundamental* limitation if a more complex controller is used. Specifically, if the model is known exactly and there are no RHP-zeros or time delays, then one may in theory extend ω_c to infinite frequency. For example, one may simply invert the plant model by placing zeros in the controller at the plant poles, and then let the controller roll off at high frequencies beyond the dynamics of the plant. However, in many practical cases the bound in (5.65) applies because we may want to use a simple controller, and also because uncertainty about the plant model often makes it difficult to place controller zeros which counteract the plant poles at high frequencies.

Remark. The *relative order* (relative degree) of the plant is sometimes used as an input–output controllability measure (e.g. Daoutidis and Kravaris, 1992). The relative order may also be defined for nonlinear plants, and it corresponds for linear plants to the pole excess of $G(s)$. For a minimum-phase plant the phase lag at infinite frequency is the relative order times $-90°$. Of course, we want the inputs directly to affect the outputs, so we want the relative order to be small. However, the practical usefulness of the relative order is rather limited since it only gives information at infinite frequency. The phase lag of $G(s)$ as a function of frequency, including the value of ω_u, provides much more information.

Another approach for quantifying the limitations of phase lags is to approximate the higher-order lags as an "effective delay" as discussed in Chapter 2; see (2.99) (PI control) and (2.100) (PID control).

5.9   Limitations imposed by unstable (RHP) poles

We now consider the limitations imposed when the plant has an unstable (RHP) pole at $s = p$. For example, the plant $G(s) = 1/(s - 3)$ has a RHP-pole at $p = 3$. We already know from the bounds on M_S and M_T, (5.15) and (5.18), that RHP-poles combined with RHP-zeros or a time delay make control difficult. The question here is: does a RHP-pole by itself pose problems in terms of control performance?

First, feedback control is required, so we need some measurement of the plant output. The reason for this is that it is impossible to stabilize a system with feedforward control – even with a perfect model that allows us to cancel perfectly the unstable pole. As discussed on page 145, we would get an internally unstable system, which eventually grows out of bounds.

Next, what problems does a RHP-pole p pose for feedback control? A good starting point for such a discussion is the fundamental constraint on the sensitivity function for internal stability, $\boxed{S(p) = 0}$. Recall that the corresponding constraint with a RHP-zero z was $S(z) = 1$, which was a problem because it is not compatible with the desire to have $|S|$ small (compared to 1) in order to have tight control (good output performance). At first, it may therefore seem that the requirement $S(p) = 0$ does not pose a problem, because it is compatible with tight control (good output performance). Actually, the main problem is at the plant input, because stabilization of an unstable plant requires feedback control with the active use of plant inputs. With feedback control, $u = KS(r - n - d_y)$, where $S = (1 + GK)^{-1}$. Note that changes in n and d_y are outside our control and therefore "unavoidable", and for an unstable plant a minimum value on $|KS|$ is also unavoidable, as derived in Section 5.3.2.

This leads to the conclusion that for an unstable plant a minimum input usage u is required. In addition, the presence of a RHP-pole imposes a lower bound on the required bandwidth and also causes an overshoot in the output signal, as summarized below:

1. **RHP-pole limitation on input usage.** *For an unstable plant, the transfer function KS (from measurement noise n or output disturbances d_y to plant input u) must satisfy, see (5.31),*

$$\|KS\|_\infty \geq |G_s^{-1}(p)| \tag{5.66}$$

   which is tight for the case of a single real RHP-pole p. A tight lower bound for a plant with multiple unstable poles is given by (5.30).

2. **RHP-pole limitation on lower bandwidth.** *To stabilize a plant, we need to react sufficiently fast, and we must require that the closed-loop bandwidth is larger than (approximately, see proof below)*

   - *$2p$, for a real RHP-pole p.*
   - *$0.67(x + \sqrt{4x^2 + 3y^2})$, for a pair of complex RHP-poles $p = x \pm jy$.*
   - *$1.15|p|$, for a pair of purely imaginary poles $p = j|p|$.*

3. **RHP-pole limitation on overshoot.** *A stable feedback system with a real RHP-pole must have an overshoot in its closed-loop response $y(t)$ to a step in the reference; see Figure 5.12(b). To quantify this overshoot y_{os}, we require a slightly different version of rise time than that defined on page 30. In accordance with Middleton (1991), we define rise time t_r as the maximum t_r for which the output signal $y(t)$ to a step r satisfies $y(t)/r \leq t/t_r \ \forall t$; see Figure 5.10.[3] Then, the step response of a system with a real RHP-pole p ($p > 0$) must satisfy (Middleton, 1991; Seron et al., 1997)*

$$y_{os} \geq y_f \frac{(pt_r - 1)e^{pt_r} + 1}{pt_r} + r \geq y_f \frac{pt_r}{2} + r \tag{5.67}$$

---

[3] The rise time t_r can also be analytically calculated as $t_r = \min_t (t\,r)/y(t)$ for $t > 0$.

where y_f is the final value of the output signal y. With integral action $y_f = r$ and a large overshoot (y_{os}) is unavoidable if the response is slow with large rise time (t_r).

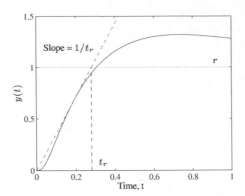

Figure 5.10: Rise time t_r according to definition $y(t)/r \le t/t_r \; \forall \, t$ for plant in (5.71) with $K_c = \tau_I = 1.25$. The straight line with slope $1/t_r$ just touches $y(t)$.

Stabilization becomes more difficult and the above bounds become worse, if the plant has a time delay or RHP-zeros located close to the RHP-poles. In essence, "the system may go unstable before we have time to react"; see also Example 5.5.

Proof of limitations on lower bandwidth: We start from the requirement that an unstable pole p requires for internal stability $T(p) = 1$. Consider that the weight $w_T(s)$ is selected such that $1/|w_T|$ is a reasonable upper bound on the complementary sensitivity function:

$$|T(j\omega)| < 1/|w_T(j\omega)| \quad \forall \omega \quad \Leftrightarrow \quad \|w_T T\|_\infty < 1$$

To satisfy this we must, since from (5.17) $\|w_T T\|_\infty \ge |w_T(p)|$, *at least* require that the weight satisfies $\boxed{|w_T(p)| < 1}$. Now consider the following weight

$$w_T(s) = \frac{s}{\omega_{BT}^*} + \frac{1}{M_T} \tag{5.68}$$

which requires that (i) T (like $|L|$) has a roll-off rate of at least 1 at high frequencies (which must be satisfied for any real system), (ii) $|T|$ is less than M_T at low frequencies, and (iii) $|T|$ drops below 1 at frequency ω_{BT}^*. The requirements on $|T|$ are shown graphically in Figure 5.11. For a real RHP-pole at $s = p$, the condition $w_T(p) < 1$ yields

$$\omega_{BT}^* > p\frac{M_T}{M_T - 1} \tag{5.69}$$

With $M_T = 2$ (reasonable robustness) this gives

$$\omega_{BT}^* > 2p \tag{5.70}$$

which proves the above bandwidth requirement.

□

Exercise 5.10 * *For purely imaginary poles located at $p = \pm j|p|$ a similar analysis of the weight (5.68) with $M_T = 2$ shows that we must at least require $\omega_{BT}^* > 1.15|p|$. Derive this bound.*

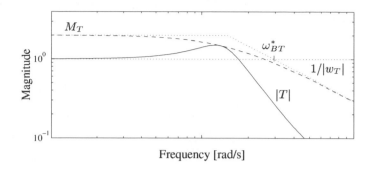

Figure 5.11: Typical complementary sensitivity, $|T|$, with upper bound $1/|w_T|$

Since $u = -KS(G_d d + n)$, the exact bounds (5.66) and (5.34) imply that stabilization may be impossible in the presence of measurement noise n or a disturbance d, since the required inputs u may be outside the saturation limit. When the input saturates, the system is practically open-loop and stabilization is impossible (see also Section 5.11.3 on page 201).

The limitations on the bandwidth and overshoot are related: to stabilize an unstable plant, we need a minimum bandwidth, which corresponds to a maximum rise time. If the rise time is too large, then control is bound to be poor. This is clearly seen from (5.67). With integral action, $y_f = r$, and the "excess" overshoot $y_{os} - r$ must exceed $\frac{pt_r}{2}r$. For example, with $t_r > 1/p$, the excess overshoot must exceed $0.5r$ (50%).

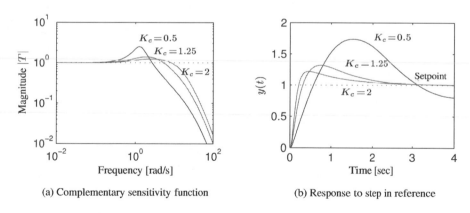

(a) Complementary sensitivity function   (b) Response to step in reference

Figure 5.12: Control of plant with RHP-pole at $p = 1$

Example 5.8 Overshoot due to RHP-pole. *Consider the following PI-controlled system, earlier used in Example 2.5 (page 30):*

$$G(s) = \frac{4}{(s-1)(0.02s+1)^2}, \quad K(s) = K_c \frac{\tau_I s + 1}{\tau_I s} \tag{5.71}$$

with $\tau_I = 1.25$. We note from the simulation in Figure 5.12(b) that as the controller becomes more aggressive (K_c increased), both the rise time and overshoot decrease. This is expected from (5.67) and also from the fact that a higher value of K_c results in higher bandwidth, but decreased peak of $|T|$; see

Figure 5.12(a). For $K_c = 2$, the rise time is 0.2 s. The resulting overshoot is 1.22, which is reasonably close to the lower bound from (5.67)

$$y_{os} \geq y_f \frac{(pt_r - 1)e^{pt_r} + 1}{pt_r} + r = 1 \cdot \frac{(0.2 - 1)e^{0.2} + 1}{0.2} + 1 = 1.11$$

It may seem that we can improve the performance by increasing K_c further. This is probably not possible, as the actual limitation due to the RHP-pole occurs at the plant input. The peak in KS increases with K_c (not shown here), so a larger value of K_c can cause saturation problems.

Combined RHP-pole and RHP-zeros. In Section 5.3 (e.g. Table 5.1 on page 177), we derived lower bounds on the peaks of important closed-loop transfer functions, and found that the combined effect of a RHP-zero z and RHP-pole p is to increase the minimum peak by a factor $\left|\frac{z+p}{z-p}\right|$. Here, we consider in some more detail the possibly conflicting bandwidth limitation imposed by having RHP-poles combined with RHP-zeros or a time delay. In order to get acceptable low-frequency performance while maintaining robustness, we have from (5.45) and (5.52) the approximate bounds $\omega_B \approx \omega_c < 0.5|z|$ for a RHP-zero and $\omega_B \approx \omega_c < 1/\theta$ for a time delay. On the other hand, for a RHP-pole we have approximately $w_B > 2|p|$. Put together we get the approximate requirements $|p| < 0.25|z|$ and $|p|\theta < 0.5$ in order to stabilize a plant while achieving acceptable low-frequency performance and robustness. The following example confirms that these requirements are reasonable.

Example 5.9 $\mathcal{H}_\infty$ **design for plant with RHP-pole and RHP-zero.** *We want to design an $\mathcal{H}_\infty$ controller for the following plant with $z = 4$ and $p = 1$:*

$$G(s) = \frac{s - 4}{(s - 1)(0.1s + 1)} \tag{5.72}$$

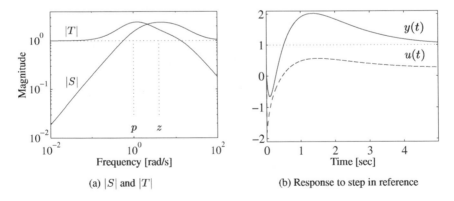

(a) $|S|$ and $|T|$ (b) Response to step in reference

Figure 5.13: $\mathcal{H}_\infty$ design for a plant with RHP-zero at $z = 4$ and RHP-pole at $p = 1$

Note that $z > p$, so from the condition on page 150 it is possible to stabilize this plant with a stable controller. Furthermore, $|p| = 0.25|z|$ so from the condition just derived it should be possible to achieve acceptable low-frequency performance and robustness. We use the S/KS design method as in Example 2.17 with input weight $w_u = 1$ and performance weight w_p in (5.49) with $A = 0$, $M = 2$, $\omega_B^* = 1$. The software gives a stable and minimum-phase controller with $\left\| \begin{bmatrix} w_pS \\ w_uKS \end{bmatrix} \right\|_\infty = 1.89$.

The corresponding sensitivity and complementary sensitivity functions, and the time response to a unit step reference change, are shown in Figure 5.13. The time response is good, taking into account the closeness of the RHP-pole and zero.

From (5.22), we have for a plant with a single real RHP-pole p and a single real RHP-zero z:

$$M_{S,\min} = M_{T,\min} = \frac{|z+p|}{|z-p|} \tag{5.73}$$

The plant in (5.72) has $z = 4$ and $p = 1$, so $\frac{|z+p|}{|z-p|} = \frac{5}{3} = 1.67$ and therefore it follows that for any controller we must at least have $\|S\|_\infty > 1.67$ and $\|T\|_\infty > 1.67$. The actual peak values for the above S/KS-design are 2.40 and 2.43, respectively.

Example 5.10 Balancing a rod. *This example is taken from Doyle et al. (1992) (also see Stein, 2003). Consider the problem of balancing a rod in the palm of one's hand. The objective is to keep the rod upright, by small hand movements, based on observing the rod either at its far end (output y_1) or the end in one's hand (output y_2). The linearized transfer functions for the two cases are*

$$G_1(s) = \frac{-g}{s^2\,(Mls^2 - (M+m)g)}; \quad G_2(s) = \frac{ls^2 - g}{s^2\,(Mls^2 - (M+m)g)}$$

Here l [m] is the length of the rod and m [kg] its mass. M [kg] is the mass of your hand and g [≈ 10 m/s^2] is the acceleration due to gravity. In both cases, the plant has three unstable poles: two at the origin and one at $p = \sqrt{\frac{(M+m)g}{Ml}}$. A short rod with a large mass gives a large value of p, and this in turn means that the system is more difficult to stabilize. For example, with $M = m$ and $l = 1$ [m] we get $p \approx 4.5$ [rad/s] and from (5.70) we desire a bandwidth of about 9 [rad/s] (corresponding to a response time of about 0.1 [s]).

If one is measuring y_1 (looking at the far end of the rod) then achieving this bandwidth is the main requirement. However, if one tries to balance the rod by looking at one's hand (y_2) there is also a RHP-zero at $z = \sqrt{\frac{g}{l}}$. If the mass of the rod is small (m/M is small), then p is close to z and stabilization is in practice impossible with any controller. Even with a large mass, stabilization is very difficult because $p > z$ whereas we would normally prefer to have the RHP-zero far from the origin and the RHP-pole close to the origin ($z > p$). So although in theory the rod may be stabilized by looking at one's hand (G_2), it seems doubtful that this is possible for a human. To quantify these problems we can use (5.73) to get

$$M_{S,\min} = M_{T,\min} = \frac{|z+p|}{|z-p|} = \frac{|1+\gamma|}{|1-\gamma|}, \quad \gamma = \sqrt{\frac{M+m}{M}}$$

Consider a light-weight rod with $m/M = 0.1$, for which we expect stabilization to be difficult. We obtain $M_{S,\min} = M_{T,\min} = 42$, and we must have $\|S\|_\infty \geq 42$ and $\|T\|_\infty \geq 42$, so poor control performance is inevitable if we try to balance the rod by looking at our hand (y_2).

The difference between the two cases, measuring y_1 and measuring y_2, highlights the importance of sensor location on the achievable performance of control.

Exercise 5.11 * *For a system with a single real RHP-zero z and N_p RHP-poles p_i and tight control at low frequencies ($A = 0$ in (5.50)) derive the following generalization of (5.52):*

$$\omega_B^* < z \left(\prod_{i=1}^{N_p} \frac{|z - p_i|}{|z + p_i|} - \frac{1}{M} \right) \tag{5.74}$$

(Hint: Use (5.13).) Note that for a plant with a single RHP-pole and RHP-zero the bound (5.74) with $M = 2$ is feasible (upper bound on ω_B^* is positive) for $p < 0.33z$. This confirms the approximate bound $p < 0.25z$ derived for stability with acceptable low-frequency performance and robustness on page 196.

5.10 Performance requirements imposed by disturbances and commands

The question we want to answer here is: how fast must the control system be in order to reject disturbances and track commands of a given magnitude? The required bandwidth varies because some plants have better "built-in" disturbance rejection capabilities than others. This may be analyzed directly by considering the appropriately scaled disturbance model, $G_d(s)$. Similarly, for tracking we may consider the magnitude R of the reference change.

Disturbance rejection. Consider a single disturbance d and assume that the reference is constant, i.e. $r = 0$. Without control the steady-state sinusoidal response is $e(\omega) = G_d(j\omega)d(\omega)$; recall (2.10). If the variables have been scaled as outlined in Section 1.4 then the worst-case disturbance at any frequency is $d(t) = \sin \omega t$, i.e. $|d(\omega)| = 1$, and the control objective is that at each frequency $|e(t)| < 1$, i.e. $|e(\omega)| < 1$. From this we can immediately conclude that

- *No control is needed if $|G_d(j\omega)| < 1$ at all frequencies (in which case the plant is said to be "self-regulated").*

If $|G_d(j\omega)| > 1$ at some frequency, then we need control (feedforward or feedback). In the following, we consider feedback control, in which case we have

$$e(s) = S(s)G_d(s)d(s) \tag{5.75}$$

The performance requirement $|e(\omega)| < 1$ for any $|d(\omega)| \leq 1$ at any frequency is satisfied if and only if

$$|SG_d(j\omega)| < 1 \quad \forall \omega \quad \Leftrightarrow \quad \|SG_d\|_\infty < 1 \tag{5.76}$$

$$\Leftrightarrow \quad \boxed{|S(j\omega)| < 1/|G_d(j\omega)| \quad \forall \omega} \tag{5.77}$$

A typical plot of $1/|G_d(j\omega)|$ is shown in Figure 5.14 (dashed line). If the plant has a RHP-zero at $s = z$, which fixes $S(z) = 1$, then using (5.14) we have the following necessary condition for satisfying $\|SG_d\|_\infty < 1$:

$$\boxed{|G_d(z)| < 1} \tag{5.78}$$

From (5.77) we also get that the frequency ω_d where $|G_d|$ crosses 1 from above yields a lower bound on the bandwidth:

$$\boxed{\omega_B > \omega_d} \quad \text{where } \omega_d \text{ is defined by } |G_d(j\omega_d)| = 1 \tag{5.79}$$

A plant with a small $|G_d|$ or a small ω_d is preferable since the need for feedback control is then less, or alternatively, given a feedback controller (which fixes S) the effect of disturbances on the output is less.

Example 5.11 *Assume that the disturbance model is $G_d(s) = k_d/(1 + \tau_d s)$ where $k_d = 10$ and $\tau_d = 100$ [seconds]. Scaling has been applied to G_d so this means that without feedback, the effect of disturbances on the outputs at low frequencies is $k_d = 10$ times larger than we desire. Thus feedback is required, and since $|G_d|$ crosses 1 at a frequency $\omega_d \approx k_d/\tau_d = 0.1$ rad/s, the minimum bandwidth requirement for disturbance rejection is $\omega_B > 0.1$ [rad/s].*

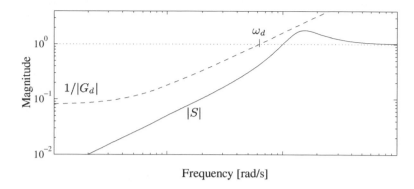

Figure 5.14: Typical performance requirement on S imposed by disturbance rejection

Remark. G_d is of high order. The actual bandwidth requirement imposed by disturbances may be higher than ω_d if $|G_d(j\omega)|$ drops with a slope steeper than -1 (on a log–log plot) just before the frequency ω_d. The reason for this is that we must, in addition to satisfying (5.77), also ensure stability with reasonable margins; so as discussed in Section 2.6.2 we cannot let the slope of $|L(j\omega)|$ around crossover be much larger than -1.

An example, in which $G_d(s)$ is of high order, is given later in Section 5.15.3 for a neutralization process. There we actually overcome the limitation on the slope of $|L(j\omega)|$ around crossover by using local feedback loops in series. We find that, although each loop has a slope -1 around crossover, the overall loop transfer function $L(s) = L_1(s)L_2(s)\cdots L_n(s)$ has a slope of about $-n$; see the example for more details. This is a case where stability is determined by each $I + L_i$ separately, but the benefits of feedback are determined by $1 + \prod_i L_i$ (also see Horowitz (1991, p. 284) who refers to lectures by Bode).

Command tracking. Assume that there are no disturbances, i.e. $d = 0$, and consider a reference change $r(t) = R\tilde{r}(t) = R\sin(\omega t)$. Since $e = Gu + G_d d - R\tilde{r}$, the same performance requirement as found for disturbances, see (5.76), applies to command tracking with G_d replaced by $-R$. Thus for acceptable control ($|e(\omega)| < 1$) we must have

$$|S(j\omega)R| < 1 \quad \forall \omega \leq \omega_r \tag{5.80}$$

where ω_r is the frequency up to which performance tracking is required.

Remark. The bandwidth requirement imposed by (5.80) depends on how sharply $|S(j\omega)|$ increases in the frequency range from ω_r (where $|S| < 1/R$) to ω_B (where $|S| \approx 1$). If $|S|$ increases with a slope of 1 then the approximate bandwidth requirement becomes $\omega_B > R\omega_r$, and if $|S|$ increases with a slope of 2 it becomes $\omega_B > \sqrt{R}\omega_r$.

5.11 Limitations imposed by input constraints

In all physical systems there are limits to the changes that can be made to the manipulated variables. In this section, we assume that the model has been scaled as outlined in Section 1.4, so that at any time we must have $|u(t)| \leq 1$. The question we want to answer is: can the expected disturbances be rejected and can we track the reference changes while maintaining $|u(t)| \leq 1$? We will consider separately the two cases of perfect control ($e = 0$) and acceptable control ($|e| < 1$). These results apply to both feedback and feedforward control.

At the end of the section we consider the additional problems encountered for unstable plants (where feedback control is required).

Remark 1 We use a frequency-by-frequency analysis and assume that at each frequency $|d(\omega)| \leq 1$ (or $|\tilde{r}(\omega)| \leq 1$). The worst-case disturbance at each frequency is $|d(\omega)| = 1$ and the worst-case reference is $r = R\tilde{r}$ with $|\tilde{r}(\omega)| = 1$.

Remark 2 Note that rate limitations, $|du/dt| \leq 1$, may also be handled by our analysis. This is done by considering du/dt as the plant input by including a term $1/s$ in the plant model $G(s)$. Alternatively we multiply the derived lower bounds on $|G|$, e.g. in (5.84), by the frequency ω. For the more general case with limitations on *both* magnitude ($|u| \leq 1$) and rate ($|du/dt| \leq \dot{u}_{\max}$), the derived lower bounds on $|G|$ should be multiplied by $\max(1, \omega/\dot{u}_{\max})$.

Remark 3 Below we require $|u| < 1$ rather than $|u| \leq 1$. This has *no* practical effect, and is used to simplify the presentation.

5.11.1 Inputs for perfect control

From (5.38) the input required to achieve perfect control ($e = 0$) is

$$u = G^{-1}r - G^{-1}G_d d \tag{5.81}$$

Disturbance rejection. With $r = 0$ and $|d(\omega)| = 1$ the requirement $|u(\omega)| < 1$ gives

$$|G^{-1}(j\omega)G_d(j\omega)| < 1 \quad \forall\omega \tag{5.82}$$

In other words, to achieve perfect control and avoid input saturation we need $|G| > |G_d|$ at all frequencies. (However, as is discussed below, we do not really need control at frequencies where $|G_d| < 1$.)

Command tracking. Next let $d = 0$ and consider the worst-case reference command which is $|r(\omega)| = R$ at all frequencies up to ω_r. To keep the inputs within their constraints we must then require from (5.81) that

$$|G^{-1}(j\omega)R| < 1 \quad \forall\omega \leq \omega_r \tag{5.83}$$

In other words, to avoid input saturation we need $|G| > R$ at all frequencies where perfect command tracking is required.

Example 5.12 *Consider a process with*

$$G(s) = \frac{40}{(5s+1)(2.5s+1)}, \quad G_d(s) = 3\frac{50s+1}{(10s+1)(s+1)}$$

From Figure 5.15 we see that $|G| < |G_d|$ for $\omega > \omega_1$, and $|G_d| < 1$ for $\omega > \omega_d$. Thus, condition (5.82) is not satisfied for $\omega > \omega_1$. However, for frequencies $\omega > \omega_d$ we do not really need control. Thus, in practice, we expect that disturbances in the frequency range between ω_1 and ω_d may cause input saturation.

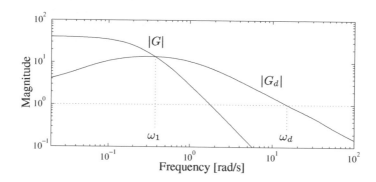

Figure 5.15: Input saturation is expected for disturbances at intermediate frequencies from ω_1 to ω_d

5.11.2 Inputs for acceptable control

For simplicity above, we assumed perfect control. However, perfect control is never really required, especially not at high frequencies, and the input magnitude required for *acceptable control* (namely $|e(j\omega)| < 1$) is somewhat smaller. For *disturbance rejection* we must then require

$$\boxed{|G| > |G_d| - 1} \quad \text{at frequencies where } |G_d| > 1 \tag{5.84}$$

Proof: Consider a "worst-case" disturbance with $|d(\omega)| = 1$. The control error is $e = y = Gu + G_dd$. Thus at frequencies where $|G_d(j\omega)| > 1$ the smallest input needed to reduce the error to $|e(\omega)| = 1$ is found when $u(\omega)$ is chosen such that the complex vectors Gu and G_dd have opposite directions. That is, $|e| = 1 = |G_dd| - |Gu|$, and with $|d| - 1$ we get $|u| - |G^{-1}|(|G_d| - 1)$, and the result follows by requiring $|u| < 1$. $\square$

Similarly, to achieve acceptable control for *command tracking* we must require

$$\boxed{|G| > |R| - 1} \quad \forall \omega \leq \omega_r \tag{5.85}$$

In summary, if we want "acceptable control" ($|e| < 1$) rather than "perfect control" ($e = 0$), then $|G_d|$ in (5.82) should be replaced by $|G_d| - 1$, and similarly, R in (5.83) should be replaced by $R - 1$. The differences are clearly small at frequencies where $|G_d|$ and $|R|$ are much larger than 1.

The requirements given by (5.84) and (5.85) are restrictions imposed on the *plant design* in order to avoid input constraints and they apply to any controller (feedback or feedforward control). If these bounds are violated at some frequency then performance will not be satisfactory (i.e. $|e(\omega)| > 1$) for a worst-case disturbance or reference occurring at this frequency.

5.11.3 Inputs for stabilization

Feedback control is required to stabilize an unstable plant. However, input constraints combined with large disturbances or noise may make stabilization difficult. To achieve $|u| < 1$ for $|d| = 1$ we must from (5.34) require (Havre and Skogestad, 2001)

$$\boxed{|G_s(p)| > |G_{d,ms}(p)|} \tag{5.86}$$

(this is for stabilization of a plant with a real RHP-pole at p). Otherwise, the input u will exceed 1 (and thus saturate) when there is a sinusoidal disturbance $d(t) = \sin \omega t$, and we may not be able to stabilize the plant.

Remark. The result in (5.86) was not available on publication of the first edition of this book (Skogestad and Postlethwaite, 1996) where we instead used the approximate, but nevertheless useful, bound

$$|G(j\omega)| > |G_d(j\omega)| \quad \forall \omega < p \tag{5.87}$$

This approximate bound is based on (5.69) where we found that we need $|T(j\omega)| \geq 1$ up to the frequency p, approximately. Since $u = KSG_d d = TG^{-1}G_d d$ this implies that we need $|u| \geq |G^{-1}G_d| \cdot |d|$ up to the frequency p, and to have $|u| \leq 1$ for $|d| = 1$ (the worst-case disturbance) we must require $|G^{-1}G_d| \leq 1$.

Example 5.13 *Consider*

$$G(s) = \frac{5}{(10s + 1)(s - 1)}, \quad G_d(s) = \frac{k_d}{(s + 1)(0.2s + 1)}, \quad k_d < 1 \tag{5.88}$$

Since $k_d < 1$ and the performance objective is $|e| < 1$, we do not really need control for disturbance rejection, but feedback control is required for stabilization, since the plant has a RHP-pole at $p = 1$. We have $|G| > |G_d|$ (i.e. $|G^{-1}G_d| < 1$) for frequencies lower than $0.5/k_d$, see Figure 5.16(a), so from the approximate bound (5.87) we do not expect problems with input constraints at low frequencies. However, at high frequencies we have $|G| < |G_d|$, and from (5.87) we must approximately require $0.5/k_d > p$, i.e. $k_d < 0.5$ to avoid problems with input saturation. This is confirmed by the exact bound in (5.86). We get

$$G_s(1) = \left. \frac{5}{(10s + 1)(s + 1)} \right|_{s=1} = 0.227, \quad G_{d,ms}(1) = \left. \frac{k_d}{(s + 1)(0.2s + 1)} \right|_{s=1} = 0.417k_d$$

and from (5.86) we must require $k_d < 0.54$ in order to avoid input saturation ($|u| < 1$) when we have sinusoidal disturbances of unit magnitude.

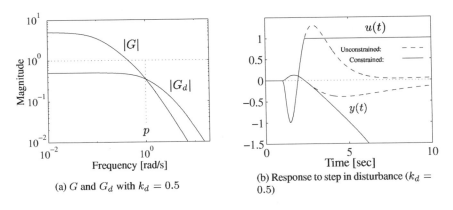

(a) G and G_d with $k_d = 0.5$

(b) Response to step in disturbance ($k_d = 0.5$)

Figure 5.16: Instability caused by input saturation for unstable plant

To check this for a particular case we select $k_d = 0.5$ and use the controller

$$K(s) = \frac{0.04}{s} \frac{(10s + 1)^2}{(0.1s + 1)^2} \tag{5.89}$$

which without constraints yields a stable closed-loop system with a gain crossover frequency, ω_c, of about 1.7. The closed-loop response to a unit step disturbance occurring after 1 second is shown in Figure 5.16(b). The stable closed-loop response when there is no input constraint is shown by the dashed line. However, we note that the input signal exceeds 1 for a short time, and when u is constrained to be within the interval $[-1, 1]$ we find indeed that the system is unstable (solid lines).

Remark. *For this example, a small reduction in the disturbance magnitude from $k_d = 0.5$ to $k_d = 0.48$ results in a stable closed-loop response in the presence of input constraints (not shown). Since $k_d = 0.54$ is the limiting value obtained from (5.86), this seems to indicate that (5.86) is a very tight condition in terms of predicting stability, but one should be careful about making such a conclusion. First, (5.86) is actually only tight for sinusoids and the simulations in this example are for a step disturbance. Second, in the example we use a particular controller, whereas (5.86) is for the "best" stabilizing controller in terms of minimizing input usage.*

For unstable plants, reference changes can also drive the system into input saturation and instability. However, this is not really a fundamental problem, because, in contrast to disturbance changes and measurement noise, one has the option of using a two degrees of-freedom controller to filter the reference signal and thus reduce the magnitude of the manipulated input.

5.12 Limitations imposed by uncertainty

The presence of uncertainty requires us to use feedback control rather than just feedforward control. The main objective of this section is to gain more insight into this statement. A further discussion is given in Section 6.10, where we consider MIMO systems.

5.12.1 Feedforward control and uncertainty

Consider feedforward control from the reference and measured disturbance (see Figure 2.5),

$$u = K_r r - K_d d \tag{5.90}$$

When applied to the nominal plant $y = Gu + G_d d$ the resulting control error is $e = y - r = -(1 - GK_r)r + (G_d - GK_d)d$. Correspondingly, for the actual plant (with model error)

$$y' = G'u + G'_d d \tag{5.91}$$

the control error is

$$e' = y' - r = -(1 - G'K_r)r + (G'_d - G'K_d)d = -S'_r r + S'_d G'_d d \tag{5.92}$$

where $S'_r \triangleq 1 - G'K_r$ and $S'_d \triangleq 1 - G'K_d G'^{-1}_d$ are the feedforward sensitivity functions. These are 1 for the case with no feedforward control, and should be less than 1 in magnitude for feedforward control to be beneficial. However, this may not be the case since any change in the process (G' and G'_d) directly propagates to a corresponding change in S'_r and S'_d and thus in the control error. This is the main problem with feedforward control.

To see this more clearly, consider the "perfect" feedforward controller $K_r = G(s)^{-1}$ and $K_d = G(s)^{-1}G_d$, which gives perfect nominal control (with $e = 0$, $S_r = 0$ and $S_d = 0$). (We must here assume that $G(s)$ is minimum-phase and stable and assume that there are no problems with input saturation.) Applying the perfect feedforward controller to the actual plant gives

$$e' = y' - r = \underbrace{\left(\frac{G'}{G} - 1\right)}_{-S_r' = \text{rel. error in } G} r - \underbrace{\left(\frac{G'/G_d'}{G/G_d} - 1\right)}_{S_d' = \text{rel. error in } G/G_d} G_d'd \qquad (5.93)$$

Thus, we find that S_r' and S_d' are equal to the (negative) relative errors in G and G/G_d, respectively. If the model error (uncertainty) is sufficiently large, such that the relative error in G/G_d is larger than 1, then $|S_d'|$ is larger than 1 and feedforward control makes this situation worse. This may quite easily happen in practice. For example, if the gain in G is increased by 33% and the gain in G_d is reduced by 33%, such that $S_d' = -\frac{G'/G}{G_d'/G_d} + 1 = -\frac{1.33}{0.67} + 1 = -2 + 1 = -1$. In words, the feedforward controller overcompensates for the disturbance, such that its negative counteracting effect is twice that of the original effect.

Another important insight from (5.93) is the following: to achieve $|e'| < 1$ for $|d| = 1$ we must require that the relative model error in G/G_d is less than $1/|G_d'|$. This requirement is unlikely to be satisfied at frequencies where $|G_d'|$ is much larger than 1 (see the following example), and this *clearly motivates the need for feedback control for "sensitive" plants where the disturbances have a large effect on the output.*

Example 5.14 *Consider disturbance rejection for a plant with*

$$G = \frac{300}{10s + 1}; \quad G_d = \frac{100}{10s + 1}$$

The objective is to keep $|y| < 1$ for $d = 1$, but notice that the disturbance gain at steady-state is 100. Nominally, the feedforward controller $K_d = G^{-1}G_d$ gives perfect control, $y = 0$. Now apply this controller to the actual process where the gains have changed by 10%

$$G' = \frac{330}{10s + 1}; \quad G_d' = \frac{90}{10s + 1}$$

From (5.93), the disturbance response in this case is

$$y' = -\left(\frac{G'/G_d'}{G/G_d} - 1\right) G_d'd = -0.22 \cdot G_d'd = \frac{-20}{10s + 1}d$$

Thus, for a step disturbance d of magnitude 1, the output y will approach -20, which is much larger than the bound $|y| < 1$. This means that we need to use feedback control, which, as discussed in the next section, is hardly affected by the above model error. Although feedforward control by itself is not sufficient for this example, it has some benefits. This is because the feedforward controller reduces the effect of the disturbance, and the minimum bandwidth requirement for feedback control is reduced from $\omega_d \approx |k_d|/\tau_d = 100/10 = 10$ rad/s (no feedforward) to about $20/10 = 2$ rad/s (with feedforward).

5.12.2 Feedback control and uncertainty

With feedback control the closed-loop response with no model error is $y - r = S(G_dd - r)$ where $S = (I + GK)^{-1}$ is the sensitivity function. With model error we get

$$y' - r = S'(G_d'd - r) \qquad (5.94)$$

where $S' = (I + G'K)^{-1}$ can be written (see (A.147)) as

$$S' = S\frac{1}{1 + ET} \tag{5.95}$$

Here $E = (G' - G)/G$ is the relative error for G, and T is the complementary sensitivity function.

From (5.94) we see that the control error is only weakly affected by model error at frequencies where feedback is effective (where $|S| \ll 1$ and $T \approx 1$). For example, if we have integral action in the feedback loop and if the feedback system with model error is stable, then $S(0) = S'(0) = 0$ and the steady-state control error is zero even with model error.

Uncertainty at crossover. Although feedback control counteracts the effect of uncertainty at frequencies where the loop gain is large, uncertainty in the crossover frequency region can result in poor performance and even instability. This may be analyzed, for example, by considering the effect of the uncertainty on the gain margin, GM $= 1/|L(j\omega_{180})|$, where ω_{180} is the frequency where $\angle L$ is $-180°$; see (2.40). Most practical controllers behave as a constant gain K_o in the crossover region, so $|L(j\omega_{180})| \approx K_o|G(j\omega_{180})|$ where $\omega_{180} \approx \omega_u$ (since the phase lag of the controller is approximately zero at this frequency; see also Section 5.8). This observation yields the following approximate rule:

- *Define ω_u as the frequency where $\angle G(j\omega_u) = -180°$. Uncertainty which keeps $|G(j\omega_u)|$ approximately constant will not change the gain margin. Uncertainty which increases $|G(j\omega_u)|$ will decrease the gain margin and may yield instability.*

This rule is useful, for example, when evaluating the effect of parametric uncertainty. This is illustrated in the following example.

Example 5.15 *Consider a stable first-order delay process, $G(s) = ke^{-\theta s}/(1 + \tau s)$, where the parameters k, τ and θ are uncertain in the sense that they may vary with operating conditions. If we assume $\tau > \theta$ then $\omega_u \approx (\pi/2)/\theta$ and we derive*

$$|G(j\omega_u)| \approx \frac{2}{\pi}k\frac{\theta}{\tau} \tag{5.96}$$

We see that to keep $|G(j\omega_u)|$ constant we want $k\frac{\theta}{\tau}$ constant. If only the delay θ increases, then $|G(j\omega_u)|$ increases and we may get instability (as we expect). However, the uncertainty in the parameters is often coupled. For example, if θ and τ increase proportionally (which is quite common in practice) such that the ratio τ/θ remains constant, then stability is not affected. In another case the steady-state gain k may change with operating point, but this may not affect stability if the ratio k/τ, which determines the high-frequency gain, is unchanged.

The above example illustrates the importance of taking into account the *structure of the uncertainty*, e.g. the coupling between the uncertain parameters. A robustness analysis which assumes the uncertain parameters to be uncorrelated is generally conservative. This is further discussed in Chapters 7 and 8.

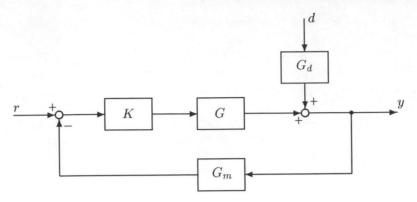

Figure 5.17: Feedback control system

5.13   Summary: controllability analysis with feedback control

We will now summarize the results of this chapter by a set of "controllability rules". We use the term "(input–output) controllability" since the bounds depend on the plant only; that is, are independent of the specific controller. Except for Rule 7, all requirements are fundamental, although some of the expressions, as seen from the derivations, are approximate (i.e. they may be off by a factor of 2 or so). However, for practical designs the bounds will need to be satisfied to get acceptable performance.

Consider the control system in Figure 5.17, where all the blocks are scalar. The model is

$$y = G(s)u + G_d(s)d; \quad y_m = G_m(s)y \tag{5.97}$$

Here $G_m(s)$ denotes the measurement transfer function and we assume $G_m(0) = 1$ (perfect steady-state measurement). The variables d, u, y and r are assumed to have been scaled as outlined in Section 1.4, and therefore $G(s)$ and $G_d(s)$ are the scaled transfer functions. Let ω_c denote the gain crossover frequency, defined as the frequency where $|L(j\omega)|$ crosses 1 from above. Let ω_d denote the frequency at which $|G_d(j\omega_d)|$ first crosses 1 from above.

The first step for controllability analysis with feedback control is to evaluate the bounds on the peaks of the different closed-loop transfer functions, i.e. S, T, KS, SG and SG_d, using formulae summarized in Table 5.1. We require that the peaks of all of these closed-loop transfer functions be small. For example, the performance requirement of keeping control error signal e small is satisfied, only if $\|S\|_\infty$ and $\|T\|_\infty$ are small. Similarly, it is necessary to ensure that $\|KS\|_\infty$ is small to avoid actuator saturation, which may destabilize the system. In addition, the following rules apply (Skogestad, 1996):

Rule 1. Speed of response to reject disturbances. *We approximately require $\omega_c > \omega_d$. More specifically, with feedback control we require $|S(j\omega)| \leq |1/G_d(j\omega)|$ $\forall \omega$. (See (5.76) and (5.79).)*

Rule 2. Speed of response to track reference changes. *We require $|S(j\omega)| \leq 1/R$ up to the frequency ω_r where tracking is required. (See (5.80).)*

Rule 3. Input constraints arising from disturbances. *For acceptable control ($|e| < 1$) we require $|G(j\omega)| > |G_d(j\omega)| - 1$ at frequencies where $|G_d(j\omega)| > 1$. For perfect control ($e = 0$) the requirement is $|G(j\omega)| > |G_d(j\omega)|$.* (See (5.82) and (5.84).)

Rule 4. Input constraints arising from setpoints. *We require $|G(j\omega)| > R - 1$ up to the frequency ω_r where tracking is required.* (See (5.85).)

Rule 5. Time delay θ in $G(s)G_m(s)$. *We approximately require $\omega_c < 1/\theta$.* (See (5.45).)

Rule 6. Tight control at low frequencies with a RHP-zero z in $G(s)G_m(s)$. *For a real RHP-zero we require $\omega_c < z/2$ and for an imaginary RHP-zero we approximately require $\omega_c < 0.86|z|$.* (See (5.52) and (5.53).)

> **Remark.** Strictly speaking, a RHP-zero only makes it impossible to have tight control in the frequency range close to the location of the RHP-zero. If we do not need tight control at low frequencies, then we may reverse the sign of the controller gain, and instead achieve tight control at higher frequencies. In this case we must for a RHP-zero z approximately require $\omega_c > 2z$. A special case is for plants with a zero at the origin; here we can achieve good transient control even though the control has no effect at steady-state.

Rule 7. Phase lag constraint. *We require in most practical cases (e.g. with PID control): $\omega_c < \omega_u$. Here the* ultimate frequency ω_u is where $\angle GG_m(j\omega_u) = -180°$. (See (5.65).)

Since time delays (Rule 5) and RHP-zeros (Rule 6) also contribute to the phase lag, one may in most practical cases combine Rules 5, 6 and 7 into the single rule: $\omega_c < \omega_u$ (Rule 7).

Rule 8. Real open-loop unstable pole in $G(s)$ at $s = p$. *We need high feedback gains to stabilize the system and we approximately require $\omega_c > 2p$.* (See (5.70).)

> *In addition, for unstable plants we need $|G_s(p)| > |G_{d,ms}(p)|$. Otherwise, the input may saturate when there are disturbances, and the plant cannot be stabilized; see (5.86).*

Most of the rules are illustrated graphically in Figure 5.18.

We have not formulated a rule to guard against model uncertainty. This is because, as given in (5.94) and (5.95), uncertainty has only a minor effect on feedback performance for SISO systems, except at frequencies where the relative uncertainty E approaches 100%, and we obviously have to detune the system. Also, since 100% uncertainty at a given frequency allows for the presence of a RHP-zero on the imaginary axis at this frequency ($G(j\omega) = 0$), it is already covered by Rule 6.

The rules are necessary conditions ("minimum requirements") to achieve acceptable control performance. They are not sufficient since among other things we have only considered one effect at a time.

The rules quantify the qualitative rules given in the introduction. For example, the rule "Control outputs that are not self-regulating" may be quantified as "Control outputs y for which $|G_d(j\omega)| > 1$ at some frequency" (Rule 1). Another important insight from Rule 1 is that a larger disturbance or a smaller specification on the control error requires faster

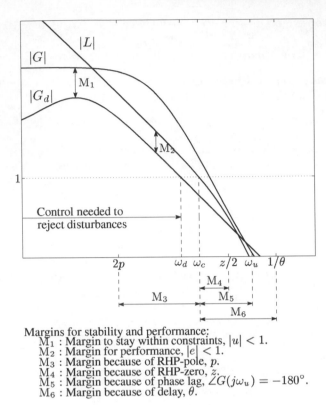

Margins for stability and performance:
M_1 : Margin to stay within constraints, $|u| < 1$.
M_2 : Margin for performance, $|e| < 1$.
M_3 : Margin because of RHP-pole, p.
M_4 : Margin because of RHP-zero, z.
M_5 : Margin because of phase lag, $\angle G(j\omega_u) = -180°$.
M_6 : Margin because of delay, θ.

Figure 5.18: Illustration of controllability requirements

response (higher bandwidth).[4] The rule "Select inputs that have a large effect on the outputs" may be quantified as "In terms of scaled variables, to avoid input saturation we must have $|G| > |G_d| - 1$ at frequencies where $|G_d| > 1$ (Rule 3), and we must have $|G| > R - 1$ at frequencies where setpoint tracking is desired (Rule 4)." The rule "Control outputs that have favourable dynamic and static characteristics" is quantified by Rule 3 ("want large gain to avoid input constraints") and Rules 4, 5 and 6 ("avoid time delay, RHP-zeros and large phase lag").

In summary, Rules 1, 2 and 8 tell us that we need high feedback gain ("fast control") in order to reject disturbances, to track setpoints and to stabilize the plant. On the other hand, Rules 5, 6 and 7 tell us that we must use low feedback gains in the frequency range where there are RHP-zeros or delays or where the plant has a lot of phase lag. We have formulated these requirements for high and low gain as bandwidth requirements. If they somehow are in conflict then the plant is not controllable and the only remedy is to introduce design modifications to the plant.

Sometimes the problem is that the disturbances are so large that the inputs saturate, or the required bandwidth is not achievable. To avoid the latter problem, we must at least require that the effect of the disturbance is less than 1 (in terms of scaled variables) at frequencies

---

[4] Another reason for preferring a large (scaled) gain from the inputs to the outputs is to be able to keep the plant close to its optimum by use of a constant setpoint policy, see "self-optimizing control" on page 390. Also note that here the scaling procedure is different than that used for controllability analysis.

beyond the bandwidth (Rule 1)

$$|G_d(j\omega)| < 1 \quad \forall \omega \geq \omega_c \tag{5.98}$$

where as found above we approximately require $\omega_c < 1/\theta$ (Rule 5), $\omega_c < z/2$ (Rule 6) and $\omega_c < \omega_u$ (Rule 7). Condition (5.98) may be used, as in the example of Section 5.15.3 below, to determine the size of equipment.

5.14 Summary: controllability analysis with feedforward control

The above controllability rules apply to feedback control, but we find that essentially the same conclusions apply to feedforward control when relevant. That is, if a plant is not controllable using feedback control, it is usually not controllable with feedforward control. A major difference, as shown below, is that a delay in $G_d(s)$ is an advantage for feedforward control ("it gives the feedforward controller more time to make the right action"). Also, a RHP-zero in $G_d(s)$ is also an advantage for feedforward control if $G(s)$ has a RHP-zero at the same location. Rules 3 and 4 on input constraints apply directly to feedforward control, but Rule 8 does not apply since unstable plants can only be stabilized by feedback control. The remaining rules in terms of performance and "bandwidth" do not apply directly to feedforward control.

Controllability can be analyzed by considering the feasibility of achieving perfect control. The feedforward controller is

$$u = K_d(s)d_m$$

where $d_m = G_{md}(s)d$ is the measured disturbance. The disturbance response with $r = 0$ becomes

$$e = Gu + G_d d = (GK_d G_{md} + G_d)d \tag{5.99}$$

(Reference tracking can be analyzed similarly by setting $G_{md} = 1$ and $G_d = -R$.)

Perfect control. From (5.99), $e = 0$ is achieved with the controller

$$K_d^{\text{perfect}} = -G^{-1}G_d G_{md}^{-1} \tag{5.100}$$

This assumes that K_d^{perfect} is stable and causal (no prediction), and so $GG_d^{-1}G_{md}$ should have no RHP-zeros and no (positive) delay. From this we find that a delay (or RHP-zero) in $G_d(s)$ is an advantage if it cancels a delay (or RHP-zero) in GG_{md}.

Ideal control. If perfect control is not possible, then one may analyze controllability by considering an "ideal" feedforward controller, K_d^{ideal}, which is (5.100) modified to be stable and causal (no prediction). The controller is ideal in that it assumes we have a perfect model. Controllability is then analyzed by using K_d^{ideal} in (5.99). An example is given below in (5.109) and (5.110) for a first-order delay process.

Model uncertainty. As discussed in Section 5.12, model uncertainty is a more serious problem for feedforward than for feedback control because there is no correction from the output measurement. For disturbance rejection, we have from (5.93) that the plant is *not* controllable with feedforward control if the relative model error for G/G_d at any frequency exceeds $1/|G_d|$. Here G_d is the scaled disturbance model. For example, if $|G_d(j\omega)| = 10$

then the error in G/G_d must not exceed 10% at this frequency. In practice, this means that feedforward control has to be combined with feedback control if the output is sensitive to the disturbance (i.e. if $|G_d|$ is much larger than 1 at some frequency).

Combined feedback and feedforward control. To analyze controllability in this case we may assume that the feedforward controller K_d has already been designed. Then from (5.99) the controllability of the remaining feedback problem can be analyzed using the rules in Section 5.13 if $G_d(s)$ is replaced by

$$\widehat{G}_d(s) = GK_dG_{md} + G_d \tag{5.101}$$

However, one must be aware that the feedforward control may be very sensitive to model error, so the benefits of feedforward may be less in practice.

Conclusion. From (5.101) we see that the primary potential benefit of feedforward control is to reduce the effect of the disturbance and make $\widehat{G}_d$ less than 1 at frequencies where feedback control is not effective due to, for example, a delay or a large phase lag in $GG_m(s)$.

5.15 Applications of controllability analysis

5.15.1 First-order delay process

Problem statement. Consider disturbance rejection for the following process:

$$G(s) = k\frac{e^{-\theta s}}{1 + \tau s}; \quad G_d(s) = k_d\frac{e^{-\theta_d s}}{1 + \tau_d s} \tag{5.102}$$

In addition there are measurement delays θ_m for the output and θ_{md} for the disturbance. All parameters have been appropriately scaled such that at each frequency $|u| < 1$, $|d| < 1$ and we want $|e| < 1$. Assume $|k_d| > 1$. Treat the two cases of (i) feedback control only, and (ii) feedforward control only, and carry out the following:

(a) For each of the eight parameters in this model explain qualitatively what value you would choose from a controllability point of view (with descriptions such as large, small, value has no effect).

(b) Give quantitative relationships between the parameters which should be satisfied to achieve controllability. Assume that appropriate scaling has been applied in such a way that the disturbance is less than 1 in magnitude, and that the input and the output are required to be less than 1 in magnitude.

Solution. (a) *Qualitative.* We want the input to have a "large, direct and fast effect" on the output, while we want the disturbance to have a "small, indirect and slow effect". By "direct" we mean without any delay or inverse response. This leads to the following conclusion. For both feedback and feedforward control we want k and τ_d large, and τ, θ and k_d small. For feedforward control we also want θ_d large (we then have more time to react), but for feedback the value of θ_d does not matter; it translates time, but otherwise has no effect. Clearly, we want θ_m small for feedback control (it is not used for feedforward), and we want θ_{md} small for feedforward control (it is not used for feedback).

(b) *Quantitative.* To stay within the input constraints ($|u| < 1$) we must require from Rule 3 that $|G(j\omega)| > |G_d(j\omega)|$ for frequencies $\omega < \omega_d$. Specifically, for both feedback and feedforward control

$$\boxed{k > k_d; \quad k/\tau > k_d/\tau_d} \tag{5.103}$$

Now consider performance where the results for feedback and feedforward control differ. (i) First consider *feedback control*. From Rule 1 we need for acceptable performance ($|e| < 1$) with disturbances

$$\omega_d \approx k_d/\tau_d < \omega_c \tag{5.104}$$

On the other hand, from Rule 5 we require for stability and performance

$$\omega_c < 1/\theta_{tot} \tag{5.105}$$

where $\theta_{tot} = \theta + \theta_m$ is the total delay around the loop. The combination of (5.104) and (5.105) yields the following requirement for controllability:

$$\boxed{\text{Feedback:} \quad \theta + \theta_m < \tau_d/k_d} \tag{5.106}$$

(ii) For *feedforward control*, any delay for the disturbance itself yields a smaller "net delay", and to have $|e| < 1$ we need "only" require

$$\boxed{\text{Feedforward:} \quad \theta + \theta_{md} - \theta_d < \tau_d/k_d} \tag{5.107}$$

Proof of (5.107): Introduce $\widehat{\theta} = \theta + \theta_{md} - \theta_d$, and consider first the case with $\widehat{\theta} \leq 0$ (so (5.107) is clearly satisfied). In this case perfect control is possible using the controller (5.100),

$$K_d^{\text{perfect}} = -G^{-1}G_dG_{md}^{-1} = -\frac{k_d}{k}\frac{1+\tau s}{1+\tau_d s}e^{\widehat{\theta}s} \tag{5.108}$$

so we can even achieve $e = 0$. Next, consider $\widehat{\theta} > 0$. Perfect control is not possible, so instead we use the "ideal" controller obtained by deleting the prediction $e^{\widehat{\theta}s}$,

$$K_d^{\text{ideal}} = -\frac{k_d}{k}\frac{1+\tau s}{1+\tau_d s} \tag{5.109}$$

From (5.99) the response with this controller is

$$e = (GK_d^{\text{ideal}}G_{md} + G_d)d = \frac{k_d e^{-\theta_d s}}{1+\tau_d s}(1 - e^{-\widehat{\theta}s})d \tag{5.110}$$

and to achieve $|e|/|d| < 1$ we must require $\frac{k_d}{\tau_d}\widehat{\theta} < 1$ (using asymptotic values and $1 - e^{-x} \approx x$ for small x) which is equivalent to (5.107). $\qquad\square$

5.15.2 Application: room heating

Consider the problem of maintaining a room at constant temperature, as discussed in Section 1.5, see Figure 1.2. Let y be the room temperature, u the heat input and d the outdoor temperature. Feedback control should be used. Let the measurement delay for temperature (y) be $\theta_m = 100$ s.

1. Is the plant controllable with respect to disturbances?
2. Is the plant controllable with respect to setpoint changes of magnitude $R = 3$ (± 3 K) when the desired response time for setpoint changes is $\tau_r = 1000$ s (17 min)?

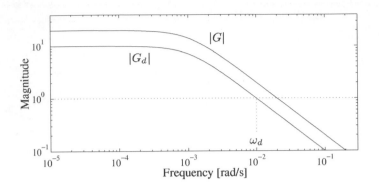

Figure 5.19: Frequency responses for room heating example

Solution. A critical part of controllability analysis is scaling. A model in terms of scaled variables was derived in (1.32)

$$G(s) = \frac{20}{1000s + 1}; \quad G_d(s) = \frac{10}{1000s + 1} \tag{5.111}$$

The frequency responses of $|G|$ and $|G_d|$ are shown in Figure 5.19.

1. Disturbances. From Rule 1 feedback control is necessary up to the frequency $\omega_d = 10/1000 = 0.01$ rad/s, where $|G_d|$ crosses 1 in magnitude ($\omega_c > \omega_d$). This is exactly the same frequency as the upper bound given by the delay, $1/\theta = 0.01$ rad/s ($\omega_c < 1/\theta$). We therefore conclude that the system is barely controllable for this disturbance. From Rule 3 no problems with input constraints are expected since $|G| > |G_d|$ at all frequencies. To support these conclusions, we design a series PID controller of the form $K(s) = K_c \frac{1+\tau_I s}{\tau_I s} \frac{\tau_D s+1}{0.1\tau_D s+1}$. With $G(s) = \frac{20e^{-100s}}{1000s+1}$, the SIMC PI tunings (page 57) for this process are $K_c = 0.25$ (scaled variables) and $\tau_I = 800$ s. This yields smooth responses, but the output peak exceeds 1.7 in response to the disturbance and the settling to the new steady-state is slow. To reduce the output peak below 1, it is necessary to increase K_c to about 0.4. Reducing τ_I from 800 s to 200 s reduces the settling time. The introduction of derivative action with $\tau_D = 60$ s gives better robustness and fewer oscillations. The final controller settings are $K_c = 0.4$ (scaled variables), $\tau_I = 200$ s and $\tau_D = 60$ s. The closed-loop simulation for a unit step disturbance (corresponding to a sudden 10 K increase in the outdoor temperature) is shown in Figure 5.20(a). The output error exceeds its allowed value of 1 for a very short time after about 100 s, but then returns quite quickly to zero. The input goes down to about -0.8 and thus remains within its allowed bound of ± 1.

2. Setpoints. The plant is controllable with respect to the desired setpoint changes. First, the delay is 100 s which is much smaller than the desired response time of 1000 s, and thus poses no problem. Second, $|G(j\omega)| \geq R = 3$ up to about $\omega_1 = 0.007$ [rad/s] which is seven times higher than the required $\omega_r = 1/\tau_r = 0.001$ [rad/s]. This means that input constraints pose no problem. In fact, we should be able to achieve response times of about $1/\omega_1 = 150$ s without reaching the input constraints. This is confirmed by the simulation in Figure 5.20(b) for a desired setpoint change $3/(150s + 1)$ using the same PID controller as above.

Exercise 5.12* *Perform closed-loop simulations with the SIMC PI controller and the proposed PID*

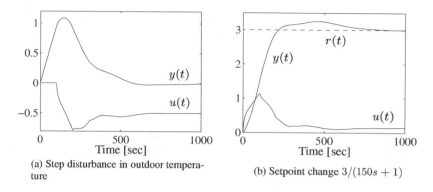

(a) Step disturbance in outdoor temperature

(b) Setpoint change $3/(150s + 1)$

Figure 5.20: PID feedback control of room heating example

controller for the room heating process. Also compute the robustness parameters (GM, PM, M_S and M_T) for the two designs.

5.15.3 Application: neutralization process

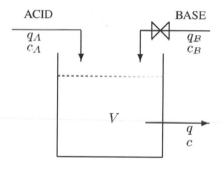

Figure 5.21: Neutralization process with one mixing tank

The following application is interesting in that it shows how the controllability analysis tools may assist the engineer in redesigning the process to make it controllable.

Problem statement. Consider the process in Figure 5.21, where a strong acid with pH $= -1$ (yes, a negative pH is possible – it corresponds to $c_{H+} = 10$ mol/l) is neutralized by a strong base (pH = 15) in a mixing tank with volume $V = 10$ m^3. We want to use feedback control to keep the pH in the product stream (output y) in the range 7 ± 1 ("salt water") by manipulating the amount of base, q_B (input u), in spite of variations in the flow of acid, q_A (disturbance d). The delay in the pH measurement is $\theta_m = 10$ s.

To achieve the desired product with pH = 7 one must exactly balance the inflow of acid (the disturbance) by the addition of base (the manipulated input). Intuitively, one might expect that the main control problem is to adjust the base accurately by means of a very accurate valve. However, as we will see, this "feedforward" way of thinking is misleading, and the main hurdle to good control is the need for very fast response times.

We take the controlled output to be the excess of acid, c [mol/l], defined as $c = c_{H+} -$

c_{OH^-}, which avoids the need to include a chemical reaction term in the model. In terms of this variable c, the control objective is to keep $|c| \leq c_{\max} = 10^{-6}$ mol/l, and the plant is a simple mixing process modelled by

$$\frac{d}{dt}(Vc) = q_A c_A + q_B c_B - qc \qquad (5.112)$$

The nominal values for the acid and base flows are $q_A^* = q_B^* = 0.005$ [m³/s] resulting in a product flow $q^* = 0.01$ [m³/s] $= 10$ [l/s]. Here superscript $*$ denotes the steady-state value. We divide each variable by its maximum deviation to get the following scaled variables:

$$y = \frac{c}{10^{-6}}; \quad u = \frac{q_B}{q_B^*}; \quad d = \frac{q_A}{0.5q_A^*} \qquad (5.113)$$

Then the appropriately scaled linear model for one tank becomes

$$G_d(s) = \frac{k_d}{1 + \tau_h s}; \quad G(s) = \frac{-2k_d}{1 + \tau_h s}; \quad k_d = 2.5 \cdot 10^6 \qquad (5.114)$$

where $\tau_h = V/q = 1000$ s is the residence time for the liquid in the tank. Note that the steady-state gain in terms of scaled variables is more than a million, so the output is extremely sensitive to both the input and the disturbance. The reason for this high gain is the much higher concentration in the two feed streams, compared to that desired in the product stream. The question is: can acceptable control be achieved?

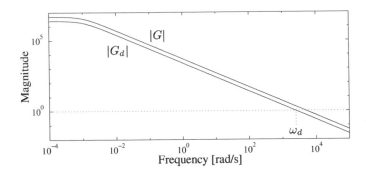

Figure 5.22: Frequency responses for the neutralization process with one mixing tank

Controllability analysis. The frequency responses of $G_d(s)$ and $G(s)$ are shown graphically in Figure 5.22. From Rule 2, input constraints do not pose a problem since $|G| = 2|G_d|$ at all frequencies. The main control problem is the high disturbance sensitivity, and from (5.104) (Rule 1) we find the frequency up to which feedback is needed

$$\omega_d \approx k_d/\tau = 2500 \text{ rad/s} \qquad (5.115)$$

This requires a response time of $1/2500 = 0.4$ milliseconds which is clearly impossible in a process control application, and is in any case much less than the measurement delay of 10 s.

Design change: multiple tanks. The only way to improve controllability is to modify the process. This is done in practice by performing the neutralization in several steps as

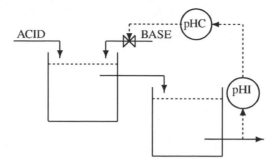

Figure 5.23: Neutralization process with two tanks and one controller

illustrated in Figure 5.23 for the case of two tanks. This is similar to playing golf where it is often necessary to use several strokes to get to the hole. With n equal mixing tanks in series the transfer function for the effect of the disturbance becomes

$$G_d(s) = k_d h_n(s); \quad h_n(s) = \frac{1}{(\frac{\tau_h}{n} s + 1)^n} \tag{5.116}$$

where $k_d = 2.5 \cdot 10^6$ is the gain for the mixing process, $h_n(s)$ is the transfer function of the mixing tanks, and τ_h is the total residence time, V_{tot}/q. The magnitude of $h_n(s)$ as a function of frequency is shown in Figure 5.24 for one to four equal tanks in series.

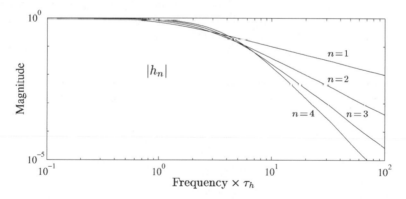

Figure 5.24: Frequency responses for n tanks in series with the same total residence time τ_h; $h_n(s) = 1/(\frac{\tau_h}{n} s + 1)^n$, $n = 1, 2, 3, 4$

From controllability Rules 1 and 5, we must at least require for acceptable disturbance rejection that

$$\boxed{|G_d(j\omega_\theta)| \le 1} \quad \omega_\theta \triangleq 1/\theta \tag{5.117}$$

where θ is the delay in the feedback loop. Thus, one purpose of the mixing tanks $h_n(s)$ is to reduce the effect of the disturbance by a factor k_d $(= 2.5 \cdot 10^6)$ at the frequency ω_θ $(= 0.1$ [rad/s]$)$, i.e. $|h_n(j\omega_\theta)| \le 1/k_d$. With $\tau_h = V_{tot}/q$ we obtain the following minimum value for the total volume for n equal tanks in series:

$$V_{tot} = q\theta n \sqrt{(k_d)^{2/n} - 1} \tag{5.118}$$

where $q = 0.01$ m^3/s. With $\theta = 10$ s we then find that the following designs have the same controllability with respect to disturbance rejection:

No. of tanks n	Total volume V_{tot} [m^3]	Volume each tank [m^3]
1	250000	250000
2	316	158
3	40.7	13.6
4	15.9	3.98
5	9.51	1.90
6	6.96	1.16
7	5.70	0.81

With one tank we need a volume corresponding to that of a supertanker to get acceptable controllability. The minimum total volume is obtained with 18 tanks of about 203 litres each – giving a total volume of 3.662 m^3. However, taking into account the additional cost for extra equipment such as piping, mixing, measurements and control, we would probably select a design with 3 or 4 tanks for this example.

Control system design. We are not quite finished yet. The condition $|G_d(j\omega_\theta)| \leq 1$ in (5.117), which formed the basis for redesigning the process, may be optimistic because it only ensures that we have $|S| < 1/|G_d|$ at the crossover frequency $\omega_B \approx \omega_c \approx \omega_\theta$. However, from Rule 1 we also require that $|S| < 1/|G_d|$, or approximately $|L| > |G_d|$, at frequencies lower than w_c, and this may be difficult to achieve since $G_d(s) = k_d h(s)$ is of order n, where n is the number of tanks. The problem is that this requires $|L|$ to drop steeply with frequency, which results in a large negative phase for L, whereas for stability and performance the slope of $|L|$ at crossover should not be steeper than -1, approximately (see Section 2.6.2).

Thus, the control system in Figure 5.23 with a single feedback controller will *not* achieve the desired performance. The solution is to install a *local feedback* control system on each tank and to add base in each tank as shown in Figure 5.25. This is another *plant design change*

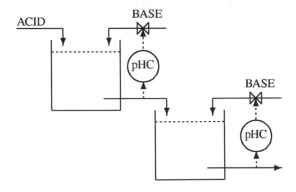

Figure 5.25: Neutralization process with two tanks and two controllers

since it requires an additional measurement and actuator for each tank. Consider the case of

n tanks in series. With n controllers the overall closed-loop response from a disturbance into the first tank to the pH in the last tank becomes

$$y = G_d \prod_{i=1}^{n} \left(\frac{1}{1 + L_i} \right) d \approx \frac{G_d}{L} d, \quad L \triangleq \prod_{i=1}^{n} L_i \tag{5.119}$$

where $G_d = \prod_{i=1}^{n} G_i$ and $L_i = G_i K_i$, and the approximation applies at low frequencies where feedback is effective.

In this case, we can design each loop $L_i(s)$ with a slope of -1 and bandwidth $\omega_c \approx \omega_\theta$, such that the overall loop transfer function L has slope $-n$ and achieves $|L| > |G_d|$ at all frequencies lower than ω_d (the size of the tanks is selected as before such that $\omega_d \approx \omega_\theta$). Thus, our analysis confirms the usual recommendation of adding base gradually and having one pH controller for each tank (McMillan, 1984, p. 208). It seems unlikely that any other control strategy can achieve a sufficiently high roll-off for $|L|$.

In summary, this application has shown how a simple controllability analysis may be used to make decisions on both the appropriate size of the equipment, and the selection of actuators and measurements for control. Our conclusions are in agreement with what is used in industry. Importantly, we arrived at these conclusions without having to design any controllers or perform any simulations. Of course, as a final test, the conclusions from the controllability analysis should be verified by simulations using a nonlinear model.

Exercise 5.13 Comparison of local feedback and cascade control. *Explain why a cascade control system with two measurements (pH in each tank) and only one manipulated input (the base flow into the first tank) will not achieve as good a performance as the control system in Figure 5.25 where we use local feedback with __two__ manipulated inputs (one for each tank).*

The following exercise further considers the use of buffer tanks for reducing quality (concentration, temperature) disturbances in chemical processes.

Exercise 5.14 * *(a) The effect of a concentration disturbance must be reduced by a factor of 100 at the frequency 0.5 rad/min. The disturbances should be dampened by use of buffer tanks and the objective is to minimize the total volume. How many tanks in series should one have? What is the total residence time?*

(b) The feed to a distillation column has large variations in concentration and the use of one buffer tank is suggested to dampen these. The effect of the feed concentration d on the product composition y is given by (scaled variables, time in minutes)

$$G_d(s) = e^{-s}/3s$$

That is, after a step in d the output y will, after an initial delay of 1 min, increase in a ramp-like fashion and reach its maximum allowed value (which is 1) after another 3 minutes. Feedback control should be used and there is an additional measurement delay of 5 minutes. What should be the residence time in the tank?

(c) Show that in terms of minimizing the total volume for buffer tanks in series, it is optimal to have buffer tanks of equal size.

(d) Is there any reason to have buffer tanks in parallel (they must not be of equal size because then one may simply combine them)?

(e) What about parallel pipes in series (pure delay). Is this a good idea?

Buffer tanks are also used in chemical processes to dampen liquid flow rate disturbances (or gas pressure disturbances). This is the topic of the following exercise.

Exercise 5.15 *Let $d_1 = q_{in}$ [m^3/s] denote a flow rate which acts as a disturbance to the process. We add a buffer tank (with liquid volume V [m^3]), and use a "slow" level controller K such that the outflow $d_2 = q_{out}$ (the "new" disturbance) is smoother than the inflow q_{in} (the "original" disturbance). The idea is to increase or decrease temporarily the liquid volume in the tank to avoid sudden changes in q_{out}. Note that the steady-state value of q_{out} must equal that of q_{in}.*

A material balance yields $V(s) = (q_{in}(s) - q_{out}(s))/s$ and with a level controller $q_{out}(s) = K(s)V(s)$ we find that

$$d_2(s) = \underbrace{\frac{K(s)}{s + K(s)}}_{h(s)} d_1(s) \qquad\qquad (5.120)$$

The design of a buffer tank for a flow rate disturbance then consists of two steps:

1. *Design the level controller $K(s)$ such that $h(s)$ has the desired shape (e.g. determined by a controllability analysis of how d_2 affects the remaining process; note that we must always have $h(0) = 1$).*

2. *Design the size of the tank (determine its volume V_{max}) such that the tank does not overflow or go empty for the expected disturbances in $d_1 = q_{in}$.*

Problem statement. *(a) Assume the inflow varies in the range $q_{in}^* \pm 100\%$ where q_{in}^* is the nominal value, and apply this stepwise procedure to two cases:*
  (i) The desired transfer function is $h(s) = 1/(\tau s + 1)$.
  (ii) The desired transfer function is $h(s) = 1/(\tau_2 s + 1)^2$.
(b) Explain why it is usually not recommended to have integral action in $K(s)$.

(c) In case (ii) one could alternatively use two tanks in series with controllers designed as in (i). Explain why this is most likely not a good solution. (Solution: The required total volume is the same, but the cost of two smaller tanks is larger than one large tank.)

5.15.4 Additional exercises

Exercise 5.16 * *What information about a plant is important for controller design, and in particular, in which frequency range is it important to know the model well? To answer this problem you may think about the following sub-problems:*

(a) Explain what information about the plant is used for Ziegler–Nichols tuning of a SISO PID controller.

(b) Is the steady-state plant gain $G(0)$ important for controller design? (As an example consider the plant $G(s) = \frac{1}{s+a}$ with $|a| \leq 1$ and design a P controller $K(s) = K_c$ such that $\omega_c = 100$. How does the controller design and the closed-loop response depend on the steady-state gain $G(0) = 1/a$?)

Exercise 5.17 *Let $G(s) = K_2 e^{-0.5s} \frac{1}{(30s+1)(Ts+1)}$, and $G_d(s) = G(s)H(s)$ where $H(s) = K_1 e^{-\theta_1 s}$. The measurement device for the output has transfer function $G_m(s) = e^{-\theta_2 s}$. The unit for time is seconds. The nominal parameter values are: $K_1 = 0.24$, $\theta_1 = 1$ [s], $K_2 = 38$, $\theta_2 = 5$ [s] and $T = 2$ [s].*

(a) Assume all variables have been appropriately scaled. Is the plant input–output controllable?

(b) What is the effect on controllability of changing one model parameter at a time in the following ways?

1. *θ_1 is reduced to 0.1 [s].*
2. *θ_2 is reduced to 2 [s].*
3. *K_1 is reduced to 0.024.*
4. *K_2 is reduced to 8.*
5. *T is increased to 30 [s].*

Exercise 5.18 * *A heat exchanger is used to exchange heat between two streams: a coolant with flow rate q (1 ± 1 kg/s) is used to cool a hot stream with inlet temperature T_0 ($100 \pm 10°C$) to the outlet* *temperature T (which should be $60 \pm 10°C$). The measurement delay for T is 3 s. The main disturbance is on T_0. The following model in terms of deviation variables is derived from heat balances:*

$$T(s) = \frac{8}{(60s + 1)(12s + 1)}q(s) + \frac{0.6(20s + 1)}{(60s + 1)(12s + 1)}T_0(s) \qquad (5.121)$$

where T and T_0 are in $°C$, q is in kg/s, and the unit for time is seconds. Derive the scaled model. Is the plant controllable with feedback control? (Solution: The delay poses no problem (performance), but the effect of the disturbance is a bit too large at high frequencies (input saturation), so the plant is not controllable.)

5.16 Conclusion

The chapter has presented a frequency domain controllability analysis for scalar systems applicable to both feedback and feedforward control. We summarized our findings in terms of eight controllability rules; see page 206. These rules are necessary conditions ("minimum requirements") to achieve acceptable control performance. They are not sufficient since among other things they only consider one effect at a time. The rules may be used to determine whether or not a given plant is controllable. The method has been applied to a pH neutralization process, and it is found that the heuristic design rules given in the literature follow directly. The key steps in the analysis are to consider disturbances and to scale the variables properly.

The tools presented in this chapter may also be used to study the effectiveness of adding extra manipulated inputs or extra measurements (cascade control). They may also be generalized to multivariable plants where directionality becomes a further crucial consideration. Interestingly, a direct generalization to decentralized control of multivariable plants is rather straightforward and involves the CLDG and the PRGA; see page 449 in Chapter 10.

6

LIMITATIONS ON PERFORMANCE IN MIMO SYSTEMS

In this chapter, we generalize the results of Chapter 5 to MIMO systems. Most of the results on fundamental limitations and controllability analysis for SISO systems also hold for MIMO systems with the additional consideration of directions. Thus, we focus on results that hold exclusively for MIMO systems or are non-trivial extensions of similar results for SISO systems. We first discuss fundamental limitations on the sensitivity and complementary sensitivity functions imposed by the presence of RHP-zeros. We then consider separately the issues of functional controllability, RHP-zeros, RHP-poles, disturbances, input constraints and uncertainty. Finally, we summarize the main steps in a procedure for analyzing the input–output controllability of MIMO plants.

6.1 Introduction

In a MIMO system, the plant gain, RHP-zeros, delays, RHP-poles and disturbances each have *directions* associated with them. This makes it more difficult to consider their effects separately, as we did in the SISO case, but we will nevertheless see that most of the SISO results can be generalized.

We will quantify the directionality of the various effects in G and G_d by their *output* directions:

- y_z: output direction of a RHP-zero, $G(z)u_z = 0 \cdot y_z$, see (4.71)
- y_p: output direction of a RHP-pole, $G(p)u_p = \infty \cdot y_p$, see (4.64)
- y_d: output direction of a disturbance, $y_d = \frac{1}{\|g_d\|_2} g_d$, see (6.42)
- u_i: i'th output direction (singular vector) of the plant, $Gv_i = \sigma_i u_i$, see (3.38)[1]

All these are $l \times 1$ vectors where l is the number of outputs. y_z and y_p are fixed complex vectors, while $y_d(s)$ and $u_i(s)$ are frequency dependent (s may here be viewed as a generalized complex frequency; in most cases $s = j\omega$). The vectors are normalized such that they have Euclidean length 1,

$$\|y_z\|_2 = 1, \quad \|y_p\|_2 = 1, \quad \|y_d(s)\|_2 = 1, \quad \|u_i(s)\|_2 = 1$$

---

[1] Note that u_i here is the i'th output singular vector, and *not* the i'th input.

Multivariable Feedback Control: Analysis and Design. Second Edition
S. Skogestad and I. Postlethwaite   © 2005, 2006 John Wiley & Sons, Ltd

We may also consider the associated input directions of G. However, these directions are usually of less interest since we are primarily concerned with the performance at the output of the plant.

The angles between the various output directions can be quantified using their inner products: $|y_z^H y_p|$, $|y_z^H y_d|$, etc. The inner product gives a number between 0 and 1, and from this we can define the angle in the first quadrant, see (A.114). For example, the output angle between a pole and a zero is

$$\phi = \cos^{-1} |y_z^H y_p|$$

where $\cos^{-1}$ denotes arccos.

We assume throughout this chapter that the models have been scaled as outlined in Section 1.4. The scaling procedure is the same as that for SISO systems, except that the scaling factors D_u, D_d, D_r and D_e are *diagonal matrices* with elements equal to the maximum change in each variable u_i, d_i, r_i and e_i. The control error in terms of scaled variables is then

$$e = y - r = Gu + G_d d - R\tilde{r}$$

where at each frequency we have $\|u(\omega)\|_{\max} \leq 1$, $\|d(\omega)\|_{\max} \leq 1$ and $\|\tilde{r}(\omega)\|_{\max} \leq 1$, and the control objective is to achieve $\|e(\omega)\|_{\max} < 1$.

Remark 1 Here $\| \cdot \|_{\max}$ is the vector infinity-norm: that is, the absolute value of the largest element in the vector. This norm is sometimes denoted $\| \cdot \|_\infty$, but this is not used here to avoid confusing it with the $\mathcal{H}_\infty$ norm of the transfer function (where the ∞ denotes the maximum over frequency rather than the maximum over the elements of the vector).

Remark 2 As for SISO systems, we see that reference changes may be analyzed as a special case of disturbances by replacing G_d by $-R$.

Remark 3 Whether various disturbances and reference changes should be considered separately or simultaneously is a matter of design philosophy. In this chapter, we mainly consider their effects separately, on the grounds that it is unlikely for several disturbances to attain their worst values simultaneously. This leads to necessary conditions for acceptable performance, which involve the elements of different matrices rather than matrix norms.

6.2   Fundamental limitations on sensitivity

6.2.1 S plus T is the identity matrix

From the identity $S + T = I$ and (A.51), we get

$$|\bar{\sigma}(S) - 1| \leq \bar{\sigma}(T) \leq \bar{\sigma}(S) + 1 \tag{6.1}$$

$$|\bar{\sigma}(T) - 1| \leq \bar{\sigma}(S) \leq \bar{\sigma}(T) + 1 \tag{6.2}$$

These can be combined to get

$$|\bar{\sigma}(S) - \bar{\sigma}(T)| \leq 1 \tag{6.3}$$

Thus, the magnitudes of $\bar{\sigma}(S)$ and $\bar{\sigma}(T)$ differ by at most 1 at a given frequency, so $\bar{\sigma}(S)$ is large if and only if $\bar{\sigma}(T)$ is large. For example, if $\bar{\sigma}(T)$ is 5 at a given frequency, then $\bar{\sigma}(S)$ must be between 4 and 6 at this frequency. The bounds (6.1) and (6.2) also show that we cannot have both S and T small (close to 0) simultaneously.

6.2.2 Interpolation constraints

RHP-zero. If $G(s)$ has a RHP-zero at z with output direction y_z, then for internal stability of the feedback system the following interpolation constraints must apply:

$$\boxed{y_z^H T(z) = 0; \quad y_z^H S(z) = y_z^H}$$
(6.4)

In words, (6.4) says that T must have a RHP-zero in the same direction as G, and that $S(z)$ has an eigenvalue of 1 corresponding to the left eigenvector y_z.

Proof of (6.4): From (4.71) there exists an output direction y_z such that $y_z^H G(z) = 0$. For internal stability, the controller cannot cancel the RHP-zero and it follows that $L = GK$ has a RHP-zero in the same direction, i.e. $y_z^H L(z) = 0$. Now $S = (I + L)^{-1}$ is stable and has no RHP-pole at $s = z$. It then follows from $T = LS$ that $y_z^H T(z) = 0$ and $y_z^H (I - S) = 0$. □

RHP-pole. If $G(s)$ has a RHP-pole at p with output direction y_p, then for internal stability the following interpolation constraints apply:

$$\boxed{S(p)y_p = 0; \quad T(p)y_p = y_p}$$
(6.5)

Proof of (6.5): The square matrix $L(p)$ has a RHP-pole at $s = p$, and if we assume that $L(s)$ has no RHP-zeros at $s = p$ then $L^{-1}(p)$ exists and from (4.75) there exists an output pole direction y_p such that

$$L^{-1}(p)y_p = 0$$
(6.6)

Since T is stable, it has no RHP-pole at $s = p$, so $T(p)$ is finite. It then follows, from $S = TL^{-1}$, that $S(p)y_p = T(p)L^{-1}(p)y_p = 0$ and $T(p) = (I - S(p))y_p = y_p$. □

Similar constraints apply to L_I, S_I and T_I, but these are in terms of the input zero and pole directions, u_z and u_p.

6.2.3 Sensitivity integrals

For SISO systems we presented several integral constraints on sensitivity (the waterbed effects). These may be generalized to MIMO systems by using the determinant or the singular values of S, see Boyd and Barratt (1991) and Freudenberg and Looze (1988). For example, the generalization of the Bode sensitivity integral in (5.5) may be written

$$\int_0^\infty \ln | \det S(j\omega)|d\omega = \sum_j \int_0^\infty \ln \sigma_j(S(j\omega))d\omega = \pi \cdot \sum_{i=1}^{N_p} Re(p_i)$$
(6.7)

For a stable $L(s)$, the integral is zero. Other generalizations are also available, see Chen (1995), Zhou et al. (1996) and Chen (2000). However, although these integral relationships are interesting, it seems difficult to derive concrete bounds on achievable performance from them.

6.3 Fundamental limitations: bounds on peaks

Based on the interpolation constraints presented in Section 6.2.2, one may derive lower bounds on various closed-loop transfer functions. The bounds are direct generalizations of those found for SISO systems, see page 172, and the comments and interpretations made for SISO systems carry over directly if we take the directions into account. The results presented in this section are from Section V in the paper by Chen (2000), unless otherwise stated. The derivations of bounds of this kind go back to the work of Zames (1981).

6.3.1 Minimum peaks for S and T

In the following, $M_{S,\min}$ and $M_{T,\min}$ denote the lowest achievable values for $\|S\|_\infty$ and $\|T\|_\infty$, respectively, using any stabilizing controller K. That is, we define

$$M_{S,\min} \triangleq \min_K \|S\|_\infty, \quad M_{T,\min} \triangleq \min_K \|T\|_\infty$$

Theorem 6.1 Sensitivity and complementary sensitivity peaks. *Consider a rational plant $G(s)$ (with no time delay). Let z_i be the N_z RHP-zeros of $G(s)$ with (unit) output zero direction vectors $y_{z,i}$. Let p_i be the N_p RHP-poles of $G(s)$ with (unit) output pole direction vectors $y_{p,i}$. Furthermore, assume that z_i and p_i are all distinct. Then we have the following tight lower bound on $\|S\|_\infty$ and $\|T\|_\infty$:*

$$M_{S,\min} = M_{T,\min} = \sqrt{1 + \bar\sigma^2 \left(Q_z^{-1/2} Q_{zp} Q_p^{-1/2} \right)} \tag{6.8}$$

where the elements of the $N_z \times N_z$ matrix Q_z, $N_p \times N_p$ matrix Q_p and $N_z \times N_p$ matrix Q_{zp} are given by Chen (2000) as

$$[Q_z]_{ij} = \frac{y_{z,i}^H y_{z,j}}{z_i + \bar z_j}, \quad [Q_p]_{ij} = \frac{y_{p,i}^H y_{p,j}}{\bar p_i + p_j}, \quad [Q_{zp}]_{ij} = \frac{y_{z,i}^H y_{p,j}}{z_i - p_j} \tag{6.9}$$

Note that (6.8) gives a tight bound for any number of RHP-poles and RHP-zeros.

Example 6.1 *Consider the SISO plant*

$$G(s) = \frac{(s-1)(s-3)}{(s-2)(s+1)^2}$$

For this plant we have $z_1 = 1$, $z_2 = 3$, $p_1 = 2$, and since this is a SISO plant, all direction vectors y_z and y_p are 1. Since we have RHP-zeros close to the RHP-pole we expect that control is fundamentally difficult. This is verified from (6.8). In Matlab, we write Qz = [1/2 1/4; 1/4 1/6]; Qp = [1/4]; Qpz = [-1 1]; msmin = sqrt(1+svd(sqrtm(inv(Qp))*Qpz*sqrtm(inv(Qz))))^2) *and find $M_{S,\min} = M_{T,\min} = 15$. This also agrees with the bound (5.23) for a SISO plant with a single RHP-pole:*

$$M_{S,\min} = M_{T,\min} = \prod_{j=1}^{N_z} \frac{|z_j + p|}{|z_j - p|} = \frac{|1+2|}{|1-2|} \cdot \frac{|3+2|}{|3-2|} = 3 \cdot 5 = 15$$

We see from the factor $\frac{y_{z,j}^H y_{p,i}}{z_j - p_i}$ in Q_{pz} that the bound will be large if we have a RHP-pole p_i close to RHP-zero z_j and with directions aligned such that $y_{z,j}^H y_{p,i}$ is not small.

Example 6.2 *Consider the MIMO plant*

$$G_\alpha(s) = \begin{bmatrix} \frac{1}{s-p} & 0 \\ 0 & \frac{1}{s+3} \end{bmatrix} \begin{bmatrix} \cos(30^o) & -\sin(30^o) \\ \sin(30^o) & \cos(30^o) \end{bmatrix} \begin{bmatrix} \frac{s-z}{0.1s+1} & 0 \\ 0 & \frac{s+2}{0.1s+1} \end{bmatrix}; \quad z = 2, p = 3 \tag{6.10}$$

which is studied in more detail in Example 6.3 (page 227). The output direction vectors corresponding to the RHP-zero at $z = 2$ and RHP-pole at $p = 3$ are, respectively,

$$y_z = \begin{bmatrix} 0.327 \\ 0.945 \end{bmatrix}, \quad y_p = \begin{bmatrix} 1 \\ 0 \end{bmatrix}$$

There is some alignment in output 1, since the RHP-zero has some effect in output 1 and the RHP-pole has all its effect in output 1. This translates into unavoidable peaks for $\bar{\sigma}(S)$ and $\bar{\sigma}(T)$. From (6.8) we get $M_{S,\min} = M_{T,\min} = 1.89$; see Matlab code in Table 6.1.

Table 6.1: **Matlab program for calculating sensitivity peak using (6.8)**

```
% G: Has distinct and at least one RHP-zero and one RHP-pole
[ptot,ztot] = pzmap(G);                                   % poles and zeros
p = ptot(find(ptot>0)); z = ztot(find(ztot>0));           % RHP poles and zeros
np = length(p); nz = length(z);
G = ss(G); [V,E] = eig(G.A); C = G.C*V;                    % output pole vectors
for i = 1:np
    Yp(:,i) = C(:,i)/norm(C(:,i));                         % pole directions
end
for i = 1:nz
    [U,S,V] = svd(evalfr(G,z(i))); Yz(:,i) = U(:,end);     % zero directions
end
Qp = (Yp'*Yp).*(1./(diag(p')*ones(np) + ones(np)*diag(p)));
Qz = (Yz'*Yz).*(1./(diag(z)*ones(nz) + ones(nz)*diag(z')));
Qzp = (Yz'*Yp).*(1./(diag(z)*ones(nz,np) - ...
        ones(nz,np)*diag(p)));
Msmin = sqrt(1+norm(sqrtm(inv(Qz))*Qzp*sqrtm(inv(Qp)))^2)
```

One RHP-pole and one RHP-zero. For a plant with one RHP-zero z and one RHP-pole p, (6.8) gives (see e.g. Chen, 2000)

$$M_{S,\min} = M_{T,\min} = \sqrt{\sin^2 \phi + \frac{|z + p|^2}{|z - p|^2} \cos^2 \phi} \tag{6.11}$$

where $\phi = \cos^{-1} |y_z^H y_p|$ is the angle between the output directions of the pole and zero. If the pole and zero are aligned such that $y_z = y_p$ and $\phi = 0$, then (6.11) simplifies to give the SISO conditions in (5.23). Conversely, if the pole and zero are orthogonal to each other, then $\phi = 90°$ and $M_{S,\min} = M_{T,\min} = 1$, and there is no additional penalty for having both a RHP-pole and a RHP-zero.

Example 6.2 continued. *For the plant in (6.10) we have $y_z^H y_p = 0.327$ which gives $\phi = \cos^{-1} 0.327 = 70.9°$. Equation (6.11) then gives $M_{S,\min} = M_{T,\min} = 1.89$, which agrees with the value obtained from (6.8).*

The bound (6.8) can be extended to include weights. With no loss of generality we assume that the weights $W_1(s)$ and $W_2(s)$ contain no RHP-poles or RHP-zeros and consider the weighted functions $W_1 S W_2$ and $W_1 T W_2$.

Theorem 6.2 Weighted sensitivity and complementary sensitivity peaks. *Consider a rational plant $G(s)$ with no time delay and no poles or zeros on the imaginary axis. Let z_i be the RHP-zeros of $G(s)$ with (unit) output zero direction vectors $y_{z,i}$. Let p_i be the RHP-poles of $G(s)$ with (unit) output pole direction vectors $y_{p,i}$. Furthermore, assume that z_i and p_i are all distinct. Define*

$$\gamma_{S,\min} \triangleq \inf_K \|W_1 S W_2\|_\infty, \quad \gamma_{T,\min} \triangleq \inf_K \|W_1 T W_2\|_\infty$$

Then

$$\gamma_{S,\min} = \lambda_{\max}^{1/2}\left(Q_{z1}^{-1/2}(Q_{z2} + Q_{zp2}^H Q_{p2}^{-1} Q_{zp2})Q_{z1}^{-1/2}\right) \qquad (6.12)$$

$$\gamma_{T,\min} = \lambda_{\max}^{1/2}\left(Q_{p2}^{-1/2}(Q_{p1} + Q_{zp1}^H Q_{z1}^{-1} Q_{zp1})Q_{p2}^{-1/2}\right) \qquad (6.13)$$

where $\lambda_{\max}$ is the largest eigenvalue and the elements of the Q-matrices are given by

$$[Q_{z1}]_{ij} = \frac{y_{z,i}^H W_1^{-1}(z_i)W_1^{-H}(z_j)y_{z,j}}{z_i + \bar{z}_j}, \quad [Q_{z2}]_{ij} = \frac{y_{z,i}^H W_2(z_i)W_2(z_j)y_{z,j}}{z_i + \bar{z}_j}$$

$$[Q_{p1}]_{ij} = \frac{y_{p,i}^H W_1^{H}(p_i)W_1(p_j)y_{p,j}}{\bar{p}_i + p_j}, \quad [Q_{p2}]_{ij} = \frac{y_{p,i}^H W_2^{-H}(p_i)W_2^{-1}(p_j)y_{p,j}}{\bar{p}_i + p_j}$$

$$[Q_{zp1}]_{ij} = \frac{y_{z,i}^H W_1^{-1}(z_i)W_1(p_j)y_{p,j}}{z_i - p_j}, \quad [Q_{zp2}]_{ij} = \frac{y_{z,i}^H W_2(z_i)W_2^{-1}(p_j)y_{p,j}}{z_i - p_j}$$

For the case with a scalar weight, we have in particular the following direct generalizations of the SISO results:

$$\|w_P S\|_\infty \geq |w_P(z)| \qquad (6.14)$$

$$\|w_T T\|_\infty \geq |w_T(p)| \qquad (6.15)$$

This shows that $\bar{\sigma}(S)$ cannot be shaped freely for a plant with a RHP-zero, and $\bar{\sigma}(T)$ cannot be shaped freely for a plant with a RHP-pole.

The bound for T in (6.8) can also be extended to include time delays at the plant output:

Theorem 6.3 Complementary sensitivity peak for plant with time delay. *Consider a plant with time delays in the output channels*

$$G_\theta(s) = \Theta(s)G(s), \quad \Theta(s) = \text{diag}\left(e^{-\theta_1 s}, \ldots, e^{-\theta_n s}\right)$$

where $G(s)$ is a rational transfer function matrix. Let z_i be the RHP-zeros of $G(s)$ with (unit) output zero direction vectors $y_{z,i}$. Let p_i be the RHP-poles of $G(s)$ with (unit) output pole direction vectors $y_{p,i}$. Note that the directions are evaluated for the plant without the time delay. Furthermore, assume that z_i and p_i are all distinct. Then we have the following tight lower bound on $\|T\|_\infty$:

$$M_{T,\min} = \lambda_{\max}\left(Q_\theta^{-1/2}(Q_p + Q_{zp}Q_z^{-1}Q_{zp}^H)Q_\theta^{-1/2}\right) \qquad (6.16)$$

where the elements of the matrices Q_z, Q_p and Q_{zp} are given in (6.9) and

$$[Q_\theta]_{ij} = \frac{y_{p,i}^H \Theta(\bar{p}_i)\Theta(p_j)y_{p,j}}{\bar{p}_i + p_j} \qquad (6.17)$$

There is *no tight* bound available for $\|S\|_\infty$ for plants with time delays. However, $\bar{\sigma}(S)$ and $\bar{\sigma}(T)$ differ by at most 1, see (6.1), and we have

$$M_{T,\min} + 1 \geq M_{S,\min} \geq M_{T,\min} - 1 \qquad (6.18)$$

where $M_{T,\min}$ is given by (6.16). An application of the bound (6.16) for a SISO plant is given in Example 5.1 (page 175).

For a time delay plant with one RHP-zero z and one RHP-pole p, similar to (6.11), we have

$$M_{T,\min} = \frac{1}{\|\Theta(p)y_p\|_2}\sqrt{\sin^2\phi + \frac{|z+p|^2}{|z-p|^2}\cos^2\phi} \qquad (6.19)$$

where $\phi = \cos^{-1}|y_z^H y_p|$ is the angle between the output directions of the pole and zero of $G(s)$.

The following example illustrates that MIMO systems may be understood from the results for SISO systems if we take directions into account.

Example 6.3 *Consider the plant*

$$G_\alpha(s) = \begin{bmatrix} \frac{1}{s-p} & 0 \\ 0 & \frac{1}{s+3} \end{bmatrix} \underbrace{\begin{bmatrix} \cos\alpha & -\sin\alpha \\ \sin\alpha & \cos\alpha \end{bmatrix}}_{U_\alpha} \begin{bmatrix} \frac{s-z}{0.1s+1} & 0 \\ 0 & \frac{s+2}{0.1s+1} \end{bmatrix}; \quad z = 2, p = 3 \qquad (6.20)$$

which has for all values of α a RHP-zero at $z = 2$ and a RHP-pole at $p = 3$. For $\alpha = 0°$ the rotation matrix $U_\alpha = I$, and the plant consists of two decoupled subsystems

$$G_0(s) = \begin{bmatrix} \frac{s-z}{(0.1s+1)(s-p)} & 0 \\ 0 & \frac{s+2}{(0.1s+1)(s+3)} \end{bmatrix}$$

Here the subsystem g_{11} has both a RHP-pole and a RHP-zero, and closed-loop performance is expected to be poor. On the other hand, there are no particular control problems related to the subsystem g_{22}. Next, consider $\alpha = 90°$ for which we have

$$U_\alpha = \begin{bmatrix} 0 & -1 \\ 1 & 0 \end{bmatrix}, \quad \text{and} \quad G_{90}(s) = \begin{bmatrix} 0 & -\frac{s+2}{(0.1s+1)(s-p)} \\ \frac{s-z}{(0.1s+1)(s+3)} & 0 \end{bmatrix}$$

and we again have two decoupled subsystems, but this time in the off-diagonal elements. The main difference, however, is that there is no interaction between the RHP-pole and RHP-zero in this case, so we expect this plant to be easier to control. For intermediate values of α we do not have decoupled subsystems, and there will be some interaction between the RHP-pole and RHP-zero.

Since in (6.20) the RHP-pole is located at the output of the plant, its output direction is fixed and we find $y_p = [1 \ 0]^T$ for all values of α. On the other hand, the RHP-zero output direction changes from $[1 \ 0]^T$ for $\alpha = 0°$ to $[0 \ 1]^T$ for $\alpha = 90°$. Thus, the angle ϕ between the pole and zero direction also varies between $0°$ and $90°$, but ϕ and α are not equal. This is seen from the table below, where we also give $M_{S,\min} = M_{T,\min}$, see (6.8) or (6.11), for four rotation angles, $\alpha = 0°, 30°, 60°$ and $90°$.

α	$0°$	$30°$	$60°$	$90°$
y_z	$\begin{matrix}1\\0\end{matrix}$	$\begin{matrix}0.33\\-0.94\end{matrix}$	$\begin{matrix}0.11\\-0.99\end{matrix}$	$\begin{matrix}0\\1\end{matrix}$
$\phi = \cos^{-1}|y_z^H y_p|$	$0°$	$70.9°$	$83.4°$	$90°$
$M_{S,\min} = M_{T,\min}$	5.0	1.89	1.15	1.0
$\|S\|_\infty$	7.00	2.60	1.59	1.98
$\|T\|_\infty$	7.40	2.76	1.60	1.31
$\gamma_{\min}(S/KS)$	9.55	3.53	2.01	1.59

The table also shows the values of $\|S\|_\infty$ and $\|T\|_\infty$ obtained by an $\mathcal{H}_\infty$ optimal S/KS design (see page 94) using the following weights:

$$W_u = I; \quad W_P = \left(\frac{s/M + \omega_B^*}{s}\right)I; \quad M = 2, \omega_B^* = 0.5 \qquad (6.21)$$

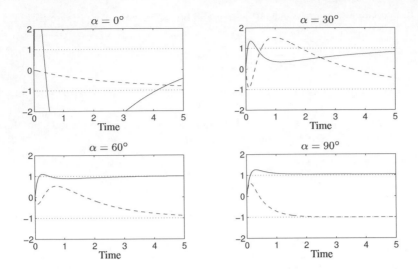

Figure 6.1: MIMO plant (6.20) with angle ϕ between RHP-pole and RHP-zero. Response to step in reference $r = [1 \ \ -1]^T$ with $\mathcal{H}_\infty$ controller for four different values of ϕ. Solid line: y_1; dashed line: y_2.

The weight W_P indicates that we require $\|S\|_\infty$ less than 2, and require tight control up to a frequency of about $\omega^*_B = 0.5 \, rad/s$. The minimum $\mathcal{H}_\infty$ norm for the overall S/KS problem is given by the value of γ in Table 6.3. The corresponding responses to a step change in the reference, $r = [1 \ \ -1]$, are shown in Figure 6.1.

Several things about the example are worth noting:

1. *We see from the simulation for $\phi = \alpha = 0°$ in Figure 6.1 that the response for y_1 is very poor. This is as expected because of the closeness of the RHP-pole and zero ($z = 2, p = 3$). The response for y_2 is also relatively sluggish, because the $\mathcal{H}_\infty$ design is only concerned with the worst-case response in y_1. The response for y_2 may therefore be made faster, if desired.*

2. *For $\phi = \alpha = 90°$ the RHP-pole and RHP-zero do not interact. From the simulation we see that y_1 (solid line) has on overshoot due to the RHP-pole, whereas y_2 (dashed line) has an inverse response due to the RHP-zero.*

3. *The lower bound $M_{S,\min} = M_{T,\min}$ on $\|S\|_\infty$ and $\|T\|_\infty$, see (6.8), is tight in the sense that there exists a controller that achieves it. This can be confirmed numerically by selecting $W_u = 0.01I$, $\omega^*_B = 0.01$ and $M = 1$. W_u and ω_B are small so the main objective is to minimize the peak of S. We find with these weights that the $\mathcal{H}_\infty$ designs for the four angles yield $\|S\|_\infty = 5.04, 1.905, 1.155, 1.005$, which are very close to $M_{S,\min}$.*

4. *The angle ϕ between the pole and zero is quite different from the rotation angle α at intermediate values between $0°$ and $90°$. This is because of the influence of the RHP-pole in output 1, which yields a strong gain in this direction, and thus tends to push the zero direction towards output 2.*

5. *For $\alpha = 0°$ we have $M_{S,\min} = M_{T,\min} = 5$ so it is clearly impossible to get $\|S\|_\infty$ less than 2, as required by the performance weight W_P. This is one reason why $\gamma_{\min} = 9.55$ is so large in this case.*

6. *The $\mathcal{H}_\infty$ optimal controller is unstable for $\alpha = 0°$ and $30°$. This is not altogether surprising, because for $\alpha = 0°$ the plant becomes two SISO systems one of which needs an unstable controller to stabilize it since $p > z$ (see condition on page 150).*

6.3.2   Minimum peaks for other closed-loop transfer functions

In this section, we provide bounds on peaks for some other closed-loop transfer functions. For motivation, we refer the reader to the discussion for SISO systems in Section 5.3.2 on page 175. The results for MIMO systems are summarized in Table 6.3.2, where we also show the performance and robustness reasons behind minimizing the peaks of different closed-loop transfer functions. We frequently make use of minimum-phase and stable versions of the plant and the disturbance models, and the details for their calculation can be found in Section A.6.

Bounds on SG. Theorem 6.2 can be used to calculate the peak value for SG with $W_1 = I$ and $W_2 = G_{ms}(s)$, where $G_{ms}(s)$ denotes the "minimum-phase, stable version" of $G(s)$. In particular, when the system has one RHP-zero z and one RHP-pole p, $\|SG\|_\infty$ must satisfy

$$\|SG\|_\infty \geq \|y_z^H G_{ms}(s)\| \sqrt{\sin^2 \widetilde{\phi} + \frac{|z + \bar{p}|^2}{|z - p|^2} \cos^2 \widetilde{\phi}} \tag{6.22}$$

where

$$\cos \widetilde{\phi} = \frac{|y_z^H G_{ms}(z) G_{ms}^{-1}(p) y_p|}{\|y_z^H G_{ms}(z)\|_2 \|G_{ms}^{-1}(p) y_p\|_2}$$

When $G(s)$ is non-square (more inputs than outputs), the "pseudo-inverse" of $G_{ms}(s)$ can be used to find bounds on $\|SG\|_\infty$.

Bounds on SG_d. In the general case, $G_d \neq G$ and we also want to keep $\|SG_d\|_\infty$ small. This case can be handled as for SG by replacing G_{ms} by $G_{d,ms}$ in (6.22), where $G_{d,ms}(s)$ denotes the "minimum-phase, stable version" of $G_d(s)$.

Bounds on KS. Glover (1986) derived the tight bound on the transfer function KS,

$$\|KS\|_\infty \geq 1/\underline{\sigma}_H(\mathcal{U}(G)^*) \tag{6.23}$$

where $\underline{\sigma}_H$ is the smallest Hankel singular value and $\mathcal{U}(G)^*$ is the mirror image of the anti-stable part of G (for a stable plant there is no lower bound).

A simpler bound is also available, since for any RHP-pole p, $\underline{\sigma}_H(\mathcal{U}(G)^*) \leq \|u_p^H G_s(p)\|_2$, where equality applies for a plant with a single real RHP-pole p. Here, u_p is the input pole direction, and G_s is the "stable version" of G with its RHP-poles mirrored into the LHP, see (5.27). This gives the bound (Havre and Skogestad, 2001)

$$\|KS\|_\infty \geq \|u_p^H G_s(p)^{-1}\|_2 \tag{6.24}$$

which is tight for the case with a single RHP-pole.

Example 6.4 *Consider the following multivariable plant:*

$$G(s) = \begin{bmatrix} \frac{s-z}{s-p} & -\frac{0.1s+1}{s-p} \\ \frac{s-z}{0.1s+1} & 1 \end{bmatrix}, \quad \text{with} \quad z = -2.5 \quad \text{and} \quad p = 2 \tag{6.25}$$

The plant G has a RHP-pole $p = 2$ (plus a LHP-zero at $z = -2.5$ which poses no limitation). The corresponding input and output pole directions are

$$u_p = \begin{bmatrix} 0.966 \\ -0.258 \end{bmatrix}, \quad y_p = \begin{bmatrix} 1 \\ 0 \end{bmatrix}$$

The stable version of $G(s)$ is given by

$$G_s(s) = \begin{bmatrix} \frac{s+2.5}{s+p} & -\frac{0.1s+1}{s+2} \\ \frac{s+2.5}{0.1s+1} & 1 \end{bmatrix}$$

The input u resulting from measurement noise n, output disturbances d_y and references r is given by (e.g. see (2.21))

$$u = KS(r - d_y - n)$$

From this it follows that we may need to bound the transfer function KS in order to avoid excessive input signals u. However, in order to stabilize the plant we must from (6.24) have that

$$\|KS\|_\infty \geq \|u_p^H G_s(p)^{-1}\|_2 = \left\| [0.966 \quad -0.258] \begin{bmatrix} 1.125 & -0.3 \\ 3.75 & 1 \end{bmatrix}^{-1} \right\|_2 = 0.859$$

The bound is tight and agrees with the value found using (6.23); see Matlab code in Table 6.2. This means, for example, that an input signal $\|u\|_2$ of magnitude 0.859 or higher is unavoidable if a sinusoidal output disturbance of magnitude $\|d_y\|_2 = 1$ hits the system at the "worst-case" frequency and direction. Havre (1998) presents more details including state-space realizations for controllers that achieve the bound.

Table 6.2: **Matlab program for calculating peak value of KS for Example 6.4**

```
% Uses Robust Control toolbox
s = tf('s');
g11=(s+2.5)/(s-2); g12=-(0.1*s+1)/(s-2); g21=(s+2.5)/(0.1*s+1); g22=1;
G = [g11 g12; g21 g22];
% Hankel singular value method (see (6.23))
[h1,h2] = hankelsv(G); % Hankel singular values
ksmin = 1/min(h2);
% Alternate method (see (6.24))
p=2;
gp=evalfr(G,p+0.0001); [U,S,V]=svd(gp); up=V(:,1); % crude method up
g11s=(s+2.5)/(s+2); g12s=-(0.1*s+1)/(s+2); g21s=(s+2.5)/(0.1*s+1); g22s=1;
Gs = [g11s g12s; g21s g22s];
ksmin1 = norm(up'*inv(evalfr(Gs,p)))
```

Exercise 6.1 *Consider the plant in (6.25), but with $z = 2.5$ so that the plant now has a RHP-zero. Compute lower bounds on $\|S\|_\infty$, $\|T\|_\infty$ and $\|KS\|_\infty$.*

Bounds on KSG_d. For arbitrary disturbances the bound (6.23) can be generalized as (Kariwala, 2004)

$$\|KSG_d\|_\infty \geq 1/\underline{\sigma}_H(\mathcal{U}(G_{d,ms}^{-1}G)^*) \tag{6.26}$$

where $\mathcal{U}(G_{d,ms}^{-1}G)^*$ is the mirror image of the anti-stable part of $G_{d,ms}^{-1}G$. Note that any unstable modes in G_d must be contained in G such that they are stabilizable with feedback control. Under the same condition, (6.24) can also be generalized to get (Havre and Skogestad, 2001)

$$\|KSG_d\|_\infty \geq |u_p^H G_s(p)^{-1} G_{d,ms}(p)| \tag{6.27}$$

which is tight for a single RHP-pole p. The bounds on the peak value of KSG_d for delay systems can be found in Kariwala (2004).

Bounds on S_I and T_I. For multivariable systems, $S_I = (I + KG)^{-1}$ is different from $S = (I + GK)^{-1}$ and similarly T_I is different from T. Thus, unlike SISO systems, the input

Table 6.3: Bounds on peaks of important closed-loop transfer functions

	Signals (see page 22)	Stability robustness (see page 303)	Special case (tight only for $N_p = 1$ and/or $N_z = 1$)	General case (including SISO)
	Want small for		*Bound on peak*	
1.	S Performance tracking ($e = -Sr$)	Multiplicative inverse output uncertainty (Δ_{io})	$\sqrt{\sin^2\phi + \frac{|z+p|^2}{|z-p|^2}\cos^2\phi}$ (6.11)	(6.8)
2.	T Performance noise ($e = -Tn$)	Multiplicative additive output uncertainty (Δ_o)	$\sqrt{\sin^2\phi + \frac{|z+p|^2}{|z-p|^2}\cos^2\phi}$ (6.11) and (6.19) for delay system	(6.8) and (6.16) for delay system
3.	KS Input usage ($u = KS(r-n)$)	Additive uncertainty (Δ_A)	$\|u_p^H G_s(p)^{-1}\|_2$ (6.24) (tight for any value of N_z)	$1/\underline{\sigma}_H(\mathcal{U}(G)^*)$ (6.23)
4.*	SG_d Performance disturbance ($e = SG_d d_u$)	$G_d = G$: Inverse uncertainty (Δ_{iA})	$\dfrac{\|y_z^H G_{d,ms}(s)\|\cdot}{\sqrt{\sin^2\tilde\phi + \frac{|z+p|^2}{|z-p|^2}\cos^2\tilde\phi}}$ (6.22)	(6.12) with $W_1 = I$ and $W_2 = G_{d,ms}$
5.*	KSG_d (T_I for $G_d = G$) Input usage disturbance ($u = KSG_d d$)	$G_d = G$: Multiplicative additive input uncertainty (Δ_I)	$|u_p^H G_s(p)^{-1}G_{d,ms}(p)|$ (6.27) (tight for any value of N_z)	$1/\underline{\sigma}_H(\mathcal{U}(G_{d,ms}^{-1}G)^*)$ (6.26)
6.	S_I Actual plant input ($(u+d_u) = S_I d_u$)	Inverse multiplicative input uncertainty (Δ_{iI})	$\sqrt{\sin^2\phi_I + \frac{|z+p|^2}{|z-p|^2}\cos^2\phi_I}$ (6.28)	(6.8) with output directions replaced by input directions (u_p, u_z)

* Special case: Input disturbance ($G_d = G$)

sensitivity function S_I and input complementary sensitivity function T_I must be considered separately from S and T, respectively. As shown in the previous section, the bounds on the peak values for S and T can be calculated using Theorem 6.1. The same result also holds for the input sensitivity S_I and input complementary sensitivity function T_I by replacing the output pole and zero directions with the corresponding input pole and zero directions. Similarly, by replacing $y_p^H y_z$ with $u_p^H u_z$, when computing ϕ, (6.11) can be used to find tight bounds on $\|S_I\|_\infty$ and $\|T_I\|_\infty$ for systems with one RHP-pole and one RHP-zero:

$$M_{S_I,\min} = M_{T_I,\min} = \sqrt{\sin^2 \phi_I + \frac{|z+p|^2}{|z-p|^2} \cos^2 \phi_I} \tag{6.28}$$

where $\cos \phi_I \triangleq |u_z^H u_p|$.

Since $T_I = KG(I + KG)^{-1} = K(I + GK)^{-1}G = KSG$ is the closed-loop transfer function from the input disturbances to the controller output (see page 69), the bound on the peak value for T_I can be alternatively calculated as a special case of (6.26) and (6.27). Note that, for a minimum-phase system, $G_{ms} = G_s$ and it follows from (6.27) that in this case $|u_p^H G_s(p)^{-1} G_{d,ms}(p)| = |u_p^H| = 1$ and we have that $\|T_I\|_\infty \geq 1$. This bound is tight for any number of unstable poles for minimum-phase systems (Kariwala, 2004).

For many practical systems, bounding one of S and S_I (or one of T and T_I) also bounds the other, but this is not true in general, as shown by the next example.

Example 6.5 *Consider the following multivariable plant:*

$$G(s) = \begin{bmatrix} \frac{s-z}{s-p} & 1 \\ \frac{0.01(s-z)}{s+10} & 0.01 \end{bmatrix} \tag{6.29}$$

The plant G has a RHP-pole at $s = p$ and a RHP-zero at $s = z$. Since the pole appears in the $(1,1)$ element and the zero only in the first column of $G(s)$, we have

$$u_p = \begin{bmatrix} 1 \\ 0 \end{bmatrix}, \quad y_p = \begin{bmatrix} 1 \\ 0 \end{bmatrix}, \quad u_z = \begin{bmatrix} 1 \\ 0 \end{bmatrix}, \quad y_z \approx \begin{bmatrix} 0.01 \\ 0.99 \end{bmatrix}$$

for all values of z and p. Note that $y_p^H y_z \approx 0.01$ and $u_p^H u_z = 1$. It follows from (6.11) and (6.28) that when p and z are located close to each other, $\|S_I\|_\infty$ and $\|T_I\|_\infty$ will be much larger than $\|S\|_\infty$ and $\|T\|_\infty$. For example, with $p = 2$ and $z = 2.1$, we have $M_{S,\min} = M_{T,\min} \approx 1$ (achievable peak values of $\|S\|_\infty$ and $\|T\|_\infty$), but from (6.28), $\|S_I\|_\infty$ and $\|T_I\|_\infty$ must be larger than 6.4.

This concludes this section on fundamental limitations. Later, in this chapter, we discuss the control implications of these results in more detail.

6.4 Functional controllability

Consider a plant $G(s)$ with l outputs and let r denote the normal rank of $G(s)$. In order to control all outputs independently we must require $r = l$ and the plant is said to be "functionally controllable". This term was introduced by Rosenbrock (1970, p. 170) for square systems. Other terms used for functional controllability are "right invertibility", "output realizability" and "output controllability". We will use the following definition:

Definition 6.1 Functional controllability. *An m-input l-output plant $G(s)$ is functionally controllable if the normal rank of $G(s)$, denoted r, is equal to the number of outputs ($r = l$); that is, if $G(s)$ has full row rank. A plant is functionally uncontrollable if $r < l$.*

The normal rank of $G(s)$ is the rank of $G(s)$ at all values of s except at a finite number of singularities (which are the zeros of $G(s)$). The minimal requirement for functional controllability is that we have at least as many inputs as outputs, i.e. $m \geq l$.

A plant is *functionally uncontrollable* if and only if $\sigma_l(G(j\omega)) = 0, \forall \omega$. As a measure of how close a plant is to being functionally uncontrollable we may therefore consider the minimum singular value $\sigma_l(G(j\omega))$. The only example of a SISO plant which is *functionally uncontrollable* is $G(s) = 0$. Similarly, a MIMO plant is functionally uncontrollable if the gain is identically zero in some output direction at all frequencies.

For strictly proper plants, $G(s) = C(sI - A)^{-1}B$, we have that $G(s)$ is *functionally uncontrollable* if $\text{rank}(B) < l$ (the system is input deficient), or if $\text{rank}(C) < l$ (the system is output deficient), or if $\text{rank}(sI - A) < l$ (fewer states than outputs). This follows since the rank of a product of matrices is less than or equal to the minimum rank of the individual matrices, see (A.36).

In most cases *functional uncontrollability* is a *structural* property of the plant; that is, it does not depend on specific parameter values, and it may often be evaluated from cause-and-effect graphs. A typical example of this is when none of the inputs u_i affect a particular output y_j which would be the case if one of the rows in $G(s)$ was identically zero. Another example is when there are fewer inputs than outputs.

If the plant is *not* functionally controllable, i.e. $r < l$, then there are $l - r$ output directions, denoted y_0, which cannot be affected. These directions will vary with frequency, and we have (analogous to the concept of a zero direction)

$$y_0^H(j\omega)G(j\omega) = 0 \tag{6.30}$$

From an SVD of $G(j\omega) = U\Sigma V^H$, the *uncontrollable output directions* $y_0(j\omega)$ are the last $l - r$ columns of $U(j\omega)$. By analyzing these directions, an engineer can then decide on whether it is acceptable to keep certain output combinations uncontrolled, or if additional actuators are needed to increase the rank of $G(s)$.

Example 6.6 *The following plant is singular and thus not functionally controllable:*

$$G(s) = \begin{bmatrix} \frac{1}{s+1} & \frac{2}{s+1} \\ \frac{2}{s+2} & \frac{4}{s+2} \end{bmatrix}$$

This is easily seen since column 2 of $G(s)$ is two times column 1. The uncontrollable output directions at low and high frequencies are, respectively,

$$y_0(0) = \frac{1}{\sqrt{2}} \begin{bmatrix} 1 \\ -1 \end{bmatrix}; \quad y_0(\infty) = \frac{1}{\sqrt{5}} \begin{bmatrix} 2 \\ -1 \end{bmatrix}$$

6.5 Limitations imposed by time delays

As for SISO systems, time delays normally introduce limitations in MIMO systems, but there are exceptions. As an example of a limitation, let θ_{ij} denote the time delay in the ij'th element of $G(s)$. Then a lower bound on the time delay for output i is given by the smallest delay in row i of $G(s)$, i.e.

$$\theta_i^{\min} = \min_j \theta_{ij}$$

This bound is obvious since $\theta_i^{\min}$ is the minimum time for any input to affect output i, and $\theta_i^{\min}$ can be regarded as a delay pinned to output i.

Holt and Morari (1985a) have derived additional bounds, but their usefulness is sometimes limited since they assume a *decoupled* closed-loop response (which is usually not desirable in terms of overall performance) and also assume infinite power in the inputs.

Exceptions. For MIMO systems we have the surprising result that an increased time delay may sometimes improve the achievable performance. As a simple example, consider the plant

$$G(s) = \begin{bmatrix} \frac{1}{s} & 1 \\ e^{-\theta s} & 1 \end{bmatrix} \tag{6.31}$$

With $\theta = 0$, the plant is singular (not functionally controllable), and controlling the two outputs independently is clearly impossible. On the other hand, for $\theta > 0$, effective feedback control is possible at high frequencies, provided the bandwidth is larger than about $1/\theta$. That is, for this example, control is easier the larger θ is. In words, the presence of the delay decouples the initial (high-frequency) response, so we can obtain tight control if the controller reacts within this initial time period. To illustrate this, we may compute the singular values of G as a function of frequency, and note that the minimum singular value is 0 at low frequencies, but increases with frequency and attains a maximum value of 1.41 at frequency π/θ.

Exercise 6.2 *Simulate the closed-loop response of the plant (6.31) for the setpoint changes $r_1 = \begin{bmatrix} 1 \\ 0 \end{bmatrix}$ and $r_2 = \begin{bmatrix} 1 \\ 1 \end{bmatrix}$ using a simple diagonal controller, $K = \frac{k}{s} I$ with $k\theta = 0.1, 1$ and 10. Plot the responses of both the inputs and outputs with $\theta = 1$. Why is control much better with r_2 as compared to r_1?*

Exercise 6.3 * *To illustrate further the above arguments, compute the sensitivity function S for the plant (6.31) and $K = \frac{k}{s} I$. Use the approximation $e^{-\theta s} \approx 1 - \theta s$ to show that at low frequencies the elements of $S(s)$ are of magnitude $1/(k\theta + 2)$. How large must k be to have acceptable performance (less than 10% offset at low frequencies)? What is the corresponding bandwidth? (Answer: Need $k > 8/\theta$. Bandwidth is equal to k.)*

Remark 1 The observant reader may have noticed that the smallest singular value of $G(s)$ in (6.31) drops to zero periodically at high frequencies, as $e^{-j\omega\theta} = 1$ for $\omega\theta = 2\pi n$, $n = 0, 1, 2, \ldots$. This will cause "ringing" irrespective of the bandwidth, as seen from the simulations.

Remark 2 The reader may also have noticed that $G(s)$ in (6.31) is singular at $s = 0$ (even with θ non-zero) and thus has a zero at $s = 0$. Therefore, a controller with integral action which cancels this zero yields an internally unstable system (e.g. the transfer function KS contains an integrator). This internal instability will manifest itself as integrating input signals that will eventually go to infinity. To "fix" these results, we may assume that the plant has an integrator in each element. Then, one of the integrators will cancel the zero at $s = 0$ and the resulting steady-state gain is finite in one direction and infinite in another. Alternatively, we may assume that $e^{-\theta s}$ is replaced by $0.99e^{-\theta s}$ so that the plant is not singular at steady-state (but it is close to singular).

Remark 3 A physical example of a model in the form of (6.31) is a distillation column where θ represents the time for a change in liquid flow at the top to reach the bottom of the column.

Exercise 6.4 *Repeat Exercise 6.2 with $e^{-\theta s}$ replaced by $0.99(1 - \frac{\theta}{2n}s)^n/(1 + \frac{\theta}{2n}s)^n$ (where $n = 2$ is the order of the Padé approximation). Also plot the elements of $S(j\omega)$ as functions of frequency for $k = 0.1/\theta$, $k = 1/\theta$ and $k = 8/\theta$. Notice that there is no ringing here as $G(s)$ is singular only at $\omega = \infty$.*

6.6   Limitations imposed by RHP-zeros

RHP-zeros are common in many practical multivariable problems. The limitations they impose are similar to those for SISO systems, although often not quite so serious because they only apply in particular directions.

For ideal ISE optimal control (the "cheap" LQR problem), the SISO result ISE $= 2/z$ from Section 5.5 can be generalized. Qiu and Davison (1993) show for a MIMO plant with RHP-zeros at z_i that the ideal ISE value (the "cheap" LQR cost function) for a step disturbance, or step reference, is directly related to $\sum_i 2/z_i$. Thus, as for SISO systems, RHP-zeros close to the origin imply poor control performance.

The limitations of a RHP-zero located at z may also be derived from the bound

$$\|w_P S(s)\|_\infty = \max_\omega |w_P(j\omega)| \cdot \bar{\sigma}(S(j\omega)) \geq |w_P(z)| \tag{6.32}$$

where $w_P(s)$ is a scalar weight. All the results derived in Section 5.7.3 for SISO systems therefore generalize if we consider the "worst" direction corresponding to the maximum singular value, $\sigma(S)$. For instance, by selecting the weight $w_P(s)$ such that we require tight control at low frequencies and a peak for $\bar{\sigma}(S)$ less than 2, we derive from (5.51) that the bandwidth (in the "worst" direction) must for a real RHP-zero satisfy $\omega_B^* < z/2$. Alternatively, if we require tight control at high frequencies, then we must from (5.57) satisfy $\omega_B^* > 2z$. The reader is also referred to Exercise 6.5, which gives the trade-off between the performances of different output for plants with a RHP-zero.

Remark 1 The use of a scalar weight $w_P(s)$ in (6.32) is somewhat restrictive. However, the assumption is less restrictive if one follows the scaling procedure in Section 1.4 and scales all outputs by their allowed variations such that their magnitudes are of approximately equal importance.

Remark 2 Note that condition (6.32) involves the maximum singular value (which is associated with the "worst" direction), and therefore the RHP-zero may not be a limitation in other directions. Furthermore, we may to some extent choose the worst direction. This is discussed next.

Exercise 6.5 * *For a system $L = GK$ with a single real RHP-zero z with input direction u_z and a diagonal performance weight matrix W_P, show that the requirement $\|W_P S\|_\infty < 1$ implies*

$$\sum_i |w_{P,i}(z)|^2 |u_{z,i}|^2 < 1 \tag{6.33}$$

If $w_{P,i}$ is given by (5.50) and $w_{P,j} = 0$, $i \neq j$ (arbitrarily poor control of all outputs other than y_i), show that tight control of y_i at low frequencies imposes the following limitation on $\omega_{B,i}^*$:

$$\omega_{B,i}^* < z\left(\frac{1}{u_{z,i}} - \frac{1}{M}\right) \tag{6.34}$$

6.6.1   Moving the effect of a RHP-zero to a specific output

In MIMO systems, one can often move the deteriorating effect of a RHP-zero to a less important output. This is possible because, although the interpolation constraint $y_z^H T(z) = 0$ imposes a certain relationship between the elements within each column of $T(s)$, the columns of $T(s)$ may still be selected independently. Let us first consider an example to motivate the results that follow. Most of the results in this section are from Holt and Morari (1985b) where further extensions can also be found.

Example 3.17 continued. *Consider the plant*

$$G(s) = \frac{1}{(0.2s + 1)(s + 1)} \begin{bmatrix} 1 & 1 \\ 1 + 2s & 2 \end{bmatrix}$$

which has a RHP-zero at $s = z = 0.5$. This is the same plant considered on page 96, where we performed some $\mathcal{H}_\infty$ controller designs. The output zero direction satisfies $y_z^H G(z) = 0$ and we find

$$y_z = \frac{1}{\sqrt{5}} \begin{bmatrix} 2 \\ -1 \end{bmatrix} = \begin{bmatrix} 0.89 \\ -0.45 \end{bmatrix}$$

Any allowable $T(s)$ must satisfy the interpolation constraint $y_z^H T(z) = 0$ in (6.4), and this imposes the following relationships between the column elements of $T(s)$:

$$2t_{11}(z) - t_{21}(z) = 0; \quad 2t_{12}(z) - t_{22}(z) = 0 \tag{6.35}$$

We will consider reference tracking $y = Tr$ and examine three possible choices for T: T_0 diagonal (a decoupled design), T_1 with output 1 perfectly controlled, and T_2 with output 2 perfectly controlled. Of course, we cannot achieve perfect control in practice, but we make the assumption to simplify our argument. In all three cases, we require perfect tracking at steady-state, i.e. $T(0) = I$.

A decoupled design has $t_{12}(s) = t_{21}(s) = 0$, and to satisfy (6.35) we then need $t_{11}(z) = 0$ and $t_{22}(z) = 0$, so the RHP-zero must be contained in both diagonal elements. One possible choice, which also satisfies $T(0) = I$, is

$$T_0(s) = \begin{bmatrix} \frac{-s+z}{s+z} & 0 \\ 0 & \frac{-s+z}{s+z} \end{bmatrix} \tag{6.36}$$

For the two designs with one output perfectly controlled we choose

$$T_1(s) = \begin{bmatrix} 1 & 0 \\ \frac{\beta_1 s}{s+z} & \frac{-s+z}{s+z} \end{bmatrix} \quad T_2(s) = \begin{bmatrix} \frac{-s+z}{s+z} & \frac{\beta_2 s}{s+z} \\ 0 & 1 \end{bmatrix}$$

The basis for the last two selections is as follows. For the output which is not perfectly controlled, the diagonal element must have a RHP-zero to satisfy (6.35), and the off-diagonal element must have an s-term in the numerator to give $T(0) = I$. To satisfy (6.35), we must then require for the two designs

$$\beta_1 = 4, \quad \beta_2 = 1$$

The RHP-zero has no effect on output 1 for design $T_1(s)$, and no effect on output 2 for design $T_2(s)$. We therefore see that it is indeed possible to move the effect of the RHP-zero to a particular output. However, we must pay for this by having to accept some interaction. We note that the magnitude of the interaction, as expressed by β_k, is largest for the case where output 1 is perfectly controlled ($\beta_1 = 4$). This is reasonable since the zero output direction $y_z = \begin{bmatrix} 0.89 & -0.45 \end{bmatrix}^T$ is mainly in the direction of output 1, so we have to "pay more" to push its effect to output 2. This was also observed in the controller designs in Section 3.6; see Figure 3.12 on page 97.

We see from the above example that by requiring a decoupled response from r to y, as in design $T_0(s)$ in (6.36), we have to accept that the multivariable RHP-zero appears as a

RHP-zero in each of the diagonal elements of $T(s)$, i.e. whereas $G(s)$ has one RHP-zero at $s = z$, $T_0(s)$ has two. In other words, requiring a decoupled response generally leads to the introduction of additional RHP-zeros in $T(s)$ which are not present in the plant $G(s)$.

We also see that we can move the effect of the RHP-zero to a particular output, but we then have to accept some interaction. This is stated more exactly in the following theorem.

Theorem 6.4 *Assume that $G(s)$ is square, functionally controllable and stable and has a single RHP-zero at $s = z$ and no RHP-pole at $s = z$. Then if the k'th element of the output zero direction is non-zero, i.e. $y_{zk} \neq 0$, it is possible to obtain "perfect" control on all outputs $j \neq k$ with the remaining output exhibiting no steady-state offset. Specifically, T can be chosen of the form*

$$T(s) = \begin{bmatrix} 1 & 0 & \cdots & 0 & 0 & 0 & \cdots & 0 \\ 0 & 1 & \cdots & 0 & 0 & 0 & \cdots & 0 \\ \vdots & \vdots & & & & & & \\ \frac{\beta_1 s}{s+z} & \frac{\beta_2 s}{s+z} & \cdots & \frac{\beta_{k-1} s}{s+z} & \frac{-s+z}{s+z} & \frac{\beta_{k+1} s}{s+z} & \cdots & \frac{\beta_n s}{s+z} \\ \vdots & & \ddots & & & & & \vdots \\ 0 & 0 & \cdots & 0 & 0 & 0 & \cdots & 1 \end{bmatrix} \tag{6.37}$$

where

$$\beta_j = -2 \frac{y_{zj}}{y_{zk}} \text{ for } j \neq k \tag{6.38}$$

Proof: It is clear that (6.37) satisfies the interpolation constraint $y_z^H T(z) = 0$; see also Holt and Morari (1985b). $\square$

The effect of moving completely the effect of a RHP-zero to output k is quantified by (6.38). We see that if the zero is not "naturally" aligned with this output, i.e. if $|y_{zk}|$ is much smaller than 1, then the interactions will be significant, in terms of yielding some $\beta_j = -2y_{zj}/y_{zk}$ much larger than 1 in magnitude. In particular, we *cannot* move the effect of a RHP-zero to an output corresponding to a zero element in y_z, which occurs frequently if we have a RHP-zero pinned to a subset of the outputs.

Exercise 6.6 * *Consider the plant*

$$G(s) = \begin{bmatrix} \alpha & 1 \\ \frac{1}{s+1} & \alpha \end{bmatrix} \tag{6.39}$$

(a) Find the zero and its output direction. (Answer: $z = \frac{1}{\alpha^2} - 1$ and $y_z = [-\alpha \ 1]^T$.)
(b) Which values of α yield a RHP-zero, and which of these values is best/worst in terms of achievable performance? (Answer: We have a RHP-zero for $|\alpha| < 1$. Best for $\alpha = 0$ with zero at infinity; if control at steady-state is required then worst for $\alpha = 1$ with zero at $s = 0$.)
(c) Suppose $\alpha = 0.1$. Which output is the most difficult to control? Illustrate your conclusion using Theorem 6.4. (Answer: Output 2 is the most difficult since the zero is mainly in that direction; we get strong interaction with $\beta = 20$ if we want to control y_2 perfectly.)

Exercise 6.7 *Repeat the above exercise for the plant*

$$G(s) = \frac{1}{s+1} \begin{bmatrix} s - \alpha & 1 \\ (\alpha + 2)^2 & s - \alpha \end{bmatrix} \tag{6.40}$$

6.7   Limitations imposed by unstable (RHP) poles

For unstable plants we *need* feedback for stabilization and a non-zero minimum value of $\|KS\|_\infty$ is also unavoidable; see (6.24). More precisely, from (6.5) the presence of an unstable pole p requires for internal stability $\boxed{T(p)y_p = y_p}$, where y_p is the output pole direction. As for SISO systems (see page 192) this imposes the following two limitations:

1. **RHP-pole limitation on input usage.** *For an unstable system, the transfer function KS (from measurement noise n or output disturbances d_y to plant inputs u) must satisfy (Havre and Skogestad, 2001)*

$$\|KS\|_\infty \geq |G_s^{-1}(p)| \tag{6.41}$$

   which is tight for the case of a single real RHP-pole p. A tighter lower bound for a system with multiple unstable poles is given by (6.23).

2. **RHP-pole limitation on bandwidth.** *To stabilize a plant, we need to react sufficiently fast, and we require that $\bar{\sigma}(T(j\omega))$ is about 1, or larger, up to the frequency $2|p|$, approximately.*

The limitation on the bandwith follows from the bound

$$\|w_T(s)T(s)\|_\infty \geq |w_T(p)|$$

and shows that a RHP-pole p imposes restrictions on $\bar{\sigma}(T)$, which are identical to those derived on $|T|$ for SISO systems in Section 5.9.

6.8   Performance requirements imposed by disturbances

For SISO systems we found that large and "fast" disturbances require tight control and a large bandwidth. The same results apply to MIMO systems, but again the issue of directions is important.

Definition 6.2 Disturbance direction. *Consider a single (scalar) disturbance and let the vector g_d represent its effect on the outputs ($y = g_d d$). The disturbance direction is defined as*

$$y_d = \frac{1}{\|g_d\|_2} g_d \tag{6.42}$$

The associated disturbance condition number is defined as

$$\gamma_d(G) = \bar{\sigma}(G)\,\bar{\sigma}(G^\dagger y_d) \tag{6.43}$$

Here $G^\dagger$ is the pseudo-inverse, which is G^{-1} for a non-singular G.

Remark. We use g_d (rather than G_d) to show that we consider a single disturbance, i.e. g_d is a vector. For a plant with many disturbances g_d is a column of the matrix G_d.

The disturbance condition number provides a measure of how a disturbance is aligned with the plant. It may vary between 1 (for $y_d = \bar{u}$) if the disturbance is in the "good" direction, and the condition number $\gamma(G) = \bar{\sigma}(G)\bar{\sigma}(G^\dagger)$ (for $y_d = \underline{u}$) if it is in the "bad" direction.

Here $\bar{u}$ and $\underline{u}$ are the output directions in which the plant has its largest and smallest gains, respectively; see Chapter 3.

In the following, let $r = 0$ and assume that the disturbance has been scaled such that at each frequency the worst-case disturbance may be selected as $|d(\omega)| = 1$. Also assume that the outputs have been scaled such that the performance objective is that at each frequency the 2-norm of the error should be less than 1, i.e. $\|e(\omega)\|_2 < 1$. With feedback control $e = Sg_d d$ and the performance objective is then satisfied if

$$\|Sg_d\|_2 = \bar{\sigma}(Sg_d) < 1 \; \forall\omega \quad \Leftrightarrow \quad \|Sg_d\|_\infty < 1 \tag{6.44}$$

For SISO systems, we used this to derive tight bounds on the sensitivity function and the loop gain: $|S| < 1/|G_d|$ and $|1 + L| > |G_d|$. A similar derivation is complicated for MIMO systems because of directions. To see this, we can use (6.42) to get the following requirement, which is equivalent to (6.44):

$$\|Sy_d\|_2 < 1/\|g_d\|_2 \; \forall\omega \tag{6.45}$$

which shows that S must be less than $1/\|g_d\|_2$ only in the direction of y_d. We can also derive bounds in terms of the singular values of S. Since g_d is a vector we have from (3.42)

$$\underline{\sigma}(S)\|g_d\|_2 \le \|Sg_d\|_2 \le \bar{\sigma}(S)\|g_d\|_2 \tag{6.46}$$

Now $\underline{\sigma}(S) = 1/\sigma(I + L)$ and $\bar{\sigma}(S) = 1/\underline{\sigma}(I + L)$, and we therefore have the requirement:

- For acceptable performance ($\|Sg_d\|_2 < 1$) we must *at least* require that $\bar{\sigma}(I + L)$ is larger than $\|g_d\|_2$ and we *may* have to require that $\underline{\sigma}(I + L)$ is larger than $\|g_d\|_2$.

Plant with RHP-zero. If $G(s)$ has a RHP-zero at $s = z$ then the performance may be poor when the disturbance is aligned with the output direction of this zero. To see this use $y_z^H S(z) = y_z^H$ and apply the maximum modulus principle to $f(s) = y_z^H Sg_d$ to get

$$\|Sg_d\|_\infty \ge |y_z^H g_d(z)| = |y_z^H y_d| \cdot \|g_d(z)\|_2 \tag{6.47}$$

To satisfy $\|Sg_d\|_\infty < 1$, we must then for a given disturbance d at least require

$$\boxed{|y_z^H g_d(z)| < 1} \tag{6.48}$$

where y_z is the direction of the RHP-zero. This provides a generalization of the SISO condition $|G_d(z)| < 1$ in (5.78). For combined disturbances, the condition is $\|y_z^H G_d(z)\|_2 < 1$.

Remark. In the above development we consider at each frequency performance in terms of $\|e\|_2$ (the 2-norm). However, the scaling procedure presented in Section 1.4 leads naturally to the vector max-norm as the way to measure signals and performance. Fortunately, this difference is not too important, and we will neglect it in the following. The reason is that for an $m \times 1$ vector a we have $\|a\|_{\max} \le \|a\|_2 \le \sqrt{m} \|a\|_{\max}$ (see (A.95)), so the values of max- and 2-norms are at most a factor of $\sqrt{m}$ apart.

Example 6.7 *Consider the following plant and disturbance models:*

$$G(s) = \frac{1}{s+2} \begin{bmatrix} s-1 & 4 \\ 4.5 & 2(s-1) \end{bmatrix}, \quad g_d(s) = \frac{6}{s+2} \begin{bmatrix} k \\ 1 \end{bmatrix}, \quad |k| \le 1 \tag{6.49}$$

It is assumed that the disturbance and outputs have been appropriately scaled, and the question is whether the plant is input–output controllable, i.e. whether we can achieve $\|Sg_d\|_\infty < 1$, for any value of $|k| \leq 1$. $G(s)$ has a RHP-zero $z = 4$ and in Example 4.13 on page 140 we have already computed the zero direction. From this we get

$$|y_z^H g_d(z)| = \left| [\, 0.83 \quad -0.55 \,] \cdot \begin{bmatrix} k \\ 1 \end{bmatrix} \right| = |0.83k - 0.55|$$

and from (6.48) we conclude that the plant is not *input–output controllable if* $|0.83k - 0.55| > 1$, *i.e. if $k < -0.54$. We cannot really conclude that the plant is controllable for $k > -0.54$ since (6.48) is only a necessary (and not sufficient) condition for acceptable performance, and there may also be other factors that determine controllability, such as input constraints which are discussed next.*

6.9 Limitations imposed by input constraints

Constraints on the manipulated variables can limit the ability to reject disturbances and track references, and to stabilize the plant. As was done for SISO plants in Chapter 5, we will consider the case of perfect control ($e = 0$) and then of acceptable control ($\|e\| \leq 1$). We derive the results for disturbances, and the corresponding results for reference tracking are obtained by replacing G_d by $-R$. The results in this section apply to both feedback and feedforward control.

Remark. For MIMO systems the choice of vector norm, $\| \cdot \|$, to measure the vector signal magnitudes at each frequency makes some difference. The vector max-norm (largest element) is the most natural choice when considering input saturation and is also the most natural in terms of our scaling procedure. However, for mathematical convenience we will also consider the vector 2-norm (Euclidean norm). In most cases, the difference between these two norms is of little practical significance.

6.9.1 Inputs for perfect control

We consider the question: can the disturbances $\|d\| \leq 1$ be rejected perfectly ($e = 0$) while maintaining $\|u\| \leq 1$? To answer this, we must quantify the set of possible disturbances and the set of allowed input signals. We will consider both the max-norm and 2-norm.

Max-norm and square plant. For a square plant the input needed for perfect disturbance rejection is $u = -G^{-1}G_d d$ (as for SISO systems). Consider a *single disturbance* (g_d is a vector). Then the worst-case disturbance is $|d(\omega)| = 1$, and we get that input saturation is avoided ($\|u\|_{max} \leq 1$) if all elements in the vector $G^{-1}g_d$ are less than 1 in magnitude; that is,

$$\|G^{-1}g_d\|_{max} \leq 1, \forall\omega$$

For *simultaneous disturbances* (G_d is a matrix), the corresponding requirement is

$$\|G^{-1}G_d\|_{i\infty} \leq 1, \forall\omega \tag{6.50}$$

where $\| \cdot \|_{i\infty}$ is the induced max-norm (induced ∞-norm, maximum row sum, see (A.106)). However, it is usually recommended in a preliminary analysis to consider one disturbance at a time, e.g. by plotting as a function of frequency the individual elements of the matrix $G^{-1}G_d$. This yields more information about which particular input is most likely to saturate and which disturbance is the most problematic.

Two-norm. We measure both the disturbance $\|d\|_2 \leq 1$ and the input in terms of the 2-norm. Assume that G has full row rank so that the outputs can be perfectly controlled. Then the smallest inputs ($\|u\|_2$) needed for perfect disturbance rejection are

$$u = -G^\dagger G_d d \tag{6.51}$$

where $G^\dagger = G^H(GG^H)^{-1}$ is the Moore–Penrose pseudo-inverse from (A.65). Then with a single disturbance we require $\|G^\dagger g_d\|_2 \leq 1$. With combined disturbances we require $\bar{\sigma}(G^\dagger G_d) \leq 1$; that is, the induced 2-norm is less than 1, see (A.107).

For *combined reference changes*, $\|\tilde{r}(\omega)\|_2 \leq 1$, the corresponding condition for perfect control with $\|u\|_2 \leq 1$ becomes $\bar{\sigma}(G^\dagger R) \leq 1$, or equivalently (see (A.63))

$$\underline{\sigma}(R^{-1}G) \geq 1, \ \forall \omega \leq \omega_r \tag{6.52}$$

where ω_r is the frequency up to which reference tracking is required. Usually R is diagonal with all elements larger than 1, and we must at least require

$$\underline{\sigma}(G(j\omega)) \geq 1, \forall \omega \leq \omega_r \tag{6.53}$$

or, more generally, we want $\underline{\sigma}(G(j\omega))$ large.

6.9.2 Inputs for acceptable control

It is possible to generalize the results applicable for SISO systems in Section 5.11.2 to MIMO systems using the singular values. The main result is summarized below and the details of the derivation can be found in the first edition of this book (Skogestad and Postlethwaite, 1996).

Let $r = 0$ and consider the response $e = Gu + G_d d$ to a disturbance d. We require $\|e\| < 1$ for any $\|d\| \leq 1$ using inputs with $\|u\| \leq 1$. We use here the max-norm, $\| \cdot \|_{\max}$ (the vector infinity-norm), for the vector signals. To simplify the problem, we consider this problem frequency by frequency and one disturbance at a time, i.e. d is a scalar and g_d a vector. The worst-case disturbance is then $|d| = 1$ and the problem at each frequency is to compute

$$U_{\min} \triangleq \min_u \|u\|_{\max} \text{ such that } \|Gu + g_d d\|_{\max} \leq 1, |d| = 1 \tag{6.54}$$

At each frequency the SVD of the plant (which may be non-square) is $G = U\Sigma V^H$. We then have that each singular value of G, $\sigma_i(G)$, must approximately satisfy

$$\boxed{\sigma_i(G) \geq |u_i^H g_d| - 1, \text{ at frequencies where } |u_i^H g_d| > 1} \tag{6.55}$$

where u_i is the i'th output singular vector of G, Note that (6.55) is approximate and is a necessary condition for achieving acceptable control.

6.9.3 Inputs for stabilization

Active use of inputs is needed to stabilize an unstable plant and from (6.24) we must require $\|KS\|_\infty \geq (1/\underline{\sigma}_H) \geq \|u_p^H G_s(p)^{-1}\|_2$ where KS is the transfer function from measurement noise and output disturbances to plant inputs, i.e. $u = -KS(d + n)$. If the required inputs exceed their constraints then stabilization is most likely not possible.

6.10 Limitations imposed by uncertainty

As discussed for SISO systems in Section 5.12, the presence of uncertainty requires us to use feedback control rather than just feedforward control. With MIMO systems there is an additional problem in that there is also uncertainty associated with the plant directionality. The main objective of this section is to introduce some simple tools, like the RGA and the condition number, which are useful in picking out plants for which one might expect sensitivity to multivariable (directional) uncertainty.

Consider the actual (uncertain) plant G' and the two-degrees of freedom controller $u' = K(r - y')K_r r$. Here K is the feedback controller and K_r the feedforward controller for references, see Figure 2.5. For simplicity, we only consider feedforward control for references, but the analysis may easily be extended to distrubances. The resulting control error e' in response to a reference change r is, see (2.28),

$$e' = y' - r = -S'S_r'r \qquad (6.56)$$

where $S = (I +' GK)^{-1}$ is the (feedback) sensitivity function and $S_r' = I - G'K_r$ is the feedforward sensitivity function. Without feedback control ($K = 0$) we have $S' = I$, and without feedforward control ($K_r = 0$) we have $S_r' = I$. For good performance ($\|e'\|_2$ small) we want $\bar\sigma(S')$ and $\bar\sigma(S_r')$ small, but this may be difficult with model uncertainty, as is discussed in more detail below; see Sections 6.10.2 (feedforward control) and 6.10.4 (feedback control). We will derive *upper bounds* on $\bar\sigma(S_r')$ and $\bar\sigma(S')$, which involve the condition numbers

$$\gamma(G) = \frac{\bar\sigma(G)}{\underline\sigma(G)}, \qquad \gamma_I^*(G) = \min_{D_I} \gamma(GD_I), \qquad (6.57)$$

The minimized condition number $\gamma_I^*(G)$ may be computed using (A.75). Similarly, we state for both feedback and feedforward control, *lower bounds* in terms of the RGA matrix of the plant.

Remark. In Chapter 8, we discuss more exact methods for analyzing performance with almost any kind of uncertainty and a given controller. This involves analyzing robust performance by use of the structured singular value. However, in this section the treatment is kept at a more elementary level as we look for results that depend on the plant only.

6.10.1 Input and output uncertainty

In practice, the difference between the true perturbed plant G' and the plant model G is caused by a number of different sources. In this section, we focus on input uncertainty and output uncertainty. In a multiplicative (relative) form, the output and input uncertainties (as in Figure 6.2) are given by[2]

Output uncertainty:	$G' = (I + E_O)G$ or $E_O = (G' - G)G^{-1}$	(6.58)
Input uncertainty:	$G' = G(I + E_I)$ or $E_I = G^{-1}(G' - G)$	(6.59)

---

[2] In this book we use Δ to represent normalized uncertainty which is norm-bounded to be less than 1, whereas $E = |\epsilon|\Delta$ is not normalized. We often use a weight $|w| = |\epsilon| = \bar\sigma(E)$ to represent the magnitude of the uncertainty.

In addition, we will for completeness consider additive uncertainty

$$G' = G + E_A \quad \text{or} \quad E_A = G' - G \tag{6.60}$$

although this is generally not a good uncertainty description because it is difficult to quantify the magnitude of E_A. If all the elements in the matrices E_I, E_O or E_A are non-zero, then we have *full-block* ("unstructured") uncertainty. However, unstructured uncertainty is often a poor (conservative) assumption for multivariable plants. We will therefore focus on diagonal input and output uncertainty, where E_I or E_O are diagonal matrices. This uncertainty is usually caused by uncertainty in the individual input or output channels. For example,

$$E_I = \text{diag}\{\epsilon_1, \epsilon_2, \ldots\} \tag{6.61}$$

where ϵ_i is the relative uncertainty in input channel i. Typically, the magnitude of ϵ_i is 0.1 or larger. It is important to stress that *diagonal input and output uncertainty* is *always* present in real systems. Of these, we will show that diagonal input uncertainty is usually the worst for control, because performance is measured at the plant output.

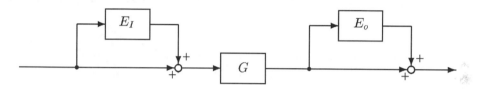

Figure 6.2: Plant with multiplicative input and output uncertainty

6.10.2   Effect of uncertainty on feedforward control

We consider here the effect of uncertainty when we use "perfect" (inverse based) feedforward control. We use the feeforward controller $u = K_r r$ and assume that the plant G is invertible so that we can select

$$K_r = G^{-1}$$

For the nominal case with no uncertainty we then achieve *perfect* control with $S_r = 0$; that is, $e = y - r = (GK_r - I)r = -S_r r = 0$. However, for the actual plant G' (with uncertainty) the control error becomes

$$e' = (G'G^{-1} - I)r = -S'_r r$$

and we get for the three sources of uncertainty

Output uncertainty:	$-S'_r = E_O$	(6.62)
Input uncertainty:	$-S'_r = GE_I G^{-1}$	(6.63)
Additive uncertainty:	$-S'_r = E_A G^{-1}$	(6.64)

For feedforward control to be effective (at a given frequency) we must require $\bar{\sigma}(S'_r) \leq 1$. We derive the following upper bounds for the three sources of uncertainty:

Output uncertainty:	$\bar{\sigma}(S'_r) = \bar{\sigma}(E_O)$	(6.65)
Input uncertainty:	$\bar{\sigma}(S'_r) \leq \bar{\sigma}(E_I)\,\gamma(G)$	(6.66)
Additive uncertainty:	$\bar{\sigma}(S'_r) \leq \bar{\sigma}(E_A)/\underline{\sigma}(G)$	(6.67)

where we have used $\bar{\sigma}(G^{-1}) = 1/\underline{\sigma}(G)$ and introduced the condition number $\gamma(G) = \bar{\sigma}(G)/\underline{\sigma}(G)$. The bounds are tight (i.e. equality can always be achieved) if we assume that any "full-block" uncertainty E_O, E_I or E_A of a given magnitude is allowed. For output uncertainty, (6.62) is identical to the result that can be derived for SISO systems (see page 204), and we must require for effective use of feeforward control that the relative output uncertainty is less than 1. For input uncertainty, the norm of the matrix GE_IG^{-1} can be a factor $\gamma(G)$ larger than the norm of E_I, and for a large $\gamma(G)$ we must require that the relative input uncertainty is much less than 1. However, inequalities (6.66) and (6.67) are generally conservative because it is not likely in practice than any full-block uncertainty of a given magnitude is possible.

Diagonal input uncertainty. We will therefore focus on diagonal input uncertainty, which always occurs in practice, and which may severely limit multivariable performance with feedforward control. In particular, we will show that

- *Feedforward control with diagonal input uncertainty is acceptable for plants with a small minimized input condition number $\gamma_I^*(G)$, see (6.68), but should be avoided for plants with large RGA elements, see (6.70).*

With diagonal input uncertainty (6.61) we may write $E_I = D_I E_I D_I^{-1}$ and $-S_r' = (GD_I)E_I(GD_I)^{-1}$ where the diagonal matrix D_I is free to be chosen. We may use this degree of freedom to make the bound on $\bar{\sigma}(S_r')$ less conservative. We have (for all diagonal E_I)

$$\bar{\sigma}(S_r') = \bar{\sigma}(GE_IG^{-1}) \leq \bar{\sigma}(E_I)\gamma_I^*(G) \tag{6.68}$$

This shows that we have <u>in</u>sensitivity to diagonal input uncertainty *if* the minimized input condition number is small. To be able to say "if and only if" we would need (6.68) to be tight (at least within some factor); that is, there should always exists a "worst-case" diagonal E_I that makes $\bar{\sigma}(S_r')$ reasonably close to the upper bound. Although this seems likely in most cases, it has not been proved to hold generally. Fortunately, we have an RGA condition that works in the opposite direction. With diagonal input uncertainty, the diagonal elements of GE_IG^{-1} are from (A.81) directly given by the corresponding row elements of the RGA

$$[GE_IG^{-1}]_{ii} = \sum_{j=1}^{n} \lambda_{ij}(G)\epsilon_j \tag{6.69}$$

The norm of a matrix is always larger than its elements, and by allowing any diagonal input uncertainty satisfying $|\epsilon_i| \leq \bar{\sigma}(E_I)$ we may select the worst-case combination of ϵ_i such that the row-sum is maximized (see remark on page 246). We then have (for some "worst-case" diagonal E_I)

$$\bar{\sigma}(S_r') = \bar{\sigma}(GE_IG^{-1}) \geq \bar{\sigma}(E_I)\|\Lambda\|_{i\infty} \tag{6.70}$$

where $\|\Lambda\|_{i\infty}$ is the induced ∞-norm (maximum row sum) of the RGA. The RGA matrix is easy to compute and independent of both input and output scalings, which make the use of condition (6.70) particularly attractive. Since diagonal input uncertainty is *always* present, we conclude from (6.63) and (6.70) that if the plant has large RGA elements then performance with feedforward control will be poor. The reverse statement is *not* true; that is, if the RGA has small elements we *cannot* conclude that the plant is insensitive to input uncertainty. This follows because we cannot from the RGA say anything about the magnitude of the off-diagonal elements of GE_IG^{-1}; see also Example 6.10.

Example 6.8 **Inverse-based control of distillation process.** *For the distillation process in (3.93) we have*

$$G(s) = \frac{1}{75s + 1} \begin{bmatrix} 87.8 & -86.4 \\ 108.2 & -109.6 \end{bmatrix}, \quad \Lambda(G) = \begin{bmatrix} 35.1 & -34.1 \\ -34.1 & 35.1 \end{bmatrix} \tag{6.71}$$

and $\gamma(G) = \gamma_I^*(G) = 141.7$. The RGA elements are large so we know that inverse-based feedforward control is sensitive to diagonal input uncertainty. With $E_I = \text{diag}\{\epsilon_1, \epsilon_2\}$ we get, for all frequencies,

$$GE_I G^{-1} = \begin{bmatrix} 35.1\epsilon_1 - 34.1\epsilon_2 & -27.7\epsilon_1 + 27.7\epsilon_2 \\ 43.2\epsilon_1 - 43.2\epsilon_2 & -34.1\epsilon_1 + 35.1\epsilon_2 \end{bmatrix} \tag{6.72}$$

The elements in the matrix $GE_I G^{-1}$ are largest when ϵ_1 and ϵ_2 have opposite signs. With a 20% error in each input channel, we may select $\epsilon_1 = 0.2$ and $\epsilon_2 = -0.2$ and find

$$GE_I G^{-1} = \begin{bmatrix} 13.8 & -11.1 \\ 17.2 & -13.8 \end{bmatrix} \tag{6.73}$$

Thus with an "ideal" feedforward controller and 20% input uncertainty, we get from (6.63) that the relative tracking error at all frequencies, including steady-state, may exceed 1000%. This demonstrates the need for feedback control. However, applying feedback control is also difficult for this plant as seen in Example 6.11.

The following example demonstrates that a large plant condition number, $\gamma(G)$, does not necessarily imply sensitivity to uncertainty even with an inverse-based controller.

Example 6.9 **Inverse-based control of distillation process, DV-model.** *In this example we consider the following distillation model given by Skogestad et al. (1988) (it is the same system as studied above but with the DV- rather than the LV-configuration for the lower control levels, see Example 10.8):*

$$G = \frac{1}{75s + 1} \begin{bmatrix} -87.8 & 1.4 \\ -108.2 & -1.4 \end{bmatrix}, \quad \Lambda(G) = \begin{bmatrix} 0.448 & 0.552 \\ 0.552 & 0.448 \end{bmatrix} \tag{6.74}$$

We have that $\|\Lambda(G(j\omega))\|_{i\infty} = 1$, $\gamma(G) \approx 70.76$ and $\gamma_I^*(G) \approx 1.11$ at all frequencies. The condition number is large, but nevertheless there is no sensitivity to diagonal input uncertainty, because $\gamma_I^*(G)$ is small. This applies to ideal inverse-based feedforward control, see (6.68), as well as to inverse-based feedback control, see (6.92) below.

Example 6.10 *For a 2×2 plant with diagonal input uncertainty we generally have*

$$GE_I G^{-1} = \begin{bmatrix} \lambda_{11}\epsilon_1 + \lambda_{12}\epsilon_2 & -\frac{g_{12}}{g_{22}}\lambda_{11}(\epsilon_1 - \epsilon_2) \\ \frac{g_{21}}{g_{11}}\lambda_{11}(\epsilon_1 - \epsilon_2) & \lambda_{21}\epsilon_1 + \lambda_{22}\epsilon_2 \end{bmatrix} \tag{6.75}$$

For example, condsider a triangular plant with $g_{12} = 0$ and with a large $|g_{21}|/|g_{11}|$,

$$G = \begin{bmatrix} 1 & 0 \\ 10 & 1 \end{bmatrix}$$

Is inverse-based feedforward control sensitive to uncertainty for this plant? $\Lambda = I$, which is small, so the lower bound (6.70) in terms of the RGA is inconclusive. The minimized input condition number for this triangular plant is $\gamma_I^* = 2|g_{21}|/|g_{11}| = 20$, which is large, so the upper bound (6.68) in terms of γ_I^* is also inconclusive. However, the system is indeed sensitive to diagonal input uncertainty, since from (6.75) the 2, 1 element of $GE_I G^{-1}$ is $(g_{21}/g_{11})(\epsilon_1 - \epsilon_2)$. For example, with 20% diagonal input uncertainty we may select $\epsilon_1 = 0.2$ and $\epsilon_2 = -0.2$ and the 2, 1 element becomes $10(0.2 + 0.2) = 4$ which is much larger than 1, and feedforward control is expected to yield poor performance with uncertainty. This motivates the use of feedback control for this plant.

Remark. Worst-case uncertainty. It is useful to know which combinations of input errors give poor performance. For an inverse-based controller (feedforward or feedback), a good indicator results if we consider GE_IG^{-1}, where $E_I = \text{diag}\{\epsilon_k\}$. If all ϵ_k have the same magnitude $|w_I| = \bar{\sigma}(E_I)$, then the largest possible magnitude of any diagonal element in GE_IG^{-1} is given by $|w_I| \cdot \|\Lambda(G)\|_{i\infty}$. To obtain this value one may select the phase of each ϵ_k such that $\angle \epsilon_k = -\angle \lambda_{ik}$, where i denotes the row of $\Lambda(G)$ with the largest elements. Also, if $\Lambda(G)$ is real (e.g. at steady-state), the signs of the ϵ_k's should be the opposite from those in the row of $\Lambda(G)$ with the largest elements.

6.10.3   Uncertainty and the benefits of feedback

To illustrate the benefits of feedback control in reducing the sensitivity to uncertainty, we consider the effect of output uncertainty on reference tracking. As a basis for comparison we first consider feedforward control.

Feedforward control. Let the nominal transfer function with feedforward control be $y = T_r r$ where $T_r = GK_r$ and K_r denotes the feedforward controller. Ideally, $T_r = I$. With model error $T'_r = G'K_r$, and the change in response is $y' - y = (T'_r - T_r)r$ where

$$T'_r - T_r = (G'G^{-1} - I)T_r = E_O T_r \tag{6.76}$$

Thus, $y' - y = E_O T_r r = E_O y$, and with feedforward control the relative control error caused by the uncertainty is equal to the relative output uncertainty.

Feedback control. With one degree-of-freedom feedback control the nominal transfer function is $y = Tr$ where $T = L(I + L)^{-1}$ is the complementary sensitivity function. Ideally, $T = I$. The change in response with model error is $y' - y = (T' - T)r$ where from (A.152)

$$T' - T = S'E_O T \tag{6.77}$$

Thus, $y' - y = S'E_O T r = S'E_O y$, and we see that

- with feedback control the effect of the uncertainty is reduced by a factor S' compared to that with feedforward control.

Thus, feedback control is much less sensitive to uncertainty than feedforward control at frequencies where feedback is effective and the elements in S' are small. However, the opposite may be true in the crossover frequency range where S' may have elements larger than 1; see Section 6.10.4.

Remark 1 For square plants, $E_O = (G' - G)G^{-1}$ and (6.77) becomes

$$\Delta T \cdot T^{-1} = S' \cdot \Delta G \cdot G^{-1} \tag{6.78}$$

where $\Delta T = T' - T$ and $\Delta G = G' - G$. Equation (6.78) provides a generalization of Bode's differential relationship (2.24) for SISO systems. To see this, consider a SISO system and let $\Delta G \to 0$. Then $S' \to S$ and we have from (6.78)

$$\frac{dT}{T} = S\frac{dG}{G} \tag{6.79}$$

Remark 2 Alternative expressions showing the benefits of feedback control are derived by introducing the inverse output multiplicative uncertainty $G' = (I - E_{iO})^{-1}G$. We then get (Horowitz and Shaked, 1975)

Feedforward control:	$T'_r - T_r = E_{iO}T'_r$	(6.80)
Feedback control:	$T' - T = SE_{iO}T'$	(6.81)

(Simple proof for square plants: switch G and G' in (6.76) and (6.77) and use $E_{iO} = (G' - G)G'^{-1}$.)

Remark 3 Another form of (6.77) is (Zames, 1981)

$$T' - T = S'(L' - L)S \qquad (6.82)$$

Conclusion. From (6.77), (6.81) and (6.82) we see that with feedback control $T' - T$ is small at frequencies where feedback is effective (i.e. S and S' are small). This is usually at low frequencies. At higher frequencies we have for real systems that L is small, so T is small, and again $T' - T$ is small. Thus with feedback, uncertainty only has a significant effect in the crossover region where S and T both have norms around 1.

6.10.4   Effect of uncertainty on feedback sensitivity peak

We demonstrated above how feedback may reduce the effect of uncertainty, but we also pointed out that uncertainty may pose limitations on achievable performance, especially at crossover frequencies. The objective in the following is to investigate how the magnitude of the sensitivity, $\bar{\sigma}(S')$, is affected by multiplicative output uncertainty and multiplicative input uncertainty given by (6.58) and (6.59), respectively. The bounds are in terms of the plant condition numbers, see (6.57), and the controller condition numbers

$$\gamma(K) = \frac{\bar{\sigma}(K)}{\underline{\sigma}(K)}, \qquad \gamma_O^*(K) = \min_{D_O} \gamma(D_O K) \qquad (6.83)$$

The minimized condition number $\gamma_O^*(K)$ may be computed using (A.76). The following factorizations of S' in terms of the nominal sensitivity S (see Appendix A.7) form the basis for the development:

Output uncertainty: $\qquad S' = S(I + E_O T)^{-1} \qquad (6.84)$

Input uncertainty: $\qquad S' = S(I + GE_I G^{-1}T)^{-1} = SG(I + E_I T_I)^{-1}G^{-1} \qquad (6.85)$

$$S' = (I + TK^{-1}E_I K)^{-1}S = K^{-1}(I + T_I E_I)^{-1}KS \qquad (6.86)$$

We assume that G and G' are stable. We also assume closed-loop stability, so that both S and S' are stable. We then get that $(I + E_O T)^{-1}$ and $(I + E_I T_I)^{-1}$ are stable. In most cases, we assume that the magnitude of the multiplicative (relative) uncertainty at each frequency can be bounded in terms of its singular value

$$\bar{\sigma}(E_I) \leq |w_I|, \qquad \bar{\sigma}(E_O) \leq |w_O| \qquad (6.87)$$

where $w_I(s)$ and $w_O(s)$ are scalar weights. Typically the uncertainty bound, $|w_I|$ or $|w_O|$, is 0.2 at low frequencies and exceeds 1 at higher frequencies.

We first state some upper bounds on $\bar{\sigma}(S')$. These are based on identities (6.84)–(6.86) and singular value inequalities (see Appendix A.3.4) of the kind

$$\bar{\sigma}((I + E_I T_I)^{-1}) = \frac{1}{\underline{\sigma}(I + E_I T_I)} \leq \frac{1}{1 - \bar{\sigma}(E_I T_I)} \leq \frac{1}{1 - \bar{\sigma}(E_I)\bar{\sigma}(T_I)} \leq \frac{1}{1 - |w_I|\bar{\sigma}(T_I)}$$

Of course these inequalities only apply if we assume $\bar{\sigma}(E_I T_I) < 1$, $\bar{\sigma}(E_I)\bar{\sigma}(T_I) < 1$ and $|w_I|\bar{\sigma}(T_I) < 1$. For simplicity, we will not state these assumptions each time.

Upper bound on $\bar{\sigma}(S')$ for output uncertainty

From (6.84), we derive

$$\bar{\sigma}(S') \leq \bar{\sigma}(S)\bar{\sigma}((I + E_O T)^{-1}) \leq \frac{\bar{\sigma}(S)}{1 - |w_O|\bar{\sigma}(T)} \tag{6.88}$$

From (6.88), we see that output uncertainty, be it diagonal or full block, poses no particular problem when performance is measured at the plant output. That is, if we have a reasonable stability margin ($\|(I + E_O T)^{-1}\|_\infty$ is not too much larger than 1), then the nominal and perturbed sensitivities do not differ very much.

Upper bounds on $\bar{\sigma}(S')$ for input uncertainty

General case (full-block or diagonal input uncertainty and any controller). From (6.85) and (6.86), we derive

$$\bar{\sigma}(S') \leq \gamma(G)\bar{\sigma}(S)\bar{\sigma}((I + E_I T_I)^{-1}) \leq \gamma(G)\frac{\bar{\sigma}(S)}{1 - |w_I|\bar{\sigma}(T_I)} \tag{6.89}$$

$$\bar{\sigma}(S') \leq \gamma(K)\bar{\sigma}(S)\bar{\sigma}((I + T_I E_I)^{-1}) \leq \gamma(K)\frac{\bar{\sigma}(S)}{1 - |w_I|\bar{\sigma}(T_I)} \tag{6.90}$$

From (6.89), we see that for a plant with a small condition number, $\gamma(G) \approx 1$, the system is insensitive to input uncertainty, irrespective of the controller. From (6.90), we have the important result that if we use a "round" controller, meaning that $\gamma(K)$ is close to 1, then the sensitivity function is *not* sensitive to input uncertainty. In many cases, (6.89) and (6.90) are not very useful because they yield unnecessarily large upper bounds.

Diagonal input uncertainty (any controller). From the first identity in (6.85) we get $S' = S(I + (GD_I)E_I(GD_I)^{-1}T)^{-1}$ and we derive, by singular value inequalities,

$$\bar{\sigma}(S') \leq \frac{\bar{\sigma}(S)}{1 - \gamma_I^*(G)|w_I|\bar{\sigma}(T)} \tag{6.91}$$

$$\bar{\sigma}(S') \leq \frac{\bar{\sigma}(S)}{1 - \gamma_O^*(K)|w_I|\bar{\sigma}(T)} \tag{6.92}$$

From (6.91), the system is insensitive to diagonal input uncertainty if $\gamma_I^*(G)$ is small, irrespective of the controller. Similarly, from (6.92) the system is insensitive to diagonal input uncertainty if $\gamma_O^*(K)$ is small, irrespective of the plant. Note that $\gamma_O^*(K) = 1$ for a diagonal controller (decentralized control), so (6.92) shows that diagonal uncertainty poses no problem with decentralized control. On the other hand, with an inverse-based (decoupling) controller of the form $K = DG^{-1}$ where D is diagonal, we have $\gamma_O^*(K) = \gamma_I^*(G)$, so decoupling control may be sensitive to diagonal input uncertainty for plants with a large $\gamma_I^*(G)$.

Lower bound on $\bar{\sigma}(S')$ for input uncertainty (including diagonal input uncertainty)

Above we derived upper bounds on $\bar{\sigma}(S')$; we will next derive a *lower* bound. A lower bound is useful because it allows us to make definite conclusions about when the plant is not input–output controllable. Importantly, the bound applies also to the special (and common) case of *diagonal* input uncertainty.

Theorem 6.5 Input uncertainty and inverse-based control. *Consider a controller $K(s) = l(s)G^{-1}(s)$ which results in a nominally decoupled response with sensitivity $S = s \cdot I$ and complementary sensitivity $T = t \cdot I$ where $t(s) = 1 - s(s)$. Suppose the plant has diagonal input uncertainty E_I of relative magnitude $|w_I(j\omega)|$ in each input channel. Then there exists a combination of input uncertainties (i.e., exists a diagonal Δ_I) such that at each frequency*

$$\bar{\sigma}(S') \geq \bar{\sigma}(S) \left(1 + \frac{|w_I t|}{1 + |w_I t|} \|\Lambda(G)\|_{i\infty} \right) \tag{6.93}$$

where $\|\Lambda(G)\|_{i\infty}$ is the maximum row sum of the RGA and $\bar{\sigma}(S) = |s|$.

The proof is given below. From (6.93), we see that with an inverse-based controller the worst-case sensitivity will be much larger than the nominal at frequencies where the plant has large RGA elements. At frequencies where control is effective ($\bar{\sigma}(S)$ is small and $|t| \approx 1$), this implies that control is not as good as expected, but it may still be acceptable. However, at crossover frequencies, where $\bar{\sigma}(S)$ and $|t| = |1 - s|$ are both close to 1, we find that $\bar{\sigma}(S')$ in (6.93) may become much larger than 1 if the plant has large RGA elements at these frequencies. The bound (6.93) applies to diagonal input uncertainty and therefore also to full-block input uncertainty (since it is a lower bound).

Proof of Theorem 6.5: (From Skogestad and Havre (1996) and Gjøsæter (1995).) Write the sensitivity function as

$$S' = (I + G'K)^{-1} = SG \underbrace{(I + E_I T_I)^{-1}}_{D} G^{-1}, \quad E_I = \text{diag}\{\epsilon_k\}, \quad S = sI \tag{6.94}$$

Since D is a diagonal matrix, we have from (6.69) that the diagonal elements of S' are given in terms of the RGA of the plant G as

$$s'_{ii} = s \sum_{k=1}^{n} \lambda_{ik} d_k; \qquad d_k = \frac{1}{1 + t\epsilon_k}; \qquad \Lambda = G \times (G^{-1})^T \tag{6.95}$$

(Note that s here is a scalar sensitivity function and not the Laplace variable.) The singular value of a matrix is larger than any of its elements, so $\bar{\sigma}(S') \geq \max_i |s'_{ii}|$, and the objective in the following is to choose a combination of input errors ϵ_k such that the worst-case $|s'_{ii}|$ is as large as possible. Consider a given output i and write each term in the sum in (6.95) as

$$\lambda_{ik} d_k = \frac{\lambda_{ik}}{1 + t\epsilon_k} = \lambda_{ik} - \frac{\lambda_{ik} t\epsilon_k}{1 + t\epsilon_k} \tag{6.96}$$

We choose all ϵ_k to have the same magnitude $|w_I(j\omega)|$, so we have $\epsilon_k(j\omega) = |w_I|e^{j\angle\epsilon_k}$. We also assume that $|t\epsilon_k| < 1$ at all frequencies[3], so that the phase of $1 + t\epsilon_k$ lies between $-90°$ and $90°$. It is then always possible to select $\angle\epsilon_k$ (the phase of ϵ_k) such that the last term in (6.96) is real and negative, and we have at each frequency, with these choices for ϵ_k,

$$\frac{s'_{ii}}{s} = \sum_{k=1}^{n} \lambda_{ik} d_k = 1 + \sum_{k=1}^{n} \frac{|\lambda_{ik}| \cdot |t\epsilon_k|}{|1 + t\epsilon_k|}$$

$$\geq 1 + \sum_{k=1}^{n} \frac{|\lambda_{ik}| \cdot |w_I t|}{1 + |w_I t|} = 1 + \frac{|w_I t|}{1 + |w_I t|} \sum_{k=1}^{n} |\lambda_{ik}| \tag{6.97}$$

---

[3] The assumption $|t\epsilon_k| < 1$ is not included in the theorem since it is actually needed for robust stability. If it does not hold we may have $\bar{\sigma}(S')$ infinite for some allowed uncertainty, and (6.93) clearly holds.

where the first equality makes use of the fact that the row elements of the RGA sum to 1 ($\sum_{k=1}^{n} \lambda_{ik} = 1$). The inequality follows since $|\epsilon_k| = |w_I|$ and $|1 + t\epsilon_k| \leq 1 + |t\epsilon_k| = 1 + |w_I t|$. This derivation holds for any i (but only for one at a time), and (6.93) follows by selecting i to maximize $\sum_{k=1}^{n} |\lambda_{ik}|$ (the maximum row sum of the RGA of G). $\qquad\qquad\qquad\qquad\qquad\qquad\qquad\qquad\qquad\qquad\qquad\qquad\qquad\square$

We next consider three examples. In the first, we consider a plant where both the RGA and $\gamma_I^*(G)$ are large. In the second, they are both small. In the third, the RGA is small, but γ_I^* is large. The first and third are sensitive to diagonal input uncertainty, whereas the second (where γ_I^* is small) is insensitive.

Example 6.11 Feedback control of distillation process. *Consider again the distillation process $G(s)$ in (6.71) which we on page 245 found to be sensitive to diagonal input uncertainty with feedforward control. For this plant we have $\|\Lambda(G(j\omega))\|_{i\infty} = 69.1$ and $\gamma(G) \approx \gamma_I^*(G) \approx 141.7$ at all frequencies.*

1. Inverse-based feedback controller. Consider the controller $K(s) = (0.7/s)G^{-1}(s)$ corresponding to the nominal sensitivity function

$$S(s) = \frac{s}{s + 0.7} I$$

The nominal response is excellent, but we found from simulations in Figure 3.14 that the closed-loop response with 20% input gain uncertainty was extremely poor (we used $\epsilon_1 = 0.2$ and $\epsilon_2 = -0.2$). The poor response is easily explained from the lower RGA bound on $\bar{\sigma}(S')$ in (6.93). With the inverse-based controller we have $l(s) = k/s$, which has a nominal phase margin of $PM = 90°$, and from (2.50) we have, at frequency ω_c, that $|s(j\omega_c)| = |t(j\omega_c)| = 1/\sqrt{2} = 0.707$. With $|w_I| = 0.2$, we then get from (6.93) that

$$\bar{\sigma}(S'(j\omega_c)) \geq 0.707 \left(1 + \frac{0.707 \cdot 0.2 \cdot 69.1}{1.14} \right) = 0.707 \cdot 9.56 = 6.76 \qquad (6.98)$$

(This is close to the peak value in (6.93) of 6.81 at frequency 0.79 rad/min.) Thus, we have that with 20% input uncertainty we may have $\|S'\|_\infty \geq 6.81$ and this explains the observed poor closed-loop performance. For comparison, the actual worst-case peak value of $\bar{\sigma}(S')$, with the inverse-based controller, is 14.5 (computed numerically using skewed-μ as discussed below). This is close to the value obtained with the uncertainty $E_I = \text{diag}\{\epsilon_1, \epsilon_2\} = \text{diag}\{0.2, -0.2\}$,

$$\|S'\|_\infty = \left\| \left(I + \frac{0.7}{s} G \begin{bmatrix} 1.2 & \\ & 0.8 \end{bmatrix} G^{-1} \right)^{-1} \right\|_\infty = 14.5$$

for which the peak occurs at 0.69 rad/min. The difference between the values 6.81 and 14.5 illustrates that the bound in terms of the RGA is generally not tight, but it is nevertheless very useful.

2. Diagonal (decentralized) feedback controller. Consider the controller

$$K_{\text{diag}}(s) = \frac{k_2(\tau s + 1)}{s} \begin{bmatrix} 1 & 0 \\ 0 & -1 \end{bmatrix}, \qquad k_2 = 2.4 \cdot 10^{-2} \, [\text{min}^{-1}]$$

The peak value for the upper bound on $\bar{\sigma}(S')$ in (6.92) is 1.26, so we are guaranteed $\|S'\|_\infty \leq 1.26$, even with 20% gain uncertainty. For comparison, the actual peak in the perturbed sensitivity function with $E_I = \text{diag}\{0.2, -0.2\}$ is $\|S'\|_\infty = 1.05$. Of course, the problem with the simple diagonal controller is that (although it is robust) even nominal performance is poor.

Remark. Relationship with the structured singular value: skewed-μ. To analyze exactly the worst-case sensitivity with a given uncertainty $|w_I|$ we may compute skewed-μ (μ^s). With reference to Section 8.11, this involves computing $\mu_{\tilde{\Delta}}(N)$ with $\tilde{\Delta} = \text{diag}(\Delta_I, \Delta_P)$ and $N = \begin{bmatrix} w_I T_I & w_I K S \\ SG/\mu^s & S/\mu^s \end{bmatrix}$ and varying μ^s until $\mu(N) = 1$. The worst-case performance at a given frequency is then $\bar{\sigma}(S') = \mu^s(N)$.

Example 6.12 *Consider the plant*

$$G(s) = \begin{bmatrix} 1 & 100 \\ 0 & 1 \end{bmatrix}$$

for which at all frequencies $\Lambda(G) = I$, $\gamma(G) = 10^4$, $\gamma^*(G) = 1.00$ and $\gamma_I^*(G) = 200$. The RGA matrix is the identity, but since $g_{12}/g_{11} = 100$ we expect from (6.75) that this plant will be sensitive to diagonal input uncertainty if we use inverse-based feedback control, $K = \frac{c}{s}G^{-1}$. This is confirmed if we compute the worst-case sensitivity function S' for $G' = G(I + w_I \Delta_I)$ where Δ_I is diagonal and $|w_I| = 0.2$. We find by computing skewed-μ, $\mu^s(N_1)$, that the peak of $\bar{\sigma}(S')$ is $\|S'\|_\infty = 20.43$.

Note that the peak is independent of the controller gain c in this case since $G(s)$ is a constant matrix. Also note that with full-block ("unstructured") input uncertainty (Δ_I is a full matrix) the worst-case sensitivity is $\|S'\|_\infty = 1021.7$.

Conclusions on input uncertainty and feedback control

Let us summarize the above findings. The following statements apply to the frequency range around crossover. By "small', we mean about 2 or smaller. By "large" we mean about 10 or larger.

1. Condition number $\gamma(G)$ or $\gamma(K)$ small: robust performance to both diagonal and full-block input uncertainty; see (6.89) and (6.90).
2. Minimized condition numbers $\gamma_I^*(G)$ or $\gamma_O^*(K)$ small: robust performance to diagonal input uncertainty; see (6.91) and (6.92). Note that a diagonal controller (decentralized control) always has $\gamma_O^*(K) = 1$.
3. RGA(G) has large elements: inverse-based controller is *not* robust to diagonal input uncertainty; see (6.93). Since diagonal input uncertainty is unavoidable in practice, the rule is never to use a decoupling controller for a plant with large RGA elements. Furthermore, a diagonal controller will most likely yield poor nominal performance for a plant with large RGA elements, so we conclude that *plants with large RGA elements are fundamentally difficult to control.*
4. $\gamma_I^*(G)$ is large while at the same time RGA(G) has small elements: cannot make any definite conclusion about the sensitivity to input uncertainty based on the bounds in this section. However, as seen in Examples 6.10 and 6.12, we may expect sensitivity to diagonal input uncertainty with inverse-based feedforward or feedback control.

6.10.5   Element-by-element uncertainty

Element-by-element uncertainty assumes independent uncertainty in the individual elements of G. This kind of uncertainty description may be questionable from a physical point of view, but it is nevertheless popular. Interestingly, the RGA matrix is a direct measure of sensitivity to element-by-element uncertainty as matrices with large RGA values become singular for small relative errors in the elements.

Theorem 6.6 *Consider a complex matrix G and let λ_{ij} denote the ij'th element in the RGA matrix of G. The matrix G becomes singular if we make a relative change $-1/\lambda_{ij}$ in its ij'th element; that is, if a single element in G is perturbed from g_{ij} to $g_{pij} = g_{ij}(1 - \frac{1}{\lambda_{ij}})$.*

The theorem is due to Yu and Luyben (1987). Our proof in Appendix A.4 is from Hovd and Skogestad (1992).

Example 6.13 *The matrix G in (6.71) is non-singular. The $1, 2$ element of the RGA is $\lambda_{12}(G) = -34.1$. Thus, the matrix G becomes singular if g_{12} is perturbed from -86.4 to*

$$g_{p12} = -86.4(1 - 1/(-34.1)) = -88.9 \tag{6.99}$$

The above theorem is an important algebraic property of the RGA, but it also has important implications for improved control:

1. **Identification.** Models of multivariable plants, $G(s)$, are often obtained by identifying one element at a time, e.g. using step responses. From Theorem 6.6 it is clear that this simple identification procedure will most likely give meaningless results (e.g. the wrong sign of the steady-state RGA) if there are large RGA elements within the bandwidth where the model is intended to be used.

2. **RHP-zeros.** Consider a plant with transfer function matrix $G(s)$. If the relative uncertainty in an element at a given frequency is larger than $|1/\lambda_{ij}(j\omega)|$ then the plant may be singular at this frequency, implying that the uncertainty allows for a RHP-zero on the $j\omega$-axis. This is of course detrimental to performance in terms of both feedforward and feedback control.

Remark. Theorem 6.6 seems to "prove" that plants with large RGA elements are fundamentally difficult to control. However, although the statement may be true (see the conclusions on page 251 based on diagonal input uncertainty, which is always present), we cannot draw this conclusion from Theorem 6.6. This is because the assumption of element-by-element uncertainty is often unrealistic from a physical point of view, since the elements are usually *coupled* in some way. For example, this is the case for the distillation column process, where the elements are coupled due to an underlying physical constraint in such a way that the model (6.71) never becomes singular, even for large changes in the transfer function matrix elements.

6.10.6 Steady-state condition for integral control

Feedback control reduces the sensitivity to model uncertainty at frequencies where the loop gains are large. With integral action in the controller we can achieve zero steady-state control error, even with large model errors, provided the sign of the plant, as expressed by $\det G(0)$, does not change. The statement applies for stable plants, or more generally for cases where the number of unstable poles in the plant does not change. The conditions are stated more exactly in the following theorem by Hovd and Skogestad (1994).

Theorem 6.7 *Let the number of open-loop unstable poles (excluding poles at $s = 0$) of $G(s)K(s)$ and $G'(s)K(s)$ be P and P', respectively. Assume that the controller K is such that GK has integral action in all channels, and that the transfer functions GK and $G'K$ are strictly proper. Then if*

$$\det G'(0)/\det G(0) \begin{cases} < 0 & \text{for } P - P' \text{ even, including zero} \\ > 0 & \text{for } P - P' \text{ odd} \end{cases} \tag{6.100}$$

at least one of the following instabilities will occur: (a) The negative feedback closed-loop system with loop gain GK is unstable. (b) The negative feedback closed-loop system with loop gain $G'K$ is unstable.

Proof: For stability of both $(I+GK)^{-1}$ and $(I+G'K)^{-1}$ we have from Lemma A.5 in Appendix A.7.3 that $\det(I + E_O T(s))$ needs to encircle the origin $P - P'$ times as s traverses the Nyquist D-contour.

Here $T(0) = I$ because of the requirement for integral action in all channels of GK. Also, since GK and $G'K$ are strictly proper, $E_O T$ is strictly proper, and hence $E_O(s)T(s) \to 0$ as $s \to \infty$. Thus, the map of $\det(I + E_O T(s))$ starts at $\det G'(0)/\det G(0)$ (for $s = 0$) and ends at 1 (for $s = \infty$). A more careful analysis of the Nyquist plot of $\det(I + E_O T(s))$ reveals that the number of encirclements of the origin will be even for $\det G'(0)/\det G(0) > 0$, and odd for $\det G'(0)/\det G(0) < 0$. Thus, if this parity (odd or even) does not match that of $P - P'$ we will get instability, and the theorem follows. $\square$

Example 6.14 *Suppose the true model of a plant is given by $G(s)$, and that by careful identification we obtain a model $G_1(s)$,*

$$G = \frac{1}{75s + 1}\begin{bmatrix} 87.8 & -86.4 \\ 108.2 & -109.6 \end{bmatrix}, \quad G_1(s) = \frac{1}{75s + 1}\begin{bmatrix} 87 & -88 \\ 109 & -108 \end{bmatrix}$$

At first glance, the identified model seems very good, but it is actually useless for control purposes since $\det G_1(0)$ has the wrong sign; $\det G(0) = -274.4$ and $\det G_1(0) = 196$ (also the RGA elements have the wrong sign; the $1, 1$ element in the RGA is -47.9 instead of $+35.1$). From Theorem 6.7 we then get that any controller with integral action designed based on the model G_1 will yield instability when applied to the plant G.

6.11 MIMO input–output controllability

We now summarize the main findings of this chapter in an analysis procedure for input–output controllability of a MIMO plant. The presence of directions in MIMO systems makes it more difficult to give a precise description of the procedure in terms of a set of rules as was done in the SISO case.

6.11.1 Controllability analysis procedure

The following procedure assumes that we have made a decision on the plant inputs and plant outputs (manipulations and measurements), and we want to analyze the model G to find out what control performance can be expected.

The procedure can also be used to assist in control structure design (the selection of inputs, outputs and control configuration), but it must then be repeated for each G corresponding to each candidate set of inputs and outputs. In some cases, the number of possibilities is so large that such an approach becomes prohibitive. Some pre-screening is then required, e.g. based on physical insight or by analyzing the "large" model, G_{all}, with all the candidate inputs and outputs included. This is briefly discussed in Section 10.4.

A typical MIMO controllability analysis may proceed as follows:

1. Scale all variables (inputs u, outputs y, disturbances d, references, r) to obtain a scaled model, $y = G(s)u + G_d(s)d$, $r = R\tilde{r}$; see Section 1.4.
2. Obtain a minimal realization.
3. Check functional controllability. To be able to control the outputs independently, we first need at least as many inputs u as outputs y. Second, we need the rank of $G(s)$ to be equal to the number of outputs, l, i.e. the minimum singular value of $G(j\omega)$, $\underline{\sigma}(G) = \sigma_l(G)$, should be non-zero (except at possible $j\omega$-axis zeros). If the plant is not functionally controllable then compute the output direction where the plant has no gain, see (6.30), to obtain insight into the source of the problem.

4. Compute the poles. For RHP (unstable) poles obtain their locations and associated directions; see (6.5). "Fast" RHP-poles far from the origin are bad.

5. Compute the zeros. For RHP-zeros obtain their locations and associated directions. Look for zeros pinned into certain outputs. "Small" RHP-zeros (close to the origin) are bad if tight performance at low frequencies is desired.

6. Calculate the bounds on different closed-loop transfer functions using the formulae summarized in Table 6.3.2. A large peak ($\gg 1$) for any of $S, T, KS, SG_d, KSG_d, S_I$ and T_I (including $G_d = G$) indicates poor closed-loop performance or poor robustness against uncertainty. Note that the peaks of KS, SG_d, KSG_d depend on the scaling of the plant and disturbance models.

7. Obtain the frequency response $G(j\omega)$ and compute the RGA matrix, $\Lambda = G \times (G^\dagger)^T$. Plants with large RGA elements at crossover frequencies are difficult to control and should be avoided. For more details about the use of the RGA see Section 3.3.6, page 81.

8. From now on scaling is critical. Compute the singular values of $G(j\omega)$ and plot them as a function of frequency. Also consider the associated input and output singular vectors.

9. The minimum singular value, $\underline{\sigma}(G(j\omega))$, is a particularly useful controllability measure. It should generally be as large as possible at frequencies where control is needed. If $\underline{\sigma}(G(j\omega)) < 1$ then we cannot (at frequency ω) make independent output changes of unit magnitude by using inputs of unit magnitude.

10. For disturbances, consider the elements of the matrix G_d. At frequencies where one or more elements is larger than 1, we need control. We get more information by considering one disturbance at a time (the columns g_d of G_d). We must require for each disturbance that S is less than $1/\|g_d\|_2$ in the disturbance direction y_d, i.e. $\|Sy_d\|_2 \le 1/\|g_d\|_2$; see (6.45). Thus, we must at least require $\underline{\sigma}(S) \le 1/\|g_d\|_2$ and we may have to require $\bar{\sigma}(S) \le 1/\|g_d\|_2$; see (6.46).

   Remark. If feedforward control is already used, then one may instead analyze $\widehat{G}_d(s) = GK_dG_{md} + G_d$ where K_d denotes the feedforward controller, see (5.101).

11. Disturbances and input saturation:

   First step. Consider the input magnitudes needed for perfect control by computing the elements in the matrix $G^\dagger G_d$. If all elements are less than 1 at all frequencies then input saturation is not expected to be a problem. If some elements of $G^\dagger G_d$ are larger than 1, then perfect control ($e = 0$) cannot be achieved at this frequency, but "acceptable" control ($\|e\|_2 < 1$) may be possible, and this may be tested in the second step.

   Second step. Check condition (6.55): that is, consider the elements of $U^H G_d$ and make sure that the elements in the i'th row are smaller than $\sigma_i(G) + 1$, at all frequencies.

12. Are the requirements compatible? Look at disturbances, RHP-poles and RHP-zeros and their associated locations and directions. For example, we must require for each disturbance and each RHP-zero that $|y_z^H g_d(z)| \le 1$; see (6.47). For combined RHP-zeros and RHP-poles see (6.8).

13. Uncertainty. If the condition number $\gamma(G)$ is small then we expect no particular problems with uncertainty. If the RGA elements are large, we expect strong sensitivity to uncertainty. For a more detailed analysis see the conclusion on page 251.

14. If decentralized control (diagonal controller) is of interest see the summary on page 449.

15. The use of the condition number and RGA are summarized separately in Section 3.3.6, page 81.

A controllability analysis may also be used to obtain initial performance weights for controller design. After a controller design one may analyze the controller by plotting, for example, its elements, singular values, RGA and condition number as a function of frequency.

6.11.2   Plant design changes

If a plant is not input–output controllable, then it must somehow be modified. Some possible modifications are listed below.

Controlled outputs. Identify the output(s) which cannot be controlled satisfactorily. Should these outputs really be controlled? Can the specifications for these be relaxed?

Manipulated inputs. If undesirable input constraints are encountered then consider replacing or moving actuators. For example, this could mean replacing a control valve with a larger one, or moving it closer to the controlled output.

If there are RHP-zeros which cause control problems then the zeros may often be eliminated by adding another input (possibly resulting in a non-square plant). This may not be possible if the zero is pinned to a particular output.

Extra measurements. If there are RHP-zeros that cause control problems, then these zeros may often be eliminated by adding extra measurements (i.e. add outputs with no associated control objective). If the effect of disturbances, or uncertainty, is large, and the dynamics of the plant are such that acceptable control cannot be achieved, then consider adding "fast local loops" based on extra measurements which are located close to the inputs and disturbances; see Section 10.6.4 and the example on page 216.

Disturbances. If the effect of disturbances is too large, then see whether the disturbance itself may be reduced. This may involve adding extra equipment to dampen the disturbances, such as a buffer tank in a chemical process or a spring in a mechanical system. In other cases, this may involve improving or changing the control of another part of the system, e.g. we may have a disturbance which is actually the manipulated input in another part of the system.

Plant dynamics and time delays. In most cases, controllability is improved by making the plant dynamics faster and by reducing time delays. An exception to this is a strongly interactive plant, where an increased dynamic lag or time delay may be helpful if it somehow "delays" the effect of the interactions; see (6.31). Another more obvious exception is for feedforward control of a measured disturbance, where a delay for the disturbance's effect on the outputs is an advantage.

Example 6.15 **Removing zeros by adding inputs.** *Consider a stable* 2×2 *plant*

$$G_1(s) = \frac{1}{(s+2)^2} \begin{bmatrix} s+1 & s+3 \\ 1 & 2 \end{bmatrix}$$

which has a RHP-zero at $s = 1$ *which limits achievable performance. The zero is not pinned to a particular output, so it will most likely disappear if we add a third manipulated input. Suppose the new plant is*

$$G_2(s) = \frac{1}{(s+2)^2} \begin{bmatrix} s+1 & s+3 & s+6 \\ 1 & 2 & 3 \end{bmatrix}$$

which indeed has no zeros. It is interesting to note that each of the three individual 2×2 *sub-plants of* $G_2(s)$ *has a RHP-zero (located at* $s = 1$, $s = 1.5$ *and* $s = 3$, *respectively).*

6.11.3   Additional exercises

The reader will be better prepared for some of these exercises following an initial reading of Chapter 10 on decentralized control. In all cases the variables are assumed to be scaled as outlined in Section 1.4.

Exercise 6.8 * *Analyze input–output controllability for*

$$G(s) = \frac{1}{s^2 + 100} \begin{bmatrix} \frac{1}{0.01s+1} & 1 \\ \frac{s+0.1}{s+1} & 1 \end{bmatrix}$$

Compute the zeros and poles, plot the RGA as a function of frequency, etc.

Exercise 6.9 *Analyze input–output controllability for*

$$G(s) = \frac{1}{(\tau s + 1)(\tau s + 1 + 2\alpha)} \begin{bmatrix} \tau s + 1 + \alpha & \alpha \\ \alpha & \tau s + 1 + \alpha \end{bmatrix}$$

where $\tau = 100$; consider two cases: (a) $\alpha = 20$, and (b) $\alpha = 2$.

   Remark. *This is a simple "two-mixing-tank" model of a heat exchanger where $u = \begin{bmatrix} T_{1in} \\ T_{2in} \end{bmatrix}$, $y = \begin{bmatrix} T_{1out} \\ T_{2out} \end{bmatrix}$ and α is the number of heat transfer units.*

Exercise 6.10 * *Let*

$$A = \begin{bmatrix} -10 & 0 \\ 0 & -1 \end{bmatrix}, B = I, C = \begin{bmatrix} 10 & 1.1 \\ 10 & 0 \end{bmatrix}, D = \begin{bmatrix} 0 & 0 \\ 0 & 1 \end{bmatrix}$$

(a) Perform a controllability analysis of $G(s)$.
(b) Let $\dot{x} = Ax + Bu + d$ and consider a unit disturbance $d = [z_1 \quad z_2]^T$. Which direction (value of z_1/z_2) gives a disturbance that is most difficult to reject (consider both RHP-zeros and input saturation)?
(c) Discuss decentralized control of the plant. How would you pair the variables?

Exercise 6.11 *Consider the following two plants. Do you expect any control problems? Could decentralized or inverse-based control be used? What pairing would you use for decentralized control?*

$$G_a(s) = \frac{1}{1.25(s + 1)(s + 20)} \begin{bmatrix} s - 1 & s \\ -42 & s - 20 \end{bmatrix}$$

$$G_b(s) = \frac{1}{(s^2 + 0.1)} \begin{bmatrix} 1 & 0.1(s - 1) \\ 10(s + 0.1)/s & (s + 0.1)/s \end{bmatrix}$$

Exercise 6.12 * *Order the following three plants in terms of their expected ease of controllability:*

$$G_1(s) = \begin{bmatrix} 100 & 95 \\ 100 & 100 \end{bmatrix}, G_2(s) = \begin{bmatrix} 100e^{-s} & 95e^{-s} \\ 100 & 100 \end{bmatrix}, G_3(s) = \begin{bmatrix} 100 & 95e^{-s} \\ 100 & 100 \end{bmatrix}$$

Remember to consider also the sensitivity to input gain uncertainty.

Exercise 6.13 *Analyze input–output controllability for*

$$G(s) = \begin{bmatrix} \frac{5000s}{(5000s+1)(2s+1)} & \frac{2(-5s+1)}{100s+1} \\ \frac{3}{5s+1} & \frac{3}{5s+1} \end{bmatrix}$$

Exercise 6.14 * *Analyze input–output controllability for*

$$G(s) = \begin{bmatrix} 100 & 102 \\ 100 & 100 \end{bmatrix}, \quad g_{d1}(s) = \begin{bmatrix} \frac{10}{s+1} \\ \frac{10}{s+1} \end{bmatrix}; \quad g_{d2} = \begin{bmatrix} \frac{1}{s+1} \\ \frac{-1}{s+1} \end{bmatrix}$$

Which disturbance is the worst?

Exercise 6.15 *(a) Analyze input–output controllability for the following three plants each of which has two inputs and one output:* $G(s) = (g_1(s) \quad g_2(s))$

(i) $g_1(s) = g_2(s) = \frac{s-2}{s+2}$.

(ii) $g_1(s) = \frac{s-2}{s+2}, \quad g_2(s) = \frac{s-2.1}{s+2.1}$.

(iii) $g_1(s) = \frac{s-2}{s+2}, \quad g_2(s) = \frac{s-20}{s+20}$.

(b) Design controllers and perform closed-loop simulations of reference tracking to complement your analysis. Consider also the input magnitudes.

Exercise 6.16 * *Find the poles and zeros and analyze input–output controllability for*

$$G(s) = \begin{bmatrix} c + (1/s) & 1/s \\ 1/s & c + (1/s) \end{bmatrix}$$

Here c is a constant, e.g. c = 1. (A similar model form is encountered for distillation columns controlled with the DB-configuration. In this case the physical reason for the model being singular at steady-state is that the sum of the two manipulated inputs is fixed at steady-state, $D + B = F$.)

Exercise 6.17 **Controllability of an FCC process.** *Consider the following 3×3 model of a fluidized catalytic cracking (FCC) process:*

$$\begin{bmatrix} y_1 \\ y_2 \\ y_3 \end{bmatrix} = G(s) \begin{bmatrix} u_1 \\ u_2 \\ u_3 \end{bmatrix}; \quad f(s) = \frac{1}{(18.8s+1)(75.8s+1)}$$

$$G(s) = f(s) \begin{bmatrix} 16.8(920s^2 + 32.4s + 1) & 30.5(52.1s+1) & 4.30(7.28s+1) \\ -16.7(75.5s+1) & 31.0(75.8s+1)(1.58s+1) & -1.41(74.6s+1) \\ 1.27(-939s+1) & 54.1(57.3s+1) & 5.40 \end{bmatrix}$$

Acceptable control of this 3×3 plant can be achieved with partial control of two outputs with input 3 in manual (not used). That is, we have a 2×2 control problem. Consider three options for the controlled outputs:

$$Y_1 = \begin{bmatrix} y_1 \\ y_2 \end{bmatrix}; \quad Y_2 = \begin{bmatrix} y_2 \\ y_3 \end{bmatrix}; \quad Y_3 = \begin{bmatrix} y_1 \\ y_2 - y_3 \end{bmatrix}$$

In all three cases, the inputs are u_1 and u_2. Assume that the third input is a disturbance ($d = u_3$).

(a) Based on the zeros of the three 2×2 plants, $G_1(s)$, $G_2(s)$ and $G_3(s)$, which choice of outputs do you prefer? Which seems to be the worst?

It may be useful to know that the zero polynomials

a	$5.75 \cdot 10^7 s^4 + 3.92 \cdot 10^7 s^3 + 3.85 \cdot 10^6 s^2 + 1.22 \cdot 10^5 s + 1.03 \cdot 10^3$
b	$4.44 \cdot 10^6 s^3 - 1.05 \cdot 10^6 s^2 - 8.61 \cdot 10^4 s - 9.43 \cdot 10^2$
c	$5.75 \cdot 10^7 s^4 - 8.75 \cdot 10^6 s^3 - 5.66 \cdot 10^5 s^2 + 6.35 \cdot 10^3 s + 1.60 \cdot 10^2$

have the following roots:

a	−0.570	−0.0529	−0.0451	−0.0132
b		0.303	−0.0532	−0.0132
c	0.199	−0.0532	0.0200	−0.0132

(b) For the preferred choice of outputs in (a) do a more detailed analysis of the expected control performance (compute poles and zeros, sketch RGA_{11}, comment on possible problems with input constraints (assume the inputs and outputs have been properly scaled), discuss the effect of the disturbance, etc.). What type of controller would you use? What pairing would you use for decentralized control?

(c) Discuss why the 3×3 plant may be difficult to control.

Remark. *This is a model of a fluidized catalytic cracking (FCC) reactor where $u = (F_s \ F_a \ k_c)^T$ represents the circulation, airflow and feed composition, and $y = (T_1 \ T_{cy} \ T_{rg})^T$ represents three temperatures. $G_1(s)$ is called the Hicks control structure and $G_3(s)$ the conventional structure. More details are found in Hovd and Skogestad (1993).*

6.12 Conclusion

We have found that most of the insights into the performance limitations of SISO systems developed in Chapter 5 carry over to MIMO systems. For RHP-zeros, RHP-poles and disturbances, the issue of directions usually makes the limitations less severe for MIMO than for SISO systems. However, the situation is usually the opposite with model uncertainty because for MIMO systems there is also uncertainty associated with plant directionality. This is an issue which is unique to MIMO systems.

We summarized on page 253 the main steps involved in an analysis of input–output controllability of MIMO plants.

7

UNCERTAINTY AND ROBUSTNESS FOR SISO SYSTEMS

In this chapter, we show how to represent uncertainty by real or complex perturbations. We also analyze robust stability (RS) and robust performance (RP) for SISO systems using elementary methods. Chapter 8 is devoted to a more general analysis and controller design for uncertain systems using the structured singular value.

7.1 Introduction to robustness

A control system is robust if it is insensitive to differences between the actual system and the model of the system which was used to design the controller. These differences are referred to as model/plant mismatch or simply model uncertainty. The key idea in the $\mathcal{H}_\infty$ robust control paradigm we use is to check whether the design specifications are satisfied even for the "worst-case" uncertainty.

Our approach is then as follows:

1. Determine the uncertainty set: find a mathematical representation of the model uncertainty ("clarify what we know about what we don't know").
2. Check robust stability (RS): determine whether the system remains stable for all plants in the uncertainty set.
3. Check robust performance (RP): if RS is satisfied, determine whether the performance specifications are met for all plants in the uncertainty set.

This approach may not always achieve optimal performance. In particular, if the worst-case plant rarely or never occurs, other approaches, such as optimizing some average performance or using adaptive control, may yield better performance. Nevertheless, the linear uncertainty descriptions presented in this book are very useful in many practical situations.

It should also be appreciated that model uncertainty is not the only concern when it comes to robustness. Other considerations include sensor and actuator failures, physical constraints, changes in control objectives, the opening and closing of loops, etc. Furthermore, if a control design is based on an optimization, then robustness problems may also be caused by the mathematical objective function not properly describing the real control problem. Also, the numerical design algorithms themselves may not be robust. However, when we refer to

Multivariable Feedback Control: Analysis and Design. Second Edition
S. Skogestad and I. Postlethwaite   © 2005, 2006 John Wiley & Sons, Ltd

robustness in this book, we mean robustness with respect to model uncertainty, and assume that a fixed (linear) controller is used.

To account for model uncertainty we will assume that the dynamic behaviour of a plant is described not by a single linear time-invariant model but by a set Π of possible linear time-invariant models, sometimes denoted as the "uncertainty set". We adopt the following notation:

Π – a set of possible perturbed plant models.

$G(s) \in \Pi$ – nominal plant model (with no uncertainty).

$G_p(s) \in \Pi$ and $G'(s) \in \Pi$ – particular perturbed plant models.

Sometimes G_p is used rather than Π to denote the uncertainty set, whereas G' always refers to a particular uncertain plant. The subscript p stands for *perturbed* or *possible* or Π (take your pick). This should not be confused with the subscript capital P, e.g. in w_P, which denotes *performance*.

We will use a "norm-bounded uncertainty description" where the set Π is generated by allowing $\mathcal{H}_\infty$ norm-bounded stable perturbations to the nominal plant $G(s)$. This corresponds to a continuous description of the model uncertainty, and there will be an infinite number of possible plants G_p in the set Π. We let E denote a perturbation which is not normalized, and let Δ denote a normalized perturbation with $\mathcal{H}_\infty$ norm less than 1.

Remark. Another strategy for dealing with model uncertainty is to approximate its effect on the feedback system by adding fictitious disturbances or noise. For example, this is the only way of handling model uncertainty within the so-called LQG approach to optimal control (see Chapter 9). Is this an acceptable strategy? In general, the answer is *no*. This is easily illustrated for linear systems where the addition of disturbances does *not* affect system stability, whereas model uncertainty combined with feedback may easily create instability.

For example, consider a plant with a nominal model $y = Gu + G_d d$, and let the perturbed plant model be $G_p = G + E$ where E represents additive model uncertainty. Then the output of the perturbed plant is

$$y = G_p u + G_d d = Gu + d_1 + d_2 \tag{7.1}$$

where y is different from what we ideally expect (namely Gu) for two reasons:

1. Uncertainty in the model ($d_1 = Eu$)
2. Signal uncertainty ($d_2 = G_d d$)

In LQG control we set $w_d = d_1 + d_2$ where w_d is assumed to be an independent variable such as white noise. Then in the design problem we may make w_d large by selecting appropriate weighting functions, but its presence will never cause instability. However, in reality $w_d = Eu + d_2$, so w_d depends on the signal u and this may cause instability in the presence of feedback when u depends on y. Specifically, the closed-loop system $(I + (G + E)K)^{-1}$ may be unstable for some $E \neq 0$. In conclusion, it may be important to take explicitly into account model uncertainty when studying feedback control.

We will next discuss some sources of model uncertainty and outline how to represent them mathematically.

7.2   Representing uncertainty

Uncertainty in the plant model may have several origins:

1. There are always parameters in the linear model which are only known approximately or are simply in error.
2. The parameters in the linear model may vary due to nonlinearities or changes in the operating conditions.
3. Measurement devices have imperfections. This may even give rise to uncertainty on the manipulated inputs, since the actual input is often measured and adjusted in a cascade manner. For example, this is often the case with valves where a flow controller is often used. In other cases, limited valve resolution may cause input uncertainty.
4. At high frequencies even the structure and the model order are unknown, and the uncertainty will always exceed 100% at some frequency.
5. Even when a very detailed model is available we may choose to work with a simpler (low-order) nominal model and represent the neglected dynamics as "uncertainty".
6. Finally, the controller implemented may differ from the one obtained by solving the synthesis problem. In this case, one may include uncertainty to allow for controller order reduction and implementation inaccuracies.

The various sources of model uncertainty mentioned above may be grouped into two main classes:

1. **Parametric (real) uncertainty.** Here the structure of the model (including the order) is known, but some of the parameters are uncertain.
2. **Dynamic (frequency-dependent) uncertainty.** Here the model is in error because of missing dynamics, usually at high frequencies, either through deliberate neglect or because of a lack of understanding of the physical process. Any model of a real system will contain this source of uncertainty.

Parametric uncertainty is quantified by assuming that each uncertain parameter is bounded within some region $[\alpha_{\min}, \alpha_{\max}]$. That is, we have parameter sets of the form

$$\alpha_p = \bar{\alpha}(1 + r_\alpha \Delta)$$

where $\bar{\alpha}$ is the mean parameter value, $r_\alpha = (\alpha_{\max} - \alpha_{\min})/(\alpha_{\max} + \alpha_{\min})$ is the relative uncertainty in the parameter, and Δ is any real scalar satisfying $|\Delta| \leq 1$.

Dynamic uncertainty is somewhat less precise and thus more difficult to quantify, but it appears that the frequency domain is particularly well suited for this class. This leads to complex perturbations which we normalize such that $\|\Delta\|_\infty \leq 1$. In this chapter, we will deal mainly with this class of perturbations.

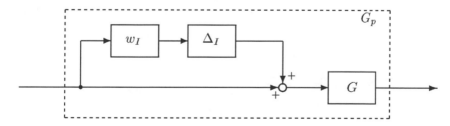

Figure 7.1: Plant with multiplicative uncertainty

In many cases, we prefer to lump the various sources of dynamic uncertainty into a *multiplicative uncertainty* of the form

$$\Pi_I: \quad G_p(s) = G(s)(1 + w_I(s)\Delta_I(s)); \quad \underbrace{|\Delta_I(j\omega)| \leq 1 \;\forall\omega}_{\|\Delta_I\|_\infty \leq 1} \qquad (7.2)$$

which may be represented by the block diagram in Figure 7.1. In (7.2), $\Delta_I(s)$ is *any* stable transfer function which at each frequency is less than or equal to 1 magnitude. Some examples of allowable $\Delta_I(s)$'s with $\mathcal{H}_\infty$ norm less than 1, $\|\Delta_I\|_\infty \leq 1$, are

$$\frac{s-z}{s+z}, \quad \frac{1}{\tau s + 1}, \quad \frac{1}{(5s+1)^3}, \quad \frac{0.1}{s^2 + 0.1s + 1}$$

The subscript I denotes "input", but for SISO systems it doesn't matter whether we consider the perturbation at the input or output of the plant, since

$$G(1 + w_I\Delta_I) = (1 + w_O\Delta_O)G \quad \text{with} \;\; \Delta_I(s) = \Delta_O(s) \;\text{and}\; w_I(s) = w_O(s)$$

Another uncertainty form, which is better suited for representing pole uncertainty, is the *inverse multiplicative uncertainty*

$$\Pi_{iI}: \quad G_p(s) = G(s)(1 + w_{iI}(s)\Delta_{iI}(s))^{-1}; \quad |\Delta_{iI}(j\omega)| \leq 1 \;\forall\omega \qquad (7.3)$$

Even with a stable $\Delta_{iI}(s)$ this form allows for uncertainty in the location of an unstable pole, and it also allows for poles crossing between the left- and right-half planes.

Parametric uncertainty is sometimes called *structured uncertainty* as it models the uncertainty in a structured manner. Analogously, lumped dynamics uncertainty is sometimes called *unstructured uncertainty*. However, one should be careful about using these terms because there can be several levels of structure, especially for MIMO systems.

Remark. Alternative approaches for describing uncertainty and the resulting performance may be considered. One approach for parametric uncertainty is to assume a probabilistic (e.g. normal) distribution of the parameters, and to consider the "average" response. This stochastic uncertainty is, however, difficult to analyze exactly.

Another approach is the multi-model approach in which one considers a finite set of alternative models. A problem with the multi-model approach is that it is not clear how to pick the set of models such that they represent the limiting ("worst-case") plants.

In this book, we will use a combination of parametric (real) uncertainty and dynamic (frequency-dependent) uncertainty. These sources can be handled within the $\mathcal{H}_\infty$ framework by allowing the perturbations to be real or complex, respectively.

7.3 Parametric uncertainty

Parametric uncertainty may be represented in the $\mathcal{H}_\infty$ framework, if we restrict the perturbations Δ to be real. Here we provide a few simple examples to illustrate this approach.

Example 7.1 Gain uncertainty. *Let the set of possible plants be*

$$G_p(s) = k_p G_0(s); \quad k_{\min} \leq k_p \leq k_{\max} \qquad (7.4)$$

where k_p is an uncertain gain and $G_0(s)$ is a transfer function with no uncertainty. By writing

$$k_p = \bar{k}(1 + r_k\Delta), \quad \bar{k} \triangleq \frac{k_{\min} + k_{\max}}{2}, \quad r_k \triangleq \frac{(k_{\max} - k_{\min})/2}{\bar{k}}, \tag{7.5}$$

where r_k is the relative magnitude of the gain uncertainty and $\bar{k}$ is the average gain, (7.4) may be rewritten as multiplicative uncertainty

$$G_p(s) = \underbrace{\bar{k}G_0(s)}_{G(s)}(1 + r_k\Delta), \quad |\Delta| \leq 1 \tag{7.6}$$

where Δ is a real scalar and $G(s)$ is the nominal plant. We see that the uncertainty in (7.6) is in the form of (7.2) with a constant multiplicative weight $w_I(s) = r_k$. The uncertainty description in (7.6) can also handle cases where the gain changes sign ($k_{\min} < 0$ and $k_{\max} > 0$) corresponding to $r_k > 1$. The usefulness of this approach is rather limited, however, since it is impossible to get any benefit from control for a plant where we can have $G_p = 0$, at least with a linear controller.

Example 7.2 **Time constant uncertainty.** *Consider a set of plants, with an uncertain time constant, given by*

$$G_p(s) = \frac{1}{\tau_p s + 1}G_0(s); \quad \tau_{\min} \leq \tau_p \leq \tau_{\max} \tag{7.7}$$

By writing $\tau_p = \bar{\tau}(1 + r_\tau\Delta)$, similar to (7.5) with $|\Delta| < 1$, the model set (7.7) can be rewritten as

$$G_p(s) = \frac{G_0}{1 + \bar{\tau}s + r_\tau\bar{\tau}s\Delta} = \underbrace{\frac{G_0}{1 + \bar{\tau}s}}_{G(s)}\frac{1}{1 + w_{iI}(s)\Delta}; \quad w_{iI}(s) = \frac{r_\tau\bar{\tau}s}{1 + \bar{\tau}s} \tag{7.8}$$

which is in the inverse multiplicative *form of (7.3). Note that it does not make physical sense for τ_p to change sign, because a value $\tau_p = 0^-$ corresponds to a pole at infinity in the RHP, and the corresponding plant would be impossible to stabilize. To represent cases in which a pole may cross between the half planes, one should instead consider parametric uncertainty in the pole itself, $1/(s+p)$, as described in (7.9).*

Example 7.3 **Pole uncertainty.** *Consider uncertainty in the parameter a in a state-space model, $\dot{y} = ay + bu$, corresponding to the uncertain transfer function $G_p(s) = b/(s - a_p)$. More generally, consider the following set of plants:*

$$G_p(s) = \frac{1}{s - a_p}G_0(s); \quad a_{\min} \leq a_p \leq a_{\max} \tag{7.9}$$

If $a_{\min}$ and $a_{\max}$ have different signs then this means that the plant can change from stable to unstable with the pole crossing through the origin (which happens in some applications). This set of plants can be written as

$$G_p = \frac{G_0(s)}{s - \bar{a}(1 + r_a\Delta)}; \quad -1 \leq \Delta \leq 1 \tag{7.10}$$

which can be exactly described by inverse multiplicative uncertainty as in (7.59) with nominal model $G = G_0(s)/(s - \bar{a})$ and

$$w_{iI}(s) = \frac{r_a\bar{a}}{s - \bar{a}} \tag{7.11}$$

The magnitude of the weight $w_{iI}(s)$ is equal to r_a at low frequencies. If r_a is larger than 1 then the plant can be both stable and unstable.

Example 7.4 Parametric zero uncertainty. *Consider zero uncertainty in the "time constant" form, as in*

$$G_p(s) = (1 + \tau_p s)G_0(s); \quad \tau_{\min} \leq \tau_p \leq \tau_{\max} \tag{7.12}$$

where the remaining dynamics $G_0(s)$ are as usual assumed to have no uncertainty. For example, let $-1 \leq \tau_p \leq 3$. Then the possible zeros $z_p = -1/\tau_p$ cross from the LHP to the RHP through infinity: $z_p \leq -1/3$ (in LHP) and $z_p \geq 1$ (in RHP). The set of plants in (7.12) may be written as multiplicative (relative) uncertainty with

$$w_I(s) = r_\tau \bar{\tau} s/(1 + \bar{\tau} s) \tag{7.13}$$

The magnitude $|w_I(j\omega)|$ is small at low frequencies, and approaches r_τ (the relative uncertainty in τ) at high frequencies. For cases with $r_\tau > 1$ we allow the zero to cross from the LHP to the RHP (through infinity).

Exercise 7.1 Parametric zero uncertainty in zero form. *Consider the following alternative form of parametric zero uncertainty:*

$$G_p(s) = (s + z_p)G_0(s); \quad z_{\min} \leq z_p \leq z_{\max} \tag{7.14}$$

which caters for zeros crossing from the LHP to the RHP through the origin (corresponding to a sign change in the steady-state gain). Show that the resulting multiplicative weight is $w_I(s) = r_z \bar{z}/(s + \bar{z})$ and explain why the set of plants given by (7.14) is entirely different from that with the zero uncertainty in "time constant" form in (7.12). Explain what the implications are for control if $r_z > 1$.

The above parametric uncertainty descriptions are mainly included to gain insight. A general procedure for handling parametric uncertainty, more suited for numerical calculations, is given by Packard (1988). Consider an uncertain state-space model

$$\dot{x} = A_p x + B_p u \tag{7.15}$$

$$y = C_p x + D_p u \tag{7.16}$$

or equivalently

$$G_p(s) = C_p(sI - A_p)^{-1}B_p + D_p \tag{7.17}$$

Assume that the underlying cause for the uncertainty is uncertainty in some real parameters $\delta_1, \delta_2, \ldots$ (these could be temperature, mass, volume, etc.), and assume in the simplest case that the state-space matrices depend linearly on these parameters, i.e.

$$A_p = A + \sum \delta_i A_i, \; B_p = B + \sum \delta_i B_i, \; C_p = C + \sum \delta_i C_i, \; D_p = D + \sum \delta_i D_i \tag{7.18}$$

where A, B, C and D model the nominal system. This description has multiple perturbations, so it cannot be represented by a single perturbation, but it should be fairly clear that we can separate out the perturbations affecting A, B, C and D, and then collect them in a large diagonal matrix Δ with the real δ_i's along its diagonal. Some of the δ_i's may have to be repeated. Also, note that seemingly nonlinear parameter dependencies may be rewritten in our standard linear block diagram form; for example, we can handle δ_1^2 (which would need δ_1 repeated), $\frac{\alpha + w_1\delta_1\delta_2}{1 + w_2\delta_2}$, etc. This is illustrated next by an example.

Example 7.5 *Assume that the linearization of a nonlinear model results in a model $y = Cu$, where $C = \delta^2$ and $|\delta| \leq 1$ in some uncertain parameter. This may be written as an upper linear fractional transformation, $F_u(M, \Delta)$, as in (A.159). To see this, define the following auxiliary variables, $y = z_1, \; z_1 = \delta x_1, \; x_1 = z_2, \; z_2 = \delta x_2$ and $x_2 = u$. Then, arrange these variables such that $[x_1 \; x_2 \; y]^T = M \cdot [z_1 \; z_2 \; u]^T$ and $[z_1 \; z_2]^T = \Delta \cdot [x_1 \; x_2]^T$ to get the desired result, where*

$$M = \begin{bmatrix} 0 & 1 & 0 \\ 0 & 0 & 1 \\ 1 & 0 & 0 \end{bmatrix} \quad \text{and} \quad \Delta = \begin{bmatrix} \delta & 0 \\ 0 & \delta \end{bmatrix}$$

Table 7.1: **Matlab program for representing repeated parametric uncertainty**

```
% Uses Robust Control toolbox
k = ureal('k',0.5,'Range',[0.4 0.6]); % Uncertain parameter
alpha = ureal('alpha',1,'Range',[0.8 1.2]);
A = [-(1+k) 0; 1 -(1+k)];
B = [(1/k -1), -1]';
C = [0 alpha];
Gp = ss(A,B,C,0);
% Use lftdata to obtain the interconnection matrix of Figure 3.23
```

The above may seem complicated. In practice, it is not, as it can be done automatically with available software. For example, in Table 7.1, we show how to generate the LFT realization for the following uncertain plant:

$$\dot{x} = \begin{bmatrix} -(1+k) & 0 \\ 1 & -(1+k) \end{bmatrix} x + \begin{bmatrix} \frac{1-k}{k} \\ -1 \end{bmatrix} u$$

$$y = \begin{bmatrix} 1 & \alpha \end{bmatrix} x$$

where $k = 0.5 + 0.1 \cdot \delta_1$, $|\delta_1| \leq 1$ and $\alpha = 1 + 0.2 \cdot \delta_2$ with $|\delta_2| \leq 1$.

7.4 Representing uncertainty in the frequency domain

In terms of quantifying uncertainty arising from unmodelled dynamics, the frequency domain approach ($\mathcal{H}_\infty$) does not seem to have much competition (when compared with other norms). In fact, Owen and Zames (1992) make the following observation:

> The design of feedback controllers in the presence of non-parametric and unstructured uncertainty ... is the *raison d'être* for $\mathcal{H}_\infty$ feedback optimization, for if disturbances and plant models are clearly parameterized then $\mathcal{H}_\infty$ methods seem to offer no clear advantages over more conventional state-space and parametric methods.

Parametric uncertainty is also often represented by complex perturbations. This has the advantage of simplifying analysis and especially controller synthesis. For example, we may simply replace the real perturbation, $-1 \leq \Delta \leq 1$, by a complex perturbation with $|\Delta(j\omega)| \leq 1$. This is of course conservative as it introduces possible plants that are not present in the original set. However, if there are several real perturbations, then the conservatism is often reduced by *lumping* these perturbations into a *single* complex perturbation. The reason for this is that with several uncertain parameters the true uncertainty region is often quite "disc-shaped", and may be more accurately represented by a single complex perturbation. This is illustrated below.

7.4.1 Uncertainty regions

To illustrate how parametric uncertainty translates into frequency domain uncertainty, consider in Figure 7.2 the Nyquist plots (or regions) generated by the following set of plants:

$$G_p(s) = \frac{k}{\tau s + 1} e^{-\theta s}, \quad 2 \leq k, \theta, \tau \leq 3 \tag{7.19}$$

Step 1. At each frequency, a *region* of complex numbers $G_p(j\omega)$ is generated by varying the three parameters in the ranges given by (7.19), see Figure 7.2. In general, these *uncertainty regions* have complicated shapes and complex mathematical descriptions, and are cumbersome to deal with in the context of control system design.

Step 2. We therefore approximate such complex regions as discs (circles) as shown in Figure 7.3, resulting in a (complex) additive uncertainty description as discussed next.

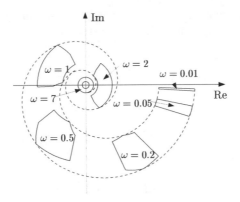

Figure 7.2: Uncertainty regions of the Nyquist plot at given frequencies. Data from (7.19).

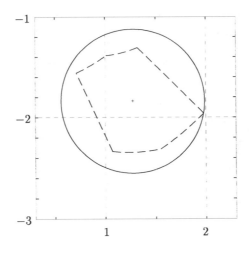

Figure 7.3: Disc approximation (solid line) of the original uncertainty region (dashed line). Plot corresponds to $\omega = 0.2$ in Figure 7.2.

Remark 1 There is no conservatism introduced in the first step when we go from a parametric uncertainty description as in (7.19) to an uncertainty region description as in Figure 7.2. This is somewhat surprising since the uncertainty regions in Figure 7.2 seem to allow for more uncertainty. For example, they allow for "jumps" in $G_p(j\omega)$ from one frequency to the next (e.g. from one corner of a region to another). Nevertheless, we derive frequency-by-frequency necessary and sufficient conditions for robust stability based on uncertainty regions in this and the next chapter. Thus, the only conservatism is in the second step where we approximate the original uncertainty region by a larger disc-shaped region as shown in Figure 7.3.

Remark 2 Exact methods do exist (using complex region mapping, e.g. see Laughlin et al. (1986)) which avoid the second conservative step. However, as already mentioned these methods are rather complex, and although they may be used in analysis, at least for simple systems, they are not really suitable for controller synthesis and will not be pursued further in this book.

Remark 3 From Figure 7.3 we see that the radius of the disc may be reduced by moving the centre (selecting another nominal model). This is discussed in Section 7.4.4.

7.4.2   Representing uncertainty regions by complex perturbations

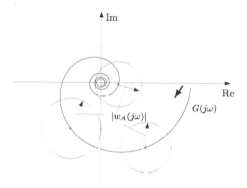

Figure 7.4: Disc-shaped uncertainty regions generated by complex additive uncertainty, $G_p = G + w_A \Delta$

We will use disc-shaped regions to represent uncertainty regions as illustrated by the Nyquist plots in Figures 7.3 and 7.4. These disc-shaped regions may be generated by additive complex norm-bounded perturbations (additive uncertainty) around a nominal plant G

$$\Pi_A : \quad G_p(s) = G(s) + w_A(s)\Delta_A(s); \quad |\Delta_A(j\omega)| \leq 1 \, \forall \omega \tag{7.20}$$

where $\Delta_A(s)$ is *any* stable transfer function which at each frequency is no larger than 1 in magnitude. How is this possible? If we consider all possible Δ_A's, then at each frequency $\Delta_A(j\omega)$ "generates" a disc-shaped region with radius 1 centred at 0, so $G(j\omega) + w_A(j\omega)\Delta_A(j\omega)$ generates at each frequency a disc-shaped region of radius $|w_A(j\omega)|$ centred at $G(j\omega)$ as shown in Figure 7.4.

In most cases $w_A(s)$ is a rational transfer function (although this need not always be the case).

One may also view $w_A(s)$ as a weight which is introduced in order to normalize the perturbation to be less than 1 in magnitude at each frequency. Thus only the magnitude of the weight matters, and in order to avoid unnecessary problems we always choose $w_A(s)$ to be stable and minimum-phase (this applies to all weights used in this book).

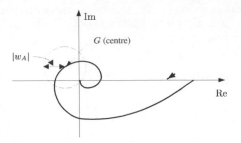

Figure 7.5: The set of possible plants includes the origin at frequencies where $|w_A(j\omega)| \geq |G(j\omega)|$, or equivalently $|w_I(j\omega)| \geq 1$

The disc-shaped regions may alternatively be represented by a *multiplicative uncertainty* description as in (7.2),

$$\Pi_I: \quad G_p(s) = G(s)(1 + w_I(s)\Delta_I(s)); \quad |\Delta_I(j\omega)| \leq 1, \forall\omega \qquad (7.21)$$

By comparing (7.20) and (7.21) we see that for SISO systems the additive and multiplicative uncertainty descriptions are equivalent if at each frequency

$$|w_I(j\omega)| = |w_A(j\omega)|/|G(j\omega)| \qquad (7.22)$$

However, multiplicative (relative) weights are often preferred because their numerical value is more informative. At frequencies where $|w_I(j\omega)| > 1$ the uncertainty exceeds 100% and the Nyquist curve may pass through the origin. This follows since, as illustrated in Figure 7.5, the radius of the discs in the Nyquist plot, $|w_A(j\omega)| = |G(j\omega)w_I(j\omega)|$, then exceeds the distance from $G(j\omega)$ to the origin. At these frequencies we do not know the phase of the plant, and we allow for zeros crossing from the LHP to the RHP. To see this, consider a frequency ω_0 where $|w_I(j\omega_0)| \geq 1$. Then there exists a $|\Delta_I| \leq 1$ such that $G_p(j\omega_0) = 0$ in (7.21); that is, there exists a possible plant with zeros at $s = \pm j\omega_0$. For this plant at frequency ω_0 the input has no effect on the output, so control has no effect. It then follows that *tight control is not possible at frequencies where* $|w_I(j\omega)| \geq 1$ (this condition is derived more rigorously in (7.43)).

7.4.3   Obtaining the weight for complex uncertainty

Consider a set Π of possible plants resulting, for example, from parametric uncertainty as in (7.19). We now want to describe this set of plants by a single (lumped) complex perturbation, Δ_A or Δ_I. This complex (disc-shaped) uncertainty description may be generated as follows:

1. Select a nominal model $G(s)$.
2. *Additive uncertainty.* At each frequency find the smallest radius $l_A(\omega)$ which includes all the possible plants Π:

$$l_A(\omega) = \max_{G_p \in \Pi} |G_p(j\omega) - G(j\omega)| \qquad (7.23)$$

If we want a rational transfer function weight, $w_A(s)$ (which may not be the case if we only want to do analysis), then it must be chosen to cover the set, so

$$|w_A(j\omega)| \geq l_A(\omega) \quad \forall\omega \qquad (7.24)$$

Usually $w_A(s)$ is of low order to simplify the controller design. Furthermore, an objective of frequency domain uncertainty is usually to represent uncertainty in a simple straightforward manner.

3. *Multiplicative (relative) uncertainty.* This is often the preferred uncertainty form, and we have

$$l_I(\omega) = \max_{G_p \in \Pi} \left| \frac{G_p(j\omega) - G(j\omega)}{G(j\omega)} \right| \tag{7.25}$$

with a rational weight

$$|w_I(j\omega)| \geq l_I(\omega), \forall \omega \tag{7.26}$$

Example 7.6 Multiplicative weight for parametric uncertainty. *Consider again the set of plants with parametric uncertainty given in (7.19)*

$$\Pi : \quad G_p(s) = \frac{k}{\tau s + 1} e^{-\theta s}, \quad 2 \leq k, \theta, \tau \leq 3 \tag{7.27}$$

We want to represent this set using multiplicative uncertainty with a rational weight $w_I(s)$. To simplify subsequent controller design we select a delay-free nominal model

$$G(s) = \frac{\bar{k}}{\bar{\tau} s + 1} = \frac{2.5}{2.5 s + 1} \tag{7.28}$$

To obtain $l_I(\omega)$ in (7.25), one may use the Matlab Robust Control toolbox command `usample`, *which gives the specified number of random plants from the uncertain set of plants. However, this command does not handle the uncertainty in time delay, thus we consider three values (2, 2.5 and 3) for each of the three parameters (k, θ, τ). (This is not, in general, guaranteed to yield the worst case as the worst case may be at the interior of the intervals.) The corresponding relative errors $|(G_p - G)/G|$ are shown as functions of frequency for the $3^3 = 27$ resulting G_p's in Figure 7.6. The curve for $l_I(\omega)$ must*

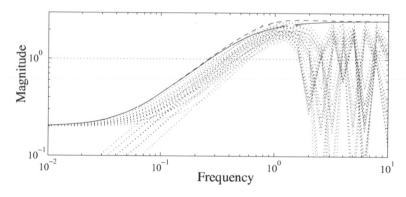

Figure 7.6: Relative errors for 27 combinations of k, τ and θ with delay-free nominal plant (dotted lines). Solid line: first-order weight $|w_{I1}|$ in (7.29). Dashed line: third order weight $|w_I|$ in (7.30).

at each frequency lie above all the dotted lines, and we find that $l_I(\omega)$ is 0.2 at low frequencies and 2.5 at high frequencies. To derive $w_I(s)$ we first try a simple first-order weight that matches this limiting behaviour:

$$w_{I1}(s) = \frac{Ts + 0.2}{(T/2.5)s + 1}, \quad T = 4 \tag{7.29}$$

As seen from the solid line in Figure 7.6, this weight gives a good fit of $l_I(\omega)$, except around $\omega = 1$ where $|w_{I1}(j\omega)|$ is slightly too small, and so this weight does not include all possible plants. To change *this so that $|w_I(j\omega)| \geq l_I(\omega)$ at all frequencies, we can multiply w_{I1} by a correction factor to lift the gain slightly at $\omega = 1$. The following works well:*

$$w_I(s) = \omega_{I1}(s) \frac{s^2 + 1.6s + 1}{s^2 + 1.4s + 1} \tag{7.30}$$

as is seen from the dashed line in Figure 7.6. The magnitude of the weight crosses 1 at about $\omega = 0.26$. This seems reasonable since we have neglected the delay in our nominal model, which by itself yields 100% uncertainty at a frequency of about $1/\theta_{max} = 0.33$ (see Figure 7.8(a) below).

An uncertainty description for the same parametric uncertainty, but with a mean-value nominal model (with delay), is given in Exercise 7.8. Parametric gain and delay uncertainty (without time constant uncertainty) are discussed further on page 272.

Remark. Pole uncertainty. In the example we represented pole (time constant) uncertainty by a multiplicative perturbation, Δ_I. We *may* even do this for unstable plants, provided the poles do not shift between the half planes and one allows $\Delta_I(s)$ to be unstable. However, if the pole uncertainty is large, and in particular if poles can cross from the LHP to the RHP, then one should use an inverse ("feedback") uncertainty representation as in (7.3).

7.4.4   Choice of nominal model

With parametric uncertainty represented as complex perturbations there are three main options for the choice of nominal model:

1. A simplified model, e.g. a low-order, delay-free model.
2. A model of mean parameter values, $G(s) = \bar{G}(s)$.
3. The central plant obtained from a Nyquist plot (yielding the smallest discs).

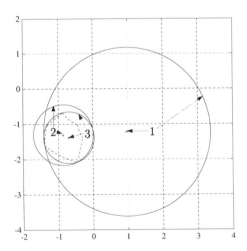

Figure 7.7: Nyquist plot of $G_p(j\omega)$ at frequency $\omega = 0.5$ (dashed region) showing complex disc approximations using three options for the nominal model:
1. Simplified nominal model with no time delay
2. Mean parameter values
3. Nominal model corresponding to the smallest radius

Option 1 usually yields the largest uncertainty region, but the model is simple and this facilitates controller design in later stages. Option 2 is probably the most straightforward choice. Option 3 yields the smallest region, but in this case a significant effort may be required to obtain the nominal model, which is usually not a rational transfer function and a rational approximation could be of very high order.

Example 7.7 *Consider again the uncertainty set (7.27) used in Example 7.6. The nominal models selected for options 1 and 2 are*

$$G_1(s) = \frac{\bar{k}}{\bar{\tau}s + 1}, \quad G_2(s) = \frac{\bar{k}}{\bar{\tau}s + 1}e^{-\bar{\theta}s}$$

For option 3 the nominal model is not rational. The Nyquist plot of the three resulting discs at frequency $\omega = 0.5$ are shown in Figure 7.7.

Remark. A similar example was studied by Wang et al. (1994), who obtained the best controller designs with option 1, although the uncertainty region is clearly much larger in this case. The reason for this is that the "worst-case region" in the Nyquist plot in Figure 7.7 corresponds quite closely to those plants with the most negative phase (at coordinates approximately equal to $(-1.5, -1.5)$). Thus, the additional plants included in the largest region (option 1) are generally easier to control and do not really matter when evaluating the worst-case plant with respect to stability or performance. In conclusion, at least for SISO plants, we find that for plants with an uncertain time delay, it is simplest and sometimes best (!) to use a delay-free nominal model, and to represent the nominal delay as additional uncertainty.

The choice of nominal model is only an issue since we are lumping several sources of parametric uncertainty into a single complex perturbation. Of course, if we use a parametric uncertainty description, based on multiple real perturbations, then we should always use the mean parameter values in the nominal model.

7.4.5   Neglected dynamics represented as uncertainty

We saw above that one advantage of frequency domain uncertainty descriptions is that one can choose to work with a simple nominal model, and represent neglected dynamics as uncertainty. We will now consider this in a little more detail. Consider a set of plants

$$G_p(s) = G_0(s)f(s)$$

where $G_0(s)$ is fixed (and certain). We want to neglect the term $f(s)$ (which may be fixed or may be an uncertain set Π_f), and represent G_p by multiplicative uncertainty with a nominal model $G = G_0$. From (7.25) we get that the magnitude of the relative uncertainty caused by neglecting the dynamics in $f(s)$ is

$$l_I(\omega) = \max_{G_p} \left| \frac{G_p - G}{G} \right| = \max_{f(s) \in \Pi_f} |f(j\omega) - 1| \qquad (7.31)$$

Three examples illustrate the procedure.

1. Neglected delay. Let $f(s) = e^{-\theta_p s}$, where $0 \le \theta_p \le \theta_{\max}$. We want to represent $G_p = G_0(s)e^{-\theta_p s}$ by a delay-free plant $G_0(s)$ and multiplicative uncertainty. Let us first consider the maximum delay, for which the relative error $|1 - e^{-j\omega\theta_{\max}}|$ is shown as a function of frequency in Figure 7.8(a). The relative uncertainty crosses 1 in magnitude at a frequency of about $1/\theta_{\max}$, reaches 2 at frequency $\pi/\theta_{\max}$ (since at this frequency $e^{j\omega\theta_{\max}} = -1$), and oscillates between 0 and 2 at higher frequencies (which corresponds to the Nyquist plot of $e^{-j\omega\theta_{\max}}$ going around and around the unit circle). Similar curves are generated for smaller values of the delay, and they also oscillate between 0 and 2 but at even higher frequencies. It then follows that if we consider all $\theta \in [0, \theta_{\max}]$ then the relative error bound is 2 at frequencies above $\pi/\theta_{\max}$, and we have

$$l_I(\omega) = \begin{cases} |1 - e^{-j\omega\theta_{\max}}| & \omega < \pi/\theta_{\max} \\ 2 & \omega \geq \pi/\theta_{\max} \end{cases} \tag{7.32}$$

Rational approximations of (7.32) are given in (7.36) and (7.37) with $r_k = 0$.

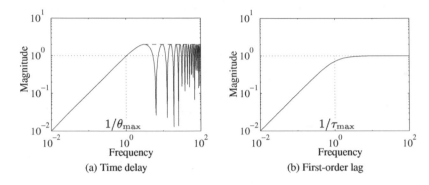

(a) Time delay   (b) First-order lag

Figure 7.8: Multiplicative uncertainty resulting from neglected dynamics

2. Neglected lag. Let $f(s) = 1/(\tau_p s + 1)$, where $0 \leq \tau_p \leq \tau_{\max}$. In this case the resulting $l_I(\omega)$, which is shown in Figure 7.8(b), can be represented by a rational transfer function with $|w_I(j\omega)| = l_I(\omega)$ where

$$w_I(s) = 1 - \frac{1}{\tau_{\max} s + 1} = \frac{\tau_{\max} s}{\tau_{\max} s + 1}$$

This weight approaches 1 at high frequencies, and the low-frequency asymptote crosses 1 at frequency $1/\tau_{\max}$.

3. Multiplicative weight for gain and delay uncertainty. Consider the following set of plants:

$$G_p(s) = k_p e^{-\theta_p s} G_0(s); \quad k_p \in [k_{\min}, k_{\max}], \quad \theta_p \in [\theta_{\min}, \theta_{\max}] \tag{7.33}$$

which we want to represent by multiplicative uncertainty and a delay-free nominal model, $G(s) = \bar{k} G_0(s)$, where $\bar{k} = \frac{k_{\min} + k_{\max}}{2}$ and $r_k = \frac{(k_{\max} - k_{\min})/2}{\bar{k}}$. Lundström (1994) derived the following exact expression for the relative uncertainty weight:

$$l_I(\omega) = \begin{cases} \sqrt{r_k^2 + 2(1 + r_k)(1 - \cos(\theta_{\max}\omega))} & \text{for } \omega < \pi/\theta_{\max} \\ 2 + r_k & \text{for } \omega \geq \pi/\theta_{\max} \end{cases} \tag{7.34}$$

where r_k is the relative uncertainty in the gain. This bound is irrational. To derive a rational weight we first approximate the delay by a first-order Padé approximation to get

$$k_{\max} e^{-\theta_{\max} s} - \bar{k} \approx \bar{k}(1 + r_k)\frac{1 - \frac{\theta_{\max}}{2}s}{1 + \frac{\theta_{\max}}{2}s} - \bar{k} = \bar{k}\frac{-\left(1 + \frac{r_k}{2}\right)\theta_{\max} s + r_k}{\frac{\theta_{\max}}{2}s + 1} \tag{7.35}$$

Since only the magnitude matters this may be represented by the following first-order weight:

$$w_I(s) = \frac{(1 + \frac{r_k}{2})\theta_{\max}s + r_k}{\frac{\theta_{\max}}{2}s + 1} \tag{7.36}$$

However, as seen from Figure 7.9, by comparing the dashed line (representing w_I) with

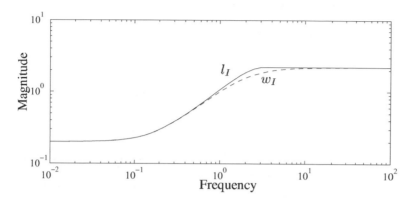

Figure 7.9: Multiplicative weight for gain and delay uncertainty in (7.33) (with $\theta_{\max} = 1$, $r_k = 0.2$)

the solid line (representing l_I), this weight w_I is somewhat optimistic (too small), especially around frequencies $1/\theta_{\max}$. To make sure that $|w_I(j\omega)| \geq l_I(\omega)$ at all frequencies we apply a correction factor and get a third-order weight

$$w_I(s) = \frac{(1 + \frac{r_k}{2})\theta_{\max}s + r_k}{\frac{\theta_{\max}}{2}s + 1} \cdot \frac{\left(\frac{\theta_{\max}}{2.363}\right)^2 s^2 + 2 \cdot 0.838 \cdot \frac{\theta_{\max}}{2.363}s + 1}{\left(\frac{\theta_{\max}}{2.363}\right)^2 s^2 + 2 \cdot 0.685 \cdot \frac{\theta_{\max}}{2.363}s + 1} \tag{7.37}$$

The improved weight $w_I(s)$ in (7.37) is not shown in Figure 7.9, but it would be almost indistinguishable from the exact bound given by the solid curve. In practical applications, it is suggested that one starts with a simple weight as in (7.36), and if it later appears important to eke out a little extra performance then one should try a higher-order weight as in (7.37).

Example 7.8 *Consider the set $G_p(s) = k_p e^{-\theta_p s} G_0(s)$ with $2 \leq k_p \leq 3$ and $2 \leq \theta_p \leq 3$. We approximate this with a nominal delay-free plant $G = \bar{k}G_0 = 2.5G_0$ and relative uncertainty. The simple first-order weight in (7.36), $w_I(s) = \frac{3.3s+0.2}{1.5s+1}$, is somewhat optimistic. To cover all the uncertainty we may use (7.37), $w_I(s) = \frac{3.3s+0.2}{1.5s+1} \cdot \frac{1.612s^2+2.128s+1}{1.612s^2+1.739s+1}$.*

7.4.6 Unmodelled dynamics uncertainty

Although we have spent a considerable amount of time on modelling uncertainty and deriving weights, we have not yet addressed the most important reason for using frequency domain ($\mathcal{H}_\infty$) uncertainty descriptions and complex perturbations, namely the incorporation of unmodelled dynamics. Of course, *unmodelled dynamics* is close to *neglected dynamics* which we have just discussed, but it is not quite the same. In unmodelled dynamics we also include unknown dynamics of unknown or even infinite order. To represent unmodelled dynamics we usually use a simple multiplicative weight of the form

$$w_I(s) = \frac{\tau s + r_0}{(\tau/r_\infty)s + 1} \tag{7.38}$$

where r_0 is the relative uncertainty at steady-state, $1/\tau$ is (approximately) the frequency at which the relative uncertainty reaches 100%, and r_∞ is the magnitude of the weight at high frequency (typically, $r_\infty \geq 2$). Based on the above examples and discussions it is hoped that the reader has now accumulated the necessary insight to select reasonable values for the parameters r_0, r_∞ and τ for a specific application. The following exercise provides further support and gives a good review of the main ideas.

Exercise 7.2[*] *Suppose that the nominal model of a plant is*

$$G(s) = \frac{1}{s+1}$$

and the uncertainty in the model is parameterized by multiplicative uncertainty with the weight

$$w_I(s) = \frac{0.125s + 0.25}{(0.125/4)s + 1}$$

Call the resulting set Π. *Now find the extreme parameter values in each of the plants (a)–(g) below so that each plant belongs to the set* Π. *All parameters are assumed to be positive. One approach is to plot* $l_I(\omega) = |G^{-1}G' - 1|$ *in (7.25) for each* G' (G_a, G_b, *etc.) and adjust the parameter in question until* l_I *just touches* $|w_I(j\omega)|$.

  (a) Neglected delay: Find the largest θ *for* $G_a = Ge^{-\theta s}$ *(Answer: 0.13).*
  (b) Neglected lag: Find the largest τ *for* $G_b = G\frac{1}{\tau s+1}$ *(Answer: 0.15).*
  (c) Uncertain pole: Find the range of a *for* $G_c = \frac{1}{s+a}$ *(Answer: 0.8 to 1.33).*
  (d) Uncertain pole (time constant form): Find the range of T *for* $G_d = \frac{1}{Ts+1}$ *(Answer: 0.7 to 1.5).*
  (e) Neglected resonance: Find the range of ζ *for* $G_e = G\frac{1}{(s/70)^2+2\zeta(s/70)+1}$ *(Answer: 0.02 to 0.8).*
  (f) Neglected dynamics: Find the largest integer m *for* $G_f = G\left(\frac{1}{0.01s+1}\right)^m$ *(Answer: 13).*

  (g) Neglected RHP-zero: Find the largest τ_z *for* $G_g = G\frac{-\tau_z s+1}{\tau_z s+1}$ *(Answer: 0.07). These results imply that a control system which meets given stability and performance requirements for all plants in* Π *is also guaranteed to satisfy the same requirements for the above plants* $G_a, G_b, \ldots, G_g$.

  (h) Repeat all of the above with a new nominal plant $G = 1/(s-1)$ *(and with everything else the same except* $G_d = 1/(Ts-1)$) *(Answers: same as above).*

Exercise 7.3 *Repeat Exercise 7.2 with a new weight,*

$$w_I(s) = \frac{s + 0.3}{(1/3)s + 1}$$

We end this section with a couple of remarks on uncertainty modelling:

1. We can usually get away with just one source of complex uncertainty for SISO systems.
2. With an $\mathcal{H}_\infty$ uncertainty description, it is possible to represent time delays (corresponding to an infinite-dimensional plant) and unmodelled dynamics of infinite order, using a nominal model and associated weights of finite order.

7.5 SISO robust stability

We have so far discussed how to represent the uncertainty mathematically. In this section, we derive conditions which will ensure that the system remains stable for all perturbations in the uncertainty set, and then in the subsequent section we study robust performance.

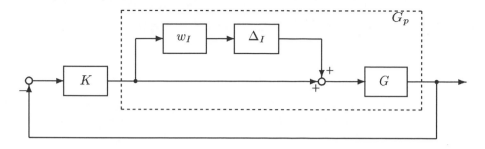

Figure 7.10: Feedback system with multiplicative uncertainty

7.5.1   RS with multiplicative uncertainty

We want to determine the stability of the uncertain feedback system in Figure 7.10 when there is multiplicative (relative) uncertainty of magnitude $|w_I(j\omega)|$. With uncertainty the loop transfer function becomes

$$L_p = G_p K = GK(1 + w_I \Delta_I) = L + w_I L \Delta_I, \quad |\Delta_I(j\omega)| \le 1, \forall \omega \qquad (7.39)$$

As always, we assume (by design) stability of the nominal closed-loop system (i.e. with $\Delta_I = 0$). For simplicity, we also assume that the loop transfer function L_p is stable. We now use the Nyquist stability condition to test for RS of the closed-loop system. We have

$$\text{RS} \quad \overset{\text{def}}{\Leftrightarrow} \quad \text{System stable } \forall L_p$$

$$\Leftrightarrow L_p \text{ should not encircle the point } -1, \quad \forall L_p \qquad (7.40)$$

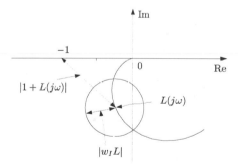

Figure 7.11: Nyquist plot of L_p for RS

1. Graphical derivation of RS condition. Consider the Nyquist plot of L_p as shown in Figure 7.11. Convince yourself that $|-1-L| = |1 + L|$ is the distance from the point -1 to the centre of the disc representing L_p, and that $|w_I L|$ is the radius of the disc. Encirclements are avoided if none of the discs cover -1, and we get from Figure 7.11

$$\text{RS} \quad \Leftrightarrow \quad |w_I L| < |1 + L|, \quad \forall \omega \qquad (7.41)$$

$$\Leftrightarrow \quad \left| \frac{w_I L}{1 + L} \right| < 1, \forall \omega \quad \Leftrightarrow \quad |w_I T| < 1, \forall \omega \tag{7.42}$$

$$\overset{\text{def}}{\Leftrightarrow} \quad \|w_I T\|_\infty < 1 \tag{7.43}$$

Note that for SISO systems $w_I = w_O$ and $T = T_I = GK(1 + GK)^{-1}$, so the condition could equivalently be written in terms of $w_I T_I$ or $w_O T$. Thus, the requirement of RS for the case with multiplicative uncertainty gives an upper bound on the complementary sensitivity:

$$\boxed{\text{RS} \Leftrightarrow |T| < 1/|w_I|, \quad \forall \omega} \tag{7.44}$$

We see that we have to detune the system (i.e. make T small) at frequencies where the relative uncertainty $|w_I|$ exceeds 1 in magnitude. Condition (7.44) is *exact* (necessary and sufficient) provided there exist uncertain plants such that at each frequency all perturbations satisfying $|\Delta(j\omega)| \leq 1$ are possible. If this is not the case, then (7.44) is only *sufficient* for RS, e.g. this is the case if the perturbation is restricted to be real, as for the parametric gain uncertainty in (7.6).

Remark. Unstable plants. The stability condition (7.43) also applies to the case when L and L_p are unstable as long as the number of RHP-poles remains the same for each plant in the uncertainty set. This follows since the nominal closed-loop system is assumed stable, so we must make sure that the perturbation does not change the number of encirclements, and (7.43) is the condition which guarantees this.

2. Algebraic derivation of RS condition. Since L_p is assumed stable, and the nominal closed loop is stable, the nominal loop transfer function $L(j\omega)$ does not encircle -1. Therefore, since the set of plants is norm-bounded, it then follows that if some L_{p1} in the uncertainty set encircles -1, then there must be another L_{p2} in the uncertainty set which goes exactly through -1 at some frequency. Thus,

$$\text{RS} \quad \Leftrightarrow \quad |1 + L_p| \neq 0, \quad \forall L_p, \forall \omega \tag{7.45}$$

$$\Leftrightarrow \quad |1 + L_p| > 0, \quad \forall L_p, \forall \omega \tag{7.46}$$

$$\Leftrightarrow \quad |1 + L + w_I L \Delta_I| > 0, \quad \forall |\Delta_I| \leq 1, \forall \omega \tag{7.47}$$

At each frequency the last condition is most easily violated (the worst case) when the complex number $\Delta_I(j\omega)$ is selected with $|\Delta_I(j\omega)| = 1$ and with phase such that the terms $(1 + L)$ and $w_I L \Delta_I$ have opposite signs (point in the opposite direction). Thus

$$\text{RS} \Leftrightarrow |1 + L| - |w_I L| > 0, \quad \forall \omega \quad \Leftrightarrow |w_I T| < 1, \quad \forall \omega \tag{7.48}$$

and we have rederived (7.43).

3. $M\Delta$-structure derivation of RS condition. This derivation is a preview of a general analysis presented in the next chapter. The reader should not be too concerned if he or she does not fully understand the details at this point. The derivation is based on applying the Nyquist stability condition to an alternative "loop transfer function" $M\Delta$ rather than L_p. The argument goes as follows. Notice that the only source of instability in Figure 7.10 is the new feedback loop created by Δ_I. If the nominal ($\Delta_I = 0$) feedback system is stable then the stability of the system in Figure 7.10 is equivalent to stability of the system in Figure 7.12, where $\Delta = \Delta_I$ and

$$M = w_I K (1 + GK)^{-1} G = w_I T \tag{7.49}$$

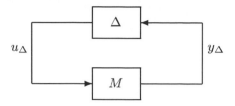

Figure 7.12: $M\Delta$-structure

is the transfer function from the output of Δ_I to the input of Δ_I. We now apply the Nyquist stability condition to the system in Figure 7.12. We assume that Δ and $M = w_I T$ are stable; the former implies that G and G_p must have the same unstable poles, the latter is equivalent to assuming nominal stability of the closed-loop system. The Nyquist stability condition then determines RS if and only if the "loop transfer function" $M\Delta$ does not encircle -1 for all Δ. Thus,

$$\text{RS} \quad \Leftrightarrow \quad |1 + M\Delta| > 0, \quad \forall\omega, \forall |\Delta| \le 1 \tag{7.50}$$

The last condition is most easily violated (the worst case) when Δ is selected at each frequency such that $|\Delta| = 1$ and the terms $M\Delta$ and 1 have opposite signs (point in the opposite direction). We therefore get

$$\text{RS} \quad \Leftrightarrow \quad 1 - |M(j\omega)| > 0, \quad \forall\omega \tag{7.51}$$

$$\Leftrightarrow \quad |M(j\omega)| < 1, \quad \forall\omega \tag{7.52}$$

which is the same as (7.43) and (7.48) since $M = w_I T$. The $M\Delta$-structure provides a very general way of handling robust stability, and we will discuss this at length in the next chapter where we will see that (7.52) is essentially a clever application of the small-gain theorem where we avoid the usual conservatism since any phase in $M\Delta$ is allowed.

Example 7.9 *Consider the following nominal plant and PI controller:*

$$G(s) = \frac{3(-2s + 1)}{(5s + 1)(10s + 1)} \quad K(s) = K_c \frac{12.7s + 1}{12.7s}$$

Recall that this is the inverse response process from Chapter 2. Initially, we select $K_c = K_{c1} = 1.13$ as suggested by the Ziegler–Nichols tuning rule. It results in a nominally stable closed-loop system. Suppose that one "extreme" uncertain plant is

$$G'(s) = 4(-3s + 1)/(4s + 1)^2 \tag{7.53}$$

For this plant the relative error $|(G' - G)/G|$ is 0.33 at low frequencies; it is 1 at about 0.1 rad/s, and it is 5.25 at high frequencies. Based on this and (7.38) we choose the following uncertainty weight:

$$w_I(s) = \frac{10s + 0.33}{(10/5.25)s + 1}$$

which closely matches this relative error. We now want to evaluate whether the system remains stable for all possible plants as given by $G_p = G(1 + w_I \Delta_I)$, where $\Delta_I(s)$ is any perturbation satisfying $\|\Delta\|_\infty \le 1$. From (7.44), we have the following necessary and sufficient condition for robust stability: $|T| < 1/|w_I| \; \forall\omega$. This condition is easy to check. Based on the nominal plant (7.53) and the given controller K_1 (with gain $K_{c1} = 1.13$), we compute $T_1 = GK_1/(1 + GK_1)$ as a function of frequency.

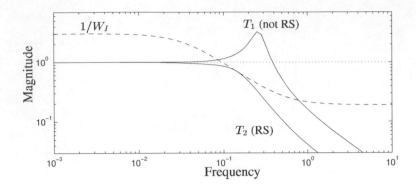

Figure 7.13: Checking robust stability with multiplicative uncertainty

From Figure 7.13, we see that $|T_1|$ exceeds $|w_I|$ over a wide frequency range, so from (7.44), we conclude that the system is <u>not</u> robustly stable.

From Figure 7.13, we notice that the worst-case frequency is $\omega = 0.26$, where $|T_1|$ is a factor of $1/0.13 = 7.7$ larger than w_I (see also Matlab code in Table 7.2, where we get `Smarg1 = 0.13`*). In other words, reducing the uncertainty weight w_I by a factor 7.7 would give stability.*

With the given uncertain plant, we need to reduce the controller gain to achieve robust stability. By trial and error, we find that reducing the gain to $K_{c2} = 0.31$ just achieves RS, as is seen from the curve for $T_2 = GK_2/(1 + GK_2)$ in Figure 7.13.

Table 7.2: **Matlab program for describing plant with complex uncertainty and analyzing RS**

```
% Uses Robust Control toolbox
G = 3*tf([-2 1],conv([5 1],[10 1]));
Wi = tf([10 0.33],[10/5.25 1]);                % Uncertainty weight
Delta = ultidyn('Delta', [1 1]);               % Dynamic uncertainty
Gp = G * (1 + Wi*Delta);
K = tf([12.7 1],[12.7 0]);
L1 = Gp*1.13*K;                                % Ziegler-Nichols Controller
T1 = feedback(L1,1);
[Smarg1,Dstab1,Report1] = robuststab(T1)       % Stability margins
L2 = Gp*1.13*K;                                % Detuned Controller
T2 = feedback(Gp*0.31*K,1);
[Smarg2,Dstab2,Report2] = robuststab(T2)
```

Remark. *For the "extreme" plant $G'(s)$ in (7.53), we find as expected that the closed-loop system is unstable with $K_{c1} = 1.13$. However, with $K_{c2} = 0.31$ the system is stable with reasonable margins (and not at the limit of instability as one might have expected); we can increase the gain by almost a factor of 2 to $K_c = 0.58$ before we get instability. This illustrates that condition (7.44) is only a* sufficient *condition for stability, and a violation of this bound does not imply instability for a specific plant G'. However, with $K_{c2} = 0.31$ there does exist another allowed complex Δ_I and a corresponding $G_p = G(1 + w_I \Delta_I)$ that yields $T_{2p} = \frac{G_p K_2}{1 + G_p K_2}$ on the limit of instability. Such a Δ_I may be identified numerically, e.g. using Matlab as shown in Table 7.2 (where the worst-case Δ_I is contained in* `Dstab2.Delta`*).*

7.5.2 Comparison with gain margin

By what factor, k_{max}, can we multiply the loop gain, $L_0 = G_0K$, before we get instability? In other words, given

$$L_p = k_pL_0; \quad k_p \in [1, k_{max}] \tag{7.54}$$

find the largest value of k_{max} such that the closed-loop system is stable.

1. Exact condition. The exact value of k_{max} (which is obtained with Δ real in (7.56)) is the gain margin (GM) from classical control. We have (recall (2.40))

$$k_{max,1} = \text{GM} = \frac{1}{|L_0(j\omega_{180})|} \tag{7.55}$$

where ω_{180} is the frequency where $\angle L_0 = -180°$.

2. Conservative condition using complex perturbation. Alternatively, represent the gain uncertainty as complex multiplicative uncertainty,

$$L_p = k_pL_0 = \bar{k}L_0(1 + r_k\Delta) \tag{7.56}$$

where

$$\bar{k} = \frac{k_{max} + 1}{2}, \quad r_k = \frac{k_{max} - 1}{k_{max} + 1} \tag{7.57}$$

Note that the nominal $L = \bar{k}L_0$ is not fixed, but depends on k_{max}. The RS condition $\|w_IT\|_\infty < 1$ (which is derived for complex Δ) with $w_I = r_k$ then gives

$$\left\| r_k \frac{\bar{k}L_0}{1 + \bar{k}L_0} \right\|_\infty < 1 \tag{7.58}$$

Here both r_k and $\bar{k}$ depend on k_{max}, and (7.58) must be solved iteratively to find $k_{max,2}$. Condition (7.58) would be exact if Δ were complex, but since it is not we expect $k_{max,2}$ to be somewhat smaller than GM.

Example 7.10 *To check this numerically consider a system with $L_0 = \frac{1}{s}\frac{-s+2}{s+2}$. We find $\omega_{180} = 2$ [rad/s] and $|L_0(j\omega_{180})| = 0.5$, and the exact factor by which we can increase the loop gain is, from (7.55), $k_{max,1} = GM = 2$. On the other hand, use of (7.58) yields $k_{max,2} = 1.78$, which as expected is less than GM = 2. This illustrates the conservatism involved in replacing a real perturbation by a complex one.*

Exercise 7.4 * *Represent the gain uncertainty in (7.54) as multiplicative complex uncertainty with nominal model $G = G_0$ (rather than $G = \bar{k}G_0$ used above).*

(a) Find w_I and use the RS condition $\|w_IT\|_\infty < 1$ to find $k_{max,3}$. Note that no iteration is needed in this case since the nominal model and thus $T = T_0$ is independent of k_{max}.

(b) One expects $k_{max,3}$ to be even more conservative than $k_{max,2}$ since this uncertainty description is not even tight when Δ is real. Show that this is indeed the case using the numerical values from Example 7.10.

7.5.3 RS with inverse multiplicative uncertainty

We will derive a corresponding RS condition for a feedback system with inverse multiplicative uncertainty (see Figure 7.14) in which

$$G_p = G(1 + w_{iI}(s)\Delta_{iI})^{-1} \tag{7.59}$$

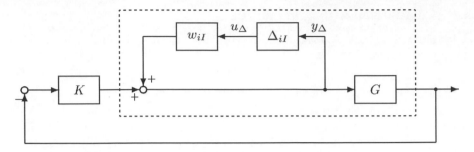

Figure 7.14: Feedback system with inverse multiplicative uncertainty

Algebraic derivation. Assume for simplicity that the loop transfer function L_p is stable, and assume stability of the nominal closed-loop system. RS is then guaranteed if encirclements by $L_p(j\omega)$ of the point -1 are avoided, and since L_p is in a norm-bounded set we have

$$\text{RS} \quad \Leftrightarrow \quad |1 + L_p| > 0, \quad \forall L_p, \forall \omega \tag{7.60}$$

$$\Leftrightarrow \quad |1 + L(1 + w_{iI}\Delta_{iI})^{-1}| > 0, \quad \forall |\Delta_{iI}| \leq 1, \forall \omega \tag{7.61}$$

$$\Leftrightarrow \quad |1 + w_{iI}\Delta_{iI} + L| > 0, \quad \forall |\Delta_{iI}| \leq 1, \forall \omega \tag{7.62}$$

The last condition is most easily violated (the worst case) when Δ_{iI} is selected at each frequency such that $|\Delta_{iI}| = 1$ and the terms $1 + L$ and $w_{iI}\Delta_{iI}$ have opposite signs (point in the opposite direction). Thus

$$\text{RS} \quad \Leftrightarrow \quad |1 + L| - |w_{iI}| > 0, \quad \forall \omega \tag{7.63}$$

$$\Leftrightarrow \quad |w_{iI}S| < 1, \quad \forall \omega \tag{7.64}$$

Remark. In this derivation we have assumed that L_p is stable, but this is not necessary as one may show by deriving the condition using the $M\Delta$-structure. Actually, the RS condition (7.64) applies even when the number of RHP-poles of G_p can change.

Control implications. From (7.64) we find that the requirement of RS for the case with inverse multiplicative uncertainty gives an upper bound on the sensitivity,

$$\boxed{\text{RS} \quad \Leftrightarrow \quad |S| < 1/|w_{iI}|, \quad \forall \omega} \tag{7.65}$$

We see that we need tight control and have to make S small at frequencies where the uncertainty is large and $|w_{iI}|$ exceeds 1 in magnitude. This may be somewhat surprising since we intuitively expect to have to detune the system (and make $S \approx 1$) when we have uncertainty, while this condition tells us to do the opposite. The reason is that this uncertainty represents pole uncertainty, and at frequencies where $|w_{iI}|$ exceeds 1 we allow for poles crossing from the LHP to the RHP (G_p becoming unstable), and we then know that we need feedback ($|S| < 1$) in order to stabilize the system.

However, $|S| < 1$ may not always be possible. In particular, assume that the plant has a RHP-zero at $s = z$. Then we have the interpolation constraint $S(z) = 1$ and we must as a prerequisite for RS, $\|w_{iI}S\|_\infty < 1$, require that $w_{iI}(z) \leq 1$ (recall the maximum modulus theorem, see (5.20)). Thus, we cannot have large pole uncertainty with $|w_{iI}(j\omega)| > 1$ (and hence the possibility of instability) at frequencies where the plant has a RHP-zero. This is consistent with the results we obtained in Section 5.3.2 (page 179).

7.6 SISO robust performance

7.6.1 SISO nominal performance in the Nyquist plot

Consider performance in terms of the weighted sensitivity function as discussed in Section 2.8.2. The condition for nominal performance (NP) is then

$$\text{NP} \quad \Leftrightarrow \quad |w_P S| < 1 \quad \forall \omega \quad \Leftrightarrow \quad |w_P| < |1 + L| \quad \forall \omega \tag{7.66}$$

Now $|1 + L|$ represents at each frequency the distance of $L(j\omega)$ from the point -1 in the Nyquist plot, so $L(j\omega)$ must be at least a distance of $|w_P(j\omega)|$ from -1. This is illustrated graphically in Figure 7.15, where we see that for NP, $L(j\omega)$ must stay outside a disc of radius $|w_P(j\omega)|$ centred on -1.

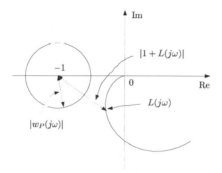

Figure 7.15: Nyquist plot of NP condition $|w_P| < |1 + L|$

7.6.2 Robust performance

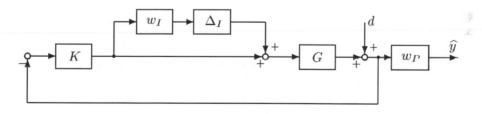

Figure 7.16: Diagram for RP with multiplicative uncertainty

For robust performance we require the performance condition (7.66) to be satisfied for *all* possible plants, i.e. including the worst-case uncertainty:

$$\text{RP} \quad \overset{\text{def}}{\Leftrightarrow} \quad |w_P S_p| < 1 \quad \forall S_p, \forall \omega \tag{7.67}$$

$$\Leftrightarrow \quad |w_P| < |1 + L_p| \quad \forall L_p, \forall \omega \tag{7.68}$$

This corresponds to requiring $|\widehat{y}/d| < 1 \ \forall \Delta_I$ in Figure 7.16, where we consider multiplicative uncertainty, and the set of possible loop transfer functions is

$$L_p = G_p K = L(1 + w_I \Delta_I) = L + w_I L \Delta_I \tag{7.69}$$

1. Graphical derivation of RP condition. Condition (7.68) is illustrated graphically by the Nyquist plot in Figure 7.17. For RP we must require that all possible $L_p(j\omega)$ stay outside a disc of radius $|w_P(j\omega)|$ centred on -1. Since L_p at each frequency stays within a disc of radius $w_I L$ centred on L, we see from Figure 7.17 that the condition for RP is that the two discs, with radii $|w_P|$ and $|w_I L|$, do not overlap. Since their centres are located a distance $|1 + L|$ apart, the RP condition becomes

$$\text{RP} \quad \Leftrightarrow \quad |w_P| + |w_I L| < |1 + L|, \quad \forall \omega \tag{7.70}$$

$$\Leftrightarrow \quad |w_P(1 + L)^{-1}| + |w_I L(1 + L)^{-1}| < 1, \quad \forall \omega \tag{7.71}$$

or in other words

$$\boxed{\text{RP} \;\Leftrightarrow\; \max_\omega \left(|w_P S| + |w_I T| \right) < 1} \tag{7.72}$$

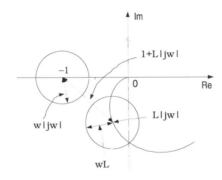

Figure 7.17: Nyquist plot of RP condition $|w_P| < |1 + L_p|$

2. Algebraic derivation of RP condition. From the definition in (7.67) we have that RP is satisfied if the worst-case (maximum) weighted sensitivity at each frequency is less than 1,

$$\text{RP} \quad \Leftrightarrow \quad \max_{S_p} |w_P S_p| < 1, \quad \forall \omega \tag{7.73}$$

(strictly speaking, max should be replaced by sup, the supremum). The perturbed sensitivity is $S_p = (I + L_p)^{-1} = 1/(1 + L + w_I L \Delta_I)$, and the worst-case (maximum) is obtained at each frequency by selecting $|\Delta_I| = 1$ such that the terms $(1 + L)$ and $w_I L \Delta_I$ (which are complex numbers) point in opposite directions. We get

$$\max_{S_p} |w_P S_p| = \frac{|w_P|}{|1 + L| - |w_I L|} = \frac{|w_P S|}{1 - |w_I T|} \tag{7.74}$$

and by substituting (7.74) into (7.73) we rederive the RP condition in (7.72).

Remarks on RP condition (7.72).

1. The RP condition (7.72) is closely approximated by the following mixed sensitivity $\mathcal{H}_\infty$ condition:

$$\left\| \begin{matrix} w_P S \\ w_I T \end{matrix} \right\|_\infty = \max_\omega \sqrt{|w_P S|^2 + |w_I T|^2} < 1 \tag{7.75}$$

To be more precise, we find from (A.96) that condition (7.75) is within a factor of at most $\sqrt{2}$ to condition (7.72). This means that for SISO systems we can closely approximate the RP condition in terms of an $\mathcal{H}_\infty$ problem, so there is little need to make use of the structured singular value. However, we will see in the next chapter that the situation can be very different for MIMO systems.

2. The RP condition (7.72) can be used to derive bounds on the loop shape $|L|$. At a given frequency we have that $|w_P S| + |w_I T| < 1$ (RP) is satisfied if (see Exercise 7.5)

$$|L| > \frac{1 + |w_P|}{1 - |w_I|} \quad \text{(at frequencies where } |w_I| < 1) \tag{7.76}$$

or if

$$|L| < \frac{1 - |w_P|}{1 + |w_I|} \quad \text{(at frequencies where } |w_P| < 1) \tag{7.77}$$

Conditions (7.76) and (7.77) may be combined over different frequency ranges. Condition (7.76) is most useful at low frequencies where generally $|w_I| < 1$ and $|w_P| > 1$ (tight performance requirement) and we need $|L|$ large. Conversely, condition (7.77) is most useful at high frequencies where generally $|w_I| > 1$ (more than 100% uncertainty), $|w_P| < 1$ and we need L small. The loop-shaping conditions (7.76) and (7.77) may in the general case be obtained numerically from μ-conditions as outlined in Remark 13 on page 311. This is discussed by Braatz et al. (1996) who derived bounds also in terms of S and T, and furthermore derived necessary bounds for RP in addition to the sufficient bounds in (7.76) and (7.77); see also Exercise 7.6.

3. The term $\mu(N_{\mathrm{RP}}) = |w_P S| + |w_I T|$ in (7.72) is the *structured singular value* (μ) for RP for this particular problem; see (8.129). We will discuss μ in much more detail in the next chapter.

4. The structured singular value μ is *not* equal to the worst-case weighted sensitivity, $\max_{S_p} |w_P S_p|$, given in (7.74) (although many people seem to think it is). The worst-case weighted sensitivity is equal to skewed-μ (μ^s) with fixed uncertainty; see Section 8.10.3. Thus, in summary we have for this particular RP problem:

$$\mu = |w_P S| + |w_I T|, \quad \mu^s = \frac{|w_P S|}{1 - |w_I T|} \tag{7.78}$$

Note that μ and μ^s are closely related since $\mu \leq 1$ if and only if $\mu^s \leq 1$.

Exercise 7.5 *Derive the loop-shaping bounds in (7.76) and (7.77) which are sufficient for $|w_P S| + |w_I T| < 1$ (RP). (Hint: Start from the RP condition in the form $|w_P| + |w_I L| < |1 + L|$ and use the facts that $|1 + L| \geq 1 - |L|$ and $|1 + L| \geq |L| - 1$.)*

Exercise 7.6 * *Also derive, from $|w_P S| + |w_I T| < 1$, the following necessary bounds for RP (which must be satisfied):*

$$|L| > \frac{|w_P| - 1}{1 - |w_I|} \quad \text{(for } |w_P| > 1 \text{ and } |w_I| < 1)$$

$$|L| < \frac{1 - |w_P|}{|w_I| - 1} \quad \text{(for } |w_P| < 1 \text{ and } |w_I| > 1)$$

(Hint: Use $|1 + L| \leq 1 + |L|$.)

Example 7.11 RP problem. *Consider RP of the SISO system in Figure 7.18, for which we have*

$$\mathrm{RP} \stackrel{\text{def}}{\Leftrightarrow} \left| \frac{\hat{y}}{d} \right| < 1, \ \forall |\Delta_u| \leq 1, \ \forall \omega; \quad w_P(s) - 0.25 + \frac{0.1}{s}; \quad w_u(s) = r_u \frac{s}{s + 1} \tag{7.79}$$

(a) Derive a condition for RP.

(b) For what values of r_u is it impossible to satisfy the RP condition?

(c) Let $r_u = 0.5$. Consider two cases for the nominal loop transfer function: (1) $GK_1(s) = 0.5/s$ and (2) $GK_2(s) = \frac{0.5}{s} \frac{1-s}{1+s}$. For each system, sketch the magnitudes of S and its performance bound as a function of frequency. Does each system satisfy RP?

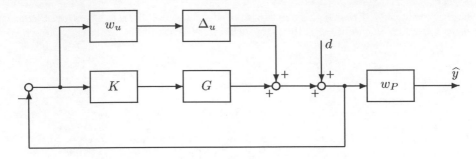

Figure 7.18: Diagram for RP in Example 7.11

Solution. *(a) The requirement for RP is* $|w_P S_p| < 1$, $\forall S_p, \forall \omega$, *where the possible sensitivities are given by*

$$S_p = \frac{1}{1 + GK + w_u \Delta_u} = \frac{S}{1 + w_u \Delta_u S} \tag{7.80}$$

The condition for RP then becomes

$$\text{RP} \quad \Leftrightarrow \quad \left| \frac{w_P S}{1 + w_u \Delta_u S} \right| < 1, \quad \forall \Delta_u, \forall \omega \tag{7.81}$$

A simple analysis shows that the worst case corresponds to selecting Δ_u *with magnitude 1 such that the term* $w_u \Delta_u S$ *is purely real and negative, and hence we have*

$$\text{RP} \quad \Leftrightarrow \quad |w_P S| < 1 - |w_u S|, \quad \forall \omega \tag{7.82}$$

$$\Leftrightarrow \quad |w_P S| + |w_u S| < 1, \quad \forall \omega \tag{7.83}$$

$$\Leftrightarrow \quad |S(jw)| < \frac{1}{|w_P(jw)| + |w_u(jw)|}, \quad \forall \omega \tag{7.84}$$

(b) Since any real system is strictly proper we have $|S| = 1$ *at high frequencies and therefore we must require* $|w_u(j\omega)| + |w_P(j\omega)| < 1$ *as* $\omega \to \infty$. *With the weights in (7.79) this is equivalent to* $r_u + 0.25 < 1$. *Therefore, we must at least require* $r_u < 0.75$ *for RP, so RP cannot be satisfied if* $r_u \geq 0.75$.

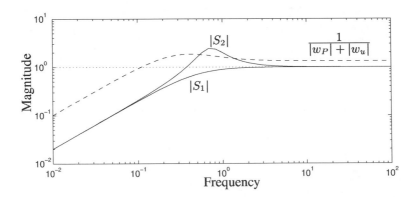

Figure 7.19: RP test

(c) Design S_1 *yields RP, while* S_2 *does not. This is seen by checking the RP condition (7.84) graphically as shown in Figure 7.19:* $|S_1|$ *has a peak of 1 while* $|S_2|$ *has a peak of about 2.45. These*

findings can also be confirmed using the Matlab commands shown in Table 7.3. We also note that `Pmarguncl.UpperBound` = 1.335, *which implies that the design* S_1 *will have RP, even if we increase the uncertainty* (w_u) *and performance requirements* (w_P) *by a factor of* 1.335. *For design* S_2, *the corresponding performance margin* (`Pmargunc2.UpperBound`) *is* 0.7, *which occurs at the frequency* $\omega = 0.801$. *This implies that the uncertainty and performance requirements must be reduced by the factor* 0.7 *at frequency* 0.801 *to achieve RP.*

Table 7.3: **Matlab program for RP analysis**

```
% Uses Robust Control toolbox
L1 = tf(0.5,[1 0]);
L2 = L1*tf([-1 1],[1 1]);
Wu = 0.5*tf([1 0],[1 1]);                    % Weights
Wp = 0.25+tf(0.1,[1 10e-6]);                 % Pole shifted for
                                             % numerical reasons
Delta = ultidyn('Delta',[1 1]);              % Dynamic uncertainty
S1 = inv(1+L1+Delta*Wu);
% Pmarg.Upperbound > 1 indicates Robust Performance
[Pmarg1,Pmarguncl,Report1] = robustperf(Wp*S1)
S2 = inv(1+L2+Delta*Wu);
[Pmarg2,Pmargunc2,Report2] = robustperf(Wp*S2)
```

7.6.3   The relationship between NP, RS and RP

Consider a SISO system with multiplicative uncertainty, and assume that the closed-loop is nominally stable (NS). The conditions for nominal performance (NP), robust stability (RS) and robust performance (RP) can then be summarized as follows:

$$\text{NP} \quad \Leftrightarrow \quad |w_P S| < 1, \forall \omega \tag{7.85}$$

$$\text{RS} \quad \Leftrightarrow \quad |w_I T| < 1, \forall \omega \tag{7.86}$$

$$\text{RP} \quad \Leftrightarrow \quad |w_P S| + |w_I T| < 1, \forall \omega \tag{7.87}$$

From this we see that a prerequisite for RP is that we satisfy NP and RS. This applies in general, both for SISO and MIMO systems and for any uncertainty. In addition, for SISO systems, if we satisfy both RS and NP, then we have at each frequency

$$|w_P S| + |w_I T| \leq 2 \max\{|w_P S|, |w_I T|\} < 2 \tag{7.88}$$

It then follows that, within a factor of at most 2, we will automatically get RP when the subobjectives of NP and RS are satisfied. Thus, RP is not a "big issue" for SISO systems, and this is probably the main reason why there is little discussion about RP in the classical control literature. On the other hand, as we will see in the next chapter, for MIMO systems we may get very poor RP even though the subobjectives of NP and RS are individually satisfied.

To satisfy RS we generally want T small, whereas to satisfy NP we generally want S small. However, we cannot make *both* S and T small at the same frequency because of the identity $S + T = 1$. This has implications for RP, since $|w_P||S| + |w_I||T| \geq \min\{|w_P|, |w_I|\}(|S| + |T|)$, where $|S| + |T| \geq |S + T| = 1$, and we derive at each frequency

$$|w_P S| + |w_I T| \geq \min\{|w_P|, |w_I|\} \tag{7.89}$$

We conclude that we *cannot have both* $|w_P| > 1$ *(i.e. good performance) and* $|w_I| > 1$ *(i.e. more than* 100% *uncertainty) at the same frequency.* One explanation for this is that at frequencies where $|w_I| > 1$ the uncertainty will allow for RHP-zeros, and we know that we cannot have tight performance in the presence of RHP-zeros.

7.6.4   The similarity between RS and RP

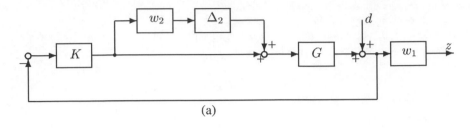

(a)

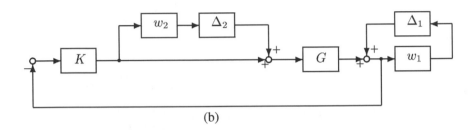

(b)

Figure 7.20: (a) RP with multiplicative uncertainty
(b) RS with combined multiplicative and inverse multiplicative uncertainty

RP may be viewed as a special case of RS (with multiple perturbations). To see this consider the following two cases as illustrated in Figure 7.20:

1. RP with multiplicative uncertainty
2. RS with combined multiplicative and inverse multiplicative uncertainty

As usual the uncertain perturbations are normalized such that $\|\Delta_1\|_\infty \leq 1$ and $\|\Delta_2\|_\infty \leq 1$. Since we use the $\mathcal{H}_\infty$ norm to define both uncertainty and performance and since the weights in Figure 7.20(a) and (b) are the same, the tests for RP and RS in cases (a) and (b), respectively, are identical. This may be argued from the block diagrams, or by simply evaluating the conditions for the two cases as shown below.

1. The condition for RP with multiplicative uncertainty was derived in (7.72), but with w_1 replaced by w_P and with w_2 replaced by w_I. We found that

$$\text{RP} \quad \Leftrightarrow \quad |w_1 S| + |w_2 T| < 1, \quad \forall \omega \qquad (7.90)$$

2. We will now derive the RS condition for the case where L_p is stable (this assumption may be relaxed if the more general $M\Delta$-structure is used, see (8.128)). We want the system to be closed-loop stable for all possible Δ_1 and Δ_2. RS is equivalent to avoiding encirclements of -1 by the Nyquist plot of L_p. That is, the distance between L_p and -1 must be larger than zero, i.e. $|1 + L_p| > 0$, and therefore

$$\text{RS} \quad \Leftrightarrow \quad |1 + L_p| > 0 \quad \forall L_p, \forall \omega \qquad (7.91)$$

$$\Leftrightarrow \quad |1 + L(1 + w_2 \Delta_2)(1 - w_1 \Delta_1)^{-1}| > 0, \ \forall \Delta_1, \forall \Delta_2, \forall \omega \qquad (7.92)$$

$$\Leftrightarrow \quad |1 + L + L w_2 \Delta_2 - w_1 \Delta_1| > 0, \ \forall \Delta_1, \forall \Delta_2, \forall \omega \qquad (7.93)$$

Here the worst case is obtained when we choose Δ_1 and Δ_2 with magnitudes 1 such that the terms $Lw_2\Delta_2$ and $w_1\Delta_1$ are in the opposite direction of the term $1 + L$. We get

$$\text{RS} \quad \Leftrightarrow \quad |1 + L| - |Lw_2| - |w_1| > 0, \quad \forall\omega \qquad (7.94)$$

$$\Leftrightarrow \quad |w_1 S| + |w_2 T| < 1, \quad \forall\omega \qquad (7.95)$$

which is the same condition as found for RP.

7.7 Additional exercises

Exercise 7.7 * *Consider a "true" plant*

$$G'(s) = \frac{3e^{-0.1s}}{(2s + 1)(0.1s + 1)^2}$$

(a) Derive and sketch the additive uncertainty weight when the nominal model is $G(s) = 3/(2s+1)$.
(b) Derive the corresponding robust stability condition.
(c) Apply this test for the controller $K(s) = k/s$ and find the values of k that yield stability. Is this condition tight?

Exercise 7.8 Uncertainty weight for a first-order model with delay. *Laughlin et al. (1987) considered the following parametric uncertainty description:*

$$G_p(s) = \frac{k_p}{\tau_p s + 1}e^{-\theta_p s}; \quad k_p \in [k_{\min}, k_{\max}], \quad \tau_p \in [\tau_{\min}, \tau_{\max}], \quad \theta_p \in [\theta_{\min}, \theta_{\max}] \qquad (7.96)$$

where all parameters are assumed positive. They chose the mean parameter values as $(\bar{k}, \bar{\theta}, \bar{\tau})$ giving the nominal model

$$G(s) = \bar{G}(s) \triangleq \frac{\bar{k}}{\bar{\tau}s + 1}e^{-\bar{\theta}s} \qquad (7.97)$$

and suggested use of the following multiplicative uncertainty weight:

$$w_{IL}(s) = \frac{k_{\max}}{\bar{k}} \cdot \frac{\bar{\tau}s + 1}{\tau_{\min}s + 1} \cdot \frac{Ts + 1}{-Ts + 1} - 1; \quad T = \frac{\theta_{\max} - \theta_{\min}}{4} \qquad (7.98)$$

(a) Show that the resulting stable and minimum-phase weight corresponding to the uncertainty description in (7.27) is

$$w_{IL}(s) = (1.25s^2 + 1.55s + 0.2)/(2s + 1)(0.25s + 1) \qquad (7.99)$$

Note that this weight cannot be compared with (7.29) or (7.30) since the nominal plant is different.
(b) Plot the magnitude of w_{IL} as a function of frequency. Find the frequency where the weight crosses 1 in magnitude, and compare this with $1/\theta_{\max}$. Comment on your answer.
(c) Find $l_I(j\omega)$ using (7.25) and compare with $|w_{IL}|$. Does the weight (7.99) and the uncertainty model (7.2) include all possible plants? (Answer: No, not quite around frequency $\omega = 5$.)

Exercise 7.9 * *Consider again the system in Figure 7.18. What kind of uncertainty might w_u and Δ_u represent?*

Exercise 7.10 Neglected dynamics. *Assume we have derived the following detailed model:*

$$G_{detail}(s) = \frac{3(-0.5s + 1)}{(2s + 1)(0.1s + 1)^2} \qquad (7.100)$$

and we want to use the simplified nominal model $G(s) = 3/(2s + 1)$ with multiplicative uncertainty. Plot $l_I(\omega)$ and approximate it by a rational transfer function $w_I(s)$.

Exercise 7.11 * **Parametric gain uncertainty.** *We showed in Example 7.1 how to represent scalar parametric gain uncertainty* $G_p(s) = k_p G_0(s)$ *where*

$$k_{\min} \le k_p \le k_{\max} \tag{7.101}$$

as multiplicative uncertainty $G_p = G(1 + w_I \Delta_I)$ *with nominal model* $G(s) = \bar{k} G_0(s)$ *and uncertainty weight* $w_I = r_k = (k_{\max} - k_{\min})/(k_{\max} + k_{\min})$. Δ_I *here is a real scalar,* $-1 \le \Delta_I \le 1$. *Alternatively, we can represent gain uncertainty as* inverse multiplicative uncertainty:

$$\Pi_{iI} : \quad G_p(s) = G(s)(1 + w_{iI}(s)\Delta_{iI})^{-1}; \quad -1 \le \Delta_{iI} \le 1 \tag{7.102}$$

with $w_{iI} = r_k$ *and* $G(s) = k_i G$ *where*

$$k_i = 2 \frac{k_{\min} k_{\max}}{k_{\max} + k_{\min}} \tag{7.103}$$

(a) *Derive (7.102) and (7.103). (Hint: The gain variation in (7.101) can be written exactly as* $k_p = k_i/(1 - r_k \Delta)$.)
(b) *Show that the form in (7.102) does not allow for* $k_p = 0$.
(c) *Discuss why (b) may be a possible advantage.*

Exercise 7.12 *The model of an industrial robot arm is as follows:*

$$G(s) = \frac{250(as^2 + 0.0001s + 100)}{s(as^2 + 0.0001(500a + 1)s + 100(500a + 1))}$$

where $a \in [0.0002, 0.002]$. *Sketch the Bode plot for the two extreme values of* a. *What kind of control performance do you expect? Discuss how you may best represent this uncertainty.*

7.8 Conclusion

In this chapter we have shown how model uncertainty for SISO systems can be represented in the frequency domain using complex norm-bounded perturbations, $\|\Delta\|_\infty \le 1$. At the end of the chapter we also discussed how to represent parametric uncertainty using real perturbations.

We showed that the requirement of robust stability for the case of multiplicative complex uncertainty imposes an upper bound on the allowed complementary sensitivity, $|w_I T| < 1, \forall \omega$. Similarly, the inverse multiplicative uncertainty imposes an upper bound on the sensitivity, $|w_{iI} S| < 1, \forall \omega$. We also derived a condition for robust performance with multiplicative uncertainty, $|w_P S| + |w_I T| < 1, \forall \omega$.

The approach in this chapter was rather elementary, and to extend the results to MIMO systems and to more complex uncertainty descriptions we need to make use of the structured singular value, μ. This is the theme of the next chapter, where we find that $|w_I T|$ and $|w_{iI} S|$ are the structured singular values for evaluating robust stability for the two sources of uncertainty in question, whereas $|w_P S| + |w_I T|$ is the structured singular value for evaluating robust performance with multiplicative uncertainty.

8

ROBUST STABILITY AND PERFORMANCE ANALYSIS FOR MIMO SYSTEMS

The objective of this chapter is to present a general method for analyzing robust stability and robust performance of MIMO systems with multiple perturbations. Our main analysis tool will be the *structured singular value*, μ. We also show how the "optimal" robust controller, in terms of minimizing μ, can be designed using DK-iteration. This involves solving a sequence of scaled $\mathcal{H}_\infty$ problems.

8.1 General control configuration with uncertainty

For useful notation and an introduction to model uncertainty, the reader is referred to Sections 7.1 and 7.2. The starting point for our robustness analysis is a system representation in which the uncertain perturbations are "pulled out" into a block-diagonal matrix,

$$\Delta = \mathrm{diag}\{\Delta_i\} = \begin{bmatrix} \Delta_1 & & & \\ & \ddots & & \\ & & \Delta_i & \\ & & & \ddots \end{bmatrix} \tag{8.1}$$

where each Δ_i represents a specific source of uncertainty, e.g. input uncertainty, Δ_I, or parametric uncertainty, δ_i, where δ_i is real. If we also pull out the controller K, we get the generalized plant P, as shown in Figure 8.1. This form is useful for controller synthesis. Alternatively, if the controller is given and we want to analyze the uncertain system, we use the $N\Delta$-structure in Figure 8.2.

In Section 3.8.8, we discussed how to find P and N for cases without uncertainty. The procedure with uncertainty is similar and is demonstrated by examples below; see Section 8.3. To illustrate the main idea, consider Figure 8.4 where it is shown how to pull out the perturbation blocks to form Δ and the nominal system N. As shown in (3.123), N is related to P and K by a lower LFT

$$N = F_l(P, K) \triangleq P_{11} + P_{12}K(I - P_{22}K)^{-1}P_{21} \tag{8.2}$$

Similarly, the uncertain closed-loop transfer function from w to z, $z = Fw$, is related to N and Δ by an upper LFT (see (3.124)),

$$F = F_u(N, \Delta) \triangleq N_{22} + N_{21}\Delta(I - N_{11}\Delta)^{-1}N_{12} \tag{8.3}$$

Multivariable Feedback Control: Analysis and Design. Second Edition
S. Skogestad and I. Postlethwaite   © 2005, 2006 John Wiley & Sons, Ltd

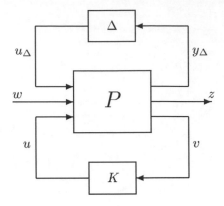

Figure 8.1: General control configuration (for controller synthesis)

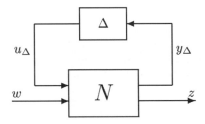

Figure 8.2: $N\Delta$-structure for robust performance analysis

To analyze robust stability of F, we can then rearrange the system into the $M\Delta$-structure of Figure 8.3 where $M = N_{11}$ is the transfer function from the output to the input of the perturbations.

8.2 Representing uncertainty

As usual, each individual perturbation is assumed to be stable and is normalized,

$$\bar{\sigma}(\Delta_i(j\omega)) \leq 1 \;\forall\omega \tag{8.4}$$

For a complex scalar perturbation we have $|\delta_i(j\omega)| \leq 1$, $\forall\omega$, and for a real scalar perturbation $-1 \leq \delta_i \leq 1$. Since from (A.49) the maximum singular value of a block-diagonal matrix is

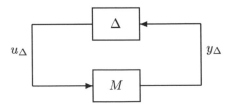

Figure 8.3: $M\Delta$-structure for robust stability analysis

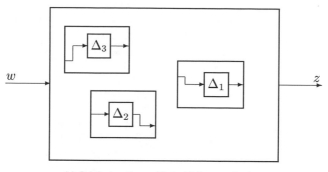

(a) Original system with multiple perturbations

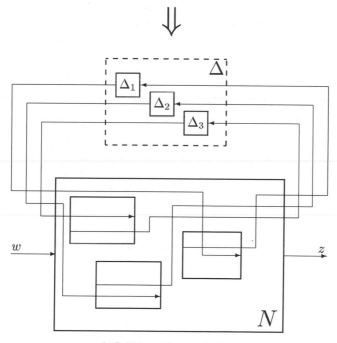

(b) Pulling out the perturbations

Figure 8.4: Rearranging an uncertain system into the $N\Delta$-structure

equal to the largest of the maximum singular values of the individual blocks, it then follows for $\Delta = \text{diag}\{\Delta_i\}$ that

$$\bar{\sigma}(\Delta_i(j\omega)) \leq 1 \; \forall \omega, \; \forall i \quad \Leftrightarrow \quad \boxed{\|\Delta\|_\infty \leq 1} \tag{8.5}$$

Note that Δ has *structure*, and therefore in the robustness analysis we do *not* want to allow all Δ such that (8.5) is satisfied. Only the subset which has the block-diagonal structure in (8.1) should be considered. In some cases the blocks in Δ may be repeated or may be real; that is, we have additional structure. For example, as shown in Example 7.5, repetition is often needed to handle parametric uncertainty.

Remark. The assumption of a stable Δ may be relaxed, but then the resulting robust stability and performance conditions will be harder to derive and more complex to state. Furthermore, if we use a suitable form for the uncertainty and allow for multiple perturbations, then we can always generate the desired class of plants with stable perturbations, so assuming Δ stable is not really a restriction.

8.2.1 Differences between SISO and MIMO systems

The main difference between SISO and MIMO systems is the concept of directions which is only relevant in the latter. As a consequence MIMO systems may experience much larger sensitivity to uncertainty than SISO systems. The following example illustrates for MIMO systems that it is sometimes critical to represent the coupling between uncertainty in different transfer function elements.

Example 8.1 **Coupling between transfer function elements.** *Consider a distillation process where at steady-state*

$$G = \begin{bmatrix} 87.8 & -86.4 \\ 108.2 & -109.6 \end{bmatrix}, \quad \Lambda = \text{RGA}(G) = \begin{bmatrix} 35.1 & -34.1 \\ -34.1 & 35.1 \end{bmatrix} \tag{8.6}$$

From the large RGA elements we know that G becomes singular for small relative changes in the individual elements. For example, from (6.99) we know that perturbing the 1, 2 element from −86.4 to −88.9 makes G singular. Since variations in the steady-state gains of ±50% or more may occur during operation of the distillation process, this seems to indicate that independent control of both outputs is impossible. However, this conclusion is incorrect since, for a distillation process, G never becomes singular. This is because the transfer function elements are coupled due to underlying physical constraints (e.g. the material balance). Specifically, for the distillation process a more reasonable description of the gain uncertainty is (Skogestad et al., 1988)

$$G_p = G + w \begin{bmatrix} \delta & -\delta \\ -\delta & \delta \end{bmatrix}, \quad |\delta| \leq 1 \tag{8.7}$$

where w in this case is a real constant, e.g. w = 50. For the numerical data above, $\det G_p = \det G$ irrespective of δ, so G_p is never singular for this uncertainty. (Note that $\det G_p = \det G$ is not generally true for the uncertainty description given in (8.7).)

Exercise 8.1 * *The uncertain plant in (8.7) may be represented in the additive uncertainty form $G_p = G + W_2 \Delta_A W_1$ where $\Delta_A = \delta$ is a single scalar perturbation. Find W_1 and W_2.*

8.2.2 Parametric uncertainty

The representation of parametric uncertainty, as discussed in Chapter 7 for SISO systems, carries straight over to MIMO systems. However, the inclusion of parametric uncertainty may be more important for MIMO plants because it offers a simple method of representing the coupling between uncertain transfer function elements. For example, the simple uncertainty description used in (8.7) originated from a parametric uncertainty description of the distillation process.

8.2.3 Unstructured uncertainty

Unstructured perturbations are often used to get a simple uncertainty model. We define *unstructured* uncertainty as the use of a "full" complex perturbation matrix Δ, usually with dimensions compatible with those of the plant, where at each frequency any $\Delta(j\omega)$ satisfying $\bar{\sigma}(\Delta(j\omega)) \leq 1$ is allowed.

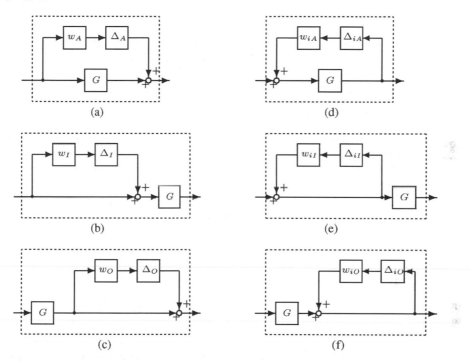

Figure 8.5: (a) Additive uncertainty, (b) multiplicative input uncertainty, (c) multiplicative output uncertainty, (d) inverse additive uncertainty, (e) inverse multiplicative input uncertainty, (f) inverse multiplicative output uncertainty

Six common forms of unstructured uncertainty are shown in Figure 8.5. In Figure 8.5(a), (b) and (c) are shown three *feedforward* forms: additive uncertainty, multiplicative input uncertainty and multiplicative output uncertainty given by

$$\Pi_A : \qquad G_p = G + E_A; \qquad E_a = w_A \Delta_a \qquad (8.8)$$

$$\Pi_I : \qquad G_p = G(I + E_I); \qquad E_I = w_I \Delta_I \qquad (8.9)$$

$$\Pi_O : \qquad G_p = (I + E_O)G; \qquad E_O = w_O \Delta_O \qquad (8.10)$$

In Figure 8.5(d), (e) and (f) are shown three *feedback* or *inverse* forms: inverse additive uncertainty, inverse multiplicative input uncertainty and inverse multiplicative output uncertainty given by

$$\Pi_{iA} : \qquad G_p = G(I - E_{iA}G)^{-1}; \qquad E_{iA} = w_{iA}\Delta_{iA} \qquad (8.11)$$

$$\Pi_{iI} : \qquad G_p = G(I - E_{iI})^{-1}; \qquad E_{iI} = w_{iI}\Delta_{iI} \qquad (8.12)$$

$$\Pi_{iO} : \qquad G_p = (I - E_{iO})^{-1}G; \qquad E_{iO} = w_{iO}\Delta_{iO} \qquad (8.13)$$

The negative sign in front of the E's does not really matter here since we assume that Δ can have any sign. Δ denotes the normalized perturbation and E the "actual" perturbation. We have used scalar weights w, so $E = w\Delta = \Delta w$, but sometimes one may want to use matrix weights, $E = W_2\Delta W_1$, where W_1 and W_2 are given transfer function matrices.

Another common form of unstructured uncertainty is coprime factor uncertainty discussed later in Section 8.6.2.

Remark. In practice, one can have several perturbations which themselves are unstructured. For example, we may have Δ_I at the input and Δ_O at the output, which may be combined into a larger perturbation, $\Delta = \text{diag}\{\Delta_I, \Delta_O\}$. However, this Δ is a block-diagonal matrix and is therefore no longer truly unstructured.

Lumping uncertainty into a single perturbation

For SISO systems, we usually lump multiple sources of uncertainty into a single complex perturbation, often in multiplicative form. This may also be done for MIMO systems, but then it makes a difference whether the perturbation is at the input or the output.

Since output uncertainty is frequently less restrictive than input uncertainty in terms of control performance (see Section 6.10.4), we first attempt to lump the uncertainty at the output. For example, a set of plants Π may be represented by *multiplicative output uncertainty* with a scalar weight $w_O(s)$ using

$$G_p = (I + w_O\Delta_O)G, \qquad \|\Delta_O\|_\infty \le 1 \qquad (8.14)$$

where, similar to (7.25),

$$l_O(\omega) = \max_{G_p \in \Pi} \bar{\sigma}\left((G_p - G)G^{-1}(j\omega)\right); \quad |w_O(j\omega)| \ge l_O(\omega) \quad \forall\omega \qquad (8.15)$$

(and we can use the pseudo-inverse if G is singular). If the resulting uncertainty weight is reasonable (i.e. it must at least be less than 1 in the frequency range where we want control), and the subsequent analysis shows that robust stability and performance may be achieved, then this lumping of uncertainty at the output is fine. If this is not the case, then one may try to lump the uncertainty at the input instead, using *multiplicative input uncertainty* with a scalar weight,

$$G_p = G(I + w_I\Delta_I), \qquad \|\Delta_I\|_\infty \le 1 \qquad (8.16)$$

where, similar to (7.25),

$$l_I(\omega) = \max_{G_p \in \Pi} \bar{\sigma}\left(G^{-1}(G_p - G)(j\omega)\right); \quad |w_I(j\omega)| \ge l_I(\omega) \quad \forall\omega \qquad (8.17)$$

However, in many cases this approach of lumping uncertainty either at the output or the input does not work well. This is because one cannot in general shift a perturbation from one location in the plant (say at the input) to another location (say the output) without introducing candidate plants which were not present in the original set. In particular, one should be careful when the plant is ill-conditioned. This is discussed next.

Moving uncertainty from the input to the output

For a scalar plant, we have $G_p = G(1+w_I\Delta_I) = (1+w_O\Delta_O)G$ and we may simply "move" the multiplicative uncertainty from the input to the output without changing the value of the weight, i.e. $w_I = w_O$. However, for multivariable plants we usually need to multiply by the condition number $\gamma(G)$ as is shown next.

Suppose the true uncertainty is represented as unstructured input uncertainty (E_I is a full matrix) in the form

$$G_p = G(I + E_I) \tag{8.18}$$

Then from (8.17), the magnitude of multiplicative input uncertainty is

$$l_I(\omega) = \max_{E_I} \bar{\sigma}(G^{-1}(G_p - G)) = \max_{E_I} \bar{\sigma}(E_I) \tag{8.19}$$

On the other hand, if we want to represent (8.18) as multiplicative output uncertainty, then from (8.15)

$$l_O(\omega) = \max_{E_I} \bar{\sigma}((G_p - G)G^{-1}) = \max_{E_I} \bar{\sigma}(GE_IG^{-1}) \tag{8.20}$$

which is much larger than $l_I(\omega)$ if the condition number of the plant is large. To see this, write $E_I = w_I\Delta_I$ where we allow any $\Delta_I(j\omega)$ satisfying $\bar{\sigma}(\Delta_I(j\omega)) \leq 1, \forall\omega$. Then at a given frequency

$$l_O(\omega) = |w_I| \max_{\Delta_I} \bar{\sigma}(G\Delta_IG^{-1}) = |w_I(j\omega)|\,\gamma(G(j\omega)) \tag{8.21}$$

Proof of (8.21): Write at each frequency $G = U\Sigma V^H$ and $G^{-1} = \tilde{U}\tilde{\Sigma}\tilde{V}^H$. Select $\Delta_I = V\tilde{U}^H$ (which is a unitary matrix with all singular values equal to 1). Then $\bar{\sigma}(G\Delta_IG^{-1}) = \bar{\sigma}(U\Sigma\tilde{\Sigma}V^H) = \bar{\sigma}(\Sigma\tilde{\Sigma}) = \bar{\sigma}(G)\bar{\sigma}(G^{-1}) = \gamma(G)$. □

Example 8.2 *Assume the relative input uncertainty is 10%, i.e. $w_I = 0.1$, and the condition number of the plant is 141.7. Then we must select $l_O = w_O = 0.1 \cdot 141.7 = 14.2$ in order to represent this as multiplicative output uncertainty (this is larger than 1 and therefore not useful for controller design).*

Also for diagonal uncertainty (E_I diagonal) we may have a similar situation. For example, if the plant has large RGA elements then the elements in GE_IG^{-1} will be much larger than those of E_I, see (A.81), making it impractical to move the uncertainty from the input to the output.

Example 8.3 *Let Π be the set of plants generated by the additive uncertainty in (8.7) with $w = 10$ (corresponding to about 10% uncertainty in each element). Then from (8.7) one plant G' in this set (corresponding to $\delta = 1$) has*

$$G' = G + \begin{bmatrix} 10 & -10 \\ -10 & 10 \end{bmatrix} \tag{8.22}$$

for which we have $l_I = \bar{\sigma}(G^{-1}(G' - G)) = 14.3$. Therefore, to represent G' in terms of input uncertainty we would need a relative uncertainty of more than 1400%. This would imply that the plant could become singular at steady-state and thus impossible to control, which we know is incorrect. Fortunately, we can instead represent this additive uncertainty as multiplicative output uncertainty (which is also generally preferable for a subsequent controller design) with $l_O = \bar{\sigma}((G' - G)G^{-1}) = 0.10$. Therefore output uncertainty works well for this particular example.

Conclusion. Ideally, we would like to lump several sources of uncertainty into a single perturbation to get a simple uncertainty description. Often an unstructured multiplicative output perturbation is used. However, from the above discussion we have learnt that we should be careful about doing this, at least for plants with a large condition number. In such cases we may have to represent the uncertainty as it occurs physically (at the input, in the elements, etc.) thereby generating several perturbations. For uncertainty associated with unstable plant poles, we should use one of the *inverse* forms in Figure 8.5.

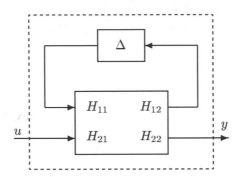

Figure 8.6: Uncertain plant, $y = G_p u$, represented by LFT, see (8.23)

Exercise 8.2 *A fairly general way of representing an uncertain plant G_p is in terms of a linear fractional transformation (LFT) of Δ as shown in Figure 8.6. Here*

$$G_p = F_u \left(\begin{bmatrix} H_{11} & H_{12} \\ H_{21} & H_{22} \end{bmatrix}, \Delta \right) = H_{22} + H_{21}\Delta(I - H_{11}\Delta)^{-1}H_{12} \qquad (8.23)$$

where $G = H_{22}$ is the nominal plant model. Obtain H for each of the six uncertainty forms in (8.8)–(8.13) using $E = W_2\Delta W_1$. (Hint for the inverse forms: $(I - W_1\Delta W_2)^{-1} = I + W_1\Delta(I - W_2W_1\Delta)^{-1}W_2$, see (3.7)–(3.9).)

Exercise 8.3 * *Obtain H in Figure 8.6 for the uncertain plant in Figure 7.20(b).*

8.2.4   Diagonal uncertainty

By "diagonal uncertainty" we mean that the perturbation is a *complex diagonal* matrix

$$\Delta(s) = \text{diag}\{\delta_i(s)\}; \quad |\delta_i(j\omega)| \leq 1, \forall \omega, \forall i \qquad (8.24)$$

(usually of the same size as the plant). For example, this is the case if Δ is diagonal in any of the six uncertainty forms in Figure 8.5. Diagonal uncertainty usually arises from a consideration of uncertainty or neglected dynamics in the individual input channels (actuators) or in the individual output channels (sensors). This type of diagonal uncertainty is *always* present, and since it has a scalar origin it may be represented using the methods presented in Chapter 7.

To make this clearer, let us consider uncertainty in the input channels. With each input u_i there is associated a separate physical system (amplifier, signal converter, actuator, valve, etc.) which based on the *controller output* signal, u_i, generates a *physical plant input m_i*

$$m_i = h_i(s)u_i \qquad (8.25)$$

The scalar transfer function $h_i(s)$ is often absorbed into the plant model $G(s)$, but for representing the uncertainty it is important to notice that it originates at the input. We can represent this actuator uncertainty as multiplicative (relative) uncertainty given by

$$h_{pi}(s) = h_i(s)(1 + w_{Ii}(s)\delta_i(s)); \quad |\delta_i(j\omega)| \leq 1, \forall \omega \qquad (8.26)$$

which after combining all input channels results in *diagonal input uncertainty* for the plant

$$G_p(s) = G(I + W_I\Delta_I); \quad \Delta_I = \text{diag}\{\delta_i\}, W_I = \text{diag}\{w_{Ii}\} \qquad (8.27)$$

Normally we would represent the uncertainty in each input or output channel using a simple weight in the form given in (7.38), namely

$$w(s) = \frac{\tau s + r_0}{(\tau/r_\infty)s + 1} \qquad (8.28)$$

where r_0 is the relative uncertainty at steady-state, $1/\tau$ is (approximately) the frequency where the relative uncertainty reaches 100%, and r_∞ is the magnitude of the weight at higher frequencies. Typically, the uncertainty $|w|$, associated with each input, is at least 10% at steady-state ($r_0 \geq 0.1$), and it increases at higher frequencies to account for neglected or uncertain dynamics (typically, $r_\infty \geq 2$).

Remark 1 The diagonal uncertainty in (8.27) originates from independent scalar uncertainty in each input channel. If we choose to represent this as *unstructured* input uncertainty (Δ_I is a full matrix) then we must realize that this will introduce non-physical couplings at the input to the plant, resulting in a set of plants which is too large, and the resulting robustness analysis may be conservative (meaning that we may incorrectly conclude that the system may not meet its specifications).

Remark 2 The claim is often made that one can easily reduce the static input gain uncertainty to significantly less than 10%, but this is not true in most cases. Consider again (8.25). A commonly suggested method to reduce the uncertainty is to measure the actual input (m_i) and employ local feedback (cascade control) to readjust u_i. As a simple example, consider a bathroom shower, in which the input variables are the flows of hot and cold water. One can then imagine measuring these flows and using cascade control so that each flow can be adjusted more accurately. However, even in this case there will be uncertainty related to the accuracy of each measurement. Note that it is *not* the absolute measurement error that yields problems, but rather the error in the sensitivity of the measurement with respect to changes (i.e. the "gain" of the sensor). For example, assume that the nominal flow in our shower is 1 l/min and we want to increase it to 1.1 l/min; that is, in terms of deviation variables we want $u = 0.1$ [l/min]. Suppose the vendor guarantees that the measurement error is less than 1%. But, even with this small absolute error, the actual flow rate may have increased from 0.99 l/min (measured value of 1 l/min is 1% higher) to 1.11 l/min (measured value of 1.1 l/min is 1% lower), corresponding to a change $u' = 0.12$ [l/min], and an input gain uncertainty of 20%.

In conclusion, diagonal input uncertainty, as given in (8.27), should always be considered because:

1. It is *always* present and a system which is sensitive to this uncertainty will not work in practice.
2. It often restricts achievable performance with multivariable control.

8.3 Obtaining P, N and M

We will now illustrate, by way of an example, how to obtain the interconnection matrices P, N and M in a given situation.

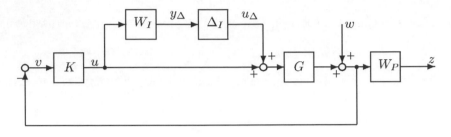

Figure 8.7: System with multiplicative input uncertainty and performance measured at the output

Example 8.4 System with input uncertainty. *Consider a feedback system with multiplicative input uncertainty Δ_I as shown in Figure 8.7. Here W_I is a normalization weight for the uncertainty and W_P is a performance weight. We want to derive the generalized plant P in Figure 8.1 which has inputs $[u_\Delta \;\; w \;\; u]^T$ and outputs $[y_\Delta \;\; z \;\; v]^T$. By writing down the equations (e.g. see Example 3.18) or simply by inspecting Figure 8.7 (remember to break the loop before and after K) we get*

$$P = \begin{bmatrix} 0 & 0 & W_I \\ W_P G & W_P & W_P G \\ -G & -I & -G \end{bmatrix} \tag{8.29}$$

It is recommended that the reader carefully derives P (as instructed in Exercise 8.4). Note that the transfer function from u_Δ to y_Δ (upper left element in P) is 0 because u_Δ has no direct effect on y_Δ (except through K). Next, we want to derive the matrix N corresponding to Figure 8.2. First, partition P to be compatible with K, i.e.

$$P_{11} = \begin{bmatrix} 0 & 0 \\ W_P G & W_P \end{bmatrix}, \; P_{12} = \begin{bmatrix} W_I \\ W_P G \end{bmatrix} \tag{8.30}$$

$$P_{21} = [-G \;\; -I], \; P_{22} = -G \tag{8.31}$$

and then find $N = F_l(P, K)$ using (8.2). We get (see Exercise 8.6)

$$N = \begin{bmatrix} -W_I KG(I + KG)^{-1} & -W_I K(I + GK)^{-1} \\ W_P G(I + KG)^{-1} & W_P(I + GK)^{-1} \end{bmatrix} \tag{8.32}$$

Alternatively, we can derive N directly from Figure 8.7 by evaluating the closed-loop transfer function from inputs $\begin{bmatrix} u_\Delta \\ w \end{bmatrix}$ to outputs $\begin{bmatrix} y_\Delta \\ z \end{bmatrix}$ (without breaking the loop before and after K). For example, to derive N_{12}, which is the transfer function from w to y_Δ, we start at the output (y_Δ) and move backwards to the input (w) using the MIMO rule in Section 3.2 (we first meet W_I, then $-K$ and we then exit the feedback loop and get the term $(I + GK)^{-1}$).

The upper left block, N_{11}, in (8.32) is the transfer function from u_Δ to y_Δ. This is the transfer function M needed in Figure 8.3 for evaluating robust stability. Thus, we have $M = -W_I KG(I + KG)^{-1} = -W_I T_I$.

Remark. Of course, deriving N from P is straightforward using available software. For example, in the Matlab Robust Control toolbox we can evaluate $N = F_l(P, K)$ using the command `N=lft(P,K)`, and with a specific Δ the perturbed transfer function $F_u(N, \Delta)$ from w to z is obtained with the command `F=lft(delta,N)`.

Exercise 8.4 * *Show in detail how P in (8.29) is derived.*

Exercise 8.5 *For the system in Figure 8.7 we see easily from the block diagram that the uncertain transfer function from w to z is $F = W_P(I + G(I + W_I\Delta_I)K)^{-1}$. Show that this is identical to $F_u(N, \Delta)$ evaluated using (8.35), where from (8.32), we have $N_{11} = -W_I T_I$, $N_{12} = -W_I KS$, $N_{21} = W_P SG$ and $N_{22} = W_P S$.*

Exercise 8.6 * *Derive N in (8.32) from P in (8.29) using the lower LFT in (8.2). You will note that the algebra is quite tedious, and that it is much simpler to derive N directly from the block diagram as described above.*

Exercise 8.7 *Derive P and N for the case when the multiplicative uncertainty is at the output rather than the input.*

Exercise 8.8 * *Find P for the uncertain system in Figure 7.18.*

Exercise 8.9 *Find P for the uncertain plant G_p in (8.23) when $w = r$ and $z = y - r$.*

Exercise 8.10 * *Find the interconnection matrix N for the uncertain system in Figure 7.18. What is M?*

Exercise 8.11 *Find the transfer function $M = N_{11}$ for studying robust stability for the uncertain plant G_p in (8.23).*

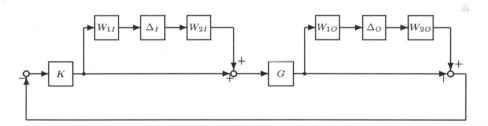

Figure 8.8: System with input and output multiplicative uncertainty

Exercise 8.12 * $M\Delta$-**structure for combined input and output uncertainties.** *Consider the block diagram in Figure 8.8 where we have both input and output multiplicative uncertainty blocks. The set of possible plants is given by*

$$G_p = (I + W_{2O}\Delta_O W_{1O})G(I + W_{2I}\Delta_I W_{1I}) \tag{8.33}$$

where $\|\Delta_I\|_\infty \le 1$ and $\|\Delta_O\|_\infty \le 1$. Collect the perturbations into $\Delta = \text{diag}\{\Delta_I, \Delta_O\}$ and re-arrange Figure 8.8 into the $M\Delta$-structure in Figure 8.3. Show that

$$M = \begin{bmatrix} W_{1I} & 0 \\ 0 & W_{1O} \end{bmatrix} \begin{bmatrix} -T_I & -KS \\ SG & -T \end{bmatrix} \begin{bmatrix} W_{2I} & 0 \\ 0 & W_{2O} \end{bmatrix} \tag{8.34}$$

8.4 Definitions of robust stability and robust performance

We have discussed how to represent an uncertain set of plants in terms of the $N\Delta$-structure in Figure 8.2. The next step is to check whether we have stability and acceptable performance for all plants in the set:

1. *Robust stability (RS) analysis*: with a given controller K we determine whether the system remains stable for all plants in the uncertainty set.
2. *Robust performance (RP) analysis*: if RS is satisfied, we determine how "large" the transfer function from exogenous inputs w to outputs z may be for all plants in the uncertainty set.

Before proceeding, we need to define performance more precisely. In Figure 8.2, w represents the exogenous inputs (normalized disturbances and references), and z the exogenous outputs (normalized errors). We have $z = F(\Delta)w$, where from (8.3)

$$F = F_u(N, \Delta) \triangleq N_{22} + N_{21}\Delta(I - N_{11}\Delta)^{-1}N_{12} \tag{8.35}$$

We will use the $\mathcal{H}_\infty$ norm to define performance and require for RP that $\|F(\Delta)\|_\infty \leq 1$ for all allowed Δ's. A typical choice is $F = w_P S_p$ (the weighted sensitivity function), where w_P is the performance weight (capital P for performance) and S_p represents the set of perturbed sensitivity functions (lower-case p for perturbed).

In terms of the $N\Delta$-structure in Figure 8.2, our requirements for stability and performance can then be summarized as follows:

$$\text{NS} \overset{\text{def}}{\Leftrightarrow} N \text{ is internally stable} \tag{8.36}$$

$$\text{NP} \overset{\text{def}}{\Leftrightarrow} \|N_{22}\|_\infty < 1; \quad \text{and NS} \tag{8.37}$$

$$\text{RS} \overset{\text{def}}{\Leftrightarrow} F = F_u(N, \Delta) \text{ is stable } \forall \Delta, \|\Delta\|_\infty \leq 1; \quad \text{and NS} \tag{8.38}$$

$$\text{RP} \overset{\text{def}}{\Leftrightarrow} \|F\|_\infty < 1, \quad \forall \Delta, \|\Delta\|_\infty \leq 1; \quad \text{and NS} \tag{8.39}$$

These definitions of RS and RP are useful only if we can test them in an efficient manner; that is, without having to search through the infinite set of allowable perturbations Δ. We will show how this can be done by introducing the structured singular value, μ, as our analysis tool. At the end of the chapter we also discuss how to synthesize controllers such that we have "optimal robust performance" by minimizing μ over the set of stabilizing controllers.

Remark 1 Important. As a prerequisite for nominal performance (NP), robust stability (RS) and robust performance (RP), we must first satisfy nominal stability (NS). This is because the frequency-by-frequency conditions can also be satisfied for unstable systems.

Remark 2 Convention for inequalities. In this book, we use the convention that the perturbations are bounded such that they are less than *or equal* to 1. This results in a stability condition with a strict inequality: for example, RS $\forall \|\Delta\|_\infty \leq 1$ if $\|M\|_\infty < 1$. (We could alternatively have bounded the uncertainty with a strict inequality, yielding the equivalent condition RS $\forall \|\Delta\|_\infty < 1$ if $\|M\|_\infty \leq 1$.)

Remark 3 Allowed perturbations. For simplicity below, we will use the shorthand notation

$$\forall \Delta \quad \text{and} \quad \max_\Delta \tag{8.40}$$

to mean "for all Δ's in the set of allowed perturbations", and "maximizing over all Δ's in the set of allowed perturbations". By *allowed perturbations* we mean that the $\mathcal{H}_\infty$ norm of Δ is less than or equal to 1, $\|\Delta\|_\infty \leq 1$, and that Δ has a specified block-diagonal structure where certain blocks may be restricted to be real. To be mathematically exact, we should replace Δ in (8.40) by $\Delta \in \mathbf{B}_\Delta$, where

$$\mathbf{B}_\Delta = \{\Delta \in \mathbf{\Delta} : \|\Delta\|_\infty \leq 1\}$$

is the set of unity norm-bounded perturbations with a given structure $\mathbf{\Delta}$. The allowed structure should also be defined, for example, by

$$\mathbf{\Delta} = \{\text{diag}\,[\delta_1 I_{r1}, \ldots, \delta_S I_{rS}, \Delta_1, \ldots, \Delta_F] : \delta_i \in \mathcal{R}, \Delta_j \in \mathcal{C}^{m_j \times m_j}\}$$

where in this case S denotes the number of real scalars (some of which may be repeated), and F the number of complex blocks. This gets rather involved. Fortunately, this amount of detail is rarely required as it is usually clear what we mean by "for all allowed perturbations" or "$\forall\Delta$".

8.5 Robust stability of the $M\Delta$-structure

Consider the uncertain $N\Delta$-system in Figure 8.2 for which the transfer function from w to z is, as in (8.35), given by

$$F_u(N,\Delta) = N_{22} + N_{21}\Delta(I - N_{11}\Delta)^{-1}N_{12} \tag{8.41}$$

Suppose that the system is nominally stable (with $\Delta = 0$); that is, N is stable (which means that the whole of N, and not only N_{22}, must be stable). We also assume that Δ is stable. We then see directly from (8.41) that the only possible source of instability is the feedback term $(I - N_{11}\Delta)^{-1}$. Thus, when we have nominal stability (NS), the stability of the system in Figure 8.2 is equivalent to the stability of the $M\Delta$-structure in Figure 8.3 where $M = N_{11}$.

We thus need to derive conditions for checking the stability of the $M\Delta$-structure. The next theorem follows from the generalized Nyquist Theorem 4.9. It applies to $\mathcal{H}_\infty$ norm-bounded Δ-perturbations, but as can be seen from the statement it also applies to any other *convex* set of perturbations (e.g. sets with other structures or sets bounded by different norms).

Theorem 8.1 Determinant stability condition (real or complex perturbations). *Assume that the nominal system $M(s)$ and the perturbations $\Delta(s)$ are stable. Consider the convex set of perturbations Δ, such that if Δ' is an allowed perturbation then so is $c\Delta'$ where c is any* __real__ *scalar such that $|c| \leq 1$. Then the $M\Delta$-system in Figure 8.3 is stable for all allowed perturbations (we have RS) if and only if*

$$Nyquist\ plot\ of\ \det(I - M\Delta(s))\ does\ not\ encircle\ the\ origin,\ \forall\Delta \tag{8.42}$$

$$\Leftrightarrow \boxed{\det(I - M\Delta(j\omega)) \neq 0, \quad \forall\omega, \forall\Delta} \tag{8.43}$$

$$\Leftrightarrow \lambda_i(M\Delta) \neq 1, \quad \forall i, \forall\omega, \forall\Delta \tag{8.44}$$

Proof: Condition (8.42) is simply the generalized Nyquist theorem (page 152) applied to a positive feedback system with a stable loop transfer function $M\Delta$.

(8.42) $\Rightarrow$ (8.43): This is obvious since by "encirclement of the origin" we also include the origin itself.

(8.42) $\Leftarrow$ (8.43) is proved by proving that not(8.42) $\Rightarrow$ not(8.43). First note that with $\Delta = 0$, $\det(I - M\Delta) = 1$ at all frequencies. Assume there exists a perturbation Δ' such that the image of $\det(I - M\Delta'(s))$ encircles the origin as s traverses the Nyquist D-contour. Because the Nyquist contour and its map are closed, there then exists another perturbation in the set, $\Delta'' = \epsilon\Delta'$, with $\epsilon \in [0, 1]$, and an ω' such that $\det(I - M\Delta''(j\omega')) = 0$.

(8.44) is equivalent to (8.43) since $\det(I - A) = \prod_i \lambda_i(I - A)$ and $\lambda_i(I - A) = 1 - \lambda_i(A)$ (see Appendix A.2.1). $\qquad\square$

The following is a special case of Theorem 8.1 which applies to complex perturbations.

Theorem 8.2 Spectral radius condition for complex perturbations. *Assume that the nominal system $M(s)$ and the perturbations $\Delta(s)$ are stable. Consider the class of perturbations, Δ, such that if Δ' is an allowed perturbation then so is $c\Delta'$ where c is any* *complex scalar such that* $|c| \leq 1$. *Then the* $M\Delta$-*system in Figure 8.3 is stable for all allowed perturbations (we have RS) if and only if*

$$\rho(M\Delta(j\omega)) < 1, \quad \forall \omega, \forall \Delta \tag{8.45}$$

or equivalently

$$\text{RS} \quad \Leftrightarrow \quad \max_{\Delta} \rho(M\Delta(j\omega)) < 1, \quad \forall \omega \tag{8.46}$$

Proof: (8.45) $\Rightarrow$ (8.43) ($\Leftrightarrow$ RS) is "obvious": it follows from the definition of the spectral radius ρ, and applies also to real Δ's.

(8.43) $\Rightarrow$ (8.45) is proved by proving that not(8.45) $\Rightarrow$ not(8.43). Assume there exists a perturbation Δ' such that $\rho(M\Delta') = 1$ at some frequency. Then $|\lambda_i(M\Delta')| = 1$ for some eigenvalue i, and there always exists another perturbation in the set, $\Delta'' = c\Delta'$, where c is a *complex* scalar with $|c| = 1$, such that $\lambda_i(M\Delta'') = +1$ (real and positive) and therefore $\det(I - M\Delta'') = \prod_i \lambda_i(I - M\Delta'') = \prod_i(1 - \lambda_i(M\Delta'')) = 0$. Finally, the equivalence between (8.45) and (8.46) is simply the definition of $\max_{\Delta}$. $\qquad\square$

Remark 1 The proof of (8.45) relies on adjusting the phase of $\lambda_i(Mc\Delta')$ using the complex scalar c and thus requires the perturbation to be complex.

Remark 2 In words, Theorem 8.2 tells us that we have stability if and only if the spectral radius of $M\Delta$ is less than 1 at all frequencies and for all allowed perturbations, Δ. The main problem here is of course that we have to test the condition for an infinite set of Δ's, and this is difficult to check numerically.

Remark 3 Theorem 8.1, which applies to both real and complex perturbations, forms the basis for the general definition of the structured singular value in (8.76).

8.6   Robust stability for complex unstructured uncertainty

In this section, we consider the special case where $\Delta(s)$ is allowed to be *any* (full) complex transfer function matrix satisfying $\|\Delta\|_{\infty} \leq 1$. This is often referred to as *unstructured uncertainty* or as *full-block complex perturbation uncertainty*.

Lemma 8.3 *Let* Δ *be the set of all complex matrices such that* $\bar{\sigma}(\Delta) \leq 1$. *Then the following holds:*

$$\max_{\Delta} \rho(M\Delta) = \max_{\Delta} \bar{\sigma}(M\Delta) = \max_{\Delta} \bar{\sigma}(\Delta)\bar{\sigma}(M) = \bar{\sigma}(M) \tag{8.47}$$

Proof: In general, the spectral radius (ρ) provides a lower bound on the spectral norm ($\bar{\sigma}$) (see (A.117)), and we have

$$\max_{\Delta} \rho(M\Delta) \leq \max_{\Delta} \bar{\sigma}(M\Delta) \leq \max_{\Delta} \bar{\sigma}(\Delta)\bar{\sigma}(M) = \bar{\sigma}(M) \tag{8.48}$$

where the second inequality in (8.48) follows since $\bar{\sigma}(AB) \leq \bar{\sigma}(A)\bar{\sigma}(B)$. Now, we need to show that we actually have equality. This will be the case if for any M there exists an allowed Δ' such that $\rho(M\Delta') = \bar{\sigma}(M)$. Such a Δ' does indeed exist if we allow Δ' to be a full matrix such that all directions in Δ' are allowed. Select $\Delta' = VU^H$ where U and V are matrices of the left and right singular vectors of $M = U\Sigma V^H$. Then $\bar{\sigma}(\Delta') = 1$ and $\rho(M\Delta') = \rho(U\Sigma V^H V U^H) = \rho(U\Sigma U^H) = \rho(\Sigma) = \bar{\sigma}(M)$. The second to last equality follows since $U^H = U^{-1}$ and the eigenvalues are invariant under similarity transformations. $\qquad\square$

Lemma 8.3 together with Theorem 8.2 directly yield the following theorem:

Theorem 8.4 RS for unstructured ("full") perturbations. *Assume that the nominal system $M(s)$ is stable (NS) and that the perturbations $\Delta(s)$ are stable. Then the $M\Delta$-system in Figure 8.3 is stable for all perturbations Δ satisfying $\|\Delta\|_\infty \leq 1$ (i.e. we have RS) if and only if*

$$\boxed{\bar{\sigma}(M(j\omega)) < 1 \quad \forall w} \qquad \Leftrightarrow \qquad \boxed{\|M\|_\infty < 1} \tag{8.49}$$

Remark 1 Condition (8.49) may be rewritten as

$$\text{RS} \Leftrightarrow \bar{\sigma}(M(j\omega))\,\bar{\sigma}(\Delta(j\omega)) < 1, \quad \forall\omega, \forall\Delta, \tag{8.50}$$

The sufficiency of (8.50) ($\Leftarrow$) also follows directly from the small-gain theorem by choosing $L = M\Delta$. The small-gain theorem applies to any operator norm satisfying $\|AB\| \leq \|A\| \cdot \|B\|$.

Remark 2 An important reason for using the $\mathcal{H}_\infty$ norm to analyze robust stability is that the stability condition in (8.50) is both necessary and sufficient. In contrast, use of the $\mathcal{H}_2$ norm, e.g. a condition like $\|M\|_2 < 1$, yields neither necessary nor sufficient conditions for stability. We do not get sufficiency since the $\mathcal{H}_2$ norm does *not* in general satisfy $\|AB\| \leq \|A\| \cdot \|B\|$; see e.g. Example 4.21.

8.6.1   Application of the unstructured RS condition

We will now present necessary and sufficient conditions for RS for each of the six single unstructured perturbations in Figure 8.5. with

$$E = W_2 \Delta W_1, \quad \|\Delta\|_\infty \leq 1 \tag{8.51}$$

To derive the matrix M, we simply "isolate" the perturbation, and determine the transfer function matrix

$$M = W_1 M_0 W_2 \tag{8.52}$$

from the output to the input of the perturbation, where M_0 for each of the six cases (disregarding some negative signs which do not affect the subsequent robustness condition) is given by

$$G_p = G + E_A : \quad M_0 = K(I + GK)^{-1} = KS \tag{8.53}$$
$$G_p = G(I + E_I) : \quad M_0 = K(I + GK)^{-1}G = T_I \tag{8.54}$$
$$G_p = (I + E_O)G : \quad M_0 = GK(I + GK)^{-1} = T \tag{8.55}$$
$$G_p = G(I - E_{iA}G)^{-1} : \quad M_0 = (I + GK)^{-1}G = SG \tag{8.56}$$
$$G_p = G(I - E_{iI})^{-1} : \quad M_0 = (I + KG)^{-1} = S_I \tag{8.57}$$
$$G_p = (I - E_{iO})^{-1}G : \quad M_0 = (I + GK)^{-1} = S \tag{8.58}$$

For example, (8.54) and (8.55) follow from the diagonal elements in the M-matrix in (8.34), and the others are derived in a similar fashion. Note that the sign of M_0 does not matter as it may be absorbed into Δ. Theorem 8.4 then yields

$$\text{RS} \quad \Leftrightarrow \quad \|W_1 M_0 W_2(j\omega)\|_\infty < 1 \tag{8.59}$$

For instance, from (8.54) and (8.59) we get for multiplicative input uncertainty with a scalar weight:

$$\text{RS} \; \forall G_p = G(I + w_I \Delta_I), \; \|\Delta_I\|_\infty \leq 1 \; \Leftrightarrow \; \|w_I T_I\|_\infty < 1 \qquad (8.60)$$

Note that the SISO condition (7.43) follows as a special case of (8.60). Similarly (7.64) follows as a special case of the inverse multiplicative output uncertainty in (8.58):

$$\text{RS} \; \forall G_p = (I - w_{iO}\Delta_{iO})^{-1}G, \; \|\Delta_{iO}\|_\infty \leq 1 \; \Leftrightarrow \; \|w_{iO}S\|_\infty < 1 \qquad (8.61)$$

In general, the unstructured uncertainty descriptions in terms of a single perturbation are not "tight" (in the sense that at each frequency all complex perturbations satisfying $\bar{\sigma}(\Delta(j\omega)) \leq 1$ may not occur in practice). Thus, the above RS conditions are often conservative. In order to get tighter conditions we must use a tighter uncertainty description in terms of a block-diagonal Δ.

8.6.2   RS for coprime factor uncertainty

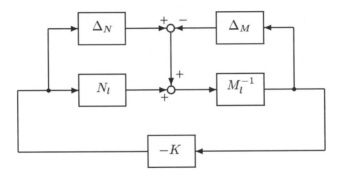

Figure 8.9: Coprime uncertainty

Robust stability bounds in terms of the $\mathcal{H}_\infty$ norm (RS $\Leftrightarrow \|M\|_\infty < 1$) are in general only tight when there is a single full perturbation block. An "exception" to this is when the uncertainty blocks enter or exit from the same location in the block diagram, because they can then be stacked on top of each other or side by side, in an overall Δ which is then a full matrix. If we norm-bound the combined (stacked) uncertainty, we then get a tight condition for RS in terms of $\|M\|_\infty$.

One important uncertainty description that falls into this category is the coprime uncertainty description shown in Figure 8.9, for which the set of plants is

$$G_p = (M_l + \Delta_M)^{-1}(N_l + \Delta_N), \quad \| [\, \Delta_N \quad \Delta_M \,] \|_\infty \leq \epsilon \qquad (8.62)$$

where $G = M_l^{-1}N_l$ is a left coprime factorization of the nominal plant, see (4.20). This uncertainty description is surprisingly general: it allows both zeros and poles to cross into the RHP, and has proved to be very useful in applications (McFarlane and Glover, 1990). Since we have no weights on the perturbations, it is reasonable to use a normalized coprime factorization of the nominal plant; see (4.25). In any case, to test for RS we can rearrange the block diagram to match the $M\Delta$-structure in Figure 8.3 with

$$\Delta = [\Delta_N \quad \Delta_M]; \quad M = - \begin{bmatrix} K \\ I \end{bmatrix} (I + GK)^{-1}M_l^{-1} \qquad (8.63)$$

We then get from Theorem 8.4

$$\text{RS} \ \forall \| \, \Delta_N \quad \Delta_M \, \|_\infty \le \epsilon \quad \Leftrightarrow \quad \|M\|_\infty < 1/\epsilon \tag{8.64}$$

The above RS result is central to the $\mathcal{H}_\infty$ loop-shaping design procedure discussed in Chapter 9.

The coprime uncertainty description provides a good "generic" uncertainty description for cases where we do not use any specific *a priori* uncertainty information. Note that the uncertainty magnitude is ϵ, so it is *not* normalized to be less than 1 in this case. This is because this uncertainty description is most often used in a controller design procedure where the objective is to maximize the magnitude of the uncertainty (ϵ) such that RS is maintained.

Remark. In (8.62) we bound the combined (stacked) uncertainty, $\|[\, \Delta_N \quad \Delta_M \,]\|_\infty \le \epsilon$, which is *not* quite the same as bounding the individual blocks, $\|\Delta_N\|_\infty \le \epsilon$ and $\|\Delta_M\|_\infty \le \epsilon$. However, from (A.46) we see that these two approaches differ at most by a factor of $\sqrt{2}$, so it is not an important issue from a practical point of view.

Exercise 8.13 * *Consider combined multiplicative and inverse multiplicative uncertainty at the output,* $G_p = (I - \Delta_{iO}W_{iO})^{-1}(I + \Delta_O W_O)G$, *where we choose to norm-bound the combined uncertainty,* $\|[\, \Delta_{iO} \quad \Delta_O \,]\|_\infty \le 1$. *Draw a block diagram of the uncertain plant, and derive a necessary and sufficient condition for RS of the closed-loop system.*

8.7 Robust stability with structured uncertainty: motivation

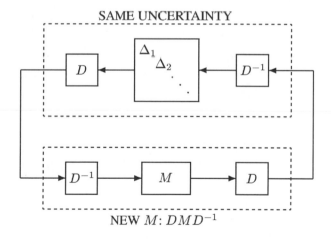

Figure 8.10: Use of block-diagonal scalings, $\Delta D = D\Delta$

Consider now the presence of structured uncertainty, where $\Delta = \text{diag}\{\Delta_i\}$ is block diagonal. To test for RS we rearrange the system into the $M\Delta$-structure and we have from (8.49)

$$\text{RS} \quad \text{if} \quad \bar{\sigma}(M(j\omega)) < 1, \forall \omega \tag{8.65}$$

We have written "if" here rather than "if and only if" since this condition is only sufficient for RS when Δ has "no structure" (full-block uncertainty). The question is whether we can take advantage of the fact that $\Delta = \text{diag}\{\Delta_i\}$ is structured to obtain an RS condition which is tighter than (8.65). One idea is to make use of the fact that stability must be independent of scaling. To this effect, we introduce the block-diagonal scaling matrix

$$D = \text{diag}\{d_i I_i\} \qquad (8.66)$$

where d_i is a scalar and I_i is an identity matrix of the same dimension as the i'th perturbation block, Δ_i. Now we rescale the inputs and outputs to M and Δ by inserting the matrices D and D^{-1} on both sides as shown in Figure 8.10. This clearly has no effect on stability. Next, note that with the chosen form for the scalings, we have for each perturbation block $\Delta_i = d_i \Delta_i d_i^{-1}$; that is, we have $\Delta = D\Delta D^{-1}$. This means that (8.65) must also apply if we replace M by DMD^{-1} (see Figure 8.10), and we have

$$\text{RS} \quad \text{if} \quad \bar{\sigma}(DMD^{-1}) < 1, \forall \omega \qquad (8.67)$$

This applies for any D in (8.66), and therefore the "most improved" (least conservative) RS condition is obtained by minimizing at each frequency the scaled singular value, and we have

$$\boxed{\text{RS} \quad \text{if} \quad \min_{D(\omega)\in\mathcal{D}} \bar{\sigma}(D(\omega)M(j\omega)D(\omega)^{-1}) < 1, \forall \omega} \qquad (8.68)$$

where $\mathcal{D}$ is the set of block-diagonal matrices whose structure is compatible to that of Δ, i.e. $\Delta D = D\Delta$. We will return with more examples of this compatibility later. Note that when Δ is a full matrix, we must select $D = dI$ and we have $\bar{\sigma}(DMD^{-1}) = \bar{\sigma}(M)$, and so as expected (8.68) is identical to (8.65). However, when Δ has structure, we get more degrees of freedom in D, and $\bar{\sigma}(DMD^{-1})$ may be significantly smaller than $\bar{\sigma}(M)$.

Remark 1 Historically, the RS condition in (8.68) directly motivated the introduction of the structured singular value, $\mu(M)$, discussed in detail in the next section. As one might guess, we have that $\mu(M) \leq \min_D \bar{\sigma}(DMD^{-1})$. In fact, for block-diagonal complex perturbations we generally have that $\mu(M)$ is very close to $\min_D \bar{\sigma}(DMD^{-1})$.

Remark 2 Other norms. Condition (8.68) is essentially a scaled version of the small-gain theorem. Thus, a similar condition applies when we use other matrix norms. The $M\Delta$-structure in Figure 8.3 is stable for all block-diagonal Δ's which satisfy $\|\Delta(j\omega)\| \leq 1, \forall w$ if

$$\min_{D(\omega)\in\mathcal{D}} \|D(\omega)M(j\omega)D(\omega)^{-1}\| < 1, \forall \omega \qquad (8.69)$$

where D as before is compatible with the block structure of Δ. Any matrix norm may be used; for example, the Frobenius norm, $\|M\|_F$, or any induced matrix norm such as $\|M\|_{i1}$ (maximum column sum), $\|M\|_{i\infty}$ (maximum row sum), or $\|M\|_{i2} = \bar{\sigma}(M)$, which is the one we will use. Although in some cases it may be convenient to use other norms, we usually prefer $\bar{\sigma}$ because for this norm we get a necessary and sufficient RS condition.

8.8   The structured singular value

The structured singular value (denoted Mu, mu, SSV or μ) is a function which provides a generalization of the singular value, $\bar{\sigma}$, and the spectral radius, ρ. We will use μ to get necessary and sufficient conditions for RS and also for RP. How is μ defined? A simple statement is:

Find the smallest structured Δ (measured in terms of $\bar{\sigma}(\Delta)$) which makes the matrix $I - M\Delta$ singular; then $\mu(M) = 1/\bar{\sigma}(\Delta)$.

Mathematically,

$$\mu(M)^{-1} \triangleq \min_{\Delta}\{\bar{\sigma}(\Delta)| \det(I - M\Delta) = 0 \text{ for structured } \Delta\} \tag{8.70}$$

Clearly, $\mu(M)$ depends not only on M but also on the allowed structure for Δ. This is sometimes shown explicitly by using the notation $\mu_{\Delta}(M)$.

Remark. For the case where Δ is "unstructured" (a full matrix), the smallest Δ which yields singularity has $\bar{\sigma}(\Delta) = 1/\bar{\sigma}(M)$, and we have $\mu(M) = \bar{\sigma}(M)$. A particular smallest Δ which achieves this is $\Delta = \frac{1}{\sigma_1}v_1 u_1^H$.

Example 8.5 **Full perturbation** (Δ *is unstructured*). *Consider*

$$M = \begin{bmatrix} 2 & 2 \\ -1 & -1 \end{bmatrix} = \begin{bmatrix} 0.894 & 0.447 \\ -0.447 & 0.894 \end{bmatrix} \begin{bmatrix} 3.162 & 0 \\ 0 & 0 \end{bmatrix} \begin{bmatrix} 0.707 & -0.707 \\ 0.707 & 0.707 \end{bmatrix}^H \tag{8.71}$$

The perturbation

$$\Delta = \frac{1}{\sigma_1}v_1 u_1^H = \frac{1}{3.162}\begin{bmatrix} 0.707 \\ 0.707 \end{bmatrix}[0.894 \quad -0.447] = \begin{bmatrix} 0.200 & -0.100 \\ 0.200 & -0.100 \end{bmatrix} \tag{8.72}$$

with $\bar{\sigma}(\Delta) = 1/\bar{\sigma}(M) = 1/3.162 = 0.316$ makes $\det(I - M\Delta) = 0$. Thus $\mu(M) = 3.162$ when Δ is a full matrix.

Note that the perturbation Δ in (8.72) is a full matrix. If we restrict Δ to be diagonal then we need a larger perturbation to make $\det(I - M\Delta) = 0$. This is illustrated next.

Example 8.5 continued. Diagonal perturbation (Δ *is structured*). *For the matrix M in (8.71), the smallest diagonal Δ which makes $\det(I - M\Delta) = 0$ is*

$$\Delta = \frac{1}{3}\begin{bmatrix} 1 & 0 \\ 0 & -1 \end{bmatrix} \tag{8.73}$$

with $\bar{\sigma}(\Delta) = 0.333$. Thus $\mu(M) = 3$ when Δ is a diagonal matrix.

The above example shows that μ depends on the structure of Δ. The following example demonstrates that μ also depends on whether the perturbation is real or complex.

Example 8.6 μ **of a scalar.** *If M is a scalar then in most cases $\mu(M) = |M|$. This follows from (8.70) by selecting $|\Delta| = 1/|M|$ such that $(1 - M\Delta) = 0$. However, this requires that we can select the phase of Δ such that $M\Delta$ is real, which is impossible when Δ is real and M has an imaginary component, so in this case $\mu(M) = 0$. In summary, we have for a scalar M*

$$\Delta \text{ complex}: \quad \mu(M) = |M| \tag{8.74}$$

$$\Delta \text{ real}: \quad \mu(M) = \begin{cases} |M| & \text{for real } M \\ 0 & \text{otherwise} \end{cases} \tag{8.75}$$

The definition of μ in (8.70) involves varying $\bar{\sigma}(\Delta)$. However, we prefer to normalize Δ such that $\bar{\sigma}(\Delta) \leq 1$. We can do this by scaling Δ by a factor k_m, and looking for the *smallest k_m* which makes the matrix $I - k_m M\Delta$ singular, and μ is then the reciprocal of this smallest k_m, i.e. $\mu = 1/k_m$. This results in the following alternative definition of μ.

Definition 8.1 Structured singular value. *Let M be a given complex matrix and let $\Delta = \mathrm{diag}\{\Delta_i\}$ denote a set of complex matrices with $\bar{\sigma}(\Delta) \leq 1$ and with a given block-diagonal structure (in which some of the blocks may be repeated and some may be restricted to be real). The real non-negative function $\mu(M)$, called the structured singular value, is defined by*

$$\mu(M) \triangleq \frac{1}{\min\{k_m|\det(I - k_m M\Delta) = 0 \text{ for structured } \Delta, \bar{\sigma}(\Delta) \leq 1\}} \qquad (8.76)$$

If no such structured Δ exists then $\mu(M) = 0$.

A value of $\mu = 1$ means that there exists a perturbation with $\bar{\sigma}(\Delta) = 1$ which is just large enough to make $I - M\Delta$ singular. A larger value of μ is "bad" as it means that a smaller perturbation makes $I - M\Delta$ singular, whereas a smaller value of μ is "good".

Exercise 8.14 *Find μ for the uncertain system in Figure 7.20(b).*

8.8.1   Remarks on the definition of μ

1. The structured singular value was introduced by Doyle (1982). At the same time (in fact, in the same issue of the same journal) Safonov (1982) introduced the *Multivariable Stability Margin* k_m for a diagonally perturbed system as the inverse of μ: that is, $k_m(M) = \mu(M)^{-1}$. In many respects, this is a more natural definition of a robustness margin. However, $\mu(M)$ has a number of other advantages, such as providing a generalization of the spectral radius, $\rho(M)$, and the spectral norm, $\bar{\sigma}(M)$.

2. The Δ corresponding to the smallest k_m in (8.76) will always have $\bar{\sigma}(\Delta) = 1$, since if $\det(I - k_m' M\Delta') = 0$ for some Δ' with $\bar{\sigma}(\Delta') = c < 1$, then $1/k_m'$ cannot be the structured singular value of M, since there exists a smaller scalar $k_m = k_m' c$ such that $\det(I - k_m M\Delta) = 0$ where $\Delta = \frac{1}{c}\Delta'$ and $\bar{\sigma}(\Delta) = 1$.

3. Note that with $k_m = 0$ we obtain $I - k_m M\Delta = I$ which is clearly non-singular. Thus, one possible way to obtain μ numerically is to start with $k_m = 0$, and gradually increase k_m until we first find an allowed Δ with $\bar{\sigma}(\Delta) = 1$ such that $(I - k_m M\Delta)$ is singular (this value of k_m is then $1/\mu$). By "allowed" we mean that Δ must have the specified block-diagonal structure and that some of the blocks may have to be real.

4. The sequence of M and Δ in the definition of μ does not matter. This follows from the identity (A.12) which yields

$$\det(I - k_m M\Delta) = \det(I - k_m \Delta M) \qquad (8.77)$$

5. In most cases M and Δ are square, but this need not be the case. If they are non-square, then we make use of (8.77) and work with either $M\Delta$ or ΔM (whichever has the lowest dimension).

The remainder of this section deals with the properties and computation of μ. Readers who are primarily interested in the practical use of μ may skip most of this material.

8.8.2   Properties of μ for real and complex Δ

Two properties of μ which hold for both real and complex perturbations Δ are:

1. $\mu(\alpha M) = |\alpha|\mu(M)$ for any *real* scalar α.
2. Let $\Delta = \mathrm{diag}\{\Delta_1, \Delta_2\}$ be a block-diagonal perturbation (in which Δ_1 and Δ_2 may have additional structure) and let M be partitioned accordingly. Then

$$\mu_\Delta(M) \geq \max\{\mu_{\Delta_1}(M_{11}), \mu_{\Delta_2}(M_{22})\} \qquad (8.78)$$

Proof: Consider $\det(I - \frac{1}{\mu}M\Delta)$ where $\mu = \mu_\Delta(M)$ and use Schur's formula in (A.14) with $A_{11} = I - \frac{1}{\mu}M_{11}\Delta_1$ and $A_{22} = I - \frac{1}{\mu}M_{22}\Delta_2$. $\square$

In words, (8.78) simply says that robustness with respect to two perturbations taken together is at least as bad as for the worst perturbation considered alone. This agrees with our intuition that we cannot improve RS by including another uncertain perturbation.

In addition, the *upper bounds* given below for complex perturbations, e.g. $\mu_\Delta(M) \leq \min_{D \in \mathcal{D}} \bar{\sigma}(DMD^{-1})$ in (8.87), also hold for real or mixed real/complex perturbations Δ. This follows because complex perturbations include real perturbations as a special case. However, the lower bounds, e.g. $\mu(M) \geq \rho(M)$ in (8.82), generally hold only for complex perturbations.

8.8.3 μ for complex Δ

When all the blocks in Δ are complex, μ may be computed relatively easily. This is discussed below and in more detail in the survey paper by Packard and Doyle (1993). The results are mainly based on the following result, which may be viewed as another definition of μ that applies for complex Δ only.

Lemma 8.5 *For complex perturbations Δ with $\bar{\sigma}(\Delta) \leq 1$:*

$$\mu(M) = \max_{\Delta, \bar{\sigma}(\Delta) \leq 1} \rho(M\Delta) \tag{8.79}$$

Proof: The lemma follows directly from the definition of μ and the equivalence between (8.43) and (8.46). $\square$

Properties of μ for complex perturbations

Most of the properties below follow easily from (8.79).

1. $\mu(\alpha M) = |\alpha|\mu(M)$ for any (*complex*) scalar α.
2. For a repeated scalar complex perturbation we have

$$\Delta = \delta I \ (\delta \text{ is a complex scalar}): \quad \mu(M) = \rho(M) \tag{8.80}$$

Proof: Follows directly from (8.79) since there are no degrees of freedom for the maximization. $\square$

3. For a full-block complex perturbation we have from (8.79) and (8.47)

$$\Delta \text{ full matrix}: \quad \mu(M) = \bar{\sigma}(M) \tag{8.81}$$

4. μ for complex perturbations is bounded by the spectral radius and the singular value (spectral norm):

$$\rho(M) \leq \mu(M) \leq \bar{\sigma}(M) \tag{8.82}$$

This follows from (8.80) and (8.81), since selecting $\Delta = \delta I$ gives the fewest degrees of freedom for the optimization in (8.79), whereas selecting Δ full gives the most degrees of freedom.

5. Consider any unitary matrix U with the same structure as Δ. Then

$$\mu(MU) = \mu(M) = \mu(UM) \qquad (8.83)$$

Proof: Follows from (8.79) by writing $MU\Delta = M\Delta'$ where $\bar{\sigma}(\Delta') = \bar{\sigma}(U\Delta) = \bar{\sigma}(\Delta)$, and so U may always be absorbed into Δ.   □

6. Consider any matrix D which commutes with Δ: that is, $\Delta D = D\Delta$. Then

$$\mu(DM) = \mu(MD) \quad \text{and} \quad \mu(DMD^{-1}) = \mu(M) \qquad (8.84)$$

Proof: $\mu(DM) = \mu(MD)$ follows from

$$\mu_\Delta(DM) = \max_\Delta \rho(DM\Delta) = \max_\Delta \rho(M\Delta D) = \max_\Delta \rho(MD\Delta) = \mu_\Delta(MD) \qquad (8.85)$$

The first equality is (8.79). The second equality applies since $\rho(AB) = \rho(BA)$ (by the eigenvalue properties in the Appendix). The key step is the third equality which applies only when $D\Delta = \Delta D$. The fourth equality again follows from (8.79).   □

7. **Improved lower bound.** Define $\mathcal{U}$ as the set of all unitary matrices U with the same block-diagonal structure as Δ. Then for complex Δ

$$\boxed{\mu(M) = \max_{U \in \mathcal{U}} \rho(MU)} \qquad (8.86)$$

Proof: The proof of this important result is given by Doyle (1982) and Packard and Doyle (1993). It follows from a generalization of the maximum modulus theorem for rational functions.   □

The result (8.86) is motivated by combining (8.83) and (8.82) to yield

$$\mu(M) \geq \max_{U \in \mathcal{U}} \rho(MU)$$

The surprise is that this is always an equality. Unfortunately, the optimization in (8.86) is not convex and so it may be difficult to use in calculating μ numerically.

8. **Improved upper bound.** Define $\mathcal{D}$ to be the set of matrices D which commute with Δ (i.e. satisfy $D\Delta = \Delta D$). Then it follows from (8.84) and (8.82) that

$$\boxed{\mu(M) \leq \min_{D \in \mathcal{D}} \bar{\sigma}(DMD^{-1})} \qquad (8.87)$$

This optimization is convex in D, i.e. has only one minimum, the global minimum; see Example 12.4 for the formulation of the optimization problem. It may be shown (Doyle, 1982) that the inequality is in fact an equality if there are three or fewer blocks in Δ. Furthermore, numerical evidence suggests that the bound is tight (within a few per cent) for four blocks or more; the worst known example to us has an upper bound which is about 15% larger than μ (Balas et al., 1993).

Some examples of D's which commute with Δ are

$$\Delta = \delta I: \quad D = \text{full matrix} \qquad (8.88)$$

$$\Delta = \text{full matrix}: \quad D = dI \qquad (8.89)$$

$$\Delta = \begin{bmatrix} \Delta_1(\text{full}) & 0 \\ 0 & \Delta_2(\text{full}) \end{bmatrix} : D = \begin{bmatrix} d_1 I & 0 \\ 0 & d_2 I \end{bmatrix} \qquad (8.90)$$

$$\Delta = \text{diag}\{\Delta_1(\text{full}), \delta_2 I, \delta_3, \delta_4\} : D = \text{diag}\{d_1 I, D_2(\text{full}), d_3, d_4\} \qquad (8.91)$$

In short, we see that the structures of Δ and D are "opposites".

9. Without affecting the optimization in (8.87), we may assume the blocks in D to be Hermitian positive definite, i.e. $D_i = D_i^H > 0$, and for scalars $d_i > 0$ (Packard and Doyle, 1993).

10. One can always simplify the optimization in (8.87) by fixing one of the scalar blocks in D equal to 1. For example, let $D = \text{diag}\{d_1, d_2, \ldots, d_n\}$, then one may without loss of generality set $d_n = 1$.

Proof: Let $D' = \frac{1}{d_n} D$ and note that $\bar{\sigma}(DMD^{-1}) = \bar{\sigma}(D'MD'^{-1})$.   □

Similarly, for cases where Δ has one or more scalar blocks, one may simplify the optimization in (8.86) by fixing one of the corresponding unitary scalars in U equal to 1. This follows from Property 1 with $|c| = 1$.

11. The following property is useful for finding $\mu(AB)$ when Δ has a structure similar to that of A or B:

$$\mu_\Delta(AB) \leq \bar{\sigma}(A)\mu_{\Delta A}(B) \tag{8.92}$$

$$\mu_\Delta(AB) \leq \bar{\sigma}(B)\mu_{B\Delta}(A) \tag{8.93}$$

Here the subscript "ΔA" denotes the structure of the matrix ΔA, and "$B\Delta$" denotes the structure of $B\Delta$.

Proof: The proof is from Skogestad and Morari (1988a). We use the fact that $\mu(AB) = \max_\Delta \rho(\Delta AB) = \max_\Delta \rho(VB)\bar{\sigma}(\Lambda)$ where $V = \Delta\Lambda/\bar{\sigma}(\Lambda)$. When we maximize over Δ, V generates a certain set of matrices with $\bar{\sigma}(V) \leq 1$. Let us *extend* this set by maximizing over *all* matrices V with $\bar{\sigma}(V) \leq 1$ and with the same structure as ΔA. We then get $\mu(AB) \leq \max_V \rho(VB)\bar{\sigma}(A) = \mu_V(B)\bar{\sigma}(A)$.   □

Some special cases of (8.92):

(a) If A is a full matrix then the structure of ΔA is a full matrix, and we simply get $\mu(AB) \leq \bar{\sigma}(A)\bar{\sigma}(B)$ (which is not a very exciting result since we always have $\mu(AB) \leq \bar{\sigma}(AB) \leq \bar{\sigma}(A)\bar{\sigma}(B)$).

(b) If Δ has the same structure as A (e.g. they are both diagonal) then

$$\boxed{\mu_\Delta(AB) \leq \bar{\sigma}(A)\mu_\Delta(B)} \tag{8.94}$$

Note: (8.94) is stated incorrectly in Doyle (1982) since it is not specified that Δ must have the same structure as A; see also Exercise 8.20 (page 313).

(c) If $\Delta = \delta I$ (i.e. Δ consists of repeated scalars), we get the spectral radius inequality $\rho(AB) \leq \bar{\sigma}(A)\mu_A(B)$. A useful special case of this is

$$\rho(M\Delta) \leq \bar{\sigma}(\Delta)\mu_\Delta(M) \tag{8.95}$$

12. A generalization of (8.92) and (8.93) is

$$\mu_\Delta(ARB) \leq \bar{\sigma}(R)\mu_{\tilde{\Delta}}^2 \begin{bmatrix} 0 & A \\ B & 0 \end{bmatrix} \tag{8.96}$$

where $\tilde{\Delta} = \text{diag}\{\Delta, R\}$. The result is proved by Skogestad and Morari (1988a).

13. The following is a further generalization of these bounds. Assume that M is an LFT of R: $M = N_{11} + N_{12}R(I - N_{22}R)^{-1}N_{21}$. The problem is to find an upper bound on R,

$\bar{\sigma}(R) \le c$, which guarantees that $\mu_\Delta(M) < 1$ when $\mu_\Delta(N_{11}) < 1$. Skogestad and Morari (1988a) show that the best upper bound is the c which solves

$$\mu_{\widetilde{\Delta}} \begin{bmatrix} N_{11} & N_{12} \\ cN_{21} & cN_{22} \end{bmatrix} = 1 \tag{8.97}$$

where $\widetilde{\Delta} = \text{diag}\{\Delta, R\}$, and c is easily computed using skewed-μ. Given the μ-condition $\mu_\Delta(M) < 1$ (for RS or RP), (8.97) may be used to derive a sufficient loop-shaping bound on a transfer function of interest, e.g. R may be S, T, L, L^{-1} or K.

Remark. In the above we have used $\min_D$. To be mathematically correct, we should have used $\inf_D$ because the set of allowed D's is not bounded and therefore the exact minimum may not be achieved (although we may get arbitrarily close). The use of $\max_\Delta$ (rather than $\sup_\Delta$) is mathematically correct since the set Δ is closed (with $\bar{\sigma}(\Delta) \le 1$).

Example 8.7 *Let*

$$M = \begin{bmatrix} a & a \\ b & b \end{bmatrix} \tag{8.98}$$

and Δ be complex 2×2 matrices. Then

$$\mu(M) = \begin{cases} \rho(M) = |a+b| & \text{for } \Delta = \delta I \\ |a| + |b| & \text{for } \Delta = \text{diag}\{\delta_1, \delta_2\} \\ \bar{\sigma}(M) = \sqrt{2|a|^2 + 2|b|^2} & \text{for } \Delta \text{ a full matrix} \end{cases} \tag{8.99}$$

Proof: For $\Delta = \delta I$, $\mu(M) = \rho(M)$ and $\rho(M) = |a+b|$ since M is singular and its non-zero eigenvalue is $\lambda_1(M) = \text{tr}(M) = a + b$. For Δ full, $\mu(M) = \bar{\sigma}(M)$ and $\bar{\sigma}(M) = \sqrt{2|a|^2 + 2|b|^2}$ since M is singular and its non-zero singular value is $\bar{\sigma}(M) = \|M\|_F$, see (A.127). For a diagonal Δ, it is interesting to consider three different proofs of the result $\mu(M) = |a| + |b|$:

(a) A direct calculation based on the definition of μ.
(b) Use of the lower "bound" in (8.86) (which is always exact).
(c) Use of the upper bound in (8.87) (which is exact here since we have only two blocks).

We will use approach (a) here and leave (b) and (c) for Exercise 8.15. We have

$$M\Delta = \begin{bmatrix} a & a \\ b & b \end{bmatrix} \begin{bmatrix} \delta_1 & \\ & \delta_2 \end{bmatrix} = \begin{bmatrix} a \\ b \end{bmatrix} [\delta_1 \quad \delta_2] = \widetilde{M}\widetilde{\Delta}$$

From (8.77) we then get

$$\det(I - M\Delta) = \det(I - \widetilde{\Delta}\widetilde{M}) = 1 - [\delta_1 \quad \delta_2]\begin{bmatrix} a \\ b \end{bmatrix} = 1 - a\delta_1 - b\delta_2$$

The smallest δ_1 and δ_2 which make this matrix singular, i.e. $1 - a\delta_1 - b\delta_2 = 0$, are obtained when $|\delta_1| = |\delta_2| = |\delta|$ and the phases of δ_1 and δ_2 are adjusted such that $1 - |a| \cdot |\delta| - |b| \cdot |\delta| = 0$. We get $|\delta| = 1/(|a| + |b|)$, and from (8.70) we have that $\mu = 1/|\delta| = |a| + |b|$. □

Exercise 8.15[*] *(continued from Example 8.7). (b) For M in (8.98) and a diagonal Δ show that $\mu(M) = |a| + |b|$ using the lower "bound" $\mu(M) = \max_U \rho(MU)$ (which is always exact). (Hint: Use $U = \text{diag}\{e^{j\phi}, 1\}$ (the blocks in U are unitary scalars, and we may fix one of them equal to 1).)*

(c) For M in (8.98) and a diagonal Δ show that $\mu(M) = |a| + |b|$ using the upper bound $\mu(M) \le \min_D \bar{\sigma}(DMD^{-1})$ (which is exact in this case since D has two "blocks").

Solution: Use $D = \text{diag}\{d, 1\}$. Since DMD^{-1} is a singular matrix we have from (A.37) that

$$\bar{\sigma}(DMD^{-1}) = \bar{\sigma}\begin{bmatrix} a & da \\ \frac{1}{d}b & b \end{bmatrix} = \sqrt{|a|^2 + |da|^2 + |b/d|^2 + |b|^2} \tag{8.100}$$

which we want to minimize with respect to d. The solution is $d = \sqrt{|b|/|a|}$ which gives $\mu(M) = \sqrt{|a|^2 + 2|ab| + |b|^2} = |a| + |b|$.

Exercise 8.16 *Let c be a complex scalar. Show that for*

$$\Delta = \text{diag}\{\Delta_1, \Delta_2\}: \quad \mu\begin{bmatrix} M_{11} & M_{12} \\ M_{21} & M_{22} \end{bmatrix} = \mu\begin{bmatrix} M_{11} & cM_{12} \\ \frac{1}{c}M_{21} & M_{22} \end{bmatrix} \quad (8.101)$$

Example 8.8 *Let M be a partitioned matrix with both diagonal blocks equal to zero. Then*

$$\mu\underbrace{\begin{bmatrix} 0 & A \\ B & 0 \end{bmatrix}}_{M} = \begin{cases} \rho(M) = \sqrt{\rho(AB)} & \text{for } \Delta = \delta I \\ \sqrt{\bar{\sigma}(A)\bar{\sigma}(B)} & \text{for } \Delta = \text{diag}\{\Delta_1, \Delta_2\}, \Delta_i \text{ full} \\ \bar{\sigma}(M) = \max\{\bar{\sigma}(A), \bar{\sigma}(B)\} & \text{for } \Delta \text{ a full matrix} \end{cases} \quad (8.102)$$

Proof: From the definition of eigenvalues and Schur's formula (A.14) we get $\lambda_i(M) = \sqrt{\lambda_i(AB)}$ and $\rho(M) = \sqrt{\rho(AB)}$ follows. For block-diagonal Δ, $\mu(M) = \sqrt{\bar{\sigma}(A)\bar{\sigma}(B)}$ follows in a similar way using $\mu(M) = \max_\Delta \rho(M\Delta) = \max_{\Delta_1,\Delta_2} \rho(A\Delta_2 B\Delta_1)$, and then realizing that we can always select Δ_1 and Δ_2 such that $\rho(A\Delta_2 B\Delta_1) = \bar{\sigma}(A)\bar{\sigma}(B)$ (recall (8.47)). $\bar{\sigma}(M) = \max\{\bar{\sigma}(A), \bar{\sigma}(B)\}$ follows since $\bar{\sigma}(M) = \sqrt{\rho(M^H M)}$ where $M^H M = \text{diag}\{B^H B, A^H A\}$. □

Exercise 8.17 *Let M be a complex 3×3 matrix and $\Delta = \text{diag}\{\delta_1, \delta_2, \delta_3\}$. Prove that*

$$M = \begin{bmatrix} a & a & a \\ b & b & b \\ c & c & c \end{bmatrix}, \quad \mu(M) = |a| + |b| + |c|$$

Exercise 8.18 * *Let a, b, c and d be complex scalars. Show that for*

$$\Delta = \text{diag}\{\delta_1, \delta_2\}: \quad \mu\begin{bmatrix} ab & ad \\ bc & cd \end{bmatrix} = \mu\begin{bmatrix} ab & ab \\ cd & cd \end{bmatrix} = |ab| + |cd| \quad (8.103)$$

Does this hold when Δ is scalar times identity, or when Δ is full? (Answers: No and No.)

Exercise 8.19 *Assume A and B are square matrices. Show by a counterexample that $\bar{\sigma}(AB)$ is not in general equal to $\bar{\sigma}(BA)$. Under what conditions is $\mu(AB) = \mu(BA)$? (Hint: Recall (8.84).)*

Exercise 8.20 * *If (8.94) were true for any structure of Δ then it would imply $\rho(AB) \leq \bar{\sigma}(A)\rho(B)$. Show by a counterexample that this is not true.*

8.9 Robust stability with structured uncertainty

Consider stability of the $M\Delta$-structure in Figure 8.3 for the case where Δ is a set of norm-bounded block-diagonal perturbations. From the determinant stability condition in (8.43) which applies to both complex and real perturbations we get

$$\text{RS} \quad \Leftrightarrow \quad \det(I - M\Delta(j\omega)) \neq 0, \quad \forall \omega, \forall \Delta, \bar{\sigma}(\Delta(j\omega)) \leq 1 \quad \forall \omega \quad (8.104)$$

A problem with (8.104) is that it is only a "yes/no" condition. To find the factor k_m by which the system is robustly stable, we scale the uncertainty Δ by k_m, and look for the smallest k_m which yields "borderline instability", namely

$$\det(I - k_m M\Delta) = 0 \quad (8.105)$$

From the definition of μ in (8.76) this value is $k_m = 1/\mu(M)$, and we obtain the following necessary and sufficient condition for robust stability.

Theorem 8.6 RS for block-diagonal perturbations (real or complex). *Assume that the nominal system M and the perturbations Δ are stable. Then the $M\Delta$-system in Figure 8.3 is stable for all allowed perturbations with $\bar{\sigma}(\Delta) \leq 1, \forall \omega$, if and only if*

$$\boxed{\mu(M(j\omega)) < 1, \quad \forall \omega} \tag{8.106}$$

Proof: $\mu(M) < 1 \Leftrightarrow k_m > 1$, so if $\mu(M) < 1$ at all frequencies the required perturbation Δ to make $\det(I - M\Delta) = 0$ is larger than 1, and the system is stable. On the other hand, $\mu(M) = 1 \Leftrightarrow k_m = 1$, so if $\mu(M) = 1$ at some frequency there does exist a perturbation with $\bar{\sigma}(\Delta) = 1$ such that $\det(I - M\Delta) = 0$ at this frequency, and the system is unstable. $\square$

Condition (8.106) for RS may be rewritten as

$$\text{RS} \quad \Leftrightarrow \quad \mu(M(j\omega))\,\bar{\sigma}(\Delta(j\omega)) < 1, \quad \forall \omega \tag{8.107}$$

which may be interpreted as a "generalized small-gain theorem" that also takes into account the *structure* of Δ.

One may argue whether Theorem 8.6 is really a theorem, or a restatement of the definition of μ. In either case, we see from (8.106) that it is trivial to check for RS provided we can compute μ.

Let us consider two examples that illustrate how we use μ to check for RS with structured uncertainty. In the first example, the structure of the uncertainty is important, and an analysis based on the $\mathcal{H}_\infty$ norm leads to the incorrect conclusion that the system is not robustly stable. In the second example the structure makes no difference.

Example 8.9 RS with diagonal input uncertainty. *Consider RS of the feedback system in*

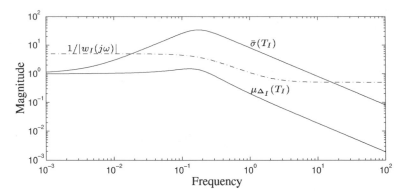

Figure 8.11: RS for diagonal input uncertainty is guaranteed since $\mu_{\Delta_I}(T_I) < 1/|w_I|$, $\forall \omega$. The use of unstructured uncertainty and $\bar{\sigma}(T_I)$ is conservative.

Figure 8.7 for the case when the multiplicative input uncertainty is diagonal. A nominal 2×2 plant and the controller (which represents PI control of a distillation process using the DV-configuration) is given by

$$G(s) = \frac{1}{\tau s + 1}\begin{bmatrix} -87.8 & 1.4 \\ -108.2 & -1.4 \end{bmatrix}; \quad K(s) = \frac{1 + \tau s}{s}\begin{bmatrix} -0.0015 & 0 \\ 0 & -0.075 \end{bmatrix} \tag{8.108}$$

(time in minutes). The controller results in a nominally stable system with acceptable performance. Assume there is complex multiplicative uncertainty in each manipulated input of magnitude

$$w_I(s) = \frac{s + 0.2}{0.5s + 1} \tag{8.109}$$

This implies a relative uncertainty of up to 20% in the low-frequency range, which increases at high frequencies, reaching a value of 1 (100% uncertainty) at about 1 rad/min. The increase with frequency allows for various neglected dynamics associated with the actuator and valve. The uncertainty may be represented as multiplicative input uncertainty as shown in Figure 8.7 where Δ_I is a diagonal complex matrix and the weight is $W_I = w_I I$ where $w_I(s)$ is a scalar. On rearranging the block diagram to match the $M\Delta$-structure in Figure 8.3 we get $M = w_I KG(I + KG)^{-1} = w_I T_I$ (recall (8.32)), and the RS condition $\mu(M) < 1$ in Theorem 8.6 yields

$$\text{RS} \Leftrightarrow \mu_{\Delta_I}(T_I) < \frac{1}{|w_I(j\omega)|} \quad \forall \omega, \quad \Delta_I = \begin{bmatrix} \delta_1 & \\ & \delta_2 \end{bmatrix} \tag{8.110}$$

This condition is shown graphically in Figure 8.11 and is seen to be satisfied at all frequencies, so the system is robustly stable. Also in Figure 8.11, $\bar{\sigma}(T_I)$ can be seen to be larger than $1/|w_I(j\omega)|$ over a wide frequency range. This shows that the system would be unstable for full-block input uncertainty (Δ_I full). However, full-block uncertainty is not reasonable for this plant, and therefore we conclude that the use of the singular value is conservative in this case. This demonstrates the need for the structured singular value.

Exercise 8.21 *Consider the same example and check for RS with full-block multiplicative output uncertainty of the same magnitude. (Solution: RS is satisfied.)*

Example 8.10 RS of spinning satellite. *Recall Motivating example no. 1 from Section 3.7.1 with the plant $G(s)$ given in (3.88) and the controller $K = I$. We want to study how sensitive this design is to multiplicative input uncertainty.*

In this case $T_I = T$, so for RS there is no difference between multiplicative input and multiplicative output uncertainty. In Figure 8.12, we plot $\mu(T)$ as a function of frequency. We find for this case that $\mu(T) = \bar{\sigma}(T)$ irrespective of the structure of the complex multiplicative perturbation (full-block, diagonal or repeated complex scalar). Since $\mu(T)$ crosses 1 at about 10 rad/s, we can tolerate more than 100% uncertainty at frequencies above 10 rad/s. At low frequencies $\mu(T)$ is about 10, so to guarantee RS we can at most tolerate 10% (complex) uncertainty at low frequencies. This confirms the results

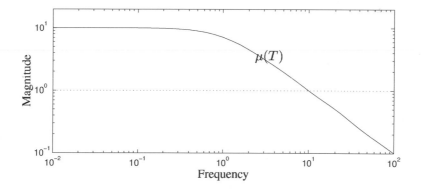

Figure 8.12: *μ-plot for spinning satellite*

from Section 3.7.1, where we found that real perturbations $\delta_1 = 0.1$ and $\delta_2 = -0.1$ yield instability. Thus, the use of complex rather than real perturbations is not conservative in this case, at least for Δ_I diagonal.

However, with repeated scalar perturbations (i.e. the uncertainty in each channel is identical) there is a difference between real and complex perturbations. With repeated real perturbations, available software (e.g. using the command mussv *with* blk = [-2 0] *in the Robust Control toolbox in Matlab) yields a peak μ-value of 1, so we can tolerate a perturbation $\delta_1 = \delta_2$ of magnitude 1 before getting instability. (This is confirmed by considering the characteristic polynomial in (3.92), from which we see that $\delta_1 = \delta_2 = -1$ yields instability.) On the other hand, with complex repeated perturbations, we have that $\mu(T) = \rho(T)$ is 10 at low frequencies, so instability may occur with a (non-physical) complex $\delta_1 = \delta_2$ of magnitude 0.1. (Indeed, from (3.92) we see that the non-physical constant perturbation $\delta_1 = \delta_2 = j0.1$ yields instability.)*

8.9.1   What do $\mu \neq 1$ and skewed-μ mean?

A value of $\mu = 1.1$ for RS means that *all* the uncertainty blocks must be decreased in magnitude by a factor 1.1 in order to guarantee stability.

But if we want to keep some of the uncertainty blocks fixed, how large can one particular source of uncertainty be before we get instability? We define this value as $1/\mu^s$, where μ^s is called *skewed-μ*. We may view $\mu^s(M)$ as a generalization of $\mu(M)$.

For example, let $\Delta = \mathrm{diag}\{\Delta_1, \Delta_2\}$ and assume we have fixed $\|\Delta_1\| \leq 1$ and we want to find how large Δ_2 can be before we get instability. The solution is to select

$$K_m = \begin{bmatrix} I & 0 \\ 0 & k_m I \end{bmatrix} \qquad (8.111)$$

and look at each frequency for the smallest value of k_m which makes $\det(I - K_m M \Delta) = 0$, and we have that skewed-μ is

$$\mu^s(M) \triangleq 1/k_m$$

Note that to compute skewed-μ we must first define which part of the perturbations is to be constant. $\mu^s(M)$ is always further from 1 than $\mu(M)$ is, i.e. $\mu^s \geq \mu$ for $\mu > 1$, $\mu^s = \mu$ for $\mu = 1$, and $\mu^s \leq \mu$ for $\mu < 1$. In practice, with available software to compute μ, we obtain μ^s by iterating on k_m until $\mu(K_m M) = 1$ where K_m may be as in (8.111). This iteration is straightforward since μ increases uniformly with k_m.

8.10   Robust performance

Robust performance (RP) means that the performance objective is satisfied for all possible plants in the uncertainty set, even the worst-case plant. We showed in Chapter 7 that for a SISO system with an $\mathcal{H}_\infty$ performance objective, the RP condition is identical to an RS condition with an additional perturbation block (!).

This also holds for MIMO systems, as illustrated by the stepwise derivation in Figure 8.13. Step B is the key step and the reader is advised to study this carefully in the treatment below. Note that the block Δ_P (where capital P denotes Performance) is always a full matrix. It is a fictitious uncertainty block representing the $\mathcal{H}_\infty$ performance specification.

8.10.1   Testing RP using μ

To test for RP, we first "pull out" the uncertain perturbations and rearrange the uncertain system into the $N\Delta$-form of Figure 8.2. Our RP requirement, as given in (8.39), is that the $\mathcal{H}_\infty$ norm of the transfer function $F = F_u(N, \Delta)$ remains less than 1 for all allowed perturbations. This may be tested exactly by computing $\mu(N)$ as stated in the following theorem.

Theorem 8.7 Robust performance. *Rearrange the uncertain system into the $N\Delta$-structure of Figure 8.13. Assume NS such that N is (internally) stable. Then*

$$\text{RP} \quad \overset{\text{def}}{\Leftrightarrow} \quad \|F\|_\infty = \|F_u(N,\Delta)\|_\infty < 1, \quad \forall \|\Delta\|_\infty \le 1 \qquad (8.112)$$

$$\Leftrightarrow \quad \boxed{\mu_{\widehat{\Delta}}(N(j\omega)) < 1, \quad \forall w} \qquad (8.113)$$

where μ is computed with respect to the structure

$$\widehat{\Delta} = \begin{bmatrix} \Delta & 0 \\ 0 & \Delta_P \end{bmatrix} \qquad (8.114)$$

and Δ_P is a full complex perturbation with the same dimensions as F^T.

Below we prove the theorem in two alternative ways, but first a few remarks:

1. Condition (8.113) allows us to test if $\|F\|_\infty < 1$ for all possible Δ's without having to test each Δ individually. Essentially, μ is defined such that it directly addresses the worst case.
2. The μ-condition for RP involves the enlarged perturbation $\widehat{\Delta} = \text{diag}\{\Delta, \Delta_P\}$. Here Δ, which itself may be a block-diagonal matrix, represents the true uncertainty, whereas Δ_P is a *full complex matrix* stemming from the $\mathcal{H}_\infty$ norm performance specification. For example, for the nominal system (with $\Delta = 0$) we get from (8.81) that $\bar{\sigma}(N_{22}) = \mu_{\Delta_P}(N_{22})$, and we see that Δ_P must be a full matrix.
3. Since $\widehat{\Delta}$ always has structure, the use of the $\mathcal{H}_\infty$ norm, $\|N\|_\infty < 1$, is generally conservative for RP.
4. From (8.78) we have that

$$\underbrace{\mu_{\widehat{\Delta}}(N)}_{\text{RP}} \ge \max\{\underbrace{\mu_\Delta(N_{11})}_{\text{RS}}, \underbrace{\mu_{\Delta_P}(N_{22})}_{\text{NP}}\} \qquad (8.115)$$

where as just noted $\mu_{\Delta_P}(N_{22}) = \bar{\sigma}(N_{22})$. Condition (8.115) implies that RS ($\mu_\Delta(N_{11}) < 1$) and NP ($\bar{\sigma}(N_{22}) < 1$) are automatically satisfied when RP ($\mu(N) < 1$) is satisfied. However, note that NS (stability of N) is not guaranteed by (8.113) and must be tested separately. (Beware! It is a common mistake to get a design with apparently great RP, but which is not nominally stable and thus is actually robustly *unstable*.)
5. For a generalization of Theorem 8.7 see the *main loop theorem* of Packard and Doyle (1993); see also Zhou et al. (1996).

Block diagram proof of Theorem 8.7

In the following, let $F = F_u(N, \Delta)$ denote the perturbed closed-loop system for which we want to test RP. The theorem is proved by the equivalence between the various block diagrams in Figure 8.13.

Step A. This is simply the definition of RP: $\|F\|_\infty < 1$.

Step B (the key step). Recall first from Theorem 8.4 that stability of the $M\Delta$-structure in Figure 8.3, where Δ is a *full* complex matrix, is equivalent to $\|M\|_\infty < 1$. From this theorem, we get that the RP condition $\|F\|_\infty < 1$ is equivalent to RS of the $F\Delta_P$-structure, where Δ_P is a *full* complex matrix.

Step C. Introduce $F = F_u(N, \Delta)$ from Figure 8.2.

Step D. Collect Δ and Δ_P into the block-diagonal matrix $\widehat{\Delta}$. Then the original RP problem is equivalent to RS of the $N\widehat{\Delta}$-structure which from Theorem 8.6 is equivalent to $\mu_{\widehat{\Delta}}(N) < 1$. $\square$

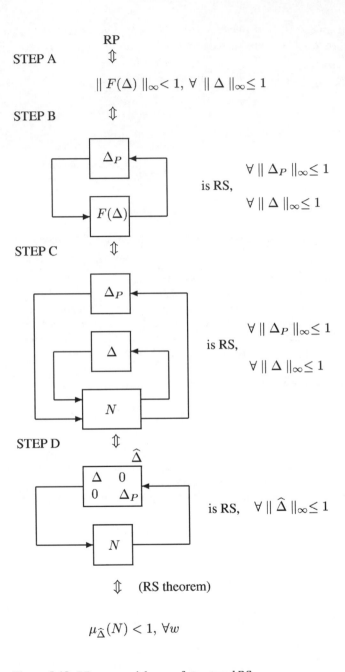

STEP A   RP
              $\Updownarrow$

$\| F(\Delta) \|_\infty < 1, \; \forall \; \| \Delta \|_\infty \leq 1$

STEP B $\Updownarrow$

$\forall \| \Delta_P \|_\infty \leq 1$
is RS,
$\forall \| \Delta \|_\infty \leq 1$

STEP C $\Updownarrow$

$\forall \| \Delta_P \|_\infty \leq 1$
is RS,
$\forall \| \Delta \|_\infty \leq 1$

STEP D $\Updownarrow$

is RS, $\forall \| \widehat{\Delta} \|_\infty \leq 1$

$\Updownarrow$ (RS theorem)

$\mu_{\widehat{\Delta}}(N) < 1, \; \forall w$

Figure 8.13: RP as a special case of structured RS

1.2 RP with input uncertainty for SISO system

a SISO system N in (8.124) is a 2×2 matrix and Δ_I and Δ_P are scalars. In this case ditions (8.116)–(8.119) become

$$\text{NS} \quad \Leftrightarrow \quad N \text{ internally stable} \Leftrightarrow S, SG, KS \text{ and } T_I \text{ are stable} \tag{8.125}$$

$$\text{NP} \quad \Leftrightarrow \quad \bar{\sigma}(N_{22}) = |w_P S| < 1, \ \forall \omega \tag{8.126}$$

$$\text{RS} \quad \Leftrightarrow \quad \mu_\Delta(N_{11}) = |w_I T_I| < 1, \ \forall \omega \tag{8.127}$$

$$\text{RP} \quad \Leftrightarrow \quad \mu_{\widehat{\Delta}}(N) = |w_P S| + |w_I T_I| < 1, \ \forall \omega \tag{8.128}$$

the RP condition (8.128) follows from (8.103); that is,

$$\mu(N) = \mu \begin{bmatrix} w_I T_I & w_I KS \\ w_P SG & w_P S \end{bmatrix} = \mu \begin{bmatrix} w_I T_I & w_I T_I \\ w_P S & w_P S \end{bmatrix} = |w_I T_I| + |w_P S| \tag{8.129}$$

we have used $T_I = KSG$. For SISO systems, $T_I = T$ and we see that (8.128) is cal to (7.72), which was derived in Chapter 7 using a simple graphical argument based Nyquist plot of $L = GK$.

optimization, in terms of weighted sensitivity with multiplicative uncertainty for a SISO , thus involves minimizing the peak value of $\mu(N) = |w_I T| + |w_P S|$. This may be using DK-iteration as outlined later in Section 8.12. A closely related problem, which er to solve both mathematically and numerically, is to minimize the peak value ($\mathcal{H}_\infty$ of the mixed sensitivity matrix

$$N_{\text{mix}} = \begin{bmatrix} w_P S \\ w_I T \end{bmatrix} \tag{8.130}$$

.96) we get that at each frequency $\mu(N) = |w_I T| + |w_P S|$ differs from $\bar{\sigma}(N_{\text{mix}}) = $ $|^2 + |w_P S|^2$ by at most a factor $\sqrt{2}$; recall (7.75). Thus, minimizing $\|N_{\text{mix}}\|_\infty$ is optimizing RP in terms of $\mu(N)$.

RP for 2×2 distillation process

r again the distillation process example from Chapter 3 (Motivating example no. 2) corresponding inverse-based controller:

$$G(s) = \frac{1}{75s + 1} \begin{bmatrix} 87.8 & -86.4 \\ 108.2 & -109.6 \end{bmatrix}; \quad K(s) = \frac{0.7}{s} G(s)^{-1} \tag{8.131}$$

roller provides a nominally decoupled system with

$$L = lI, \ S = \epsilon I \text{ and } T = tI \tag{8.132}$$

$$l = \frac{0.7}{s}, \quad \epsilon = \frac{1}{1 + l} = \frac{s}{s + 0.7}, \quad t = 1 - \epsilon = \frac{0.7}{s + 0.7} = \frac{1}{1.43s + 1}$$

used ϵ for the nominal sensitivity in each loop to distinguish it from the Laplace . Recall from Figure 3.14 that this controller gave an excellent nominal response, e response with 20% gain uncertainty in each input channel was extremely poor. We

---

Algebraic proof of Theorem 8.7

The definition of μ gives at each frequency

$$\mu_{\widehat{\Delta}}(N(j\omega)) < 1 \Leftrightarrow \det(I - N(j\omega)\widehat{\Delta}(j\omega)) \neq 0, \forall \widehat{\Delta}, \bar{\sigma}(\widehat{\Delta}(j\omega)) \leq 1$$

By Schur's formula in (A.14) we have

$$\det(I - N\widehat{\Delta}) = \det \begin{bmatrix} I - N_{11}\Delta & -N_{12}\Delta_P \\ -N_{21}\Delta & I - N_{22}\Delta_p \end{bmatrix}$$

$$= \det(I - N_{11}\Delta) \cdot \det[I - N_{22}\Delta_P - N_{21}\Delta(I - N_{11}\Delta)^{-1} N_{12}\Delta_P]$$

$$= \det(I - N_{11}\Delta) \cdot \det[I - (N_{22} + N_{21}\Delta(I - N_{11}\Delta)^{-1} N_{12})\Delta_P]$$

$$= \det(I - N_{11}\Delta) \cdot \det(I - F_u(N, \Delta)\Delta_P)$$

Since this expression should not be zero, both terms must be non-zero at each frequency, i.e.

$$\det(I - N_{11}\Delta) \neq 0 \ \forall \Delta \ \Leftrightarrow \ \mu_\Delta(N_{11}) < 1, \quad \forall \omega \quad (\text{RS})$$

and for all Δ

$$\det(I - F\Delta_P) \neq 0 \ \forall \Delta_P \Leftrightarrow \mu_{\Delta_P}(F) < 1 \Leftrightarrow \bar{\sigma}(F) < 1, \quad \forall \omega \quad (\text{RP definition})$$

Theorem 8.7 is proved by reading the above lines in the opposite direction. Note that it is not necessary to test for RS separately as it follows as a special case of the RP requirement. $\quad\square$

8.10.2 Summary of μ-conditions for NP, RS and RP

First, we rearrange the uncertain system into the $N\Delta$-structure of Figure 8.2, where the block-diagonal perturbations satisfy $\|\Delta\|_\infty \leq 1$. Then we introduce

$$F = F_u(N, \Delta) = N_{22} + N_{21}\Delta(I - N_{11}\Delta)^{-1} N_{12}$$

and let the performance requirement (RP) be $\|F\|_\infty \leq 1$ for all allowable perturbations. It then follows that

$$\text{NS} \quad \Leftrightarrow \quad N \text{ (internally) stable} \tag{8.116}$$

$$\text{NP} \quad \Leftrightarrow \quad \bar{\sigma}(N_{22}) = \mu_{\Delta_P} < 1, \ \forall \omega, \text{ and NS} \tag{8.117}$$

$$\text{RS} \quad \Leftrightarrow \quad \mu_\Delta(N_{11}) < 1, \ \forall \omega, \text{ and NS} \tag{8.118}$$

$$\text{RP} \quad \Leftrightarrow \quad \mu_{\widehat{\Delta}}(N) < 1, \ \forall \omega, \ \widehat{\Delta} = \begin{bmatrix} \Delta & 0 \\ 0 & \Delta_P \end{bmatrix}, \text{ and NS} \tag{8.119}$$

Here Δ is a block-diagonal matrix (its detailed structure depends on the uncertainty we are representing), whereas Δ_P is always a full complex matrix representing the $\mathcal{H}_\infty$ performance specification. Δ_P does not need to be a square matrix. Note that nominal NS must be tested separately in all cases.

Although the structured singular value is not a norm, it is sometimes convenient to refer to the peak μ-value as the "Δ-norm". For a stable rational transfer matrix $H(s)$, with an associated block structure Δ, we therefore define

$$\|H(s)\|_\Delta \triangleq \max_\omega \mu_\Delta(H(j\omega))$$

For a nominally stable system we then have

$$\text{NP} \Leftrightarrow \|N_{22}\|_\infty < 1, \quad \text{RS} \Leftrightarrow \|N_{11}\|_\Delta < 1, \quad \text{RP} \Leftrightarrow \|N\|_{\widehat{\Delta}} < 1$$

8.10.3   Worst-case performance and skewed-μ

Assume we have a system for which the peak μ-value for RP is 1.1. What does this mean? The definition of μ tells us that our RP requirement would be satisfied exactly if we reduced *both* the performance requirement *and* the uncertainty by a factor of 1.1. So μ does *not* directly give us the worst-case performance, i.e. $\max_\Delta \bar{\sigma}(F(\Delta))$, as one might have expected.

To find the worst-case weighted performance for a given uncertainty, one needs to keep the magnitude of the perturbations fixed ($\bar{\sigma}(\Delta) \leq 1$); that is, we must compute skewed-μ of N as discussed in Section 8.9.1. We have, in this case,

$$\max_{\bar{\sigma}(\Delta) \leq 1} \bar{\sigma}(F_u(N, \Delta)(j\omega)) = \mu^s(N(j\omega)) \tag{8.120}$$

To find μ^s numerically, we scale the performance part of N by a factor $k_m = 1/\mu^s$ and iterate on k_m until $\mu = 1$. That is, at each frequency skewed-μ is the value $\mu^s(N)$ which solves

$$\mu(K_m N) = 1, \quad K_m = \begin{bmatrix} I & 0 \\ 0 & 1/\mu^s \end{bmatrix} \tag{8.121}$$

Note that μ underestimates how bad or good the actual worst-case performance is. This follows because $\mu^s(N)$ is always further from 1 than $\mu(N)$.

Remark. The corresponding worst-case perturbation may be obtained as follows. First compute the worst-case performance at each frequency using skewed-μ. At the frequency where $\mu^s(N)$ has its peak, we may extract the corresponding worst-case perturbation generated by the software, and then find a stable, all-pass transfer function that matches this. In the Matlab Robust Control toolbox, the single command `robustperf` combines these steps: `[perfmarg,perfmargunc] = robustperf(lft(Delta,N));`.

8.11   Application: robust performance with input uncertainty

We will now consider in some detail the case of multiplicative input uncertainty with performance defined in terms of weighted sensitivity, as illustrated in Figure 8.14. The performance requirement is then

$$\text{RP} \stackrel{\text{def}}{\Leftrightarrow} \|w_P(I + G_p K)^{-1}\|_\infty < 1, \quad \forall G_p \tag{8.122}$$

where the set of plants is given by

$$G_p = G(I + w_I \Delta_I), \quad \|\Delta_I\|_\infty \leq 1 \tag{8.123}$$

Here $w_P(s)$ and $w_I(s)$ are scalar weights, so the performance objective is the same for all the outputs, and the uncertainty is the same for all inputs. We will mostly assume that Δ_I is diagonal, but we will also consider the case when Δ_I is a full matrix. This problem is excellent for illustrating the robustness analysis of uncertain multivariable systems. It should be noted, however, that although the problem setup in (8.122) and (8.123) is fine for analyzing a given controller, it is less suitable for controller synthesis. For example, the problem formulation does not penalize directly the outputs from the controller.

In this section, we will:

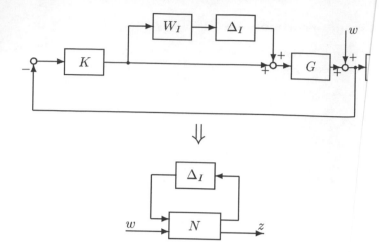

Figure 8.14: RP of system with input uncertainty

1. Find the interconnection matrix N for this problem.
2. Consider the SISO case, so that useful connections can be ma[de with the] previous chapter.
3. Consider a multivariable distillation process for which we [saw in the] simulations in Chapter 3 that a decoupling controller is sensi[tive to] input gains. We will find that μ for RP is indeed much larger [for this] controller.
4. Find some simple bounds on μ for this problem and discus[s the resulting] number.
5. Make comparisons with the case where the uncertainty is loca[ted...]

8.11.1   Interconnection matrix

On rearranging the system into the $N\Delta$-structure, as shown i[n] (8.32),

$$N = \begin{bmatrix} w_I T_I & w_I K S \\ w_P S G & w_P S \end{bmatrix}$$

where $T_I = KG(I+KG)^{-1}$, $S = (I+GK)^{-1}$. For simplicit[y we omit the] signs in the 1,1 and 1,2 blocks of N, since $\mu(N) = \mu(UN)$ w[ith... see] (8.83).

For a given controller K we can now test for NS, NP, RS [and RP] with

$$\hat{\Delta} = \begin{bmatrix} \Delta_I & 0 \\ 0 & \Delta_P \end{bmatrix}$$

Here $\Delta = \Delta_I$ may be a full or diagonal matrix (depending on [...]) the fictitious perturbation matrix Δ_P, representing the $\mathcal{H}_\infty$ [...] always a full matrix.

will now confirm these findings by a μ-analysis. To this effect we use the following weights for uncertainty and performance:

$$w_I(s) = \frac{s + 0.2}{0.5s + 1}; \quad w_P(s) = \frac{s/2 + 0.05}{s} \tag{8.133}$$

With reference to (7.36) we see that the weight $w_I(s)$ may approximately represent a 20% gain error and a neglected time delay of 0.9 min. $|w_I(j\omega)|$ levels off at 2 (200% uncertainty) at high frequencies. With reference to (2.105) we see that the performance weight $w_P(s)$ specifies integral action, a closed-loop bandwidth of about 0.05 [rad/min] (which is relatively slow in the presence of an allowed time delay of 0.9 min) and a maximum peak for $\bar{\sigma}(S)$ of $M_S = 2$.

We now test for NS, NP, RS and RP. Note that Δ_I is a diagonal matrix in this example.

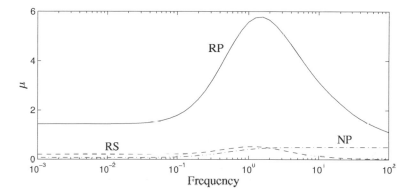

Figure 8.15: μ-plots for distillation process with decoupling controller

NS With G and K as given in (8.131) we find that S, SG, KS and T_I are stable, so the system is nominally stable.

NP With the decoupling controller we have

$$\sigma(N_{22}) = \bar{\sigma}(w_P S) = \left| \frac{s/2 + 0.05}{s + 0.7} \right|$$

and we see from the dashed-dot line in Figure 8.15 that the NP condition is easily satisfied: $\bar{\sigma}(w_P S)$ is small at low frequencies ($0.05/0.7 = 0.07$ at $\omega = 0$) and approaches $1/2 = 0.5$ at high frequencies.

RS Since in this case $w_I T_I = w_I T$ is a scalar times the identity matrix, we have, independent of the structure of Δ_I, that

$$\mu_{\Delta_I}(w_I T_I) = |w_I t| = \left| 0.2 \frac{5s + 1}{(0.5s + 1)(1.43s + 1)} \right|$$

and we see from the dashed line in Figure 8.15 that RS is easily satisfied. The peak value of $\mu_{\Delta_I}(M)$ over frequency is $\|M\|_{\Delta_I} = 0.53$. This means that we may increase the uncertainty by a factor of $1/0.53 = 1.89$ before the worst-case uncertainty yields instability. That is, we can tolerate about 38% gain uncertainty and a time delay of about 1.7 min before we get instability.

RP   Although our system has good robustness margins (RS easily satisfied) and excellent NP
we know from the simulations in Figure 3.14 that RP is poor. This is confirmed by the
μ-curve for RP in Figure 8.15 which was computed numerically using $\mu_{\widehat{\Delta}}(N)$ with N
as in (8.124), $\widehat{\Delta} = \text{diag}\{\Delta_I, \Delta_P\}$ and $\Delta_I = \text{diag}\{\delta_1, \delta_2\}$. The peak value is close to
6, meaning that even with six times less uncertainty, the weighted sensitivity will be
about six times larger than we require. The peak of the actual worst-case weighted
sensitivity with uncertainty blocks of magnitude 1, which may be computed using
skewed-μ, is for comparison 44.93.

The Matlab Robust Control toolbox commands to generate Figure 8.15 are given in Table 8.1.

In general, μ with unstructured uncertainty (Δ_I full) is larger than μ with structured
uncertainty (Δ_I diagonal). However, for our particular plant and controller in (8.131) it
appears from numerical calculations, and by use of (8.136) below, that they are the same.
Of course, this is not generally true, as is confirmed in the following exercise.

Exercise 8.22* *Consider the plant $G(s)$ in (8.108) which is ill-conditioned with $\gamma(G) = 70.8$ at all
frequencies (but note that the RGA elements of G are all about 0.5). With an inverse-based controller
$K(s) = \frac{0.7}{s}G(s)^{-1}$, compute μ for RP with both diagonal and full-block input uncertainty using the
weights in (8.133). The value of μ is much smaller in the former case.*

8.11.4   RP and the condition number

In this subsection, we consider the relationship between μ for RP and the condition number
of the plant or of the controller. We consider *unstructured* multiplicative input uncertainty
(i.e. Δ_I is a full matrix) and performance measured in terms of weighted sensitivity.

Any controller. Let N be given as in (8.124). Then

$$\overbrace{\mu_{\widetilde{\Delta}}(N)}^{\text{RP}} \leq [\overbrace{\bar{\sigma}(w_I T_I)}^{\text{RS}} + \overbrace{\bar{\sigma}(w_P S)}^{\text{NP}}](1 + \sqrt{k}) \tag{8.134}$$

where k is the condition number of either the plant or the controller (the smallest one should
be used):

$$k = \gamma(G) \quad \text{or} \quad k = \gamma(K) \tag{8.135}$$

Proof of (8.134): Since Δ_I is a full matrix, (8.87) yields

$$\mu(N) = \min_d \bar{\sigma} \begin{bmatrix} N_{11} & dN_{12} \\ d^{-1}N_{21} & N_{22} \end{bmatrix}$$

where from (A.47)

$$\bar{\sigma} \begin{bmatrix} w_I T_I & dw_I KS \\ d^{-1}w_P SG & w_P S \end{bmatrix} \leq \bar{\sigma}(w_I T_I \begin{bmatrix} I & dG^{-1} \end{bmatrix}) + \bar{\sigma}(w_P S \begin{bmatrix} d^{-1}G & I \end{bmatrix})$$

$$\leq \bar{\sigma}(w_I T_I)\underbrace{\bar{\sigma}(I \quad dG^{-1})}_{\leq 1+|d|\bar{\sigma}(G^{-1})} + \bar{\sigma}(w_P S)\underbrace{\bar{\sigma}(d^{-1}G \quad I)}_{\leq 1+|d^{-1}|\bar{\sigma}(G)}$$

and selecting $d = \sqrt{\frac{\bar{\sigma}(G)}{\bar{\sigma}(G^{-1})}} = \sqrt{\gamma(G)}$ gives

$$\mu(N) \leq [\bar{\sigma}(w_I T_I) + \bar{\sigma}(w_P S)] (1 + \sqrt{\gamma(G)})$$

A similar derivation may be performed using $SG = K^{-1}T_I$ to derive the same expression but with
$\gamma(K)$ instead of $\gamma(G)$.   □

Table 8.1: Matlab program for μ-analysis (generates Figure 8.15)

```
% Uses the Robust Control toolbox
G0=[87.8 -86.4; 108.2 -109.6];
G=tf([1],[75 1])*G0;
G=minreal(ss(G));
%
% Inverse-based controller
%
Kinv=0.7*tf([75 1],[1 1e-5])*inv(G0);
%
% Weights
%
Wp=0.5*tf([10 1],[10 1e-5])*eye(2);
Wi=tf([1 0.2],[0.5 1])*eye(2);
%
% Generalized plant P
%
systemnames = 'G Wp Wi';
inputvar = '[ydel(2); w(2) ; u(2)]';
outputvar = '[Wi ; Wp ; -G-w]';
input_to_G = '[u+ydel]';
input_to_Wp = '[G+w]';
input_to_Wi = '[u]';
sysoutname = 'P';
cleanupsysic= 'yes'; sysic;
%
N=lft(P,Kinv);
omega = logspace(-3,3,61); Nf=frd(N,omega);
%
% mu for RP
%
blk=[1 1; 1 1; 2 2];
[mubnds,muinfo]=mussv(Nf,blk,'c');
muRP=mubnds(:,1); [muRPinf,muRPw] = norm(muRP,inf);          % (ans = 5.7726)
%
% Worst case weighted sensitivity
%
delta = [ultidyn('del1',[1 1]) 0;0 ultidyn('del2',[1 1])];
Np = lft(delta,N); %Perturbed model
opt = wcgopt('ABadThreshold',100);
Npw = wcgain(Np,opt);                                        % (ans = 44.98 for
%                                                               delta = 1)
% mu for RS
%
Nrs=Nf(1:2,1:2); % Picking out WiTi
[mubnds,muinfo]=mussv(Nrs,[1 1; 1 1],'c');
muRS=mubnds(:,1); [muRSinf,muRSw]=norm(muRS,inf)            % (ans = 0.5242)
%
% mu for NS (=max. singular value of Nnp)
%
Nnp=Nf(3:4,3:4); % Picking out wP*Si
[mubnds,muinfo]=mussv(Nnp,[1 1;1 1],'c');
muNS=mubnds(:,1); [muNSinf,muNSw]=norm(muNS,inf)           % (ans = 0.500)
bodemag(muRP,'',muRS,'--',muNS,'-.',omega)
```

From (8.134) we see that with a "round" controller, i.e. one with $\gamma(K) = 1$, there is less sensitivity to uncertainty (but it may be difficult to achieve NP in this case). On the other hand, we would expect μ for RP to be large if we used an inverse-based controller for a plant with a large condition number, since then $\gamma(K) = \gamma(G)$ is large. This is confirmed by (8.136) below.

Example 8.11 *For the distillation process studied above, we have $\gamma(G) = \gamma(K) = 141.7$ at all frequencies, and at frequency $w = 1$ rad/min the upper bound given by (8.134) becomes $(0.52 + 0.41)(1 + \sqrt{141.7}) = 13.1$. This is higher than the actual value of $\mu(N)$ which is 5.56, which illustrates that the bound in (8.134) is generally not tight.*

Inverse-based controller. With an inverse-based controller (resulting in the nominal decoupled system (8.132)) and unstructured input uncertainty, it is possible to derive an analytic expression for μ for RP with N as in (8.124):

$$\mu_{\tilde{\Delta}}(N) = \sqrt{|w_P\epsilon|^2 + |w_I t|^2 + |w_P\epsilon| \cdot |w_I t| \left(\gamma(G) + \frac{1}{\gamma(G)}\right)} \qquad (8.136)$$

where ϵ is the nominal sensitivity and $\gamma(G)$ is the condition number of the plant. We see that for plants with a large condition number, μ for RP increases approximately in proportion to $\sqrt{\gamma(G)}$.

Proof of (8.136): The proof originates from Stein and Doyle (1991). The upper μ-bound in (8.87) with $D = \text{diag}\{dI, I\}$ yields

$$\mu(N) = \min_d \bar{\sigma} \begin{bmatrix} w_I tI & w_I t(dG)^{-1} \\ w_P\epsilon(dG) & w_P\epsilon I \end{bmatrix} = \min_d \bar{\sigma} \begin{bmatrix} w_I tI & w_I t(d\Sigma)^{-1} \\ w_P\epsilon(d\Sigma) & w_P\epsilon I \end{bmatrix}$$

$$= \min_d \max_i \bar{\sigma} \begin{bmatrix} w_I t & w_I t(d\sigma_i)^{-1} \\ w_P\epsilon(d\sigma_i) & w_P\epsilon \end{bmatrix}$$

$$= \min_d \max_i \sqrt{|w_P\epsilon|^2 + |w_I t|^2 + |w_P\epsilon d\sigma_i|^2 + |w_I t(d\sigma_i)^{-1}|^2}$$

We have used here the SVD of $G = U\Sigma V^H$ at each frequency, and the fact that $\bar{\sigma}$ is unitary invariant. σ_i denotes the i'th singular value of G. The expression is minimized by selecting at each frequency $d = |w_I t|/(|w_P\epsilon|\bar{\sigma}(G)\underline{\sigma}(G))$, see (8.100), and hence the desired result. For more details see Zhou et al. (1996, pp. 293–295). □

Example 8.12 *For the distillation column example studied above, we have at frequency $\omega = 1$ rad/min, $|w_P\epsilon| = 0.41$ and $|w_I t| = 0.52$, and since $\gamma(G) = 141.7$ at all frequencies, (8.136) yields $\mu(N) = \sqrt{0.17 + 0.27 + 30.51} = 5.56$ which agrees with the plot in Figure 8.15.*

Worst-case performance (any controller)

We next derive relationships between worst-case performance and the condition number. Suppose that at each frequency the worst-case sensitivity is $\bar{\sigma}(S')$. We then have that the worst-case weighted sensitivity is equal to skewed-μ:

$$\max_{S_p} \bar{\sigma}(w_P S_p) = \bar{\sigma}(w_P S') = \mu^s(N)$$

Now, recall that in Section 6.10.4 we derived a number of upper bounds on $\bar{\sigma}(S')$, and referring back to (6.89) we find

$$\bar{\sigma}(S') \leq \gamma(G)\frac{\bar{\sigma}(S)}{1 - \bar{\sigma}(w_I T_I)} \qquad (8.137)$$

A similar bound involving $\gamma(K)$ applies. We then have

$$\mu^s(N) = \bar{\sigma}(w_P S') \le k \frac{\bar{\sigma}(w_P S)}{1 - \bar{\sigma}(w_I T_I)} \tag{8.138}$$

where k as before denotes the condition number of *either* the plant *or* the controller (preferably the smallest). Equation (8.138) holds for any controller and for any structure of the uncertainty (including Δ_I unstructured).

Remark 1 In Section 6.10.4, we derived tighter upper bounds for cases when Δ_I is restricted to be diagonal and when we have a decoupling controller. In (6.93), we also derived a lower bound in terms of the RGA.

Remark 2 Since $\mu^s = \mu$ when $\mu = 1$, we may, from (8.134), (8.138) and expressions similar to (6.91) and (6.92), derive the following *sufficient* (conservative) tests for RP ($\mu(N) < 1$) with unstructured input uncertainty (any controller):

$$\text{RP} \quad \Leftarrow \quad [\bar{\sigma}(w_P S) + \bar{\sigma}(w_I T_I)](1 + \sqrt{k}) < 1, \ \forall \omega$$
$$\text{RP} \quad \Leftarrow \quad k\bar{\sigma}(w_P S) + \bar{\sigma}(w_I T_I) < 1, \quad \forall \omega$$
$$\text{RP} \quad \Leftarrow \quad \bar{\sigma}(w_P S) + k\bar{\sigma}(w_I T) < 1, \quad \forall \omega$$

where k denotes the condition number of *either* the plant *or* the controller (the smallest being the most useful).

Example 8.13 *For the distillation process, the upper bound given by (8.138) at $\omega = 1$ rad/min is* $141.7 \cdot 0.41/(1 - 0.52) = 121$. *This is higher than the actual peak value of $\mu^s = \max_{S_p} \bar{\sigma}(w_P S_p)$, which as found earlier is 44.9 (at frequency 1.2 rad/min), and demonstrates that these bounds are not generally tight.*

8.11.5 Comparison with output uncertainty

Consider output multiplicative uncertainty of magnitude $w_O(j\omega)$. In this case, we get the interconnection matrix

$$N = \begin{bmatrix} w_O T & w_O T \\ w_P S & w_P S \end{bmatrix} \tag{8.139}$$

and for any structure of the uncertainty $\mu(N)$ is bounded as follows:

$$\bar{\sigma}\begin{bmatrix} w_O T \\ w_P S \end{bmatrix} \le \overbrace{\mu(N)}^{\text{RP}} \le \sqrt{2}\,\underbrace{\overbrace{\bar{\sigma}\begin{bmatrix} w_O T \\ w_P S \end{bmatrix}}^{\text{RS}}}_{\text{NP}} \tag{8.140}$$

This follows since the uncertainty and performance blocks both enter at the output (see Section 8.6.2) and from (A.46) the difference between bounding the combined perturbations, $\bar{\sigma}[\Delta_O \ \ \Delta_P]$, and individual perturbations, $\bar{\sigma}(\Delta_O)$ and $\bar{\sigma}(\Delta_P)$, is at most a factor of $\sqrt{2}$. Thus, in this case we "automatically" achieve RP (at least within $\sqrt{2}$) if we have satisfied separately the subobjectives of NP and RS. This confirms our findings from Section 6.10.4 that multiplicative output uncertainty poses no particular problem for performance. It also implies that for practical purposes we may optimize RP with output uncertainty by minimizing the $\mathcal{H}_\infty$ norm of the stacked matrix $\begin{bmatrix} w_O T \\ w_P S \end{bmatrix}$.

Exercise 8.23 *Consider the RP problem with weighted sensitivity and multiplicative output uncertainty. Derive the interconnection matrix N for (1) the conventional case with $\widehat{\Delta} = \mathrm{diag}\{\Delta, \Delta_P\}$, and (2) the stacked case when $\widehat{\Delta} = [\,\Delta \quad \Delta_P\,]$. Use this to prove (8.140).*

8.12 μ-synthesis and DK-iteration

The structured singular value μ is a very powerful tool for the analysis of RP with a given controller. However, one may also seek to find the controller that minimizes a given μ-condition: this is the μ-synthesis problem.

8.12.1 DK-iteration

At present there is no direct method to synthesize a μ-optimal controller. However, for complex perturbations a method known as DK-iteration is available. It combines $\mathcal{H}_\infty$ synthesis and μ-analysis, and often yields good results. The starting point is the upper bound (8.87) on μ in terms of the scaled singular value

$$\mu(N) \leq \min_{D \in \mathcal{D}} \bar{\sigma}(DND^{-1})$$

The idea is to find the controller that minimizes the peak value over frequency of this upper bound, namely

$$\min_K (\min_{D \in \mathcal{D}} \|DN(K)D^{-1}\|_\infty) \tag{8.141}$$

by alternating between minimizing $\|DN(K)D^{-1}\|_\infty$ with respect to either K or D (while holding the other fixed). To start the iterations, one selects an initial stable rational transfer matrix $D(s)$ with appropriate structure. The identity matrix is often a good initial choice for D provided the system has been reasonably scaled for performance. The DK-iteration then proceeds as follows:

1. K**-step.** Synthesize an $\mathcal{H}_\infty$ controller for the scaled problem, $\min_K \|DN(K)D^{-1}\|_\infty$ with fixed $D(s)$.
2. D**-step.** Find $D(j\omega)$ to minimize at each frequency $\bar{\sigma}(DND^{-1}(j\omega))$ with fixed N.
3. Fit the magnitude of each element of $D(j\omega)$ to a stable and minimum-phase transfer function $D(s)$ and go to step 1.

The iteration may continue until satisfactory performance is achieved, $\|DND^{-1}\|_\infty < 1$, or until the $\mathcal{H}_\infty$ norm no longer decreases. One fundamental problem with this approach is that although each of the minimization steps (K-step and D-step) are convex, joint convexity is *not* guaranteed. Therefore, the iterations may converge to a local optimum. However, practical experience suggests that the method works well in most cases.

The order of the controller resulting from each iteration is equal to the number of states in the plant $G(s)$ plus the number of states in the weights plus twice the number of states in $D(s)$. For most cases, the true μ-optimal controller is not rational, and will thus be of infinite order, but because we use a finite-order $D(s)$ to approximate the D-scales, we get a controller of finite (but often high) order. The true μ-optimal controller would have a flat μ-curve (as a function of frequency), except at infinite frequency where μ generally has to approach a fixed value independent of the controller (because $L(j\infty) = 0$ for real systems). However, with a finite-order controller we will generally not be able (and it may not be desirable) to extend the flatness to infinite frequencies.

The DK-iteration depends heavily on optimal solutions for steps 1 and 2, and also on good fits in step 3, preferably by a transfer function of low order. One reason for preferring a low-order fit is that this reduces the order of the $\mathcal{H}_\infty$ problem, which usually improves the numerical properties of the $\mathcal{H}_\infty$ optimization (step 1) and also yields a controller of lower order. In some cases the iterations converge slowly, and it may be difficult to judge whether the iterations are converging or not. One may even experience the μ-value increasing. This may be caused by numerical problems or inaccuracies (e.g. the upper bound μ-value in step 2 being higher than the $\mathcal{H}_\infty$ norm obtained in step 1), or by a poor fit of the D-scales. In any case, if the iterations converge slowly, then one may consider going back to the initial problem and rescaling the inputs and outputs.

In the K-step (step 1) where the $\mathcal{H}_\infty$ controller is synthesized, it is often desirable to use a slightly suboptimal controller (e.g. with an $\mathcal{H}_\infty$ norm, γ, which is 5% higher than the optimal value, $\gamma_{\min}$). This yields a blend of $\mathcal{H}_\infty$ and $\mathcal{H}_2$ optimality with a controller which usually has a steeper high-frequency roll-off than the $\mathcal{H}_\infty$ optimal controller.

8.12.2 Adjusting the performance weight

Recall that if μ at a given frequency is different from 1, then the interpretation is that at this frequency we can tolerate $1/\mu$-times more uncertainty and satisfy our performance objective with a margin of $1/\mu$. In μ-synthesis, the designer will usually adjust some parameter(s) in the performance or uncertainty weights until the peak μ-value is close to 1. Sometimes the uncertainty is fixed, and we effectively optimize worst-case performance by adjusting a parameter in the performance weight. For example, consider the performance weight

$$w_P(s) = \frac{s/M + \omega_B^*}{s + \omega_B^* A}$$ (8.142)

where we want to keep M constant and find the highest achievable bandwidth frequency ω_B^*. The optimization problem becomes

$$\max |\omega_B^*| \quad \text{such that} \quad \mu(N) < 1, \forall \omega$$ (8.143)

where N, the interconnection matrix for the RP problem, depends on ω_B^*. This may be implemented as an outer loop around the DK-iteration.

8.12.3 Fixed structure controller

Sometimes it is desirable to find a low order controller with a given structure, e.g. a decentralized PID controller. This may be achieved by numerical optimization where μ is minimized with respect to the controller parameters. The problem here is that the optimization is not generally convex in the parameters. Sometimes it helps to switch the optimization between minimizing the peak of μ (i.e. $\|\mu\|_\infty$) and minimizing the integral square deviation of μ away from k (i.e. $\|\mu(j\omega) - k\|_2$) where k usually is close to 1. The latter is an attempt to "flatten out" μ.

8.12.4   Example: μ-synthesis with DK-iteration

We will consider again the case of multiplicative input uncertainty and performance defined in terms of weighted sensitivity, as discussed in detail in Section 8.11. We noted there that this setup is fine for analysis, but less suitable for controller synthesis, as it does not explicitly penalize the outputs from the controller. Nevertheless we will use it here as an example of μ-synthesis because of its simplicity. The resulting controller will have very large gains at high frequencies and should not be used directly for implementation. In practice, one can add extra roll-off to the controller (which should work well because the system should be robust with respect to uncertain high-frequency dynamics), or one may consider a more complicated problem setup (see Section 13.4).

With this caution in mind, we proceed with the problem description. Again, we use the model of the simplified distillation process

$$G(s) = \frac{1}{75s + 1}\begin{bmatrix} 87.8 & -86.4 \\ 108.2 & -109.6 \end{bmatrix} \tag{8.144}$$

The uncertainty weight $w_I I$ and performance weight $w_P I$ are given in (8.133), and are shown graphically in Figure 8.16. The objective is to minimize the peak value of $\mu_{\tilde{\Delta}}(N)$, where N is given in (8.124) and $\tilde{\Delta} = \mathrm{diag}\{\Delta_I, \Delta_P\}$. We will consider diagonal input uncertainty (which is always present in any real problem), so Δ_I is a 2×2 diagonal matrix. Δ_P is a full 2×2 matrix representing the performance specification. Note that we have only three complex uncertainty blocks, so $\mu(N)$ is equal to the upper bound $\min_D \bar{\sigma}(DND^{-1})$ in this case.

We will now use DK-iteration in an attempt to obtain the μ-optimal controller for this example. The appropriate commands for the Matlab Robust Control toolbox are listed in Table 8.2. The Matlab Robust Control toolbox contains commands that "automate" the DK-iteration (listed at the bottom of Table 8.2), but we use a "manual" approach here, as this yields more insight.

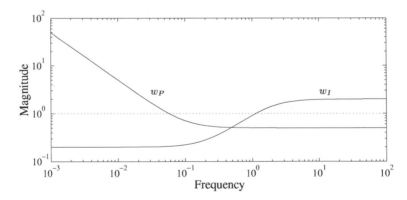

Figure 8.16: Uncertainty and performance weights. Notice that there is a frequency range ("window") where both weights are less than 1 in magnitude.

First the generalized plant P as given in (8.29) is constructed. It includes the plant model, the uncertainty weight and the performance weight, but not the controller which is to be designed (note that $N = F_l(P, K)$). Then the block structure is defined; it consists of two 1×1 blocks to represent Δ_I and a 2×2 block to represent Δ_P. The scaling matrix D for

Table 8.2: **Matlab program to perform** DK-**iteration**

```
% Uses the Robust Control toolbox
G0 = [87.8 -86.4; 108.2 -109.6];                        % Distillation
dyn = tf(1,[75 1]); G=dyn*eye(2)*G0;                    % process.
%
% Weights.
%
Wp = 0.5*tf([10 1],[10 1.e-5])*eye(2);                  % Approximated
Wi = tf([1 0.2],[0.5 1])*eye(2);                        % integrator.
%
% Generalized plant P. %
systemnames = 'G Wp Wi';
inputvar = '[udel(2); w(2) ; u(2)]';
outputvar = '[Wi; Wp; -G-w]';
input_to_G = '[u+udel]';
input_to_Wp = '[G+w]'; input_to_Wi = '[u]';
sysoutname = 'P'; cleanupsysic = 'yes';
sysic;
P = minreal(ss(P));
%
% Initialize.
%
omega = logspace(-3,3,61);
blk = [1 1; 1 1; 2 2];
nmeas = 2; nu = 2; d0 = 1;
D = append(d0,d0,tf(eye(2)),tf(eye(2)));                % Initial scaling.
%
% START ITERATION.
%
% STEP 1: Find H-infinity optimal controller
% with given scalings:
%
[K,Nsc,gamma,info] = hinfsyn(D*P*inv(D),nmeas,nu,....
                 'method','lmi','Tolgam',1e-3);
Nf = frd(lft(P,K),omega);
%
% STEP 2: Compute mu using upper bound:
%
[mubnds,Info] = mussv(Nf,blk,'c');
bodemag(mubnds(1,1),omega);
murp = norm(mubnds(1,1),inf,1e-6);
%
% STEP 3: Fit resulting D-scales:
%
[dsysl,dsysr] = mussvunwrap(Info);
dsysl = dsysl/dsysl(3,3);
d1 = fitfrd(genphase(dsysl(1,1)),4);                    % Choose 4th order.
%
% GOTO STEP 1 (unless satisfied with murp).
%
% Alternatively use automatic software
%
% Delta = [ultidyn('D_1',[1 1]) 0;0 ultidyn('D_2',[1 1])];  % Diagonal uncertainty.
% Punc = lft(Delta,P);
% opt = dkitopt('FrequencyVector',omega);
% [K,clp,bnd,dkinfo] = dksyn(Punc,nmeas,nu,opt);
```

DND^{-1} then has the structure $D = \text{diag}\{d_1, d_2, d_3 I_2\}$ where I_2 is a 2×2 identity matrix, and we may set $d_3 = 1$. As initial scalings we select $d_1^0 = d_2^0 = 1$. P is then scaled with the matrix $\text{diag}\{D, I_2\}$ where I_2 is associated with the inputs and outputs from the controller (we do not want to scale the controller).

Iteration no. 1. Step 1: With the initial scalings, $D^0 = I$, the $\mathcal{H}_\infty$ software (see Table 8.2) produced a five-state controller with an $\mathcal{H}_\infty$ norm of $\gamma = 1.1798$. Step 2: The upper μ-bound gave the μ-curve shown as curve "Iter. 1" in Figure 8.17, corresponding to a peak value of μ=1.1798. Step 3: The frequency-dependent $d_1(\omega)$ and $d_2(\omega)$ from step 2 were each fitted using a fourth-order transfer function. $d_1(w)$ and the fitted fourth-order transfer function (dotted line) are shown in Figure 8.18 and labelled "Iter. 1". The fit is very good, except at higher frequencies. At low frequencies, it is hard to distinguish the two curves. d_2 is not shown because it was found that $d_1 \approx d_2$ (indicating that the worst-case full-block Δ_I is in fact diagonal).

Iteration no. 2. Step 1: With the 8-state scaling $D^1(s)$ the $\mathcal{H}_\infty$ software gave a 21-state controller and $\|D^1 N (D^1)^{-1}\|_\infty = 1.0274$. Step 2: This controller gave a peak value of μ of 1.0272. Step 3: The resulting scalings D^2 were only slightly changed from the previous iteration as can be seen from $d_1^2(\omega)$ labelled "Iter. 2" in Figure 8.18.

Iteration no. 3. Step 1: With the scalings $D^2(s)$ the $\mathcal{H}_\infty$ norm was only slightly reduced from 1.0274 to 1.0208. Since the improvement was small and since the value was very close to the desired value of 1, it was decided to stop the iterations. The resulting controller with 21 states (denoted K_3 in the following) gives a peak μ-value of 1.0205.

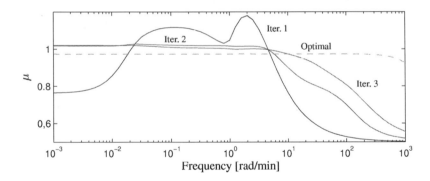

Figure 8.17: Change in μ during DK-iteration

Analysis of μ-"optimal" controller K_3

The final μ-curves for NP, RS and RP with controller K_3 are shown in Figure 8.19. The objectives of RS and NP are easily satisfied. Furthermore, the peak μ-value of 1.0205 with controller K_3 is only slightly above 1, so the performance specification $\bar\sigma(w_P S_p) < 1$ is almost satisfied for all possible plants. To confirm this we considered the nominal plant and six perturbed plants

$$G_i'(s) = G(s) E_{Ii}(s)$$

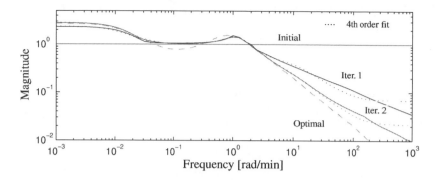

Figure 8.18: Change in D-scale d_1 during DK-iteration

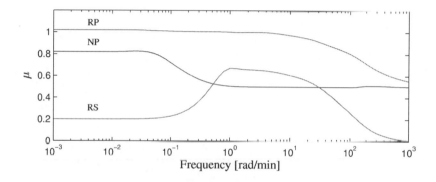

Figure 8.19: μ-plots with μ-"optimal" controller K_3

where $E_{Ii} = I + w_I \Delta_I$ is a diagonal transfer function matrix representing input uncertainty (with nominal $E_{I0} = I$). Recall that the uncertainty weight is

$$w_I(s) = \frac{s + 0.2}{0.5s + 1}$$

which is 0.2 in magnitude at low frequencies. Thus, the following input gain perturbations are allowable:

$$E_{I1} = \begin{bmatrix} 1.2 & 0 \\ 0 & 1.2 \end{bmatrix}, \ E_{I2} = \begin{bmatrix} 0.8 & 0 \\ 0 & 1.2 \end{bmatrix}, \ E_{I3} = \begin{bmatrix} 1.2 & 0 \\ 0 & 0.8 \end{bmatrix}, \ E_{I4} = \begin{bmatrix} 0.8 & 0 \\ 0 & 0.8 \end{bmatrix}$$

These perturbations do not make use of the fact that $w_I(s)$ increases with frequency. Two allowed dynamic perturbations for the diagonal elements in $w_I \Delta_I$ are

$$\epsilon_1(s) = \frac{-s + 0.2}{0.5s + 1}, \quad \epsilon_2(s) = -\frac{s + 0.2}{0.5s + 1}$$

corresponding to elements in E_{I_i} of

$$f_1(s) = 1 + \epsilon_1(s) = 1.2\frac{-0.417s + 1}{0.5s + 1}, \quad f_2(s) = 1 + \epsilon_2(s) = 0.8\frac{-0.633s + 1}{0.5s + 1}$$

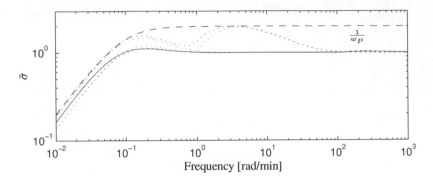

Figure 8.20: Perturbed sensitivity functions $\bar{\sigma}(S')$ using μ-"optimal" controller K_3. Dotted lines: plants $G'_i, i = 1, 6$. Solid line: nominal plant G. Dashed line: inverse of performance weight.

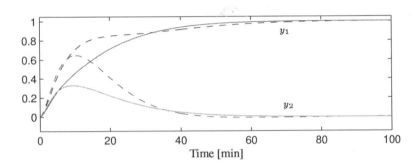

Figure 8.21: Setpoint response for μ-"optimal" controller K_3. Solid line: nominal plant. Dashed line: uncertain plant G'_3.

so let us also consider

$$E_{I5} = \begin{bmatrix} f_1(s) & 0 \\ 0 & f_1(s) \end{bmatrix}, \ E_{I6} = \begin{bmatrix} f_2(s) & 0 \\ 0 & f_1(s) \end{bmatrix}$$

The maximum singular value of the sensitivity, $\bar{\sigma}(S'_i)$, is shown in Figure 8.20 for the nominal and six perturbed plants, and is seen to be almost below the bound $1/|w_I(j\omega)|$ for all seven cases ($i = 0, 6$) illustrating that RP is almost satisfied. The sensitivity for the nominal plant is shown by the solid line, and the others with dotted lines. At low frequencies the worst-case corresponds closely to a plant with gains 1.2 and 0.8, such as G'_2, G'_3 or G'_6. Overall, the worst case of these six plants seems to be $G'_6 = GE_{I6}$, which has $\bar{\sigma}(S')$ close to the bound at low frequencies, and has a peak of about 2.003 (above the allowed bound of 2) at 3.5 rad/min.

To find the "true" worst-case performance and plant we used the Matlab Robust Control toolbox command `robustperf` as explained in Section 8.10.3 on page 320. This gives a worst-case performance of $\max_{S_p} \|w_P S_p\|_\infty = 1.0205$, and the sensitivity function for the corresponding worst-case plant $G'_{wc}(s) = G(s)(I + w_I(s)\Delta_{wc}(s))$ found with the software has a peak value of $\bar{\sigma}(S_p)$ of about 1.0979 at 0.02 rad/min. It may seem surprising that $\|S_p\|_\infty$ is much smaller than the sensitivity peak for the perturbed plants considered earlier; however, note that $G'_{wc}(s)$ is the worst-case plant with respect to the peak value of $\|w_P S_p\|_\infty$ and not $\|S_p\|_\infty$.

Remark. The "worst-case" plant is not unique, and there are many plants which yield a worst-case performance of $\max_{S_p} \|w_P S_p\|_\infty = 1.037$. For example, it is likely that we could find plants which were more consistently "worse" at all frequencies than the one shown by the dotted lines in Figure 8.20.

The time responses of y_1 and y_2 to a filtered setpoint change in y_1, $r_1 = 1/(5s + 1)$, are shown in Figure 8.21 both for the nominal case (solid line) and for 20% input gain uncertainty (dashed line) using the plant $G'_3 = GE_3$ (which we know is one of the worst plants). The responses are interactive, but show no strong sensitivity to the uncertainty. The responses with uncertainty are seen to be much better than those with the inverse-based controller studied earlier and shown in Figure 3.14.

Remarks on the μ-synthesis example.

1. By trial and error, and many long nights, Petter Lundström was able to reduce the peak μ-value for RP for this problem down to about $\mu_{opt} = 0.974$ (Lundström, 1994). The resulting design produces the curves labelled *optimal* in Figures 8.17 and 8.18. The corresponding controller, K_{opt}, may be synthesized using $\mathcal{H}_\infty$ synthesis with the following third-order D-scales:

$$d_1(s) = d_2(s) = 2\frac{(0.001s + 1)(s + 0.25)(s + 0.054)}{((s + 0.67)^2 + 0.56^2)(s + 0.013)}, \quad d_3 = 1 \tag{8.145}$$

2. Note that the optimal controller K_{opt} for this problem has an SVD form. That is, let $G = U\Sigma V^H$, then $K_{opt} = VK_sU^H$ where K_s is a diagonal matrix. This arises because in this example U and V are constant matrices. For more details see Hovd (1992) and Hovd et al. (1997).
3. For this particular plant it appears that the worst-case full-block input uncertainty is a diagonal perturbation, so we might as well have used a full matrix for Δ_I. But this does *not* hold in general.
4. The $\mathcal{H}_\infty$ software may encounter numerical problems if $P(s)$ has poles on the $j\omega$-axis. This is the reason why in the Matlab code we have moved the integrators (in the performance weights) slightly into the LHP
5. The initial choice of scaling $D = I$ gave a good design for this plant with an $\mathcal{H}_\infty$ norm of about 1.18. This scaling worked well because the inputs and outputs had been scaled to be of unit magnitude. For a comparison, consider the original model in Skogestad et al. (1988) which was in terms of *unscaled* physical variables:

$$G_{\text{unscaled}}(s) = \frac{1}{75s + 1}\begin{bmatrix} 0.878 & -0.864 \\ 1.082 & -1.096 \end{bmatrix} \tag{8.146}$$

Equation (8.146) has all its elements 100 times smaller than in the scaled model (8.144). Therefore, using this model should give the same optimal μ-value but with controller gains 100 times larger. However, starting the DK-iteration with $D = I$ works very poorly in this case. The first iteration yields an $\mathcal{H}_\infty$ norm of 14.9 (step 1) resulting in a peak μ-value of 5.2 (step 2). Subsequent iterations yield with third- and fourth-order fits of the D-scales the following peak μ-values: 2.92, 2.22, 1.87, 1.67, 1.58, 1.53, 1.49, 1.46, 1.44, 1.42. At this point (after 11 iterations) the μ-plot is fairly flat up to 10 [rad/min] and one may be tempted to stop the iterations. However, we are still far away from the optimal value which we know is less than 1. This demonstrates the importance of good initial D-scales, which is related to scaling the plant model properly.
6. We used the stepwise procedure for DK-iteration primarily for insight. Matlab Robust Control toolbox command dksyn provides an automated version of this procedure, which is shown at the bottom of Table 8.2. For the distillation process example, dksyn yields a 26-state controller with a μ value of 1.094 in four iterations, which is inferior compared to the "manual" stepwise procedure.

Exercise 8.24 * *Explain why the optimal μ-value would not change if in the model (8.144) we changed the time constant of 75 [min] to another value. Note that the μ-iteration itself would be affected.*

8.13 Further remarks on μ

8.13.1 Further justification for the upper bound on μ

For complex perturbations, the scaled singular value $\bar{\sigma}(DND^{-1})$ is a tight upper bound on $\mu(N)$ in most cases, and minimizing the upper bound $\|DND^{-1}\|_{\infty}$ forms the basis for the DK-iteration. However, $\|DND^{-1}\|_{\infty}$ is also of interest in its own right. The reason for this is that when all uncertainty blocks are full and complex, the upper bound provides a necessary and sufficient condition for robustness to *arbitrary-slow time-varying linear uncertainty* (Poolla and Tikku, 1995). On the other hand, the use of μ assumes the uncertain perturbations to be *time invariant*. In some cases, it can be argued that slowly time-varying uncertainty is more useful than constant perturbations, and therefore that it is better to minimize $\|DND^{-1}\|_{\infty}$ instead of $\mu(N)$. In addition, by considering how $D(\omega)$ varies with frequency, one can find bounds on the allowed time variations in the perturbations.

Another interesting fact is that the use of *constant* D-scales (D is *not* allowed to vary with frequency) provides a necessary and sufficient condition for robustness to *arbitrary-fast time-varying linear uncertainty* (Shamma, 1994). It may be argued that such perturbations are unlikely in a practical situation. Nevertheless, we see that if we can get an acceptable controller design using constant D-scales, then we know that this controller will work very well even for rapid changes in the plant model. Another advantage of constant D-scales is that the computation of μ is then straightforward and may be solved using LMIs, see Example 12.4.

8.13.2 Real perturbations and the mixed μ-problem

We have not discussed in any detail the analysis and design problems which arise with real or, more importantly, mixed real and complex perturbations.

The current algorithms, implemented in the Matlab μ-toolbox, employ a generalization of the upper bound $\bar{\sigma}(DMD^{-1})$, where in addition to D-matrices, which exploit the block-diagonal structure of the perturbations, there are G-matrices, which exploit the structure of the real perturbations. The G-matrices (which should not be confused with the plant transfer function $G(s)$) have real diagonal elements at locations where Δ is real and have zeros elsewhere. The algorithm in the μ-toolbox makes use of the following result from Young et al. (1992): if there exist a $\beta > 0$, a D and a G with the appropriate block-diagonal structure such that

$$\bar{\sigma}\left((I + G^2)^{-\frac{1}{4}}\left(\frac{1}{\beta}DMD^{-1} - jG\right)(I + G^2)^{-\frac{1}{4}}\right) \leq 1 \tag{8.147}$$

then $\mu(M) \leq \beta$. For more details, the reader is referred to Young (1993).

There is also a corresponding DGK-iteration procedure for synthesis (Young, 1994). The practical implementation of this algorithm is, however, difficult, and a very high-order fit may be required for the G-scales. An alternative approach which involves solving a series of scaled DK-iterations is given by Tøffner-Clausen et al. (1995).

8.13.3 Computational complexity

It has been established that the computational complexity of computing μ has a combinatoric (non-polynomial or "*NP*-hard") growth with the number of parameters involved (Braatz et al., 1994), even for purely complex perturbations (Toker and Ozbay, 1998).

This does not mean, however, that practical algorithms are not possible, and we have described practical algorithms for computing upper bounds of μ for cases with complex, real or mixed real/complex perturbations.

As mentioned on page 310, the upper bound $\bar{\sigma}(DMD^{-1})$ for complex perturbations is generally tight, whereas the present upper bounds for mixed perturbations (see (8.147)) may be arbitrarily conservative.

There also exist a number of lower bounds for computing μ. Most of these involve generating a perturbation which makes $I - M\Delta$ singular, see e.g. Young and Doyle (1997).

8.13.4 Discrete case

It is also possible to use μ for analyzing RP of discrete time systems (Packard and Doyle, 1993). Consider a discrete time system

$$x_{k+1} = Ax_k + Bu_k, \quad y_k = Cx_k + Du_k$$

The corresponding discrete transfer function matrix from u to y is $N(z) = C(zI - A)^{-1}B + D$. First, note that the $\mathcal{H}_\infty$ norm of a discrete transfer function is

$$\|N\|_\infty \triangleq \max_{|z|\geq 1} \bar{\sigma}(C(zI - A)^{-1}B + D)$$

This follows since evaluation on the $j\omega$-axis in the continuous case is equivalent to the unit circle ($|z| = 1$) in the discrete case. Second, note that $N(z)$ may be written as an LFT in terms of $1/z$,

$$N(z) = C(zI - A)^{-1}B + D = F_u\left(H, \frac{1}{z}I\right); \quad H = \begin{bmatrix} A & B \\ C & D \end{bmatrix} \qquad (8.148)$$

Thus, by introducing $\delta_z = 1/z$ and $\Delta_z = \delta_z I$ we have from the main loop theorem of Packard and Doyle (1993) (which generalizes Theorem 8.7) that $\|N\|_\infty < 1$ (NP) if and only if

$$\mu_{\widehat{\Delta}}(H) < 1, \quad \widehat{\Delta} = \text{diag}\{\Delta_z, \Delta_P\} \qquad (8.149)$$

where Δ_z is a matrix of repeated complex scalars, representing the discrete "frequencies", and Δ_P is a full complex matrix, representing the singular value performance specification. Thus, we see that the search over frequencies in the frequency domain is avoided, but at the expense of a complicated μ-calculation. The condition in (8.149) is also referred to as the *state-space μ-test*.

Condition (8.149) only considers nominal performance (NP). However, note that in this case nominal stability (NS) follows as a special case (and thus does not need to be tested separately), since when $\mu_{\widehat{\Delta}}(H) \leq 1$ (NP) we have from (8.78) that $\mu_{\Delta_z}(A) = \rho(A) < 1$, which is the well-known stability condition for discrete systems.

We can also generalize the treatment to consider RS and RP. In particular, since the state-space matrices are contained explicitly in H in (8.148), it follows that the discrete time formulation is convenient if we want to consider parametric uncertainty in the state-space matrices. This is discussed by Packard and Doyle (1993). However, this results in real perturbations, and the resulting μ-problem which involves repeated complex perturbations (from the evaluation of z on the unit circle), a full-block complex perturbation (from the performance specification), and real perturbations (from the uncertainty), is difficult to solve numerically both for analysis and in particular for synthesis. For this reason the discrete time formulation is little used in practical applications.

8.14   Conclusion

In this chapter and the last we have discussed how to represent uncertainty and how to analyze its effect on stability (RS) and performance (RP) using the structured singular value μ as our main tool.

To analyze robust stability (RS) of an uncertain system we make use of the $M\Delta$-structure (Figure 8.3) where M represents the transfer function for the "new" feedback part generated by the uncertainty. From the small-gain theorem,

$$\text{RS} \quad \Leftarrow \quad \bar{\sigma}(M) < 1 \quad \forall \omega \tag{8.150}$$

which is tight (necessary and sufficient) for the special case where at each frequency *any* complex Δ satisfying $\bar{\sigma}(\Delta) \leq 1$ is allowed. More generally, the tight condition is

$$\text{RS} \quad \Leftrightarrow \quad \mu(M) < 1 \quad \forall \omega \tag{8.151}$$

where $\mu(M)$ is the structured singular value $\mu(M)$. The calculation of μ makes use of the fact that Δ has a given block-diagonal structure, where certain blocks may also be real (e.g. to handle parametric uncertainty).

We defined robust performance (RP) as $\|F_u(N, \Delta)\|_\infty < 1$ for all allowed Δ's. Since we used the $\mathcal{H}_\infty$ norm in both the representation of uncertainty and the definition of performance, we found that RP could be viewed as a special case of RS, and we derived

$$\text{RP} \quad \Leftrightarrow \quad \mu(N) < 1 \quad \forall \omega \tag{8.152}$$

where μ is computed with respect to the block-diagonal structure $\text{diag}\{\Delta, \Delta_P\}$. Here Δ represents the uncertainty and Δ_P is a fictitious full uncertainty block representing the $\mathcal{H}_\infty$ performance bound.

It should be noted that there are two main approaches to getting a robust design:

1. We aim to make the system robust to some "general" class of uncertainty which we do not explicitly model. For SISO systems the classical gain and phase margins and the peaks of S and T provide useful general robustness measures. For MIMO systems, normalized coprime factor uncertainty provides a good general class of uncertainty, and the associated Glover–McFarlane $\mathcal{H}_\infty$ loop-shaping design procedure, see Chapter 9, has proved itself very useful in applications.
2. We explicitly model and quantify the uncertainty in the plant and aim to make the system robust to this specific uncertainty. This second approach has been the focus of the preceding two chapters. Potentially, it yields better designs, but it may require a much larger effort in terms of uncertainty modelling, especially if parametric uncertainty is considered. Analysis and, in particular, synthesis using μ can be very involved.

In applications, it is therefore recommended to start with the first approach, at least for design. The robust stability and performance are then analyzed in simulations and using the structured singular value; for example, by considering first simple sources of uncertainty such as multiplicative input uncertainty. One then iterates between design and analysis until a satisfactory solution is obtained.

Practical μ-analysis

We end the chapter by providing a few recommendations on how to use the structured singular value μ in practice.

1. Because of the effort involved in deriving detailed uncertainty descriptions, and the subsequent complexity in synthesizing controllers, the rule is to "start simple" with a crude uncertainty description, and then to see whether the performance specifications can be met. Only if they can't, should one consider more detailed uncertainty descriptions such as parametric uncertainty (with real perturbations).
2. The use of μ implies a worst-case analysis, so one should be careful about including too many sources of uncertainty, noise and disturbances – otherwise it becomes very unlikely for the worst case to occur, and the resulting analysis and design may be unnecessarily conservative.
3. There is always uncertainty with respect to the inputs and outputs, so it is generally "safe" to include *diagonal* input and output uncertainty. The relative (multiplicative) form is very convenient in this case.
4. μ is most commonly used for analysis. If μ is used for synthesis, then we recommend that you keep the uncertainty fixed and adjust the parameters in the performance weight until μ is close to 1.

9

CONTROLLER DESIGN

In this chapter, we present practical procedures for multivariable controller design which are relatively straightforward to apply and which, in our opinion, have an important role to play in industrial control.

For industrial systems which are either SISO or loosely coupled, the classical loop-shaping approach to control system design as described in Section 2.6 has been successfully applied. But for truly multivariable systems it has only been in the last two decades, or so, that reliable generalizations of this classical approach have emerged.

9.1 Trade-offs in MIMO feedback design

The shaping of multivariable transfer functions is based on the idea that a satisfactory definition of gain (range of gain) for a matrix transfer function is given by the singular values of the transfer function. By multivariable transfer function shaping, therefore, we mean the shaping of singular values of appropriately specified transfer functions such as the loop transfer function or possibly one or more closed-loop transfer functions. This methodology for controller design is central to the practical procedures described in this chapter.

In February 1981, the *IEEE Transactions on Automatic Control* published a Special Issue on Linear Multivariable Control Systems, the first six papers of which were on the use of singular values in the analysis and design of multivariable feedback systems. The paper by Doyle and Stein (1981) was particularly influential: it was primarily concerned with the fundamental question of how to achieve the benefits of feedback in the presence of unstructured uncertainty, and through the use of singular values it showed how the classical loop-shaping ideas of feedback design could be generalized to multivariable systems. To see how this was done, consider the one degree-of-freedom configuration shown in Figure 9.1. The plant G and controller K interconnection is driven by reference commands r, output disturbances d and measurement noise n. y are the outputs to be controlled, and u are the control signals. In terms of the sensitivity function $S = (I + GK)^{-1}$ and the closed-loop transfer function $T = GK(I + GK)^{-1} = I - S$, we have the following important relationships:

$$y(s) = T(s)r(s) + S(s)d(s) - T(s)n(s) \qquad (9.1)$$
$$u(s) = K(s)S(s)\left[r(s) - n(s) - d(s)\right] \qquad (9.2)$$

These relationships determine several closed-loop objectives, in addition to the requirement that K stabilizes G, namely:

Multivariable Feedback Control: Analysis and Design. Second Edition
S. Skogestad and I. Postlethwaite   © 2005, 2006 John Wiley & Sons, Ltd

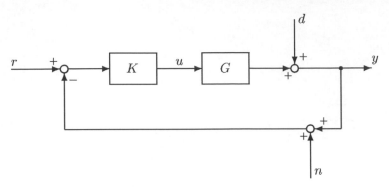

Figure 9.1: One degree-of-freedom feedback configuration

1. For *disturbance rejection* make $\bar{\sigma}(S)$ small.
2. For *noise attenuation* make $\bar{\sigma}(T)$ small.
3. For *reference tracking* make $\bar{\sigma}(T) \approx \underline{\sigma}(T) \approx 1$.
4. For *input usage (control energy) reduction* make $\bar{\sigma}(KS)$ small.

If the unstructured uncertainty in the plant model G is represented by an additive perturbation, i.e. $G_p = G + \Delta$, then from (8.53), a further closed-loop objective is:

5. For *robust stability* in the presence of an additive perturbation make $\bar{\sigma}(KS)$ small.

Alternatively, if the uncertainty is modelled by a multiplicative output perturbation such that $G_p = (I + \Delta)G$, then from (8.55), we have:

6. For *robust stability* in the presence of a multiplicative output perturbation make $\bar{\sigma}(T)$ small.

The closed-loop requirements 1 to 6 cannot all be satisfied simultaneously. Feedback design is therefore a trade-off over frequency of conflicting objectives. This is not always as difficult as it sounds because the frequency ranges over which the objectives are important can be quite different. For example, disturbance rejection is typically a low-frequency requirement, while noise mitigation is often only relevant at higher frequencies.

In classical *loop shaping*, it is the magnitude of the open-loop transfer function $L = GK$ which is shaped, whereas the above design requirements are all in terms of closed-loop transfer functions. However, recall from (3.51) that

$$\underline{\sigma}(L) - 1 \le \frac{1}{\bar{\sigma}(S)} \le \underline{\sigma}(L) + 1 \tag{9.3}$$

from which we see that $\bar{\sigma}(S) \approx 1/\underline{\sigma}(L)$ at frequencies where $\underline{\sigma}(L)$ is much larger than 1. It also follows that at the bandwidth frequency (where $1/\bar{\sigma}(S(j\omega_B)) = \sqrt{2} = 1.41$), we have $\underline{\sigma}(L(j\omega_B))$ between 0.41 and 2.41. Furthermore, from $T = L(I + L)^{-1}$ it follows that $\bar{\sigma}(T) \approx \bar{\sigma}(L)$ at frequencies where $\bar{\sigma}(L)$ is small. Thus, over specified frequency ranges, it is relatively easy to approximate the closed-loop requirements by the following open-loop objectives:

1. For *disturbance rejection* make $\underline{\sigma}(GK)$ large; valid for frequencies at which $\underline{\sigma}(GK) \gg 1$.

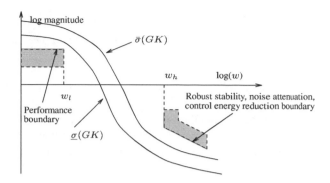

Figure 9.2: Design trade-offs for the multivariable loop transfer function GK

2. For *noise attenuation* make $\bar{\sigma}(GK)$ small; valid for frequencies at which $\bar{\sigma}(GK) \ll 1$.
3. For *reference tracking* make $\underline{\sigma}(GK)$ large; valid for frequencies at which $\underline{\sigma}(GK) \gg 1$.
4. For *input usage (control energy) reduction* make $\bar{\sigma}(K)$ small; valid for frequencies at which $\bar{\sigma}(GK) \ll 1$.
5. For *robust stability to an additive perturbation* make $\bar{\sigma}(K)$ small; valid for frequencies at which $\bar{\sigma}(GK) \ll 1$.
6. For *robust stability to a multiplicative output perturbation* make $\bar{\sigma}(GK)$ small; valid for frequencies at which $\bar{\sigma}(GK) \ll 1$.

Typically, the open-loop requirements 1 and 3 are valid and important at low frequencies, $0 \leq \omega \leq \omega_l \leq \omega_B$, while 2, 4, 5 and 6 are conditions which are valid and important at high frequencies, $\omega_B \leq \omega_h \leq \omega \leq \infty$, as illustrated in Figure 9.2. From this we see that at frequencies where we want high gains (at low frequencies) the "worst-case" direction is related to $\underline{\sigma}(GK)$, whereas at frequencies where we want low gains (at high frequencies) the "worst-case" direction is related to $\bar{\sigma}(GK)$.

Exercise 9.1 * *Show that the closed-loop objectives 1 to 6 can be approximated by the open-loop objectives 1 to 6 at the specified frequency ranges.*

From Figure 9.2, it follows that the control engineer must design K so that $\bar{\sigma}(GK)$ and $\underline{\sigma}(GK)$ avoid the shaded regions. That is, for good performance, $\underline{\sigma}(GK)$ must be made to lie above a performance boundary for all ω up to ω_l, and for robust stability $\bar{\sigma}(GK)$ must be forced below a robustness boundary for all ω above ω_h. To shape the singular values of GK by selecting K is a relatively easy task, but to do this in a way which also guarantees closed-loop stability is in general difficult. Closed-loop stability cannot be determined from open-loop singular values.

For SISO systems, it is clear from Bode's work (1945) that closed-loop stability is closely related to open-loop gain and phase near the crossover frequency ω_c, where $|GK(j\omega_c)| = 1$. In particular, the roll-off rate from high to low gain at crossover is limited by phase requirements for stability, and in practice this corresponds to a roll-off rate less than $40\ dB/decade$ (slope -2 on log–log plot); see Section 2.6.2. An immediate consequence of this is that there is a lower limit to the difference between ω_h and ω_l in Figure 9.2.

For MIMO systems a similar gain–phase relationship holds in the crossover frequency region, but this is in terms of the eigenvalues of GK and results in a limit on the roll-off rate of the magnitude of the eigenvalues of GK, not the singular values (Doyle and Stein, 1981).

The stability constraint is therefore even more difficult to handle in multivariable loop shaping than it is in classical loop shaping. To overcome this difficulty Doyle and Stein (1981) proposed that the loop shaping should be done with a controller that was already known to guarantee stability. They suggested that an LQG controller could be used in which the regulator part is designed using a "sensitivity recovery" procedure of Kwakernaak (1969) to give desirable properties (gain and phase margins) in GK. They also gave a dual "robustness recovery" procedure for designing the filter in an LQG controller to give desirable properties in KG. Recall that KG is not in general equal to GK, which implies that stability margins vary from one break point to another in a multivariable system. Both of these loop transfer recovery (LTR) procedures are discussed below after first describing traditional LQG control.

9.2 LQG control

Optimal control, building on the optimal filtering work of Wiener in the 1940's, reached maturity in the 1960's with what we now call linear quadratic Gaussian or LQG control. Its development coincided with large research programmes and considerable funding in the United States and the former Soviet Union on space-related problems. These were problems, such as rocket manoeuvring with minimum fuel consumption, which could be well defined and easily formulated as optimization problems. Aerospace engineers were particularly successful at applying LQG, but when other control engineers attempted to use LQG on everyday industrial problems a different story emerged. Accurate plant models were frequently not available and the assumption of white noise disturbances was not always relevant or meaningful to practising control engineers. As a result LQG designs were sometimes not robust enough to be used in practice. In this section, we will describe the LQG problem and its solution, we will discuss its robustness properties, and we will describe procedures for improving robustness. Many textbooks consider this topic in far greater detail; we recommend Anderson and Moore (1989) and Kwakernaak and Sivan (1972).

9.2.1 Traditional LQG and LQR problems

In traditional LQG control, it is assumed that the plant dynamics are linear and known, and that the measurement noise inputs and disturbance signals (process noise) are stochastic with known statistical properties. That is, we have a plant model

$$\dot{x} = Ax + Bu + w_d \qquad (9.4)$$
$$y = Cx + Du + w_n \qquad (9.5)$$

where for simplicity we set $D = 0$ (see Remark 2 on page 347). w_d and w_n are the disturbance (process noise) and measurement noise respectively, which are usually assumed to be uncorrelated zero-mean Gaussian stochastic processes with constant power spectral density matrices W and V respectively. That is, w_d and w_n are white noise processes with covariances

$$E\left\{w_d(t)w_d(\tau)^T\right\} = W\delta(t-\tau) \qquad (9.6)$$
$$E\left\{w_n(t)w_n(\tau)^T\right\} = V\delta(t-\tau) \qquad (9.7)$$

and

$$E\left\{w_d(t)w_n(\tau)^T\right\} = 0, \ E\left\{w_n(t)w_d(\tau)^T\right\} = 0 \tag{9.8}$$

where E is the expectation operator and $\delta(t - \tau)$ is a delta function.

The LQG control problem is to find the optimal control $u(t)$ which minimizes

$$J = E\left\{\lim_{T \to \infty} \frac{1}{T} \int_0^T \left[x^T Q x + u^T R u\right] dt\right\} \tag{9.9}$$

where Q and R are appropriately chosen constant weighting matrices (design parameters) such that $Q = Q^T \geq 0$ and $R = R^T > 0$. The name LQG arises from the use of a Linear model, an integral Quadratic cost function, and Gaussian white noise processes to model disturbance signals and noise.

The solution to the LQG problem, known as the separation theorem or certainty equivalence principle, is surprisingly simple and elegant. It consists of first determining the optimal controller for a deterministic linear quadratic regulator (LQR) problem: namely, the above LQG problem without w_d and w_n. It happens that the solution to this problem can be written in terms of the simple state feedback law

$$u(t) = -K_r x(t) \tag{9.10}$$

where K_r is a constant matrix which is easy to compute and is clearly independent of W and V, the statistical properties of the plant noise. Note that (9.10) requires that x is measured and available for feedback, which is not generally the case. This difficulty is overcome by the next step, where we find an optimal estimate $\hat{x}$ of the state x, so that $E\left\{[x - \hat{x}]^T [x - \hat{x}]\right\}$ is minimized. The optimal state estimate is given by a Kalman filter and is independent of Q and R. The required solution to the LQG problem is then found by replacing x by $\hat{x}$, to give $u(t) = -K_r \hat{x}(t)$. We therefore see that the LQG problem and its solution can be separated into two distinct parts, as illustrated in Figure 9.3.

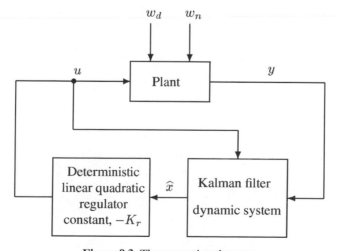

Figure 9.3: The separation theorem

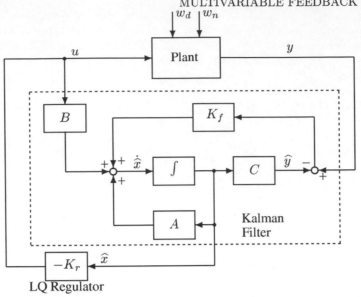

Figure 9.4: The LQG controller and noisy plant

We will now give the equations necessary to find the optimal state feedback matrix K_r and the Kalman filter.

Optimal state feedback. The LQR problem, where all the states are known, is the deterministic initial value problem: given the system $\dot{x} = Ax + Bu$ with a non-zero initial state $x(0)$, find the input signal $u(t)$ which takes the system to the zero state ($x = 0$) in an optimal manner, i.e. by minimizing the deterministic cost

$$J_r = \int_0^\infty (x(t)^T Q x(t) + u(t)^T R u(t)) dt \tag{9.11}$$

The optimal solution (for any initial state) is $u(t) = -K_r x(t)$, where

$$K_r = R^{-1} B^T X \tag{9.12}$$

and $X = X^T \geq 0$ is the unique positive semi-definite solution of the algebraic Riccati equation

$$A^T X + XA - XBR^{-1}B^T X + Q = 0 \tag{9.13}$$

Kalman filter. The Kalman filter has the structure of an ordinary state estimator or observer, as shown in Figure 9.4, with

$$\dot{\hat{x}} = A\hat{x} + Bu + K_f(y - C\hat{x}) \tag{9.14}$$

The optimal choice of K_f, which minimizes $E\left\{[x - \hat{x}]^T [x - \hat{x}]\right\}$, is given by

$$K_f = YC^T V^{-1} \tag{9.15}$$

where $Y = Y^T \geq 0$ is the unique positive semi-definite solution of the algebraic Riccati equation

$$YA^T + AY - YC^T V^{-1} CY + W = 0 \tag{9.16}$$

LQG: combined optimal state estimation and optimal state feedback. The LQG control problem is to minimize J in (9.9). The structure of the LQG controller is illustrated in Figure 9.4; its transfer function, from y to u (i.e. assuming positive feedback), is easily shown to be given by

$$K_{\text{LQG}}(s) \stackrel{s}{=} \left[\begin{array}{c|c} A - BK_r - K_fC & K_f \\ \hline -K_r & 0 \end{array} \right]$$

$$= \left[\begin{array}{c|c} A - BR^{-1}B^TX - YC^TV^{-1}C & YC^TV^{-1} \\ \hline -R^{-1}B^TX & 0 \end{array} \right] \qquad (9.17)$$

It has the same degree (number of poles) as the plant.

Remark 1 The optimal gain matrices K_f and K_r exist, and the LQG-controlled system is internally stable, provided the systems with state-space realizations $(A, B, Q^{\frac{1}{2}})$ and $(A, W^{\frac{1}{2}}, C)$ are stabilizable and detectable.

Remark 2 If the plant model is bi-proper, with a non-zero D-term in (9.5), then the Kalman filter equation (9.14) has the extra term $-K_fDu$ on the right hand side, and the A-matrix of the LQG controller in (9.17) has the extra term $+K_fDK_r$.

Exercise 9.2 *For the plant and LQG controller arrangement of Figure 9.4, show that the closed-loop dynamics are described by*

$$\frac{d}{dt} \left[\begin{array}{c} x \\ x - \hat{x} \end{array} \right] = \left[\begin{array}{cc} A - BK_r & BK_r \\ 0 & A - K_fC \end{array} \right] \left[\begin{array}{c} x \\ x - \hat{x} \end{array} \right] + \left[\begin{array}{cc} I & 0 \\ I & -K_f \end{array} \right] \left[\begin{array}{c} w_d \\ w_n \end{array} \right]$$

This shows that the closed-loop poles are simply the union of the poles of the deterministic LQR system (eigenvalues of $A - BK_r$) and the poles of the Kalman filter (eigenvalues of $A - K_fC$). It is exactly as we would have expected from the separation theorem.

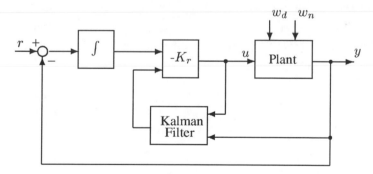

Figure 9.5: LQG controller with integral action and reference input

For the LQG controller, as shown in Figure 9.4, it is not easy to see where to position the reference input r, and how integral action may be included, if desired. One strategy is illustrated in Figure 9.5. Here the control error $r - y$ is integrated and the regulator K_r is designed for the plant augmented with the integrator states.

Example 9.1 LQG design with integral action for inverse response process. *The standard LQG design procedure does not give a controller with integral action, so we will use the setup in Figure 9.5*

and augment the plant $G(s)$ with an integrator before designing the state feedback regulator. The plant is the SISO inverse response process $G(s) = \frac{3(-2s+1)}{(5s+1)(10s+1)}$ in (2.31), which was studied extensively in Chapter 2. For the objective function $J = \int (x^T Q x + u^T R u) dt$, we choose Q such that only the integrated state $y - r$ is weighted, and we choose the input weight $R = 1$. (Only the ratio between Q and R matters and reducing R yields a faster response.) The Kalman filter is set up so that we do not estimate the integrated states. For the noise weights we select $W = wI$ (process noise directly on the states) with $w = 1$, and we choose $V = 1$ (measurement noise). (Only the ratio between w and V matters and reducing V yields a faster response.) The Matlab file in Table 9.1 was used to design the LQG controller. The resulting closed-loop response is shown in Figure 9.6. The response is good and very similar to the loop-shaping design in Figure 2.20 (page 45).

Table 9.1: Matlab commands to generate LQG controller in Example 9.1

```
% Uses the Control toolbox
G = tf(3*[-2 1],conv([5 1],[10 1]));    % inverse response process
[a,b,c,d] = ssdata(G);
% Model dimensions:
p = size(c,1);                          % no. of outputs (y)
[n,m] = size(b);                        % no. of states and inputs (u)
Znm=zeros(n,m); Zmm=zeros(m,m);
Znn=zeros(n,n); Zmn=zeros(m,n);
% 1) Design state feedback regulator
A = [a Znm;-c Zmm]; B = [b;-d];         % augment plant with integrators
Q=[Znn Znm;Zmn eye(m,m)];              % weight on integrated error
R=eye(m);                               % input weight
Kr=lqr(A,B,Q,R);                        % optimal state-feedback regulator
Krp=Kr(1:m,1:n);Kri=Kr(1:m,n+1:n+m);   % extract integrator and state feedbacks
% 2) Design Kalman filter                % don't estimate integrator states
Bnoise = eye(n);                        % process noise model (Gd)
W = eye(n); V = 1*eye(m);               % process and measurement noise weight
Estss = ss(a,[b Bnoise],c,[0 0 0]);
[Kess, Ke] = kalman(Estss,W,V);         % Kalman filter gain
% 3) Form overall controller
Ac=[Zmm Zmn;-b*Kri a-b*Krp-Ke*c];      % integrators included
Bcr = [eye(m); Znm]; Bcy = [-eye(m); Ke];
Cc = [-Kri -Krp]; Dcr = Zmm; Dcy = Zmm;
Klqg2 = ss(Ac,[Bcr Bcy],Cc,[Dcr Dcy]); % Final 2-DOF controller from [r y]' to u
Klqg = ss(Ac,-Bcy,Cc,-Dcy);            % Feedback part of controller from -y to u
% Simulation
sys1 = feedback(G*Klqg,1); step(sys1,50);  % 1-DOF simulation
sys = feedback(G*Klqg2,1,2,1,+1);      % 2-DOF simulation
sys2 = sys*[1; 0]; hold; step(sys2,50);
```

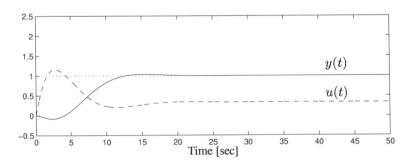

Figure 9.6: LQG design (Klqg2 in Table 9.1) for inverse response process. Closed-loop response to unit step in reference r.

Remark. *We just noted the similarity between the responses in Figure 9.6 (LQG) and Figure 2.20* *(loop shaping). However, note that the loop-shaping controller is a one degree-of-freedom controller, whereas the LQG controller is actually a two degrees-of-freedom controller (Klqg2 in Table 9.1). As seen from Figure 9.5, the reference change is not sent directly to the Kalman filter and this avoids the "derivative or proportional kick". For our specific example, the step response (not shown) of the one degree-of-freedom LQG controller $u = K_{\mathrm{LQG}}(r - y)$ (Klqg in Table 9.1) is significantly worse with an overshoot in y of about 40%. This large overshoot translates into much poorer robustness margins for LQG than the loop-shaping design; see Exercise 9.4. Also, the disturbance rejection is poorer for the LQG design.*

Exercise 9.3 *Derive the equations used in the Matlab file in Table 9.1.*

Exercise 9.4 *Compare the robustness of the loop-shaping and LQG designs in Examples 2.8 and 9.1, respectively, by computing the gain and phase margins (GM and PM) and the sensitivity peaks (M_S and M_T). (Note that robustness is given by the feedback loop, and is thus the same for the LQG controllers* Klqg *and* Klqg2 *defined in Table 9.1.)*

	GM	PM	M_S	M_T
Solution:				
Loop-shaping	2.92	53.9°	1.11	1.75
LQG	1.83	37.4°	1.63	2.39

9.2.2 Robustness properties

For an LQG-controlled system with a combined Kalman filter and LQR control law there are no guaranteed stability margins. This was brought starkly to the attention of the control community by Doyle (1978) (in a paper entitled "Guaranteed Margins for LQG Regulators" with a very compact abstract which simply states "There are none"). He showed, by example, that there exist LQG combinations with arbitrarily small-gain margins.

However, for an LQR-controlled system (i.e. assuming all the states are available and no stochastic inputs) it is well known (Kalman, 1964; Safonov and Athans, 1977) that, if the weight R is chosen to be diagonal, the sensitivity function $S = (I + K_r(sI - A)^{-1}B)^{-1}$ satisfies the Kalman inequality

$$\bar{\sigma}(S(j\omega)) \leq 1, \ \forall w \tag{9.18}$$

From this it can be shown that the system will have a gain margin equal to infinity, a gain reduction margin (lower gain margin) equal to 0.5, and a (minimum) phase margin of 60° in each plant input control channel. This means that in the LQR-controlled system $u = -K_r x$, a complex perturbation diag $\{k_i e^{j\theta_i}\}$ can be introduced at the plant inputs without causing instability provided

(i) $\theta_i = 0$ and $0.5 \leq k_i \leq \infty$, $i = 1, 2, \ldots, m$

or

(ii) $k_i = 1$ and $|\theta_i| \leq 60°$, $i = 1, 2, \ldots, m$

where m is the number of plant inputs. For a single-input plant, the above shows that the Nyquist diagram of the open-loop regulator transfer function $K_r(sI - A)^{-1}B$ will always lie outside the unit circle with centre -1. This was first shown by Kalman (1964), and is illustrated in Example 9.2 below.

Example 9.2 LQR design of a first-order process. *Consider a first-order process $G(s) = 1/(s-a)$ with the state-space realization*

$$\dot{x}(t) = ax(t) + u(t), \quad y(t) = x(t)$$

so that the state is directly measured. For a non-zero initial state the cost function to be minimized is

$$J_r = \int_0^\infty (x^2 + Ru^2)dt$$

The algebraic Riccati equation (9.13) becomes $(A = a, B = 1, Q = 1)$

$$aX + Xa - XR^{-1}X + 1 = 0 \quad \Leftrightarrow \quad X^2 - 2aRX - R = 0$$

which, since $X \geq 0$, *gives* $X = aR + \sqrt{(aR)^2 + R}$. *The optimal control is given by* $u = -K_r x$ *where from (9.12)*

$$K_r = X/R = a + \sqrt{a^2 + 1/R}$$

and we get the closed-loop system

$$\dot{x} = ax + u = -\sqrt{a^2 + 1/R}\, x$$

The closed-loop pole is located at $s = -\sqrt{a^2 + 1/R} < 0$. *Thus, the root locus for the optimal closed-loop pole with respect to* R *starts at* $s = -|a|$ *for* $R = \infty$ *(infinite weight on the input) and moves to* $-\infty$ *along the real axis as* R *approaches zero. Note that the root locus is identical for stable* $(a < 0)$ *and unstable* $(a > 0)$ *plants* $G(s)$ *with the same value of* $|a|$. *In particular, for* $a > 0$ *we see that the minimum input energy needed to stabilize the plant (corresponding to* $R = \infty$) *is obtained with the input* $u = -2|a|x$, *which moves the pole from* $s = a$ *to its mirror image at* $s = -a$.

For R *small ("cheap control") the gain crossover frequency of the loop transfer function* $L = GK_r = K_r/(s - a)$ *is given approximately by* $\omega_c \approx \sqrt{1/R}$. *Note also that* $L(j\omega)$ *has a roll-off of* -1 *(−20 dB/decade) at high frequencies, which is a general property of LQR designs. Furthermore, the Nyquist plot of* $L(j\omega)$ *avoids the unit disc centred on the critical point* -1, *i.e.* $|S(j\omega)| = 1/|1 + L(j\omega)| \leq 1$ *at all frequencies. This is obvious for the stable plant with* $a < 0$ *since* $K_r > 0$ *and then the phase of* $L(j\omega)$ *varies from* $0°$ *(at zero frequency) to* $-90°$ *(at infinite frequency). The surprise is that it is also true for the unstable plant with* $a > 0$ *even though the phase of* $L(j\omega)$ *varies from* $-180°$ *to* $-90°$.

Consider now the Kalman filter shown earlier in Figure 9.4. Notice that it is itself a feedback system. Arguments dual to those employed for the LQR-controlled system can then be used to show that, if the power spectral density matrix V is chosen to be diagonal, then at the input to the Kalman gain matrix K_f there will be an infinite gain margin, a gain reduction margin of 0.5 and a minimum-phase margin of 60°. Consequently, for a single-output plant, the Nyquist diagram of the open-loop filter transfer function $C(sI - A)^{-1}K_f$ will lie outside the unit circle with centre at -1.

An LQR-controlled system has good stability margins at the plant inputs, and a Kalman filter has good stability margins at the inputs to K_f, so why are there no guarantees for LQG control? To answer this, consider the LQG controller arrangement shown in Figure 9.7. The loop transfer functions associated with the labelled points 1 to 4 are respectively

$$\begin{aligned}
L_1(s) &= K_r \left[\Phi(s)^{-1} + BK_r + K_f C\right]^{-1} K_f C\Phi(s)B \\
&= -K_{\text{LQG}}(s)G(s) & (9.19) \\
L_2(s) &= -G(s)K_{\text{LQG}}(s) & (9.20) \\
L_3(s) &= K_r\Phi(s)B \text{ (regulator transfer function)} & (9.21) \\
L_4(s) &= C\Phi(s)K_f \text{ (filter transfer function)} & (9.22)
\end{aligned}$$

where

$$\Phi(s) \triangleq (sI - A)^{-1} \tag{9.23}$$

$K_{\text{LQG}}(s)$ is as in (9.17) and $G(s) = C\Phi(s)B$ is the plant model.

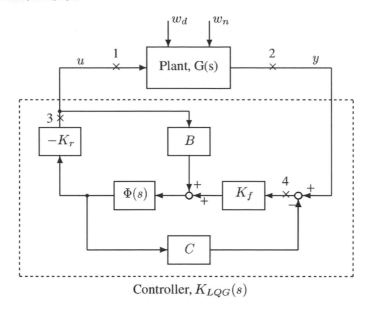

Figure 9.7: LQG-controlled plant

Remark. $L_3(s)$ and $L_4(s)$ are surprisingly simple. For $L_3(s)$ the reason is that after opening the loop at point 3 the error dynamics (point 4) of the Kalman filter are not excited by the plant inputs; in fact they are uncontrollable from u.

Exercise 9.5 *Derive the expressions for $L_1(s)$, $L_2(s)$, $L_3(s)$ and $L_4(s)$, and explain why $L_4(s)$ (like $L_3(s)$) has such a simple form.*

At points 3 and 4 we have the guaranteed robustness properties of the LQR system and the Kalman filter respectively. But at the actual input and output of the plant (points 1 and 2) where we are most interested in achieving good stability margins, we have complex transfer functions which in general give no guarantees of satisfactory robustness properties. Notice also that points 3 and 4 are effectively inside the LQG controller which has to be implemented, most likely as software, and so we have good stability margins where they are not really needed and no guarantees where they are.

Fortunately, for a minimum-phase plant procedures developed by Kwakernaak (1969) and Doyle and Stein (1979; 1981) show how, by a suitable choice of parameters, either $L_1(s)$ can be made to tend asymptotically to $L_3(s)$ or $L_2(s)$ can be made to approach $L_4(s)$. These procedures are considered next.

9.2.3 Loop transfer recovery (LTR) procedures

For full details of the recovery procedures, we refer the reader to the original communications (Kwakernaak, 1969; Doyle and Stein, 1979; Doyle and Stein, 1981) or to the tutorial paper by Stein and Athans (1987). We will only give an outline of the major steps here, since we will argue later that the procedures are somewhat limited for practical control system design. For a more recent appraisal of LTR, we recommend the Special Issue of the *International*

Journal of Robust and Nonlinear Control, edited by Niemann and Stoustrup (1995).

The LQG loop transfer function $L_2(s)$ can be made to approach $C\Phi(s)K_f$, with its guaranteed stability margins, if K_r in the LQR problem is designed to be large using the sensitivity recovery procedure of Kwakernaak (1969). It is necessary to assume that the plant model $G(s)$ is minimum-phase and that it has at least as many inputs as outputs.

Alternatively, the LQG loop transfer function $L_1(s)$ can be made to approach $K_r\Phi(s)B$ by designing K_f in the Kalman filter to be large using the robustness recovery procedure of Doyle and Stein (1979). Again, it is necessary to assume that the plant model $G(s)$ is minimum-phase, but this time it must have at least as many outputs as inputs.

The procedures are dual and therefore we will only consider recovering robustness at the plant output. That is, we aim to make $L_2(s) = G(s)K_{\mathrm{LQG}}(s)$ approximately equal to the Kalman filter transfer function $C\Phi(s)K_f$.

First, we design a Kalman filter whose transfer function $C\Phi(s)K_f$ is desirable. This is done, in an iterative fashion, by choosing the power spectral density matrices W and V so that the minimum singular value of $C\Phi(s)K_f$ is large enough at low frequencies for good performance and its maximum singular value is small enough at high frequencies for robust stability, as discussed in Section 9.1. Notice that W and V are being used here as design parameters and their associated stochastic processes are considered to be fictitious. In tuning W and V we should be careful to choose V as diagonal and $W = (BS)(BS)^T$, where S is a scaling matrix which can be used to balance, raise, or lower the singular values. When the singular values of $C\Phi(s)K_f$ are thought to be satisfactory, loop transfer recovery is achieved by designing K_r in an LQR problem with $Q = C^T C$ and $R = \rho I$, where ρ is a scalar. As ρ tends to zero $G(s)K_{\mathrm{LQG}}$ tends to the desired loop transfer function $C\Phi(s)K_f$.

Much has been written on the use of LTR procedures in multivariable control system design. But as methods for multivariable loop shaping they are limited in their applicability and sometimes difficult to use. Their main limitation is to minimum-phase plants. This is because the recovery procedures work by cancelling the plant zeros, and a cancelled non-minimum-phase zero would lead to instability. The cancellation of lightly damped zeros is also of concern because of undesirable oscillations at these modes during transients. A further disadvantage is that the limiting process ($\rho \to 0$) which brings about full recovery also introduces high gains which may cause problems with unmodelled dynamics. Because of the above disadvantages, the recovery procedures are not usually taken to their limits ($\rho \to 0$) to achieve full recovery, but rather a set of designs is obtained (for small ρ) and an acceptable design is selected. The result is a somewhat ad-hoc design procedure in which the singular values of a loop transfer function, $G(s)K_{\mathrm{LQG}}(s)$ or $K_{\mathrm{LQG}}(s)G(s)$, are indirectly shaped. A more direct and intuitively appealing method for multivariable loop shaping will be given in Section 9.4.

9.3 $\mathcal{H}_2$ and $\mathcal{H}_\infty$ control

Motivated by the shortcomings of LQG control, there was a significant shift in the 1980's towards $\mathcal{H}_\infty$ optimization for robust control. This development originated from the influential work of Zames (1981), although an earlier use of $\mathcal{H}_\infty$ optimization in an engineering context can be found in Helton (1976). Zames argued that the poor robustness properties of LQG could be attributed to the integral criterion in terms of the $\mathcal{H}_2$ norm, and he also criticized the representation of uncertain disturbances by white noise processes as often unrealistic.

As the $\mathcal{H}_\infty$ theory developed, however, the two approaches of $\mathcal{H}_2$ and $\mathcal{H}_\infty$ control were seen to be more closely related than originally thought, particularly in the solution process; see for example Glover and Doyle (1988) and Doyle et al. (1989). In this section, we will begin with a general control problem formulation into which we can cast all $\mathcal{H}_2$ and $\mathcal{H}_\infty$ optimizations of practical interest. The general $\mathcal{H}_2$ and $\mathcal{H}_\infty$ problems will be described along with some specific and typical control problems. It is not our intention to describe in detail the mathematical solutions, since efficient, commercial software for solving such problems is readily available. Rather we seek to provide an understanding of some useful problem formulations, which might then be used by the reader, or modified to suit his or her application.

9.3.1   General control problem formulation

There are many ways in which feedback design problems can be cast as $\mathcal{H}_2$ and $\mathcal{H}_\infty$ optimization problems. It is very useful therefore to have a standard problem formulation into which any particular problem may be manipulated. Such a general formulation is afforded by the general configuration shown in Figure 9.8 and discussed earlier in Chapter 3. The system

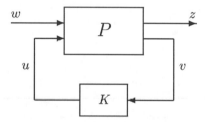

Figure 9.8: General control configuration

of Figure 9.8 is described by

$$\begin{bmatrix} z \\ v \end{bmatrix} = P(s) \begin{bmatrix} w \\ u \end{bmatrix} = \begin{bmatrix} P_{11}(s) & P_{12}(s) \\ P_{21}(s) & P_{22}(s) \end{bmatrix} \begin{bmatrix} w \\ u \end{bmatrix} \tag{9.24}$$

$$u = K(s)v \tag{9.25}$$

with a state-space realization of the generalized plant P given by

$$P \stackrel{s}{=} \left[\begin{array}{c|cc} A & B_1 & B_2 \\ \hline C_1 & D_{11} & D_{12} \\ C_2 & D_{21} & D_{22} \end{array} \right] \tag{9.26}$$

The signals are: u the control variables, v the measured variables, w the exogenous signals such as disturbances w_d and commands r, and z the so-called "error" signals which are to be minimized in some sense to meet the control objectives. As shown in (3.114) the closed-loop transfer function from w to z is given by the linear fractional transformation

$$z = F_l(P, K)w \tag{9.27}$$

where

$$F_l(P, K) = P_{11} + P_{12}K(I - P_{22}K)^{-1}P_{21} \tag{9.28}$$

$\mathcal{H}_2$ and $\mathcal{H}_\infty$ control involve the minimization of the $\mathcal{H}_2$ and $\mathcal{H}_\infty$ norms of $F_l(P, K)$ respectively. We will consider each of them in turn.

First some remarks about the algorithms used to solve such problems. The most general, widely available and widely used algorithms for $\mathcal{H}_2$ and $\mathcal{H}_\infty$ control problems are based on the state-space solutions in Glover and Doyle (1988) and Doyle et al. (1989). It is worth mentioning again that the similarities between $\mathcal{H}_2$ and $\mathcal{H}_\infty$ theory are most clearly evident in the aforementioned algorithms. For example, both $\mathcal{H}_2$ and $\mathcal{H}_\infty$ require the solutions to two Riccati equations, they both give controllers of state dimension equal to that of the generalized plant P, and they both exhibit a separation structure in the controller already seen in LQG control. An algorithm for $\mathcal{H}_\infty$ control problems is summarized in Section 9.3.4.

The following assumptions are typically made in $\mathcal{H}_2$ and $\mathcal{H}_\infty$ problems:

(A1) (A, B_2, C_2) is stabilizable and detectable.

(A2) D_{12} and D_{21} have full rank.

(A3) $\begin{bmatrix} A - j\omega I & B_2 \\ C_1 & D_{12} \end{bmatrix}$ has full column rank for all ω.

(A4) $\begin{bmatrix} A - j\omega I & B_1 \\ C_2 & D_{21} \end{bmatrix}$ has full row rank for all ω.

(A5) $D_{11} = 0$ and $D_{22} = 0$.

Assumption (A1) is required for the existence of stabilizing controllers K, and assumption (A2) is sufficient to ensure the controllers are proper and hence realizable. Assumptions (A3) and (A4) ensure that the optimal controller does not try to cancel poles or zeros on the imaginary axis which would result in closed-loop instability. Assumption (A5) is conventional in $\mathcal{H}_2$ control. $D_{11} = 0$ makes P_{11} strictly proper. Recall that $\mathcal{H}_2$ is the set of strictly proper stable transfer functions. The assumption $D_{22} = 0$ simplifies the formulae in the $\mathcal{H}_2$ algorithms and is made without loss of generality, since a substitution $K_D = K(I + D_{22}K)^{-1}$ gives the controller, when $D_{22} \neq 0$ (Zhou et al., 1996, p. 317). In $\mathcal{H}_\infty$, neither $D_{11} = 0$, nor $D_{22} = 0$, are required but they do significantly simplify the algorithm formulae. If they are not zero, an equivalent $\mathcal{H}_\infty$ problem can be constructed in which they are; see Safonov et al. (1989) and Green and Limebeer (1995). For simplicity, it is also sometimes assumed that D_{12} and D_{21} are given by

(A6) $D_{12} = \begin{bmatrix} 0 \\ I \end{bmatrix}$ and $D_{21} = \begin{bmatrix} 0 & I \end{bmatrix}$.

This can be achieved, without loss of generality, by a scaling of u and v and a unitary transformation of w and z; see for example Maciejowski (1989). In addition, for simplicity of exposition, the following additional assumptions are sometimes made:

(A7) $D_{12}^T C_1 = 0$ and $B_1 D_{21}^T = 0$.

(A8) (A, B_1) is stabilizable and (A, C_1) is detectable.

Assumption (A7) is common in $\mathcal{H}_2$ control, e.g. in LQG where there are no cross-terms in the cost function ($D_{12}^T C_1 = 0$), and the process noise and measurement noise are uncorrelated ($B_1 D_{21}^T = 0$). Notice that if (A7) holds then (A3) and (A4) may be replaced by (A8).

Whilst the above assumptions may appear daunting, most sensibly posed control problems will meet them. Therefore, if the software (e.g. the Robust Control toolbox of Matlab) complains, then it probably means that your control problem is not well formulated and you should think again.

Lastly, it should be said that $\mathcal{H}_\infty$ algorithms, in general, find a suboptimal controller. That is, for a specified γ a stabilizing controller is found for which $\|F_l(P,K)\|_\infty < \gamma$. If an optimal controller is required then the algorithm can be used iteratively, reducing γ until the minimum is reached within a given tolerance. In general, to find an optimal $\mathcal{H}_\infty$ controller is numerically and theoretically complicated. This contrasts significantly with $\mathcal{H}_2$ theory, in which the optimal controller is unique and can be found from the solution of just two Riccati equations.

9.3.2 $\mathcal{H}_2$ optimal control

The standard $\mathcal{H}_2$ optimal control problem is to find a stabilizing controller K which minimizes

$$\|F(s)\|_2 = \sqrt{\frac{1}{2\pi} \int_{-\infty}^{\infty} \mathrm{tr}\left[F(j\omega)F(j\omega)^H\right] d\omega}; \quad F \triangleq F_l(P,K) \qquad (9.29)$$

For a particular problem the generalized plant P will include the plant model, the interconnection structure, and the designer-specified weighting functions. This is illustrated for the LQG problem in the next subsection.

As discussed in Section 4.10.1 and noted in Tables A.1 and A.2 on page 540, the $\mathcal{H}_2$ norm can be given different deterministic interpretations. It also has the following stochastic interpretation. Suppose in the general control configuration that the exogenous input w is white noise of unit intensity. That is,

$$E\left\{w(t)w(\tau)^T\right\} = I\delta(t - \tau) \qquad (9.30)$$

The expected power in the error signal z is then given by

$$E\left\{\lim_{T\to\infty} \frac{1}{2T} \int_{-T}^{T} z(t)^T z(t) dt\right\} \qquad (9.31)$$

$$= \mathrm{tr}\, E\left\{z(t)z(t)^T\right\}$$

$$= \frac{1}{2\pi} \int_{-\infty}^{\infty} \mathrm{tr}\left[F(j\omega)F(j\omega)^H\right] d\omega$$

(by Parseval's theorem)

$$= \|F\|_2^2 = \|F_l(P,K)\|_2^2 \qquad (9.32)$$

Thus, by minimizing the $\mathcal{H}_2$ norm, the output (or error) power of the generalized system, due to a unit intensity white noise input, is minimized; we are minimizing the root-mean-square (rms) value of z.

9.3.3   LQG: a special $\mathcal{H}_2$ optimal controller

An important special case of $\mathcal{H}_2$ optimal control is the LQG problem described in Section 9.2.1. For the stochastic system

$$
\dot{x} \;=\; Ax + Bu + w_d \tag{9.33}
$$
$$
y \;=\; Cx + w_n \tag{9.34}
$$

where

$$
E\left\{ \begin{bmatrix} w_d(t) \\ w_n(t) \end{bmatrix} [\, w_d(\tau)^T \;\; w_n(\tau)^T \,] \right\} = \begin{bmatrix} W & 0 \\ 0 & V \end{bmatrix} \delta(t - \tau) \tag{9.35}
$$

The LQG problem is to find $u = K(s)y$ such that

$$
J = E\left\{ \lim_{T \to \infty} \frac{1}{T} \int_0^T [x^T Q x + u^T R u]\, dt \right\} \tag{9.36}
$$

is minimized with $Q = Q^T \geq 0$ and $R = R^T > 0$.

This problem can be cast as an $\mathcal{H}_2$ optimization in the general framework in the following manner. Define an error signal z as

$$
z = \begin{bmatrix} Q^{\frac{1}{2}} & 0 \\ 0 & R^{\frac{1}{2}} \end{bmatrix} \begin{bmatrix} x \\ u \end{bmatrix} \tag{9.37}
$$

and represent the stochastic inputs w_d, w_n as

$$
\begin{bmatrix} w_d \\ w_n \end{bmatrix} = \begin{bmatrix} W^{\frac{1}{2}} & 0 \\ 0 & V^{\frac{1}{2}} \end{bmatrix} w \tag{9.38}
$$

where w is a white noise process of unit intensity. Then the LQG cost function is

$$
J = E\left\{ \lim_{T \to \infty} \frac{1}{T} \int_0^T z(t)^T z(t)\, dt \right\} = \|F_l(P, K)\|_2^2 \tag{9.39}
$$

where

$$
z(s) = F_l(P, K) w(s) \tag{9.40}
$$

and the generalized plant P is given by

$$
P = \begin{bmatrix} P_{11} & P_{12} \\ P_{21} & P_{22} \end{bmatrix} \stackrel{s}{=} \left[\begin{array}{c|ccc|c} A & W^{\frac{1}{2}} & 0 & B \\ \hline Q^{\frac{1}{2}} & 0 & 0 & 0 \\ 0 & 0 & 0 & R^{\frac{1}{2}} \\ \hline C & 0 & V^{\frac{1}{2}} & 0 \end{array} \right] \tag{9.41}
$$

The above formulation of the LQG problem is illustrated in the general setting in Figure 9.9. With the standard assumptions for the LQG problem, application of the general $\mathcal{H}_2$ formulae (Doyle et al., 1989) to this formulation gives the familiar LQG optimal controller as in (9.17).

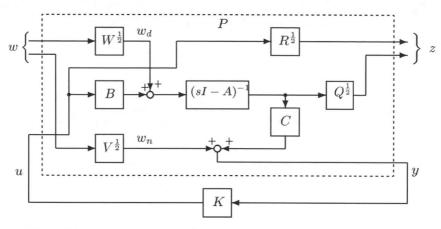

Figure 9.9: The LQG problem formulated in the general control configuration

9.3.4 $\mathcal{H}_\infty$ optimal control

With reference to the general control configuration of Figure 9.8, the standard $\mathcal{H}_\infty$ optimal control problem is to find all stabilizing controllers K which minimize

$$\|F_l(P, K)\|_\infty = \max_\omega \bar\sigma(F_l(P, K)(j\omega)) \tag{9.42}$$

As discussed in Section 4.10.2 the $\mathcal{H}_\infty$ norm has several interpretations in terms of performance. One is that it minimizes the peak of the maximum singular value of $F_l(P(j\omega), K(j\omega))$. It also has a time domain interpretation as the induced (worst-case) 2-norm. Let $z = F_l(P, K)w$, then

$$\|F_l(P, K)\|_\infty = \max_{w(t)\neq 0} \frac{\|z(t)\|_2}{\|w(t)\|_2} \tag{9.43}$$

where $\|z(t)\|_2 = \sqrt{\int_0^\infty \sum_i |z_i(t)|^2 dt}$ is the 2-norm of the vector signal.

In practice, it is usually not necessary to obtain an optimal controller for the $\mathcal{H}_\infty$ problem, and it is often computationally (and theoretically) simpler to design a suboptimal one (i.e. one close to the optimal ones in the sense of the $\mathcal{H}_\infty$ norm). Let $\gamma_{\min}$ be the minimum value of $\|F_l(P, K)\|_\infty$ over all stabilizing controllers K. Then the $\mathcal{H}_\infty$ suboptimal control problem is: given a $\gamma > \gamma_{\min}$, find all stabilizing controllers K such that

$$\|F_l(P, K)\|_\infty < \gamma$$

This can be solved efficiently using the algorithm of Doyle et al. (1989), and by reducing γ iteratively, an optimal solution is approached. The algorithm is summarized below with all the simplifying assumptions.

General $\mathcal{H}_\infty$ algorithm. *For the general control configuration of Figure 9.8 described by (9.24)–(9.26), with assumptions (A1) to (A8) in Section 9.3.1, there exists a stabilizing controller $K(s)$ such that $\|F_l(P, K)\|_\infty < \gamma$ if and only if*

(i) $X_\infty \geq 0$ *is a solution to the algebraic Riccati equation*

$$A^T X_\infty + X_\infty A + C_1^T C_1 + X_\infty(\gamma^{-2} B_1 B_1^T - B_2 B_2^T)X_\infty = 0 \tag{9.44}$$

such that Re $\lambda_i \left[A + (\gamma^{-2}B_1B_1^T - B_2B_2^T)X_\infty \right] < 0, \forall i;$ *and*

(ii) $Y_\infty \geq 0$ *is a solution to the algebraic Riccati equation*

$$AY_\infty + Y_\infty A^T + B_1 B_1^T + Y_\infty (\gamma^{-2}C_1^T C_1 - C_2^T C_2)Y_\infty = 0 \qquad (9.45)$$

such that Re $\lambda_i \left[A + Y_\infty(\gamma^{-2}C_1^T C_1 - C_2^T C_2) \right] < 0, \forall i;$ *and*

(iii) $\rho(X_\infty Y_\infty) < \gamma^2$

All such controllers are then given by $K = F_l(K_c, Q)$ where

$$K_c(s) \overset{s}{=} \left[\begin{array}{c|cc} A_\infty & -Z_\infty L_\infty & Z_\infty B_2 \\ \hline F_\infty & 0 & I \\ -C_2 & I & 0 \end{array} \right] \qquad (9.46)$$

$$F_\infty = -B_2^T X_\infty, \ L_\infty = -Y_\infty C_2^T, \ Z_\infty = (I - \gamma^{-2}Y_\infty X_\infty)^{-1} \qquad (9.47)$$

$$A_\infty = A + \gamma^{-2}B_1 B_1^T X_\infty + B_2 F_\infty + Z_\infty L_\infty C_2 \qquad (9.48)$$

and $Q(s)$ is any stable proper transfer function such that $\|Q\|_\infty < \gamma$. For $Q(s) = 0$, we get

$$K(s) = K_{c_{11}}(s) = -F_\infty(sI - A_\infty)^{-1}Z_\infty L_\infty \qquad (9.49)$$

This is called the "central" controller and has the same number of states as the generalized plant $P(s)$. The central controller can be separated into a state estimator (observer) of the form

$$\dot{\hat{x}} = A\hat{x} + B_1 \underbrace{\gamma^{-2}B_1^T X_\infty \hat{x}}_{\widehat{w}_{\text{worst}}} + B_2 u + Z_\infty L_\infty(C_2\hat{x} - y) \qquad (9.50)$$

and a state feedback

$$u = F_\infty \hat{x} \qquad (9.51)$$

Upon comparing the observer in (9.50) with the Kalman filter in (9.14) we see that it contains an additional term $B_1\widehat{w}_{\text{worst}}$, where $\widehat{w}_{\text{worst}}$ can be interpreted as an estimate of the worst-case disturbance (exogenous input). Note that for the special case of $\mathcal{H}_\infty$ loop shaping this extra term is not present. This is discussed in Section 9.4.4.

γ-**iteration.** If we desire a controller that achieves $\gamma_{\min}$, to within a specified tolerance, then we can perform a bisection on γ until its value is sufficiently accurate. The above result provides a test for each value of γ to determine whether it is less than $\gamma_{\min}$ or greater than $\gamma_{\min}$.

Given all the assumptions (A1) to (A8), the above is the most simple form of the general $\mathcal{H}_\infty$ algorithm. For the more general situation, where some of the assumptions are relaxed, the reader is referred to the original source (Glover and Doyle, 1988). In practice, we would expect a user to have access to commercial software such as Matlab and its toolboxes.

In Section 2.8, we distinguished between two methodologies for $\mathcal{H}_\infty$ controller design: the transfer function shaping approach and the signal-based approach. In the former, $\mathcal{H}_\infty$ optimization is used to shape the singular values of specified transfer functions over frequency. The maximum singular values are relatively easy to shape by forcing them to lie below user-defined bounds, thereby ensuring desirable bandwidths and roll-off rates. In the signal-based approach, we seek to minimize the energy in certain error signals given a set of exogenous input signals. The latter might include the outputs of perturbations representing uncertainty, as well as the usual disturbances, noise and command signals. Both of these two approaches will be considered again in the remainder of this section. In each case we will examine a particular problem and formulate it in the general control configuration.

A difficulty that sometimes arises with $\mathcal{H}_\infty$ control is the selection of weights such that the $\mathcal{H}_\infty$ optimal controller provides a good trade-off between conflicting objectives in various frequency ranges. Thus, for practical designs it is sometimes recommended to perform only a few iterations of the $\mathcal{H}_\infty$ algorithm. The justification for this is that the initial design, after one iteration, is similar to an $\mathcal{H}_2$ design which does trade off over various frequency ranges. Therefore stopping the iterations before the optimal value is achieved gives the design an $\mathcal{H}_2$ flavour which may be desirable.

9.3.5 Mixed-sensitivity $\mathcal{H}_\infty$ control

Mixed-sensitivity is the name given to transfer function shaping problems in which the sensitivity function $S = (I + GK)^{-1}$ is shaped along with one or more other closed-loop transfer functions such as KS or the complementary sensitivity function $T = I - S$. Earlier in this chapter, by examining a typical one degree-of-freedom configuration, Figure 9.1, we saw quite clearly the importance of S, KS and T.

Suppose, therefore, that we have a regulation problem in which we want to reject a disturbance d entering at the plant output and it is assumed that the measurement noise is relatively insignificant. Tracking is not an issue and therefore for this problem it makes sense to shape the closed-loop transfer functions S and KS in a one degree-of-freedom setting. Recall that S is the transfer function between d and the output, and KS the transfer function between d and the control signals. It is important to include KS as a mechanism for limiting the size and bandwidth of the controller, and hence the control energy used. The size of KS is also important for robust stability with respect to uncertainty modelled as additive plant perturbations; see (8.53) on page 303.

The disturbance d is typically a low-frequency signal, and therefore it will be successfully rejected if the maximum singular value of S is made small over the same low frequencies. To do this we could select a scalar low-pass filter $w_1(s)$ with a bandwidth equal to that of the disturbance, and then find a stabilizing controller that minimizes $\|w_1 S\|_\infty$. This cost function alone is not very practical. It focuses on just one closed-loop transfer function and for plants without RHP-zeros the optimal controller has infinite gains. In the presence of a non-minimum-phase zero, the stability requirement will indirectly limit the controller gains, but it is far more useful in practice to minimize

$$\left\| \begin{bmatrix} w_1 S \\ w_2 KS \end{bmatrix} \right\|_\infty \tag{9.52}$$

where $w_2(s)$ is a scalar high-pass filter with a crossover frequency approximately equal to that of the desired closed-loop bandwidth.

In general, the scalar weighting functions $w_1(s)$ and $w_2(s)$ can be replaced by matrices $W_1(s)$ and $W_2(s)$. This can be useful for systems with channels of quite different bandwidths when diagonal weights are recommended, but anything more complicated is usually not worth the effort.

Remark. Note that we have outlined here an alternative way of selecting the weights from that in

Example 2.17 and Section 3.5.7. There $W_1 = W_P$ was selected with a crossover frequency equal to that of the desired closed-loop bandwidth and $W_2 = W_u$ was selected as a constant, usually $W_u = I$.

To see how this mixed-sensitivity problem can be formulated in the general setting, we can imagine the disturbance d as a single exogenous input and define an error signal $z = [\, z_1^T \quad z_2^T \,]^T$, where $z_1 = W_1 y$ and $z_2 = -W_2 u$, as illustrated in Figure 9.10. It is

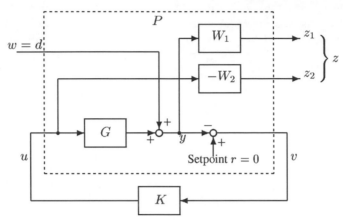

Figure 9.10: S/KS mixed-sensitivity optimization in standard form (regulation)

not difficult from Figure 9.10 to show that $z_1 = W_1 S w$ and $z_2 = W_2 K S w$ as required, and to determine the elements of the generalized plant P as

$$P_{11} = \begin{bmatrix} W_1 \\ 0 \end{bmatrix} \quad P_{12} = \begin{bmatrix} W_1 G \\ -W_2 \end{bmatrix}$$
$$P_{21} = -I \qquad P_{22} = -G$$

(9.53)

where the partitioning is such that

$$\begin{bmatrix} z_1 \\ z_2 \\ \hline v \end{bmatrix} = \begin{bmatrix} P_{11} & P_{12} \\ P_{21} & P_{22} \end{bmatrix} \begin{bmatrix} w \\ u \end{bmatrix}$$

(9.54)

and

$$F_l(P, K) = \begin{bmatrix} W_1 S \\ W_2 K S \end{bmatrix}$$

(9.55)

Another interpretation can be put on the S/KS mixed-sensitivity optimization as shown in the standard control configuration of Figure 9.11. Here we consider a tracking problem. The exogenous input is a reference command r, and the error signals are $z_1 = -W_1 e = W_1(r-y)$ and $z_2 = W_2 u$. As in the regulation problem of Figure 9.10, we have in this tracking problem $z_1 = W_1 S w$ and $z_2 = W_2 K S w$. An example of the use of S/KS mixed-sensitivity minimization is given in Chapter 13, where it is used to design a rotorcraft control law. In this helicopter problem, you will see that the exogenous input w is passed through a weight W_3 before it impinges on the system. W_3 is chosen to weight the input signal and not directly to shape S or KS. This signal-based approach to weight selection is the topic of the next subsection.

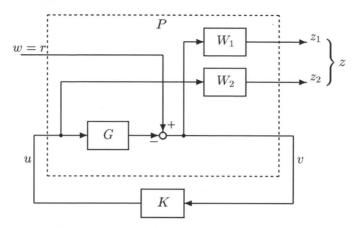

Figure 9.11: S/KS mixed-sensitivity minimization in standard form (tracking)

Another useful mixed-sensitivity optimization problem, again in a one degree-of-freedom setting, is to find a stabilizing controller which minimizes

$$\left\| \begin{bmatrix} W_1 S \\ W_2 T \end{bmatrix} \right\|_\infty \tag{9.56}$$

The ability to shape T is desirable for tracking problems and noise attenuation. It is also important for robust stability with respect to multiplicative perturbations at the plant output. The S/T mixed-sensitivity minimization problem can be put into the standard control configuration as shown in Figure 9.12. The elements of the corresponding generalized plant

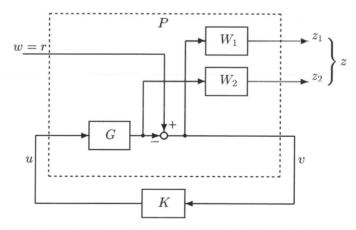

Figure 9.12: S/T mixed-sensitivity optimization in standard form

P are

$$P_{11} = \begin{bmatrix} W_1 \\ 0 \end{bmatrix} \quad P_{12} = \begin{bmatrix} -W_1 G \\ W_2 G \end{bmatrix} \tag{9.57}$$
$$P_{21} = I \qquad P_{22} = -G$$

Exercise 9.6 * *For the cost function*

$$
\left\| \begin{bmatrix} W_1 S \\ W_2 T \\ W_3 KS \end{bmatrix} \right\|_\infty
\tag{9.58}
$$

formulate a standard problem, draw the corresponding control configuration and give expressions for the generalized plant P.

The shaping of closed-loop transfer functions as described above with the "stacked" cost functions becomes difficult with more than two functions. With two, the process is relatively easy. The bandwidth requirements on each are usually complementary and simple, stable, low-pass and high-pass filters are sufficient to carry out the required shaping and trade-offs. We stress that the weights W_i in mixed-sensitivity $\mathcal{H}_\infty$ optimal control must all be stable. If they are not, assumption (A1) in Section 9.3.1 is not satisfied, and the general $\mathcal{H}_\infty$ algorithm is not applicable. Therefore if we wish, for example, to emphasize the minimization of S at low frequencies by weighting with a term including integral action, we would have to approximate $\frac{1}{s}$ by $\frac{1}{s+\epsilon}$, where $\epsilon \ll 1$. This is exactly what was done in Example 2.17. Similarly one might be interested in weighting KS with a non-proper weight to ensure that K is small outside the system bandwidth. But the standard assumptions preclude such a weight. The trick here is to replace a non-proper term such as $(1 + \tau_1 s)$ by $(1 + \tau_1 s)/(1 + \tau_2 s)$ where $\tau_2 \ll \tau_1$. A useful discussion of the tricks involved in using "unstable" and "non-proper" weights in $\mathcal{H}_\infty$ control can be found in Meinsma (1995).

For more complex problems, information might be given about several exogenous signals in addition to a variety of signals to be minimized and classes of plant perturbations to be robust against. In this case, the mixed-sensitivity approach is not general enough and we are forced to look at more advanced techniques such as the signal-based approach considered next.

9.3.6   Signal-based $\mathcal{H}_\infty$ control

The signal-based approach to controller design is very general and is appropriate for multivariable problems in which several objectives must be taken into account simultaneously. In this approach, we define the plant and possibly the model uncertainty, we define the class of external signals affecting the system and we define the norm of the error signals we want to keep small. The focus of attention has moved to the size of signals and away from the size and bandwidth of selected closed-loop transfer functions.

Weights are used to describe the expected or known frequency content of exogenous signals and the desired frequency content of error signals. Weights are also used if a perturbation is used to model uncertainty, as in Figure 9.13, where G is the nominal model, W is a weighting function that captures the relative model fidelity over frequency, and Δ represents unmodelled dynamics usually normalized via W so that $\|\Delta\|_\infty < 1$; see Chapter 8 for more details. As in mixed-sensitivity $\mathcal{H}_\infty$ control, the weights in signal-based $\mathcal{H}_\infty$ control need to be stable and proper for the general $\mathcal{H}_\infty$ algorithm to be applicable.

LQG control is a simple example of the signal-based approach, in which the exogenous signals are assumed to be stochastic (or alternatively impulses in a deterministic setting) and the error signals are measured in terms of the 2-norm. As we have already seen, the weights Q and R are constant, but LQG can be generalized to include frequency-dependent weights on the signals leading to what is sometimes called Wiener–Hopf design, or simply $\mathcal{H}_2$ control.

When we consider a system's response to persistent sinusoidal signals of varying

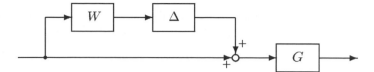

Figure 9.13: Multiplicative dynamic uncertainty model

frequency, or when we consider the induced 2-norm between the exogenous input signals and the error signals, we are required to minimize the $\mathcal{H}_\infty$ norm. In the absence of model uncertainty, there does not appear to be an overwhelming case for using the $\mathcal{H}_\infty$ norm rather than the more traditional $\mathcal{H}_2$ norm. However, when uncertainty is addressed, as it always should be, $\mathcal{H}_\infty$ is clearly the more natural approach using component uncertainty models as in Figure 9.13.

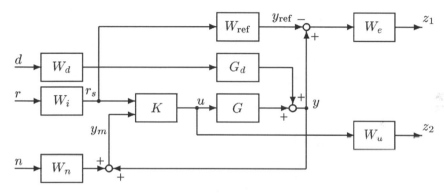

Figure 9.14: A signal-based $\mathcal{H}_\infty$ control problem

A typical problem using the signal-based approach to $\mathcal{H}_\infty$ control is illustrated in the interconnection diagram of Figure 9.14. G and G_d are nominal models of the plant and disturbance dynamics, and K is the controller to be designed. The weights W_d, W_i and W_n may be constant or dynamic and describe the relative importance and/or frequency content of the disturbances, setpoints and noise signals. The weight W_{ref} is a desired closed-loop transfer function between the weighted setpoint r_s and the actual output y. The weights W_e and W_u reflect the desired frequency content of the error $(y - y_{\text{ref}})$ and the control signals u, respectively. The problem can be cast as a standard $\mathcal{H}_\infty$ optimization in the general control configuration by defining

$$
w = \begin{bmatrix} d \\ r \\ n \end{bmatrix} \quad z = \begin{bmatrix} z_1 \\ z_2 \end{bmatrix}
$$

$$
v = \begin{bmatrix} r_s \\ y_m \end{bmatrix} \quad u = u
$$

(9.59)

in the general setting of Figure 9.8.

Suppose we now introduce a multiplicative dynamic uncertainty model at the input to the plant as shown in Figure 9.15. The problem we now want to solve is: find a stabilizing controller K such that the $\mathcal{H}_\infty$ norm of the transfer function between w and z is less than 1 for all Δ, where $\|\Delta\|_\infty < 1$. We have assumed in this statement that the signal weights have normalized the 2-norm of the exogenous input signals to unity. This problem is a non-standard $\mathcal{H}_\infty$ optimization. It is a robust performance problem for which the μ-synthesis procedure, outlined in Chapter 8, can be applied. Mathematically, we require the structured

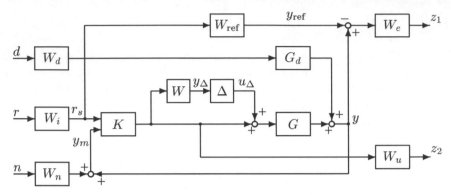

Figure 9.15: An $\mathcal{H}_\infty$ robust performance problem

singular value

$$\mu(M(j\omega)) < 1, \forall \omega \tag{9.60}$$

where M is the transfer function matrix between

$$\begin{bmatrix} d \\ r \\ n \\ \delta \end{bmatrix} \quad \text{and} \quad \begin{bmatrix} z_1 \\ z_2 \\ \epsilon \end{bmatrix} \tag{9.61}$$

and the associated block-diagonal perturbation has two blocks: a fictitious performance block between $[\, d^T \quad r^T \quad n^T \,]^T$ and $[\, z_1^T \quad z_2^T \,]^T$, and an uncertainty block Δ between u_Δ and y_Δ. Whilst the structured singular value is a useful analysis tool for assessing designs, μ-synthesis is sometimes difficult to use and often too complex for the practical problem at hand. In its full generality, the μ-synthesis problem has not yet been solved mathematically; where solutions exist the controllers tend to be of very high order; the algorithms may not always converge; and design problems are sometimes difficult to formulate directly.

For many industrial control problems, a design procedure is required which offers more flexibility than mixed-sensitivity $\mathcal{H}_\infty$ control, but is not as complicated as μ-synthesis. For simplicity, it should be based on classical loop-shaping ideas and it should not be limited in its applications like LTR procedures. In the next section, we present such a controller design procedure.

9.4 $\mathcal{H}_\infty$ loop-shaping design

The loop-shaping design procedure described in this section is based on $\mathcal{H}_\infty$ robust stabilization combined with classical loop shaping, as proposed by McFarlane and Glover

(1990). It is essentially a two-stage design process. First, the open-loop plant is augmented by pre- and post-compensators to give a desired shape to the singular values of the open-loop frequency response. This could be based on an initial controller design. Then the resulting shaped plant (initial loop shape) is robustly stabilized ("robustified") with respect to the quite general class of coprime factor uncertainty using $\mathcal{H}_\infty$ optimization.

An important advantage is that no problem-dependent uncertainty modelling, or weight selection, is required in this second step.

We will begin the section with a description of the $\mathcal{H}_\infty$ robust stabilization problem (Glover and McFarlane, 1989). This is a particularly nice problem because it does not require γ-iteration for its solution, and explicit formulae for the corresponding controllers are available. The formulae are relatively simple and so will be presented in full.

Following this, a step-by-step procedure for $\mathcal{H}_\infty$ loop-shaping design is presented. This systematic procedure has its origin in the PhD thesis of Hyde (1991) and has since been successfully applied to several industrial problems. The procedure synthesizes what is in effect a single degree-of-freedom controller. This can be a limitation if there are stringent requirements on command following. However, as shown by Limebeer et al. (1993), the procedure can be extended by introducing a second degree of freedom in the controller and formulating a standard $\mathcal{H}_\infty$ optimization problem which allows one to trade off robust stabilization against closed-loop model matching. We will describe this two degrees-of-freedom extension and further show that such controllers have a special observer-based structure which can be taken advantage of in controller implementation.

9.4.1 Robust stabilization

For multivariable systems, classical gain and phase margins are unreliable indicators of robust stability when defined for each channel (or loop), taken one at a time, because simultaneous perturbations in more than one loop are not then catered for; see the spinning satellite example in Chapter 3 (page 98). More general perturbations like $\text{diag}\{k_i\}$ and $\text{diag}\{e^{j\theta_i}\}$, as discussed in Section 9.2.2, are required to capture the uncertainty, but even these are limited. It is now common practice, as seen in Chapter 8, to model uncertainty by stable norm-bounded dynamic (complex) matrix perturbations. With a single perturbation, the associated robustness test is in terms of the maximum singular value of a closed-loop transfer function. Use of a single stable perturbation, however, restricts the plant and perturbed plant models to have either the same number of unstable poles or the same number of unstable (RHP) zeros. To overcome this, two stable perturbations can be used, one on each of the factors in a coprime factorization of the plant, as shown in Section 8.6.2. Although this uncertainty description seems unrealistic and less intuitive than the others, it is in fact quite general, and for our purposes it leads to a very useful $\mathcal{H}_\infty$ robust stabilization problem. Before presenting the problem, we will first recall the uncertainty model given in (8.62).

We will consider the stabilization of a plant G which has a normalized left coprime factorization (as discussed in Section 4.1.5)

$$G = M^{-1}N \tag{9.62}$$

where we have dropped the subscript from M and N for simplicity. A perturbed plant model G_p can then be written as

$$G_p = (M + \Delta_M)^{-1}(N + \Delta_N) \tag{9.63}$$

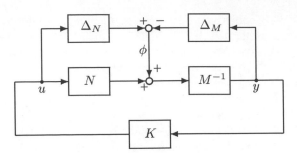

Figure 9.16: $\mathcal{H}_\infty$ robust stabilization problem

where Δ_M, Δ_N are stable unknown transfer functions which represent the uncertainty in the nominal plant model G. The objective of robust stabilization it to stabilize not only the nominal model G, but a family of perturbed plants defined by

$$G_p = \left\{ (M + \Delta_M)^{-1}(N + \Delta_N) : \|[\Delta_N \quad \Delta_M]\|_\infty < \epsilon \right\} \qquad (9.64)$$

where $\epsilon > 0$ is then the stability margin. To maximize this stability margin is the problem of robust stabilization of normalized coprime factor plant descriptions as introduced and solved by Glover and McFarlane (1989).

For the perturbed feedback system of Figure 9.16, as already derived in (8.64), the stability property is robust if and only if the nominal feedback system is stable and

$$\gamma_K \triangleq \left\| \begin{bmatrix} K \\ I \end{bmatrix} (I - GK)^{-1} M^{-1} \right\|_\infty \le \frac{1}{\epsilon} \qquad (9.65)$$

Notice that γ_K is the $\mathcal{H}_\infty$ norm from ϕ to $\begin{bmatrix} u \\ y \end{bmatrix}$ and $(I - GK)^{-1}$ is the sensitivity function for this positive feedback arrangement.

The lowest achievable value of γ_K and the corresponding maximum stability margin ϵ are given by Glover and McFarlane (1989) as

$$\gamma_{min} = \epsilon_{max}^{-1} = \left\{ 1 - \|[N \; M]\|_H^2 \right\}^{-\frac{1}{2}} = (1 + \rho(XZ))^{\frac{1}{2}} \qquad (9.66)$$

where $\| \cdot \|_H$ denotes Hankel norm, ρ denotes the spectral radius (maximum eigenvalue), and for a minimal state-space realization (A, B, C, D) of G, Z is the unique positive definite solution to the algebraic Riccati equation

$$(A - BS^{-1}D^T C)Z + Z(A - BS^{-1}D^T C)^T - ZC^T R^{-1}CZ + BS^{-1}B^T = 0 \quad (9.67)$$

where

$$R = I + DD^T, \quad S = I + D^T D$$

and X is the unique positive definite solution of the following algebraic Riccati equation:

$$(A - BS^{-1}D^T C)^T X + X(A - BS^{-1}D^T C) - XBS^{-1}B^T X + C^T R^{-1}C = 0 \qquad (9.68)$$

Notice that the formulae simplify considerably for a strictly proper plant, i.e. when $D = 0$.

A controller (the "central" controller in McFarlane and Glover (1990)) which guarantees that

$$\left\| \begin{bmatrix} K \\ I \end{bmatrix} (I - GK)^{-1} M^{-1} \right\|_\infty \leq \gamma \qquad (9.69)$$

for a specified $\gamma > \gamma_{\min}$, is given by

$$K \overset{s}{=} \left[\begin{array}{c|c} A + BF + \gamma^2 (L^T)^{-1} Z C^T (C + DF) & \gamma^2 (L^T)^{-1} Z C^T \\ \hline B^T X & -D^T \end{array} \right] \qquad (9.70)$$

$$F = -S^{-1}(D^T C + B^T X) \qquad (9.71)$$

$$L = (1 - \gamma^2)I + XZ \qquad (9.72)$$

The Matlab function `coprimeunc`, listed in Table 9.2, can be used to generate the controller in (9.70). It is important to emphasize that since we can compute $\gamma_{\min}$ from (9.66) we get an explicit solution by solving just two Riccati equations (`care`) and avoid the γ-iteration needed to solve the general $\mathcal{H}_\infty$ problem.

Table 9.2: Matlab function to generate the $\mathcal{H}_\infty$ controller in (9.70)

```
% Uses Control toolbox
function [Ac,Bc,Cc,Dc,gammin]=coprimeunc(a,b,c,d,gamrel)
%
% Finds the controller which optimally ''robustifies'' a given shaped plant
% in terms of tolerating maximum coprime uncertainty.
%
% INPUTS:
% a,b,c,d: State-space description of (shaped) plant.
% gamrel: gamma used is gamrel*gammin  (typical gamrel-1.1)
%
% OUTPUTS:
% Ac,Bc,Cc,Dc: "Robustifying" controller (positive feedback).
%
S = eye(size(d'*d))+d'*d;
R = eye(size(d*d'))+d*d';
Rinv = inv(R);Sinv=inv(S);
A1 = (a-b*Sinv*d'*c); R1 = S; B1 = b; Q1 = c'*Rinv*c;
[X,XAMP,G] = care(A1,B1,Q1,R1);
A2 = A1'; Q2 = b*Sinv*b'; B2 = c'; R2 = R;
[Z,ZAMP,G] = care(A2,B2,Q2,R2);
% optimal gamma
XZ = X*Z; gammin = sqrt(1+max(oig(XZ)));
% Use higher gamma
gam = gamrel*gammin; gam2 = gam*gam; gamconst = (1-gam2)*eye(size(XZ));
Lc = gamconst + XZ; Li = inv(Lc'); Fc = -Sinv*(d'*c+b'*X);
Ac = a + b*Fc + gam2*Li*Z*c'*(c+d*Fc);
Bc = gam2*Li*Z*c';
Cc = b'*X;
Dc = -d';
```

Remark 1 An example of the use of `coprimeunc` is given in Example 9.3 below.

Remark 2 Notice that, if $\gamma = \gamma_{\min}$ in (9.70), then $L = -\rho(XZ)I + XZ$, which is singular, and thus (9.70) cannot be implemented. If for some unusual reason the truly optimal controller is required, then this problem can be resolved using a descriptor system approach, the details of which can be found in Safonov et al. (1989).

Remark 3 Alternatively, from Glover and McFarlane (1989), all controllers achieving $\gamma = \gamma_{\min}$ are given by $K = UV^{-1}$, where U and V are stable, (U, V) is a right coprime factorization of K, and

U, V satisfy

$$\left\| \begin{bmatrix} -N^* \\ M^* \end{bmatrix} + \begin{bmatrix} U \\ V \end{bmatrix} \right\|_\infty = \|[N \ \ M]\|_H \qquad (9.73)$$

The determination of U and V is a Nehari extension problem: that is, a problem in which an unstable transfer function $R(s)$ is approximated by a stable transfer function $Q(s)$, such that $\|R + Q\|_\infty$ is minimized, the minimum being $\|R^*\|_H$. A solution to this problem is given in Glover (1984).

Exercise 9.7 *Formulate the $\mathcal{H}_\infty$ robust stabilization problem in the general control configuration of Figure 9.8, and determine a transfer function expression and a state-space realization for the generalized plant P.*

9.4.2   A systematic $\mathcal{H}_\infty$ loop-shaping design procedure

Robust stabilization alone is not much use in practice because the designer is not able to specify any performance requirements. To do this McFarlane and Glover (1990) proposed pre- and post-compensating the plant to shape the open-loop singular values prior to robust stabilization of the "shaped" plant.

If W_1 and W_2 are the pre- and post-compensators respectively, then the shaped plant (initial loop shape) G_s is given by

$$G_s = W_2 G W_1 \qquad (9.74)$$

as shown in Figure 9.17. The controller K_s is synthesized by solving the robust stabilization

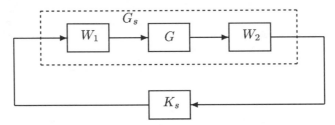

Figure 9.17: The shaped plant and controller

problem of section 9.4.1 for the shaped plant G_s with a normalized left coprime factorization $G_s = M_s^{-1} N_s$. The feedback controller for the plant G is then $K = W_1 K_s W_2$. The above procedure contains all the essential ingredients of classical loop shaping, and can easily be implemented using the formulae already presented and reliable algorithms in, for example, Matlab.

We first present a simple SISO example, where $W_2 = 1$, and we select W_1 to get acceptable disturbance rejection. We will afterwards present a systematic procedure for selecting the weights W_1 and W_2.

Example 9.3 Glover–McFarlane $\mathcal{H}_\infty$ loop shaping for the disturbance process. *Consider the disturbance process in (2.62) which was studied in detail in Chapter 2:*

$$G(s) = \frac{200}{10s + 1} \frac{1}{(0.05s + 1)^2}, \quad G_d(s) = \frac{100}{10s + 1} \qquad (9.75)$$

We want as good disturbance rejection as possible, and the gain crossover frequency w_c for the final design should be about 10 rad/s.

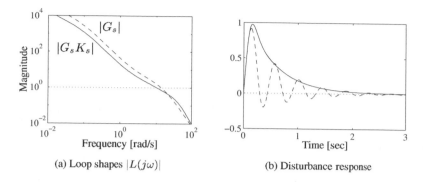

(a) Loop shapes $|L(j\omega)|$ (b) Disturbance response

Figure 9.18: Glover–McFarlane loop-shaping design for the disturbance process. Dashed line: initial "shaped" design, G_s. Solid line: "robustified" design, $G_s K_s$.

In Example 2.10 we argued that for acceptable disturbance rejection with minimum input usage, the loop shape ("shaped plant") $|G_s| = |GW_1|$ should be similar to $|G_d|$, so $|W_1| = |G^{-1}G_d|$ is desired. Then after neglecting the high-frequency dynamics in $G(s)$ this yields an initial weight $W_1 = 0.5$. To improve the performance at low frequencies we add integral action, and we also add a phase-advance term $s + 2$ to reduce the slope for L from -2 at lower frequencies to about -1 at crossover. Finally, to make the response a little faster we multiply the gain by a factor 2 to get the weight

$$W_1 = \frac{s+2}{s} \tag{9.76}$$

This yields a shaped plant $G_s = GW_1$ with a gain crossover frequency of 13.7 rad/s, and the magnitude of $G_s(j\omega)$ is shown by the dashed line in Figure 9.18(a). The response to a unit step in the disturbance response is shown by the dashed line in Figure 9.18(b), and, as may be expected, the response with the "controller" $K = W_1$ is too oscillatory.

We now "robustify" this design so that the shaped plant tolerates as much $\mathcal{H}_\infty$ coprime factor uncertainty as possible. This may be done with Matlab using either the command ncfsyn *in the Robust Control toolbox or using the function* coprimeunc *given in Table 9.2. Here the shaped plant $G_s = GW_1$ has state-space matrices A, B, C and D, and the function returns the "robustifying" positive feedback controller K_s with state-space matrices A_c, B_c, C_c and D_c. In general, K_s has the same number of poles (states) as G_s.* qamrel *is the value of γ relative to $\gamma_{\min}$, and was in our case selected as 1.1. The returned variable* gammin *($\gamma_{\min}$) is the inverse of the magnitude of coprime uncertainty we can tolerate before we get instability. We want $\gamma_{\min} \geq 1$ as small as possible, and we usually require that $\gamma_{\min}$ is less than 4, corresponding to 25% allowed coprime uncertainty.*

By applying this to our example we get $\gamma_{\min} = 2.34$ and an overall controller $K = W_1 K_s$ with five states (G_s, and thus K_s, has four states, and W_1 has one state). The corresponding loop shape $|G_s K_s|$ is shown by the solid line in Figure 9.18(a). We see that the change in the loop shape is small, and we note with interest that the slope around crossover is somewhat gentler. This translates into better margins: the gain margin (GM) is improved from 1.62 (for G_s) to 3.48 (for $G_s K_s$), and the phase margin (PM) is improved from $13.2°$ to $51.5°$. The gain crossover frequency w_c is reduced slightly from 13.7 to 10.3 rad/s. The corresponding disturbance response is shown in Figure 9.18(b) and is seen to be much improved.

Remark. The response with the controller $K = W_1 K_s$ is quite similar to that of the loop-shaping controller $K_3(s)$ designed in Chapter 2 (see curves L_3 and y_3 in Figure 2.24). The response for reference tracking with controller $K = W_1 K_s$ is not shown; it is also very similar to that with K_3 (see Figure 2.26), but it has a slightly smaller overshoot of 21% rather than 24%. To reduce this overshoot we would need to use a two degrees-of-freedom controller.

Exercise 9.8 * *Design an $\mathcal{H}_\infty$ loop-shaping controller for the disturbance process in (9.75) using the weight W_1 in (9.76), i.e. generate plots corresponding to those in Figure 9.18. Next, repeat the design with $W_1 = 2(s+3)/s$ (which results in an initial G_s which would yield closed-loop instability with $K_c = 1$). Compute the gain and phase margins and compare the disturbance and reference responses. In both cases find ω_c and use (2.45) to compute the maximum delay that can be tolerated in the plant before instability arises.*

Skill is required in the selection of the weights (pre- and post-compensators W_1 and W_2), but experience on real applications has shown that robust controllers can be designed with relatively little effort by following a few simple rules. An excellent illustration of this is given in the thesis of Hyde (1991) who worked with Glover on the robust control of VSTOL (Vertical and/or Short Take-Off and Landing) aircraft. Their work culminated in a successful flight test of $\mathcal{H}_\infty$ loop-shaping control laws implemented on a Harrier jump-jet research vehicle at the former UK Defence Research Agency (now QinetiQ), Bedford, in 1993. The $\mathcal{H}_\infty$ loop-shaping procedure has also been extensively studied and worked on by Postlethwaite and Walker (1992) in their work on advanced control of high-performance helicopters, also for the UK DRA (now QinetiQ) at Bedford. This application is discussed in detail in the helicopter case study in Section 13.2. More recently, $\mathcal{H}_\infty$ loop shaping has been tested in flight on a Bell 205 fly-by-wire helicopter; see Postlethwaite et al. (1999), Smerlas et al. (2001), Prempain and Postlethwaite (2004), Postlethwaite et al. (2005).

Based on these, and other, studies, it is recommended that the following systematic procedure is followed when using $\mathcal{H}_\infty$ loop shaping design:

1. Scale the plant outputs and inputs. This is very important for most design procedures and is sometimes forgotten. In general, scaling improves the conditioning of the design problem, it enables meaningful analysis to be made of the robustness properties of the feedback system in the frequency domain, and for loop shaping it can simplify the selection of weights. There are a variety of methods available including normalization with respect to the magnitude of the maximum or average value of the signal in question. Scaling with respect to maximum values is important if the controllability analysis of earlier chapters is to be used. However, if one is to go straight to a design the following variation has proved useful in practice:

   (a) The outputs are scaled such that equal magnitudes of cross-coupling into each of the outputs is equally undesirable.

   (b) Each input is scaled by a given percentage (say 10%) of its expected range of operation. That is, the inputs are scaled to reflect the relative actuator capabilities. An example of this type of scaling is given in the aero-engine case study of Chapter 13.

2. Order the inputs and outputs so that the plant is as diagonal as possible. The relative gain array can be useful here. The purpose of this pseudo-diagonalization is to ease the design of the pre- and post-compensators which, for simplicity, will be chosen to be diagonal.

Next, we discuss the selection of weights to obtain the shaped plant $G_s = W_2 G W_1$ where

$$W_1 = W_p W_a W_g \tag{9.77}$$

3. Select the elements of diagonal pre- and post-compensators W_p and W_2 so that the singular values of $W_2 G W_p$ are desirable. This would normally mean high gain at low

frequencies, roll-off rates of approximately 20 dB/decade (a slope of about -1) at the desired bandwidth(s), with higher rates at high frequencies. Some trial and error is involved here. W_2 is usually chosen as a constant, reflecting the relative importance of the outputs to be controlled and the other measurements being fed back to the controller. For example, if there are feedback measurements of two outputs to be controlled and a velocity signal, then W_2 might be chosen to be diag$[1, 1, 0.1]$, where 0.1 is in the velocity signal channel. W_p contains the dynamic shaping. Integral action, for low-frequency performance; phase advance for reducing the roll-off rates at crossover; and phase lag to increase the roll-off rates at high frequencies should all be placed in W_p if desired. The weights should be chosen so that no unstable hidden modes are created in G_s.

4. *Optional*: Align the singular values at a desired bandwidth using a further constant weight W_a cascaded with W_p. This is effectively a constant decoupler and should not be used if the plant is ill-conditioned in terms of large RGA elements (see Section 6.10.4). The align algorithm of Kouvaritakis (1974) is recommended (see file `align.m` available at the book's home page).

5. *Optional*: Introduce an additional gain matrix W_g cascaded with W_a to provide control over actuator usage. W_g is diagonal and is adjusted so that actuator rate limits are not exceeded for reference demands and typical disturbances on the scaled plant outputs. This requires some trial and error.

6. Robustly stabilize the shaped plant $G_s = W_2 G W_1$, where $W_1 = W_p W_a W_g$, using the formulae of the previous section. First, calculate the maximum stability margin $\epsilon_{\max} = 1/\gamma_{\min}$. If the margin is too small, $\epsilon_{\max} < 0.25$, then go back to step 4 and modify the weights. Otherwise, select $\gamma > \gamma_{\min}$, by about 10%, and synthesize a suboptimal controller using (9.70). There is usually no advantage to be gained by using the optimal controller. When $\epsilon_{\max} > 0.25$ (respectively $\gamma_{\min} < 4$) the design is usually successful. In this case, at least 25% coprime factor uncertainty is allowed, and we also find that the shape of the open-loop singular values will not have changed much after robust stabilization. A small value of $\epsilon_{\max}$ indicates that the chosen singular value loop shapes are incompatible with robust stability requirements. That the loop shapes do not change much following robust stabilization if γ is small (ϵ large) is justified theoretically in McFarlane and Glover (1990).

7. Analyze the design and if all the specifications are not met make further modifications to the weights.

8. Implement the controller. The configuration shown in Figure 9.19 has been found useful when compared with the conventional setup in Figure 9.1. This is because the references

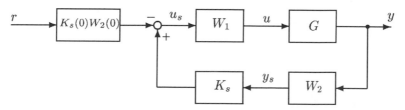

Figure 9.19: A practical implementation of the loop-shaping controller

do not directly excite the dynamics of K_s, which can result in large amounts of overshoot (classical derivative kick). The constant prefilter ensures a steady-state gain of 1 between

r and y, assuming integral action in W_1 or G.

It has recently been shown (Glover et al., 2000) that the stability margin $\epsilon_{\max} = 1/\gamma_{\min}$, here defined in terms of coprime factor perturbations, can be interpreted in terms of simultaneous gain and phase margins in all the plant's inputs and outputs, when the $\mathcal{H}_\infty$ loop-shaping weights W_1 and W_2 are diagonal. The derivation of these margins is based on the gap metric (Georgiou and Smith, 1990) and the ν-gap metric (Vinnicombe, 1993) measures for uncertainty. A discussion of these measures lies outside the scope of this book, but the interested reader is referred to the excellent book on the subject by Vinnicombe (2001) and the paper by Glover et al. (2000).

We will conclude this subsection with a summary of the advantages offered by the above $\mathcal{H}_\infty$ loop-shaping design procedure:

- It is relatively easy to use, being based on classical loop-shaping ideas.
- There exists a closed formula for the $\mathcal{H}_\infty$ optimal cost $\gamma_{\min}$, which in turn corresponds to a maximum stability margin $\epsilon_{\max} = 1/\gamma_{\min}$.
- No γ-iteration is required in the solution.
- Except for special systems, ones with all-pass factors, there are no pole–zero cancellations between the plant and controller (Sefton and Glover, 1990; Tsai et al., 1992). Pole–zero cancellations are common in some other $\mathcal{H}_\infty$ control problems, like the S/T-problem in (9.56), and are a problem when the plant has lightly damped modes.

Exercise 9.9 *First a definition and some useful properties.*

Definition: A stable transfer function matrix $H(s)$ is inner if $H^* H = I$, and co-inner if $HH^* = I$. The operator H^* is defined as $H^*(s) = H^T(-s)$.

Properties: The $\mathcal{H}_\infty$ norm is invariant under right multiplication by a co-inner function and under left multiplication by an inner function.

Equipped with the above definition and properties, show for the shaped $G_s = M_s^{-1} N_s$ that the matrix $[\, M_s \quad N_s \,]$ is co-inner and hence that the $\mathcal{H}_\infty$ loop-shaping cost function

$$\left\| \begin{bmatrix} K_s \\ I \end{bmatrix} (I - G_s K_s)^{-1} M_s^{-1} \right\|_\infty \tag{9.78}$$

is equivalent to

$$\left\| \begin{bmatrix} K_s S_s & K_s S_s G_s \\ S_s & S_s G_s \end{bmatrix} \right\|_\infty \tag{9.79}$$

where $S_s = (I - G_s K_s)^{-1}$. This shows that the problem of finding a stabilizing controller to minimize the four-block cost function (9.79) has an exact solution.

Whilst it is highly desirable, from a computational point of view, to have exact solutions for $\mathcal{H}_\infty$ optimization problems, such problems are rare. We are fortunate that the above robust stabilization problem is also one of great practical significance.

9.4.3 Two degrees-of-freedom controllers

Many control design problems possess two degrees of freedom: on the one hand, measurement or feedback signals; and on the other, commands or references. Sometimes, one degree of freedom is left out of the design, and the controller is driven (for example) by an error signal, i.e. the difference between a command and the output. But in cases where stringent time domain specifications are set on the output response, a one degree-of-freedom structure may not be sufficient. A general two degrees-of-freedom feedback control scheme is depicted in Figure 9.20. The commands and feedbacks enter the controller separately and are independently processed.

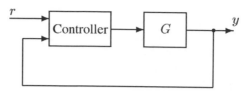

Figure 9.20: General two degrees-of-freedom feedback control scheme

The $\mathcal{H}_\infty$ loop-shaping design procedure of McFarlane and Glover is a one degree-of-freedom design, although as we showed in Figure 9.19 a simple constant prefilter can easily be implemented for steady-state accuracy. For many tracking problems, however, this will not be sufficient and a dynamic two degrees-of-freedom design is required. In Hoyle et al. (1991) and Limebeer et al. (1993) a two degrees-of-freedom extension of the Glover–McFarlane procedure was proposed to enhance the model-matching properties of the closed loop. With this the feedback part of the controller is designed to meet robust stability and disturbance rejection requirements in a manner similar to the one degree-of-freedom loop-shaping design procedure except that only a pre-compensator weight W is used. It is assumed that the measured outputs and the outputs to be controlled are the same, although this assumption can be removed as shown later. An additional prefilter part of the controller is then introduced to force the response of the closed-loop system to follow that of a specified model, T_{ref}, often called the reference model. Both parts of the controller are synthesized by solving the design problem illustrated in Figure 9.21.

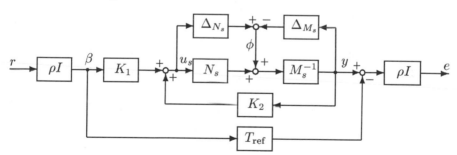

Figure 9.21: Two degrees-of-freedom $\mathcal{H}_\infty$ loop-shaping design problem

The design problem is to find the stabilizing controller $K = [K_1 \quad K_2]$ for the shaped plant $G_s = GW_1$, with a normalized coprime factorization $G_s = M_s^{-1} N_s$, which minimizes the $\mathcal{H}_\infty$ norm of the transfer function between the signals $[r^T \quad \phi^T]^T$ and $[u_s^T \quad y^T \quad e^T]^T$ as defined in Figure 9.21. The problem is easily cast into the general control configuration and solved suboptimally using standard methods and γ-iteration. We will show this later.

The control signal to the shaped plant u_s is given by

$$u_s = [K_1 \quad K_2] \begin{bmatrix} \beta \\ y \end{bmatrix} \tag{9.80}$$

where K_1 is the prefilter, K_2 is the feedback controller, β is the scaled reference, and y is the measured output. The purpose of the prefilter is to ensure that

$$\|(I - G_s K_2)^{-1} G_s K_1 - T_{\text{ref}}\|_\infty \leq \gamma \rho^{-2} \tag{9.81}$$

T_{ref} is the desired closed-loop transfer function selected by the designer to introduce time domain specifications (desired response characteristics) into the design process; and ρ is a scalar parameter that the designer can increase to place more emphasis on model matching in the optimization at the expense of robustness.

From Figure 9.21 and a little bit of algebra, we have that

$$\begin{bmatrix} u_s \\ y \\ e \end{bmatrix} = \begin{bmatrix} \rho(I - K_2 G_s)^{-1} K_1 & K_2(I - G_s K_2)^{-1} M_s^{-1} \\ \rho(I - G_s K_2)^{-1} G_s K_1 & (I - G_s K_2)^{-1} M_s^{-1} \\ \rho^2 \left[(I - G_s K_2)^{-1} G_s K_1 - T_{\text{ref}}\right] & \rho(I - G_s K_2)^{-1} M_s^{-1} \end{bmatrix} \begin{bmatrix} r \\ \phi \end{bmatrix} \tag{9.82}$$

In the optimization, the $\mathcal{H}_\infty$ norm of this block matrix transfer function is minimized.

Notice that the (1,2) and (2,2) blocks taken together are associated with robust stabilization and the (3,1) block corresponds to model matching. In addition, the (1,1) and (2,1) blocks help to limit actuator usage and the (3,2) block is linked to the performance of the loop. For $\rho = 0$, the problem reverts to minimizing the $\mathcal{H}_\infty$ norm of the transfer function between ϕ and $[u_s^T \quad y^T]^T$, namely, the robust stabilization problem, and the two degrees-of-freedom controller reduces to an ordinary $\mathcal{H}_\infty$ loop-shaping controller.

To put the two degrees-of-freedom design problem into the standard control configuration, we can define a generalized plant P by

$$\begin{bmatrix} u_s \\ y \\ e \\ \hline \beta \\ y \end{bmatrix} = \begin{bmatrix} P_{11} & P_{12} \\ P_{21} & P_{22} \end{bmatrix} \begin{bmatrix} r \\ \phi \\ \hline u_s \end{bmatrix} \tag{9.83}$$

$$= \begin{bmatrix} 0 & 0 & I \\ 0 & M_s^{-1} & G_s \\ -\rho^2 T_{\text{ref}} & \rho M_s^{-1} & \rho G_s \\ \hline \rho I & 0 & 0 \\ 0 & M_s^{-1} & G_s \end{bmatrix} \begin{bmatrix} r \\ \phi \\ u_s \end{bmatrix} \tag{9.84}$$

Further, if the shaped plant G_s and the desired stable closed-loop transfer function T_{ref} have the following state-space realizations

$$G_s \stackrel{s}{=} \left[\begin{array}{c|c} A_s & B_s \\ \hline C_s & D_s \end{array} \right] \tag{9.85}$$

$$T_{\text{ref}} \stackrel{s}{=} \left[\begin{array}{c|c} A_r & B_r \\ \hline C_r & D_r \end{array} \right] \tag{9.86}$$

then P may be realized by

$$\left[\begin{array}{cc|ccc|c} A_s & 0 & 0 & (B_s D_s^T + Z_s C_s^T)R_s^{-1/2} & B_s \\ 0 & A_r & B_r & 0 & 0 \\ \hline 0 & 0 & 0 & 0 & I \\ C_s & 0 & 0 & R_s^{1/2} & D_s \\ \rho C_s & -\rho^2 C_r & -\rho^2 D_r & \rho R_s^{1/2} & \rho D_s \\ \hline 0 & 0 & \rho I & 0 & 0 \\ C_s & 0 & 0 & R_s^{1/2} & D_s \end{array}\right] \qquad (9.87)$$

and used in standard $\mathcal{H}_\infty$ algorithms (Doyle et al., 1989) to synthesize the controller K. Note that $R_s = I + D_s D_s^T$, and Z_s is the unique positive definite solution to the generalized Riccati equation (9.67) for G_s. Matlab commands to synthesize the controller are given in Table 9.3.

Table 9.3: **Matlab commands to synthesize the $\mathcal{H}_\infty$ two degrees-of-freedom controller in (9.80)**

```
% Uses Robust Control toolbox
%
% INPUTS: Shaped plant Gs
%         Reference model Tref
%
% OUTPUT: Two degrees-of-freedom controller K
%
% Coprime factorization of Gs
%
[As,Bs,Cs,Ds] = ssdata(balreal(Gs));
[Ar,Br,Cr,Dr] = ssdata(Tref);
[nr,nr] = size(Ar); [lr,mr] = size(Dr);
[ns,ns] = size(As); [ls,ms] = size(Ds);
Rs = eye(ls)+Ds*Ds'; Ss = eye(ms)+Ds'*Ds;
A = (As - Bs*inv(Ss)*Ds'*Cs);
B=sqrtm(Cs'*inv(Rs)*Cs);
Q=Bs*inv(Ss)*Bs';
[Zs,ZAMP,G,REP]=care(A,B,Q);
%
% Choose rho=1 (Designer's choice) and
% build the generalized plant P in (9.87)
%
rho=1;
A = blkdiag(As,Ar);
B1 = [zeros(ns,mr) ((Bs*Ds')+(Zs*Cs'))*inv(sqrtm(Rs));
        Br zeros(nr,ls)];
B2 = [Bs;zeros(nr,ms)];
C1 = [zeros(ms,ns+nr);Cs zeros(ls,nr);rho*Cs -rho*rho*Cr];
C2 = [zeros(mr,ns+nr);Cs zeros(ls,nr)];
D11 = [zeros(ms,mr+ls);zeros(ls,mr) sqrtm(Rs);-rho*rho*Dr rho*sqrtm(Rs)];
D12 = [eye(ms);Ds;rho*Ds];
D21 = [rho*eye(mr) zeros(mr,ls);zeros(ls,mr) sqrtm(Rs)];
D22 = [zeros(mr,ms);Ds];
B = [B1 B2]; C = [C1;C2]; D = [D11 D12;D21 D22];
P = ss(A,B,C,D);
% Alternative: Use sysic to generate P from Figure 9.21
% but may get extra states, since states from Gs may enter twice.
%
% Gamma iterations to obtain H-infinity controller
%
[l1,m2] = size(D12); [l2,m1] = size(D21);
nmeas = l2; ncon = m2; gmin = 1; gmax = 5; gtol = 0.01;
[K,Gnclp, gam] = hinfsyn(P,nmeas,ncon,'GMIN',gmin,'GMAX',gmax,...
                'TOLGAM',gtol,'DISPLAY','on');
```

Remark 1 We stress that we aim here to minimize the $\mathcal{H}_\infty$ norm of the entire transfer function in (9.82). An alternative problem would be to minimize the $\mathcal{H}_\infty$ norm form r to e subject to an upper bound on $\|[\,\Delta_{N_s} \quad \Delta_{M_S}\,]\|_\infty$. This problem would involve the structured singular value, and the optimal controller could be obtained from solving a series of $\mathcal{H}_\infty$ optimization problems using DK-iteration; see Section 8.12.

Remark 2 Extra measurements. In some cases, a designer has more plant outputs available as measurements than can (or even need) to be controlled. These extra measurements can often make the design problem easier (e.g. velocity feedback) and therefore when beneficial should be used by the feedback controller K_2. This can be accommodated in the two degrees-of-freedom design procedure by introducing an output selection matrix W_o. This matrix selects from the output measurements y only those which are to be controlled and hence included in the model-matching part of the optimization. In Figure 9.21, W_o is introduced between y and the summing junction. In the optimization problem, only the equation for the error e is affected, and in the realization (9.87) for P one simply replaces ρC_s by $\rho W_o C_s$, $\rho R_s^{1/2}$ by $\rho W_o R_s^{1/2}$ and ρD_s by $\rho W_o D_s$ in the fifth row. For example, if there are four feedback measurements and only the first three are to be controlled, then

$$W_o = \begin{bmatrix} 1 & 0 & 0 & 0 \\ 0 & 1 & 0 & 0 \\ 0 & 0 & 1 & 0 \end{bmatrix} \tag{9.88}$$

Remark 3 Steady-state gain matching. The command signals r can be scaled by a constant matrix W_i to make the closed-loop transfer function from r to the controlled outputs $W_o y$ match the desired model T_{ref} exactly at steady-state. This is not guaranteed by the optimization which aims to minimize the ∞-norm of the error. The required scaling is given by

$$W_i \triangleq \left[W_o (I - G_s(s) K_2(s))^{-1} G_s(s) K_1(s) \right]^{-1} T_{\text{ref}}(s)\big|_{s=0} \tag{9.89}$$

Recall that $W_o = I$ if there are no extra feedback measurements beyond those that are to be controlled. The resulting controller is $K = [\,K_1 W_i \quad K_2\,]$.

We will conclude this subsection with a summary of the main steps required to synthesize a two degrees-of-freedom $\mathcal{H}_\infty$ loop-shaping controller.

1. Design a one degree-of-freedom $\mathcal{H}_\infty$ loop-shaping controller using the procedure of Section 9.4.2, but without a post-compensator weight W_2. Hence W_1.
2. Select a desired closed-loop transfer function T_{ref} between the commands and controlled outputs.
3. Set the scalar parameter ρ to a small value greater than 1; something in the range 1 to 3 will usually suffice.
4. For the shaped plant $G_s = GW_1$, the desired response T_{ref}, and the scalar parameter ρ, solve the standard $\mathcal{H}_\infty$ optimization problem defined by P in (9.87) to a specified tolerance to get $K = [\,K_1 \quad K_2\,]$. Remember to include W_o in the problem formulation if extra feedback measurements are to be used.
5. Replace the prefilter K_1 by $K_1 W_i$ to give exact model matching at steady-state.
6. Analyze and, if required, redesign making adjustments to ρ and possibly W_1 and T_{ref}.

The final two degrees-of-freedom $\mathcal{H}_\infty$ loop-shaping controller is illustrated in Figure 9.22

9.4.4   Observer-based structure for $\mathcal{H}_\infty$ loop-shaping controllers

$\mathcal{H}_\infty$ designs exhibit a separation structure in the controller. As seen from (9.50) and (9.51) the controller has an observer/state feedback structure, but the observer is non-standard, having a

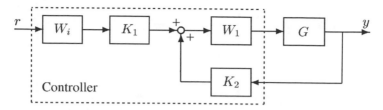

Figure 9.22: Two degrees-of-freedom $\mathcal{H}_\infty$ loop-shaping controller

disturbance term (a "worst" disturbance) entering the observer state equations. For $\mathcal{H}_\infty$ loop-shaping controllers, whether of the one or two degrees-of-freedom variety, this extra term is not present. The clear structure of $\mathcal{H}_\infty$ loop-shaping controllers has several advantages:

- It is helpful in describing a controller's function, especially to one's managers or clients who may not be familiar with advanced control.
- It lends itself to implementation in a gain-scheduled scheme, as shown by Hyde and Glover (1993).
- It offers computational savings in digital implementations and some multi-mode switching schemes, as shown in Samar (1995).

We will present the controller equations, for both one and two degrees-of-freedom $\mathcal{H}_\infty$ loop-shaping designs. For simplicity we will assume the shaped plant is strictly proper, with a stabilizable and detectable state-space realization

$$G_s \overset{s}{=} \left[\begin{array}{c|c} A_s & B_s \\ \hline C_s & 0 \end{array} \right] \tag{9.90}$$

In this case, as shown in Sefton and Glover (1990), the single degree-of-freedom $\mathcal{H}_\infty$ loop-shaping controller can be realized as an observer for the shaped plant plus a state feedback control law. The equations are

$$\dot{\widehat{x}}_s = A_s\widehat{x}_s + H_s(C_s\widehat{x}_s - y_s) + B_s u_s \tag{9.91}$$

$$u_s = \bar{K}_s\widehat{x}_s \tag{9.92}$$

where $\widehat{x}_s$ is the observer state, u_s and y_s are respectively the input and output of the shaped plant, and

$$H_s = -Z_s C_s^T \tag{9.93}$$

$$\bar{K}_s = -B_s^T \left[I - \gamma^{-2}I - \gamma^{-2}X_s Z_s \right]^{-1} X_s \tag{9.94}$$

where Z_s and X_s are the appropriate solutions to the generalized algebraic Riccati equations for G_s given in (9.67) and (9.68).

In Figure 9.23, an implementation of an observer-based $\mathcal{H}_\infty$ loop-shaping controller is shown in block diagram form. The same structure was used by Hyde and Glover (1993) in their VSTOL design which was scheduled as a function of aircraft forward speed.

Walker (1996) has shown that the two degrees-of-freedom $\mathcal{H}_\infty$ loop-shaping controller also has an observer-based structure. He considers a stabilizable and detectable plant

$$G_s \overset{s}{=} \left[\begin{array}{c|c} A_s & B_s \\ \hline C_s & 0 \end{array} \right] \tag{9.95}$$

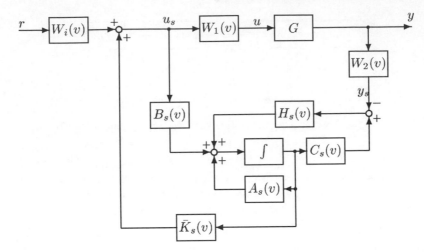

Figure 9.23: An implementation of an $\mathcal{H}_\infty$ loop-shaping controller for use when gain scheduling against a variable v

and a desired closed-loop transfer function

$$T_{\text{ref}} \stackrel{s}{=} \left[\begin{array}{c|c} A_r & B_r \\ \hline C_r & 0 \end{array} \right] \tag{9.96}$$

in which case the generalized plant $P(s)$ in (9.87) simplifies to

$$P \stackrel{s}{=} \left[\begin{array}{cc|ccc} A_s & 0 & 0 & Z_s C_s^T & B_s \\ 0 & A_r & B_r & 0 & 0 \\ \hline 0 & 0 & 0 & 0 & I \\ C_s & 0 & 0 & I & 0 \\ \rho C_s & -\rho^2 C_r & 0 & \rho I & 0 \\ \hdashline 0 & 0 & \rho I & 0 & 0 \\ C_s & 0 & 0 & I & 0 \end{array} \right] \stackrel{\triangle}{=} \left[\begin{array}{c|cc} A & B_1 & B_2 \\ \hline C_1 & D_{11} & D_{12} \\ C_2 & D_{21} & D_{22} \end{array} \right] \tag{9.97}$$

Walker (1996) then shows that a stabilizing controller $K = \begin{bmatrix} K_1 & K_2 \end{bmatrix}$ satisfying $\|F_l(P, K)\|_\infty < \gamma$ exists if, and only if,

(i) $\gamma > \sqrt{1 + \rho^2}$, and

(ii) $X_\infty \geq 0$ is a solution to the algebraic Riccati equation

$$X_\infty A + A^T X_\infty + C_1^T C_1 - \bar{F}^T (\bar{D}^T \bar{J} \bar{D}) \bar{F} = 0 \tag{9.98}$$

such that Re $\lambda_i \begin{bmatrix} A - B\bar{F} \end{bmatrix} < 0\ \forall i$, where

$$\bar{F} = (\bar{D}^T \bar{J} \bar{D})^{-1} (\bar{D}^T \bar{J} C + B^T X_\infty) \tag{9.99}$$

$$\bar{D} = \begin{bmatrix} D_{11} & D_{12} \\ I_w & 0 \end{bmatrix} \tag{9.100}$$

$$\bar{J} = \begin{bmatrix} I_z & 0 \\ 0 & -\gamma^2 I_w \end{bmatrix} \tag{9.101}$$

where I_z and I_w are unit matrices of dimensions equal to those of the error signal z, and exogenous input w, respectively, in the standard configuration.

Notice that this $\mathcal{H}_\infty$ controller depends on the solution to just one algebraic Riccati equation, not two. This is a characteristic of the two degrees-of-freedom $\mathcal{H}_\infty$ loop-shaping controller (Hoyle et al., 1991).

Walker (1996) further shows that if (i) and (ii) are satisfied, then a stabilizing controller $K(s)$ satisfying $\|F_l(P, K)\|_\infty < \gamma$ has the following equations:

$$\dot{\hat{x}}_s = A_s \hat{x}_s + H_s (C_s \hat{x}_s - y_s) + B_s u_s \tag{9.102}$$

$$\dot{x}_r = A_r x_r + B_r r \tag{9.103}$$

$$u_s = -B_s^T X_{\infty 11} \hat{x}_s - B_s^T X_{\infty 12} x_r \tag{9.104}$$

where $X_{\infty 11}$ and $X_{\infty 12}$ are elements of

$$X_\infty = \begin{bmatrix} X_{\infty 11} & X_{\infty 12} \\ X_{\infty 21} & X_{\infty 22} \end{bmatrix} \tag{9.105}$$

which has been partitioned conformably with

$$A = \begin{bmatrix} A_s & 0 \\ 0 & A_r \end{bmatrix} \tag{9.106}$$

and H_s is as in (9.93).

The structure of this controller is shown in Figure 9.24, where the state feedback gain matrices F_s and F_r are defined by

$$F_s \triangleq B_s^T X_{\infty 11} \qquad F_r \triangleq B_s^T X_{\infty 12} \tag{9.107}$$

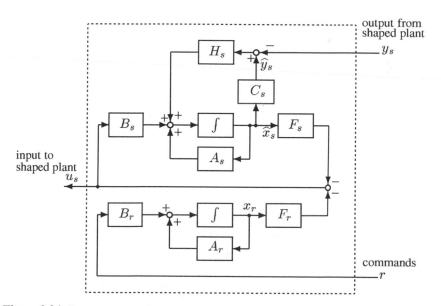

Figure 9.24: Structure of the two degrees-of-freedom $\mathcal{H}_\infty$ loop-shaping controller

The controller consists of a state observer for the shaped plant G_s, a model of the desired closed-loop transfer function T_{ref} (without C_r) and a state feedback control law that uses both the observer and reference-model states.

As in the one degree-of-freedom case, this observer-based structure is useful in gain scheduling. The reference-model part of the controller is also nice because it is often the same at different design operating points and so may not need to be changed at all during a scheduled operation of the controller. Likewise, parts of the observer may not change; for example, if the weight $W_1(s)$ is the same at all the design operating points. Therefore whilst the structure of the controller is comforting in the familiarity of its parts, it also has some significant advantages when it comes to implementation.

9.4.5 Implementation issues

Discrete time controllers. For implementation purposes, discrete time controllers are usually required. These can be obtained from a continuous time design using a bilinear transformation from the s-domain to the z-domain, but there can be advantages in being able to design directly in discrete time. In Samar (1995) and Postlethwaite et al. (1995), observer-based state-space equations are derived directly in discrete time for the two degrees-of-freedom $\mathcal{H}_\infty$ loop-shaping controller and successfully applied to an aero-engine. This application was on a real engine, a Spey engine, which is a Rolls-Royce two-spool reheated turbofan that was housed at the UK Defence Research Agency (now QinetiQ), Pyestock. As this was a real application, a number of important implementation issues needed to be addressed. Although these are outside the general scope of this book, they will be briefly mentioned now.

Anti-windup. In $\mathcal{H}_\infty$ loop shaping the pre-compensator weight W_1 would normally include integral action in order to reject low-frequency disturbances acting on the system. However, in the case of actuator saturation the integrators continue to integrate their input and hence cause windup problems. An anti-windup scheme is therefore required on the weighting function W_1. One approach is to implement the weight W_1 in its *self-conditioned* or *Hanus* form. Let the weight W_1 have a realization

$$
W_1 \overset{s}{=} \left[\begin{array}{c|c} A_w & B_w \\ \hline C_w & D_w \end{array} \right]
\tag{9.108}
$$

and let u be the input to the plant actuators and u_s the input to the shaped plant. Then $u = W_1 u_s$. When implemented in Hanus form, the expression for u becomes (Hanus et al., 1987)

$$
u = \left[\begin{array}{c|cc} A_w - B_w D_w^{-1} C_w & 0 & B_w D_w^{-1} \\ \hline C_w & D_w & 0 \end{array} \right] \left[\begin{array}{c} u_s \\ u_a \end{array} \right]
\tag{9.109}
$$

where u_a is the actual plant input; that is, the measurement at the output of the actuators which therefore contains information about possible actuator saturation. The situation is illustrated in Figure 9.25, where the actuators are each modelled by a unit gain and a saturation. The Hanus form prevents windup by keeping the states of W_1 consistent with the actual plant input at all times. When there is no saturation $u_a = u$, the dynamics of W_1 remain unaffected and (9.109) simplifies to (9.108). But when $u_a \neq u$ the dynamics are inverted and driven by u_a so that the states remain consistent with the actual plant input u_a. Notice that such an implementation requires W_1 to be invertible and minimum-phase. A more general approach to anti-windup is given in Section 12.4.

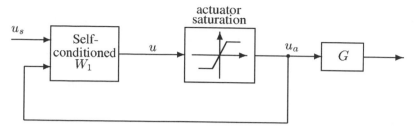

Figure 9.25: Self-conditioned weight W_1

Exercise 9.10 * *Show that the Hanus form of the weight W_1 in (9.109) simplifies to (9.108) when there is no saturation, i.e. when $u_a = u$.*

Bumpless transfer. In the aero-engine application, a multi-mode switched controller was designed. This consisted of three controllers, each designed for a different set of engine output variables, which were switched between depending on the most significant outputs at any given time. To ensure smooth transition from one controller to another – bumpless transfer – it was found useful to condition the reference models and the observers in each of the controllers. Thus when on-line, the observer state evolves according to an equation of the form (9.102) but when off-line the state equation becomes

$$\dot{\widehat{x}}_s = A_s\widehat{x}_s + H_o(C_s\widehat{x}_s - y_s) + B_s u_{as} \tag{9.110}$$

where u_{as} is the actual input to the shaped plant governed by the on-line controller. The reference model with state feedback given by (9.103) and (9.104) is not invertible and therefore cannot be self-conditioned. However, in discrete time the optimal control also has a feed-through term from r which gives a reference model that can be inverted. Consequently, in the aero-engine example the reference models for the three controllers were each conditioned so that the inputs to the shaped plant from the off-line controller followed the actual shaped plant input u_{as} given by the on-line controller. For a more recent treatment of bumpless transfer see Turner and Walker (2000).

Satisfactory solutions to implementation issues such as those discussed above are crucial if advanced control methods are to gain wider acceptance in industry. We have tried to demonstrate here that the observer-based structure of the $\mathcal{H}_\infty$ loop-shaping controller is helpful in this regard.

9.5 Conclusion

We have described several methods and techniques for controller design, but our emphasis has been on $\mathcal{H}_\infty$ loop shaping which is easy to apply and in our experience works very well in practice. It combines classical loop-shaping ideas (familiar to most practising engineers) with an effective method for robustly stabilizing the feedback loop. For complex problems, such as unstable plants with multiple gain crossover frequencies, it may not be easy to decide on a desired loop shape. In this case, we would suggest doing an initial LQG design (with simple weights) and using the resulting loop shape as a reasonable one to aim for in $\mathcal{H}_\infty$ loop shaping.

An alternative to $\mathcal{H}_\infty$ loop shaping is a standard $\mathcal{H}_\infty$ design with a "stacked" cost function such as in S/KS mixed-sensitivity optimization. In this approach, $\mathcal{H}_\infty$ optimization is used to shape two or sometimes three closed-loop transfer functions. However, with more functions the shaping becomes increasingly difficult for the designer.

In other design situations where there are several performance objectives (e.g. on signals, model following and model uncertainty), it may be more appropriate to follow a signal-based $\mathcal{H}_2$ or $\mathcal{H}_\infty$ approach. But again the problem formulations become so complex that the designer has little direct influence on the design.

After a design, the resulting controller should be analyzed with respect to robustness and tested by nonlinear simulation. For the former, we recommend μ-analysis as discussed in Chapter 8, and if the design is not robust, then the weights will need modifying in a redesign. Sometimes one might consider synthesizing a μ-optimal controller, but this complexity is rarely necessary in practice. Moreover, one should be careful about combining controller synthesis and analysis into a single step. The following quote from Rosenbrock (1974) illustrates the dilemma:

> In synthesis the designer specifies in detail the properties which his system must have, to the point where there is only one possible solution. ...The act of specifying the requirements in detail implies the final solution, yet has to be done in ignorance of this solution, which can then turn out to be unsuitable in ways that were not foreseen.

Therefore, control system design usually proceeds iteratively through the steps of modelling, control structure design, controllability analysis, performance and robustness weights selection, controller synthesis, control system analysis and nonlinear simulation. Rosenbrock (1974) makes the following observation:

> Solutions are constrained by so many requirements that it is virtually impossible to list them all. The designer finds himself threading a maze of such requirements, attempting to reconcile conflicting demands of cost, performance, easy maintenance, and so on. A good design usually has strong aesthetic appeal to those who are competent in the subject.

10

CONTROL STRUCTURE DESIGN

Most (if not all) available control theories assume that a control structure is given at the outset. They therefore fail to answer some basic questions, which a control engineer regularly meets in practice. Which variables should be controlled, which variables should be measured, which inputs should be manipulated, and which links should be made between them? The objective of this chapter is to describe the main issues involved in control structure design and to present some of the quantitative methods available, for example, for selection of controlled variables and for decentralized control.

10.1 Introduction

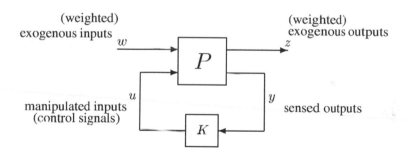

Figure 10.1: General control configuration

In much of this book, we consider the general control problem formulation shown in Figure 10.1, where the *controller design* problem is to

- Find a stabilizing controller K, which, based on the information in y, generates a control signal u, which counteracts the influence of w on z, thereby minimizing the closed-loop norm from w to z.

We presented different techniques for controller design in Chapters 2, 8 and 9. However, if we go back to Chapter 1 (page 1), then we see that controller design is only one step, step 9, in the overall process of designing a control system. In this chapter, we are concerned with the structural decisions of *control structure design*, which are the steps necessary to get to Figure 10.1:

Multivariable Feedback Control: Analysis and Design. Second Edition
S. Skogestad and I. Postlethwaite   © 2005, 2006 John Wiley & Sons, Ltd

Step 4 on page 1: The selection of controlled outputs (a set of variables which are to be controlled to achieve a set of specific objectives).
See Sections 10.2 and 10.3: *What are the variables z in Figure 10.1?*

Step 5 on page 1: The selection of manipulated inputs and measurements (sets of variables which can be manipulated and measured for control purposes).
See Section 10.4: *What are the variable sets u and y in Figure 10.1?*

Step 6 on page 1: The selection of a *control configuration* (a structure of interconnecting measurements/commands and manipulated variables).
See Sections 10.5 and 10.6: *What is the structure of K in Figure 10.1; that is, how should we "pair" the variable sets u and y?*

The distinction between the words control *structure* and control *configuration* may seem minor, but note that it is significant within the context of this book. The *control structure* (or control strategy) refers to all structural decisions included in the design of a control system (steps 4, 5 and 6). On the other hand, the *control configuration* refers only to the structuring (decomposition) of the controller *K* itself (step 6) (also called the measurement/manipulation partitioning or input/output pairing). Control configuration issues are discussed in more detail in Section 10.5. The selection of controlled outputs, manipulations and measurements (steps 4 and 5 combined) is sometimes called *input/output selection*.

One important reason for decomposing the control system into a specific *control configuration* is that it may allow for simple tuning of the subcontrollers without the need for a detailed plant model describing the dynamics and interactions in the process. Multivariable centralized controllers can always outperform decomposed (decentralized) controllers, but this performance gain must be traded off against the cost of obtaining and maintaining a sufficiently detailed plant model and the additional hardware.

The number of possible control structures shows a combinatorial growth, so for most systems a careful evaluation of all alternative control structures is impractical. Fortunately, we can often obtain a reasonable choice of controlled outputs, measurements and manipulated inputs from physical insight. In other cases, simple controllability measures as presented in Chapters 5 and 6 may be used for quickly evaluating or screening alternative control structures. Additional tools are presented in this chapter.

From an engineering point of view, the decisions involved in designing a complete control system are taken sequentially: first, a "top-down" selection of controlled outputs, measurements and inputs (steps 4 and 5) and then a "bottom-up" design of the control system (in which step 6, the selection of the control configuration, is the most important decision). However, the decisions are closely related in the sense that one decision directly influences the others, so the procedure may involve iteration. Skogestad (2004a) has proposed a procedure for control structure design for complete chemical plants, consisting of the following *structural* decisions:

"Top-down" (mainly step 4)

(i) Identify operational constraints and identify a scalar cost function *J* that characterizes optimal operation.

(ii) Identify degrees of freedom (manipulated inputs *u*) and in particular identify the ones that affect the cost *J* (in process control, the cost *J* is usually determined by the steady-state).

(iii) Analyze the solution of optimal operation for various disturbances, with the aim of finding primary controlled variables ($y_1 = z$) which, when kept constant, indirectly minimize the cost ("self-optimizing control"). (Section 10.3)

(iv) Determine where in the plant to set the production rate.

"Bottom-up" (steps 5 and 6)

(v) *Regulatory/base control layer*: Identify additional variables to be measured and controlled (y_2), and suggest how to pair these with manipulated inputs. (Section 10.4)

(vi) *"Advanced"/supervisory control layer* configuration: Should it be decentralized or multivariable? (Sections 10.5.1 and 10.6)

(vii) *On-line optimization layer*: Is this needed or is a constant setpoint policy sufficient ("self-optimizing control")? (Section 10.3)

Except for decision (iv), which is specific to process control, this procedure may be applied to any control problem.

Control structure design was considered by Foss (1973) in his paper entitled "Critique of chemical process control theory" where he concluded by challenging the control theoreticians of the day to close the gap between theory and applications in this important area. Control structure design is clearly important in the chemical process industry because of the complexity of these plants, but the same issues are relevant in most other areas of control where we have large-scale systems. In the late 1980's Carl Nett (Nett, 1989; Nett and Minto, 1989) gave a number of lectures based on his experience of aero-engine control at General Electric, under the title "A quantitative approach to the selection and partitioning of measurements and manipulations for the control of complex systems". He noted that increases in controller complexity unnecessarily outpace increases in plant complexity, and that the objective should be to

> minimize control system complexity subject to the achievement of accuracy specifications in the face of uncertainty.

Balas (2003) recently surveyed the status of flight control. He states, with reference to the *Boeing* company, that "the key to the control design is selecting the variables to be regulated and the controls to perform regulation" (steps 4 and 5). Similarly, the first step in *Honeywell*'s procedure for controller design is "the selection of controlled variables (CVs) for performance and robustness" (step 4).

Surveys on control structure design and input–output selection are given by Van de Wal (1994) and Van de Wal and de Jager (2001), respectively. A review of control structure design in the chemical process industry (plantwide control) is given by Larsson and Skogestad (2000). The reader is referred to Chapter 5 (page 164) for an overview of the literature on input–output controllability analysis.

10.2   Optimal operation and control

The overall control objective is to maintain acceptable operation (in terms of safety, environmental impact, load on operators, and so on) while keeping the operating conditions close to economically optimal. In Figure 10.2, we show three different implementations for optimization and control:

(a) Open-loop optimization

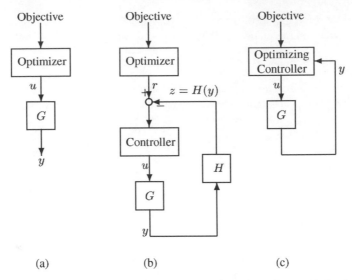

Figure 10.2: Different structures for optimization and control. (a) Open-loop optimization. (b) Closed-loop implementation with separate control layer. (c) Integrated optimization and control.

(b) Closed-loop implementation with separate control layer

(c) Integrated optimization and control ("optimizing control")

Structure (a) with open-loop optimization is usually not acceptable because of model error and unmeasured disturbances. Theoretically, optimal performance is obtained with the *centralized optimizing controller* in structure (c), which combines the functions of optimization and control in one layer. All control actions in such an ideal control system would be perfectly coordinated and the control system would use on-line dynamic optimization based on a nonlinear dynamic model of the complete plant instead of, for example, infrequent steady-state optimization. However, this solution is normally not used for a number of reasons, including: the cost of modelling, the difficulty of controller design, maintenance and modification, robustness problems, operator acceptance, and the lack of computing power.

In practice, the *hierarchical control system* in Figure 10.2(b) is used, with different tasks assigned to each layer in the hierarchy. In the simplest case we have two layers:

- *optimization layer* – computes the desired optimal reference commands r (outside the scope of this book)
- *control layer* – implements the commands to achieve $z \approx r$ (the focus of this book).

The optimization tends to be performed *open-loop* with limited use of feedback. On the other hand, the control layer is mainly based on *feedback* information. The optimization is often based on nonlinear steady-state models, whereas linear dynamic models are mainly used in the control layer (as we do throughout the book).

Additional layers are possible, as is illustrated in Figure 10.3 which shows a typical control hierarchy for a complete chemical plant. Here the control layer is subdivided into two layers: *supervisory control* ("advanced control") and *regulatory control* ("base control"). We have also included a scheduling layer above the optimization layer. Similar hierarchies

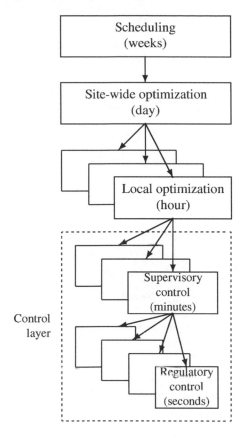

Figure 10.3: Typical control system hierarchy in a chemical plant

are found in control systems for most applications, although the time constants and names of the layers may be different. Note that we have not included any functions related to logic control (startup/ shutdown) and safety systems. These are of course important, but need not be considered during normal operation.

In general, the information flow in such a control hierarchy is based on the upper layer sending *setpoints* (references, commands) to the layer below, and the lower layer reporting back any problems in achieving this. There is usually a *time scale separation* between the upper layers and the lower layers as indicated in Figure 10.3. The slower upper layer controls variables that are more important from an overall (long time scale) point of view, using as degrees of freedom the setpoints for the faster lower layer. The lower layer should take care of fast (high-frequency) disturbances and keep the system reasonably close to its optimum in the fast time scale. To reduce the need for frequent setpoint changes, we should control variables that require small setpoint changes, and this observation is the basis for Section 10.3 which deals with selecting controlled variables.

With a "reasonable" time scale separation between the layers, typically a factor of five or more in terms of closed-loop response time, we have the following advantages:

1. The stability and performance of a lower (faster) layer is not much influenced by the presence of upper (slow) layers because the frequency of the "disturbance" from the upper layer is well inside the bandwidth of the lower layer.
2. With the lower (faster) layers in place, the stability and performance of the upper (slower) layers do not depend much on the specific controller settings used in the lower layers because they only effect high frequencies outside the bandwidth of the upper layers.

More generally, there are two ways of partitioning the control system:

Vertical (hiearchical) decomposition. This is the decomposition just discussed which usually results from a time scale difference between the various control objectives ("decoupling in time"). The controllers are normally designed sequentially, starting with the fast layers, and then cascaded (series interconnected) in a hierarchical manner.

Horizontal decomposition. This is used when the plant is "decoupled in space", and normally involves a set of independent decentralized controllers. Decentralized control is discussed in more detail in Section 10.6 (page 429).

Remark 1 In accordance with Lunze (1992) we have purposely used the word *layer* rather than *level* for the hierarchical decomposition of the control system. The somewhat subtle difference is that in a *multilevel* system all units contribute to satisfying the same goal, whereas in a *multilayer* system the different units have different local objectives (which preferably contribute to the overall goal). Multilevel systems have been studied in connection with the solution of optimization problems.

Remark 2 The tasks within any layer can be performed by humans (e.g. manual control), and the interaction and task sharing between the automatic control system and the human operators are very important in most cases, e.g. an aircraft pilot. However, these issues are outside the scope of this book.

Remark 3 As noted above, we may also decompose the control layer, and from now on when we talk about control configurations, hierarchical decomposition and decentralization, we generally refer to the control layer.

Remark 4 A fourth possible strategy for optimization and control, not shown in Figure 10.2, is (d) *extremum-seeking control*. Here the model-based block in Figure 10.2(c) is replaced by an "experimenting" controller, which, based on measurements of the cost J, perturbs the input in order to seek the extremum (minimum) of J; see e.g. Ariyur and Krstic (2003) for details. The main disadvantage with this strategy is that a fast and accurate on-line measurement of J is rarely available.

10.3   Selection of primary controlled outputs

We are concerned here with the selection of controlled outputs (controlled variables, CVs). This involves selecting the variables z to be controlled at given reference values, $z \approx r$, where r is set by some higher layer in the control hierarchy. Thus, the selection of controlled outputs (for the control layer) is usually intimately related to the hierarchical structuring of the control system shown in Figure 10.2(b). The aim of this section is to provide systematic methods for selecting controlled variables. Until recently, this has remained an unsolved problem. For example, Fisher et al. (1985) state that "Our current approach to control of a complete plant is to solve the optimal steady-state problem on-line, and then use the results of this analysis to fix the setpoints of selected controlled variables. There is no available procedure for selecting this set of controlled variables, however. Hence experience and intuition still plays a major role in the design of control systems."

The important variables in this section are:

- u – degrees of freedom (inputs)
- z – primary ("economic") controlled variables
- r – reference value (setpoint) for z
- y – measurements, process information (often including u)

In the general case, the controlled variables are selected as functions of the measurements, $z = H(y)$. For example, z can be a linear combination of measurements, i.e. $z = Hy$. In many cases, we select individual measurements as controlled variables and H is a "selection matrix" consisting of ones and zeros. Normally, we select as many controlled variables as the number of available degrees of freedom, i.e. $n_z = n_u$.

The controlled variables z are often not important variables in themselves, but are controlled in order to achieve some overall operational objective. A reasonable question is then: why not forget the whole thing about selecting controlled variables, and instead directly adjust the manipulated variables u? The reason is that an open-loop implementation usually fails because we are not able to adjust to changes (disturbances d) and errors (in the model). The following example illustrates the issues.

Example 10.1 Cake baking. *The overall goal is to make a cake which is well baked inside and has a nice exterior. The manipulated input for achieving this is the heat input, $u = Q$ (and we will assume that the duration of the baking is fixed, e.g. at 15 minutes).*

(a) If we had never baked a cake before, and if we were to construct the oven ourselves, we might consider directly manipulating the heat input to the oven, possibly with a watt-meter measurement. However, this open-loop *implementation would not work well, as the optimal heat input depends strongly on the particular oven we use, and the operation is also sensitive to disturbances; for example, opening the oven door or whatever else might be in the oven. In short, the open-loop implementation is sensitive to uncertainty.*

(b) An effective way of reducing the uncertainty is to use feedback. Therefore, in practice we use a closed-loop *implementation where we control the oven temperature ($z = T$) using a thermostat. The temperature setpoint $r = T_s$ is found from a cook book (which plays the role of the "optimizer"). The (a) open-loop and (b) closed-loop implementations of the cake baking process are illustrated in Figure 10.2.*

The key question is: what variables z should we control? In many cases, it is clear from a physical understanding of the process what these are. For example, if we are considering heating or cooling a room, then we should select the room temperature as the controlled variable z. Furthermore, we generally control variables that are optimally at their constraints (limits). For example, we make sure that the air conditioning is on maximum if we want to cool down our house quickly. In other cases, it is less obvious what to control, because the overall control objective may not be directly associated with keeping some variable constant.

To get an idea of the issues involved, we will consider some simple examples. Let us first consider two cases where implementation is obvious because the optimal strategy is to keep variables at their constraints.

Example 10.2 Short-distance (100 m) running. *The objective is to minimize the time T of the race ($J = T$). The manipulated input (u) is the muscle power. For a well-trained runner, the optimal solution lies at the constraint $u = u_{max}$. Implementation is then easy: select $z = u$ and $r = u_{max}$ or alternatively "run as fast as possible".*

Example 10.3 Driving from A to B. *Let y denote the speed of the car. The objective is to minimize the time T of driving from A to B or, equivalently, to maximize the speed (y), i.e. $J = -y$. If we are driving on a straight and clear road, then the optimal solution is always to stay on the speed limit constraint (y_{max}). Implementation is then easy: use a feedback scheme (cruise control) to adjust the engine power (u) such that we are at the speed limit; that is, select $z = y$ and $r = y_{max}$.*

In the next example, the optimal solution does not lie at a constraint and the selection of the controlled variable is not obvious.

Example 10.4 Long-distance running. *The objective is to minimize the time T of the race ($J = T$), which is achieved by maximizing the average speed. It is clear that running at maximum input power is not a good strategy. This would give a high speed at the beginning, but a slower speed towards the end, and the average speed will be lower. A better policy would be to keep constant speed ($z = y_1 =$ speed). The optimization layer (e.g. the trainer) will then choose an optimal setpoint r for the speed, and this is implemented by the control layer (the runner). Alternative strategies, which may work better in a hilly terrain, are to keep a constant heart rate ($z = y_2 =$ heart rate) or a constant lactate level ($z = y_3 =$ lactate level).*

10.3.1 Self-optimizing control

Recall that the title of this section is selection of primary controlled outputs. In the cake baking process, we select the *oven temperature* as the controlled output z in the control layer. It is interesting to note that controlling the oven temperature in itself has no direct relation to the overall goal of making a well-baked cake. So why do we select the oven temperature as a controlled output? We now want to outline an approach for answering questions of this kind. Two distinct questions arise:

1. What variables z should be selected as the controlled variables?
2. What is the optimal reference value (z_{opt}) for these variables?

The second problem is one of optimization and is extensively studied (but not in this book). Here we want to gain some insight into the first problem which has been much less studied. We make the following *assumptions*:

1. The overall goal can be quantified in terms of a scalar cost function J.
2. For a given disturbance d, there exists an optimal value $u_{opt}(d)$ (and corresponding value $z_{opt}(d)$), which minimizes the cost function J.
3. The reference values r for the controlled outputs z are kept constant, i.e. r is independent of the disturbances d. Typically, some average value is selected, e.g. $r = z_{opt}(\bar{d})$.

In the following, we assume that the optimally constrained variables are already controlled at their constraints ("active constraint control") and consider the "remaining" unconstrained problem with controlled variables z and remaining unconstrained degrees of freedom u.

The system behaviour is a function of the independent variables u and d, so we may formally write $J = J(u, d)$.[1] For a given disturbance d the optimal value of the cost function

---

[1] Note that the cost J is usually not a simple function of u and d, but is rather given by some implied relationship such as

$$\min_{u,x} J = J_0(u, x, d) \quad \text{s.t.} \quad f(x, u, d) = 0$$

where $\dim f = \dim x$ and $f(x, u, d) = 0$ represents the model equations. Formally eliminating the internal state variables x gives the problem $\min_u J(u, d)$.

is

$$J_{\mathrm{opt}}(d) \triangleq J(u_{\mathrm{opt}}(d), d) = \min_{u} J(u, d) \tag{10.1}$$

Ideally, we want $u = u_{\mathrm{opt}}(d)$. However, this will not be achieved in practice and we have a loss $L = J(u, d) - J_{\mathrm{opt}}(d) > 0$.

We consider the simple feedback policy in Figure 10.2(b), where we attempt to keep z constant. Note that the open-loop implementation is included as a special case by selecting $z = u$. The aim is to adjust u automatically, if necessary, when there is a disturbance d such that $u \approx u_{\mathrm{opt}}(d)$. This effectively turns the complex optimization problem into a simple feedback problem. The goal is to achieve "self-optimizing control" (Skogestad, 2000):

> *Self-optimizing control is when we can achieve an acceptable loss with constant setpoint values for the controlled variables without the need to reoptimize when disturbances occur.*

Remark. In Chapter 5, we introduced the term self-regulation, which is when acceptable dynamic control performance can be obtained with constant manipulated variables (u). Self-optimizing control is a direct generalization to the layer above where we can achieve acceptable (economic) performance with constant controlled variables (z).

The concept of self-optimizing control is inherent in many real-life scenarios including (Skogestad, 2004b):

- The *central bank* attempts to optimize the welfare of the country (J) by keeping a constant inflation rate (z) by varying the interest rate (u).
- The *long-distance runner* may attempt to minimize the total running time ($J = T$) by keeping a constant heart rate ($z = y_1$) or constant lactate level ($z = y_2$) by varying the muscle power (u).
- A driver attempts to minimize the fuel consumption and engine wear (J) by keeping a constant engine rotation speed (z) by varying the gear position (u).

The presence of self-optimizing control is also evident in biological systems, which have no capacity for solving complex on-line optimization problems. Here, self-optimizing control policies are the only viable solution and have developed by evolution. In business systems, the primary ("economic") controlled variables are called key performance indicators (KPIs) and their optimal values are obtained by analyzing successful businesses ("benchmarking").

The idea of self-optimizing control is further illustrated in Figure 10.4, where we see that there is a loss if we keep a constant value for the controlled variable z, rather than reoptimizing when a disturbance moves the process away from its nominal optimal operating point (denoted $\bar{d}$).

An ideal self-optimizing variable would be the gradient of the Lagrange function for the optimization problem, which should be zero. However, a direct measurement of the gradient (or a closely related variable) is rarely available, and computing the gradient generally requires knowing the value of unmeasured disturbances. We will now outline some approaches for selecting the controlled variables z. Although a model is used to find z, note that the goal of self-optimizing control is to eliminate the need for on-line model-based optimization.

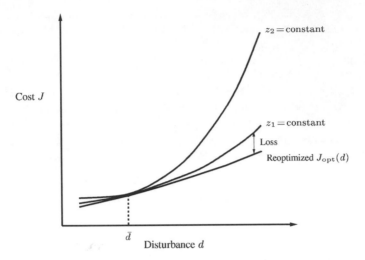

Figure 10.4: Loss imposed by keeping constant setpoint for the controlled variable. In this case z_1 is a better "self-optimizing" controlled variable than z_2.

10.3.2   Selecting controlled outputs: local analysis

We use here a local second-order accurate analysis of the loss function. From this, we derive the useful minimum singular value rule, and an exact local method; see Halvorsen et al. (2003) for further details. Note that this is a local analysis, which may be misleading; for example, if the optimum point of operation is close to infeasibility.

Consider the loss $L = J(u,d) - J_{opt}(d)$, where d is a fixed (generally non-zero) disturbance. We here make the following additional assumptions:

1. The cost function J is smooth, or more precisely twice differentiable.
2. As before, we assume that the optimization problem is unconstrained. If it is optimal to keep some variable at a constraint, then we assume that this is implemented ("active constraint control") and consider the remaining unconstrained problem.
3. The dynamics of the problem can be neglected when evaluating the cost; that is, we consider steady-state control and optimization.
4. We control as many variables z as there are available degrees of freedom, i.e. $n_z = n_u$.

For a fixed d we may then express $J(u,d)$ in terms of a Taylor series expansion in u around the optimal point. We get

$$
J(u,d) = J_{opt}(d) + \underbrace{\left(\frac{\partial J}{\partial u}\right)_{opt}^{T}}_{=0}(u - u_{opt}(d))
$$

$$
+ \frac{1}{2}(u - u_{opt}(d))^{T}\underbrace{\left(\frac{\partial^2 J}{\partial u^2}\right)_{opt}}_{=J_{uu}}(u - u_{opt(d)}) + \cdots \qquad (10.2)
$$

We will neglect terms of third order and higher (which assumes that we are reasonably close to the optimum). The second term on the right hand side in (10.2) is zero at the optimal point for an unconstrained problem. Equation (10.2) quantifies how a non-optimal input $u - u_{\text{opt}}$ affects the cost function. To study how this relates to output selection we use a linearized model of the plant

$$z = Gu + G_d d \tag{10.3}$$

where G and G_d are the steady-state gain matrix and disturbance model respectively. For a fixed d, we have $z - z_{\text{opt}} = G(u - u_{\text{opt}})$. If G is invertible we then get

$$u - u_{\text{opt}} = G^{-1}(z - z_{\text{opt}}) \tag{10.4}$$

Note that G is a square matrix, since we have assumed that $n_z = n_u$. From (10.2) and (10.4) we get the second-order accurate approximation

$$L = J - J_{\text{opt}} \approx \frac{1}{2}(z - z_{\text{opt}})^T G^{-T} J_{uu} G^{-1}(z - z_{\text{opt}}) \tag{10.5}$$

where the term $J_{uu} = (\partial^2 J / \partial u^2)_{\text{opt}}$ is independent of z. Alternatively, we may write

$$L = \frac{1}{2}\|\tilde{z}\|_2^2 \tag{10.6}$$

where $\tilde{z} = J_{uu}^{1/2} G^{-1}(z - z_{\text{opt}})$. These expressions for the loss L yield considerable insight. Obviously, we would like to select the controlled outputs z such that $z - z_{\text{opt}}$ is zero. However, this is not possible in practice because of (1) varying disturbances d and (2) implementation error e associated with control of z. To see this more clearly, we write

$$z - z_{\text{opt}} = z - r + r - z_{\text{opt}} = e + e_{\text{opt}}(d) \tag{10.7}$$

where

$$\text{Optimization error}: \quad e_{\text{opt}}(d) \triangleq r - z_{\text{opt}}(d)$$

$$\text{Implementation error}: \quad e \triangleq z - r$$

First, we have an optimization error $e_{\text{opt}}(d)$ because the algorithm (e.g. the cook book for cake baking) gives a desired r which is different from the optimal $z_{\text{opt}}(d)$. Second, we have a control or implementation error e because control is not perfect; either because of poor control performance or because of an incorrect measurement (steady-state bias) n^z. If we have integral action in the controller, then the steady-state control error is zero, and we have

$$e = n^z$$

If z is directly measured then n^z is its measurement error. If z is a combination of several measurements y, $z = Hy$, see Figure 10.2(b), then $n^z = Hn^y$, where n^y is the vector of measurement errors for the measurements y.

In most cases, the errors e and $e_{\text{opt}}(d)$ can be assumed independent. The maximum value of $|z - z_{\text{opt}}|$ for the expected disturbances and implementation errors, which we call the "expected optimal span", is then

$$\text{span}(z) = \max_{d,e} |z - z_{\text{opt}}| = \max_d |e_{\text{opt}}(d)| + \max_e |e| \tag{10.8}$$

Example 10.1 Cake baking continued. *Let us return to the question: why select the oven temperature as a controlled output? We have two alternatives: a closed-loop implementation with $z = T$ (the oven temperature) and an open-loop implementation with $z = u = Q$ (the heat input). From experience, we know that the optimal oven temperature T_{opt} is largely independent of disturbances and is almost the same for any oven. This means that we may always specify the same oven temperature, say $r = T_s = 190°C$, as obtained from the cook book. On the other hand, the optimal heat input Q_{opt} depends strongly on the heat loss, the size of the oven, etc., and may vary between, say, 100 W and 5000 W. A cook book would then need to list a different value of $r = Q_s$ for each kind of oven and would in addition need some correction factor depending on the room temperature, how often the oven door is opened, etc. Therefore, we find that it is much easier to get $e_{opt} = T_s - T_{opt}$ [°C] small than to get $e_{opt} = Q_s - Q_{opt}$ [W] small. Thus, the main reason for controlling the oven temperature is to minimize the optimization error.* In addition, the control error e is expected to be much smaller when controlling temperature.

From (10.5) and (10.7), we conclude that we should select the controlled outputs z such that:

1. G^{-1} is small (i.e. G is large); the choice of z should be such that the inputs have a large effect on z.
2. $e_{opt}(d) = r - z_{opt}(d)$ is small; the choice of z should be such that its optimal value $z_{opt}(d)$ depends only weakly on the disturbances (and other changes).
3. $e = z - r$ is small; the choice of z should be such that it is easy to keep the control or implementation error e small.
4. G^{-1} is small, which implies that G should not be close to singular. For cases with two or more controlled variables, the variables should be selected such that they are independent of each other.

By proper scaling of the variables, these four requirements can be combined into the "maximize minimum singular value rule" as discussed next.

10.3.3   Selecting controlled outputs: maximum scaled gain method

We here derive a very simple method for selecting controlled variables in terms of the steady-state gain matrix G from inputs u (unconstrained degrees of freedom) to outputs z (candidate controlled variables).

Scalar case. In many cases we only have one unconstrained degree of freedom (u is a scalar and we want to select one z to control). Introduce the scaled gain from u to z:

$$G' = G/\text{span}(z)$$

Note form (10.8) that $\text{span}(z) = \max_{d,e} |z - z_{opt}|$ includes both the optimization (setpoint) error and the implementation error. Then, from (10.5), the maximum expected loss imposed by keeping z constant is

$$L_{max} = \max_{d,e} L = \frac{|J_{uu}|}{2} \left(\frac{\max_{d,e} |z - z_{opt}|}{G} \right)^2 = \frac{|J_{uu}|}{2} \frac{1}{|G'|^2} \tag{10.9}$$

Here $|J_{uu}|$, the Hessian of the cost function, is independent of the choice for z. From (10.9), we then get that *the "scaled gain" $|G'|$ should be maximized to minimize the loss.* Note that the loss decreases with the square of the scaled gain. For an application, see Example 10.6 on page 398.

Multivariable case. Here u and z are vectors. Introduce the scaled outputs $z' \triangleq S_1 z$ and the scaled plant $G' = S_1 G$. Similar to the scalar case we scale with respect to the span,

$$S_1 = \text{diag}\{\frac{1}{\text{span}(z_i)}\} \qquad (10.10)$$

where

$$\text{span}(z_i) = \max_{d,e}|z_i - z_{i,\text{opt}}| = \max_d e_{i,\text{opt}}(d) + \max_e |e_i|$$

From (10.6), we have $L = \frac{1}{2}\|\tilde{z}\|_2^2$ where $\tilde{z} = J_{uu}^{1/2} G^{-1}(z - z_{\text{opt}})$. Introducing the scaled outputs gives $\tilde{z} = J_{uu}^{1/2} G'^{-1}(z' - z'_{\text{opt}})$. With the assumed scaling, the individual scaled output deviations $z_i' - z_{i,\text{opt}}'$ are less than 1 in magnitude. However, the variables z_i are generally correlated, so any combinations of deviations with magnitudes less than 1 may not possible. For example, the optimal values of both z_1 and z_2 may change in the same direction when there is a disturbance. Nevertheless, we will here assume that the expected output deviations are uncorrelated by making the following assumption:

A1 The variations in $z_i' - z_{i_\text{opt}}'$ are uncorrelated, or more precisely, the "worst-case" combination of output deviations $z_i' - z_{i_\text{opt}}'$, with $\|z' - z'_{\text{opt}}\|_2 = 1$, can occur in practice. Here $z' = S_1 z$ denotes the scaled outputs.

The reason for using the vector 2-norm, and not the max-norm, is mainly for mathematical comvenience. With assumption A1 and (A.104), we then have from (10.6) that the maximum (worst-case) loss is

$$L_{max} = \max_{\|z'-z'_{\text{opt}}\|_2 \leq 1} \frac{\|\tilde{z}\|_2}{2} = \frac{1}{2}\bar{\sigma}^2(J_{uu}^{1/2} G'^{-1}) = \frac{1}{2}\frac{1}{\underline{\sigma}^2(G' J_{uu}^{-1/2})} \qquad (10.11)$$

where $G' = S_1 G$ and the last equality follows from (A.40). The result may be stated as follows

> **Maximum gain (minimum singular value) rule.** *Let G denote the steady-state gain matrix from inputs u (unconstrained degrees of freedom) to outputs z (candidate controlled variables). Scale the outputs using S_1 in (10.10) and assume that A1 holds. Then to minimize the steady-state loss select controlled variables z that maximize $\underline{\sigma}(S_1 G J_{uu}^{-1/2})$.*

The rule may stated as minimizing the scaled minimum singular value, $\underline{\sigma}(G')$, of the scaled gain matrix $G' = S_1 G S_2$, where the output scaling matrix S_1 has the inverse of the spans along its diagonal, whereas the input "scaling" is generally a full matrix, $S_2 = J_{uu}^{-1/2}$. This important result was first presented in the first edition of this book (Skogestad and Postlethwaite, 1996) and proven in more detail by Halvorsen et al. (2003).

Example 10.5 *The aero-engine application in Chapter 13 (page 500) provides a nice illustration of output selection. There the overall goal is to operate the engine optimally in terms of fuel consumption, while at the same time staying safely away from instability. The optimization layer is a look-up table, which gives the optimal parameters for the engine at various operating points. Since the engine at steady-state has three degrees of freedom we need to specify three variables to keep the engine approximately at the optimal point, and six alternative sets of three outputs are given in Table 13.3.2 (page 503). For the scaled variables, the value of $\underline{\sigma}(G'(0))$ is 0.060, 0.049, 0.056, 0.366, 0.409 and 0.342 for the six alternative sets. Based on this, the first three sets are eliminated. The final choice is then based on other considerations including controllability.*

Remark 1 In the maximum gain rule, the objective function and the magnitudes of the disturbances and measurement noise enter indirectly through the scaling S_1 of the outputs z. To obtain $S_1 = \text{diag}\{\frac{1}{\text{span}(z_i)}\}$ we need to obtain for each candidate output $\text{span}(z_i) = \max_d |e_{i,\text{opt}}(d)| + \max |e_i|$. The second contribution to the span is simply the expected measurement error, which is the measurement error plus the control error. The first contribution, $e_{i,\text{opt}}$, may be obtained from a (nonlinear) model as follows: Compute the optimal values of the unconstrained z for the expected disturbances (with optimally constrained variables fixed). This yields a "look-up" table of z_{opt} for various expected disturbance combinations. From this data obtain for each candidate output, the expected variation in its optimal value, $e_{i_{\text{opt}}} = (z_{i_{\text{opt,max}}} - z_{i_{\text{opt,min}}})/2$.

Remark 2 Our desire to have $\underline{\sigma}(G')$ large for output selection is *not* related to the desire to have $\underline{\sigma}(G)$ large to avoid input constraints as discussed in Section 6.9. In particular, the scalings, and thus the matrix G', are different for the two cases.

Remark 3 We have in our derivation assumed that the nominal operating point is optimal. However, it can be shown that the results are independent of the operating point, provided we are in the region where the cost can be approximated by a quadratic function as in (10.2) (Alstad, 2005). Thus, it is equally important to select the right controlled variables when we are nominally non-optimal.

Exercise 10.1 *Recall that the maximum gain rule requires that the minimum singular value of the (scaled) gain matrix be maximized. It is proposed that the loss can simply be minimized by selecting the controlled variables as $z = \beta y$, where β is a large number. Show that such a scaling does not affect the selection of controlled variables using the singular value method.*

10.3.4   Selecting controlled outputs: exact local method

The maximum gain rule is based on assumption A1 on page 395, which may not hold for some cases with more than one controlled variable ($n_z = n_u > 1$). This is pointed out by Halvorsen et al. (2003), who derived the following exact local method.

Let the diagonal matrix W_d contain the magnitudes of expected disturbances and the diagonal matrix W_e contain the expected implementation errors associated with the individual controlled variables. We assume that the combined disturbance and implementation error vector has norm 1, $\left\| \begin{bmatrix} d' \\ e' \end{bmatrix} \right\|_2 = 1$. Then, it may be shown that the worst-case loss is (Halvorsen et al., 2003)

$$\max_{\left\| \begin{bmatrix} d' \\ e' \end{bmatrix} \right\|_2 \leq 1} L = \frac{1}{2}\bar{\sigma}([\, M_d \quad M_e\,])^2 \tag{10.12}$$

where

$$M_d = J_{uu}^{1/2} \left(J_{uu}^{-1} J_{ud} - G^{-1} G_d \right) W_d \tag{10.13}$$

$$M_e = J_{uu}^{1/2} G^{-1} W_e \tag{10.14}$$

Here $J_{uu} = \left(\partial^2 J / \partial u^2 \right)_{\text{opt}}$, $J_{ud} = \left(\partial^2 J / \partial u \partial d \right)_{\text{opt}}$ and the scaling enters through the weights W_d and W_e.

10.3.5 Selecting controlled outputs: direct evaluation of cost

The local methods presented in Sections 10.3.2-10.3.4 are very useful. However, in many practical examples nonlinear effects are important. In particular, the local methods may not be able to detect feasibility problems. For example, in marathon running, selecting a control strategy based on constant speed may be good locally (for small disturbances). However, if we encounter a steep hill (a large disturbance), then operation may not be feasible, because the selected reference value may be too high. In such cases, we may need to use a "brute force" direct evaluation of the loss and feasibility for alternative sets of controlled variables. This is done by solving the nonlinear equations, and evaluating the cost function J for various selected disturbances d and control errors e, assuming $z = r + e$ where r is kept constant (Skogestad, 2000). Here r is usually selected as the optimal value for the nominal disturbance, but this may not be the best choice and its value may also be found by optimization ("optimal back-off") (Govatsmark, 2003). The set of controlled outputs with smallest worst-case or average value of J is then preferred. This approach may be time consuming because the solution of the nonlinear equations must be repeated for each candidate set of controlled outputs.

10.3.6 Selecting controlled outputs: measurement combinations

We have so far selected z as a subset of the available measurements y. More generally, we may consider *combinations* of the measurements. We will restrict ourselves to *linear* combinations

$$z = Hy \tag{10.15}$$

where y now denotes all the available measurements, including the inputs u used by the control system. The objective is to find the measurement combination matrix H.

Optimal combination. Write the linear model in terms of the measurements y as $y = G^y u + G_d^y d$. Locally, the optimal linear combination is obtained by minimizing $\bar{\sigma}([M_d \quad M_e])$ in (10.12) with $W_e = HW_{n^y}$, where W_{n^y} contains the expected measurement errors associated with the individual measured variables; see Halvorsen et al. (2003). Note that H enters (10.12) indirectly, since $G = HG^y$ and $G_d = HG_d^y$ depend on H. However, (10.12) is a nonlinear function of H and numerical search-based methods need to be used.

Null space method. A simpler method for finding H is the *null space method* proposed by Alstad and Skogestad (2004), where we neglect the implementation error, i.e., $M_e = 0$ in (10.14). Then, a constant setpoint policy ($z = r$) is optimal if $z_{\text{opt}}(d)$ is independent of d, that is, when $z_{\text{opt}} = 0 \cdot d$ in terms of deviation variables. Note that the optimal values of the individual measurements y_{opt} still depend on d and we may write

$$y_{\text{opt}} = Fd \tag{10.16}$$

where F denotes the *optimal* sensitivity of y with respect to d. We would like to find $z = Hy$ such that $z_{\text{opt}} = Hy_{\text{opt}} = HFd = 0 \cdot d$ for all d. To satisfy this, we must require

$$HF = 0 \tag{10.17}$$

or that H lies in the left null space of F. This is always possible, provided $n_y \geq n_u + n_d$. This is because the null space of F has dimension $n_y - n_d$ and to make $HF = 0$, we must require that $n_z = n_u < n_y - n_d$. It can be shown that when (10.17) holds, $M_d = 0$. If there are too many disturbances, i.e. $n_y < n_u + n_d$, then one should select only the important disturbances (in terms of economics) or combine disturbances with a similar effect on y (Alstad, 2005).

In the presence of implementation errors, even when (10.17) holds such that $M_d = 0$, the loss can be large due to non-zero M_e. Therefore, the null space method does not guarantee that the loss L using a combination of measurements will be less than using the individual measurements. One practical approach is to select first the candidate measurements y, whose sensitivity to the implementation error is small (Alstad, 2005).

10.3.7 Selecting controlled outputs: examples

The following example illustrates the simple "maximize scaled gain rule" (mimimum singular value method).

Example 10.6 Cooling cycle. *A simple cooling cycle or heat pump consists of a compressor (where work W_s is supplied and the pressure is increased to p_h), a high-pressure condenser (where heat is supplied to the surroundings at high temperature), an expansion valve (where the fluid is expanded to*

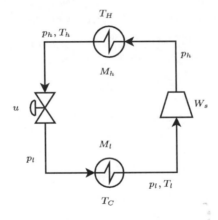

Figure 10.5: Cooling cycle

a lower pressure p_l such that the temperature drops) and a low-pressure evaporator (where heat is removed from the surroundings at low temperature); see Figure 10.5. The compressor work is indirectly set by the amount of heating or cooling, which is assumed given. We consider a design with a flooded evaporator where there is no super-heating. In this case, the expansion valve position (u) remains as an unconstrained degree of freedom, and should be adjusted to minimize the work supplied, $J = W_s$. The question is: what variable should we control?

Seven alternative controlled variables are considered in Table 10.1. The data is for an ammonia cooling cycle, and we consider Δy_{opt} for a small disturbance of 0.1 K in the hot surroundings ($d_1 = T_H$). We do not consider implementation errors. Details are given in Jensen and Skogestad (2005). From (10.9), it follows that it may be useful to compute the scaled gain $G' = G/\text{span}(z(d_i))$ for the various disturbances d_i and look for controlled variables z with a large value of $|G'|$. From a physical point of view, two obvious candidate controlled variables are the high and low pressures (p_h and p_l). However, these appear to be poor choices with scaled gains $|G'|$ of 126 and 0, respectively. The

Table 10.1: Local "maximum gain" analysis for selecting controlled variable for cooling cycle

Variable (y)	$\Delta z_{opt}(d_1)$	$G = \frac{\Delta z}{\Delta u}$	$|G'| = \frac{|G|}{|\Delta z_{opt}(d_1)|}$
Condenser pressure, p_h [Pa]	3689	-464566	126
Evaporator pressure, p_l [Pa]	-167	0	0
Temperature at condenser exit, T_h [K]	0.1027	316	3074
Degree of sub-cooling, $T_h - T^{\text{sat}}(p_h)$ [K]	-0.0165	331	20017
Choke valve opening, u	8.0×10^{-4}	1	1250
Liquid level in condenser, M_h [m^3]	6.7×10^{-6}	-1.06	157583
Liquid level in evaporator, M_l [m^3]	-1.0×10^{-5}	1.05	105087

zero gain is because we assume a given cooling duty $Q_C = UA(T_l - T_C)$ and further assume saturation $T_l = T^{\text{sat}}(p_l)$. Keeping p_l constant is then infeasible when, for example, there are disturbances in T_C. Other obvious candidates are the temperatures at the exit of the heat exchangers, T_h and T_l. However, the temperature T_l at the evaporator exit is directly related to p_l (because of saturation) and also has a zero gain. The open-loop policy with a constant valve position u has a scaled gain of 1250, and the temperature at the condenser exit (T_h) has a scaled gain of 3074. Even more promising is the degree of subcooling at the condenser exit with a scaled gain of 20017. Note that the loss decreases in proportion to $|G'|^2$, so the increase in the gain by a factor $20017/1250 = 16.0$ when we change from constant choke valve opening ("open-loop") to constant degree of subcooling, corresponds to a decrease in the loss (at least for small perturbations) by a factor $16.0^2 = 256$. Finally, the best single measurements seem to be the amount of liquid in the condenser and evaporator, M_h and M_l, with scaled gains of 157583 and 105087, respectively. Both these strategies are used in actual heat pump systems. A "brute force" evaluation of the cost for a (large) disturbance in the surrounding temperature ($d_1 = T_H$) of about 10 K, confirms the linear analysis, except that the choice $z = T_h$ turns out to be infeasible. The open-loop policy with constant valve position ($z = u$) increases the compressor work by about 10%, whereas the policy with a constant condenser level ($z = M_h$) has an increase of less than 0.003%. Similar results hold for a disturbance in the cold surroundings ($d_2 = T_C$). Note that the implementation error was not considered, so the actual losses will be larger.

The next simple example illustrates the use of different methods for selection of controlled variables.

Example 10.7 Selection of controlled variables. *As a simple example, consider a scalar unconstrained problem, with the cost function $J = (u - d)^2$, where nominally $d^* = 0$. For this problem we have three candidate measurements,*

$$y_1 = 0.1(u - d); \quad y_2 = 20u; \quad y_3 = 10u - 5d$$

We assume the disturbance and measurement noises are of unit magnitude, i.e. $|d| \leq 1$ and $|n_i^y| \leq 1$. For this problem, we always have $J_{opt}(d) = 0$ corresponding to

$$u_{opt}(d) = d, \quad y_{1,opt}(d) = 0, \quad y_{2,opt}(d) = 20d \quad \text{and} \quad y_{3,opt}(d) = 5d$$

For the nominal case with $d^* = 0$, we thus have $u_{opt}(d^*) = 0$ and $y_{opt}(d^*) = 0$ for all candidate controlled variables and at the nominal operating point we have $J_{uu} = 2$, $J_{ud} = -2$. The linearized models for the three measured variables are

$$
\begin{array}{lll}
y_1: & G_1^y = 0.1, & G_{d1}^y = -0.1 \\
y_2: & G_2^y = 20, & G_{d2}^y = 0 \\
y_3: & G_3^y = 10, & G_{d3}^y = -5
\end{array}
$$

Let us first consider selecting one of the individual measurements as a controlled variable. We have

$$\text{Case 1:} \quad z = y_1, \quad G = G_1^y$$
$$\text{Case 2:} \quad z = y_2, \quad G = G_2^y$$
$$\text{Case 3:} \quad z = y_3, \quad G = G_3^y$$

The losses for this example can be evaluated analytically, and we find for the three cases

$$L_1 = (10e_1)^2; \quad L_2 = (0.05e_2 - d)^2; \quad L_3 = (0.1e_3 - 0.5d)^2$$

(For example, with $z = y_3$, we have $u = (y_3 + 5d)/10$ and with $z = n_3^y$, we get $L_3 = (u - d)^2 = (0.1n_3^y + 0.5d - d)^2$.) With $|d| \leq 1$ and $|n_i^y| \leq 1$, the worst-case losses (with $|d| = 1$ and $|n_i^y| = 1$) are $L_1 = 100$, $L_2 = 1.05^2 = 1.1025$ and $L_3 = 0.6^2 = 0.36$, and we find that $z = y_3$ is the best overall choice for self-optimizing control and $z = y_1$ is the worst. We note that $z = y_1$ is perfectly self-optimizing with respect to disturbances, but has the highest loss. This highlights the importance of considering the implementation error when selecting controlled variables. Next, we compare the three different methods discussed earlier in this section.

A. *Maximum scaled gain (singular value rule): For the three choices of controlled variables we have without scaling $|G_1| = \underline{\sigma}(G_1) = 0.1$, $\underline{\sigma}(G_2) = 20$ and $\underline{\sigma}(G_3) = 10$. This indicates that z_2 is the best choice, but this is only correct with no disturbances. Let us now follow the singular value procedure.*

1. *The input is scaled by the factor $1/\sqrt{(\partial^2 J/\partial u^2)_{\text{opt}}} = 1/\sqrt{2}$ such that a unit deviation in each input from its optimal value has the same effect on the cost function J.*

2. *To find the optimum setpoint error, first note that $u_{\text{opt}}(d) = d$. Substituting $d = 1$ (the maximum disturbance) and $u = u_{\text{opt}} = 1$ (the optimal input) into the defining expressions for the candidate measurements, then gives $e_{\text{opt},1} = 0.1(u - d) = 0$, $e_{\text{opt},2} = 20u = 20$ and $e_{\text{opt},3} = 10u - 5d = 5$. Alternatively, one may use the expression (Halvorsen et al., 2003) $e_{\text{opt},i} = (G_i^y J_{uu}^{-1} J_{ud} - G_{di}^y)\Delta d$. Note that only the magnitude of $e_{\text{opt},i}$ matters.*

3. *For each candidate controlled variable the implementation error is assumed to be $n^z = 1$.*

4. *The expected variation ("span") for $z = y_1$ is $|e_{\text{opt},i}| + |n_1^y| = 0 + 1 = 1$. Similarly, for $z = y_2$ and $z = y_3$, the spans are $20 + 1 = 21$ and $5 + 1 = 6$, respectively.*

5. *The scaled gain matrices and the worst-case losses are*

$$z = y_1: \quad |G_1'| = \tfrac{1}{1} \cdot 0.1/\sqrt{2} = 0.071; \quad L_1 = \tfrac{1}{2|G'|^2} = 100$$
$$z = y_2: \quad |G_2'| = \tfrac{1}{21} \cdot 20/\sqrt{2} = 0.67; \quad L_2 = \tfrac{1}{2|G'|^2} = 1.1025$$
$$z = y_3: \quad |G_3'| = \tfrac{1}{6} \cdot 10/\sqrt{2} = 1.18; \quad L_3 = \tfrac{1}{2|G'|^2} = 0.360$$

We note from the computed losses that the singular value rule (= maximize scaled gain rule) suggests that we should control $z = y_3$, which is the same as found with the "exact" procedure. The losses are also identical.

B. *Exact local method: In this case, we have $W_d = 1$ and $W_{e_i} = 1$ and for y_1*

$$M_d = \sqrt{2}\left(2^{-1} \cdot (-2) - 0.1^{-1} \cdot (-0.1)\right) \cdot 1 = 0 \quad \text{and} \quad M_e = \sqrt{2} \cdot 0.1^{-1} \cdot 1 = 10\sqrt{2}$$

$$L_1 = \frac{\bar{\sigma}([\,M_d \quad M_e\,])2}{2} = \frac{1}{2}(\bar{\sigma}(0 \quad 10\sqrt{2})) = 100$$

Similarly, we find with z_2 and z_3

$$L_2 = \frac{1}{2}(\bar{\sigma}(-\sqrt{2} \quad \sqrt{2}/20)) = 1.0025 \quad \text{and} \quad L_3 = \frac{1}{2}(\bar{\sigma}(-\sqrt{2}/2 \quad \sqrt{2}/10)) = 0.26$$

Thus, the exact local method also suggests selecting $z = y_3$ as the controlled variable. The reason for the slight difference from the "exact" nonlinear losses is that we assumed d and n^y individually to be less than 1 in the exact nonlinear method, whereas in the exact linear method we assumed that the combined 2-norm of d and n^y was less than 1.

C. *Combinations of measurements: We now want to find the best combination $z = Hy$. In addition to y_1, y_2 and y_3, we also include the input u in the set y, i.e.*

$$y = [\, y_1 \quad y_2 \quad y_3 \quad u \,]^T$$

We assume that the implementation error for u is 1, i.e. $n^u = 1$. We then have $W_n^y = I$, where W_n^y is a 4×4 matrix. Furthermore, we have

$$G^y = [\, 0.1 \quad 20 \quad 10 \quad 1 \,]^T \qquad G_d^y = [\, -0.1 \quad 0 \quad -5 \quad 0 \,]^T$$

Optimal combination. We wish to find H such that $\bar{\sigma}([\, M_d \quad M_e \,])$ in (10.12) is minimized, where $G = HG^y$, $G_d = HG_d^y$, $W_e = HW_n^y$, $J_{uu} = 2$, $J_{ud} = -2$ and $W_d = 1$. Numerical optimization yields $H_{\mathrm{opt}} = [\, 0.0209 \quad -0.2330 \quad 0.9780 \quad -0.0116 \,]$; that is, the optimal combination of the three measurements and the manipulated input u is

$$z = 0.0209y_1 - 0.23306y_2 + 0.9780y_3 - 0.0116u$$

We note, as expected, that the most important contribution to z comes from the variable y_3. The loss is $L = 0.0405$, so it is reduced by a factor 6 compared to the previous best case ($L = 0.26$) with $z = y_3$.

Null space method. In the null space method we find the optimal combination without implementation error. This first step is to find the optimal sensitivity with respect to the disturbances. Since $u_{\mathrm{opt}} = d$, we have

$$\Delta y_{opt} = F\Delta d = G^y \Delta u_{\mathrm{opt}} + G_d^y \Delta d = \underbrace{(G^y + G_d^y)}_{F} \Delta d$$

and thus the optimal sensitivity is

$$F = [\, 0 \quad 20 \quad 5 \quad 1 \,]^T$$

To have zero loss with respect to disturbances we need to combine at least $n_u + n_d = 1 + 1 = 2$ measurements. Since we have four candidate measurements, there are an infinite number of possible combinations, but for simplicity of the control system, we prefer to combine only two measurements. To reduce the effect of implementation errors, it is best to combine measurements y with a large gain, provided they contain different information about u and d. More precisely, we should maximize $\underline{\sigma}([\, G^y \quad G_d^y \,])$. From this we find that measurements 2 and 3 are the best, with $\underline{\sigma}([\, G^y \quad G_d^y \,]) = \underline{\sigma} \begin{bmatrix} 20 & 0 \\ 10 & -5 \end{bmatrix} = 4.45$. To find the optimal combination we use $HF = 0$ or

$$20h_2 + 5h_3 = 0$$

Setting $h_2 = 1$ gives $h_3 = -4$, and the optimal combination is $z = y_2 - 4y_3$ or (normalizing the 2-norm of H to 1):

$$z = -0.2425y_2 + 0.9701y_3$$

The resulting loss when including the implementation error is $L = 0.0425$. We recommend the use of this solution, because the loss is only marginally higher (0.0425 instead of 0.0405) than that obtained using the optimal combination of all four measurements.

Maximizing scaled gain for combined measurements. For the scalar case, the "maximize scaled gain rule" can also be used to find the best combination. Consider a linear combination of measurements 2 and 3, $z = h_2y_2 + h_3y_3$. The gain from u to z is $G = h_2G_2^y + h_3G_3^y$. The span for z, $\mathrm{span}(z) = |e_{\mathrm{opt},z}| + |e_z|$, is obtained by combining the individual spans

$$e_{\mathrm{opt},z} = h_2e_{\mathrm{opt},2} + h_3e_{\mathrm{opt},3} = h_2f_2 + h_3f_3 = 20h_2 + 5h_3$$

and $|e_z| = h_2|e_2| + h_3|e_3|$. *If we assume that the combined implementation errors are 2-norm bounded,* $\left\| \begin{bmatrix} e_2 \\ e_3 \end{bmatrix} \right\|_2 \leq 1$, *then the worst-case implementation error for z is* $|e_z| = \left\| \begin{bmatrix} h_2 \\ h_3 \end{bmatrix} \right\|_2$. *The resulting scaled gain that should be maximized in magnitude is*

$$G' = \frac{G}{\text{span}} = \frac{h_2 G_2^y + h_3 G_3^y}{|h_2 e_{\text{opt},2} + h_3 e_{\text{opt},3}| + |e_z|} \tag{10.18}$$

The expression (10.18) gives considerable insight into the selection of a good measurement combination. We should select H (i.e. h_2 and h_3) in order to maximize $|G'|$. The null space method corresponds to selecting H such that $e_{\text{opt}} = h_2 e_{\text{opt},2} + h_3 e_{\text{opt},3} = 0$. This gives $h_2 = -0.2425$ and $h_3 = 0.9701$, and $|e_z| = \left\| \begin{bmatrix} h_2 \\ h_3 \end{bmatrix} \right\|_2 = 1$. The corresponding scaled gain is

$$G' = \frac{-20 \cdot 0.2425 + 10 \cdot 0.9701}{0 + 1} = -4.851$$

with a loss $L = \alpha/(2|G'|^2) = 0.0425$ (as found above). (The factor $\alpha = J_{uu} = 2$ is included because we did not scale the inputs when obtaining G'.)

Some additional examples can be found in Skogestad (2000), Halvorsen et al. (2003), Skogestad (2004b) and Govatsmark (2003).

Exercise 10.2 * *Suppose that we want to minimize the LQG-type objective function, $J = x^2 + ru^2$, $r > 0$, where the steady-state model of the system is*

$$x + 2u - 3d = 0$$

$$y_1 = 2x, \quad y_2 = 6x - 5d, \quad y_3 = 3x - 2d$$

Which measurement would you select as a controlled variable for $r = 1$? How does your conclusion change with variation in r? Assume unit implementation error for all measurements.

Exercise 10.3 *In Exercise 10.2, how would your conclusions change when u (open-loop implementation policy) is also included as a candidate controlled variable? First, assume the implementation error for u is unity. Repeat the analysis, when the implementation error for u and each of the measurements is 10.*

10.3.8 Selection of controlled variables: summary

When the optimum coincides with constraints, optimal operation is achieved by controlling the active constraints. It is for the remaining unconstrained degrees of freedom that the selection of controlled variables is a difficult issue.

The most common "unconstrained case" is when there is only a single unconstrained degree of freedom. The rule is then to select a controlled variable such that the (scaled) gain is maximized.

Scalar rule: "maximize scaled gain $|G'|$"

- G = unscaled gain from u to z
- Scaled gain $G' = G/\text{span}$
- span = optimal range ($|e_{\text{opt}}|$) + implementation error ($|e|$)

In words, this "maximize scaled gain rule" may be expressed as follows:

Select controlled variables z with a large controllable range compared to their sum of optimal variation and implementation error. Here

- controllable range = range which may be reached by varying the inputs (as given by the steady-state gain)
- optimal variation: due to disturbance (at steady-state)
- implementation error = sum of control error and measurement error (at steady-state)

For cases with more than one unconstrained degree of freedom, we use the gain in the most difficult direction as expressed by the minimum singular value.

General **"maximum gain" rule:** *"maximize the (scaled) minimum singular value $\underline{\sigma}(G')$ (at steady-state)", where $G' = S_1 G S_2$ and $S_2 = J_{uu}^{-1/2}$ (see page 395 for details).*

We have written "at steady-state" because the cost usually depends on the steady state, but more generally it could be replaced by "at the bandwidth frequency of the layer above (which adjusts the setpoints for z)".

10.4   Regulatory control layer

In this section, we are concerned with the regulatory control layer. This is at the bottom of the control hierarchy and the objective of this layer is generally to "stabilize" the process and facilitate smooth operation. It is *not* to optimize objectives related to profit, which is done at higher layers. Usually, this is a decentralized control system of "low complexity" which keeps a set of measurements at given setpoints. The regulatory control layer is usually itself hierarchical, consisting of cascaded loops. If there are "truly" unstable modes (RHP-poles) then these are usually stabilized first. Then, we close loops to "stabilize" the system in the more general sense of keeping the states within acceptable bounds (avoiding drift), for which the key issue is local disturbance rejection.

The most important issues for regulatory control are what to measure and what to manipulate. Some simple rules for these are given on page 405. A fundamental issue is whether the introduction of a separate regulatory control layer imposes an inherent performance loss in terms of control of the primary variables z. Interestingly, the answer is "no" provided the regulatory controller does not contain RHP-zeros, and provided the layer above has full access to changing the reference values in the regulatory control layer (see Theorem 10.2 on page 416).

10.4.1   Objectives of regulatory control

Some more specific objectives of the regulatory control layer may be:

O1. Provide sufficient quality of control to enable a trained operator to keep the plant running safely without use of the higher layers in the control system.

This sharply reduces the need for providing costly backup systems for the higher layers of the control hierarchy in case of failures.

O2. Allow for simple decentralized (local) controllers (in the regulatory layer) that can be tuned on-line.

O3. Take care of "fast" control, such that acceptable control is achievable using "slow" control in the layer above.

O4. Track references (setpoints) set by the higher layers in the control hierarchy.

The setpoints of the lower layers are often the manipulated variables for the higher levels in the control hierarchy, and we want to be able to change these variables as directly and with as little interaction as possible. Otherwise, the higher layer will need a model of the dynamics and interactions of the outputs from the lower layer.

O5. Provide for local disturbance rejection.

This follows from O4, since we want to be able to keep the controlled variables in the regulatory control system at their setpoints.

O6. Stabilize the plant (in the mathematical sense of shifting RHP-poles to the LHP).

O7. Avoid "drift" so that the system stays within its "linear region" which allows the use of linear controllers.

O8. Make it possible to use simple (at least in terms of dynamics) models in the higher layers.

We want to use relatively simple models because of reliability and the costs involved in obtaining and maintaining a detailed dynamic model of the plant, and because complex dynamics will add to the computational burden on the higher-layer control system.

O9. Do not introduce unnecessary performance limitations for the remaining control problem.

The "remaining control problem" is the control problem as seen from the higher layer which has as manipulated inputs the setpoints to the lower-level control system and the possible "unused" manipulated inputs. By "unnecessary" we mean limitations (e.g. RHP-zeros, large RGA elements, strong sensitivity to disturbances) that do not exist in the original problem formulation.

10.4.2　Selection of variables for regulatory control

For the following discussion, it is useful to divide the outputs y into two classes:

- y_1 – (locally) uncontrolled outputs (for which there is an associated control objective)
- y_2 – (locally) measured and controlled outputs (with reference value r_2)

By "locally" we mean here "in the regulatory control layer". Thus, the variables y_2 are the selected controlled variables in the regulatory control layer. We also subdivide the available manipulated inputs u in a similar manner:

- u_1 – (locally) unused inputs (this set may be empty)

- u_2 – (locally) used inputs for control of y_2 (usually $n_{u_2} = n_{y_2}$)

We will study the regulatory control layer, but a similar subdivision and analysis could be performed for any control layer. The variables y_1 are sometimes called "primary" outputs, and the variables y_2 "secondary" outputs. Note that y_2 is the controlled variable (CV) in the control layer presently considered. Typically, you can think of y_1 as the variables we would really like to control and y_2 as the variables we control locally to make control of y_1 easier.

The regulatory control layer should assist in achieving the overall operational goals, so if the "economic" controlled variables z are known, then we should include them in y_1. In other cases, if the objective is to stop the system from "drifting" away from its steady-state, then the variables y_1 could be a weighted subset of the system states; see the discussion on page 418.

The most important issues for regulatory control are:

1. What should we control (what is the variable set y_2)?
2. What should we select as manipulated variables (what is the variable set u_2) and how should it be paired with y_2?

The pairing issue arises because we aim at using decentralized SISO control, if at all possible. In many cases, it is "clear" from physical considerations and experience what the variables y_2 are (see the distillation example below for a typical case). However, we have put the word "clear" in quotes, because it may sometimes be useful to question the conventional control wisdom.

We will below, see (10.28), derive transfer functions for "partial control", which are useful for a more exact analysis of the effects of various choices for y_2 and u_2. However, we will first present some simple rules that may be useful for reducing the number of alternatives that could be studied. This is important in order to avoid a combinatorial growth in possibilities. For a plant where we want to select m from M candidate inputs u, and l from L candidate measurements y, the number of possibilities is

$$\binom{L}{l}\binom{M}{m} = \frac{L!}{l!(L-l)!}\,\frac{M!}{m!(M-m)!} \tag{10.19}$$

A few examples: for $m = l = 1$ and $M = L = 2$ the number of possibilities is 4; for $m = l = 2$ and $M = L = 4$ it is 36; and for $m = M$, $l = 5$ and $L = 100$ (selecting 5 measurements out of 100 possible) there are 75287520 possible combinations.

It is useful to distinguish between two main cases:

1. **Cascade and indirect control.** The variables y_2 are controlled solely to assist in achieving good control of the "primary" outputs y_1. In this case r_2 (sometimes denoted $r_{2,u}$) is usually "free" for use as manipulated inputs (MVs) in the layer above for the control of y_1.
2. **Decentralized control (using sequential design).** The variables y_2 are important in themselves. In this case, their reference values r_2 (sometimes denoted $r_{2,d}$) are usually not available for the control of y_1, but rather act as disturbances to the control of y_1.

Rules for selecting y_2. Especially for the first case (cascade and indirect control), the following rules may be useful for identifying candidate controlled variables y_2 in the regulatory control layer:

1. y_2 should be easy to measure.

2. Control of y_2 should "stabilize" the plant.
3. y_2 should have good controllability; that is, it has favourable dynamics for control.
4. y_2 should be located "close" to the manipulated variable u_2 (as a consequence of rule 3, because for good controllability we want a small effective delay; see page 57).
5. The (scaled) gain from u_2 to y_2 should be large.

In words, the last rule says that the controllable range for y_2 (which may be reached by varying the inputs u_2) should be large compared to its expected variation (span). It is a restatement of the maximum gain rule presented on page 395 for selecting primary ("economic") controlled variables z. The rule follows because we would like to control variables y_2 that contribute to achieving optimal operation. For the scalar case, we should maximize the gain $|G'_{22}| = |G_{22}|/\text{span}(y_2)$, where G_{22} is the unscaled transfer function from u_2 to y_2, and span(y_2) is the sum of the optimal variation and the implementation error for y_2. For cases with more than one output, the "gain" is given by the minimum singular value, $\underline{\sigma}(G'_{22})$. The scaled gain (including the optimal variation and implementation error) should be evaluated for constant u_1 and approximately at the bandwidth frequency of the control layer immediately above (which adjust the references r_2 for y_2).

Rules for selecting u_2. To control y_2, we select a subset u_2 of the available manipulated inputs u. Similar considerations as for y_2 apply to the choice of candidate manipulated variables u_2:

1. Select u_2 so that controllability for y_2 is good; that is, u_2 has a "large" and "direct" effect on y_2. Here "large" means that the gain is large, and "direct" means good dynamics with no inverse response and a small effective delay.
2. Select u_2 to maximize the magnitude of the (scaled) gain from u_2 to y_2.
3. Avoid using variables u_2 that may saturate.

The last item is the only "new" requirement compared to what we stated for selecting y_2. By "saturate" we mean that the desired value of the input u_2 exceeds a physical constraint; for example, on its magnitude or rate. The last rule applies because, when an input saturates, we have effectively lost control, and reconfiguration may be required. Preferably, we would like to minimize the need for reconfiguration and its associated logic in the regulatory control layer, and rather leave such tasks for the upper layers in the control hierarchy.

Example 10.8 Regulatory control for distillation column: basic layer. *The overall control problem for the distillation column in Figure 10.6 has five manipulated inputs*

$$u = [\,L \quad V \quad D \quad B \quad V_T\,]^T$$

These are all flows [mol/s]: reflux L, boilup V, distillate D, bottom flow B, and overhead vapour (cooling) V_T. What to control (y) is yet to be decided.

Overall objective. From a steady-state (and economic) point of view, the column has only three degrees of freedom[2] With pressure also controlled, there are two remaining steady-state degrees of freedom, and we want to identify the economic controlled variables $y_1 = z$ associated with these. To do this, we define the cost function J and minimize it for various disturbances, subject to the constraints, which include specifications on top composition (x_D) and bottom composition (x_B), together with upper and lower bounds on the flows. In most cases, the optimal solution lies at the constraints. A very

---

[2] A distillation column has two fewer steady-state than dynamic degrees of freedom, because the integrating condenser and reboiler levels, which need to be controlled to stabilize the process, have no steady-state effect.

common situation is that both top and bottom composition optimally lie at their specifications ($y_{D,\min}$ and $x_{B,\max}$). We generally choose to control active constraints and then have

$$y_1 = z = [\,x_D \quad x_B\,]^T$$

Regulatory control: selection of y_2. We need to stabilize the two integrating modes associated with the liquid holdups (levels) in the condenser and reboiler of the column (M_D and M_B [mol]). In addition, we normally have tight control of pressure (p), because otherwise the (later) control of temperature and composition becomes more difficult. In summary, we decide to control the following three variables in the regulatory control layer:

$$y_2 = [\,M_D \quad M_B \quad p\,]^T$$

Note that these three variables are important to control in themselves.

Overall control problem. In summary, we have now identified five variables that we want to control

$$y = \underbrace{[x_D \quad x_B}_{y_1} \quad \underbrace{M_D \quad M_B \quad p]^T}_{y_2}$$

The resulting overall 5×5 control problem from u to y can be approximated as (Skogestad and Morari, 1987a):

$$
\begin{bmatrix} x_D \\ x_B \\ M_D \\ M_B \\ M_V(p) \end{bmatrix}
=
\begin{bmatrix}
g_{11}(s) & g_{12}(s) & 0 & 0 & 0 \\
g_{21}(s) & g_{22}(s) & 0 & 0 & 0 \\
-1/s & 0 & -1/s & 0 & 0 \\
g_L(s)/s & -1/s & 0 & -1/s & 0 \\
0 & 1/(s+k_p) & 0 & 0 & -1/(s+k_p)
\end{bmatrix}
\begin{bmatrix} L \\ V \\ D \\ B \\ V_T \end{bmatrix}
\tag{10.20}
$$

In addition, there are high-frequency dynamics (delays) associated with the inputs (valves) and outputs (measurements). For control purposes it is very important to include the transfer function $g_L(s)$, which represents the liquid flow dynamics from the top to the bottom of the column, $\Delta L_B = g_L(s)\Delta L$. For control purposes, it may be approximated by a delay, $g_L(s) = e^{-\theta_L s}$. $g_L(s)$ also enters into the transfer function $g_{21}(s)$ from L to x_B, and by this decouples the distillation column dynamics at high frequencies. The overall plant model in (10.20) usually has no inherent control limitations caused by RHP-zeros, but the plant has two poles at the origin (from the integrating liquid levels, M_D and M_B), and also one pole close to the origin ("almost integrating") in $G_{\mathrm{LV}} = \begin{bmatrix} g_{11} & g_{12} \\ g_{21} & g_{22} \end{bmatrix}$ originating from the internal recycle in the column. These three modes need to be "stabilized". In addition, for high-purity separations, there is a potential control problem in that the G_{LV}-subsystem is strongly coupled at steady-state, e.g. resulting in large elements in the RGA matrices for G_{LV} and also for the overall 5×5 plant, but fortunately the system is decoupled at high frequency because of the liquid flow dynamics represented by $g_L(s)$. Another complication is that composition measurements (y_1) are often expensive and unreliable.

Regulatory control: selection of u_2. As already mentioned, the distillation column is first stabilized by closing three decentralized SISO loops for level and pressure, $y_2 = [\,M_D \quad M_B \quad p\,]^T$. These loops usually interact weakly with each other and may be tuned independently. However, there exist many possible choices for u_2 (and thus for u_1). For example, the condenser holdup tank (M_D) has one inlet flow (V_T) and two outlet flows (L and D), and any one of these flows, or a combination, may be used effectively to control M_D. By convention, each choice ("configuration") of u_2 used for controlling level and pressure is named by the inputs u_1 left for composition control. For example, the "LV-configuration" used in many examples in this book refers to a partially controlled system where $u_2 = [\,D \quad B \quad V_T\,]^T$ is used to control levels and pressure (y_2) in the regulatory layer, and we are left with

$$u_1 = [\,L \quad V\,]^T$$

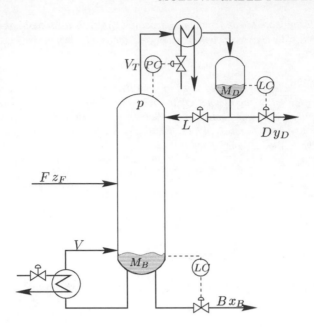

Figure 10.6: Distillation column controlled with the LV-configuration

to control composition (y_1). The LV-configuration is known to be strongly interactive at steady-state, as can been seen from the large steady-state RGA elements; see (3.94) on page 100. On the other hand, the LV-configuration is good from the point of view that it is the only configuration where control of y_1 (using u_1) is nearly independent of the tuning of the level controllers (K_2). This is quite important, because we normally want "slow" (smooth control) rather than tight control of the levels (M_D and M_B). This may give undesirable interactions from the regulatory control layer (y_2) into the primary control layer (y_1). However, this is avoided with the LV-configuration.

Another configuration is the DV-configuration where $u_2 = \begin{bmatrix} L & B & V_T \end{bmatrix}^T$ is used to control levels and pressure, and we are left with

$$u_1 = \begin{bmatrix} D & V \end{bmatrix}^T$$

to control compositions. If we were only concerned with controlling the condenser level (M_D) then this choice would be better for cases with difficult separations where $L/D \gg 1$. This is because to avoid saturation in u_2 we would like to use the largest flow (in this case $u_2 = L$) to control condenser level (M_D). In addition for this case, the steady-state interactions from u_1 to y_1, as expressed by the RGA, are generally much less; see (6.74) on page 245. However, a disadvantage with the DV-configuration is that the effect of u_1 on y_1 depends strongly on the tuning of K_2. This is not surprising, since using D to control x_D corresponds to pairing on $g_{31} = 0$ in (10.20), and D (u_1) therefore only has an effect on x_D (y_1) when the level loop (from $u_2 = L$ to $y_2 = M_D$) has been closed.

There are also many other possible configurations (choices for the two inputs in u_1); with five inputs there are ten alternative configurations. Furthermore, one often allows for the possibility of using ratios between flows, e.g. L/D, as possible degrees of freedom in u_1, and this sharply increases the number of alternatives. However, for all these configurations, the effect of u_1 on y_1 depends on the tuning of K_2, which is undesirable. This is one reason why the LV-configuration is used most in practice. In the next section, we discuss how closing a "fast" temperature loop may improve the controllability of the LV-configuration.

In the above example, the variables y_2 were important variables in themselves. In the following example, the variable y_2 is controlled to assist in the control of the primary variables y_1.

Example 10.9 Regulatory control for distillation column: temperature control. *We will assume that we have closed the three basic control loops for liquid holdup (M_D, M_B) and pressure (p) using the LV-configuration, see Example 10.8, and we are left with a 2×2 control problem with*

$$u = [\, L \quad V \,]^T$$

(reflux and boilup) and

$$y_1 = [\, x_D \quad x_B \,]^T$$

(product compositions). A controllability analysis of the model $G_{LV}(s)$ from u to y_1 shows that there is (1) an almost integrating mode, and (2) strong interactions. The integrating mode results in high sensitivity to disturbances at lower frequencies. The control implication is that we need to close a "stabilizing" loop. A closer analysis of the interactions (e.g. a plot of the RGA elements as a function of frequency) shows that they are much smaller at high frequencies. The physical reason for this is that L and x_D are at the top of the column, and V and x_B at the bottom, and since it takes some time (θ_L) for a change in L to reach the bottom, the high-frequency response is decoupled. The control implication is that the interactions may be avoided by closing a loop with a closed-loop response time less than about θ_L.

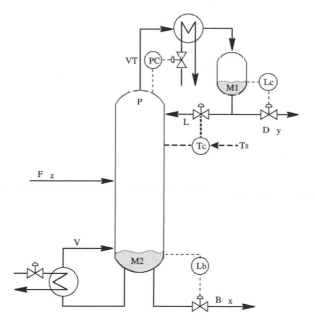

Figure 10.7: Distillation column with LV-configuration and regulatory temperature loop

It turns out that closing <u>one</u> fast loop may take care of both stabilization and reducing interactions. The issue is then which loop to close. The most obvious choice is to close one of the composition loops (y_1). However, there is usually a time delay involved in measuring composition (x_D and x_B), and the measurement may be unreliable. On the other hand, the temperature T is a good indicator of composition and is easy to measure. The preferred solution is therefore to close a fast temperature loop somewhere along the column. This loop will be implemented as part of the regulatory control system.

We have two available manipulated variables u, so temperature may be controlled using reflux L or boilup V. We choose reflux L here (see Figure 10.7) because it is more likely that boilup V will reach its maximum value, and input saturation is not desired in the regulatory control layer. In terms of the notation presented above, we then have a SISO regulatory loop with

$$y_2 = T; \quad u_2 = L$$

and $u_1 = V$. The "primary" composition control layer adjusts the temperature setpoint $r_2 = T_s$ for the regulatory layer. Thus, for the primary layer we have

$$y_1 = [\, x_D \quad x_B \,]^T; \quad u = [\, u_1 \quad r_2 \,]^T = [\, V \quad T_s \,]^T$$

The issue is to find which temperature T in the column to control, and for this we may use the "maximum gain rule". The objective is to maximize the scaled gain $|G'_{22}(j\omega)|$ from $u_2 = L$ to $y_2 = T$. Here, $|G'_{22}| = |G_{22}|/\text{span}$ where G_{22} is the unscaled gain and span = optimal range ($|e_{opt}|$) + implementation error ($|e|$) for the selected temperature. The gain should be evaluated at approximately the bandwidth frequency of the composition layer that adjusts the setpoint $r_2 = T_s$. For this application, we assume that the primary layer is relatively slow, such that we can evaluate the gain at steady-state, i.e. $\omega = 0$.

In Table 10.2, we show the normalized temperatures $y_2 = x$, unscaled gain, optimal variation for the two disturbances, implementation error, and the resulting span and scaled gain for measurements located at stages 1 (reboiler), 5, 10, 15, 21 (feed stage), 26, 31, 36 and 41 (condenser). The gains are also plotted as a function of stage number in Figure 10.8. The largest scaled gain of about 88 is achieved when the temperature measurement is located at stage 15 from the bottom. However, this is below the feed stage and it takes some time for the change in reflux ($u_2 = L$), which enters at the top, to reach this stage. Thus, for dynamic reasons it is better to place the measurement in the top part of the column; for example, at stage 27 where the gain has a "local" peak of about 74.

Table 10.2: Evaluation of scaled gain $|G'_{22}|$ for alternative temperature locations (y_2) for distillation example. Span $= |\Delta y_{2,\text{opt}}(d_1)| + |\Delta y_{2,\text{opt}}(d_2)| + e_{y_2}$. Scaled gain $|G'_{22}| = |G_{22}|/\text{span}$.

Stage	Nominal value y_2	Unscaled G_{22}	$\Delta y_{2,\text{opt}}(d_1)$	$\Delta y_{2,\text{opt}}(d_2)$	e_{y_2}	span(y_2)	Scaled $|G'_{22}|$
1	0.0100	1.0846	0.0077	0.0011	0.05	0.0588	18.448
5	0.0355	3.7148	0.0247	0.0056	0.05	0.0803	46.247
10	0.1229	10.9600	0.0615	0.0294	0.05	0.1408	77.807
15	0.2986	17.0030	0.0675	0.0769	0.05	0.1944	87.480
21	0.4987	9.6947	-0.0076	0.0955	0.05	0.1532	63.300
26	0.6675	14.4540	-0.0853	0.0597	0.05	0.1950	74.112
31	0.8469	10.5250	-0.0893	0.0130	0.05	0.1524	69.074
36	0.9501	4.1345	-0.0420	-0.0027	0.05	0.0947	43.646
41	0.9900	0.8754	-0.0096	-0.0013	0.05	0.0609	14.376

Remarks to example.

1. *We use data for "column A" (see Section 13.4) which has 40 stages. This column separates a binary mixture, and for simplicity we assume that the temperature T on stage i is directly given by the mole fraction of the light component, $T_i = x_i$. This can be regarded as a "normalized" temperature which ranges from 0 in the bottom to 1 in the top of the column. The implementation error is assumed to be the same on all stages, namely $e_{y_2} = 0.05$ (and with a temperature difference between the two components of 13.5 K, this corresponds to an implementation error of ± 0.68 K). The disturbances are a 20% increase in feed rate F ($d_1 = 0.2$) and a change from 0.5 to 0.6 in feed mole fraction z_F ($d_2 = 0.1$).*

2. *The optimal variation* $(\Delta y_{2,\text{opt}}(d))$ *is often obtained from a detailed steady-state model, but it was generated here from the linear model. For any disturbance d we have in terms of deviation variables (we omit the Δ's)*

$$y_1 = G_{12}u_2 + G_{d1}d$$

$$y_2 = G_{22}u_2 + G_{d2}d$$

The optimal strategy is to have the product compositions constant; that is, $y_1 = \begin{bmatrix} x_D & x_B \end{bmatrix}^T = 0$. However, since $u_2 = L$ is a scalar, this is not possible. The best solution in a least squares sense (minimize $\|y_1\|_2$) is found by using the pseudo-inverse, $u_2^{\text{opt}} = -G_{12}^{\dagger}G_{d1}d$. The resulting optimal change in the temperature $y_2 = T$ is then

$$y_2^{\text{opt}} = (-G_{22}G_{12}^{\dagger}G_{d1} + G_{d2})d \qquad (10.21)$$

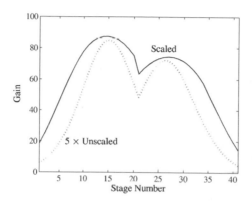

Figure 10.8: Scaled ($|G_{22}'|$) and unscaled ($|G_{22}|$) gains for alternative temperature locations for the distillation example

3. *As seen from the solid and dashed lines in Figure 10.8, the local peaks of the unscaled and scaled gains occur at stages 26 and 27, respectively. Thus, scaling does not affect the final conclusion much in this case. However, if we were to set the implementation error e to zero, then the maximum scaled gain would be at the bottom of the column (stage 1).*
4. *We made the choice $u_2 = L$ to avoid saturation in the boilup V in the regulatory control layer. However, if saturation is not a problem, then the other alternative $u_2 = V$ may be better. A similar analysis with $u_2 = V$ gives a maximum scaled gain of about 100 is obtained with the temperature measured at stage 14.*

In summary, the overall 5×5 distillation control problem may be solved by first designing a 4×4 "stabilizing" (regulatory) controller K_2 for levels, pressure and temperature

$$y_2 = \begin{bmatrix} M_D & M_B & p & T \end{bmatrix}^T, \quad u_2 = \begin{bmatrix} D & B & V_T & L \end{bmatrix}^T$$

and then designing a 2×2 "primary" controller K_1 for composition control

$$y_1 = \begin{bmatrix} x_D & x_B \end{bmatrix}, \quad u_1 = \begin{bmatrix} V & T_s \end{bmatrix}$$

Alternatively, we may interchange L and V in u_1 and u_2. The temperature sensor (T) should be located at a point with a large scaled gain.

We have discussed some simple rules and tools ("maximum gain rule") for selecting the variables in the regulatory control layer. The regulatory control layer is usually itself hierarchical, consisting of a layer for stabilization of unstable modes (RHP-poles) and a layer for "stabilization" in terms of disturbance rejection. Next, we introduce pole vectors and partial control, which are more specific tools for addressing the issues of stabilization and disturbance rejection.

10.4.3   Stabilization: pole vectors

Pole vectors are useful for selecting inputs and outputs for stabilization of unstable modes (RHP-poles) when input usage is an issue. An important advantage is that the selection of inputs is treated separately from the selection of outputs and hence we avoid the combinatorial issue. The main disadvantage is that the theoretical results only hold for cases with a *single* RHP-pole, but applications show that the tool is more generally useful.

The issue is: which outputs (measurements) and inputs (manipulations) should be used for stabilization? We should clearly avoid saturation of the inputs, because this makes the system effectively open-loop and stabilization is then impossible. A reasonable objective is therefore to minimize the input usage required for stabilization. In addition, this choice also minimizes the "disturbing" effect that the stabilization layer has on the remaining control problem.

Recall that $u = -KS(r + n - d)$, so input usage is minimized when the norm of KS is minimal. We will consider both the $\mathcal{H}_2$ and $\mathcal{H}_\infty$ norms.

Theorem 10.1 (Input usage for stabilization) *For a rational plant with a single unstable mode p, the minimal $\mathcal{H}_2$ and $\mathcal{H}_\infty$ norms of the transfer function KS are given as (Havre and Skogestad, 2003; Kariwala, 2004)*

$$\min_K \|KS\|_2 = \frac{(2p)^{3/2} \cdot |q^T t|}{\|u_p\|_2 \cdot \|y_p\|_2} \tag{10.22}$$

$$\min_K \|KS\|_\infty = \frac{2p \cdot |q^T t|}{\|u_p\|_2 \cdot \|y_p\|_2} \tag{10.23}$$

Here u_p and y_p denote the input and output pole vectors (see page 127), respectively, and t and q are the right and left eigenvectors of the state matrix A, satisfying $At = pt$ and $q^T A = q^T p$.

Theorem 10.1 applies to plants with any number of RHP-zeros and to both multivariable (MIMO) and single-loop (SISO) control. In the SISO case, u_p and y_p are the elements in the pole vectors, $u_{p,j}$ and $y_{p,i}$, corresponding to the selected input (u_j) and output (y_i). Notice that the term ($q^T t$) is independent of the selected inputs and outputs, u_j and y_i. Thus, for a single unstable mode and SISO control:

> *The input usage required for stabilization is minimized by selecting the output y_i (measurement) and input u_j (manipulation) corresponding to the largest elements in the output and input pole vectors (y_p and u_p), respectively* (see also Remark 2 on page 137).

This choice maximizes the (state) controllability and observability of the unstable mode. Note that the selections of measurement y_i and input u_j are performed *independently*. The above result is for unstable poles. However, Havre (1998) shows that the input requirement for pole placement is minimized by selecting the output and input corresponding to the largest elements in the y_p and u_p, respectively. This property also holds for LHP-poles, and shows that pole vectors may also be useful when we want to move stable poles.

Exercise 10.4[*] *Show that for a system with a single unstable pole, (10.23) represents the least achievable value of $\|KS\|_\infty$. (Hint: Rearrange (5.31) on page 178 using the definition of pole vectors.)*

When the plant has *multiple* unstable poles, the pole vectors associated with a specific RHP-pole give a measure of input usage required to move this RHP-pole assuming that the other RHP-poles are unchanged. This is of course unrealistic; nevertheless, the pole vector approach can be used by stabilizing one source of instability at a time. That is, first an input and an output are selected considering one real RHP-pole or a pair of complex RHP-poles and a stabilizing controller is designed. Then, the pole vectors are recomputed for the partially controlled system and another set of variables is selected. This process is repeated until all the modes are stabilized. This process results in a sequentially designed decentralized controller and has been useful in several practical applications, as demonstrated by the next example.

Example 10.10 Stabilization of Tennessee Eastman process. *The Tennessee Eastman chemical process (Downs and Vogel, 1993) was introduced as a challenge problem to test methods for control structure design.[3] The process has 12 manipulated inputs and 41 candidate measurements, of which we consider 11 here; see Havre (1998) for details on the selection of these variables and scaling. The model has six unstable poles at the operating point considered, $p = [0 \quad 0.001 \quad 0.023 \pm j0.156 \quad 3.066 \pm j5.079]$. The absolute values of the output and input pole vectors are*

$$|Y_p| = \begin{bmatrix} 0.000 & 0.001 & 0.041 & 0.112 \\ 0.000 & 0.004 & 0.169 & 0.065 \\ 0.000 & 0.000 & 0.013 & 0.366 \\ 0.000 & 0.001 & 0.051 & 0.410 \\ 0.009 & 0.581 & 0.488 & 0.316 \\ 0.000 & 0.001 & 0.041 & 0.115 \\ 1.605 & 1.192 & 0.754 & 0.131 \\ 0.000 & 0.001 & 0.039 & 0.108 \\ 0.000 & 0.001 & 0.038 & 0.217 \\ 0.000 & 0.001 & 0.055 & \mathbf{1.485} \\ 0.000 & 0.002 & 0.132 & 0.272 \end{bmatrix} \quad |U_p|^T = \begin{bmatrix} 6.815 & 6.909 & 2.573 & 0.964 \\ 6.906 & 7.197 & 2.636 & 0.246 \\ 0.148 & 1.485 & 0.768 & 0.044 \\ 3.973 & 11.550 & 5.096 & 0.470 \\ 0.012 & 0.369 & 0.519 & 0.356 \\ 0.507 & 0.077 & 0.066 & 0.033 \\ 0.135 & 1.850 & 1.682 & 0.110 \\ 22.006 & 0.049 & 0.000 & 0.000 \\ 0.007 & 0.054 & 0.010 & 0.013 \\ 0.247 & 0.708 & 1.501 & \mathbf{2.021} \\ 0.109 & 0.976 & 1.447 & 0.753 \\ 0.033 & 0.095 & 0.201 & 0.302 \end{bmatrix}$$

where we have combined pole vectors corresponding to a complex eigenvalue into a single column. The individual columns of $|Y_p|$ and individual rows of $|U_p|$ correspond to the poles at 0, 0.001, $0.023 \pm j0.156$ and $3.066 \pm j5.079$, respectively.

When designing a stabilizing control system, we normally start by stabilizing the "most unstable" (fastest) pole, i.e. complex poles at $3.066 \pm j5.079$ in this case. From the pole vectors, this mode is most easily stabilized by use of u_{10} and y_{10}. A PI controller, with proportional gain of -0.05 and integral time of 300 minutes, is designed for this loop. This simple controller stabilizes the complex unstable poles at $3.066 \pm j5.079$ and also at $0.023 \pm j0.156$. This is reasonable since the pole vectors show that the modes at $0.023 \pm j0.156$ are observable and controllable through y_{10} and u_{10}, respectively. For stabilizing the integrating modes, the pole vectors can be recomputed to select two additional inputs and outputs; see Havre (1998) for details.

Note that the different choices of inputs and outputs for stabilization have different effects on the controllability of the stabilized system. Thus, in some cases, variable selection using pole vectors may need to be repeated a few times before a satisfactory solution is obtained. An alternative approach is to use the method by Kariwala (2004), which also handles the case of multiple unstable modes directly, but is more involved than the simple pole-vector-based method.

---

[3] Simulink and Matlab models for the Tennessee Eastman process are available from Professor Larry Ricker at the University of Washington (easily found using a search engine).

Exercise 10.5 * *For systems with multiple unstable poles, the variables can be selected sequentially using the pole vector approach by stabilizing one real pole or a pair of complex poles at a time. Usually, the selected variable does not depend on the controllers designed in the previous steps. Verify this for each of the following two systems:*

$$G_1(s) = Q(s) \cdot \begin{bmatrix} 10 & 2 & 1 \\ 12 & 1.5 & 5.01 \end{bmatrix} \quad G_2(s) = Q(s) \cdot \begin{bmatrix} 10 & 2 & 1 \\ 12 & 1 & 1.61 \end{bmatrix}$$

$$Q(s) = \begin{bmatrix} 1/(s-1) & 0 \\ 0 & 1/(s-0.5) \end{bmatrix}$$

(Hint: Use simple proportional controllers for stabilization of p = 1 and evaluate the effect of change of controller gain on pole vectors in the second iteration.)

10.4.4   Local disturbance rejection: partial control

Let y_1 denote the primary variables, and y_2 the locally controlled variables. We start by deriving the transfer functions for y_1 for the *partially controlled system* when y_2 is controlled. We also partition the inputs u into the sets u_1 and u_2, where the set u_2 is used to control y_2. The model $y = Gu$ may then be written[4]

$$y_1 = G_{11}u_1 + G_{12}u_2 + G_{d1}d \tag{10.24}$$

$$y_2 = G_{21}u_1 + G_{22}u_2 + G_{d2}d \tag{10.25}$$

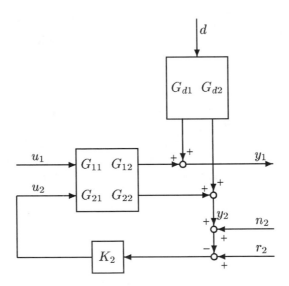

Figure 10.9: Partial control

Now assume that feedback control

$$u_2 = K_2(r_2 - y_{2,m})$$

---

[4] We may assume that any stabilizing loops have already been closed, so for the model $y = Gu$, G includes the stabilizing controller and u includes any "free" setpoints to the stabilizing layer below.

is used for the secondary subsystem involving u_2 and y_2, see Figure 10.9, where $y_{2,m} = y_2 + n_2$ is the measured value of y_2. By eliminating u_2 and y_2, we then get the following model for the resulting partially controlled system from u_1, r_2, d and n_2 to y_1:

$$y_1 = \underbrace{\left(G_{11} - G_{12}K_2(I + G_{22}K_2)^{-1}G_{21}\right)}_{P_u} u_1$$

$$+ \underbrace{\left(G_{d1} - G_{12}K_2(I + G_{22}K_2)^{-1}G_{d2}\right)}_{P_d} d$$

$$+ \underbrace{G_{12}K_2(I + G_{22}K_2)^{-1}}_{P_r}(r_2 - n_2) \qquad (10.26)$$

Note that P_d, the *partial disturbance gain*, is the disturbance gain for a system under partial control. P_u is the effect of u_1 on y_1 with y_2 controlled. In many cases, the set u_1 is empty because there are no extra inputs. In such cases, r_2 is probably available for control of y_1, and P_r gives the effect of r_2 on y_1. In other cases, r_2 may be viewed as a disturbance for the control of y_1.

In the following discussion, we assume that the control of y_2 is fast compared to the control of y_1. This results in a time scale separation between these layers, which simplifies controller design. To obtain the resulting model we may let $K_2 \to \infty$ in (10.26). Alternatively, we may solve for u_2 in (10.25) to get

$$u_2 = -G_{22}^{-1}G_{d2}d - G_{22}^{-1}G_{21}u_1 + G_{22}^{-1}y_2 \qquad (10.27)$$

We have assumed that G_{22} is square and invertible, otherwise we can use a least squares solution by replacing G_{22}^{-1} by the pseudo inverse, $G_{22}^{\dagger}$. On substituting (10.27) into (10.24) and assuming $y_2 \approx r_2 - n_2$ ("perfect" control), we get

$$y_1 \approx \underbrace{(G_{11} - G_{12}G_{22}^{-1}G_{21})}_{P_u} u_1 + \underbrace{(G_{d1} - G_{12}G_{22}^{-1}G_{d2})}_{P_d} d + \underbrace{G_{12}G_{22}^{-1}}_{P_r}\underbrace{(r_2 - n_2)}_{y_2} \qquad (10.28)$$

The advantage of the approximation (10.28) over (10.26) is that it is independent of K_2, but we stress that it is useful only at frequencies where y_2 is tightly controlled.

Remark 1 Relationships similar to those given in (10.28) have been derived by many authors, e.g. see the work of Manousiouthakis et al. (1986) on block relative gains and the work of Haggblom and Waller (1988) on distillation control configurations.

Remark 2 Equation (10.26) may be rewritten in terms of linear fractional transformations (page 543). For example, the transfer function from u_1 to y_1 is

$$F_l(G, -K_2) = G_{11} - G_{12}K_2(I + G_{22}K_2)^{-1}G_{21} \qquad (10.29)$$

Exercise 10.6 *The block diagram in Figure 10.11 below shows a cascade control system where the primary output y_1 depends directly on the extra measurement y_2, so $G_{12} = G_1G_2$, $G_{22} = G_2$, $G_{d1} = [\, I \quad G_1\,]$ and $G_{d2} = [0 \quad I]$. Assume tight control of y_2. Show that $P_d = [\, I \quad 0\,]$ and $P_r = G_1$ and discuss the result. Note that P_r is the "new" plant as it appears with the inner loop closed.*

The selection of secondary variables y_2 depends on whether u_1 or r_2 (or any) are available for control of y_1. Next, we consider in turn each of the three cases that may arise.

1. Cascade control system

Cascade control is a special case of partial control, where we use u_2 to control (tightly) the secondary outputs y_2, and r_2 replaces u_2 as a degree of freedom for controlling y_1. We would like to avoid the introduction of additional (new) RHP-zeros, when closing the secondary loops. The next theorem shows that this is not a problem.

Theorem 10.2 (RHP-zeros due to closing of secondary loop) *Assume that* $n_{y_1} = n_{u_1}$ $+ n_{u_2}$ *and* $n_{y_2} = n_{r_2} = n_{u_2}$ *(see Figure 10.9). Let the plant* $G = \begin{bmatrix} G_{11} & G_{12} \\ G_{21} & G_{22} \end{bmatrix}$ *and the secondary loop* $(S_2 = (I + G_{22}K_2)^{-1})$ *be stable. Then the partially controlled plant*

$$P_{\mathrm{CL}} = [\, G_{11} - G_{12}K_2S_2G_{21} \quad G_{12}K_2S_2 \,] \tag{10.30}$$

from $[u_1 \ r_2]$ *to* y_1 *in (10.26) has* <u>*no*</u> *additional RHP-zeros (that are not present in the open-loop plant* $[\, G_{11} \quad G_{12} \,]$ *from* $[u_1 \ u_2]$ *to* y_1*) if*

1. r_2 *is available for control of* y_1*, and*
2. K_2 *is minimum-phase.*

Proof: Under the dimensional and stability assumptions, P_{CL} is a stable and square transfer function matrix. Thus, the RHP-zeros of P_{CL} are the points in RHP where $\det(P_{\mathrm{CL}}(s)) = 0$ (also see Remark 4 on page 141). Using Schur's formula in (A.14),

$$\det(P_{\mathrm{CL}}) = \det(M) \cdot \det(S_2)$$

where

$$M = \left[\begin{array}{cc|c} G_{11} & 0 & G_{12}K_2 \\ G_{21} & -I & I + G_{22}K_2 \end{array} \right]$$

with the partitioning as shown above. By exchanging the columns of M, we have

$$
\begin{aligned}
\det(M) &= (-1)^n \det\left(\left[\begin{array}{cc|c} G_{11} & G_{12}K_2 & 0 \\ G_{21} & I + G_{22}K_2 & -I \end{array} \right] \right) \\
&= \det\left([\, G_{11} \quad G_{12}K_2 \,] \right) \\
&= \det\left([\, G_{11} \quad G_{12} \,] \right) \det\left(\left[\begin{array}{cc} I & 0 \\ 0 & K_2 \end{array} \right] \right) \\
&= \det\left([\, G_{11} \quad G_{12} \,] \right) \cdot \det(K_2)
\end{aligned}
$$

The second equality follows since the rearranged matrix is block triangular and $\det(-I) = (-1)^n$. Then, putting everything together, we have that

$$\det(P_{\mathrm{CL}}) = \det\left([\, G_{11} \quad G_{12} \,] \right) \cdot \det(K_2) \cdot \det(S_2)$$

Although the RHP-poles of K_2 appear as RHP-zeros of S_2 due to the interpolation constraints, these zeros are cancelled by K_2 and thus $\det(K_2) \cdot \det(S_2)$ evaluated at RHP-poles of K_2 is non-zero. Therefore, when r_2 is available for control of y_1 and K_2 is minimum-phase, the RHP-zeros of P_{CL} are the same as the RHP-zeros of $[\, G_{11} \quad G_{12} \,]$ and the result follows. When u_1 is empty, the transfer matrix from r_2 to y_1 is given as $G_{12}K_2(I + G_{22}K_2)^{-1}$ and thus K_2 being minimum-phase implies that the secondary loop does not introduce any additional RHP-zeros. A somewhat more restrictive version of this theorem was proven by Larsson (2000). The proof here is due to V. Kariwala. Note that the assumptions on the dimensions of y_1 and u_2 are made for simplicity of the proof and the conclusions of Theorem 10.2 still hold when these assumptions are relaxed. $\qquad\square$

For a stable plant G, the controller K_2 can usually be chosen to be minimum-phase. Then, Theorem 10.2 implies that whenever r_2 is available for control of y_1, closing the secondary loops does not introduce additional RHP-zeros. However, note that closing secondary loops *may* make the system more sensitive to disturbances if the action of the secondary (inner) loop "overcompensates" and thereby makes the system more sensitive to the disturbance. As an example consider a plant with $G_{d1} = 1, G_{12} = 1, G_{22} = -0.1$ and $G_{d2} = 1$. Then with tight control of y_2, the disturbance gain for y_1 increases by a factor 9, from $G_{d1} = 1$ to $P_d = G_{d1} - G_{12}G_{22}^{-1}G_{d2} = 9$. In summary, it follows that we should select secondary variables for cascade control such that the input–output controllability of the "new" partially controlled plant $P_{CL} = [\, G_{11} - G_{12}K_2S_2G_{21} \quad G_{12}K_2S_2 \,] = [\, P_u \quad P_r \,]$ with disturbance model P_d is better than that of the "original" plant $[\, G_{11} \quad G_{12} \,]$ with disturbance model G_{d1}. In particular, this requires that

1. $\underline{\sigma}([\, P_u \quad P_r \,])$ (or $\underline{\sigma}(P_r)$, if u_1 is empty) is large at low frequencies.
2. $\bar{\sigma}([\, P_d \quad -P_r \,])$ is small and at least smaller than $\bar{\sigma}(G_{d1})$. In particular, this argument applies at higher frequencies. Note that P_r measures the effect of measurement noise n_2 on y_1.
3. To ensure that u_2 has enough power to reject the local disturbances d and track r_2, based on (10.27), we require that $\bar{\sigma}(G_{22}^{-1}G_{d2}) < 1$ and $\bar{\sigma}(G_{22}^{-1}) < 1$. Here, we have assumed that the inputs have been scaled as outlined in Section 1.4.

Remark 1 The above recommendations for selection of secondary variables are stated in terms of singular values, but the choice of norm is usually of secondary importance. The minimization of $\bar{\sigma}([\, P_d \quad -P_r \,])$ arises if $\left\| \begin{bmatrix} d \\ n_2 \end{bmatrix} \right\|_2 \leq 1$ and we want to minimize $\|y_1\|_2$.

Remark 2 By considering the cost function $J = \min_{d,n_2} y_1^T y_1$, the selection of secondary variables for disturbance rejection using the objectives outlined above is closely related to the concept of self-optimizing control discussed in Section 10.3.

2. Sequentially designed decentralized control system

When r_2 is *not* available for control of y_1, we have a sequentially designed decentralized controller. Here the variables y_2 are important in themselves and we first design a controller K_2 to control the subset y_2. With this controller K_2 in place (a partially controlled system), we may then design a controller K_1 for the remaining outputs.

In this case, secondary loops can introduce "new" RHP-zeros in the partially controlled system P_u. For example, this is likely to happen if we pair on negative RGA elements (Shinskey, 1967; 1996); see Example 10.22 (page 447). Such zeros, however, can be moved to high frequencies (beyond the bandwidth), if it is possible to tune the inner (secondary) loop sufficiently fast (Cui and Jacobsen, 2002).

In addition, based on the general objectives for variable selection, we require that $\underline{\sigma}(P_u)$ instead of $\underline{\sigma}([\, P_u \quad P_r \,])$ be large. The other objectives for secondary variable selection are the same as for cascade control and are therefore not repeated here.

3. Indirect control

Indirect control is when neither r_2 nor u_1 are available for control of y_1. The objective is to minimize $J = \|y_1 - r_1\|$, but we assume that we cannot measure y_1. Instead we hope that y_1

is indirectly controlled by controlling y_2. With perfect control of y_2, as before

$$y_1 = P_d d + P_r(r_2 - n_2)$$

With $n_2 = 0$ and $d = 0$ this gives $y_1 = G_{12}G_{22}^{-1}r_2$, so r_2 must be chosen such that

$$r_2 = G_{22}G_{12}^{-1}r_1 \qquad (10.31)$$

The control error in the primary output is then

$$y_1 - r_1 = P_d d - P_r n_2 \qquad (10.32)$$

To minimize $J = \|y_1 - r_1\|$ we should therefore (as for the two other cases) select the controlled outputs y_2 such that $\|P_d d\|$ and $\|P_r n_2\|$ are small or, in terms of singular values, $\bar{\sigma}([\begin{matrix} P_d & -P_r \end{matrix}])$ is small. The problem of indirect control is closely related to that of cascade control. The main difference is that in *cascade control* we also measure and control y_1 in an outer loop; so in cascade control we need $\|[\begin{matrix} P_d & P_r \end{matrix}]\|$ small only at frequencies outside the bandwidth of the outer control loop (involving y_1).

Remark 1 In some cases, this measurement selection problem involves a trade-off between wanting $\|P_d\|$ small (wanting a strong correlation between measured outputs y_2 and "primary" outputs y_1) and wanting $\|P_r\|$ small (wanting the effect of control errors (measurement noise) to be small). For example, this is the case in a distillation column when we use temperatures inside the column (y_2) for indirect control of the product compositions (y_1). For a high-purity separation, we cannot place the measurement close to the column end due to sensitivity to measurement error ($\|P_r\|$ becomes large), and we cannot place it far from the column end due to sensitivity to disturbances ($\|P_d\|$ becomes large); see also Example 10.9 (page 409).

Remark 2 Indirect control is related to the idea of *inferential control* which is commonly used in the process industry. However, with inferential control the idea is usually to use the measurement of y_2 to estimate (infer) y_1 and then to control this estimate rather than controlling y_2 directly, e.g. see Stephanopoulos (1984). However, there is no universal agreement on these terms, and Marlin (1995) uses the term inferential control to mean indirect control as discussed above.

Optimal "stabilizing" control in terms of minimizing drift

A primary objective of the regulatory control system is to "stabilize" the plant in terms of minimizing its steady-state drift from a nominal operating point. To quantify this, let w represent the variables in which we would like to avoid drift; for example, w could be the weighted states of the plant. For now let y denote the available measurements and u the manipulated variables to be used for stabilizing control. The problem is: to minimize the drift, which variables c should be controlled (at constant setpoints) by u? We assume linear measurement combinations,

$$c = Hy \qquad (10.33)$$

and that we control as many variables as the number of degrees of freedom, $n_c = n_u$. The linear model is

$$w = G^w u + G_d^w d = \widetilde{G}^w \begin{bmatrix} u \\ d \end{bmatrix}$$

$$y = G^y u + G_d^y d = \widetilde{G}^y \begin{bmatrix} u \\ d \end{bmatrix}$$

With perfect regulatory control ($c = 0$), the closed-loop response from d to w is

$$w = P_d^w d; \quad P_d^w = G_d^w - G^w (HG^y)^{-1} HG_d^y$$

Since generally $n_w > n_u$, we do not have enough degrees of freedom to make $w = 0$ ("zero drift"). Instead, we seek the least squares solution that minimizes $\|w\|_2$. In the absence of implementation error, an explicit solution, which also minimizes $\|P_d^w\|_2$, is

$$H = (G^w)^T \widetilde{G}^w (\widetilde{G}^y)^\dagger \tag{10.34}$$

where we have assumed that we have enough measurements, $n_y \geq n_u + n_d$.

Proof of (10.34): We want to minimize

$$J = \|w\|_2^2 = u^T (G^w)^T G^w u + d^T (G_d^w)^T G_d^w d + 2u^T (G^w)^T G_d^w d$$

Then,

$$dJ/du = 2(G^w)^T G^w u + 2(G^w)^T G_d^w d = 2(G^w)^T \widetilde{G}^w \begin{bmatrix} u \\ d \end{bmatrix}$$

An ideal "self-optimizing" variable is $c = dJ/du$, as then $c = 0$ is always optimal with zero loss (in the absence of implementation error). Now, $c = Hy = H\widetilde{G}^y \begin{bmatrix} u \\ d \end{bmatrix}$, so to get $c = dJ/du$, we would like

$$H\widetilde{G}^y = (G^w)^T \widetilde{G}^w \tag{10.35}$$

(the factor 2 does not matter). Since $n_y \geq n_u + n_d$, (10.35) has an infinite number of solutions, and the one using the right inverse of $\widetilde{G}^y$ is given by (10.34). It can be shown that the use of the right inverse is optimal in terms of minimizing the effect of the (until now neglected) implementation error on w, provided the measurements (y) have been normalized (scaled) with respect to their expected measurement error (n^y) (Alstad, 2005, p. 52). The result (10.34) was originally proved by Hori et al. (2005), but this proof is due to V. Kariwala. □

H computed from (10.34) will be dynamic (frequency-dependent), but for practical purposes, we recommend that it is evaluated at the closed-loop bandwidth frequency of the outer loop that adjusts the setpoints for r. In most cases. it is acceptable to use the steady-state matrices.

Example 10.11 Combination of measurements for minimizing drift of distillation column. *We consider the distillation column (column "A") with the LV-configuration and use the same data as in Example 10.9 (page 409). The objective is to minimize the steady-state drift of the 41 composition variables (w = states) due to variations in the feed rate and feed composition by controlling a combination of the available temperature measurements. We have $u = L$, $n_u = 1$ and $n_d = 2$ and we need at least $n_u + n_d = 1 + 2 = 3$ measurements to achieve zero loss (see null space method, page 397). We select three temperature measurements (y) at stages 15, 20 and 26. One reason for not selecting the measurements located at the column ends is their sensitivity to implementation error, see Example 10.9. By ignoring the implementation error, the optimal combination of variables that minimizes $\|P_d^w(0)\|_2$ is, from (10.34),*

$$c = 0.719 T_{15} - 0.018 T_{20} + 0.694 T_{26}$$

When c is controlled perfectly at $c_s = 0$, this gives $\bar{\sigma}(P_d^w(0)) = 0.363$. This is significantly smaller than $\bar{\sigma}(G_d^w(0)) = 9.95$, which is the "open-loop" deviation of the state variables due to the disturbances. We have not considered the effect of implementation error so far. Similar to (10.28), it can be shown that the effect of implementation error on w is given by $\bar{\sigma}(G_w(G_y)^\dagger)$. With an implementation error of 0.05 in the individual temperature measurements, we get $\bar{\sigma}(G_w(G_y)^\dagger) = 0.135$, which is small.

10.5   Control configuration elements

In this section, we discuss in more detail some of the control configuration elements mentioned above. We assume that the measurements y, manipulations u and controlled outputs z are fixed. The available synthesis theories presented in this book result in a multivariable controller K which connects all available measurements/commands (y) with all available manipulations (u),

$$u = Ky \tag{10.36}$$

However, such a "big" (full) controller may not be desirable. By control configuration selection we mean the partitioning of measurements/commands and manipulations within the control layer. More specifically, we define

> **Control configuration.** *The restrictions imposed on the overall controller K by decomposing it into a set of local controllers (subcontrollers, units, elements, blocks) with predetermined links and with a possibly predetermined design sequence where subcontrollers are designed locally.*

In a conventional feedback system, a typical restriction on K is to use a one degree-of-freedom controller (so that we have the same controller for r and $-y$). Obviously, this limits the achievable performance compared to that of a two degrees-of-freedom controller. In other cases, we may use a two degrees-of-freedom controller, but we may impose the restriction that the feedback part of the controller (K_y) is first designed locally for disturbance rejection, and then the prefilter (K_r) is designed for command tracking. In general, this will limit the achievable performance compared to a simultaneous design (see also the remark on page 111). Similar arguments apply to other cascade schemes.

Some elements used to build up a specific control configuration are:

- Cascade controllers
- Decentralized controllers
- Feedforward elements
- Decoupling elements
- Selectors

These are discussed in more detail below, and in the context of the process industry in Shinskey (1967, 1996) and Balchen and Mumme (1988). First, some definitions:

> **Decentralized control** *is when the control system consists of independent feedback controllers which interconnect a subset of the output measurements/commands with a subset of the manipulated inputs. These subsets should not be used by any other controller.*

This definition of decentralized control is consistent with its use by the control community. In decentralized control, we may rearrange the ordering of measurements/commands and manipulated inputs such that the feedback part of the overall controller K in (10.36) has a fixed block-diagonal structure.

> **Cascade control** *arises when the output from one controller is the input to another.* This is broader than the conventional definition of cascade control which is that the output from one controller is the reference command (setpoint) to another. In addition, in cascade control, it is usually assumed that the inner loop (K_2) is much faster than the outer loop (K_1).

Feedforward elements *link measured disturbances to manipulated inputs.*

Decoupling elements *link one set of manipulated inputs ("measurements") with another set of manipulated inputs. They are used to improve the performance of decentralized control systems, and are often viewed as feedforward elements (although this is not correct when we view the control system as a whole) where the "measured disturbance" is the manipulated input computed by another decentralized controller.*

Selectors *are used to select for control, depending on the conditions of the system, a subset of the manipulated inputs or a subset of the outputs.*

In addition to restrictions on the structure of K, we may impose restrictions on the way, or rather in which *sequence*, the subcontrollers are designed. For most decomposed control systems we design the controllers sequentially, starting with the "fast" or "inner" or "lower-layer" control loops in the control hierarchy. Since cascade and decentralized control systems depend more strongly on feedback rather than models as their source of information, it is usually more important (relative to centralized multivariable control) that the fast control loops are tuned to respond quickly.

In this section, we discuss cascade controllers and selectors, and in the following section, we consider decentralized diagonal control. Let us first give some justification for using such "suboptimal" configurations rather than directly designing the overall controller K.

10.5.1 Why use simplified control configurations?

Decomposed control configurations can be quite complex, see for example Figure 10.13 (page 427), and it may therefore be both simpler and better in terms of control performance to set up the controller design problem as an optimization problem and let the computer do the job, resulting in a centralized multivariable controller as used in other chapters of this book.

If this is the case, why are simplified parameterizations (e.g. PID) and control configurations (e.g. cascade and decentralized control) used in practice? There are a number of reasons, but the most important one is probably the cost associated with obtaining good plant models, which are a prerequisite for applying multivariable control. On the other hand, with cascade and decentralized control the controllers are usually tuned one at a time with a minimum of modelling effort, sometimes even *on-line* by selecting only a few parameters (e.g., the gain and integral time constant of a PI controller). Thus:

- *A fundamental reason for applying cascade and decentralized control is to save on modelling effort.*

Other *benefits* of cascade and decentralized control may include the following:

- easy for operators to understand
- ease of tuning because the tuning parameters have a direct and "localized" effect
- insensitive to uncertainty, e.g. in the input channels
- failure tolerance and the possibility of taking individual control elements into or out of service
- few control links and the possibility for simplified (decentralized) implementation
- reduced computation load

The latter two benefits are becoming less relevant as the cost of computing power is reduced. Based on the above discussion, the main challenge is to find a *control configuration* which allows the (sub)controllers to be tuned independently based on a minimum of model information (the pairing problem). For industrial problems, the number of possible pairings is usually very high, but in most cases physical insight and simple tools, such as the RGA, can be helpful in reducing the number of options to a manageable number. To be able to tune the controllers independently, we must require that the loops interact only to a limited extent. For example, one desirable property is that the steady-state gain from u_i to y_i in an "inner" loop (which has already been tuned) does not change too much as outer loops are closed. For decentralized diagonal control the RGA is a useful tool for addressing this pairing problem (see page 450).

Remark. We just argued that the main advantage of applying cascade and decentralized control is that the controllers can be tuned on-line and this saves on the modelling effort. However, in our theoretical treatment we need a model, for example, to decide on a control configuration. This seems to be a contradiction, but note that the model required for selecting a configuration may be more "generic" and does not need to be modified for each particular application. Thus, if we have found a good control configuration for one particular applications, then it is likely that it will work well also for similar applications.

10.5.2   Cascade control systems

We want to illustrate how a control system which is decomposed into subcontrollers can be used to solve multivariable control problems. For simplicity, we use SISO controllers here of the form

$$u_i = K_i(s)(r_i - y_i) \tag{10.37}$$

where $K_i(s)$ is a scalar. Note that whenever we close a SISO control loop we lose the corresponding input, u_i, as a degree of freedom, but at the same time the reference, r_i, becomes a new degree of freedom.

It may look like it is not possible to handle non-square systems with SISO controllers. However, since the input to the controller in (10.37) is a reference minus a measurement, we can cascade controllers to make use of extra measurements or extra inputs. A *cascade control structure* results when either of the following two situations arise:

- The reference r_i is an output from another controller (typically used for the case of an extra measurement y_i), see Figure 10.10(a). This is *conventional cascade control*.
- The "measurement" y_i is an output from another controller (typically used for the case of an extra manipulated input u_j, e.g. in Figure 10.10(b) where u_2 is the "measurement" for controller K_1). This cascade scheme where the "extra" input u_2 is used to improve the dynamic response, but is reset to a desired "mid-range" target value on a longer time scale, is referred to as *input resetting* (also known as *mid-ranging* or valve position control).

10.5.3   Extra measurements: cascade control

In many cases, we make use of extra measurements y_2 (*secondary outputs*) to provide local disturbance rejection and linearization, or to reduce the effects of measurement noise. For example, velocity feedback is frequently used in mechanical systems, and local flow cascades

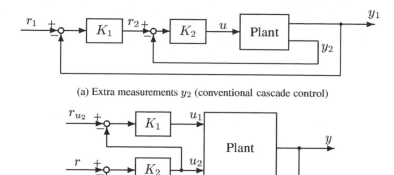

(a) Extra measurements y_2 (conventional cascade control)

(b) Extra inputs u_2 (input resetting)

Figure 10.10: Cascade implementations

are used in process systems. For distillation columns, it is usually recommended to close an inner temperature loop ($y_2 = T$), see Example 10.9.

A typical implementation with two cascaded SISO controllers is shown in Figure 10.10(a) where

$$r_2 = K_1(s)(r_1 - y_1) \tag{10.38}$$

$$u = K_2(s)(r_2 - y_2) \tag{10.39}$$

u is the manipulated input, y_1 the controlled output (with an associated control objective r_1) and y_2 the extra measurement. Note that the output r_2 from the slower *primary* controller K_1 is not a manipulated *plant* input, but rather the reference input to the faster *secondary* (or slave) controller K_2. For example, cascades based on measuring the actual manipulated variable (in which case $y_2 = u_m$) are commonly used to reduce uncertainty and nonlinearity at the plant input.

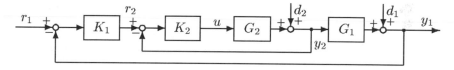

Figure 10.11: Common case of cascade control where the primary output y_1 depends directly on the extra measurement y_2

In the general case, y_1 and y_2 in Figure 10.10(a) are not directly related to each other, and this is sometimes referred to as *parallel cascade control*. However, it is common to encounter the situation in Figure 10.11 where y_1 depends directly on y_2. This is a special case of Figure 10.10(a) with "Plant" $= \begin{bmatrix} G_1 G_2 \\ G_2 \end{bmatrix}$, and it is considered further in Example 10.12 and Exercise 10.7.

Remark. Centralized (parallel) implementation. Alternatively, we may use a centralized implementation $u = K(r - y)$ where K is a 2-input 1-output controller. This gives

$$u = K_{11}(s)(r_1 - y_1) + K_{12}(s)(r_2 - y_2) \tag{10.40}$$

where in most cases $r_2 = 0$ (since we do not have a degree of freedom to control y_2). With $r_2 = 0$ in (10.40) the relationship between the centralized and cascade implementations is $K_{11} = K_2 K_1$ and $K_{12} = K_2$.

An advantage with the cascade implementation is that it more clearly decouples the design of the two controllers. It also shows that r_2 is not a degree of freedom at higher layers in the control system. Finally, it allows for integral action in both loops (whereas usually only K_{11} would have integral action in (10.40)). On the other hand, a centralized implementation is better suited for direct multivariable synthesis; see the velocity feedback for the helicopter case study in Section 13.2.

When should we use cascade control? With reference to the special (but common) case of conventional cascade control shown in Figure 10.11, Shinskey (1967, 1996) states that the principal advantages of cascade control are:

1. Disturbances arising within the secondary loop (before y_2 in Figure 10.11) are corrected by the secondary controller before they can influence the primary variable y_1.
2. Phase lag existing in the secondary part of the process (G_2 in Figure 10.11) is reduced measurably by the secondary loop. This improves the speed of response of the primary loop.
3. Gain variations in the secondary part of the process are overcome within its own loop.

Morari and Zafiriou (1989) conclude, again with reference to Figure 10.11, that the use of an extra measurement y_2 is useful under the following circumstances:

(a) The disturbance d_2 (entering before the measurement y_2) is significant and G_1 is non-minimum-phase – e.g. G_1 contains an effective time delay [see Example 10.12].
(b) The plant G_2 has considerable uncertainty associated with it – e.g. G_2 has a poorly known nonlinear behaviour – and the inner loop serves to remove the uncertainty.

In terms of design, they recommended that K_2 is first designed to minimize the effect of d_2 on y_1 (with $K_1 = 0$) and then K_1 is designed to minimize the effect of d_1 on y_1.

An example where local feedback control is required to counteract the effect of high-order lags is given for a neutralization process in Figure 5.25 on page 216. The benefits of local feedback are also discussed by Horowitz (1991).

Exercise 10.7 *We want to derive the above conclusions (a) and (b) from an input–output controllability analysis, and also explain (c) why we may choose to use cascade control if we want to use simple controllers (even with $d_2 = 0$).*

Outline of solution: (a) Note that if G_1 is minimum-phase, then the input–output controllability of G_2 and $G_1 G_2$ are in theory the same, and for rejecting d_2 there is no fundamental advantage in measuring y_1 rather than y_2. (b) The inner loop $L_2 = G_2 K_2$ removes the uncertainty if it is sufficiently fast (high-gain feedback). It yields a transfer function $(I + L_2)^{-1} L_2$ which is close to I at frequencies where K_1 is active. (c) In most cases, such as when PID controllers are used, the practical closed-loop bandwidth is limited approximately by the frequency w_u, where the phase of the plant is $-180°$ (see Section 5.8 on page 191), so an inner cascade loop may yield faster control (for rejecting d_1 and tracking r_1) if the phase of G_2 is less than that of $G_1 G_2$.

Tuning of cascaded PID controllers using the SIMC rules. Recall the SIMC PID procedure presented on page 57, where the idea is to tune the controllers such that the resulting transfer function from r to y is $T \approx \frac{e^{-\theta s}}{\tau_c s + 1}$. Here, θ is the effective delay in G (from u to y) and τ_c is a tuning parameter with $\tau_c = \theta$ being selected for fast (and still robust) control. Let us apply this approach to the cascaded system in Figure 10.11. The inner loop (K_2) is tuned based on G_2. We then get $y_2 = T_2 r_2$, where $T_2 \approx \frac{e^{-\theta_2 s}}{\tau_{c2} s + 1}$ and θ_2 is the effective delay in G_2. Since the inner loop is fast (θ_2 and τ_{c2} are small), its response may be approximated as a pure time delay for the tuning of the slower outer loop (K_1),

$$T_2 \approx 1 \cdot e^{-(\theta_2 + \tau_{c2})s} \tag{10.41}$$

The resulting model for tuning of the outer loop (K_1) is then

$$\widetilde{G}_1 = G_1 T_2 \approx G_1 e^{-(\theta_2 + \tau_{c2})s} \tag{10.42}$$

and the PID tuning parameters for K_1 are easily obtained using the SIMC rules. For a "fast response" from r_2 to y_2 in the inner loop, the SIMC-rule is to select $\tau_{c2} = \theta_2$. However, this may be unnecessarily fast and to improve robustness we may want to select a larger τ_{c2}. Its value will not affect the outer loop, provided $\tau_{c2} < \tau_{c1}/5$ approximately, where τ_{c1} is the response time in the outer loop.

Example 10.12 *Consider the closed-loop system in Figure 10.11, where*

$$G_1 = \frac{(-0.6s + 1)}{(6s + 1)} e^{-s} \quad \text{and} \quad G_2 = \frac{1}{(6s + 1)(0.4s + 1)}$$

We first consider the case where we only use the primary measurement (y_1), i.e. design the controller based on $G = G_1 G_2$. Using the half rule on page 57, we find that the effective delay is $\theta_1 = 6/2 + 0.4 + 0.6 + 1 = 5$, and using the SIMC tuning rules on page 57, a PI controller is designed with $K_c = 0.9$ and $\tau_I = 9$. The closed-loop response of the system to step changes of magnitude 1 in the setpoint (at $t = 0$) and of magnitude 6 in disturbance d_2 (at $t = 50$) is shown in Figure 10.12. From the dashed line, we see that the closed-loop disturbance rejection is poor.

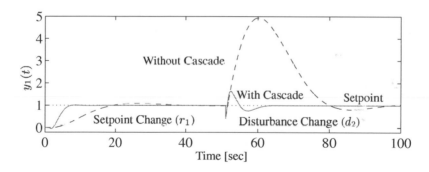

Figure 10.12: Improved control performance with cascade control (solid) as compared to single-loop control (dashed)

Next, to improve disturbance rejection, we make use of the measurement y_2 in a cascade implementation as shown in Figure 10.11. First, the PI controller for the inner loop is designed based on G_2. The effective delay is $\theta_2 = 0.2$. For "fast control" the SIMC rule (page 57) is to use $\tau_{c2} = \theta_2$. However, since this is an inner loop, where tight control is not critical, we choose $\tau_{c2} = 2\theta_2 = 0.4$, which gives somewhat less aggressive settings with $K_{c2} = 10.33$ and $\tau_{I2} = 2.4$. The PI controller for the outer loop is next designed with the inner loop closed. From (10.41), the transfer function for the inner loop is approximated as a delay of $\tau_{c2} + \theta_2 = 0.6$ giving $\widetilde{G}_1 = G_1 e^{-0.6s} = \frac{(-0.6s + 1)}{(6s + 1)} e^{-1.6s}$. Thus, for the outer loop, the effective delay is $\theta_1 = 0.6 + 1.6 = 2.2$ and with $\tau_{c1} = \theta_1 = 2.2$ ("fast *control"), the resulting SIMC PI tunings are $K_{c1} = 1.36$ and $\tau_{I1} = 6$. From Figure 10.12, we note that the cascade controller greatly improves the rejection of d_2. The speed of the setpoint tracking is also improved, because the local control (K_2) reduces the effective delay for control of y_1.*

Exercise 10.8 *To illustrate the benefit of using inner cascades for high-order plants, consider Figure 10.11 and a plant $G = G_1 G_2 G_3 G_4 G_5$ with*

$$G_1 = G_2 = G_3 = G_4 = G_5 = \frac{1}{s+1}$$

Consider the following two cases:

(a) Measurement of y_1 only, i.e. $G = \frac{1}{(s+1)^5}$.

(b) Four additional measurements available (y_2, y_3, y_4, y_5) on outputs of G_1, G_2, G_3 and G_4.

For case (a) design a PID controller and for case (b) use five simple proportional controllers with gains with gains 10 (innermost loop), 5, 2, 1 and 0.5 (outer loop) (note that the gain has to be smaller in the outer loop to avoid instability caused by the effective delay in the inner loop). For case (b) also try using a PI controller in the outer loop to avoid the steady-state offset. Compare the responses to disturbances entering before G_1 (at $t = 0$), G_2 ($t = 20$), G_3 ($t = 40$), G_4 ($t = 60$), G_5 ($t = 80$), and for a setpoint change ($t = 100$)".

10.5.4   Extra inputs

In some cases, we have more manipulated inputs than controlled outputs. These may be used to improve control performance. Consider a plant with a single controlled output y and two manipulated inputs u_1 and u_2. Sometimes u_2 is an extra input which can be used to improve the fast (transient) control of y, but if it does not have sufficient power or is too costly to use for long-term control, then after a while it is reset to some desired value ("ideal resting value").

Cascade implementation (input resetting). An implementation with two cascaded SISO controllers is shown in Figure 10.10(b). We let input u_2 take care of the fast control and u_1 the long-term control. The fast control loop is then

$$u_2 = K_2(s)(r - y) \tag{10.43}$$

The objective of the other slower controller is then to use input u_1 to reset input u_2 to its desired value r_{u_2}:

$$u_1 = K_1(s)(r_{u_2} - y_1), \quad y_1 = u_2 \tag{10.44}$$

and we see that the output u_2 from the fast controller K_2 is the "measurement" y_1 for the slow controller K_1.

In process control, the cascade implementation with input resetting often involves *valve position control*, because the extra input u_2, usually a valve, is reset to a desired position by the outer cascade.

Centralized (parallel) implementation. Alternatively, we may use a centralized implementation $u = K(r - y)$ where K is a 1-input 2-output controller. This gives

$$u_1 = K_{11}(s)(r - y), \quad u_2 = K_{21}(s)(r - y) \tag{10.45}$$

Here two inputs are used to control one output, so to get a unique steady-state for the inputs u_1 and u_2 we usually let K_{11} have integral control, whereas K_{21} does not. Then $u_2(t)$ will only be used for transient (fast) control and will return to zero (or more precisely to its desired value r_{u_2}) as $t \to \infty$. With $r_{u_2} = 0$ the relationship between the centralized and cascade implementation is $K_{11} = -K_1 K_2$ and $K_{21} = K_2$.

Comparison of cascade and centralized implementations. The cascade implementation in Figure 10.10(b) has the advantage, compared to the centralized (parallel) implementation, of decoupling the design of the two controllers. It also shows more clearly that r_{u_2}, the reference for u_2, may be used as a degree of freedom at higher layers in the control system. Finally, we can have integral action in both K_1 and K_2, but note that the gain of K_1 should be negative (if effects of u_1 and u_2 on y are both positive).

Exercise 10.9 * *Draw the block diagrams for the two centralized (parallel) implementations corresponding to Figure 10.10.*

Exercise 10.10 *Derive the closed-loop transfer functions for the effect of r on y, u_1 and u_2 in the cascade input resetting scheme of Figure 10.10(b). As an example use $G = [\,G_{11} \quad G_{12}\,] = [\,1 \quad 1\,]$ and use integral action in both controllers, $K_1 = -1/s$ and $K_2 = 10/s$. Show that input u_2 is reset at steady-state.*

10.5.5 Extra inputs and outputs

In some cases performance may be improved with local control loops involving both extra manipulated inputs and extra measurements. However, as always, the improvement must be traded off against the cost of the extra actuators, measurements and control system.

Example 10.13 **Two layers of cascade control.** *Consider the system in Figure 10.13 with two manipulated plant inputs (u_2 and u_3), one controlled output (y_1, which should be close to r_1) and two measured variables (y_1 and y_2). Input u_2 has a more direct effect on y_1 than does input u_3 (since there is a large delay in $G_3(s)$). Input u_2 should only be used for transient control as it is desirable that it remains close to $r_3 = r_{u_2}$. The extra measurement y_2 is closer than y_1 to the input u_2 and may be useful for detecting disturbances (not shown) affecting G_1.*

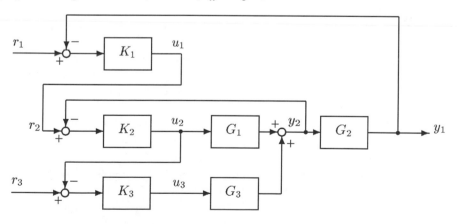

Figure 10.13: Control configuration with two layers of cascade control

In Figure 10.13, controllers K_1 and K_2 are cascaded in a conventional manner, whereas controllers K_2 and K_3 are cascaded to achieve input resetting. The "input" u_1 is not a (physical) plant input, but it

does play the role of an input (manipulated variable) as seen from the controller K_1. The corresponding equations are

$$u_1 = K_1(s)(r_1 - y_1) \tag{10.46}$$

$$u_2 = K_2(s)(r_2 - y_2), \quad r_2 = u_1 \tag{10.47}$$

$$u_3 = K_3(s)(r_3 - y_3), \quad y_3 = u_2 \tag{10.48}$$

Controller K_1 controls the primary output y_1 at its reference r_1 by adjusting the "input" u_1, which is the reference value for y_2. Controller K_2 controls the secondary output y_2 using input u_2. Finally, controller K_3 manipulates u_3 slowly in order to reset input u_2 to its desired value r_3.

Typically, the controllers in a cascade system are tuned one at a time starting with the fastest loop. For example, for the control system in Figure 10.13 we would probably tune the three controllers in the order K_2 (inner cascade using fast input), K_3 (input resetting using slower input), and K_1 (final adjustment of y_1).

Exercise 10.11 * **Process control application.** *A practical case of a control system like the one in Figure 10.13 is in the use of a pre-heater to keep a reactor temperature y_1 at a given value r_1. In this case, y_2 may be the outlet temperature from the pre-heater, u_2 the bypass flow (which should be reset to r_3, say 10% of the total flow), and u_3 the flow of heating medium (steam). Process engineering students: Make a process flowsheet with instrumentation lines (not a block diagram) for this heater/reactor process.*

10.5.6   Selectors

Split-range control for extra inputs. We considered above the case where the primary input is "slow", and an extra input is added to improve the dynamic performance. For economic reasons or to avoid saturation the extra input is reset to a desired "mid-range" target value on a longer time scale (input resetting or mid-ranging). Another situation is when the primary input may saturate, and an extra input is added to maintain control of the output. In this case, the control range is often split such that, for example, u_1 is used for control when $y \in [y_{min}, y_1]$, and u_2 is used when $y \in [y_1, y_{max}]$.

    Selectors for too few inputs. A completely different situation occurs if there are too few inputs. Consider the case with one input (u) and several outputs ($y_1, y_2, \ldots$). In this case, we cannot control all the outputs independently, so we either need to control all the outputs in some average manner, or we need to make a choice about which outputs are the most important to control. Selectors or logic switches are often used for the latter. *Auctioneering selectors* are used to decide to control one of several similar outputs. For example, such a selector may be used to adjust the heat input (u) to keep the maximum temperature ($\max_i y_i$) in a fired heater below some value. *Override selectors* are used when several controllers compute the input value, and we select the smallest (or largest) as the input. For example, this is used in a heater where the heat input (u) normally controls temperature (y_1), except when the pressure (y_2) is too large and pressure control takes over.

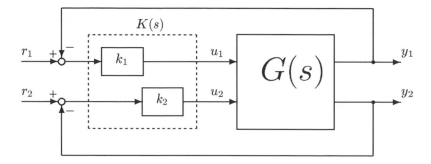

Figure 10.14: Decentralized diagonal control of a 2×2 plant

10.6 Decentralized feedback control

10.6.1 Introduction

We have already discussed, in the previous sections on control configurations, the use of decentralized control, but here we consider it in more detail. To this end, we assume in this section that $G(s)$ is a square plant which is to be controlled using a diagonal controller (see Figure 10.14)

$$K(s) = \text{diag}\{k_i(s)\} = \begin{bmatrix} k_1(s) & & & \\ & k_2(s) & & \\ & & \ddots & \\ & & & k_m(s) \end{bmatrix} \qquad (10.49)$$

This is the problem of decentralized (or diagonal) feedback control.

It may seem like the use of decentralized control seriously limits the achievable control performance. However, often the performance loss is small, partly because of the benefits of high-gain feedback. For example, it can be proved theoretically (Zames and Bensoussan, 1983) that with decentralized control one may achieve perfect control of all outputs, provided the plant has no RHP-zeros that limit the use of high feedback gains. Furthermore, for a stable plant $G(s)$ (also with RHP-zeros), it is possible to use integral control in all channels (to achieve perfect steady-state control) if and only if $G(0)$ is non-singular (Campo and Morari, 1994). Both these conditions are also required with full multivariable control. Nevertheless, for "interactive" plants and finite bandwidth controllers, there is a performance loss with decentralized control because of the interactions caused by non-zero off-diagonal elements in G. The interactions may also cause stability problems. A key element in decentralized control is therefore to select good "pairings" of inputs and outputs, such that the effect of the interactions is minimized.

The design of decentralized control systems typically involves two steps:

1. The choice of pairings (control configuration selection).
2. The design (tuning) of each controller, $k_i(s)$.

The optimal solution to this problem is very difficult mathematically. First, the number of pairing options in step 1 is $m!$ for an $m \times m$ plant and thus increases *exponentially* with the size of the plant. Second, the optimal controller in step 2 is in general of infinite order and may be non-unique. In step 2, there are three main approaches:

Fully coordinated design. All the diagonal controller elements $k_i(s)$ are designed simultaneously based on the complete model $G(s)$. This is the theoretically optimal approach for decentralized control, but it is not commonly used in practice. First, as just mentioned, the design problem is very difficult. Second, it offers few of the "normal" benefits of decentralized control (see page 421), such as ease of tuning, reduced modelling effort, and good failure tolerance. In fact, since a detailed dynamic model is required for the design, an optimal coordinated decentralized design offers few benefits compared to using a "full" multivariable controller which is easier to design and has better performance. The exception is situations where multivariable control cannot be used, for example, when centralized cooordination is difficult for geographical reasons. We do not address the optimal coordinated design of decentralized controllers in this book, and the reader is referred to the literature (e.g. Sourlas and Manousiouthakis, 1995) for more details.

Independent design. Each controller element $k_i(s)$ is designed based on the corresponding diagonal element of $G(s)$, such that each individual loop is stable. Possibly, there is some consideration of the off-diagonal interactions when tuning each loop. This approach is the main focus in the remaining part of this chapter. It is used when it is desirable that we have *integrity* where the individual parts of the system (including each loop) can operate independently. The pairing rules on page 450 can be used to obtain pairings for independent design. In short the rules are to (1) pair on RGA elements close to 1 at crossover frequencies, (2) pair on positive steady-state RGA elements, and (3) pair on elements that impose minimal bandwidth limitations (e.g., small delay). The first and second rules are to avoid that the interactions cause instability. The third rule follows because we for good performance want to use high-gain feedback, but we require stable individual loops. For many interactive plants, it is not possible to find a set of pairing satisfying all the three rules.

Sequential design. The controllers are designed sequentially, one at a time, with the previously designed ("inner") controllers implemented. This has the important advantage of reducing each design to a scalar (SISO) problem, and is well suited for on-line tuning. The sequential design approach can be used for interactive problems where the independent design approach does not work, provided it is acceptable to have "slow" control of some output so that we get a difference in the closed-loop response times of the outputs. One then starts by closing the fast "inner" loops (involving the outputs with the fastest desired response times), and continues by closing the slower "outer" loops. The main disadvantage with this approach is that failure tolerance is not guaranteed when the inner loops fail (integrity). In particular, the individual loops are not guaranteed to be stable. Furthermore, one has to decide on the order in which to close the loops.

The effective use of a decentralized controller requires some element of decoupling. Loosely speaking, *independent design* is used when the system is decoupled in space ($G(s)$ is close to diagonal), whereas *sequential design* is used when the system outputs can be decoupled in time.

The analysis of *sequentially designed* decentralized control systems may be performed using the results on partial control presented earlier in this chapter. For example, after closing the inner loops (from u_2 to y_2), the transfer function for the remaining outer system (from u_1 to y_1) is $P_u = \left(G_{11} - G_{12}K_2(I + G_{22}K_2)^{-1}G_{21}\right)$; see (10.26). Notice that in the general case we need to take into account the details of the controller K_2. However, when there is a time scale separation between the layers with the fast loops (K_2) being closed first, then we may for the design of K_1 assume $K_2 \to \infty$ ("perfect control of y_2"), and the transfer function for the remaining "slow" outer system becomes $P_u = G_{11} - G_{12}G_{22}^{-1}G_{21}$; see (10.28). The advantages of the time scale separation for sequential design of decentralized controllers (with fast "inner" and slow "outer" loops), are the same as those for hierarchical cascade control (with fast "lower" and slow "upper" layers) as listed on page 387. Examples of sequential design are given in Example 10.15 (page 433) and in Section 10.6.6 (page 446).

The relative gain array (RGA) is a very useful tool for decentralized control. It is defined as $\Lambda = G \times (G^{-1})^T$, where $\times$ denotes element-by-element multiplication. It is recommended to read the discussion about the "original interpretation" of the RGA on page 83, before continuing. Note in particular from (3.56) that each RGA element represents the ratio between the open-loop (g_{ij}) and "closed-loop" ($\widehat{g}_{ij}$) gains for the corresponding input-output pair, $\lambda_{ij} = g_{ij}/\widehat{g}_{ij}$. By "closed-loop" here we mean "partial control with the other outputs perfectly controlled". Intuitively, we would like to pair on elements with $\lambda_{ij}(s)$ close to 1, because then the transfer function from u_j to y_i is unaffected by closing the other loops.

Remark. We assume in this section that the decentralized controllers $k_i(s)$ are scalar. The treatment may be generalized to block-diagonal controllers by, for example, introducing tools such as the block relative gain; e.g., see Manousiouthakis et al. (1986) and Kariwala et al. (2003).

10.6.2   Introductory examples

To provide some insight into decentralized control and to motivate the material that follows we start with some simple 2×2 examples. We assume that the outputs y_1 and y_2 have been scaled so that the allowable control errors ($e_i = y_i - r_i$), $i = 1, 2$ are approximately between 1 and -1. We design the decentralized controller to give first-order responses with time constant τ_i in each of the individual loops, that is, $y_i = \frac{1}{\tau_i s + 1} r_i$. For simplicity, the plants have no dynamics, and the individual controllers are then simple integral controllers $k_i(s) = \frac{1}{g_{ii}} \frac{1}{\tau_i s}$; see the IMC design procedure on page 54. To make sure that we do not use aggressive control, we use (in all simulations) a "real" plant, where we add a delay of 0.5 time units in each output, i.e. $G_{\text{sim}} = Ge^{-0.5s}$. This delay is not included in the analytic expressions, e.g. (10.52), in order to simplify our discussion, but it is included for simulation and tuning. With a delay of 0.5 we should, for stability and acceptable robustness, select $\tau_i \geq 1$; see the SIMC rule for "fast but robust" control on page 57. In all simulations we drive the system with reference changes of $r_1 = 1$ at $t = 0$ and $r_2 = 1$ at $t = 20$.

Example 10.14 Diagonal plant. *Consider the simplest case of a diagonal plant*

$$G = \begin{bmatrix} g_{11} & g_{12} \\ g_{21} & g_{22} \end{bmatrix} = \begin{bmatrix} 1 & 0 \\ 0 & 1 \end{bmatrix} \tag{10.50}$$

with RGA $= I$. The off-diagonal elements are zero, so there are no interactions and decentralized control with diagonal pairings is obviously optimal.

   Diagonal pairings. The controller

$$K = \begin{bmatrix} \frac{1}{\tau_1 s} & 0 \\ 0 & \frac{1}{\tau_2 s} \end{bmatrix} \tag{10.51}$$

gives nice decoupled first-order responses

$$y_1 = \frac{1}{\tau_1 s + 1} r_1 \quad \text{and} \quad y_2 = \frac{1}{\tau_2 s + 1} r_2 \tag{10.52}$$

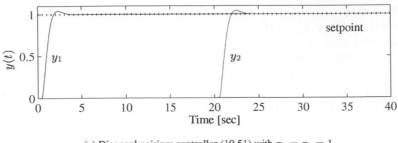

(a) Diagonal pairing; controller (10.51) with $\tau_1 = \tau_2 = 1$

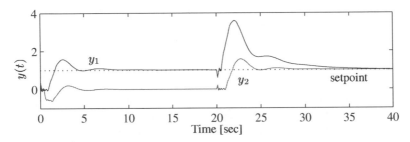

(b) Off-diagonal pairing; plant (10.53) and controller (10.54)

Figure 10.15: Decentralized control of diagonal plant (10.50)

as illustrated in Figure 10.15(a) for the case with $\tau_1 = \tau_2 = 1$.

Off-diagonal pairings. When considering pairings other than diagonal, we recommend to first permute the inputs such that the paired elements are along the diagonal. For the off-diagonal pairing, we use the permuted inputs

$$u_1^* = u_2 \; , \; u_2^* = u_1$$

corresponding to the permuted plant (denoted with *)

$$G^* = G \begin{bmatrix} 0 & 1 \\ 1 & 0 \end{bmatrix}^T = \begin{bmatrix} g_{12} & g_{11} \\ g_{22} & g_{21} \end{bmatrix} = \begin{bmatrix} 0 & 1 \\ 1 & 0 \end{bmatrix} \tag{10.53}$$

This corresponds to pairing on two zero elements, $g_{11}^* = 0$ and $g_{22}^* = 0$, and we cannot use independent or sequential controller design. A coordinated (simultaneous) controller design is required and after some trial and error we arrived at the following design

$$K^*(s) = \begin{bmatrix} \dfrac{-(0.5s+0.1)}{s} & 0 \\ 0 & \dfrac{(0.5s+2)}{s} \end{bmatrix} \tag{10.54}$$

Performance is of course quite poor as is illustrated in Figure 10.15(b), but it is nevertheless workable (surprisingly!).

Remark. The last example, where a diagonal plant is controlled using the off-diagonal pairings, is quite striking. A simple physical example is the control of temperatures in two unrelated rooms, say one located in the UK (Ian's office) and one in Norway (Sigurd's office). The setup is then that Ian gets a measurement of Sigurd's room temperature, and based on this adjusts the heating in his room (in the UK). Similarly, Sigurd gets a measurement of

Ian's room temperature, and based on this adjusts the heating in his room (in Norway). As shown in Figure 10.15(b), such a ridiculous setup (with $g_{11} = 0$ and $g_2 = 0$) is actually workable because of the "hidden" feedback loop going through the off-diagonal elements and the controllers ($k_1 k_2 g_{12} g_{21}$ is nonzero), provided one is able to tune the controllers k_1 and k_2 (which is not trivial – as seen it requires a negative sign in one of the controllers). Two lessons from this example are that (1) decentralized control can work for almost any plant, and (2) the fact that we have what seems to be acceptable closed-loop performance does not mean that we are using the best pairing.

Exercise 10.12 *Consider in more detail the off-diagonal pairings for the diagonal plant in the example above. (i) Explain why it is necessary to use a negative sign in (10.54). (ii) Show that the plant (10.53) cannot be stabilized by a pure integral action controller of the form $K^*(s) = \mathrm{diag}(\frac{k_i}{s})$.*

Example 10.15 **One-way interactive (triangular) plant.** *Consider*

$$G = \begin{bmatrix} 1 & 0 \\ 5 & 1 \end{bmatrix} \tag{10.55}$$

for which

$$G^{-1} = \begin{bmatrix} 1 & 0 \\ -5 & 1 \end{bmatrix} \quad \text{and} \quad \mathrm{RGA} = \begin{bmatrix} 1 & 0 \\ 0 & 1 \end{bmatrix}$$

The RGA matrix is identity, which suggests that the diagonal pairings are best for this plant. However, we see that there is a large interaction ($g_{21} = 5$) from u_1 to y_2, which, as one might expect, implies poor performance with decentralized control. Note that this is not a fundamental control limitation as the decoupling controller $K(s) = \frac{1}{s}\begin{bmatrix} 1 & 0 \\ -5 & 1 \end{bmatrix}$ gives nice decoupled responses, identical to those shown in Figure 10.15 (but the decoupler may be sensitive to uncertainty; see Exercise 10.13).

Diagonal pairings using independent design. If we use independent design based on the paired (diagonal) elements only (without considering the interactions caused by $g_{21} = 5 \neq 0$), then the controller becomes

$$K = \begin{bmatrix} \frac{1}{\tau_1 s} & 0 \\ 0 & \frac{1}{\tau_2 s} \end{bmatrix} \tag{10.56}$$

with $\tau_1 = \tau_2 = 1$ (assuming a 0.5 time delay). However, a closer analysis shows that the closed-loop response with the controller (10.56) becomes

$$y_1 = \frac{1}{\tau_1 s + 1} r_1 \tag{10.57}$$

$$y_2 = \frac{5\tau_2 s}{(\tau_1 s + 1)(\tau_2 s + 1)} r_1 + \frac{1}{\tau_2 s + 1} r_2 \tag{10.58}$$

If we plot the interaction term from r_1 to y_2 as a function of frequency, then we find that for $\tau_1 = \tau_2$ it has a peak value of about 2.5. Therefore, with this controller the response for y_2 is not acceptable when we make a change in r_1. To keep this peak below 1, we need to select $\tau_1 \geq 5\tau_2$, approximately. This is illustrated in Figure 10.16(a) where we have selected $\tau_1 = 5$ and $\tau_2 = 1$. Thus, to keep $|e_2| \leq 1$, we must accept slow control of y_1.

Remark. *The performance problem was not detected from the RGA matrix, because it only measures two-way interactions. However, it may be detected from the "Performance RGA" matrix (PRGA), which for our plant with unity diagonal elements is equal to G^{-1}. As discussed on page 438, a large element in a row of PRGA indicates that fast control is needed to get acceptable reference tracking. Thus, the 2, 1 element in G^{-1} of magnitude 5, confirms that control of y_2 must be about 5 times faster than that of y_1.*

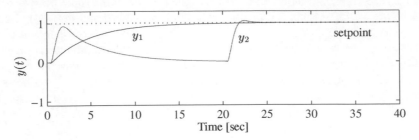

(a) Diagonal pairing; controller (10.56) with $\tau_1 = 5$ and $\tau_2 = 1$

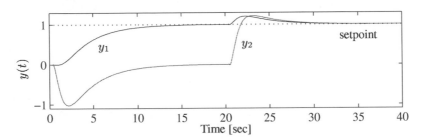

(b) Off-diagonal pairing; plant (10.59) and controller (10.60) with $\tau_1 = 5$ and $\tau_2 = 1$

Figure 10.16: Decentralized control of triangular plant (10.55)

Off-diagonal pairings using sequential design. The permuted plant is

$$G^* = G \begin{bmatrix} 0 & 1 \\ 1 & 0 \end{bmatrix}^T = \begin{bmatrix} 0 & 1 \\ 1 & 5 \end{bmatrix} \tag{10.59}$$

This corresponds to pairing on a zero element $g_{11}^* = 0$. This pairing is not *acceptable if we use the independent design approach, because u_1^* has no effect on y_1 so "loop 1" does not work by itself. However, with the sequential design approach, we may first close the loop around y_2 (on the element $g_{22}^* = 5$). With the IMC design approach, the controller becomes $k_2^*(s) = 1/(g_{22}^* \tau_2 s) = 1/(5\tau_2 s)$ and with this loop closed, u_1^* does have an effect on y_1. Assuming tight control of y_2 gives (using the expression for "perfect" partial control in (10.28))*

$$y_1 = \left(g_{11}^* - \frac{g_{12}^* g_{21}^*}{g_{22}^*} \right) u_1^* = -\frac{1}{5} u_1^*$$

The controller for the pairing u_1^*-y_1 becomes $k_1^*(s) = 1/(g_{11}^* \tau_1 s) = -5/(\tau_1 s)$ and thus

$$K^* = \begin{bmatrix} \frac{-5}{\tau_1 s} & 0 \\ 0 & \frac{1}{5\tau_2 s} \end{bmatrix} \tag{10.60}$$

The response with $\tau_1 = 5$ and $\tau_2 = 1$ is shown in Figure 10.16(b). We see that performance is only slightly worse than with the diagonal pairings. However, more seriously, we have the problem that if control of y_2 fails, e.g. because $u_2^* = u_1$ saturates, then we also lose control of y_1 (in addition, we get instability with y_2 drifting away, because of the integral action for y_1). The situation is particularly bad in this case because of the pairing on a zero element, but the dependence on faster (inner) loops being in service is a general problem with sequential design.

Exercise 10.13 . *Redo the simulations in Example 10.15 with 20% diagonal input uncertainty.* *Specifically, add a block* $\begin{bmatrix} 1.2 & 0 \\ 0 & 0.8 \end{bmatrix}$ *between the plant and the controller. Also simulate with the* *decoupler* $K(s) = \frac{1}{s}\begin{bmatrix} 1 & 0 \\ -5 & 1 \end{bmatrix}$ *which is expected to be particularly sensitive to uncertainty (why? –* *see conclusions on page 251 and note that* $\gamma_I^*(G) = 10$ *for this plant).*

Example 10.16 **Two-way interactive plant.** *Consider the plant*

$$G = \begin{bmatrix} 1 & g_{12} \\ 5 & 1 \end{bmatrix} \tag{10.61}$$

for which

$$G^{-1} = \frac{1}{1-5g_{12}}\begin{bmatrix} 1 & -g_{12} \\ -5 & 1 \end{bmatrix} \quad \text{and} \quad \text{RGA} = \frac{1}{1-5g_{12}}\begin{bmatrix} 1 & -5g_{12} \\ -5g_{12} & 1 \end{bmatrix}$$

The control properties of this plant depend on the parameter g_{12}. *The plant is singular* ($\det(G) = 1 - 5g_{12} = 0$) *for* $g_{12} = 0.2$, *and in this case independent control of both outputs is impossible,* *whatever the controller. We will examine the diagonal pairings using the independent design controller*

$$K = \begin{bmatrix} \frac{1}{\tau_1 s} & 0 \\ 0 & \frac{1}{\tau_2 s} \end{bmatrix} \tag{10.62}$$

The individual loops are stable with responses $y_1 = \frac{1}{(\tau_1 s + 1)}r_1$ *and* $y_2 = \frac{1}{(\tau_2 s + 1)}r_2$, *respectively. With* *both loops closed, the response is* $y = GK(I + GK)^{-1}r = Tr$, *where*

$$T = \frac{1}{(\tau_1 s + 1)(\tau_2 s + 1) - 5g_{12}}\begin{bmatrix} \tau_2 s + 1 - 5g_{12} & g_{12}\tau_1 s \\ 5\tau_2 s & \tau_1 s + 1 - 5g_{12} \end{bmatrix}$$

We see that $T(0) = I$, *so we have perfect steady-state control, as is expected with integral action.* *However, the interactions as expressed by the term* $5g_{12}$ *may yield instability, and we find that the* *system is closed-loop unstable for* $g_{12} > 0.2$. *This is also expected because the diagonal RGA elements* *are negative for* $g_{12} > 0.2$, *indicating a gain change between the open-loop* (g_{ii}) *and closed-loop* ($\widehat{g}_{ii}$) *transfer functions, which is incompatible with integral action. Thus, for* $g_{12} > 0.2$, *the off-diagonal* *pairings must be used if we want to use an independent design (with stable individual loops).*

We will now consider three cases, (a) $g_{12} = 0.17$, (b) $g_{12} = -0.2$ *and* (c) $g_{12} = -1$, *each with the* *same controller (10.62) with* $\tau_1 = 5$ *and* $\tau_2 = 1$. *Because of the large interactions given by* $g_{21} = 5$, *we need to control* y_2 *faster than* y_1.

(a) $\underline{g_{12} = 0.17}$. *In this case,*

$$G^{-1} = \begin{bmatrix} 6.7 & -1.1 \\ -33.3 & 6.7 \end{bmatrix} \quad \text{and} \quad \text{RGA} = \begin{bmatrix} 6.7 & -5.7 \\ -5.7 & 6.7 \end{bmatrix}$$

The large RGA elements indicate strong interactions. Furthermore, recall from (3.56) that the *RGA gives the ratio of the open-loop and (partially) closed-loop gains,* $g_{ij}/\widehat{g}_{ij}$. *Thus, in terms of* *decentralized control, the large positive RGA elements indicate that* $\widehat{g}_{ij}$ *is small and the loops will* *tend to counteract each other by reducing the effective loop gain. This is confirmed by simulations* *in Figure 10.17(a).*

(b) $\underline{g_{12} = -0.2}$. *In this case,*

$$G^{-1} = \begin{bmatrix} 0.5 & 0.1 \\ -2.5 & 0.5 \end{bmatrix} \quad \text{and} \quad \text{RGA} = \begin{bmatrix} 0.5 & 0.5 \\ 0.5 & 0.5 \end{bmatrix}$$

The RGA elements of 0.5 indicate quite strong interactions and show that the interaction increases *the effective gain. This is confirmed by the closed-loop responses in Figure 10.17(b).*

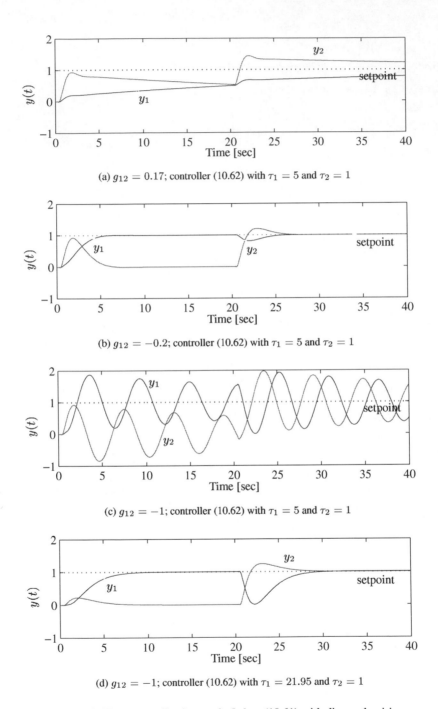

(a) $g_{12} = 0.17$; controller (10.62) with $\tau_1 = 5$ and $\tau_2 = 1$

(b) $g_{12} = -0.2$; controller (10.62) with $\tau_1 = 5$ and $\tau_2 = 1$

(c) $g_{12} = -1$; controller (10.62) with $\tau_1 = 5$ and $\tau_2 = 1$

(d) $g_{12} = -1$; controller (10.62) with $\tau_1 = 21.95$ and $\tau_2 = 1$

Figure 10.17: Decentralized control of plant (10.61) with diagonal pairings

(c) $g_{12} = -1$. *In this case,*

$$G^{-1} = \begin{bmatrix} 0.17 & 0.17 \\ -0.83 & 0.17 \end{bmatrix} \quad \text{and} \quad \text{RGA} = \begin{bmatrix} 0.17 & 0.83 \\ 0.83 & 0.17 \end{bmatrix}$$

The RGA indicates clearly that the off-diagonal pairings are preferable. Nevertheless, we will consider the diagonal pairings with $\tau_1 = 5$ and $\tau_2 = 1$ (as before). The response is poor as seen in Figure 10.17(c). The closed-loop system is stable, but very oscillatory. This is not surprising as the diagonal RGA elements of 0.17 indicate that the interactions increase the effective loop gains by a factor 6 ($= 1/0.17$). To study this in more detail, we write the closed-loop polynomial in standard form

$$(\tau_1 s + 1)(\tau_2 s + 1) - 5g_{12} = \tau^2 s^2 + 2\tau\zeta s + 1$$

with

$$\tau = \sqrt{\frac{\tau_1 \tau_2}{1 - 5g_{12}}} \quad \text{and} \quad \zeta = \frac{1}{2}\frac{\tau_1 + \tau_2}{\sqrt{\tau_1 \tau_2}}\frac{1}{\sqrt{1 - 5g_{12}}}$$

We note that we get oscillations ($0 < \zeta < 1$), when g_{12} is negative and large. For example, $g_{12} = -1$, $\tau_1 = 5$ and $\tau_2 = 1$ gives $\zeta = 0.55$. Interestingly, we see from the expression for ζ that the oscillations may be reduced by selecting τ_1 and τ_2 to be more different. This follows because $\frac{1}{2}\frac{\tau_1 + \tau_2}{\sqrt{\tau_1 \tau_2}}$ is the ratio between the arithmetic and geometric means, which is larger the more different τ_1 and τ_2 are. Indeed, with $g_{12} = -1$ we find that oscillations can be eliminated ($\zeta = 1$) by selecting $\tau_1 = 21.95\tau_2$. This is confirmed by the simulations in Figure 10.17(d). The response is surprisingly good taking into account that we are using the wrong pairings.

Exercise 10.14 *Design decentralized controllers for the 3×3 plant $G(s) = G(0)e^{-0.5s}$ where $G(0)$ is given by (10.80). Try both the diagonal pairings and the pairings corresponding to positive steady-state RGA elements, i.e. $G^* = G\begin{bmatrix} 0 & 1 & 0 \\ 1 & 0 & 0 \\ 0 & 0 & 1 \end{bmatrix}^T$.*

The above examples show that in many cases we can achieve quite good performance with decentralized control, even for interactive plants. However, decentralized controller design is more difficult for such plants, and this, in addition to the possibility for improved performance, favours the use of multivariable control for interactive plants.

With the exception of Section 10.6.6, the focus in the rest of this chapter is on *independently designed* decentralized control systems, which cannot be analyzed using the expressions for partial control presented earlier in (10.28). We present tools for pairing selections (step 1) and for analyzing the stability and performance of decentralized control systems based on independent design. Readers who are primarily interested in applications of decentralized control may want to go directly to the summary in Section 10.6.8 (page 449).

10.6.3 Notation and factorization of sensitivity function

$G(s)$ denotes a square $m \times m$ plant with elements g_{ij}. With a particular choice of pairings we can rearrange the columns or rows of $G(s)$ such that the paired elements are along the diagonal of $G(s)$. We then have that the controller $K(s)$ is diagonal (diag$\{k_i\}$). We introduce

$$\widetilde{G} \triangleq \text{diag}\{g_{ii}\} = \begin{bmatrix} g_{11} & & & \\ & g_{22} & & \\ & & \ddots & \\ & & & g_{mm} \end{bmatrix} \tag{10.63}$$

as the matrix consisting of the diagonal elements of G. The loop transfer function in loop i is denoted $L_i = g_{ii}k_i$, which is also equal to the i'th diagonal element of $L = GK$.

$$\widetilde{S} \triangleq (I + \widetilde{G}K)^{-1} = \text{diag}\left\{\frac{1}{1 + g_{ii}k_i}\right\} \quad \text{and} \quad \widetilde{T} = I - \widetilde{S} \qquad (10.64)$$

contain the sensitivity and complementary sensitivity functions for the individual loops. Note that $\widetilde{S}$ is *not* equal to the matrix of diagonal elements of $S = (I + GK)^{-1}$.

With decentralized control, the interactions are given by the off-diagonal elements $G - \widetilde{G}$. The interactions can be normalized with respect to the diagonal elements and we define

$$E \triangleq (G - \widetilde{G})\widetilde{G}^{-1} \qquad (10.65)$$

The "magnitude" of the matrix E is commonly used as an "interaction measure". We will show that $\mu(E)$ (where μ is the structured singular value) is the best (least conservative) measure, and will define "generalized diagonal dominance" to mean $\mu(E) < 1$. To derive these results we make use of the following important factorization of the "overall" sensitivity function $S = (I + GK)^{-1}$ with all loops closed,

$$\underbrace{S}_{\text{overall}} = \underbrace{\widetilde{S}}_{\text{individual loops}} \underbrace{(I + E\widetilde{T})^{-1}}_{\text{interactions}} \qquad (10.66)$$

Equation (10.66) follows from (A.147) with $G = \widetilde{G}$ and $G' = G$. The reader is encouraged to confirm that (10.66) is correct, because most of the important results for stability and performance using independent design may be derived from this expression.

A related factorization which follows from (A.148) is

$$S = \widetilde{S}(I - E_S\widetilde{S})^{-1}(I - E_S) \qquad (10.67)$$

where

$$E_S = (G - \widetilde{G})G^{-1} \qquad (10.68)$$

(10.67) may be rewritten as

$$S = (I + \widetilde{S}(\Gamma - I))^{-1}\widetilde{S}\Gamma \qquad (10.69)$$

where Γ is the performance relative gain array (PRGA),

$$\Gamma(s) \triangleq \widetilde{G}(s)G^{-1}(s) \qquad (10.70)$$

Γ is a normalized inverse of the plant. Note that $E_S = I - \Gamma$ and $E = \Gamma^{-1} - I$. In Section 10.6.7 we discuss in more detail the use of the PRGA.

These factorizations are particularly useful for analyzing decentralized control systems based on *independent design*, because the basis is then the individual loops with transfer function $\widetilde{S}$.

10.6.4 Stability of decentralized control systems

We consider the independent design procedure and assume that (a) the plant G is stable and (b) each individual loop is stable by itself ($\widetilde{S}$ and $\widetilde{T}$ are stable). Assumption (b) is the basis for independent design. Assumption (a) is also required for independent design because we want to be able to take any loop(s) out of service and remain stable, and this is not possible if the plant is unstable.

To achieve stability of the overall system with all loops closed, we must require that the interactions do not cause instability. We use the expressions for S in (10.66) and (10.69) to derive conditions for this.

Theorem 10.3 *With assumptions (a) and (b), the overall system is stable (S is stable):*
(i) if and only if $(I + E\widetilde{T})^{-1}$ is stable, where $E = (G - \widetilde{G})\widetilde{G}^{-1}$,
(ii) if and only if $\det(I + E\widetilde{T}(s))$ does not encircle the origin as s traverses the Nyquist D-contour,
(iii) if

$$\rho(E\widetilde{T}(j\omega)) < 1, \forall \omega \tag{10.71}$$

(iv) (and (10.71) is satisfied) if

$$\bar{\sigma}(\widetilde{T}) = \max_i |\widetilde{t}_i| < 1/\mu(E) \quad \forall \omega \tag{10.72}$$

The structured singular value $\mu(E)$ is computed with respect to a diagonal structure (of $\widetilde{T}$).

Proof: (Grosdidier and Morari, 1986) (ii) follows from the factorization $S = \widetilde{S}(I + E\widetilde{T})^{-1}$ in (10.66) and the generalized Nyquist theorem in Lemma A.5 (page 543). (iii) Condition (10.71) follows from the spectral radius stability condition in (4.110). (iv) The least conservative way to split up $\rho(E\widetilde{T})$ is to use the structured singular value. From (8.92) we have $\rho(E\widetilde{T}) \leq \mu(E)\bar{\sigma}(T)$ and (10.72) follows.   □

Theorem 10.4 *With assumptions (a) and (b) and also assuming that that G and $\widetilde{G}$ have no RHP-zeros, the overall system is stable (S is stable):*
(i) if and only if $(I - E_S\widetilde{S}(s))^{-1}$ is stable, where $E_S = (G - \widetilde{G})G^{-1}$,
(ii) if and only if $\det(I - E_S\widetilde{S})$ does not encircle the origin as s traverses the Nyquist D-contour,
(iii) if

$$\rho(E_S\widetilde{S}(j\omega)) < 1, \forall \omega \tag{10.73}$$

(iv) (and (10.73) is satisfied) if

$$\bar{\sigma}(\widetilde{S}) = \max_i |\widetilde{s}_i| < 1/\mu(E_S) \quad \forall \omega \tag{10.74}$$

The structured singular value $\mu(E_S)$ is computed with respect to a diagonal structure (of $\widetilde{S}$).

Proof: The proof is similar to that of Theorem 10.3. We need to assume no RHP-zeros in order to get (i).   □

Remark. The μ-conditions (10.72) and (10.74) for (nominal) stability of the decentralized control system can be generalized to include robust stability and robust performance; see equations (31a-b) in Skogestad and Morari (1989).

In both the above Theorems, (i) and (ii) are necessary and sufficient conditions for stability, whereas the spectral radius condition (iii) is weaker (only sufficient) and the μ-condition condition (iv) is even weaker. Nevertheless, the use of μ is the least conservative way of "splitting up" the spectral radius ρ in condition (iii).

Equation (10.72) is easy to satisfy at high frequencies, where generally $\bar{\sigma}(\widetilde{T}) \to 0$. Similarly, (10.74) is usually easy to satisfy at low frequencies since $\bar{\sigma}(\widetilde{S}(0)) = 0$ for systems with integral control (no steady-state offset). Unfortunately, the two conditions cannot be combined over different frequency ranges (Skogestad and Morari, 1989). Thus, to guarantee stability we need to satisfy one of the conditions over the whole frequency range.

Since (10.72) is generally most difficult to satisfy at low frequencies, where usually $\bar{\sigma}(\widetilde{T}) \approx 1$, this gives rise to the following pairing rule:

- *Prefer pairings with $\mu(E) < 1$ ("diagonal dominance") at frequencies within the closed-loop bandwidth.*

Let Λ denote the RGA of G. For an $n \times n$ plant $\lambda_{ii}(0) > 0.5 \; \forall i$ is a necessary condition for $\mu(E(0)) < 1$ (diagonal dominance at steady state) (Kariwala et al., 2003). This gives the following pairing rule: *Prefer pairing on steady-state RGA elements larger than 0.5 (because otherwise we can never have $\mu(E(0)) < 1$).*

Since (10.74) is generally most difficult to satisfy at high frequencies where $\bar{\sigma}(\widetilde{S}) \approx 1$, and since encirclement of the origin of $\det(I - E_S\widetilde{S}(s))$ is most likely to occur at frequencies up to crossover, this gives rise to the following pairing rule:

- *Prefer pairings with $\mu(E_S) < 1$ ("diagonal dominance") at crossover frequencies.*

Gershgorin bounds. An alternative to splitting up $\rho(E\widetilde{T})$ using μ, is to use Gershgorin's theorem, see page 519. From (10.71) we may then derive (Rosenbrock, 1974) sufficient conditions for overall stability, either in terms of the rows of G,

$$|\tilde{t}_i| < |g_{ii}| / \sum_{j\neq i} |g_{ij}| \quad \forall i, \forall \omega \tag{10.75}$$

or, alternatively, in terms of the columns,

$$|\tilde{t}_i| < |g_{ii}| / \sum_{j\neq i} |g_{ji}| \quad \forall i, \forall \omega \tag{10.76}$$

This gives the important insight that it is preferable to pair on large elements in G, because then the sum of the off-diagonal elements, $\sum_{j\neq i} |g_{ij}|$ and $\sum_{j\neq i} |g_{ji}|$, is small. The "Gershgorin bounds", which should be small, are the inverse of the right hand sides in (10.75) and (10.76),

The Gershgorin conditions (10.75) and (10.76), are complementary to the μ-condition in (10.72). Thus, the use of (10.72) is not always better (less conservative) than (10.75) and (10.76). It is true that the *smallest* of the $i = 1, \ldots m$ upper bounds in (10.75) or (10.76) is always smaller (more restrictive) than $1/\mu(E)$ in (10.72). However, (10.72) imposes the *same* bound on $|\tilde{t}_i|$ for each loop, whereas (10.75) and (10.76) give *individual* bounds, some of which may be less restrictive than $1/\mu(E)$.

Diagonal dominance. Although "diagonal dominance" is a matrix property, its definition has been motivated by control, where, loosely speaking, diagonal dominance means that the interactions will not introduce instability. Originally, for example in the Inverse Nyquist Array method of Rosenbrock (1974), diagonal dominance was defined in terms of the Gershgorin bounds, resulting in the conditions $\|E\|_{i1} < 1$ ("column dominance") and $\|E\|_{i\infty} < 1$ ("row dominance"), where $E = (G - \widetilde{G})\widetilde{G}^{-1}$. However, stability is scaling independent, and by "optimally" scaling the plant using DGD^{-1}, where the scaling matrix D is diagonal, one obtains from these conditions that the matrix G is (generalized) diagonally dominant if $\rho(|E|) < 1$; see (A.128). Here $\rho(|E|)$ is the Perron root of E. An even less restrictive definition of diagonal dominance is obtained by starting from the stability condition in terms of $\mu(E)$ in (10.72). This leads us to propose the improved definition below.

Definition 10.1 A matrix G is generalized diagonally dominant if and only if $\mu(E) < 1$.

Here the term "generalized diagonally dominant" means "can be scaled to be diagonally dominant". Note that we always have $\mu(E) \leq \rho(|E|)$, so the use of μ is less restrictive than the Perron root. Also note that $\mu(E) = 0$ for a triangular plant.[5] It is also possible to use $\mu(E_s)$ as measure of diagonal dominance, and we then have that a matrix is generalized diagonally dominant if $\mu(E) < 1$ or if $\mu(E_S) < 1$.

Example 10.17 *Consider the following plant where we pair on its diagonal elements:*

$$G = \begin{bmatrix} -5 & 1 & 2 \\ 4 & 2 & -1 \\ -3 & -2 & 6 \end{bmatrix}; \quad \widetilde{G} = \begin{bmatrix} -5 & 0 & 0 \\ 0 & 2 & 0 \\ 0 & 0 & 6 \end{bmatrix}; \quad E = (G - \widetilde{G})\widetilde{G}^{-1} = \begin{bmatrix} 0 & 0.5 & 0.33 \\ -0.8 & 0 & -0.167 \\ 0.6 & -1 & 0 \end{bmatrix}$$

The μ-interaction measure is $\mu(E) = 0.9189$, so the plant is diagonally dominant. From (10.72), stability of the individual loops $\tilde{t}_i$ guarantees stability of the overall closed-loop system, provided we keep the individual peaks of $|\tilde{t}_i|$ less than $1/\mu(E) = 1.08$. This allows for integral control with $\tilde{t}(0) = 1$. Note that it is not possible in this case to conclude from the Gershgorin bounds in (10.75) and (10.76) that the plant is diagonally dominant, because the 2, 2 element of G $(= 2)$ is smaller than both the sum of the off-diagonal elements in row 2 $(- 5)$ and in column 2 $(- 3)$.

Iterative RGA. An iterative computation of the RGA, $\Lambda^k(G)$, gives a permuted identity matrix that corresponds to the (permuted) generalized diagonal dominant pairing, if it exists (Johnson and Shapiro, 1986, Theorem 2) (see also page 88). Note that the iterative RGA avoids the combinatorial problem of testing all pairings, as is required when computing $\mu(E)$ or the RGA number. Thus, we may use the iterative RGA to find a promising pairing, and check for diagonal dominance using $\mu(E)$.

Exercise 10.15 *For the plant in Example 10.17 check that the iterative RGA converges to the diagonally dominant pairings.*

Example 10.18 RGA number. *The RGA number, $\|\Lambda - I\|_{\text{sum}}$, is commonly used as a measure of diagonal dominance, but unfortunately for 4×4 plants or larger, a small RGA number does not guarantee diagonal dominance. To illustrate this, consider the matrix G = [1 1 0 0; 0 0.1 1 1; 1 1 0.1 0; 0 0 1 1]. It has has RGA= I, but $\mu(E) = \mu(E_S) = 10.9$ so it is far from diagonally dominant.*

Triangular plants. Overall stability is trivially satisfied for a triangular plant as described in the theorem below.

Theorem 10.5 *Suppose the plant $G(s)$ is stable and upper or lower triangular (at all frequencies), and is controlled by a diagonal controller. Then the overall system is stable if and only if the individual loops are stable.*

---

[5] A triangular plant may have large off-diagonal elements, but it can be scaled to be diagonal. For example $\begin{bmatrix} d_1 & 0 \\ 0 & d_2 \end{bmatrix} \begin{bmatrix} g_{11} & 0 \\ g_{21} & g_{22} \end{bmatrix} \begin{bmatrix} 1/d_1 & 0 \\ 0 & 1/d_2 \end{bmatrix} = \begin{bmatrix} g_{11} & 0 \\ \frac{d_2}{d_1} g_{12} & g_{22} \end{bmatrix}$ which approaches $\begin{bmatrix} g_{11} & 0 \\ 0 & g_{22} \end{bmatrix}$ for $|d_1| \gg |d_2|$.

Proof: For a triangular plant G, $E = (G - \widetilde{G})\widetilde{G}^{-1}$ is triangular with all diagonal elements zero, so it follows that all eigenvalues of $E\widetilde{T}$ are zero. Thus $\det(I + E\widetilde{T}(s)) = 1$ and from (ii) in Theorem 10.3 the interactions can not cause instability.   □

Because of interactions, there may not exists pairings such that the plant is triangular at low frequencies. Fortunately, in practice it is sufficient for stability that the plant is triangular at crossover frequencies, and we have:

Triangular pairing rule. *To achieve stability with decentralized control, prefer pairings such that at frequencies ω around crossover, the rearranged plant matrix $G(j\omega)$ (with the paired elements along the diagonal) is close to triangular.*

Derivation of triangular pairing rule. The derivation is based on Theorem 10.4. From the spectral radius stability condition in (10.74) the overall system is stable if $\rho(\widetilde{S}E_S(j\omega)) < 1$, $\forall\omega$. At low frequencies, this condition is usually satisfied because $\widetilde{S}$ is small. At higher frequencies, where $\widetilde{S} = \mathrm{diag}\{\widetilde{s}_i\} \approx I$, (10.74) may be satisfied if $G(j\omega)$ is close to triangular. This is because E_S and thus $\widetilde{S}E_S$ are then close to triangular, with diagonal elements close to zero, so the eigenvalues of $\widetilde{S}E_S(j\omega)$ are close to zero. Thus (10.74) is satisfied and we have stability of S. The use of Theorem 10.4 assumes that G and $\widetilde{G}$ have no RHP-zeros, but in practice the result also holds for plants with RHP-zeros provided they are located beyond the crossover frequency range.   □

Remark. Triangular plant, RGA= I and stability. An important RGA-property is that the RGA of a triangular plant is always the identity matrix ($\Lambda = I$) or equivalently the RGA number is zero; see property 4 on page 527. In the first edition of this book (Skogestad and Postlethwaite, 1996), we incorrectly claimed that the reverse is also true; that is, an identity RGA matrix ($\Lambda(G) = I$) implies that G is triangular. Then, in the first printing of the second edition we incorrectly claimed that it holds for 3×3 systems or smaller, but actually it holds only for 2×2 systems or smaller as illustrated by the following 3×3 counterexample (due to Vinay Kariwala):

$$G = \begin{bmatrix} g_{11} & 0 & 0 \\ g_{21} & g_{22} & g_{23} \\ g_{31} & 0 & g_{33} \end{bmatrix} \tag{10.77}$$

has RGA= I in all cases (for any nonzero value of the indicated entries g_{ij}), but G is not triangular. On the other hand, note that this G is diagonally dominant since $\mu(E) = 0$ in all cases. However, more generally RGA= I does not imply diagonal dominance as illustrated by the following 4×4 matrix [6]

$$G = \begin{bmatrix} 1 & 1 & 0 & 0 \\ 0 & \alpha & 1 & 1 \\ 1 & 1 & \beta & 0 \\ 0 & 0 & 1 & 1 \end{bmatrix} \tag{10.78}$$

which has RGA= I for any nonzero value of α and β, but G is not triangular and not always diagonal dominant. For example, $\mu(E) = 3.26$ (not diagonally dominant) for $\alpha = \beta = 0.4$. Also, for this plant stability of the individual loops does not necessarily give overall stability. For example, $\widetilde{T} = \frac{1}{\tau s+1} I$ (stable individual loops) gives instability (T unstable) with $\alpha = \beta$ when $|\alpha| = |\beta| < 0.4$. Therefore, RGA= I and stable individual loops do *not* generally guarantee overall stability (it is *not* a sufficient stability condition). Nevertheless, it is clear that we would *prefer* to have RGA= I, because otherwise the plant cannot be triangular. Thus, from the triangular pairing rule we have that it is desirable to select pairings such that the RGA is close to the identity matrix in the crossover region.

---

[6] (10.78) is a generalization of a counterexample given by Johnson and Shapiro (1986). On our book's home page a physical mixing process is given with a transfer function of this form.

10.6.5 Integrity and negative RGA elements

A desirable property of a decentralized control system is that it has *integrity*, that is, the closed-loop system should remain stable as subsystem controllers are brought in and out of service or when inputs saturate. Mathematically, the system possesses integrity if it remains stable when the controller K is replaced by $\mathbb{E}K$ where $\mathbb{E} = \text{diag}\{\epsilon_i\}$ and ϵ_i may take on the values of $\epsilon_i = 0$ or $\epsilon_i = 1$.

An even stronger requirement ("complete detunability") is when it is required that the system remains stable as the gain in various loops is reduced (detuned) by an arbitrary factor, i.e. ϵ_i may take any value between 0 and 1, $0 \leq \epsilon_i \leq 1$. *Decentralized integral controllability* (DIC) is concerned with whether complete detunability is *possible* with *integral control*.

Definition 10.2 Decentralized integral controllability (DIC). *The plant $G(s)$ (corresponding to a given pairing with the paired elements along its diagonal) is DIC if there exists a stabilizing decentralized controller with integral action in each loop such that each individual loop may be detuned independently by a factor ϵ_i ($0 \leq \epsilon_i \leq 1$) without introducing instability.*

Note that DIC considers the *existence* of a controller, so it depends only on the plant G and the chosen pairings. The steady-state RGA provides a very useful tool to test for DIC, as is clear from the following result which was first proved by Grosdidier et al. (1985).

Theorem 10.6 Steady-state RGA and DIC. *Consider a stable square plant G and a diagonal controller K with integral action in all elements, and assume that the loop transfer function GK is strictly proper. If a pairing of outputs and manipulated inputs corresponds to a negative steady-state relative gain, then the closed-loop system has at least one of the following properties:*
(a) The overall closed-loop system is unstable.
(b) The loop with the negative relative gain is unstable by itself.
(c) The closed-loop system is unstable if the loop with the negative relative gain is opened (broken).

This can be summarized as follows:

$$\text{A stable (reordered) plant } G(s) \text{ is DIC only if } \lambda_{ii}(0) \geq 0 \text{ for all } i. \qquad (10.79)$$

Proof: Use Theorem 6.7 on page 252 and select $G' = \text{diag}\{g_{ii}, G^{ii}\}$. Since $\det G' = g_{ii} \det G^{ii}$ and from (A.78) $\lambda_{ii} = \frac{g_{ii} \det G^{ii}}{\det G}$ we have $\det G' / \det G = \lambda_{ii}$ and Theorem 10.6 follows. $\square$

Each of the three possible instabilities in Theorem 10.6 resulting from pairing on a negative value of $\lambda_{ij}(0)$ is undesirable. The worst case is (a) when the overall system is unstable, but situation (c) is also highly undesirable as it will imply instability if the loop with the negative relative gain somehow becomes inactive, e.g. due to input saturation. Situation (b) is unacceptable if the loop in question is intended to be operated by itself, or if all the other loops may become inactive, e.g. due to input saturation.

The RGA is a very efficient tool because it does not have to be recomputed for each possible choice of pairing. This follows since any permutation of the rows and columns of G results in the same permutation in the RGA of G. To achieve DIC one has to pair on a positive RGA(0) element in each row and column, and therefore one can often eliminate many candidate pairings by a simple glance at the RGA matrix. This is illustrated by the following examples:

Example 10.19 *Consider a* 3 × 3 *plant with*

$$G(0) = \begin{bmatrix} 10.2 & 5.6 & 1.4 \\ 15.5 & -8.4 & -0.7 \\ 18.1 & 0.4 & 1.8 \end{bmatrix} \quad \text{and} \quad \Lambda(0) = \begin{bmatrix} 0.96 & \mathbf{1.45} & -1.41 \\ \mathbf{0.94} & -0.37 & 0.43 \\ -0.90 & -0.07 & \mathbf{1.98} \end{bmatrix} \quad (10.80)$$

For a 3 × 3 *plant there are six possible pairings, but from the steady-state RGA we see that there is only one positive element in column 2 ($\lambda_{12} = 1.45$), and only one positive element in row 3 ($\lambda_{33} = 1.98$), and therefore there is only one possible pairing with all RGA elements positive ($u_1 \leftrightarrow y_2$, $u_2 \leftrightarrow y_1$, $u_3 \leftrightarrow y_3$). Thus, if we require to pair on the positive RGA elements, we can from a quick glance at the steady-state RGA eliminate five of the six pairings.*

Example 10.20 *Consider the following plant and RGA:*

$$G(0) = \begin{bmatrix} 0.5 & 0.5 & -0.004 \\ 1 & 2 & -0.01 \\ -30 & -250 & 1 \end{bmatrix} \quad \text{and} \quad \Lambda(0) = \begin{bmatrix} -1.56 & -2.19 & 4.75 \\ 3.12 & 4.75 & -6.88 \\ -0.56 & -1.56 & 3.12 \end{bmatrix} \quad (10.81)$$

From the RGA, we see that it is impossible to rearrange the plant such that all diagonal RGA elements are positive. Consequently, this plant is not DIC for any choice of pairings.

Example 10.21 *Consider the following plant and RGA:*

$$G(s) = \frac{(-s+1)}{(5s+1)^2} \begin{bmatrix} 1 & -4.19 & -25.96 \\ 6.19 & 1 & -25.96 \\ 1 & 1 & 1 \end{bmatrix} \quad \text{and} \quad \Lambda(G) = \begin{bmatrix} 1 & 5 & -5 \\ -5 & 1 & 5 \\ 5 & -5 & 1 \end{bmatrix}$$

Note that the RGA is constant, independent of frequency. Only two of the six possible pairings give positive steady-state RGA elements (see pairing rule 2 on page 450): (a) the (diagonal) pairing on all $\lambda_{ii} = 1$ and (b) the pairing on all $\lambda_{ii} = 5$. Intuitively, one may expect pairing (a) to be the best since it corresponds to pairing on RGA elements equal to 1. However, the RGA matrix is far from identity, and the RGA number, $\|\Lambda - I\|_{\mathrm{sum}}$, is 30 for both pairings. Also, none of the pairings are diagonally dominant as $\mu(E) = 8.84$ for pairing (a) and $\mu(E) = 1.25$ for the pairing (b). These are larger than 1, so none of the two alternatives satisfy pairing rule 1 discussed on page 450, and we are led to conclude that decentralized control should not be used for this plant.

Hovd and Skogestad (1992) confirm this conclusion by designing PI controllers for the two cases. They found pairing (a) corresponding to $\lambda_{ii} = 1$ to be significantly worse than (b) with $\lambda_{ii} = 5$, in agreement with the values for $\mu(E)$. They also found the achievable closed-loop time constants to be 1160 and 220, respectively, which in both cases is very slow compared to the RHP-zero which has a time constant of 1.

Exercise 10.16 *Use the method of "iterative RGA" (page 88) on the model in Example 10.21, and confirm that it results in "recommending" the pairing on $\lambda_{ii} = 5$, which indeed was found to be the best choice based on $\mu(E)$ and the simulations. (This is partly good luck, because the proven theoretical result for iterative RGA only holds for a generalized diagonally dominant matrix.)*

Exercise 10.17* *(a) Assume that the* 4 × 4 *matrix in (A.83) represents the steady-state model of a plant. Show that 20 of the 24 possible pairings can be eliminated by requiring DIC. (b) Consider the* 3 × 3 *FCC process in Exercise 6.17 on page 257. Show that five of the six possible pairings can be eliminated by requiring DIC.*

Remarks on DIC and RGA.

1. DIC was introduced by Skogestad and Morari (1988b) who also give necessary and sufficient conditions for testing DIC. A detailed survey of conditions for DIC and other related properties is given by Campo and Morari (1994).

2. DIC is also closely related to D-stability, see papers by Yu and Fan (1990) and Campo and Morari (1994). The theory of D-stability provides necessary and sufficient conditions (except in a few special cases, such as when the determinant of one or more of the submatrices is zero).

3. Unstable plants are not DIC. The reason for this is that with all $\epsilon_i = 0$ we are left with the uncontrolled plant G, and the system will be (internally) unstable if $G(s)$ is unstable.

4. For $\epsilon_i = 0$ we assume that the integrator of the corresponding SISO controller has been removed, otherwise the integrator would yield internal instability.

5. For 2×2 and 3×3 plants we have even tighter RGA conditions for DIC than (10.79). For 2×2 plants (Skogestad and Morari, 1988b)

$$\text{DIC} \quad \Leftrightarrow \quad \lambda_{11}(0) > 0 \tag{10.82}$$

For 3×3 plants with positive diagonal RGA elements of $G(0)$ and of $G^{ii}(0), i = 1, 2, 3$ (its three principal submatrices), we have (Yu and Fan, 1990)

$$\text{DIC} \quad \Leftrightarrow \quad \sqrt{\lambda_{11}(0)} + \sqrt{\lambda_{22}(0)} + \sqrt{\lambda_{33}(0)} \geq 1 \tag{10.83}$$

(Strictly speaking, as pointed out by Campo and Morari (1994), we do not have equivalence for the case when $\sqrt{\lambda_{11}(0)} + \sqrt{\lambda_{22}(0)} + \sqrt{\lambda_{33}(0)}$ is identically equal to 1, but this has little practical significance.)

6. One cannot in general expect tight conditions for DIC in terms of the RGA (i.e. for 4×4 systems or higher). The reason for this is that the RGA essentially only considers "corner values", $\epsilon_i = 0$ or $\epsilon_i = 1$, for the detuning factor, that is, it tests for integrity. This is clear from the fact that $\lambda_{ii} = \frac{g_{ii} \det G^{ii}}{\det G}$, where G corresponds to $\epsilon_i = 1$ for all i, g_{ii} corresponds to $\epsilon_i = 1$ with the other $\epsilon_k = 0$, and G^{ii} corresponds to $\epsilon_i = 0$ with the other $\epsilon_k = 1$. A more complete integrity ("corner-value") result is given next.

7. **Determinant condition for integrity (DIC).** The following condition is concerned with whether it is possible to design a decentralized controller for the plant such that the system possesses *integrity*, which is a prerequisite for having DIC. *Assume without loss of generality that the signs of the rows or columns of G have been adjusted such that all diagonal elements of $G(0)$ are positive, i.e. $g_{ii}(0) \geq 0$. Then one may compute the determinant of $G(0)$ and all its principal submatrices (obtained by deleting rows and corresponding columns in $G(0)$), which should all have the same sign for integrity.* This determinant condition follows by applying Theorem 6.7 to all possible combinations of $\epsilon_i = 0$ or 1 as illustrated in the proof of Theorem 10.6.

8. The Niederlinski index of a matrix G is defined as

$$N_I(G) = \det G / \Pi_i g_{ii} \tag{10.84}$$

A simple way to test the determinant condition for integrity, which is a necessary condition for DIC, is to require that the Niederlinski index of $G(0)$ and the Niederlinski indices of all the principal submatrices $G^{ii}(0)$ of $G(0)$ are positive.

The original result of Niederlinski, which involved only testing N_I of $G(0)$, obviously yields less information than the determinant condition as does the use of the sign of the RGA elements. This is because the RGA element is $\lambda_{ii} = \frac{g_{ii} \det G^{ii}}{\det G}$, so we may have cases where two negative determinants result in a positive RGA element. Nevertheless, the RGA is usually the preferred tool because it does not have to be recomputed for each pairing. Let us first consider an example where the Niederlinski index is inconclusive:

$$G_1(0) = \begin{bmatrix} 10 & 0 & 20 \\ 0.2 & 1 & -1 \\ 11 & 12 & 10 \end{bmatrix} \quad \text{and} \quad \Lambda(G_1(0)) = \begin{bmatrix} 4.58 & 0 & -3.58 \\ 1 & -2.5 & 2.5 \\ -4.58 & 3.5 & 2.08 \end{bmatrix}$$

Since one of the diagonal RGA elements is negative, we conclude that this pairing is *not* DIC. On the other hand, $N_I(G_1(0)) = 0.48$ (which is positive), so Niederlinski's original condition is inconclusive. However, the N_I of the three principal submatrices $\begin{bmatrix} 10 & 0 \\ 0.2 & 1 \end{bmatrix}$, $\begin{bmatrix} 10 & 20 \\ 11 & 10 \end{bmatrix}$ and $\begin{bmatrix} 1 & -1 \\ 12 & 10 \end{bmatrix}$ are $1, -1.2$ and 2.2, and since one of these is negative, the determinant condition correctly tells us that we do not have DIC.

For this 4×4 example the RGA is inconclusive:

$$G_2(0) = \begin{bmatrix} 8.72 & 2.81 & 2.98 & -15.80 \\ 6.54 & -2.92 & 2.50 & -20.79 \\ -5.82 & 0.99 & -1.48 & -7.51 \\ -7.23 & 2.92 & 3.11 & 7.86 \end{bmatrix} \quad \text{and} \quad \Lambda(G_2(0)) = \begin{bmatrix} 0.41 & 0.47 & -0.06 & 0.17 \\ -0.20 & 0.45 & 0.32 & 0.44 \\ 0.40 & 0.08 & 0.17 & 0.35 \\ 0.39 & 0.001 & 0.57 & 0.04 \end{bmatrix}$$

All the diagonal RGA values are positive, so it is inconclusive when it comes to DIC. However, the Niederlinski index of the gain matrix is negative, $N_I(G_2(0)) = -18.65$, and we conclude that this pairing is not DIC (further evaluation of the 3×3 and 2×2 submatrices is not necessary in this case).

9. The above results, including the requirement that we should pair on positive RGA elements, give *necessary* conditions for DIC. If we assume that the controllers have integral action, then $T(0) = I$, and we can derive from (10.72) that a *sufficient condition for DIC* is that G is generalized diagonally dominant at steady-state, i.e.

$$\mu(E(0)) < 1$$

This is proved by Braatz (1993, p. 154). Since the requirement is only sufficient for DIC, it cannot be used to eliminate designs.

10. If the plant has $j\omega$-axis poles, e.g. integrators, it is recommended that, prior to the RGA analysis, these are moved slightly into the LHP (e.g. by using very low-gain feedback). This will have no practical significance for the subsequent analysis.

11. Since Theorem 6.7 applies to unstable plants, we may also easily extend Theorem 10.6 to unstable plants (and in this case one may actually desire to pair on a negative RGA element). This is shown in Hovd and Skogestad (1994). Alternatively, one may first implement a stabilizing controller and then analyze the partially controlled system as if it were the plant $G(s)$.

10.6.6 RHP-zeros and RGA: reasons for avoiding negative RGA elements with sequential design

So far we have considered decentralized control based on independent design, where we require that the individual loops are stable and that we do not get instability as loops are closed or taken out of service. This led to the integrity (DIC) result of avoiding pairing on negative RGA elements at steady state. However, if we use sequential design, then the "inner" loops should *not* be taken out of service, and one may even end up with loops that are unstable by themselves (if the inner loops were to be removed). Nevertheless, for sequential design we find that it is also generally undesirable to pair on negative RGA elements, and the purpose of this section is primarily to illustrate this, by using some results that link the RGA and RHP-zeros.

Bristol (1966) claimed that negative values of $\lambda_{ii}(0)$ imply the presence of RHP-zeros, but did not provide any proof. However, it is indeed true as illustrated by the following two theorems.

Theorem 10.7 *(Hovd and Skogestad, 1992) Consider a transfer function matrix $G(s)$ with no zeros or poles at $s = 0$. Assume that $\lim_{s \to \infty} \lambda_{ij}(s)$ is finite and different from zero. If $\lambda_{ij}(j\infty)$ and $\lambda_{ij}(0)$ have different signs then at least one of the following must be true:*
(a) The element $g_{ij}(s)$ has a RHP-zero.

(b) The overall plant $G(s)$ has a RHP-zero.
(c) The subsystem with input j and output i removed, $G^{ij}(s)$, has a RHP-zero.

Theorem 10.8 *(Grosdidier et al., 1985) Consider a stable transfer function matrix $G(s)$ with elements $g_{ij}(s)$. Let $\hat{g}_{ij}(s)$ denote the closed-loop transfer function between input u_j and output y_i with all the other outputs under integral control. Assume that: (i) $g_{ij}(s)$ has no RHP-zeros, (ii) the loop transfer function GK is strictly proper, (iii) all other elements of $G(s)$ have equal or higher pole excess than $g_{ij}(s)$. We then have:*
If $\lambda_{ij}(0) < 0$, then for $\hat{g}_{ij}(s)$ the number of RHP-poles plus RHP-zeros is odd.

Note that $\hat{g}_{ij}(s)$ in Theorem 10.8 is the same as the transfer function P_u from u_1 to y_1 for the partially controlled system in (10.26).

Sequential design and RHP-zeros. We design and implement the diagonal controller by tuning and closing one loop at a time in a sequential manner. Assume that we end by pairing on a *negative* steady-state RGA element, $\lambda_{ij}(0) < 0$, and that the corresponding element $g_{ij}(s)$ has no RHP-zero. Then we have the following implications:

(a) If we have integral action (as we normally have), then we will get a RHP-zero in $\hat{g}_{ij}(s)$ which will limit the performance in the "final" output y_i (follows from Theorem 10.8). However, the performance limitation is less if the inner loop is tuned sufficiently fast (Cui and Jacobsen, 2002), see also Example 10.22.

(b) If $\lambda_{ij}(\infty)$ is positive (it is usually close to 1, see pairing rule 1), then irrespective of integral action, we have a RHP-zero in $G^{ij}(s)$, which will also limit the performance in the *other* outputs (follows from Theorem 10.7).

In conclusion, for performance we should avoid ending up by pairing on a negative RGA element.

Example 10.22 **Negative RGA element and RHP-zeros.** *Consider a plant with*

$$G(s) = \frac{1}{s+10} \begin{bmatrix} 4 & 4 \\ 2 & 1 \end{bmatrix} \quad \Lambda(s) = \begin{bmatrix} -1 & 2 \\ 2 & -1 \end{bmatrix}$$

Note that the RGA is independent of frequency for this plant, so $\lambda_{11}(0) = \lambda_\infty = 1$. We want to illustrate that pairing on negative RGA elements gives performance problems. We start by closing the loop from u_1 to y_1 with a controller $u_1 = k_{11}(s)(r_1 - y_1)$. For the partially controlled system, the resulting transfer function from u_2 to y_2 ("outer loop") is

$$\hat{g}_{22}(s) = g_{22}(s) - \frac{k_{11}(s)g_{21}(s)g_{12}(s)}{1 + g_{11}(s)k_{11}(s)}$$

With an integral controller $k_{11}(s) = K_I/s$, we find, as expected from Theorem 10.8, that

$$\hat{g}_{22}(s) = \frac{s^2 + 10s - 4K_I}{(s+10)(s^2 + 10s + 4K_I)}$$

always has a RHP-zero. For large values of K_I, the RHP-zero moves further away, and is less limiting in terms of performance for the outer loop. With a proportional controller, $k_{11}(s) = K_c$, we find that

$$\hat{g}_{22}(s) = \frac{s + 10 - 4K_c}{(s+10)(s + 10 + 4K_c)}$$

has a zero at $4K_c - 10$. For $K_c < 2.5$, the zero is in the LHP, but it crosses into the RHP, when K_c exceeds 2.5. For large values of K_c, the RHP-zero moves further away, and does not limit the performance in the outer loop in practice. The worst value is $K_c = 2.5$, where we have a zero at the origin and the steady-state gain $\hat{g}_{22}(0)$ changes sign.

10.6.7 Performance of decentralized control systems

Consider again the factorization

$$S = (I + \widetilde{S}(\Gamma - I))^{-1}\widetilde{S}\Gamma$$

in (10.69) where $\Gamma = \widetilde{G}G^{-1}$ is the performance relative gain array (PRGA), The diagonal elements of the PRGA matrix are equal to the diagonal elements of the RGA, $\gamma_{ii} = \lambda_{ii}$, and this is the reason for its name. Note that the off-diagonal elements of the PRGA depend on the relative scaling on the outputs, whereas the RGA is scaling independent. On the other hand, the PRGA also measures one-way interaction, whereas the RGA only measures two-way interaction. At frequencies where feedback is effective ($\widetilde{S} \approx 0$), (10.69) yields $S \approx \widetilde{S}\Gamma$ Thus, large elements in the PRGA (Γ) (compared to 1 in magnitude) mean that the interactions "slow down" the overall response and cause performance to be worse than for the individual loops. On the other hand, small PRGA elements (compared to 1 in magnitude) mean that the interactions actually improve performance at this frequency.

We will also make use of the related closed-loop disturbance gain (CLDG) matrix, defined as

$$\widetilde{G}_d(s) \triangleq \Gamma(s)G_d(s) = \widetilde{G}(s)G^{-1}(s)G_d(s) \qquad (10.85)$$

The CLDG depends on both output and disturbance scaling.

In the following, we consider performance in terms of the control error

$$e = y - r = Gu + G_d d - r \qquad (10.86)$$

Suppose the system has been scaled as outlined in Section 1.4, such that at each frequency:

1. Each disturbance is less than 1 in magnitude, $|d_k| < 1$.
2. Each reference change is less than the corresponding diagonal element in R, $|r_j| < R_j$.
3. For each output the acceptable control error is less than 1, $|e_i| < 1$.

Single disturbance. Consider a single disturbance, in which case G_d is a vector, and let g_{di} denote the i'th element of G_d. Let $L_i = g_{ii}k_i$ denote the loop transfer function in loop i. Consider frequencies where feedback is effective so $\widetilde{S}\Gamma$ is small (and (10.89) is valid). Then for acceptable disturbance rejection ($|e_i| < 1$) with decentralized control, we must require for each loop i,

$$|1 + L_i| > |\widetilde{g}_{di}| \qquad (10.87)$$

which is the same as the SISO condition (5.77) except that G_d is replaced by the CLDG, $\widetilde{g}_{di}$. In words, $\widetilde{g}_{di}$ gives the "apparent" disturbance gain as seen from loop i when the system is controlled using decentralized control.

Single reference change. We can similarly address a change in reference for output j of magnitude R_j and consider frequencies where feedback is effective (and (10.89) is valid). Then for acceptable reference tracking ($|e_i| < 1$) we must require for each loop i

$$|1 + L_i| > |\gamma_{ij}| \cdot |R_j| \qquad (10.88)$$

which is the same as the SISO condition (5.80) except for the PRGA factor, $|\gamma_{ij}|$. In other words, when the other loops are closed the response in loop i gets slower by a factor $|\gamma_{ii}|$. Consequently, for *performance* it is desirable to have *small* elements in Γ, at least at frequencies where feedback is effective. However, at frequencies close to crossover, stability is the main issue, and since the diagonal elements of the PRGA and RGA are equal, we usually prefer to have $\gamma_{ii} = \lambda_{ii}$ close to 1 (see pairing rule 1 on page 450).

Proofs of (10.87) and (10.88): At frequencies where feedback is effective, $\widetilde{S}$ is small, so

$$I + \widetilde{S}(\Gamma - I) \approx I \tag{10.89}$$

and from (10.69) we have

$$S \approx \widetilde{S}\Gamma \tag{10.90}$$

The closed-loop response then becomes

$$e = SG_d d - Sr \approx \widetilde{S}\widetilde{G}_d d - \widetilde{S}\Gamma r \tag{10.91}$$

and the response in output i to a single disturbance d_k and a single reference change r_j is

$$e_i \approx \widetilde{s}_i \widetilde{g}_{dik} d_k - \widetilde{s}_i \gamma_{ik} r_k \tag{10.92}$$

where $\widetilde{s}_i = 1/(1 + g_{ii}k_i)$ is the sensitivity function for loop i by itself. Thus, to achieve $|e_i| < 1$ for $|d_k| = 1$ we must require $|\widetilde{s}_i \widetilde{g}_{dik}| < 1$ and (10.87) follows. Similarly, to achieve $|e_i| < 1$ for $|r_j| = |R_j|$ we must require $|s_i \gamma_{ik} R_j| < 1$ and (10.88) follows. Also note that $|s_i \gamma_{ik}| < 1$ will imply that assumption (10.89) is valid. Since R usually has all of its elements larger than 1, in most cases (10.89) will be automatically satisfied if (10.88) is satisfied, so we normally need not check assumption (10.89). □

Remark 1 Relation (10.90) may also be derived from (10.66) by assuming $\widetilde{T} \approx I$ which yields $(I + E\widetilde{T})^{-1} \approx (I + E)^{-1} = \Gamma$.

Remark 2 Consider a particular disturbance with model g_d. Its effect on output i with no control is g_{di}, and the ratio between $\widetilde{g}_{di}$ (the CLDG) and g_{di} is the *relative disturbance gain* (RDG) (β_i) of Stanley et al. (1985) (see also Skogestad and Morari (1987b)):

$$\beta_i \triangleq \widetilde{g}_{di}/g_{di} = [\widetilde{G}G^{-1}g_d]_i/[g_d]_i \tag{10.93}$$

Thus β_i, which is scaling independent, gives the *change* in the effect of the disturbance caused by decentralized control. It is desirable to have β_i small, as this means that the interactions are such that they reduce the apparent effect of the disturbance, such that one does not need high gains $|L_i|$ in the individual loops.

10.6.8 Summary: pairing selection and controllability analysis for decentralized control

When considering decentralized diagonal control of a plant, one should first check that the plant is controllable with any controller, see Section 6.11.

If the plant is unstable, then it recommended that a lower-layer stabilizing controller is first implemented, at least for the "fast" unstable modes. The pole vectors (page 412) are useful in selecting which inputs and outputs to use for stabilizing control. Note that some unstable plants are not stabilizable with a *diagonal* controller. This happens if the unstable modes belong to the "decentralized fixed modes", which are the modes unaffected by diagonal feedback control (e.g. Lunze (1992)). A simple example is a triangular plant where the unstable mode appears only in the off-diagonal elements, but here the plant can be stabilized by changing the pairings.

10.6.9   Independent design

We first consider the case of independent design, where the controller elements are designed based on the diagonal (paired) elements of the plant such that individual loops are stable.

The first step is to determine if one can find a good set of input–output pairs bearing in mind the following three pairing rules:

> **Pairing rule 1. RGA at crossover frequencies.** *Prefer pairings such that the rearranged system, with the selected pairings along the diagonal, has an RGA matrix close to identity at frequencies around the closed-loop bandwidth.*

To help in identifying the pairing with RGA closest to identity, one may, at the bandwidth frequency, compute the iterative RGA, $\Lambda^k(G)$; see Exercise 10.6.4 on page 441.

Pairing rule 1 is to ensure that we have diagonal dominance where interactions from other loops do not cause instability. Actually, pairing rule 1 does not ensure this, see the Remark on page 442, and to ensure stability we may instead require that the rearranged plant is triangular at crossover frequencies. However, the RGA is simple and only requires one computation, and since (a) all triangular plants have RGA $= I$ and (b) there is at most one choice of pairings with RGA $= I$ at crossover frequencies, we do nothing wrong in terms of missing good pairing alternatives by following pairing rule 1. To check for diagonal dominance of a promising pairing (with RGA $= I$) one may subsequently compute $\mu(E_S) = \mu(\text{PRGA} - I)$ to check if it is smaller than 1 at crossover frequencies.

> **Pairing rule 2.** *For a stable plant avoid pairings that correspond to negative steady-state RGA elements, $\lambda_{ij}(0) < 0$.*

This rule follows because we require integrity (DIC) with independent design (page 443), and also because we would like to avoid the introduction of RHP-zeros with sequential design (page 446).

Remark. Even if we have $\lambda_{ii}(0) = 1$ and $\lambda_{ii}(\infty) = 1$ for all i, this does not necessarily mean that the diagonal pairing is the best, even for a 2×2 plant. The reason for this is that the behaviour at "intermediate" bandwidth frequencies is more important. This was illustrated in Example 3.11, where we found from the frequency-dependent RGA in Figure 3.8 (page 86) that the off-diagonal pairing is preferable, because it has RGA close to identity at the bandwidth frequencies.

> **Pairing rule 3.** *Prefer a pairing ij where g_{ij} puts minimal restrictions on the achievable bandwidth. Specifically, the effective delay θ_{ij} in $g_{ij}(s)$ should be small.*

This rule favours pairing on variables physically "close to each other", which makes it easier to use high-gain feedback and satisfy (10.87) and (10.88), while at the same time achieving stability in each loop. It is also consistent with the desire that $\Lambda(j\omega)$ is close to I at crossover frequencies. Pairing rule 3 implies that we should avoid pairing on elements with high order, a time delay or a RHP-zero, because these result in an increased effective delay; see page 58. Goodwin et al. (2005) discuss performance limitations of independent design, in particular when pairing rule 3 is violated.

When a reasonable choice of pairings has been found (if possible), one should rearrange G to have the paired elements along the diagonal and perform a controllability analysis as follows.

1. Compute the PRGA ($\Gamma = \tilde{G}G^{-1}$) and CLDG ($\tilde{G}_d = \Gamma G_d$), and plot these as functions of frequency. For systems with many loops, it is best to perform the analysis one loop at a time. That is, for each loop i, plot $|\tilde{g}_{dik}|$ for each disturbance k and plot $|\gamma_{ij}|$ for each reference j (assuming here for simplicity that each reference is of unit magnitude). For performance, see (10.88) and(10.87), we need $|1 + L_i|$ to be larger than each of these

$$\text{Performance}: \quad |1 + L_i| > \max_{k,j}\{|\tilde{g}_{dik}|, |\gamma_{ij}|\} \tag{10.94}$$

To achieve stability of the individual loops one must analyze $g_{ii}(s)$ to ensure that the bandwidth required by (10.94) is achievable. Note that RHP-zeros in the diagonal elements may limit achievable decentralized control, whereas they may not pose any problems for a multivariable controller. Since with decentralized control we usually want to use simple controllers, the achievable bandwidth in each loop will be limited by the effective delay θ_{ij} in $g_{ij}(s)$.
2. In general, see rule 5.13 on page 207, one may check for constraints by considering the elements of $G^{-1}G_d$ and making sure that they do not exceed 1 in magnitude within the frequency range where control is needed. Equivalently, one may plot $|g_{ii}|$ for each loop i, and the requirement is then

$$\text{To avoid input constraints}: \quad |g_{ii}| > |\tilde{g}_{dik}|, \quad \forall k \tag{10.95}$$

at frequencies where $|\tilde{g}_{dik}|$ is larger than 1 (this follows since $\tilde{G}_d = \tilde{G}G^{-1}G_d$). This provides a direct generalization of the requirement $|G| > |G_d|$ for SISO systems. The advantage of (10.95) compared to using $G^{-1}G_d$ is that we can limit ourselves to frequencies where control is needed to reject the disturbance (where $|\tilde{g}_{dik}| > 1$).

If the plant is not controllable with any choice of pairings, then one may consider another pairing choice and go back to step 1. Most likely this will not help, and one would need to consider decentralized sequential design, or multivariable control.

If the chosen pairing *is* controllable then the analysis based on (10.94) tells us directly how large the loop gain $|L_i| = |g_{ii}k_i|$ must be, and this can be used as a basis for designing the controller $k_i(s)$ for loop i.

10.6.10 Sequential design

Sequential design may be applied when it is not possible to find a suitable set of pairings for independent design using the above three pairing rules. For example, with sequential design one may choose to pair on an element with $g_{ii} = 0$ (and $\lambda_{ii} = 0$), which violates both pairing rules 1 and 3. One then relies on the interactions to achieve the desired performance, as loop i by itself has no effect. This was illustrated for the case with off-diagonal pairings in Example 10.15 on page 434. Another case with pairing on a zero element is in distillation control when the LV-configuration is *not* used, see Example 10.8. One may also in some cases pair on negative steady-state RGA elements, although we have established that to avoid introducing RHP-zeros one should avoid closing a loop on a negative steady-state RGA (see page 447).

The procedure and rules for independent design can be used as a starting point for finding good pairings for sequential design. With sequential design, one also has to decide the order in which the loops are closed, and one generally starts by closing the fast loops. This favours starting with a pairing where g_{ij} has good controllability, including a large gain and a small effective delay. One may also consider the disturbance gain to find which outputs need to be tightly controlled. After closing one loop, one needs to obtain the transfer function for the resulting partially controlled system, see (10.28), and then redo the analysis in order to select the next pairing, and so on.

Example 10.23 Application to distillation process. *In order to demonstrate the use of the frequency-dependent RGA and CLDG for evaluation of expected diagonal control performance, we again consider the distillation process used in Example 10.8. The LV -configuration is used; that is, the manipulated inputs are reflux L (u_1) and boilup V (u_2). The outputs are the product compositions y_D (y_1) and x_B (y_2). Disturbances in feed flow rate F (d_1) and feed composition z_F (d_2) are included in the model. The disturbances and outputs have been scaled such that a magnitude of 1 corresponds to a change in F of 20%, a change in z_F of 20%, and a change in x_B and y_D of 0.01 mole fraction units. The five state dynamic model is given in Section 13.4.*

Initial controllability analysis. *$G(s)$ is stable and has no RHP-zeros. The plant and RGA matrix at steady-state are*

$$G(0) = \begin{bmatrix} 87.8 & -86.4 \\ 108.2 & -109.6 \end{bmatrix} \quad \Lambda(0) = \begin{bmatrix} 35.1 & -34.1 \\ -34.1 & 35.1 \end{bmatrix} \tag{10.96}$$

The RGA elements are much larger than 1 and indicate a plant that is fundamentally difficult to control (recall property C1, page 89). Fortunately, the flow dynamics partially decouple the response at higher frequencies, and we find that $\Lambda(j\omega) \approx I$ at frequencies above about 0.5 rad/min. Therefore if we can achieve sufficiently fast control, the large steady-state RGA elements may be less of a problem.

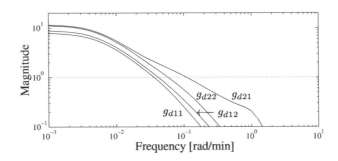

Figure 10.18: Disturbance gains $|g_{dik}|$ for assessing the effect of disturbance k on output i

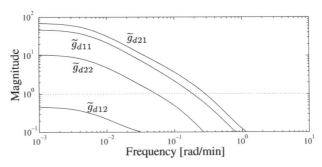

Figure 10.19: Closed-loop disturbance gains $|\widetilde{g}_{dik}|$ for assessing the effect of disturbance k on output i

The steady-state effect of the two disturbances is given by

$$G_d(0) = \begin{bmatrix} 7.88 & 8.81 \\ 11.72 & 11.19 \end{bmatrix} \tag{10.97}$$

and the magnitudes of the elements in $G_d(j\omega)$ are plotted as functions of frequency in Figure 10.18. From this plot the two disturbances seem to be equally difficult to reject with magnitudes larger than 1 up to a frequency of about 0.1 rad/min. We conclude that control is needed up to 0.1 rad/min. The magnitude of the elements in $G^{-1}G_d(j\omega)$ (not shown) are all less than 1 at all frequencies (at least up to 10 rad/min), and so it will be assumed that input constraints pose no problem.

Choice of pairings. *The selection of u_1 to control y_1 and u_2 to control y_2 corresponds to pairing on positive elements of $\Lambda(0)$ and $\Lambda(j\omega) \approx I$ at high frequencies. This seems sensible, and is used in the following.*

Analysis of decentralized control. *The elements in the CLDG and PRGA matrices are shown as functions of frequency in Figures 10.19 and 10.20. At steady-state we have*

$$\Gamma(0) = \begin{bmatrix} 35.1 & -27.6 \\ -43.2 & 35.1 \end{bmatrix}, \quad \widetilde{G}_d(0) = \Gamma(0)G_d(0) = \begin{bmatrix} -47.7 & -0.40 \\ 70.5 & 11.7 \end{bmatrix} \tag{10.98}$$

In this particular case, the off-diagonal elements of RGA (Λ) and PRGA (Γ) are quite similar. We note that $\widetilde{G}_d(0)$ is very different from $G_d(0)$, and this also holds at higher frequencies. For disturbance 1 (first column in $\widetilde{G}_d$) we find that the interactions increase the apparent effect of the disturbance, whereas they reduce the effect of disturbance 2, at least on output 1.

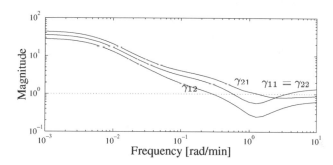

Figure 10.20: PRGA elements $|\gamma_{ij}|$ for effect of reference j on output i

We now consider one loop at a time to find the required bandwidth. For loop 1 (output 1) we consider γ_{11} and γ_{12} for references, and $\widetilde{g}_{d11}$ and $\widetilde{g}_{d12}$ for disturbances. Disturbance 1 is the most difficult, and we need $|1+L_1| > |\widetilde{g}_{d11}|$ at frequencies where $|\widetilde{g}_{d11}|$ is larger than 1, which is up to about 0.2 rad/min. The magnitudes of the PRGA elements are somewhat smaller than $|\widetilde{g}_{d11}|$ (at least at low frequencies), so reference tracking will be achieved if we can reject disturbance 1. From $\widetilde{g}_{d12}$ we see that disturbance 2 has almost no effect on output 1 under feedback control.

Also, for loop 2 we find that disturbance 1 is the most difficult, and from $\widetilde{g}_{d12}$ we require a loop gain larger than 1 up to about 0.3 rad/min. A bandwidth of about 0.2 to 0.3 rad/min in each loop is required for rejecting disturbance 1, and should be achievable in practice.

Observed control performance. *To check the validity of the above results we designed two single-loop PI controllers:*

$$k_1(s) = 0.261\frac{1+3.76s}{3.76s}; \quad k_2(s) = -0.375\frac{1+3.31s}{3.31s} \tag{10.99}$$

The loop gains, $L_i = g_{ii}k_i$, with these controllers are larger than the closed-loop disturbance gains, $|\delta_{ik}|$, at frequencies up to crossover. Closed-loop simulations with these controllers are shown in Figure 10.21. The simulations confirm that disturbance 2 is more easily rejected than disturbance 1.

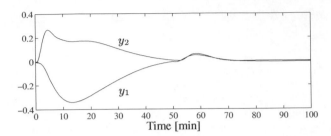

Figure 10.21: Decentralized PI control. Responses to a unit step in d_1 at $t = 0$ and a unit step in d_2 at $t = 50$ min.

In summary, there is an excellent agreement between the controllability analysis and the simulations, as has also been confirmed by a number of other examples.

10.6.11   Conclusions on decentralized control

In this section, we have derived a number of conditions for the stability, e.g. (10.72) and (10.79), and performance, e.g. (10.87) and (10.88), of decentralized control systems. The conditions may be useful in determining appropriate pairings of inputs and outputs and the sequence in which the decentralized controllers should be designed. Recall, however, that in many practical cases decentralized controllers are tuned off-line, and sometimes on-line, using local models. In such cases, the conditions may be used in an input–output controllability analysis to determine the viability of decentralized control.

Some exercises which include a controllability analysis of decentralized control are given at the end of Chapter 6.

10.7   Conclusion

Control structure design is very important in applications, but it has traditionally received little attention in the control community. In this chapter, we have discussed the issues involved, and we have provided some results and rules, dos and don'ts, which we believe will be helpful in practice. However, there is still a need for improved tools and theory in this important area.

11

MODEL REDUCTION

This chapter describes methods for reducing the order of a plant or controller model. We place considerable emphasis on reduced order models obtained by residualizing the less controllable and observable states of a balanced realization. We also present the more familiar methods of balanced truncation and optimal Hankel norm approximation.

11.1 Introduction

Modern controller design methods such as $\mathcal{H}_\infty$ and LQG produce controllers of order at least equal to that of the plant, and usually higher because of the inclusion of weights. These control laws may be too complex with regards to practical implementation and simpler designs are then sought. For this purpose, one can either reduce the order of the plant model prior to controller design, or reduce the controller in the final stage, or both.

The central problem we address is: given a high-order linear time-invariant stable model G, find a low-order approximation G_a such that the infinity ($\mathcal{H}_\infty$ or $\mathcal{L}_\infty$) norm of the difference, $\|G-G_a\|_\infty$, is small. By model order, we mean the dimension of the state vector in a minimal realization. This is sometimes called the McMillan degree.

So far in this book we have only been interested in the infinity ($\mathcal{H}_\infty$) norm of stable systems. But the error $G - G_a$ may be unstable and the definition of the infinity norm needs to be extended to unstable systems. $\mathcal{L}_\infty$ defines the set of rational functions which have no poles on the imaginary axis, it includes $\mathcal{H}_\infty$, and its norm (like $\mathcal{H}_\infty$) is given by $\|G\|_\infty = \sup_w \bar{\sigma}\left(G(jw)\right)$.

We will describe three main methods for tackling this problem: balanced truncation, balanced residualization and optimal Hankel norm approximation. Each method gives a stable approximation and a guaranteed bound on the error in the approximation. We will further show how the methods can be employed to reduce the order of an *unstable* model G. All these methods start from a special state-space realization of G referred to as balanced. We will describe this realization, but first we will show how the techniques of truncation and residualization can be used to remove the high-frequency or fast modes of a state-space realization.

11.2 Truncation and residualization

Let (A, B, C, D) be a minimal realization of a stable system $G(s)$, and partition the state vector x, of dimension n, into $\begin{bmatrix} x_1 \\ x_2 \end{bmatrix}$ where x_2 is the vector of $n - k$ states which we wish to remove. With appropriate partitioning of A, B and C, the state-space equations become

$$
\begin{aligned}
\dot{x}_1 &= A_{11}x_1 + A_{12}x_2 + B_1 u \\
\dot{x}_2 &= A_{21}x_1 + A_{22}x_2 + B_2 u \\
y &= C_1 x_1 + C_2 x_2 + Du
\end{aligned}
\tag{11.1}
$$

11.2.1 Truncation

A k'th-order truncation of the realization $G \stackrel{s}{=} (A, B, C, D)$ is given by $G_a \stackrel{s}{=} (A_{11}, B_1, C_1, D)$. The truncated model G_a is equal to G at infinite frequency, $G(\infty) = G_a(\infty) = D$, but apart from this there is little that can be said in the general case about the relationship between G and G_a. If, however, A is in Jordan form then it is easy to order the states so that x_2 corresponds to high-frequency or fast modes. This is discussed next.

Modal truncation. For simplicity, assume that A has been diagonalized so that

$$
A = \begin{bmatrix} \lambda_1 & 0 & \cdots & 0 \\ 0 & \lambda_2 & \cdots & 0 \\ \vdots & \vdots & \ddots & \vdots \\ 0 & 0 & \cdots & \lambda_n \end{bmatrix} \quad B = \begin{bmatrix} b_1^T \\ b_2^T \\ \vdots \\ b_n^T \end{bmatrix} \quad C = \begin{bmatrix} c_1 & c_2 & \cdots & c_n \end{bmatrix} \tag{11.2}
$$

Then, if the λ_i's are ordered so that $|\lambda_1| < |\lambda_2| < \cdots$, the fastest modes are removed from the model after truncation. The difference between G and G_a following a k'th-order model truncation is given by

$$
G - G_a = \sum_{i=k+1}^{n} \frac{c_i b_i^T}{s - \lambda_i} \tag{11.3}
$$

and therefore

$$
\|G - G_a\|_\infty \leq \sum_{i=k+1}^{n} \frac{\bar{\sigma}(c_i b_i^T)}{|Re(\lambda_i)|} \tag{11.4}
$$

It is interesting to note that the error depends on the residues $c_i b_i^T$ as well as the λ_i's. The distance of λ_i from the imaginary axis is therefore by itself not a reliable indicator of whether the associated mode should be included in the reduced order model or not.

An advantage of modal truncation is that the poles of the truncated model are a subset of the poles of the original model and therefore retain any physical interpretation they might have, e.g. the phugoid mode in aircraft dynamics.

11.2.2 Residualization

In truncation, we discard all the states and dynamics associated with x_2. Suppose that instead of this we simply set $\dot{x}_2 = 0$, i.e. we *residualize* x_2, in the state-space equations. One can then solve for x_2 in terms of x_1 and u, and back substitution of x_2 then gives

$$
\dot{x}_1 = (A_{11} - A_{12}A_{22}^{-1}A_{21})x_1 + (B_1 - A_{12}A_{22}^{-1}B_2)u \tag{11.5}
$$

$$y \;=\; (C_1 - C_2 A_{22}^{-1} A_{21}) x_1 + (D - C_2 A_{22}^{-1} B_2) u \tag{11.6}$$

Let us assume A_{22} is invertible and define

$$A_r \;\stackrel{\triangle}{=}\; A_{11} - A_{12} A_{22}^{-1} A_{21} \tag{11.7}$$

$$B_r \;\stackrel{\triangle}{=}\; B_1 - A_{12} A_{22}^{-1} B_2 \tag{11.8}$$

$$C_r \;\stackrel{\triangle}{=}\; C_1 - C_2 A_{22}^{-1} A_{21} \tag{11.9}$$

$$D_r \;\stackrel{\triangle}{=}\; D - C_2 A_{22}^{-1} B_2 \tag{11.10}$$

The reduced order model $G_a(s) \stackrel{s}{=} (A_r, B_r, C_r, D_r)$ is called a residualization of $G(s) \stackrel{s}{=} (A, B, C, D)$. Usually (A, B, C, D) will have been put into Jordan form, with the eigenvalues ordered so that x_2 contains the fast modes. Model reduction by residualization is then equivalent to *singular perturbational approximation*, where the derivatives of the fastest states are allowed to approach zero with some parameter ϵ. An important property of residualization is that it preserves the steady-state gain of the system, $G_a(0) = G(0)$. This should be no surprise since the residualization process sets derivatives to zero, which are zero anyway at steady-state. But it is in stark contrast to truncation which retains the system behaviour at infinite frequency. This contrast between truncation and residualization follows from the simple bilinear relationship $s \rightarrow \frac{1}{s}$ which relates the two (e.g. Liu and Anderson, 1989).

It is clear from the discussion above that truncation is to be preferred when accuracy is required at high frequencies, whereas residualization is better for low-frequency modelling.

Both methods depend to a large extent on the original realization and we have suggested the use of the Jordan form. A better realization, with many useful properties, is the balanced realization which will be considered next.

11.3   Balanced realizations

In words only: a *balanced realization* is an asymptotically stable minimal realization in which the controllability and observability Gramians are equal and diagonal.

More formally: let (A, B, C, D) be a minimal realization of a stable, rational transfer function $G(s)$, then (A, B, C, D) is called *balanced* if the solutions to the following Lyapunov equations

$$AP + PA^T + BB^T \;=\; 0 \tag{11.11}$$

$$A^T Q + QA + C^T C \;=\; 0 \tag{11.12}$$

are $P = Q = \mathrm{diag}(\sigma_1, \sigma_2, \ldots, \sigma_n) \stackrel{\triangle}{=} \Sigma$, where $\sigma_1 \geq \sigma_2 \geq \cdots \geq \sigma_n > 0$. P and Q are the controllability and observability Gramians, also defined by

$$P \;\stackrel{\triangle}{=}\; \int_0^\infty e^{At} BB^T e^{A^T t} dt \tag{11.13}$$

$$Q \;\stackrel{\triangle}{=}\; \int_0^\infty e^{A^T t} C^T C e^{At} dt \tag{11.14}$$

Σ is therefore simply referred to as the Gramian of $G(s)$. The σ_i's are the ordered Hankel singular values of $G(s)$, more generally defined as $\sigma_i \triangleq \lambda_i^{\frac{1}{2}}(PQ), i = 1, \ldots, n$. Notice that $\sigma_1 = \|G\|_H$, the Hankel norm of $G(s)$.

Any minimal realization of a stable transfer function can be balanced by a simple state similarity transformation, and routines for doing this are available in Matlab. For further details on computing balanced realizations, see Laub et al. (1987). Note that balancing does not depend on D.

So what is so special about a balanced realization? In a balanced realization the value of each σ_i is associated with a state x_i of the balanced system. And the size of σ_i is a relative measure of the contribution that x_i makes to the input–output behaviour of the system; also see the discussion on page 161. Therefore if $\sigma_1 \gg \sigma_2$, then the state x_1 affects the input–output behaviour much more than x_2, or indeed any other state because of the ordering of the σ_i. After balancing a system, each state is just as controllable as it is observable, and a measure of a state's joint observability and controllability is given by its associated Hankel singular value. This property is fundamental to the model reduction methods in the remainder of this chapter which work by removing states having little effect on the system's input–output behaviour.

11.4   Balanced truncation and balanced residualization

Let the balanced realization (A, B, C, D) of $G(s)$ and the corresponding Σ be partitioned compatibly as

$$A = \begin{bmatrix} A_{11} & A_{12} \\ A_{21} & A_{22} \end{bmatrix}, \quad B = \begin{bmatrix} B_1 \\ B_2 \end{bmatrix}, \quad C = \begin{bmatrix} C_1 & C_2 \end{bmatrix} \tag{11.15}$$

$$\Sigma = \begin{bmatrix} \Sigma_1 & 0 \\ 0 & \Sigma_2 \end{bmatrix} \tag{11.16}$$

where $\Sigma_1 = \mathrm{diag}(\sigma_1, \sigma_2, \ldots, \sigma_k)$, $\Sigma_2 = \mathrm{diag}(\sigma_{k+1}, \sigma_{k+2}, \ldots, \sigma_n)$ and $\sigma_k > \sigma_{k+1}$.

Balanced truncation. The reduced order model given by (A_{11}, B_1, C_1, D) is called a *balanced truncation* of the full-order system $G(s)$. This idea of balancing the system and then discarding the states corresponding to small Hankel singular values was first introduced by Moore (1981). A balanced truncation is also a balanced realization (Pernebo and Silverman, 1982), and the $\mathcal{H}_\infty$ norm of the error between $G(s)$ and the reduced order system is bounded by twice the sum of the last $n - k$ Hankel singular values, i.e. twice the trace of Σ_2 or simply "twice the sum of the tail" (Glover, 1984; Enns, 1984). For the case of repeated Hankel singular values, Glover (1984) shows that each repeated Hankel singular value is to be counted only once in calculating the sum.

A precise statement of the bound on the approximation error is given in Theorem 11.1 below.

Useful algorithms that compute balanced truncations without first computing a balanced realization have been developed by Tombs and Postlethwaite (1987) and Safonov and Chiang (1989). These still require the computation of the observability and controllability Gramians, which can be a problem if the system to be reduced is of very high order. In such cases the technique of Jaimoukha et al. (1992), based on computing approximate solutions to Lyapunov equations, is recommended.

Balanced residualization. In balanced truncation above, we discarded the least controllable and observable states corresponding to Σ_2. In balanced residualization, we simply set to zero the derivatives of all these states. The method was introduced by Fernando and Nicholson (1982) who called it a singular perturbational approximation of a balanced system. The resulting balanced residualization of $G(s)$ is (A_r, B_r, C_r, D_r) as given by the formulae (11.7)–(11.10).

Liu and Anderson (1989) have shown that balanced residualization enjoys the same error bound as balanced truncation. An alternative derivation of the error bound, more in the style of Glover (1984), is given by Samar et al. (1995). A precise statement of the error bound is given in the following theorem.

Theorem 11.1 *Let $G(s)$ be a stable rational transfer function with Hankel singular values $\sigma_1 > \sigma_2 > \cdots > \sigma_N$ where each σ_i has multiplicity r_i and let $G_a^k(s)$ be obtained by truncating or residualizing the balanced realization of $G(s)$ to the first $(r_1 + r_2 + \cdots + r_k)$ states. Then*

$$\|G(s) - G_a^k(s)\|_\infty \leq 2(\sigma_{k+1} + \sigma_{k+2} + \cdots + \sigma_N) \qquad (11.17)$$

The following two exercises are to emphasize that (i) balanced residualization preserves the steady-state gain of the system and (ii) balanced residualization is related to balanced truncation by the bilinear transformation $s \to s^{-1}$.

Exercise 11.1 * *The steady-state gain of a full-order balanced system (A, B, C, D) is $D - CA^{-1}B$. Show, by algebraic manipulation, that this is also equal to $D_r - C_r A_r^{-1} B_r$, the steady-state gain of the balanced residualization given by (11.7)–(11.10).*

Exercise 11.2 *Let $G(s)$ have a balanced realization $\left[\begin{array}{c|c} A & B \\ \hline C & D \end{array} \right]$, then*

$$\left[\begin{array}{c|c} A^{-1} & A^{-1}B \\ \hline -CA^{-1} & D - CA^{-1}B \end{array} \right]$$

is a balanced realization of $H(s) \triangleq G(s^{-1})$, and the Gramians of the two realizations are the same.

1. *Write down an expression for a balanced truncation $H_t(s)$ of $H(s)$.*
2. *Apply the reverse transformation $s^{-1} \to s$ to $H_t(s)$, and hence show that $G_r(s) \triangleq H_t(s^{-1})$ is a balanced residualization of $G(s)$ as defined by (11.7)–(11.10).*

11.5   Optimal Hankel norm approximation

In this approach to model reduction, the problem that is directly addressed is the following: given a stable model $G(s)$ of order (McMillan degree) n, find a reduced order model $G_h^k(s)$ of degree k such that the Hankel norm of the approximation error, $\|G(s) - G_h^k(s)\|_H$, is minimized.

The Hankel norm of any stable transfer function $E(s)$ is defined as

$$\|E(s)\|_H \triangleq \rho^{\frac{1}{2}}(PQ) \qquad (11.18)$$

where P and Q are the controllability and observability Gramians of $E(s)$. It is also the maximum Hankel singular value of $E(s)$. So in the optimization we seek an error which is in some sense closest to being completely unobservable and completely uncontrollable, which seems sensible. A more detailed discussion of the Hankel norm was given in Section 4.10.4 (page 160).

The Hankel norm approximation problem has been considered by many but especially Glover (1984). In Glover (1984) a complete treatment of the problem is given, including a closed-form optimal solution and a bound on the infinity norm of the approximation error. The infinity norm bound is of particular interest because it is better than that for balanced truncation and residualization.

The theorem below gives a particular construction for optimal Hankel norm approximations of square stable transfer functions.

Theorem 11.2 *Let $G(s)$ be a stable, square, transfer function $G(s)$ with Hankel singular values $\sigma_1 \geq \sigma_2 \geq \cdots \geq \sigma_k \geq \sigma_{k+1} = \sigma_{k+2} = \cdots = \sigma_{k+l} > \sigma_{k+l+1} \geq \cdots \geq \sigma_n > 0$, then an optimal Hankel norm approximation of order k, $G_h^k(s)$, can be constructed as follows.*

Let (A, B, C, D) be a balanced realization of $G(s)$ with the Hankel singular values reordered so that the Gramian matrix is

$$
\begin{aligned}
\Sigma &= \operatorname{diag}\left(\sigma_1, \sigma_2, \cdots, \sigma_k, \sigma_{k+l+1}, \cdots, \sigma_n, \sigma_{k+1}, \cdots, \sigma_{k+l}\right) \\
&\triangleq \operatorname{diag}\left(\Sigma_1, \sigma_{k+1} I\right)
\end{aligned}
\tag{11.19}
$$

Partition (A, B, C, D) to conform with Σ:

$$
A = \begin{bmatrix} A_{11} & A_{12} \\ A_{21} & A_{22} \end{bmatrix}, \; B = \begin{bmatrix} B_1 \\ B_2 \end{bmatrix}, \; C = \begin{bmatrix} C_1 & C_2 \end{bmatrix}
\tag{11.20}
$$

Define $(\widehat{A}, \widehat{B}, \widehat{C}, \widehat{D})$ by

$$
\widehat{A} \triangleq \Gamma^{-1}\left(\sigma_{k+1}^2 A_{11}^T + \Sigma_1 A_{11} \Sigma_1 - \sigma_{k+1} C_1^T U B_1^T\right)
\tag{11.21}
$$

$$
\widehat{B} \triangleq \Gamma^{-1}\left(\Sigma_1 B_1 + \sigma_{k+1} C_1^T U\right)
\tag{11.22}
$$

$$
\widehat{C} \triangleq C_1 \Sigma_1 + \sigma_{k+1} U B_1^T
\tag{11.23}
$$

$$
\widehat{D} \triangleq D - \sigma_{k+1} U
\tag{11.24}
$$

where U is a unitary matrix satisfying

$$
B_2 = -C_2^T U
\tag{11.25}
$$

and

$$
\Gamma \triangleq \Sigma_1^2 - \sigma_{k+1}^2 I
\tag{11.26}
$$

The matrix $\widehat{A}$ has k "stable" eigenvalues (in the open LHP); the remaining ones are in the open RHP. Then

$$
G_h^k(s) + F(s) \stackrel{s}{=} \left[\begin{array}{c|c} \widehat{A} & \widehat{B} \\ \hline \widehat{C} & \widehat{D} \end{array}\right]
\tag{11.27}
$$

where $G_h^k(s)$ is a stable optimal Hankel norm approximation of order k, and $F(s)$ is an anti-stable (all poles in the open RHP) transfer function of order $n - k - l$. The Hankel norm of the error between G and the optimal approximation G_h^k is equal to the $(k+1)$'th Hankel singular value of G:

$$
\|G - G_h^k\|_H = \sigma_{k+1}(G)
\tag{11.28}
$$

Remark 1 The $k + 1$'th Hankel singular value is generally not repeated, but the possibility is included in the theory for completeness.

Remark 2 The order k of the approximation can be selected either directly, or indirectly by choosing the "cut-off" value σ_k for the included Hankel singular values. In the latter case, one often looks for large "gaps" in the relative magnitude, σ_k/σ_{k+1}.

Remark 3 There is an infinite number of unitary matrices U satisfying (11.25); one choice is $U = -C_2(B_2^T)^\dagger$.

Remark 4 If $\sigma_{k+1} = \sigma_n$, i.e. only the smallest Hankel singular value is deleted, then $F = 0$, otherwise $(\widehat{A}, \widehat{B}, \widehat{C}, \widehat{D})$ has a non-zero anti-stable part and G_h^k has to be separated from F.

Remark 5 When the order k is chosen to be zero, G_h^k is a constant matrix and $(\widehat{A}, \widehat{B}, \widehat{C}, \widehat{D} - G_h^k) = F(s)$, which is entirely anti-stable. In this case, $\|G(s) - G_h^k(s)\|_H = \|G(s) - F(s)\|_{\mathcal{L}_\infty} = \|G^T(-s) - F^T(-s)\|_{\mathcal{L}_\infty}$. The last inequality follows since the $\mathcal{L}_\infty$ norm of a system is equal to the $\mathcal{L}_\infty$ norm of its mirror image across the imaginary axis. This special case can be interpreted as approximating a stable system by an unstable system or an unstable system by a stable one. The latter problem is alternatively known as the *Nehari extension problem*, which was used extensively in the early solutions of $\mathcal{H}_\infty$ optimal controller design problems (Francis, 1987); also see the robust stabilization problem on page 368.

Remark 6 For non-square systems, an optimal Hankel norm approximation can be obtained by first augmenting $G(s)$ with zero to form a square system. For example, if $G(s)$ is flat, define $\bar{G}(s) \triangleq \begin{bmatrix} G(s) \\ 0 \end{bmatrix}$ which is square, and let $\bar{G}_h(s) = \begin{bmatrix} G_1(s) \\ G_2(s) \end{bmatrix}$ be a k'th-order optimal Hankel norm approximation of $\bar{G}(s)$ such that $\|\bar{G}(s) - \bar{G}_h(s)\|_H = \sigma_{k+1}(\bar{G}(s))$. Then

$$\sigma_{k+1}(G(s)) \leq \|G - G_1\|_H \leq \|\bar{G} - \bar{G}_h\|_H = \sigma_{k+1}(\bar{G}) = \sigma_{k+1}(G)$$

Consequently, this implies that $\|G - G_1\|_H = \sigma_{k+1}(G)$ and $G_1(s)$ is an optimal Hankel norm approximation of $G(s)$.

Remark 7 The Hankel norm of a system does not depend on the D-matrix in the system's state-space realization. The choice of the D-matrix in G_h^k is therefore arbitrary except when $F = 0$, in which case it is equal to $\widehat{D}$.

Remark 8 The infinity norm does depend on the D-matrix, and therefore the D-matrix of G_h^k can be chosen to reduce the infinity norm of the approximation error (without changing the Hankel norm). Glover (1984) showed that through a particular choice of D, called D_o, the following bound could be obtained:

$$\|G - G_h^k - D_o\|_\infty \leq \sigma_{k+1} + \delta \tag{11.29}$$

where

$$\delta \triangleq \|F - D_o\|_\infty \leq \sum_{i=1}^{n-k-l} \sigma_i(F(-s)) \leq \sum_{i=1}^{n-k-l} \sigma_{i+k+l}(G(s)) \tag{11.30}$$

This results in an infinity norm bound on the approximation error, $\delta \leq \sigma_{k+l+1} + \cdots + \sigma_n$, which is equal to the "sum of the tail" or less since the Hankel singular value σ_{k+1}, which may be repeated, is only included once. Recall that the bound for the error in balanced truncation and balanced residualization is *twice* the "sum of the tail".

11.6   Reduction of unstable models

Balanced truncation, balanced residualization and optimal Hankel norm approximation only apply to stable models. In this section we will briefly present two approaches for reducing the order of an unstable model.

11.6.1   Stable part model reduction

Enns (1984) and Glover (1984) proposed that the unstable model could first be decomposed into its stable and anti-stable parts. Namely

$$G(s) = G_u(s) + G_s(s) \tag{11.31}$$

where $G_u(s)$ has all its poles in the closed RHP and $G_s(s)$ has all its poles in the open LHP. Balanced truncation, balanced residualization or optimal Hankel norm approximation can then be applied to the stable part $G_s(s)$ to find a reduced order approximation $G_{sa}(s)$. This is then added to the anti-stable part to give

$$G_a(s) = G_u(s) + G_{sa}(s) \tag{11.32}$$

as an approximation to the full-order model $G(s)$.

11.6.2   Coprime factor model reduction

The coprime factors of an unstable transfer function $G(s)$ are stable, and therefore we could reduce the order of these factors using balanced truncation, balanced residualization or optimal Hankel norm approximation, as proposed in the following scheme (McFarlane and Glover, 1990):

- Let $G(s) = M^{-1}(s)N(s)$, where $M(s)$ and $N(s)$ are stable left coprime factors of $G(s)$.
- Approximate $[N\ M]$ of degree n by $[N_a\ M_a]$ of degree $k < n$, using balanced truncation, balanced residualization or optimal Hankel norm approximation.
- Realize the reduced order transfer function $G_a(s)$, of degree k, by $G_a(s) = M_a^{-1}N_a$.

A dual procedure could be written down based on a right coprime factorization of $G(s)$.

For related work in this area, we refer the reader to Anderson and Liu (1989) and Meyer (1987). In particular, Meyer (1987) has derived the following result:

Theorem 11.3 *Let (N, M) be a normalized left coprime factorization of $G(s)$ of degree n. Let $[N_a\ M_a]$ be a degree k balanced truncation of $[N\ M]$ which has Hankel singular values $\sigma_1 \geq \sigma_2 \geq \cdots \geq \sigma_k > \sigma_{k+1} \geq \cdots \geq \sigma_n > 0$. Then (N_a, M_a) is a normalized left coprime factorization of $G_a = M_a^{-1}N_a$, and $[N_a\ M_a]$ has Hankel singular values $\sigma_1, \sigma_2, \ldots, \sigma_k$.*

Exercise 11.3* *Is Theorem 11.3 true, if we replace balanced truncation by balanced residualization?*

11.7   Model reduction using Matlab

The commands in Table 11.1 from the Matlab Robust Control toolbox may be used to perform model reduction for both stable and unstable systems. Note that most reduction commands in

Matlab automatically separate out the unstable part and then add it to the stable part after its reduction.

Table 11.1: Matlab commands for model reduction

```
% Uses Robust Control toolbox
% Remove fast stable modes
p=pole(sys);
sysd=canon(sys);                    % Diagonalize the system
elim=(abs(p)>tol) & (real(p)<0);    % and identify fast stable modes
syst=modred(sysd,elim,'t');         % then: Truncate fast modes.
sysr=modred(sysd,elim);             % or: Residualize fast modes.
% Balanced model reduction
% Works for stable modes, so use k > number of unstable modes
n=size(sys.A,1);
sysbt=balancmr(sys,k);              % kth order balanced truncation.
sysbr=modred(balreal(sys),k+1:n);   % or: kth order balanced residualization.
sysbh=hankelmr(sys,k);              % or: kth order optimal Hankel norm approx.
% Using coprime factors (works also for unstable modes)
nu=size(sys,2);
sysct=ncfmr(sys,k);                 % balanced truncation of coprime factors.
[sysc,cinfo]=ncfmr(sys,n);          % or: obtain coprime factors of system
syscr=modred(cinfo.GL,k+1:n);       % residualize.
syscrm=minreal(inv(syscr(:,nu+1...  % and obtain kth order model.
          :end))*syscr(:,1:nu));
sysch=hankelmr(cinfo.GL,k);         % or: optimal Hankel norm approximation.
syschm=minreal(inv(sysch(:,nu+1...  % and obtain kth order model.
          :end))*sysch(:,1:nu));
```

11.8 Two practical examples

In this section, we make comparisons between the three main model reduction techniques presented by applying them to two practical examples. The first example is on the reduction of a plant model and the second considers the reduction of a two degrees-of-freedom controller. Our presentation is similar to that in Samar et al. (1995).

11.8.1 Reduction of a gas turbine aero-engine model

For the first example, we consider the reduction of a stable model of a Rolls-Royce Spey gas turbine engine. This engine will be considered again in Chapter 13. The model has 3 inputs, 3 outputs, and 15 states. Inputs to the engine are fuel flow, variable nozzle area and an inlet guide vane with a variable angle setting. The outputs to be controlled are the high-pressure compressor's spool speed, the ratio of the high-pressure compressor's outlet pressure to engine inlet pressure, and the low-pressure compressor's exit Mach number measurement. The model describes the engine at 87% of maximum thrust with sea-level static conditions. The Hankel singular values for the 15-state model are listed in Table 11.2. Recall that the $\mathcal{L}_\infty$ error bounds after reduction are "twice the sum of the tail" for balanced residualization and balanced truncation and the "sum of the tail" for optimal Hankel norm approximation. Based on this we decided to reduce the model to 6 states.

Figure 11.1 shows the singular values (not Hankel singular values) of the reduced and full-order models plotted against frequency for the residualized, truncated and optimal Hankel norm approximated cases respectively. The D-matrix used for optimal Hankel norm approximation is such that the error bound given in (11.29) is met. It can be seen that

Table 11.2: Hankel singular values of the gas turbine aero-engine model

1) 2.0005e+01	6) 6.2964e-01	11) 1.3621e-02
2) 4.0464e+00	7) 1.6689e-01	12) 3.9967e-03
3) 2.7546e+00	8) 9.3407e-02	13) 1.1789e-03
4) 1.7635e+00	9) 2.2193e-02	14) 3.2410e-04
5) 1.2965e+00	10) 1.5669e-02	15) 3.3073e-05

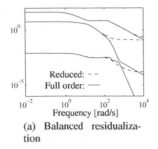 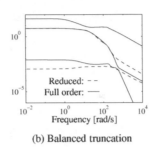 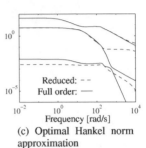

(a) Balanced residualization

(b) Balanced truncation

(c) Optimal Hankel norm approximation

Figure 11.1: Singular values for model reductions of the aero-engine from 15 to 6 states

the residualized system matches perfectly at steady-state. The singular values of the error system $(G - G_a)$, for each of the three approximations, are shown in Figure 11.2(a). The

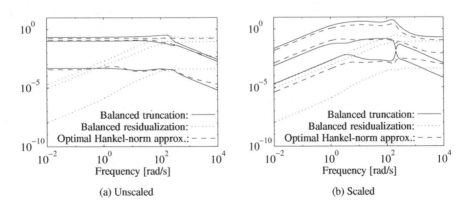

(a) Unscaled

(b) Scaled

Figure 11.2: Singular values for the scaled and unscaled error systems

$\mathcal{H}_\infty$ norm of the error system is computed to be 0.295 for balanced residualization and occurs at 208 rad/s; the corresponding error norms for balanced truncation and optimal Hankel norm approximation are 0.324 and 0.179 occurring at 169 rad/s and 248 rad/s, respectively. The theoretical upper bounds for these error norms are 0.635 (twice the sum of the tail) for residualization and truncation, and 0.187 (using (11.29)) for optimal Hankel norm approximation respectively. It should be noted that the plant under consideration is desired to have a closed-loop bandwidth of around 10 rad/s. The error around this frequency, therefore, should be as small as possible for good controller design. Figure 11.2(a) shows that the error for balanced residualization is the smallest in this frequency range.

Steady-state gain preservation. It is sometimes desirable to have the steady-state gain of the reduced plant model the same as the full-order model. For example, this is the case if we want to use the model for feedforward control. The truncated and optimal Hankel norm approximated systems do not preserve the steady-state gain and have to be scaled, i.e. the model approximation G_a is replaced by $G_a W_s$, where $W_s = G_a(0)^{-1}G(0)$, G being the full-order model. The scaled system no longer enjoys the bounds guaranteed by these methods and $\|G - G_a W_s\|_\infty$ can be quite large as is shown in Figure 11.2(b). Note that the residualized system does not need scaling, and the error system for this case has been shown again only for ease of comparison. The $\mathcal{H}_\infty$ norms of these errors are computed and are found to degrade to 5.71 (at 151 rad/s) for the scaled truncated system and 2.61 (at 168.5 rad/s) for the scaled optimal Hankel norm approximated system. The truncated and Hankel norm approximated systems are clearly worse after scaling since the errors in the critical frequency range around crossover become large despite the improvement at steady state. Hence residualization is to be preferred over these other techniques whenever good low-frequency matching is desired.

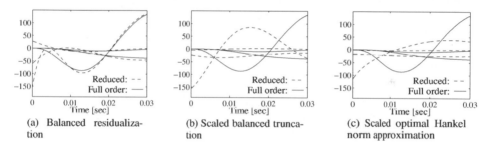

(a) Balanced residualization

(b) Scaled balanced truncation

(c) Scaled optimal Hankel norm approximation

Figure 11.3: Aero-engine: impulse responses (second input)

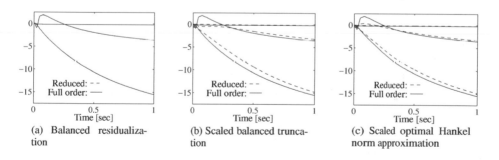

(a) Balanced residualization

(b) Scaled balanced truncation

(c) Scaled optimal Hankel norm approximation

Figure 11.4: Aero-engine: step responses (second input)

Impulse and step responses from the second input to all the outputs for the three reduced systems (with the truncated and optimal Hankel norm approximated systems scaled) are shown in Figures 11.3 and 11.4, respectively. The responses for the other inputs were found to be similar. The simulations confirm that the residualized model's response is closer to the full-order model's response.

11.8.2   Reduction of an aero-engine controller

We now consider reduction of a stable two degrees-of-freedom H_∞ loop-shaping controller. The plant for which the controller is designed is the full-order gas turbine engine model described in Section 11.8.1 above.

A robust controller was designed using the procedure outlined in Section 9.4.3; see Figure 9.21 which describes the design problem. $T_{\text{ref}}(s)$ is the desired closed-loop transfer function, ρ is a design parameter, $G_s = M_S^{-1} N_s$ is the shaped plant and $(\Delta_{N_s}, \Delta_{M_S})$ are perturbations on the normalized coprime factors representing uncertainty. We denote the actual closed-loop transfer function (from β to y) by $T_{y\beta}$.

The controller $K = [K_1 \ K_2]$, which excludes the loop-shaping weight W_1 (which includes 3 integral action states), has 6 inputs (because of the two degrees-of-freedom structure), 3 outputs and 24 states. It has not been scaled (i.e. the steady-state value of $T_{y\beta}$ has not been matched to that of T_{ref} by scaling the prefilter). It is reduced to 7 states in each of the cases that follow.

Let us first compare the magnitude of $T_{y\beta}$ with that of the specified model T_{ref}. By magnitude, we mean singular values. These are shown in Figure 11.5(a). The infinity norm of

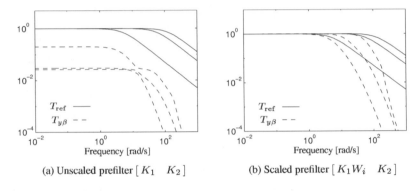

(a) Unscaled prefilter $[\ K_1 \quad K_2\ ]$ (b) Scaled prefilter $[\ K_1 W_i \quad K_2\ ]$

Figure 11.5: Singular values of T_{ref} (solid) and $T_{y\beta}$ (dashed)

the difference $T_{y\beta} - T_{\text{ref}}$ is computed to be 0.974 and occurs at 8.5 rad/s. Note that we have $\rho = 1$ and the γ achieved in the $\mathcal{H}_\infty$ optimization is 2.32, so that $\|T_{y\beta} - T_{\text{ref}}\|_\infty \le \gamma \rho^{-2}$ as required; see (9.81). The prefilter is now scaled so that $T_{y\beta}$ matches T_{ref} exactly at steady-state, i.e. we replace K_1 by $K_1 W_i$ where $W_i = T_{y\beta}(0)^{-1} T_{\text{ref}}(0)$. It is argued by Hoyle et al. (1991) that this scaling produces better model matching at all frequencies, because the $\mathcal{H}_\infty$ optimization process has already given $T_{y\beta}$ the same magnitude frequency response shape as the model T_{ref}. The scaled transfer function is shown in Figure 11.5(b), and the infinity norm of the difference $(T_{y\beta} - T_{\text{ref}})$ computed to be 1.44 (at 46 rad/s). It can be seen that this scaling has not degraded the infinity norm of the error significantly as was claimed by Hoyle et al. (1991). To ensure perfect steady-state tracking the controller is always scaled in this way. We are now in a position to discuss ways of reducing the controller. We will look at the following two approaches:

1. The scaled controller $[\ K_1 W_i \quad K_2\ ]$ is reduced. A balanced residualization of this controller preserves the controller's steady-state gain and would not need to be scaled again. Reductions via truncation and optimal Hankel norm approximation techniques,

however, lose the steady-state gain. The prefilters of these reduced controllers would therefore need to be rescaled to match $T_{\text{ref}}(0)$.

2. The full-order controller $[\,K_1 \quad K_2\,]$ is directly reduced without first scaling the prefilter. In this case, scaling is done after reduction.

We now consider the first approach. A balanced residualization of $[\,K_1 W_i \quad K_2\,]$ is obtained. The theoretical upper bound on the $\mathcal{H}_\infty$ norm of the error (twice the sum of the tail) is 0.698, i.e.

$$\| K_1 W_i - (K_1 W_i)_a \quad K_2 - K_{2a} \|_\infty \leq 0.698 \tag{11.33}$$

where the subscript a refers to the low-order approximation. The actual error norm is computed to be 0.365. $T_{y\beta}$ for this residualization is computed and its magnitude plotted in Figure 11.6(a). The $\mathcal{H}_\infty$ norm of the difference $(T_{y\beta} - T_{\text{ref}})$ is computed to be 1.44 (at 43

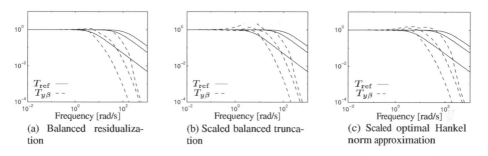

(a) Balanced residualization

(b) Scaled balanced truncation

(c) Scaled optimal Hankel norm approximation

Figure 11.6: Singular values of T_{ref} and $T_{y\beta}$ for reduced $[\,K_1 W_i \quad K_2\,]$

rad/s). This value is very close to that obtained with the full-order controller $[\,K_1 W_i \quad K_2\,]$, and so the closed-loop response of the system with this reduced controller is expected to be very close to that with the full-order controller. Next $[\,K_1 W_i \quad K_2\,]$ is reduced via balanced truncation. The bound given by (11.33) still holds. The steady-state gain, however, falls below the adjusted level, and the prefilter of the truncated controller is thus scaled. The bound given by (11.33) can no longer be guaranteed for the prefilter (it is in fact found to degrade to 3.66), but it holds for $K_2 - K_{2a}$. Singular values of T_{ref} and $T_{y\beta}$ for the scaled truncated controller are shown in Figure 11.6(b). The infinity norm of the difference is computed to be 1.44 and this maximum occurs at 46 rad/s. Finally $[\,K_1 W_i \quad K_2\,]$ is reduced by optimal Hankel norm approximation. The following error bound is theoretically guaranteed:

$$\| K_1 W_i - (K_1 W_i)_a \quad K_2 - K_{2a} \|_\infty \leq 0.189 \tag{11.34}$$

Again the reduced prefilter needs to be scaled and the above bound can no longer be guaranteed; it actually degrades to 1.87. Magnitude plots of $T_{y\beta}$ and T_{ref} are shown in Figure 11.6(c), and the infinity norm of the difference is computed to be 1.43 and occurs at 43 rad/s.

It has been observed that both balanced truncation and optimal Hankel norm approximation cause a lowering of the system steady-state gain. In the process of adjustment of these steady-state gains, the infinity norm error bounds are destroyed. In the case of our two degrees-of-freedom controller, where the prefilter has been optimized to give closed-loop responses within a tolerance of a chosen ideal model, large deviations may be incurred. Closed-loop

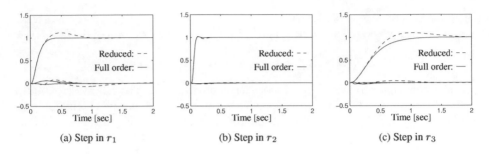

(a) Step in r_1 (b) Step in r_2 (c) Step in r_3

Figure 11.7: Closed-loop step responses: $[\, K_1 W_i \quad K_2 \,]$ balanced residualized

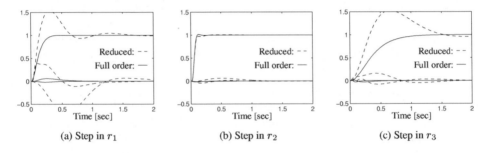

(a) Step in r_1 (b) Step in r_2 (c) Step in r_3

Figure 11.8: Closed-loop step responses: $[\, K_1 W_i \quad K_2 \,]$ balanced truncated

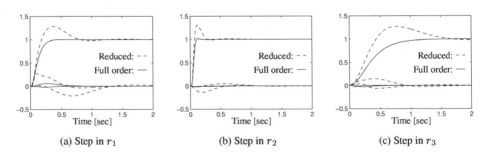

(a) Step in r_1 (b) Step in r_2 (c) Step in r_3

Figure 11.9: Closed-loop step responses: $[\, K_1 W_i \quad K_2 \,]$ optimal Hankel norm approximated and rescaled

responses for the three reduced controllers discussed above are shown in Figures 11.7, 11.8 and 11.9.

It is seen that the residualized controller performs much closer to the full-order controller and exhibits better performance in terms of interactions and overshoots. It may not be possible to use the other two reduced controllers if the deviation from the specified model becomes larger than the allowable tolerance, in which case the number of states by which the controller is reduced would probably have to be reduced. It should also be noted from (11.33) and (11.34) that the guaranteed bound for $K_2 - K_{2a}$ is lowest for optimal Hankel norm approximation.

Let us now consider the second approach. The controller $[K_1 \quad K_2]$ obtained from the H_∞ optimization algorithm is reduced directly. The theoretical upper bound on the error for balanced residualization and truncation is

$$\| K_1 - K_{1a} \quad K_2 - K_{2a} \|_\infty \leq 0.165 \qquad (11.35)$$

The residualized controller retains the steady-state gain of $[K_1 \quad K_2]$. It is therefore scaled with the same W_i as was required for scaling the prefilter of the full-order controller. Singular values of T_{ref} and $T_{y\beta}$ for this reduced controller are shown in Figure 11.10(a), and the infinity norm of the difference was computed to be 1.50 at 44 rad/s. $[K_1 \quad K_2]$ is next

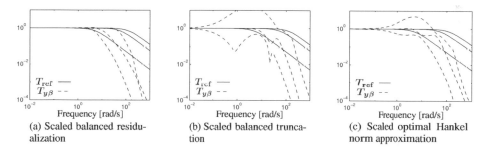

(a) Scaled balanced residualization

(b) Scaled balanced truncation

(c) Scaled optimal Hankel norm approximation

Figure 11.10: Singular values of T_{ref} and $T_{y\beta}$ for reduced $[K_1 \quad K_2]$

truncated. The steady-state gain of the truncated controller is lower than that of $[K_1 \quad K_2]$, and it turns out that this has the effect of reducing the steady-state gain of $T_{y\beta}$. Note that the steady-state gain of $T_{y\beta}$ is already less than that of T_{ref} (Figure 11.5). Thus in scaling the prefilter of the truncated controller, the steady-state gain has to be pulled up from a lower level as compared with the previous (residualized) case. This causes greater degradation at other frequencies. The infinity norm of $(T_{y\beta} - T_{ref})$ in this case is computed to be 25.3 and occurs at 3.4 rad/s (see Figure 11.10(b)). Finally $[K_1 \quad K_2]$ is reduced by optimal Hankel norm approximation. The theoretical bound given in (11.29) is computed and found to be 0.037, i.e. we have

$$\| K_1 - K_{1a} \quad K_2 - K_{2a} \|_\infty \leq 0.037 \qquad (11.36)$$

The steady-state gain falls once more in the reduction process, and again a larger scaling is required. Singular value plots for $T_{y\beta}$ and T_{ref} are shown in Figure 11.10(c). $\|T_{y\beta} - T_{ref}\|_\infty$ is computed to be 4.5 and occurs at 5.1 rad/s.

Some closed-loop step response simulations are shown in Figures 11.11, 11.12 and 11.13. It can be seen that the truncated and Hankel norm approximated systems have deteriorated

to an unacceptable level. Only the residualized system maintains an acceptable level of
performance.

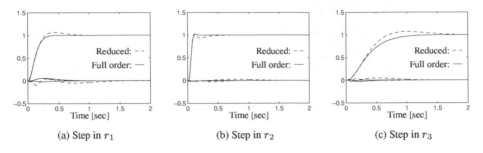

(a) Step in r_1 (b) Step in r_2 (c) Step in r_3

Figure 11.11: Closed-loop step responses: $[\,K_1 \quad K_2\,]$ balanced residualized and scaled

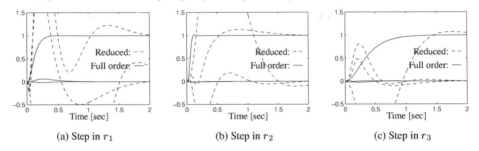

(a) Step in r_1 (b) Step in r_2 (c) Step in r_3

Figure 11.12: Closed-loop step responses: $[\,K_1 \quad K_2\,]$ balanced truncated and scaled

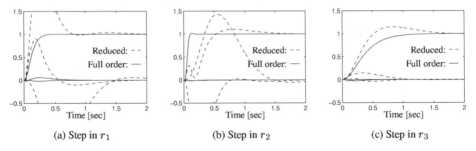

(a) Step in r_1 (b) Step in r_2 (c) Step in r_3

Figure 11.13: Closed-loop step responses: $[\,K_1 \quad K_2\,]$ optimal Hankel norm approximated and scaled

We have seen that the first approach yields better model matching, though at the expense
of a larger infinity norm bound on $K_2 - K_{2a}$ (compare (11.33) and (11.35), or (11.34) and
(11.36)). We have also seen how the scaling of the prefilter in the first approach gives poorer
performance for the truncated and optimal Hankel norm approximated controllers, relative to
the residualized one.

In the second case, all the reduced controllers need to be scaled, but a "larger" scaling is
required for the truncated and optimal Hankel norm approximated controllers. There appears
to be no formal proof of this observation. It is, however, intuitive in the sense that controllers
reduced by these two methods yield poorer model matching at steady-state as compared with
that achieved by the full-order controller. A larger scaling is therefore required for them
than is required by the full-order or residualized controllers. In any case, this larger scaling

gives poorer model matching at other frequencies, and only the residualized controller's performance is deemed acceptable.

11.9 Conclusion

We have presented and compared three main methods for model reduction based on balanced realizations: balanced truncation, balanced residualization and optimal Hankel norm approximation.

Residualization, unlike truncation and optimal Hankel norm approximation, preserves the steady-state gain of the system, and, like truncation, it is simple and computationally inexpensive. It is observed that truncation and optimal Hankel norm approximation perform better at high frequencies, whereas residualization performs better at low and medium frequencies, i.e. up to the critical frequencies. Thus for plant model reduction, where models are not accurate at high frequencies to start with, residualization would seem to be a better option. Further, if the steady-state gains are to be kept unchanged, truncated and optimal Hankel norm approximated systems require scaling, which may result in large errors. In such a case, too, residualization would be a preferred choice.

Frequency-weighted model reduction has been the subject of numerous papers over the past few years. The idea is to emphasize frequency ranges where better matching is required. This, however, has been observed to have the effect of producing larger errors (greater mismatching) at other frequencies (Anderson, 1986; Enns, 1984). In order to get good steady-state matching, a relatively large weight would have to be used at steady-state, which would cause poorer matching elsewhere. The choice of weights is not straightforward, and an error bound is available only for weighted Hankel norm approximation. The computation of the bound is also not as easy as in the unweighted case (Anderson and Liu, 1989). Balanced residualization can, in this context, be seen as a reduction scheme with implicit low- and medium-frequency weighting.

For controller reduction, we have shown in a two degrees-of-freedom example the importance of scaling and steady-state gain matching.

In general, steady-state gain matching may not be crucial, but the matching should usually be good near the desired closed-loop bandwidth. Balanced residualization has been seen to perform close to the full-order system in this frequency range. Good approximation at high frequencies may also sometimes be desired. In such a case, using truncation or optimal Hankel norm approximation with appropriate frequency weightings may yield better results.

Finally, for controller reduction it is important that any subsequent loss in closed-loop performance is minimized. This problem has been addressed by Goddard (1995).

12

LINEAR MATRIX INEQUALITIES

This chapter gives an introduction to the use of linear matrix inequalities (LMIs) in the numerical solution of some important control problems. LMI problems are defined and tools described for transforming such problems into suitable formats for solution. The chapter ends with a case study on anti-windup compensator synthesis.

12.1 Introduction to LMI problems

LMIs are matrix inequalities which are linear (or affine) in a set of *matrix* variables. Many problems in control theory can be stated in terms of LMIs and their existence can be traced back over 100 years to the work of Lyapunov. However, until relatively recently, there were few (if any) routines available to solve LMIs numerically. During the past 10–15 years, the development of sophisticated numerical routines has made it possible to solve LMIs in a reasonably efficient manner. These routines exploit the convexity of LMI problems in order to obtain reliable numerical calculations.

From a control engineering perspective, one of the main attractions of LMIs is that they can be used to solve problems which involve several matrix variables, and, moreover, different structures can be imposed on these matrix variables. Another attractive feature of LMI methods is that they are flexible, so it is often relatively straightforward to pose a variety of problems as LMI problems, amenable to LMI methods. Furthermore, in many cases the use of LMIs can remove restrictions associated with conventional methods and aid their extension to more general scenarios. Often LMI methods can be applied in instances where conventional methods either fail or struggle to find a solution.

Another advantage of LMIs, at least in a pedagogical sense, is that they are able to *unite* many previous results in a common framework. This can enable one to obtain additional insight into established areas. Some important controller design problems, which have been shown to be solvable using LMIs, include: $\mathcal{H}_\infty$ controller design (Gahinet and Apkarian, 1994; Iwasaki and Skelton, 1994), $\mathcal{H}_2$ controller design (e.g. Sato and Liu, 1999), mixed $\mathcal{H}_2/\mathcal{H}_\infty$ optimal controller design (e.g. Khargonekar and Rotea, 1991), pole placement (e.g. Chilali and Gahinet, 1996) and robust model predictive control (e.g. Kothare et al., 1996). This list is by no means exhaustive and we encourage the reader to refer to the book by Boyd et al. (1994) for an overview of control-theoretic problems that can be posed as LMIs and for a more complete exposure to LMIs.

Multivariable Feedback Control: Analysis and Design. Second Edition
S. Skogestad and I. Postlethwaite   © 2005, 2006 John Wiley & Sons, Ltd

The main contents of this chapter are available as a University of Leicester technical report (Turner et al., 2004) which draws heavily on the material contained within the book by Boyd et al. (1994) and the Matlab LMI control toolbox by Gahinet et al. (1995). Note that the LMI control toolbox is now part of the Robust Control toolbox in Matlab.

12.1.1   Fundamental LMI properties

A notion central to the understanding of matrix inequalities is *definiteness*. In particular, a real square matrix Q is defined to be *positive definite* if

$$x^T Q x > 0 \quad \forall x \neq 0 \tag{12.1}$$

and Q is said to be *positive semi-definite* if

$$x^T Q x \geq 0 \quad \forall x \tag{12.2}$$

It is common practice to write $Q > 0$ ($Q \geq 0$) to indicate that Q is positive (semi-)definite. Likewise, a matrix $P = -Q$ is said to be *negative (semi-)definite* if Q is positive (semi-)definite and to indicate negative (semi-)definiteness, we write $P < 0$ ($P \leq 0$).

Notice that any real square matrix Q can be written as

$$Q = \left(\frac{Q + Q^T}{2} \right) + \left(\frac{Q - Q^T}{2} \right) \tag{12.3}$$

where the first term on the right hand side of (12.3) is symmetric and the second term is skew-symmetric. A property of a skew-symmetric matrix is that its associated quadratic function is always zero and therefore

$$x^T Q x = x^T \left(\frac{Q + Q^T}{2} \right) x \tag{12.4}$$

It then follows that Q is positive definite, if the symmetric matrix $(Q+Q^T)$ is positive definite. A consequence of this is that Q is positive definite, if all the eigenvalues of $(Q + Q^T)$ are positive.

If Q is a complex matrix, it is said to be positive definite, if $x^H Q x > 0$ for any non-zero x; and Q will then be Hermitian. In this chapter, however, we are largely interested in real matrices and real-valued LMIs as discussed below.

The basic structure of an LMI is

$$F(x) = F_0 + \sum_{i=1}^{m} x_i F_i > 0 \tag{12.5}$$

where $x \in \mathbb{R}^m$ is a variable and F_0, F_i are given constant symmetric real matrices. The representation (12.5) may seem restrictive, as we have not allowed for cases where some of the matrices F_i are complex Hermitian or the LMI is non-strict having the form $F(x) \geq 0$. However, complex-valued LMIs can be easily turned into real-valued LMIs; see Exercise 12.1. Similarly, it is also possible to convert any "feasible" non-strict LMI to the strict LMI form in (12.5); see Boyd et al. (1994).

The basic LMI problem – the *feasibility* problem – is to find x such that inequality (12.5) holds. Note that $F(x) > 0$ in (12.5) describes an *affine* relationship in terms of the variable

x. Normally the variable x, which we are interested in, is composed of one or many matrices whose columns have been "stacked" as a vector. That is,

$$F(x) = F(X_1, X_2, \ldots, X_n) \tag{12.6}$$

where $X_i \in \mathbb{R}^{q_i \times p_i}$ is a matrix, $\sum_{i=1}^{n} q_i \times p_i = m$, and the columns of all the matrix variables are stacked up to form a single vector variable.

Hence, from now on, we will consider functions of the form

$$F(X_1, X_2, \ldots, X_n) = F_0 + G_1 X_1 H_1 + G_2 X_2 H_2 + \ldots \tag{12.7}$$

$$= F_0 + \sum_{i=1}^{n} G_i X_i H_i > 0 \tag{12.8}$$

where F_0, G_i, H_i are given matrices and the X_i are the matrix variables which we seek.

Exercise 12.1 * *Let Q be a Hermitian matrix ($Q = Q^H$) having the form $Q = Q_R + jQ_I$. Show that $Q > 0$ if and only if*

$$\begin{bmatrix} Q_R & Q_I \\ -Q_I & Q_R \end{bmatrix} > 0 \tag{12.9}$$

12.1.2   Systems of LMIs

In general, we are frequently faced with **LMI** constraints of the form

$$F_1(X_1, \ldots, X_n) \;\; > \;\; 0 \tag{12.10}$$

$$\vdots$$

$$F_p(X_1, \ldots, X_n) \;\; > \;\; 0 \tag{12.11}$$

where

$$F_j(X_1, \ldots, X_n) = F_{0j} + \sum_{i=1}^{n} G_{ij} X_i H_{ij} \tag{12.12}$$

However, it is easily seen that, by defining $\widetilde{F}_0, \widetilde{G}_i, \widetilde{H}_i, \widetilde{X}_i$ as

$$\widetilde{F}_0 \;\; = \;\; \mathrm{diag}(F_{01}, \ldots, F_{0p}) \tag{12.13}$$

$$\widetilde{G}_i \;\; = \;\; \mathrm{diag}(G_{i1}, \ldots, G_{ip}) \tag{12.14}$$

$$\widetilde{H}_i \;\; = \;\; \mathrm{diag}(H_{i1}, \ldots, H_{ip}) \tag{12.15}$$

$$\widetilde{X}_i \;\; = \;\; \mathrm{diag}(X_i, \ldots X_i) \tag{12.16}$$

we actually have the inequality

$$F_{\mathrm{big}}(X_1, \ldots, X_n) \triangleq \widetilde{F}_0 + \sum_{i=1}^{n} \widetilde{G}_i \widetilde{X}_i \widetilde{H}_i > 0 \tag{12.17}$$

That is, we can represent a (big) system of LMIs as a single LMI. Therefore, we do not distinguish a single LMI from a system of LMIs; they are the same mathematical entity. We may also encounter systems of LMIs of the form

$$F_1(X_1, \ldots, X_n) > 0 \tag{12.18}$$

$$F_2(X_1, \ldots, X_n) > F_3(X_1, \ldots, X_n) \tag{12.19}$$

Again, it is easy to see that this can be written in the same form as inequality (12.17) above. For the remainder of the chapter we do not distinguish between LMIs which can be written as above, or those which are in the more generic form of inequality (12.17).

Notation. It is standard to let X_i denote the generic LMI variables. In the examples that follow in this chapter, we will use the notation more commonly associated with the specific problem. For instance, in Example 12.1, $P = X_1$, and in Example 12.6, P and Q are the LMI variables X_1 and X_2, respectively.

12.2   Types of LMI problems

The term "LMI problem" is rather vague and in fact there are several sub-groups of LMI problems including LMI feasibility problems, linear objective minimization problems and generalized eigenvalue problems. These three problems will be described below in the same way that they are separated in the Matlab LMI toolbox. Note that by "LMI problem" we normally mean solving an optimization problem or an eigenvalue problem with LMI constraints.

12.2.1   LMI feasibility problems

These are simply problems for which we seek a *feasible* solution $\{X_1, \ldots, X_n\}$ such that

$$F(X_1, \ldots, X_n) > 0 \tag{12.20}$$

We are not interested in the optimality of the solution, only in finding a solution, which may not be unique.

Example 12.1 **Determining stability of a linear system.** *Consider an autonomous linear system*

$$\dot{x} = Ax \tag{12.21}$$

then the Lyapunov LMI problem for proving stability of this system ($\mathrm{Re}\{\lambda_i(A)\} < 0$, $\forall i$) *is to find a $P > 0$ such that (see e.g. Boyd et al., 1994, p. 20)*

$$A^T P + PA < 0 \tag{12.22}$$

This is an LMI feasibility problem in $P > 0$. However, given any $P > 0$ which satisfies this, it is obvious that any matrix from the set

$$\mathcal{P} = \{\beta P : \text{ scalar } \beta > 0\} \tag{12.23}$$

also solves the problem. In fact, as will be seen later in the anti-windup case study, the matrix P forms part of a Lyapunov function for the linear system. Further, note that the LMI (12.22) and the requirement $P > 0$ can be combined into a single LMI as

$$\begin{bmatrix} A^T P + PA & 0 \\ 0 & -P \end{bmatrix} < 0 \tag{12.24}$$

Matlab code for solving this problem is given in Table 12.1.

Table 12.1: MATLAB program for determining stability in Example 12.1

```
% Uses MATLAB Robust Control toolbox
% A: n × n state matrix
setlmis([])
P = lmivar(1,[size(A,1) 1]);               % Specify structure and size of P
Lyap = newlmi
% Only the terms above the diagonal need to be specified:
lmiterm([Lyap 1 1 P],1,A,'s')              % AP + P'A < 0
lmiterm([Lyap 1 2 0],1)                     % 0
lmiterm([Lyap 2 2 P],-1,1)                  % P > 0
LMIsys = getlmis;                           % Obtain the system of LMIs
[tmin,xfeas] = feasp(LMIsys);               % Solve the feasibility problem
% Feasible (A is stable) iff tmin < 0
```

12.2.2 Linear objective minimization problems

These problems are also called eigenvalue problems. They involve the minimization (or maximization) of some *linear scalar* function, $\alpha(.)$, of the matrix variables, subject to LMI constraints:

$$\min \alpha(X_1, \ldots, X_n) \tag{12.25}$$

$$\text{s.t.} \quad F(X_1, \ldots, X_n) > 0 \tag{12.26}$$

where we have used the abbreviation "s.t." to mean "such that". In this case, we are therefore trying to optimize some quantity whilst ensuring some LMI constraints are satisfied. Actually, $\alpha(.)$ does not need to be a linear function but the problem should be convex. Examples, where $\alpha(.)$ is not linear, can be found in Boyd et al. (1994) and in some LMI software.

Example 12.2 Calculating the $\mathcal{H}_\infty$ norm of a linear system. *Consider a linear system*

$$\dot{x} = Ax + Bw \tag{12.27}$$

$$z = Cx + Dw \tag{12.28}$$

then the problem of finding the $\mathcal{H}_\infty$ norm of the transfer function matrix T_{zw} from w to z is equivalent to the following optimization problem in $P > 0$ (see e.g. Gahinet and Apkarian, 1994):

$$\min \gamma \tag{12.29}$$

$$\text{s.t.} \quad \begin{bmatrix} A^T P + PA & PB & C^T \\ B^T P & -\gamma I & D^T \\ C & D & -\gamma I \end{bmatrix} < 0 \tag{12.30}$$

Note that although $\gamma > 0$ is unique, the uniqueness of $P > 0$ is, in general, not guaranteed. The LMI problem (12.29)–(12.30) can be easily solved using Matlab, as shown in Table 12.2. Also note that we have here chosen to write the two LMIs (12.30) and $P > 0$ separately and not combined, as in (12.24).

12.2.3 Generalized eigenvalue problems

The generalized eigenvalue problem, or GEVP, is slightly different to the preceding problem in the sense that the objective of the optimization problem is not actually convex, but *quasi-convex*. However, the methods used to solve such problems are similar. Specifically a GEVP is formulated as

Table 12.2: MATLAB program for calculating $\mathcal{H}_\infty$ norm in Example 12.2

```
% Uses MATLAB Robust Control toolbox
% [A,B,C,D]: State-space realization
n = size(A,1)
setlmis([])
P = lmivar(1,[size(A,1) 1]);              %Specify structure and size of P
gamma = lmivar(1,[1 1]);
HinfLMI = newlmi                          % LMI # 1
lmiterm([HinfLMI 1 1 P],1,A,'s')          % AP + P'A
lmiterm([HinfLMI 1 2 P],1,B)              % PB
lmiterm([HinfLMI 1 3 0],C')               % C'
lmiterm([HinfLMI 2 2 gamma],-1,1)         % -gamma*I
lmiterm([HinfLMI 2 3 0],D')               % D'
lmiterm([HinfLMI 3 3 gamma],-1,1)         % -gamma*I
Ppos = newlmi                             % LMI # 2
lmiterm([Ppos 1 1 P],-1,1)                % P > 0
LMIsys = getlmis;                         % Obtain the system of LMIs
c = mat2dec(LMIsys,zeros(n),1);           % Vector c in c'x
options = [1e-5,0,0,0,0];                 % Relative accuracy of solution
[normhinf,xopt] = mincx(LMIsys,c, options); % Solve minimization problem
```

$$\min \lambda \tag{12.31}$$

$$\text{s.t.} \quad F_1(X_1,\ldots,X_n) - \lambda F_2(X_1,\ldots,X_n) < 0 \tag{12.32}$$

$$F_2(X_1,\ldots,X_n) > 0 \tag{12.33}$$

$$F_3(X_1,\ldots,X_n) > 0 \tag{12.34}$$

The first two lines are equivalent to minimizing the largest "generalized" eigenvalue of the matrix pencil $F_1(X_1,\ldots,X_n) - \lambda F_2(X_1,\ldots,X_n)$. In some cases, a GEVP problem can be reduced to a linear objective minimization problem, through an appropriate change of variables.

Example 12.3 **Bounding the decay rate of a linear system.** *A good example of a GEVP is given by Boyd et al. (1994). Given a stable linear system $\dot{x} = Ax$, the decay rate is the largest α such that*

$$\|x(t)\| \le e^{-\alpha t}\beta\|x(0)\| \quad \forall x(t) \tag{12.35}$$

where β is a constant. If we choose $V(x) = x^T P x > 0$ as a Lyapunov function for the system and ensure that $\dot{V}(x) \le -2\alpha V(x)$ it is easily shown that the system will have a decay rate of at least α. Hence, the problem of finding the decay rate could be posed as the optimization problem in $P > 0$

$$\min -\alpha \tag{12.36}$$

$$\text{s.t.} \quad A^T P + PA + 2\alpha P \le 0 \tag{12.37}$$

This problem is a GEVP with the functions

$$F_1(P) = A^T P + PA \tag{12.38}$$

$$F_2(P) = 2P \tag{12.39}$$

Example 12.4 **Calculating upper bound on μ.** *Consider the problem of calculating the upper bound on the structured singular value, μ in (8.87), given as*

$$\mu_{\text{up}}(M) = \min_{D \in \mathcal{D}} \bar{\sigma}(DMD^{-1}) \tag{12.40}$$

where $\mathcal{D}$ is the set of matrices D which commute with the uncertainty block Δ (i.e. satisfy $D\Delta = \Delta D$). This bound is tight for complex Δ with three or fewer blocks. Due to the presence of the inverse term, the optimization problem is difficult to solve in its original form; however, it can be transformed into an equivalent LMI problem. To see this note that

$$\bar{\sigma}(DMD^{-1}) < \gamma \Leftrightarrow \rho(D^{-H}M^H D^H DMD^{-1}) < \gamma^2 \tag{12.41}$$

$$\Leftrightarrow D^{-H}M^H D^H DMD^{-1} - \gamma^2 I < 0 \Leftrightarrow M^H PM - \gamma^2 P < 0 \tag{12.42}$$

where we have introduced $P = D^H D$. Note that $P > 0$ and in addition has the structure of D, i.e. $P \in \mathcal{D}$. Now, μ_{up} can be found by solving the following optimization problem:

$$\min \ \gamma^2 \tag{12.43}$$

$$\text{s.t.} \quad M^H PM - \gamma^2 P < 0 \tag{12.44}$$

which is a GEVP with the functions

$$F_1(P) = M^H PM \tag{12.45}$$

$$F_2(P) = P \tag{12.46}$$

A Matlab program for solving the GEVP (12.43)–(12.44) with structured Δ is shown in Table 12.3.

Table 12.3: **MATLAB program for calculating upper bound on μ in Example 12.4**

```
% Uses MATLAB Robust Control toolbox
% Here: M is 4 × 4 real matrix
% Here: Structured Delta with a full 2 × 2 block and a scalar 2 × 2 block
setlmis([])
P = lmivar(1,[2 0;2 1]);                    % Specify P to commute with Delta
gamma = lmivar(1,[1 1]);
Ppos = newlmi;                              % LMI # 2
lmiterm([-Ppos 1 1 P],1,1)                  % P > 0
MuupLMI = newlmi;                           % LMI # 1
lmiterm([MuupLMI 1 1 P],M',M)              % F₁(P) = M'PM
lmiterm([-MuupLMI 1 1 P],1,1)              % -F₂(P) = -P
LMIsys = getlmis;                           % Obtain the system of LMIs
[gmin,xopt] = gevp(LMIsys,1);               % Solve the GEVP problem
muup = sqrt(gmin)                           % Upper bound on μ
```

Exercise 12.2 Let $M = \begin{bmatrix} -1 & -1 \\ 3 & 3 \end{bmatrix}$ and compute $\mu(M)$ for (i) $\Delta = \delta \cdot I$ (scalar 2×2 block), (ii) $\Delta = \mathrm{diag}(\delta_1, \delta_2)$ (two 1×1 blocks) and (iii) $\Delta = $ full 2×2 block using the Matlab program in Table 12.3. Verify with (8.99). (Solution: $\mu(M) = 2, 4$ and $\sqrt{20} = 4.47$.)

12.3 Tricks in LMI problems

Although many control problems can be cast as LMI problems, a substantial number of these need to be manipulated before they are in a suitable LMI problem format. Fortunately, there are a number of common tools or "tricks" which can be used to transform problems into suitable LMI forms. Some of the more useful ones are described below.

12.3.1 Change of variables

Many control problems can be posed in the form of a set of nonlinear matrix inequalities; that is, the inequalities are nonlinear in the matrix variables we seek. However, by defining new variables it is sometimes possible to "linearize" the nonlinear inequalities, hence making them solvable by LMI methods.

Example 12.5 State feedback control synthesis problem. *Consider the problem of finding a matrix $F \in \mathbb{R}^{m \times n}$ such that the matrix $A + BF \in \mathbb{R}^{n \times n}$ has all of its eigenvalues in the open left-half complex plane. By the theory of Lyapunov equations (see Zhou et al., 1996), this is equivalent to finding a matrix F and a positive definite matrix $P \in \mathbb{R}^{n \times n}$ such that the following inequality holds:*

$$(A + BF)^T P + P(A + BF) \quad < \quad 0 \tag{12.47}$$

or

$$A^T P + PA + F^T B^T P + PBF \quad < \quad 0 \tag{12.48}$$

This problem is not in LMI form due to the terms which contain products of F and P – these terms are 'nonlinear' and as there are products of two variables, they are said to be "bilinear". If we multiply either side of (12.48) by $Q := P^{-1}$ (which does not change the definiteness of the expression since $\mathrm{rank}(P) = \mathrm{rank}(Q) = n$) we obtain

$$QA^T + AQ + QF^T B^T + BFQ < 0 \tag{12.49}$$

This is a new matrix inequality in the variables $Q > 0$ and F, but it is still nonlinear. To rectify this, we simply define a second new variable $L = FQ$ giving

$$QA^T + AQ + L^T B^T + BL < 0 \tag{12.50}$$

We now have an LMI feasibility problem in the new variables $Q > 0$ and $L \in \mathbb{R}^{m \times n}$. Once this LMI has been solved we can recover a suitable state feedback matrix as $F = LQ^{-1}$ and our Lyapunov variable as $P = Q^{-1}$. Hence, by making a change of variables we have obtained an LMI from a nonlinear matrix inequality.

The key fact to consider when making a change of variables is the assurance that the original variables can be recovered and that they are not over-determined. Notice also that multiplication by Q above is an example of a congruence transformation as considered in the next section.

Exercise 12.3* *With reference to Example 12.2, formulate the problem of finding the worst-case (maximum) gain of each of the uncertain systems*

$$G_1(s) = \frac{k}{s + \tau}; G_2(s) = \frac{k}{\tau s + 1} \tag{12.51}$$

as LMI problems. Verify your results with the Robust Control toolbox command wcgain *using numerical values $2 \le k, \tau \le 3$.*

12.3.2 Congruence transformation

For a given positive definite matrix $Q \in \mathbb{R}^{n \times n}$, we know that, for another real matrix $W \in \mathbb{R}^{n \times n}$ such that $\mathrm{rank}(W) = n$, the following inequality holds:

$$W Q W^T > 0 \tag{12.52}$$

In other words, *definiteness* of a matrix is invariant under pre- and postmultiplication by a full rank real matrix, and its transpose, respectively. The process of transforming $Q > 0$ into (12.52) using a real full rank matrix is called a "congruence transformation". It is very useful for "removing" bilinear terms in matrix inequalities and is often used, in conjunction with a change of variables, to make a bilinear matrix inequality *linear*. Often W is chosen to have a diagonal structure.

Example 12.6 **Making a bilinear matrix inequality linear.** *Consider*

$$Q = \begin{bmatrix} A^T P + P A & P B F + C^T V \\ \star & -2V \end{bmatrix} < 0 \tag{12.53}$$

where the matrices $P > 0, V > 0$ and F (definiteness not specified) are the matrix variables and the remaining matrices are constant. The $\star$ in the bottom left entry of the matrix denotes the term required to make the expression symmetric (in this case, $\star = F^T B^T P^T + V^T C$) and will be used frequently hereafter. Notice that this inequality is bilinear in the variables P and F which occur in the $(1, 2)$ and $(2, 1)$ blocks of the matrix $Q \in \mathbb{R}^{(n+l) \times (n+l)}$. However, if we choose the matrix

$$W = \begin{bmatrix} P^{-1} & 0 \\ 0 & V^{-1} \end{bmatrix} \in \mathbb{R}^{(n+l) \times (n+l)} \tag{12.54}$$

which is full rank ($\mathrm{rank}(W) = n + l$) by virtue of the invertibility of P and V (which exist as the matrices are positive definite), then calculating $W Q W^T$ gives

$$W Q W^T = \begin{bmatrix} P^{-1} A^T + A P^{-1} & B F V^{-1} + P^{-1} C^T \\ \star & -2V^{-1} \end{bmatrix} < 0 \tag{12.55}$$

Hence, in the new variables $X = P^{-1}, U = V^{-1}$ and $L = F V^{-1}$ we have a linear *matrix inequality*

$$W Q W^T = \begin{bmatrix} X A^T + A X & B L + X C^T \\ \star & -2U \end{bmatrix} \tag{12.56}$$

Notice that the original variables can be recovered by inverting X and U.

12.3.3 Schur complement

The main use of the Schur complement is to transform quadratic matrix inequalities into LMIs, or at least as a step in this direction. Schur's complement formula says that the following statements are equivalent:

$$(i) \qquad \Phi = \begin{bmatrix} \Phi_{11} & \Phi_{12} \\ \Phi_{12}^T & \Phi_{22} \end{bmatrix} < 0$$

$$(ii) \qquad \Phi_{22} < 0$$
$$\Phi_{11} - \Phi_{12} \Phi_{22}^{-1} \Phi_{12}^T < 0$$

A non-strict form involving a Moore–Penrose pseudo-inverse also exists if Φ is only negative semi-definite; see Boyd et al. (1994).

Example 12.7 Making a quadratic inequality linear. *Consider the LQR-type matrix inequality (Riccati inequality)*

$$A^T P + PA + PBR^{-1}B^T P + Q < 0 \tag{12.57}$$

where $P > 0$ is the matrix variable and the other matrices are constant with $Q, R > 0$. This inequality can be used to minimize the cost function (seen in Chapter 9)

$$J = \int_0^\infty (x^T Q x + u^T R u) dt \tag{12.58}$$

If we now define

$$\Phi_{11} := A^T P + PA + Q \tag{12.59}$$
$$\Phi_{12} := PB \tag{12.60}$$
$$\Phi_{22} := -R \tag{12.61}$$

and use the Schur complement identities we can transform our Riccati inequality into

$$\begin{bmatrix} A^T P + PA + Q & PB \\ \star & -R \end{bmatrix} < 0 \tag{12.62}$$

In other words, we have transformed a quadratic matrix inequality into an LMI.

Exercise 12.4 *Verify that, for a given complex matrix A, the constraint $\bar{\sigma}(A) < \gamma$ can be posed as an LMI. Is it possible to represent the constraint $\underline{\sigma}(A) > \gamma$ as an LMI?*

12.3.4 The S-procedure

The S-procedure is essentially a method which enables one to combine several quadratic inequalities into one single inequality (generally with some conservatism). There are many instances in control engineering when we would like to ensure that a single quadratic function of $x \in \mathbb{R}^m$ is such that

$$F_0(x) \leq 0; \quad F_0(x) \triangleq x^T A_0 x + 2b_0 x + c_0 \tag{12.63}$$

whenever certain other quadratic functions are positive semi-definite, i.e. when

$$F_i(x) \geq 0 \quad F_i(x) \triangleq x^T A_i x + 2b_0 x + c_0, \quad i \in \{1, 2, \ldots, q\} \tag{12.64}$$

To illustrate the S-procedure, consider $i = 1$, for simplicity. That is, we would like to ensure $F_0(x) \leq 0$ for all x such that $F_1(x) \geq 0$. Now, if there exists a positive (or zero) scalar, τ, such that

$$F_{\text{aug}}(x) \triangleq F_0(x) + \tau F_1(x) \leq 0 \quad \forall x \quad \text{s.t.} \quad F_1(x) \geq 0 \tag{12.65}$$

it follows that our goal is achieved. To see this, note that $F_{\text{aug}}(x) \leq 0$ implies that $F_0(x) \leq 0$ if $\tau F_1(x) \geq 0$ because $F_0(x) \leq F_{\text{aug}}(x)$ if $F_1(x) \geq 0$. Thus, extending this idea to q inequality constraints we have that

$$F_0(x) \leq 0 \quad \text{whenever} \quad F_i(x) \geq 0 \tag{12.66}$$

holds if

$$F_0(x) + \sum_{i=1}^q \tau_i F_i(x) \leq 0, \quad \tau_i \geq 0 \tag{12.67}$$

In general the S-procedure is conservative; inequality (12.67) implies inequality (12.66), but not vice versa. When $q = 1$, however, the S-procedure is non-conservative. The usefulness of the S-procedure is in the possibility of including the τ_i's as variables in an LMI problem.

Example 12.8 Combining quadratic constraints to yield an LMI. *An instructive example, from Boyd et al. (1994), involves finding a matrix variable $P > 0$ such that*

$$\begin{bmatrix} x \\ z \end{bmatrix}^T \begin{bmatrix} A^T P + PA & PB \\ B^T P & 0 \end{bmatrix} \begin{bmatrix} x \\ z \end{bmatrix} < 0 \qquad (12.68)$$

whenever $x \neq 0$ and z satisfy the constraint

$$z^T z \leq x^T C^T C x \qquad (12.69)$$

Note that inequality (12.69) is equivalent to

$$(x^T C^T C x - z^T z) \geq 0 \qquad (12.70)$$

or

$$\begin{bmatrix} x \\ z \end{bmatrix}^T \begin{bmatrix} C^T C & 0 \\ 0 & -I \end{bmatrix} \begin{bmatrix} x \\ z \end{bmatrix} \geq 0 \qquad (12.71)$$

The two quadratic constraints (12.68) and (12.71) can thus be combined with the S-procedure to yield the LMI

$$\begin{bmatrix} A^T P + PA + \tau C^T C & PB \\ B^T P & -\tau I \end{bmatrix} < 0 \qquad (12.72)$$

in the variables $P > 0$ and $\tau \geq 0$.

12.3.5 The projection lemma and Finsler's lemma

In some types of control problems, particularly those seeking dynamic controllers, we encounter inequalities of the form

$$\Psi(X) + G(X)\Lambda H^T(X) + H(X)\Lambda^T G^T(X) < 0 \qquad (12.73)$$

where X and Λ are the matrix variables and $\Psi(.), G(.), H(.)$ are (normally affine) functions of X but not of Λ.

In Gahinet and Apkarian (1994), it is proved that inequality (12.73) is satisfied, for some X, if and only if

$$\begin{cases} W_{G(X)}^T \Psi(X) W_{G(X)} & < \quad 0 \\ W_{H(X)}^T \Psi(X) W_{H(X)} & < \quad 0 \end{cases} \qquad (12.74)$$

where $W_{G(X)}$ and $W_{H(X)}$ are matrices with columns which form bases for the null spaces of $G(X)$ and $H(X)$ respectively. Alternatively, $W_{G(X)}$ and $W_{H(X)}$ are sometimes called *orthogonal complements* of $G(X)$ and $H(X)$ respectively. Note that

$$W_{G(X)} G(X) = 0, \quad W_{H(X)} H(X) = 0 \qquad (12.75)$$

The main point of this result (referred to as Gahinet and Apkarian's projection lemma) is that it enables one to transform a matrix inequality, which is a, not necessarily linear, function of *two* variables, into two inequalities which are functions of just *one* variable. This has two useful consequences:

(i) It can facilitate the derivation of an LMI.

(ii) There are fewer variables for computation.

Finsler (1937) also proved that inequality (12.73) is equivalent to two inequalities

$$
\begin{cases}
\Psi(X) - \sigma G(X)G(X)^T & < 0 \\
\Psi(X) - \sigma H(X)H(X)^T & < 0
\end{cases}
\tag{12.76}
$$

for some real σ. In other words, inequalities (12.74) and (12.76) are equivalent. This result is often referred to as *Finsler's lemma*.

Example 12.5 (State feedback) continued. *Consider again the state feedback synthesis problem of finding $P > 0$ and F such that*

$$
(A + BF)^T P + P(A + BF) < 0
\tag{12.77}
$$

Using the change of variables described earlier in Example 12.5, we can change this problem into that of finding $Q > 0$ and L such that

$$
QA^T + AQ + L^T B^T + BL < 0
\tag{12.78}
$$

If we choose to eliminate the variable L using the projection lemma we get

$$
\begin{cases}
W_B^T (AQ + QA^T)W_B & < \quad 0, \quad Q > 0 \\
W_I^T (AQ + QA^T)W_I & < \quad 0, \quad Q > 0
\end{cases}
\tag{12.79}
$$

However, as W_I is a matrix whose columns span the null space of the identity matrix which is $\mathcal{N}(I) = \{0\}$, the above equation simply reduces to

$$
W_B^T (AQ + QA^T)W_B < 0, \quad Q > 0
\tag{12.80}
$$

which is an LMI problem.
   Alternatively, using Finsler's lemma we get

$$
\begin{cases}
AQ + QA^T - \sigma BB^T & < \quad 0, \quad Q > 0 \\
AQ + QA^T - \sigma I & < \quad 0, \quad Q > 0
\end{cases}
\tag{12.81}
$$

However, we can neglect the second inequality because if we can find a σ satisfying the first inequality, we can always find one which satisfies the second.

Notice that the use of both the projection lemma and Finsler's lemma effectively reduces our original LMI problem into two separate ones: the first LMI problem involves the calculation of $Q > 0$; the second involves the back substitution of Q into the original problem in order for us to find L (and then F). The reader is, however, cautioned against the possibility of ill-conditioning in this two-step approach. For some problems, normally those with large numbers of variables, X can be poorly conditioned, which can hinder the numerical determination of Λ from (12.73).

12.4 Case study: anti-windup compensator synthesis

Linear controllers can be very effective at controlling real plants until they encounter actuator saturation, which can cause the behaviour of the system to deteriorate dramatically, or even become unstable. To limit this loss of performance special compensators called *anti-windup compensators* are added which take action when the control signal saturates. As the anti-windup compensator is inactive for large periods of time, conventional linear methods are not always useful for designing such a compensator. However, as we will discover, LMIs can play an important part in this design.

Anti-windup was also discussed in Section 9.4.5, where the Hanus scheme was briefly introduced. The approach below is more general and rigorous.

12.4.1 Representing anti-windup compensators

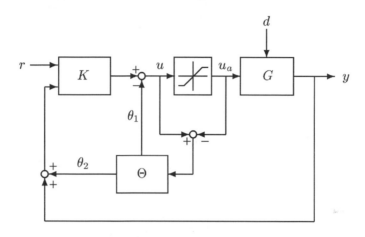

Figure 12.1: Generic anti-windup scheme

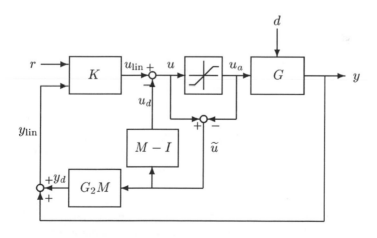

Figure 12.2: Conditioning with $M(s)$

A generic anti-windup compensator is depicted in Figure 12.1. The plant $G(s) = [G_1(s) \quad G_2(s)]$ is assumed to be stable (to enable global results to be obtained – see Turner and Postlethwaite (2004) for more detail about this). $G_1(s)$ represents the disturbance

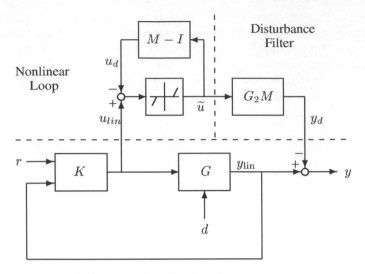

Nominal Linear Transfer Function

Figure 12.3: Equivalent representation of conditioning with $M(s)$

feedforward part of the plant and therefore is the transfer function from the disturbance $d(s)$ to the output $y(s)$. Similarly $G_2(s)$ represents the feedback part of the plant and therefore is the transfer function from the actual control input $u_a(s)$ to the output $y(s)$. Only $G_2(s)$ plays a part in anti-windup synthesis and its state-space realization is given by

$$G_2(s) \overset{s}{=} \left[\begin{array}{c|c} A_p & B_p \\ \hline C_p & D_p \end{array} \right] \qquad (12.82)$$

$K(s)$ is the linear controller, which we assume has been designed such that its closed loop interconnection with $G(s)$ is stable, in the absence of saturation, and such that some linear performance specifications have been satisfied.

The anti-windup compensator, $\Theta(s)$, adds extra signals to the controller input and output when control signal saturation occurs. By choosing $\Theta(s)$ in different ways, the closed-loop properties during and following saturation are influenced. Figure 12.2 shows the closed-loop system, when $\Theta(s)$ is parameterized in terms of the transfer function $M(s)$. An interesting choice of $M(s)$ is $M(s) = I$. In this case, the anti-windup solution is similar to the internal model control scheme discussed by Campo and Morari (1990). However, this is not always a good solution, especially when $G_2(s)$ has lightly damped modes (Weston and Postlethwaite, 2000). As shown later, better solutions can be obtained by choosing $M(s)$ as a coprime factor of $G_2(s)$.

From the identity

$$\mathrm{Dz}(u) = u - \mathrm{sat}(u) \qquad (12.83)$$

where $\mathrm{Dz}(.)$ and $\mathrm{sat}(.)$ represent the deadzone and saturation functions respectively, it can be proven that Figures 12.2 and 12.3 are equivalent (Weston and Postlethwaite, 2000). Figure 12.3 is convenient to analyze the stability and performance of the system and, in particular, it can be seen that, providing the nominal linear closed loop is stable, overall stability is governed by the stability of the *nonlinear loop*. Moreover, the performance of the system can be measured by the "size" of the map from u_{lin} to y_d. This map, call it $\mathcal{T}_p$, governs how much the linear output is perturbed by the saturation of the control signal. Hence, it would be useful to minimize the size of the norm of this – nonlinear – operator. For more information on the motivation behind this see, for example, Turner and Postlethwaite (2004) and Turner et al. (2003).

12.4.2 Lyapunov stability

The stability of nonlinear systems is more difficult to ascertain than that of linear systems. A sufficient (but not necessary) condition was given by Lyapunov; see, for example, Khalil (1996).

Theorem 12.1 Lyapunov's theorem *Given a positive definite function $V(x) > 0\ \forall x \neq 0$ and an autonomous system $\dot{x} = f(x)$, then the system $\dot{x} = f(x)$ is stable if*

$$\dot{V}(x) = \frac{\partial V}{\partial x} f(x) < 0 \quad \forall x \neq 0 \tag{12.84}$$

As our anti-windup system is nonlinear due to the presence of the saturation function, we will use Lyapunov's theorem to establish stability.

12.4.3 $\mathcal{L}_2$ gain

In linear systems, the $\mathcal{H}_\infty$ norm is equivalent to the maximum root mean square or rms energy gain of the system. The equivalent measure for nonlinear systems is the so-called $\mathcal{L}_2$ gain, which is a bound on the rms energy gain. Specifically a *nonlinear* system with input $u(t)$ and output $y(t)$ is said to have an $\mathcal{L}_2$ gain of γ if

$$\|y\|_2 < \gamma\|u\|_2 + \beta \tag{12.85}$$

where β is a positive constant and $\|(.)\|_2$ denotes the standard 2-norm-in-time ($\mathcal{L}_2$ norm) of a vector. Thus the $\mathcal{L}_2$ gain of a system can be taken as a measure of the size of the output a system exhibits relative to the size of its input.

12.4.4 Sector boundedness

The saturation function is defined as

$$\text{sat}(u) = [\text{sat}_1(u_1), \ldots, \text{sat}_m(u_m)]^T \tag{12.86}$$

and $\text{sat}_i(u_i) = \text{sign}(u_i) \times \min\{|u_i|, \bar{u}_i\}$, $\bar{u}_i > 0\ \forall i \in \{1, \ldots, m\}$, where $\bar{u}_i$ is the i'th saturation limit. From this, the deadzone function can be defined as

$$\text{Dz}(u) = u - \text{sat}(u) \tag{12.87}$$

It is easy to verify that the saturation function, $\text{sat}_i(u_i)$, satisfies the following inequality:

$$u_i\text{sat}_i(u_i) \geq \text{sat}_i^2(u_i) \tag{12.88}$$

or

$$\mathrm{sat}_i(u_i)[u_i - \mathrm{sat}_i(u_i)]w_i \geq 0 \tag{12.89}$$

for some $w_i > 0$. Collecting this inequality for all i we can write

$$\mathrm{sat}(u)^T W[u - \mathrm{sat}(u)] \geq 0 \tag{12.90}$$

for some diagonal $W > 0$. Similarly it follows that

$$\mathrm{Dz}(u)^T W[u - \mathrm{Dz}(u)] \geq 0 \tag{12.91}$$

for some diagonal $W > 0$. We will make use of this inequality in the derivation of our anti-windup compensator synthesis equations.

12.4.5 Full-order anti-windup compensators

The term "full-order" anti-windup compensators has a similar meaning to the term "full-order" $\mathcal{H}_\infty$ controller; that is, the compensator is of order equal to the plant. We will confine our attention to full-order anti-windup compensator synthesis. For a treatment of low-order and static anti-windup synthesis, see Turner and Postlethwaite (2004).

Assume that we factorize $G_2(s) = N(s)M(s)^{-1}$, i.e. the anti-windup parameter $M(s)$ is chosen as part of a coprime factorization of $G_2(s)$; for example, see Section 4.1.5 or Zhou et al. (1996). In this case, the operator $\mathcal{T}_p : u_{lin} \mapsto y_d$ is given by

$$\mathcal{T}_p \triangleq \begin{cases} \dot{x}_p &= (A_p + B_p F)x_p + B_p \tilde{u} \\ u_d &= F x_p \\ y_d &= (C_p + D_p F)x_p + D_p \tilde{u} \\ \tilde{u} &= \mathrm{Dz}(u_{lin} - u_d) \end{cases} \tag{12.92}$$

The matrix F determines the coprime factorization of $G_2(s)$, which in turn influences the performance of the anti-windup compensator. Hence, our goal in full-order anti-windup synthesis is to find an appropriate matrix F such that the closed-loop performance in the presence of saturation is good.

12.4.6 Anti-windup synthesis

We would like to choose F (and therefore $M(s)$) such that $\mathcal{T}_p$ is internally stable with sufficiently small $\mathcal{L}_2$ gain. It can be verified see (see Turner and Postlethwaite, 2004) that if we choose a Lyapunov function $V(x) = x_p^T P x_p > 0$ and ensure that

$$\dot{V}(x) + y_d^T y_d - \gamma^2 u_{lin}^T u_{lin} < 0 \tag{12.93}$$

then the operator $\mathcal{T}_p$ is indeed internally stable with an $\mathcal{L}_2$ gain of γ. Therefore, using the expression for $\mathcal{T}_p$ we can write inequality (12.93) as

$$z^T \begin{bmatrix} \bar{A}^T P + P\bar{A} + \bar{C}^T\bar{C} & PB_p + \bar{C}^T D_p & 0 \\ \star & D_p^T D_p & 0 \\ \star & \star & -\gamma^2 I \end{bmatrix} z < 0 \tag{12.94}$$

where

$$\bar{A} = A_p + B_p F \tag{12.95}$$

$$\bar{C} = C_p + D_p F \tag{12.96}$$

$$z = \begin{bmatrix} x_p^T & \tilde{u}^T & u_{\text{lin}}^T \end{bmatrix}^T \tag{12.97}$$

However, from the sector boundedness of the deadzone we also have that

$$2\tilde{u}^T W[u_{lin} - Fx_p - \tilde{u}] \geq 0 \tag{12.98}$$

We will use the S-procedure to combine inequalities (12.94) and (12.98). First note that inequality (12.98) may be written as

$$z^T \begin{bmatrix} 0 & -F^T W & 0 \\ \star & -2W & W \\ \star & \star & 0 \end{bmatrix} z \geq 0 \tag{12.99}$$

We have added the factor of 2 into inequality (12.98) so that inequality (12.99) can be written in a tidier fashion; without this factor of 2, there would be several factors of $1/2$ present. Using the S-procedure described earlier, we can combine inequality (12.94) with (12.99) to obtain

$$\begin{bmatrix} \bar{A}^T P + P\bar{A} + \bar{C}^T \bar{C} & PB_p + \bar{C}^T D_p - F^T W\tau & 0 \\ \star & -2W\tau + D_p^T D_p & W\tau \\ \star & \star & -\gamma^2 I \end{bmatrix} < 0 \tag{12.100}$$

Notice that τ *only* appears adjacent to W, so we can define a new variable $V = W\tau$ and use this from now on. Applying the Schur complement we obtain

$$\begin{bmatrix} \bar{A}^T P + P\bar{A} & PB_p - F^T V & 0 & \bar{C}^T \\ \star & -2V & V & D_p^T \\ \star & \star & -\gamma I & 0 \\ \star & \star & \star & -\gamma I \end{bmatrix} < 0 \tag{12.101}$$

Next, using the congruence transformation $\text{diag}(P^{-1}, V^{-1}, I, I)$ we obtain

$$\begin{bmatrix} P^{-1}A_p^T + A_p P^{-1} + P^{-1}F^T B_p^T + B_p F P^{-1} & B_p V^{-1} - P^{-1} F^T \\ \star & -2V^{-1} \\ \star & \star \\ \star & \star \end{bmatrix}$$

$$\begin{matrix} 0 & P^{-1}C_p^T + P^{-1}F^T D_p^T \\ I & V^{-1}D_p^T \\ -\gamma I & 0 \\ \star & -\gamma I \end{matrix} \Bigg] \tag{12.102}$$

Finally, defining new variables $Q = P^{-1}, U = V^{-1}, L = QF$ we get

$$\begin{bmatrix} QA_p^T + A_p Q + L^T B_p^T + B_p L & B_p U - QF^T & 0 & QC_p^T + L^T D_p^T \\ \star & -2U & I & U D_p^T \\ \star & \star & -\gamma I & 0 \\ \star & \star & \star & -\gamma I \end{bmatrix} < 0$$

which is now an LMI in $Q > 0$, $U > 0$ and diagonal, $\gamma > 0$ and L. To obtain F we can thus compute $F = Q^{-1}L$, which allows us to construct our anti-windup compensator.

For applications of these and similar formulae see Turner and Postlethwaite (2004) and Herrmann et al. (2003a; 2003b).

12.5 Conclusion

In recent years, efficient interior-point algorithms have been developed to solve convex LMI optimization problems of the type presented in this chapter. We have described the main (generic) LMI problems in control and the tools and tricks required to transform them into formats that can readily take advantage of the algorithms now available, especially in Matlab. In the examples, we have only used Matlab code, as we have throughout the book. Alternative LMI software is available and in this context we would like to mention YALMIP (`http://control.ee.ethz.ch/~joloef/yalmip.php`), which is particularly useful for interfacing with the free solvers available. By including this chapter, we have attempted to give the essential ingredients for developing an understanding of the power and usefulness of LMIs. More details can be found in Boyd et al. (1994). A cautionary note is that the complexity of LMI computations is high, and certainly higher, for example, than solving a Riccati equation in a conventional approach. Nevertheless, the LMI approach opens the way to solving problems that conventional methods cannot.

13

CASE STUDIES

In this chapter, we present three case studies which illustrate a number of important practical issues, namely: weights selection in $\mathcal{H}_\infty$ mixed-sensitivity design, disturbance rejection, output selection, two degrees-of-freedom $\mathcal{H}_\infty$ loop-shaping design, ill-conditioned plants, μ-analysis and μ-synthesis.

13.1   Introduction

The complete design process for an industrial control system will normally include the following steps:

1. *Plant modelling:* to determine a mathematical model of the plant either from experimental data using identification techniques, or from physical equations describing the plant dynamics, or a combination of these.
2. *Plant input–output controllability analysis:* to discover what closed-loop performance can be expected and what inherent limitations there are to "good" control, and to assist in deciding upon an initial control structure and maybe an initial selection of performance weights.
3. *Control structure design:* to decide on which variables to be manipulated and measured and which links should be made between them.
4. *Controller design:* to formulate a mathematical design problem which captures the engineering design problem and to synthesize a corresponding controller.
5. *Control system analysis:* to assess the control system by analysis and simulation against the performance specifications or the designer's expectations.
6. *Controller implementation:* to implement the controller, almost certainly in software for computer control, taking care to address important issues such as anti-windup and bumpless transfer.
7. *Control system commissioning:* to bring the controller on-line, to carry out on-site testing and to implement any required modifications before certifying that the controlled plant is fully operational.

In this book, we have focused on steps 2, 3, 4 and 5, and in this chapter we will present three case studies which demonstrate many of the ideas and practical techniques which can be used in these steps. The case studies are not meant to produce the "best" controller for the application considered but rather are used here to illustrate a particular technique from the book.

In case study 1, a helicopter control law is designed for the rejection of atmospheric turbulence. The gust disturbance is modelled as an extra input to an S/KS $\mathcal{H}_\infty$

mixed-sensitivity design problem. Results from nonlinear simulations indicate significant improvement over a standard S/KS design. For more information on the applicability of $\mathcal{H}_\infty$ control to advanced helicopter flight, the reader is referred to Walker and Postlethwaite (1996) who describe the design and ground-based piloted simulation testing of a high-performance helicopter flight control system. The first flight test results are given in Postlethwaite et al. (1999).

Case study 2 illustrates the application and usefulness of the two degrees-of-freedom $\mathcal{H}_\infty$ loop-shaping approach by applying it to the design of a robust controller for a high-performance aero-engine. Nonlinear simulation results are shown. Efficient and effective tools for control structure design (input–output selection) are also described and applied to this problem. This design work on the aero-engine has been further developed and forms the basis of a multi-mode controller which has been implemented and successfully tested on a Rolls-Royce Spey engine test facility at the former UK Defence Research Agency (now QinetiQ), Pyestock (Samar, 1995).

The final case study is concerned with the control of an idealized distillation column. A very simple plant model is used, but it is sufficient to illustrate the difficulties of controlling ill-conditioned plants and the adverse effects of model uncertainty. The structured singular value μ is seen to be a powerful tool for robustness analysis.

Case studies 1, 2 and 3 are based on papers by Postlethwaite et al. (1994), Samar and Postlethwaite (1994) and Skogestad et al. (1988), respectively.

13.2 Helicopter control

This case study is used to illustrate how weights can be selected in $\mathcal{H}_\infty$ mixed-sensitivity design, and how this design problem can be modified to improve disturbance rejection properties.

13.2.1 Problem description

In this case study, we consider the design of a controller to reduce the effects of atmospheric turbulence on helicopters. The reduction of the effects of gusts is very important in reducing a pilot's workload, and enables aggressive manoeuvres to be carried out in poor weather conditions. Also, as a consequence of decreased buffeting, the airframe and component lives are lengthened and passenger comfort is increased.

The design of rotorcraft flight control systems, for robust stability and performance, has been studied over a number of years using a variety of methods including: H_∞ optimization (Yue and Postlethwaite, 1990; Postlethwaite and Walker, 1992); eigenstructure assignment (Manness and Murray-Smith, 1992; Samblancatt et al., 1990); sliding mode control (Foster et al., 1993); and H_2 design (Takahashi, 1993). These early H_∞ controller designs were particularly successful (Walker et al., 1993), and have proved themselves in piloted simulations. These designs have used frequency information about the disturbances to limit the system sensitivity but in general there has been no explicit consideration of the effects of atmospheric turbulence. Therefore by incorporating practical knowledge about the disturbance characteristics, and how they affect the real helicopter, improvements to the overall performance should be possible. We will demonstrate this below.

The nonlinear helicopter model we will use for simulation purposes was developed at the former Defence Research Agency (now QinetiQ), Bedford (Padfield, 1981) and is known as the Rationalized Helicopter Model (RHM). A turbulence generator module has recently been included in the RHM and this enables controller designs to be tested on-line for their disturbance rejection properties. It should be noted that the model of the gusts affects the helicopter equations in a complicated fashion and is self-contained in the code of the RHM. For design purposes we will imagine that the gusts affect the model in a much simpler manner.

We will begin by repeating the design of Yue and Postlethwaite (1990) which used an S/KS $\mathcal{H}_\infty$ mixed-sensitivity problem formulation without explicitly considering atmospheric turbulence. We will then, for the purposes of design, represent gusts as a perturbation in the velocity states of the helicopter model and include this disturbance as an extra input to the S/KS design problem. The resulting controller is seen to be substantially better at rejecting atmospheric turbulence than the earlier standard S/KS design. More recent references on the application of H_∞ optimization to helicopter flight control, including flight tests, are given in the conclusions, Section 13.2.6.

13.2.2   The helicopter model

The aircraft model used in our work is representative of the Westland Lynx, a twin-engined multi-purpose military helicopter, approximately 9000 lbs (4000 kg) gross weight, with a four-blade semi-rigid main rotor. The unaugmented aircraft is unstable, and exhibits many of the cross-couplings characteristic of a single main-rotor helicopter. In addition to the basic rigid body, engine and actuator components, the model also includes second order rotor flapping and coning modes for off-line use. The model has the advantage that essentially the same code can be used for a real-time piloted simulation as for a workstation-based off-line handling qualities assessment.

The equations governing the motion of the helicopter are complex and difficult to formulate with high levels of precision. For example, the rotor dynamics are particularly difficult to model. A robust design methodology is therefore essential for high-performance helicopter control. The starting point for this study was to obtain an eighth-order differential equation

Table 13.1: Helicopter state vector

State	Description
θ	Pitch attitude
ϕ	Roll attitude
p	Roll rate (body-axis)
q	Pitch rate (body-axis)
ξ	Yaw rate
v_x	Forward velocity
v_y	Lateral velocity
v_z	Vertical velocity

modelling the small-perturbation rigid motion of the aircraft about hover. The corresponding state-space model is

$$\dot{x} = Ax + Bu \tag{13.1}$$

$$y \;=\; Cx \tag{13.2}$$

where the matrices A, B and C for the appropriately scaled system are available over the Internet as described in the preface. The eight-state rigid-body vector x is given in Table 13.2.2. The model is open-loop unstable with a pair of complex RHP-poles located at 0.23 ± 0.55 rad/s. The outputs consist of four primary controlled outputs

- Heave velocity $\dot{H}$
- Pitch attitude θ
- Roll attitude ϕ
- Heading rate $\dot{\psi}$

$\left.\right\} y_1$

together with two additional (body-axis) measurements

- Roll rate p
- Pitch rate q

$\left.\right\} y_2$

The controller (or pilot in manual control) generates four blade angle demands which are effectively the helicopter inputs, since the actuators (which are typically modelled as first-order lags) are modelled as unity gains in this study. The blade angles are

- main rotor collective
- longitudinal cyclic
- lateral cyclic
- tail rotor collective

$\left.\right\} u$

The action of each of these blade angles can be briefly described as follows. The main rotor collective changes all the blades of the main rotor by an equal amount and so roughly speaking controls lift. The longitudinal and lateral cyclic inputs change the main rotor blade angles differently thereby tilting the lift vector to give longitudinal and lateral motion, respectively. The tail rotor is used to balance the torque generated by the main rotor, and so stops the helicopter spinning around; it is also used to give lateral motion. This description, which assumes the helicopter inputs and outputs are decoupled, is useful to get a feeling of how a helicopter works but the dynamics are actually highly coupled. They are also unstable, and about some operating points exhibit non-minimum-phase characteristics.

We are interested in the design of *full-authority controllers*, which means that the controller has total control over the blade angles of the main and tail rotors, and is interposed between the pilot and the actuation system. It is normal in conventional helicopters for the controller to have only limited authority leaving the pilot to close the loop for much of the time (manual control). With a full-authority controller, the pilot merely provides the reference commands.

One degree-of-freedom controllers as shown in Figure 13.1 are to be designed. Notice that in the standard one degree-of-freedom configuration the pilot reference commands r_1 are augmented by a zero vector because of the rate feedback signals. These zeros indicate that there are no *a priori* performance specifications on $y_2 = [p \quad q]^T$.

13.2.3 $\mathcal{H}_\infty$ mixed-sensitivity design

We will consider the $\mathcal{H}_\infty$ mixed-sensitivity design problem illustrated in Figure 13.2. It can be viewed as a tracking problem as previously discussed in Chapter 9 (see Figure 9.11), but with an additional weight W_3. W_1 and W_2 are selected as loop-shaping weights whereas W_3

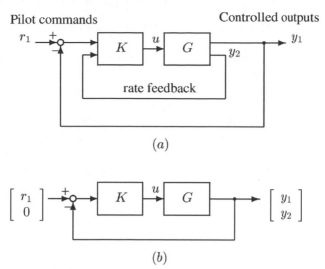

Figure 13.1: Helicopter control structure (a) as implemented, (b) in the standard one degree-of-freedom configuration

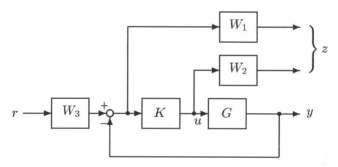

Figure 13.2: S/KS mixed-sensitivity minimization

is signal-based. The optimization problem is to find a stabilizing controller K to minimize the cost function

$$\left\| \begin{bmatrix} W_1 S W_3 \\ W_2 K S W_3 \end{bmatrix} \right\|_\infty \qquad (13.3)$$

This cost was also considered by Yue and Postlethwaite (1990) in the context of helicopter control. Their controller was successfully tested on a piloted flight simulator at DRA Bedford and so we propose to use the same weights here. The design weights W_1, W_2 and W_3 were selected as

$$
\begin{aligned}
W_1 &= \operatorname{diag}\left\{ 0.5\frac{s+12}{s+0.012},\ 0.89\frac{s+2.81}{s+0.005},\ 0.89\frac{s+2.81}{s+0.005}, \right. \\
&\qquad \left. 0.5\frac{s+10}{s+0.01},\ \frac{2s}{(s+4)(s+4.5)},\ \frac{2s}{(s+4)(s+4.5)} \right\} \qquad (13.4) \\
W_2 &= 0.5\frac{s+0.0001}{s+10}I_4 \qquad (13.5)
\end{aligned}
$$

$$W_3 \quad = \quad \mathrm{diag}\,\{1,\, 1,\, 1,\, 1,\, 0.1,\, 0.1\} \tag{13.6}$$

The reasoning behind these selections of Yue and Postlethwaite (1990) is summarized below.

Selection of $W_1(s)$ (performance weight): For good tracking accuracy in each of the controlled outputs the sensitivity function is required to be small. This suggests forcing integral action into the controller by selecting an s^{-1} shape in the weights associated with the controlled outputs. It was not thought necessary to have exactly zero steady-state errors and therefore these weights were given a finite gain of 500 at low frequencies. (Notice that a pure integrator cannot be included in W_1 anyway, since the standard $\mathcal{H}_\infty$ optimal control problem would not then be well posed in the sense that the corresponding generalized plant P could not then be stabilized by the feedback controller K.) In tuning W_1 it was found that a finite attenuation at high frequencies was useful in reducing overshoot. Therefore, high-gain low-pass filters were used in the primary channels to give accurate tracking up to about 6 rad/s. The presence of unmodelled rotor dynamics around 10 rad/s limits the bandwidth of W_1. With four inputs to the helicopter, we can only expect to control four outputs independently. Because of the rate feedback measurements the sensitivity function S is a 6×6 matrix and therefore two of its singular values (corresponding to p and q) are always close to 1 across all frequencies. All that can be done in these channels is to improve the disturbance rejection properties around crossover, 4 to 7 rad/s, and this was achieved using second-order band-pass filters in the last two elements of W_1.

Selection of $W_2(s)$ (input weight): The same first-order high-pass filter is used in each channel with a corner frequency of 10 rad/s to limit input magnitudes at high-frequencies and thereby limit the closed-loop bandwidth. The high-frequency gain of W_2 can be increased to limit fast actuator movement. The low-frequency gain of W_2 was set to approximately -100 dB to ensure that the cost function is dominated by W_1 at low frequencies.

Selection of $W_3(s)$ (setpoint filter): W_3 is a weighting on the reference input r. It is chosen to be a constant matrix with unity weighting on each of the output commands and a weighting of 0.1 on the fictitious rate demands. The reduced weighting on the rates (which are not directly controlled) enables some disturbance rejection on these outputs, without them significantly affecting the cost function. The main aim of W_3 is to force equally good tracking of each of the primary signals.

For the controller designed using the above weights, the singular value plots of S and KS are shown in Figure 13.3(a) and (b). These have the general shapes and bandwidths designed for and, as already mentioned, the controlled system performed well in piloted simulation. The effects of atmospheric turbulence will be illustrated later after designing a second controller in which disturbance rejection is explicitly included in the design problem.

13.2.4   Disturbance rejection design

In the design below we will assume that the atmospheric turbulence can be modelled as gust velocity components that perturb the helicopter's velocity states v_x, v_y and v_z by $d = [\,d_1 \quad d_2 \quad d_3\,]^T$ as in the following equations. The disturbed system is therefore expressed as

$$\dot{x} \quad = \quad Ax + A \begin{bmatrix} 0 \\ d \end{bmatrix} + Bu \tag{13.7}$$

$$y \quad = \quad Cx \tag{13.8}$$

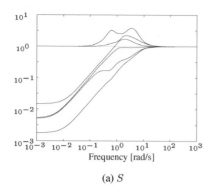
(a) S

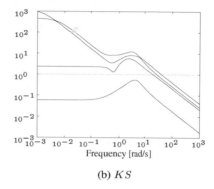
(b) KS

Figure 13.3: Singular values of S and KS (S/KS design)

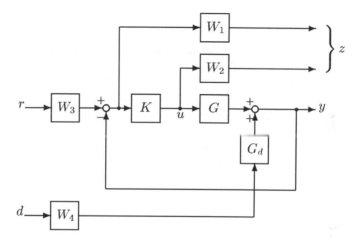

Figure 13.4: Disturbance rejection design

We define $B_d \triangleq$ columns 6, 7 and 8 of A. Then we have

$$\dot{x} = Ax + Bu + B_d d \qquad (13.9)$$
$$y = Cx \qquad (13.10)$$

which in transfer function terms can be expressed as

$$y = G(s)u + G_d(s)d \qquad (13.11)$$

where $G(s) = C(sI - A)^{-1}B$, and $G_d(s) = C(sI - A)^{-1}B_d$. The design problem we will solve is illustrated in Figure 13.4. The optimization problem is to find a stabilizing controller K that minimizes the cost function

$$\left\| \begin{bmatrix} W_1 S W_3 & -W_1 S G_d W_4 \\ W_2 K S W_3 & -W_2 K S G_d W_4 \end{bmatrix} \right\|_\infty \qquad (13.12)$$

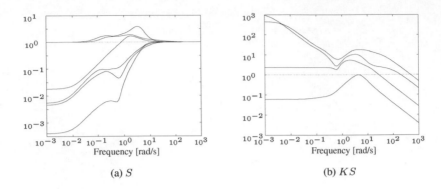

(a) S (b) KS

Figure 13.5: Singular values of S and KS (disturbance rejection design)

which is the $\mathcal{H}_\infty$ norm of the transfer function from $\begin{bmatrix} r \\ d \end{bmatrix}$ to z. This is easily cast into the general control configuration and solved using standard software. Notice that if we set W_4 to zero the problem reverts to the S/KS mixed-sensitivity design of the previous subsection. To synthesize the controller we used the same weights W_1, W_2 and W_3 as in the S/KS design, and selected $W_4 = \alpha I$, with α a scalar parameter used to emphasize disturbance rejection. After a few iterations we finalized on $\alpha = 30$. For this value of α, the singular value plots of S and KS, see Figure 13.5(a) and (b), are quite similar to those of the S/KS design, but as we will see in the next subsection there is a significant improvement in the rejection of gusts. Also, since G_d shares the same dynamics as G, and W_4 is a constant matrix, the degree of the disturbance rejection controller is the same as that for the S/KS design.

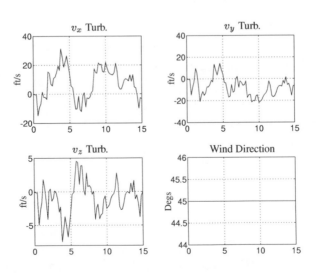

Figure 13.6: Velocity components of turbulence (time in seconds)

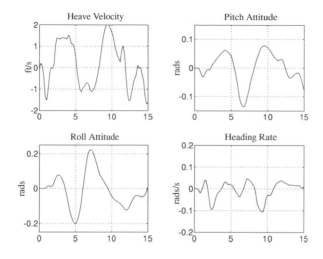

Figure 13.7: Response to turbulence of the S/KS design (time in seconds)

13.2.5 Comparison of disturbance rejection properties of the two designs

To compare the disturbance rejection properties of the two designs we simulated both controllers on the RHM nonlinear helicopter model equipped with a statistical discrete gust model for atmospheric turbulence (Dahl and Faulkner, 1979). With this simulation facility, gusts cannot be generated at hover and so the nonlinear model was trimmed at a forward flight speed of 20 knots (at an altitude of 100 ft (30 m)), and the effect of turbulence on the four controlled outputs observed. Recall that both designs were based on a linearized model about hover and therefore these tests at 20 knots also demonstrate the robustness of the controllers. Tests were carried out for a variety of gusts, and in all cases the disturbance rejection design was significantly better than the S/KS design.

In Figure 13.6, we show a typical gust generated by the RHM. The effects of this on the controlled outputs are shown in Figures 13.7 and 13.8 for the S/KS design and the disturbance rejection design, respectively. Compared with the S/KS design, the disturbance rejection controller practically halves the turbulence effect on heave velocity, pitch attitude and roll attitude. The change in the effect on heading rate is small.

13.2.6 Conclusions

The two controllers designed were of the same degree and had similar frequency domain properties. But by incorporating knowledge about turbulence activity into the second design, substantial improvements in disturbance rejection were achieved. The reduction of the turbulence effects by a half in heave velocity, pitch attitude and roll attitude indicates the possibility of a significant reduction in a pilot's workload, allowing more aggressive manoeuvres to be carried out with greater precision. Passenger comfort and safety would also be increased.

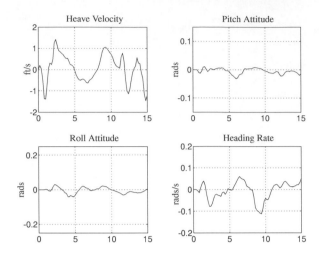

Figure 13.8: Response to turbulence of the disturbance rejection design (time in seconds)

The study was primarily meant to illustrate the ease with which information about disturbances can be beneficially included in controller design. The case study also demonstrated the selection of weights in $\mathcal{H}_\infty$ mixed-sensitivity design. To read how the $\mathcal{H}_\infty$ methods have been successfully used and tested in flight on a Bell 205 fly-by-wire helicopter, see Postlethwaite et al. (1999), Smerlas et al. (2001), Prempain and Postlethwaite (2004) and Postlethwaite et al. (2005). A series of flight tests carried out in 2004 resulted in level 1 (the highest) handling qualities ratings for all manoeuvres tested. There results were still to be written up, when this book went to press. For more flight control examples and illustrations of the usefulness of robust multivariable control, see Bates and Postlethwaite (2002).

13.3   Aero-engine control

In this case study, we apply a variety of tools to the problem of output selection, and illustrate the application of the two degrees-of-freedom $\mathcal{H}_\infty$ loop-shaping design procedure.

13.3.1   Problem description

This case study explores the application of advanced control techniques to the problem of control structure design and robust multivariable controller design for a high-performance gas turbine engine. The engine under consideration is the Spey engine which is a Rolls-Royce two-spool reheated turbofan, used to power modern military aircraft. The engine has two compressors: a low-pressure (LP) compressor or fan, and a high-pressure (HP) or core compressor as shown in Figure 13.9. The high-pressure flow at the exit of the core compressor is combusted and allowed to expand partially through the HP and LP turbines which drive the two compressors. The flow finally expands to atmospheric pressure at the nozzle exit, thus producing thrust for aircraft propulsion. The efficiency of the engine and the thrust produced

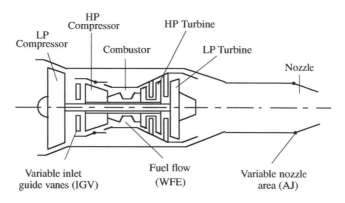

Figure 13.9: Schematic of the aero-engine

depends on the pressure ratios generated by the two compressors. If the pressure ratio across a compressor exceeds a certain maximum, it may no longer be able to hold the pressure head generated and the flow will tend to reverse its direction. This happens in practice, with the flow actually going negative, but it is only a momentary effect. When the back pressure has cleared itself, positive flow is re-established but, if flow conditions do not change, the pressure builds up causing flow reversal again. Thus the flow surges back and forth at high frequency, the phenomenon being referred to as *surge*. Surging causes excessive aerodynamic pulsations which are transmitted through the whole machine and must be avoided at all costs. However, for higher performance and greater efficiency the compressors must also be operated close to their surge lines. The primary aim of the control system is thus to control engine thrust whilst regulating compressor surge margins. But these engine parameters, namely thrust and the two compressor surge margins, are not directly measurable. There are, however, a number of measurements available which represent these quantities, and our first task is to choose from the available measurements, the ones that are in some sense better for control purposes. This is the problem of output selection as discussed in Chapter 10.

The next step is the design of a robust multivariable controller which provides satisfactory performance over the entire operating range of the engine. Since the aero-engine is a highly nonlinear system, it is normal for several controllers to be designed at different operating points and then to be scheduled across the flight envelope. Also in an aero-engine there are a number of parameters, apart from the ones being primarily controlled, that are to be kept within specified safety limits, e.g. the turbine blade temperature. The number of parameters to be controlled and/or limited exceeds the number of available inputs, and hence all these parameters cannot be controlled independently at the same time. The problem can be tackled by designing a number of scheduled controllers, each for a different set of output variables, which are then switched between, depending on the most significant limit at any given time. The switching is usually done by means of lowest-wins or highest-wins gates, which serve to propagate the output of the most suitable controller to the plant input. Thus, a switched gain-scheduled controller can be designed to cover the full operating range and all possible configurations. In Postlethwaite et al. (1995) a digital multi-mode scheduled controller is designed for the Spey engine under consideration here. In their study gain scheduling was not required to meet the design specifications. Below we will describe the design of a robust controller for the primary engine outputs using the two degrees-of-freedom $\mathcal{H}_\infty$ loop-shaping approach. The same methodology was used in the design of Postlethwaite et al. (1995) which was successfully implemented and tested on the Spey engine.

13.3.2   Control structure design: output selection

The Spey engine has three inputs, namely fuel flow (WFE), a nozzle with a variable area (AJ), and inlet guide vanes with a variable angle setting (IGV):

$$u = [\,\text{WFE}\quad \text{AJ}\quad \text{IGV}\,]^T$$

In this study, there are six output measurements available,

$$y_{\text{all}} = [\,\text{NL}\quad \text{OPR1}\quad \text{OPR2}\quad \text{LPPR}\quad \text{LPEMN}\quad \text{NH}\,]^T$$

as described below. For each one of the six output measurements, a look-up table provides its desired optimal value (setpoint) as a function of the operating point. However, with three inputs we can only control three outputs independently so the first question we face is: which three?

Engine thrust (one of the parameters to be controlled) can be defined in terms of the LP compressor's spool speed (NL), the ratio of the HP compressor's outlet pressure to engine inlet pressure (OPR1), or the engine overall pressure ratio (OPR2). We will choose from these three measurements the one that is best for control:

- Engine thrust: Select one of NL, OPR1 and OPR2 (outputs 1, 2 and 3).

Similarly, the surge margin of the LP compressor can be represented by either the LP compressor's pressure ratio (LPPR) or the LP compressor's exit Mach number measurement (LPEMN), and a selection between the two has to be made:

- Surge margin: Select one of LPPR and LPEMN (outputs 4 and 5).

In this study we will not consider control of the HP compressor's surge margin, or other configurations concerned with the limiting of engine temperatures. Our third output will be the HP compressor's spool speed (NH), which it is also important to maintain within safe limits. (NH is actually the HP spool speed made dimensionless by dividing by the square root of the total inlet temperature and scaled so that it is a percentage of the maximum spool speed at a standard temperature of 288.15 K.)

- Spool speed: Select NH (output 6).

We have now subdivided the available outputs into three subsets, and decided to select one output from each subset. This gives rise to the six candidate output sets as listed in Table 13.3.2.

We now apply some of the tools given in Chapter 10 for tackling the output selection problem. It is emphasized at this point that a good physical understanding of the plant is very important in the context of this problem, and some measurements may have to be screened beforehand on practical grounds. A 15-state linear model of the engine (derived from a nonlinear simulation at 87% of maximum thrust) will be used in the analysis that follows. The model is available over the Internet (as described in the Preface), along with actuator dynamics which result in a plant model of 18 states for controller design. The nonlinear model used in this case study was provided by the UK Defence Research Agency (now QinetiQ) at Pyestock with the permission of Rolls-Royce Military Aero Engines Ltd.

Scaling. Some of the tools we will use for control structure selection are dependent on the scalings employed. Scaling the inputs and the candidate measurements, therefore, is vital before comparisons are made and can also improve the conditioning of the problem for design purposes. We use the method of scaling described in Section 9.4.2. The outputs are scaled such that equal magnitudes of cross-coupling into each of the outputs are equally undesirable. We have chosen to scale the thrust-related outputs such that one unit of each scaled measurement represents 7.5% of maximum thrust. A unit step demand on each of these scaled outputs would thus correspond to a demand of 7.5% (of maximum) in thrust. The surge-margin-related outputs are scaled so that one unit corresponds to 5% surge margin. If the controller designed provides an interaction of less than 10% between the scaled outputs (for unit reference steps), then we would have 0.75% or less change in thrust for a step demand of 5% in surge margin, and a 0.5% or less change in surge margin for a 7.5% step demand in thrust. The final output NH (which is already a scaled variable) was further scaled (divided by 2.2) so that a unit change in NH corresponds to a 2.2% change in NH. The inputs are scaled by 10% of their expected ranges of operation.

Table 13.2: RHP zeros and minimum singular value for the six candidate output sets

Set no.	Candidate controlled outputs	RHP zeros < 100 rad/s	$\underline{\sigma}(G(0))$
1	NL, LPPR, NH (1, 4, 6)	none	0.060
2	OPR1, LPPR, NH (2, 4, 6)	none	0.049
3	OPR2, LPPR, NH (3, 4, 6)	30.9	0.056
4	NL, LPEMN, NH (1, 5, 6)	none	0.366
5	OPR1, LPEMN, NH (2, 5, 6)	none	0.409
6	OPR2, LPEMN, NH (3, 5, 6)	27.7	0.392

Steady-state model. With these scalings the steady-state model $y_{all} = G_{all}u$ (with all the candidate outputs included) and the corresponding non-square RGA matrix, $\Lambda = G_{all} \times G_{all}^{\dagger T}$, are given by

$$G_{all} = \begin{bmatrix} 0.696 & -0.046 & -0.001 \\ 1.076 & -0.027 & 0.004 \\ 1.385 & 0.087 & -0.002 \\ 11.036 & 0.238 & -0.017 \\ -0.064 & -0.412 & 0.000 \\ 1.474 & -0.093 & 0.983 \end{bmatrix} \quad \Lambda(G_{all}) = \begin{bmatrix} 0.009 & 0.016 & 0.000 \\ 0.016 & 0.008 & -0.000 \\ 0.006 & 0.028 & -0.000 \\ 0.971 & -0.001 & 0.002 \\ -0.003 & 0.950 & 0.000 \\ 0.002 & -0.000 & 0.998 \end{bmatrix} \quad (13.13)$$

and the singular value decomposition of $G_{all}(0) = U_0 \Sigma_0 V_0^H$ is

$$U_0 = \begin{bmatrix} 0.062 & 0.001 & -0.144 & -0.944 & -0.117 & -0.266 \\ 0.095 & 0.001 & -0.118 & -0.070 & -0.734 & 0.659 \\ 0.123 & -0.025 & 0.133 & -0.286 & 0.640 & 0.689 \\ 0.977 & -0.129 & -0.011 & 0.103 & -0.001 & -0.133 \\ -0.006 & 0.065 & -0.971 & 0.108 & 0.195 & 0.055 \\ 0.131 & 0.989 & 0.066 & -0.000 & 0.004 & -0.004 \end{bmatrix}$$

$$\Sigma_0 = \begin{bmatrix} 11.296 & 0 & 0 \\ 0 & 0.986 & 0 \\ 0 & 0 & 0.417 \\ 0 & 0 & 0 \\ 0 & 0 & 0 \\ 0 & 0 & 0 \end{bmatrix} \qquad V_0 = \begin{bmatrix} 1.000 & -0.007 & -0.021 \\ 0.020 & -0.154 & 0.988 \\ 0.010 & 0.988 & 0.154 \end{bmatrix}$$

The six row sums of the RGA matrix are

$$\Lambda_\Sigma = [0.025 \quad 0.023 \quad 0.034 \quad 0.972 \quad 0.947 \quad 1.000]^T$$

and from (A.85) this indicates that we should select outputs 4, 5 and 6 (corresponding to the three largest elements) in order to maximize the projection of the selected outputs onto the space corresponding to the three non-zero singular values. However, this selection is not one of our six candidate output sets because there is no output directly related to engine thrust (outputs 1, 2 and 3).

We now proceed with a more detailed input–output controllability analysis of the six candidate output sets. In the following, $G(s)$ refers to the transfer function matrix for the effect of the three inputs on the selected three outputs.

Minimum singular value. In Chapter 10, we showed that a reasonable criterion for selecting controlled outputs y is to make $\|G^{-1}(y - y_{\text{opt}})\|$ small (page 395), in particular at steady-state. Here $y - y_{\text{opt}}$ is the deviation in y from its optimal value. At steady-state this deviation arises mainly from errors in the (look-up table) setpoint due to disturbances and unknown variations in the operating point. If we assume that, with the scalings given above, the magnitude $|(y - y_{\text{opt}})_i|$ is similar (close to 1) for each of the six outputs, then we should select a set of outputs such that the elements in $G^{-1}(0)$ are small, or alternatively, such that $\underline{\sigma}(G(0))$ is as large as possible (minimum singular value rule; see page 395). In Table 13.3.2 we have listed $\underline{\sigma}(G(0))$ for the six candidate output sets. We conclude that we can eliminate sets 1, 2 and 3, and consider only sets 4, 5 and 6. For these three sets we find that the value of $\underline{\sigma}(G(0))$ is between 0.366 and 0.409 which is only slightly smaller than $\underline{\sigma}(G_{\text{all}}(0)) = 0.417$.

Remark. The three eliminated sets all include output 4, LPPR. Interestingly, this output is associated with the largest element in the gain matrix $G_{\text{all}}(0)$ of 11.0, and is thus also associated with the largest singular value (as seen from the first column of U). This illustrates that the preferred choice is often not associated with $\bar{\sigma}(G)$.

Right-half plane zeros. RHP-zeros limit the achievable performance of a feedback loop by limiting the open-loop gain–bandwidth product. They can be a cause of concern, particularly if they lie within the desired closed-loop bandwidth. Also, choosing different outputs for feedback control can give rise to different numbers of RHP-zeros at different locations. The choice of outputs should be such that a minimum number of RHP-zeros are encountered, and should be as far removed from the imaginary axis as possible.

Table 13.3.2 shows the RHP-zeros slower than 100 rad/s for all combinations of prospective output variables. The closed-loop bandwidth requirement for the aero-engine is approximately 10 rad/s. RHP-zeros close to this value or smaller (closer to the origin) will, therefore, cause problems and should be avoided. It can be seen that the variable OPR2 introduces (relatively) slow RHP-zeros. It was observed that these zeros move closer to the origin at higher thrust levels. Thus sets 3 and 6 are unfavourable for closed-loop control. This along with the minimum singular value analysis leaves us with sets 4 and 5 for further consideration

Relative gain array (RGA). We here consider the RGAs of the candidate square transfer function matrices $G(s)$ with three outputs,

$$\Lambda(G(s)) = G(s) \times G^{-T}(s) \tag{13.14}$$

In Section 3.4, it is argued that the RGA provides useful information for the analysis of input–output controllability and for the pairing of inputs and outputs. Specifically input and output variables should be paired so that the diagonal elements of the RGA are as close as possible to unity. Furthermore, if the plant has large RGA elements and an inverting controller is used, the closed-loop system will have little robustness in the face of diagonal input uncertainty. Such a perturbation is quite common due to uncertainty in the actuators. Thus we want Λ to have small elements and for diagonal dominance we want $\Lambda - I$ to be small. These two objectives can be combined in the single objective of a small RGA number, defined as

$$\text{RGA number} \triangleq \|\Lambda - I\|_{\text{sum}} = \sum_{i=j} |1 - \lambda_{ij}| + \sum_{i \neq j} |\lambda_{ij}| \tag{13.15}$$

The lower the RGA number, the more preferred is the control structure. Before calculating the RGA number over frequency we rearranged the output variables so that the steady-state RGA matrix was as close as possible to the identity matrix.

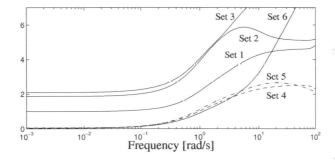

Figure 13.10: RGA numbers

The RGA numbers for the six candidate output sets are shown in Figure 13.10. As in the minimum singular value analysis above, we again see that sets 1, 2 and 3 are less favourable. Once more, sets 4 and 5 are the best but too similar to allow a decisive selection.

Hankel singular values. Notice that sets 4 and 5 differ only in one output variable, NL in set 4 and OPR1 in set 5. Therefore, to select between them we next consider the Hankel singular values of the two transfer functions between the three inputs and output NL and output OPR1, respectively. Hankel singular values reflect the joint controllability and observability of the states of a balanced realization (as described in Section 11.3). Recall that the Hankel singular values are invariant under state transformations but they do depend on scaling.

Figure 13.11 shows the Hankel singular values of the two transfer functions for outputs NL and OPR1, respectively. The Hankel singular values for OPR1 are larger, which indicates that OPR1 has better state controllability and observability properties than NL. In other words, output OPR1 contains more information about the system internal states than output NL. It therefore seems to be preferable to use OPR1 for control purposes rather than NL, and hence (in the absence of no other information) set 5 is our final choice.

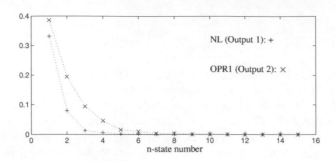

Figure 13.11: Hankel singular values

13.3.3 A two degrees-of-freedom $\mathcal{H}_\infty$ loop-shaping design

The design procedure given in Section 9.4.3 will be used to design a two degrees-of-freedom $\mathcal{H}_\infty$ loop-shaping controller for the 3-input 3-output plant G. An 18-state linear plant model G (including actuator dynamics) is available over the Internet. It is based on scaling, output selection and input–output pairing as described below. To summarize, the selected outputs (set 5) are

- y_1 : engine inlet pressure, OPR1
- y_2 : LP compressor's exit Mach number measurement, LPEMN
- y_3 : HP compressor's spool speed, NH

and the corresponding inputs are

- u_1 : fuel flow, WFE
- u_2 : nozzle area, AJ
- u_3 : inlet guide vane angle, IGV

The corresponding steady-state $(s = 0)$ model and RGA matrix is

$$G = \begin{bmatrix} 1.076 & -0.027 & 0.004 \\ -0.064 & -0.412 & 0.000 \\ 1.474 & -0.093 & 0.983 \end{bmatrix}, \quad \Lambda(G) = \begin{bmatrix} 1.002 & 0.004 & -0.006 \\ 0.004 & 0.996 & -0.000 \\ -0.006 & -0.000 & 1.006 \end{bmatrix} \quad (13.16)$$

Pairing of inputs and outputs. The pairing of inputs and outputs is important because it makes the design of the prefilter easier in a two degrees-of-freedom control configuration and simplifies the selection of weights. It is of even greater importance if a decentralized control scheme is to be used, and gives insight into the working of the plant. In Chapter 10, it is argued that negative entries on the principal diagonal of the steady-state RGA should be avoided and that the outputs in G should be (re)arranged such that the RGA is close to the identity matrix. For the selected output set, we see from (13.16) that no rearranging of the outputs is needed. That is, we should pair OPR1, LPEMN and NH with WFE, AJ and IGV, respectively.

$\mathcal{H}_\infty$ **loop-shaping design.** We follow the design procedure given in Section 9.4.3. In steps 1 to 3 we discuss how pre- and post-compensators are selected to obtain the desired shaped plant (loop shape) $G_s = W_2 G W_1$ where $W_1 = W_p W_a W_b$. In steps 4 to 6 we present the subsequent $\mathcal{H}_\infty$ design.

1. The singular values of the plant are shown in Figure 13.12(a) and indicate a need for extra low-frequency gain to give good steady-state tracking and disturbance rejection. The

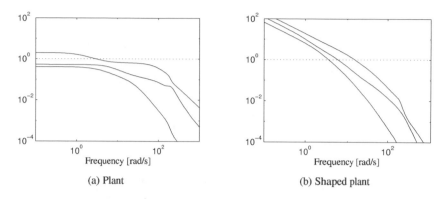

(a) Plant (b) Shaped plant

Figure 13.12: Singular values for plant and shaped plant

pre-compensator weight is chosen as simple integrators, i.e. $W_p = \frac{1}{s} I_3$, and the post-compensator weight is selected as $W_2 = I_3$.

2. $W_2 G W_p$ is next aligned at 7 rad/s. The align gain W_a (used in front of W_p) is the approximate real inverse of the shaped system at the specified frequency. The crossover is thus adjusted to 7 rad/s in order to give a closed-loop bandwidth of approximately 10 rad/s. Alignment should not be used if the plant is ill-conditioned with large RGA elements at the selected alignment frequency. In our case the RGA elements are small (see Figure 13.10) and hence alignment is not expected to cause problems.

3. An additional gain W_g is used in front of the align gain to give some control over actuator usage. W_g is adjusted so that the actuator rate limits are not exceeded for reference and disturbance steps on the scaled outputs. By some trial and error, W_g is chosen to be $\mathrm{diag}(1, 2.5, 0.3)$. This indicates that the second actuator (AJ) is made to respond at higher rates whereas the third actuator (IGV) is made slower. The shaped plant now becomes $G_s = G W_1$ where $W_1 = W_p W_a W_g$. Its singular values are shown in Figure 13.12(b).

4. $\gamma_{\min}$ in (9.66) for this shaped plant is found to be 2.3 which indicates that the shaped plant is compatible with robust stability.

5. ρ is set to 1 and the reference model T_{ref} is chosen as

$$T_{\mathrm{ref}} = \mathrm{diag} \left\{ \frac{1}{0.018s + 1}, \frac{1}{0.008s + 1}, \frac{1}{0.2s + 1} \right\}$$

The third output NH is thus made slower than the other two in following reference inputs.

6. The standard $\mathcal{H}_\infty$ optimization defined by P in (9.87) is solved. γ iterations are performed and a slightly suboptimal controller achieving $\gamma = 2.9$ is obtained. Moving closer to optimality introduces very fast poles in the controller which, if the controller is to be discretized, would ask for a very high sample rate. Choosing a slightly suboptimal controller alleviates this problem and also improves on the H_2 performance. The prefilter is finally scaled to achieve perfect steady-state model matching. The controller (with the weights W_1 and W_2) has 27 states.

13.3.4   Analysis and simulation results

Step responses of the linear controlled plant model are shown in Figure 13.13. The decoupling

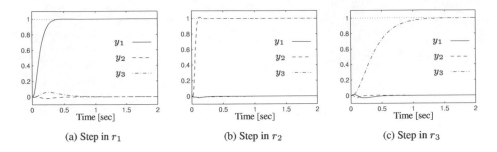

(a) Step in r_1 (b) Step in r_2 (c) Step in r_3

Figure 13.13: Reference step responses

is good with less than 10% interactions. Although not shown here, the control inputs were analyzed and the actuator signals were found to lie within specified limits. Responses to disturbance steps on the outputs were also seen to meet the problem specifications. Notice that because there are two degrees of freedom in the controller structure, the reference to output and disturbance to output transfer functions can be given different bandwidths.

The robustness properties of the closed-loop system are now analyzed. Figure 13.14(a) shows the singular values of the sensitivity function. The peak value is less than 2 (actually

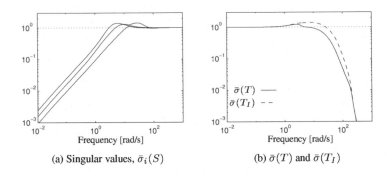

(a) Singular values, $\bar{\sigma}_i(S)$ (b) $\bar{\sigma}(T)$ and $\bar{\sigma}(T_I)$

Figure 13.14: Sensitivity and complementary sensitivity functions

it is $1.44 = 3.2$ dB), which is considered satisfactory. Figure 13.14(b) shows the maximum singular values of $T = (I - GW_1K_2)^{-1}GW_1K_2$ and $T_I = (I - W_1K_2G)^{-1}W_1K_2G$. Both of these have small peaks and go to zero quickly at high frequencies. From Section 9.2.2, this indicates good robustness with respect to both multiplicative output and multiplicative input plant perturbations.

Nonlinear simulation results are shown in Figure 13.15. Reference signals are given to each of the scaled outputs simultaneously. The solid lines show the references, and the dashed-dot lines, the outputs. It can be seen that the controller exhibits good performance with low interactions.

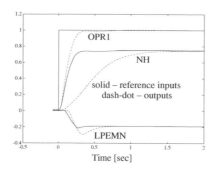

Figure 13.15: Nonlinear simulation results

13.3.5 Conclusions

The case study has demonstrated the ease with which the two degrees-of-freedom $\mathcal{H}_\infty$ loop-shaping design procedure can be applied to a complex engineering system. Some tools for control structure design have also been usefully applied to the aero-engine example. We stress that a good control structure selection is very important. It results in simpler controllers and, in general, a simpler design exercise.

13.4 Distillation process

A typical distillation column is shown in Figure 10.6 on page 408. The overall 5×5 control problem is discussed in Example 10.8 (page 406) and you are advised to read this first. The commonly used LV- and DV-configurations, which are discussed below, are partially controlled systems where three loops for liquid level and pressure have already been closed.

For a general discussion on distillation column control, the reader is also referred to Shinskey (1984), Skogestad and Morari (1987a) and the survey paper by Skogestad (1997).

We have throughout the book studied a particular high-purity binary distillation column with 40 theoretical stages (39 ideal trays and a reboiler) plus a total condenser. This is "column A" in Skogestad et al. (1990). The feed is an equimolar liquid mixture of two components with a relative volatility of 1.5. The pressure p is assumed constant (perfect control of p using V_T as an input). The operating variables (e.g. reflux and boilup rates) are such that we nominally have 99% purity for each product (y_D and x_B). The nominal holdups on all stages, including the reboiler and condenser, are $M_i^*/F = 0.5$ min. The liquid flow dynamics, which are important for control, are modelled by a simple linear relationship, $L_i(t) = L_i^* + (M_i(t) - M_i^*)/\tau_L$, where $\tau_L = 0.063$ min (the same value is used on all trays). No actuator or measurement dynamics are included. This results in a model with 82 states. This distillation process is difficult to control because of strong interactions between the two product compositions. More information, including steady-state profiles along the column, is available over the Internet.

The complete linear distillation column model with 4 inputs (L, V, D, B), 4 outputs (y_D, x_B, M_D, M_B), 2 disturbances (F, z_F) and 82 states is available over the Internet. The states are the mole fractions and liquid holdups on each of the 41 stages. By closing the two level loops (M_D and M_B) this model may be used to generate the model for any configuration (LV, DV, etc.). The Matlab commands for generating the LV-, DV- and DB-configurations are given in Table 13.3.

A 5-state LV-model, obtained by model reducing the above 82-state model, is given on page 513. This model is also available over the Internet.

Table 13.3: Matlab program for generating model of various distillation configurations

```
% Uses Matlab Robust control toolbox
% G4: State-space model (4 inputs, 2 disturbances, 4 outputs, 82 states)
% Level controllers using D and B (P-controllers; bandwidth = 10 rad/min):
Kd = 10; Kb = 10;
% Now generate the LV-configuration from G4 using sysic:
systemnames = 'G4 Kd Kb';
inputvar = '[L(1); V(1); d(2)]';
outputvar = '[G4(1);G4(2)]';
input_to_G4 = '[L; V; Kd; Kb; d ]';
input_to_Kd = '[G4(3)]';
input_to_Kb = '[G4(4)]';
sysoutname ='Glv';
cleanupsysic='yes'; sysic;
%
% Modifications needed to generate DV-configuration:
Kl = 10; Kb = 10;
systemnames = 'G4 Kl Kb';
inputvar = '[D(1); V(1); d(2)]';
input_to_G4 = '[Kl; V; D; Kb; d ]';
input_to_Kl = '[G4(3)]';
input_to_Kb = '[G4(4)]';
sysoutname ='Gdv';
%
% Modifications needed to generate DB-configuration:
Kl = 10; Kv = 10;
systemnames = 'G4 Kl Kv';
inputvar = '[D(1); B(1); d(2)]';
input_to_G4 = '[Kl; Kv; D; B; d ]';
input_to_Kl = '[G4(3)]';
input_to_Kv = '[G4(4)]';
sysoutname ='Gdb';
```

This distillation process has been used as an illustrative example throughout the book, and so to avoid unnecessary repetition we will simply summarize what has been done and refer to the many exercises and examples for more details. The steady-state properties of the model, including the choice of temperature measurement, are discussed in Examples 10.8 and 10.9.

13.4.1 Idealized LV-model

The following idealized model of the distillation process, originally from Skogestad et al. (1988), has been used in examples throughout the book:

$$G(s) = \frac{1}{75s + 1}\begin{bmatrix} 87.8 & -86.4 \\ 108.2 & -109.6 \end{bmatrix} \tag{13.17}$$

The inputs are the reflux (L) and boilup (V), and the controlled outputs are the top and bottom product compositions (y_D and x_B). This is a very crude model of the distillation process, but it provides an excellent example of an ill-conditioned process where control is difficult, primarily due to the presence of input uncertainty.

We refer the reader to the following places in the book where the model (13.17) is used:

Example 3.5 (page 78): SVD analysis. The singular values are plotted as a function of frequency in Figure 3.7(b) on page 80.

Example 3.6 (page 79): Discussion of the physics of the process and the interpretation of directions.

Example 3.14 (page 89): The condition number, $\gamma(G)$, is 141.7, and the 1, 1 element of the RGA, $\lambda_{11}(G)$, is 35.1 (at all frequencies).

Motivating example no. 2 (page 100): Introduction to robustness problems with inverse-based controller using simulation with 20% input uncertainty.

Exercise 3.10 (page 103): Design of robust SVD controller.

Exercise 3.11 (page 103): Combined input and output uncertainty for inverse-based controller.

Exercise 3.12 (page 103): Attempt to "robustify" an inverse-based design using McFarlane–Glover $\mathcal{H}_\infty$ loop-shaping procedure.

Example 6.8 (page 245): Sensitivity to input uncertainty with feedforward control (RGA).

Example 6.11 (page 250): Sensitivity to input uncertainty with inverse-based controller, sensitivity peak (RGA).

Example 6.14 (page 253): Sensitivity to element-by-element uncertainty (relevant for identification).

Example 8.1 (page 292): Coupling between uncertainty in transfer function elements.

Example in Section 8.11.3 (page 322): μ for robust performance which explains poor performance in Motivating example no. 2.

Example in Section 8.12.4 (page 330): Design of μ-optimal controller using DK-iteration.

In addition, the reader is referred to the first edition of this book (Skogestad and Postlethwaite, 1996) for an example on the magnitude of inputs for rejecting disturbances (in feed rate and feed composition) at steady state.

The model in (13.17) has also been the basis for two benchmark problems.

Original benchmark problem. The original control problem was formulated by Skogestad et al. (1988) as a bound on the weighted sensitivity with frequency-bounded input uncertainty. The optimal solution to this problem is provided by the one degree-of-freedom μ-optimal controller given in the example in Section 8.12.4 where a peak μ-value of 0.974 (Remark 1 on page 335) was obtained.

Revised CDC benchmark problem. The original problem formulation is unrealistic in that there is no bound on the input magnitudes. Furthermore, the bounds on performance and uncertainty are given in the frequency domain (in terms of weighted $\mathcal{H}_\infty$ norm), whereas many engineers feel that time domain specifications are more realistic. Limebeer (1991) therefore suggested the following CDC specifications. The set of plants Π is defined by

$$
\tilde{G}(s) = \frac{1}{75s + 1} \begin{bmatrix} 0.878 & -0.864 \\ 1.082 & -1.096 \end{bmatrix} \begin{bmatrix} k_1 e^{-\theta_1 s} & 0 \\ 0 & k_2 e^{-\theta_2 s} \end{bmatrix}
$$
$$
k_i \in \begin{bmatrix} 0.8 & 1.2 \end{bmatrix}, \quad \theta_i \in \begin{bmatrix} 0 & 1.0 \end{bmatrix} \tag{13.18}
$$

In physical terms this means 20% gain uncertainty and up to 1 minute delay in each input channel. The specification is to achieve for every plant $\widetilde{G} \in \Pi$:

S1: Closed-loop stability.

S2: For a unit step demand in channel 1 at $t = 0$ the plant output y_1 (tracking) and y_2 (interaction) should satisfy:

- $y_1(t) \geq 0.9$ for all $t \geq 30$ min
- $y_1(t) \leq 1.1$ for all t
- $0.99 \leq y_1(\infty) \leq 1.01$
- $y_2(t) \leq 0.5$ for all t
- $-0.01 \leq y_2(\infty) \leq 0.01$

The same corresponding requirements hold for a unit step demand in channel 2.

S3: $\bar{\sigma}(K_y \widetilde{S}) < 0.316, \ \forall \omega$

S4: $\bar{\sigma}(\widetilde{G} K_y) < 1$ for $\omega \geq 150$

Note that a two degrees-of-freedom controller may be used and K_y then refers to the feedback plant of the controller. In practice, specification S4 is indirectly satisfied by S3. Note that the uncertainty description $G_p = G(I + \omega_I \Delta_I)$ with $w_I = \frac{s+0.2}{0.5s+1}$ (as used in the examples in the book) only allows for about 0.9 minute time delay error. To get a weight $w_I(s)$ which includes the uncertainty in (13.18) we may use the procedure described on page 272, i.e. (7.36) or (7.37) with $r_k = 0.2$ and $\theta_{max} = 1$.

Several designs have been presented which satisfy the specifications for the CDC problem in (13.18). For example, a two degrees-of-freedom $\mathcal{H}_\infty$ loop-shaping design is given by Limebeer et al. (1993), and an extension of this by Whidborne et al. (1994). A two degrees-of-freedom μ-optimal design is presented by Lundström et al. (1999).

13.4.2   Detailed LV-model

In the book we have also used a 5-state dynamic model of the distillation process which includes liquid flow dynamics (in addition to the composition dynamics) as well as disturbances. This 5-state model was obtained from model reduction of the detailed model with 82 states. The steady-state gains for the two disturbances are given in (10.97).

The 5-state model is similar to (13.17) at low frequencies, but the model is much less interactive at higher frequencies. The physical reason for this is that the liquid flow dynamics decouple the response and make $G(j\omega)$ upper triangular at higher frequencies. The effect is illustrated in Figure 13.16 where we show the singular values and the magnitudes of the RGA elements as functions of frequency. As a comparison, the RGA element $\lambda_{11}(G) = 35.1$ at all frequencies (and not just at steady-state) for the simplified model in (13.17). The implication is that control at crossover frequencies is easier than expected from the simplified model (13.17).

Applications based on the 5-state model are found in:

Example 10.9 (page 409): Selection of secondary (temperature) measurement for improving controllability of primary (composition) variables.

Example in Section 10.23 (page 452): Controllability analysis of decentralized control.

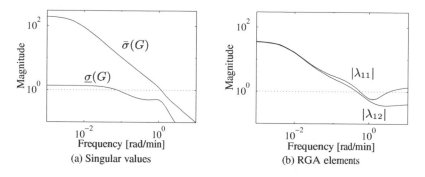

Figure 13.16: Detailed 5-state model of distillation column

Details on the 5-state model. A state-space realization is

$$G(s) \overset{s}{=} \left[\begin{array}{c|c} A & B \\ \hline C & 0 \end{array}\right], \quad G_d(s) \overset{s}{=} \left[\begin{array}{c|c} A & B_d \\ \hline C & 0 \end{array}\right] \tag{13.19}$$

where

$$A = \begin{bmatrix} -.005131 & 0 & 0 & 0 & 0 \\ 0 & -.07366 & 0 & 0 & 0 \\ 0 & 0 & -.1829 & 0 & 0 \\ 0 & 0 & 0 & -.4620 & .9895 \\ 0 & 0 & 0 & -.9895 & -.4620 \end{bmatrix}, \quad B = \begin{bmatrix} -.629 & .624 \\ .055 & -.172 \\ .030 & -.108 \\ -.186 & -.139 \\ -1.23 & -.056 \end{bmatrix}$$

$$C = \begin{bmatrix} .7223 & .5170 & .3386 & .1633 & .1121 \\ -.8913 & .4728 & .9876 & .8425 & .2186 \end{bmatrix}, \quad B_d = \begin{bmatrix} -0.062 & -0.067 \\ 0.131 & 0.040 \\ 0.022 & -0.106 \\ -0.188 & 0.027 \\ -0.045 & 0.014 \end{bmatrix}$$

Scaling. The model is scaled such that a magnitude of 1 corresponds to the following: 0.01 mole fraction units for each output (y_D and x_B), the nominal feed flow rate for the two inputs (L and V) and a 20% change for each disturbance (feed rate F and feed composition z_F). Notice that the steady-state gains computed with this model are slightly different from the ones used in the examples.

Remark. A similar dynamic LV-model, but with 8 states, is given by Green and Limebeer (1995), who also design an $\mathcal{H}_\infty$ loop-shaping controller.

Exercise 13.1 * *Repeat the μ-optimal design based on DK-iteration in Section 8.12.4 using the model (13.19).*

13.4.3   Idealized DV-model

Finally, we have also made use of an idealized model for the DV-configuration:

$$G(s) = \frac{1}{75s + 1}\begin{bmatrix} -87.8 & 1.4 \\ -108.2 & -1.4 \end{bmatrix} \tag{13.20}$$

In this case the condition number $\gamma(G) = 70.8$ is still large, but the RGA elements are small (about 0.5).

Example 6.9 (page 245): Bounds on the sensitivity peak show that an inverse-based controller is robust with respect to diagonal input uncertainty.

Example 8.9 (page 314): μ for robust stability with a diagonal controller is computed. The difference between diagonal and full-block input uncertainty is significant.

Remark. In practice, the DV-configuration may not be as favourable as indicated by these examples, because the level controller is not perfect as was assumed when deriving (13.20).

13.4.4   Further distillation case studies

The full distillation model, which is available over the Internet, may form the basis for several case studies (projects). These could include input–output controllability analysis, controller design, robustness analysis and closed-loop simulation. The following cases may be considered:

1. Model with four inputs and four outputs
2. LV-configuration (studied extensively in this book)
3. DV-configuration (see previous page)
4. DB-configuration (see also Exercise 6.16, page 257)

The models in the latter three cases are generated from the 4×4 model by closing two level loops (see the Matlab file in Table 13.3) to get a partially controlled plant with two inputs and two outputs (in addition to the two disturbances).

Remark. For the DV- and DB-configurations the resulting model depends quite strongly on the tuning of the level loops, so one may consider separately the two cases of tight level control (e.g. $K = 10$, as in Table 13.3) or loosely tuned level control (e.g. $K = 0.2$ corresponding to a time constant of 5 min). Level control tuning may also be considered as a source of uncertainty. The models do not include actuator or measurement dynamics, which may also be considered as a source of uncertainty.

13.5   Conclusion

The case studies in this chapter have served to demonstrate the usefulness and ease of application of many of the techniques discussed in the book. Realistic problems have been considered but the idea has been to illustrate the techniques rather than to provide "optimal" solutions.

For the helicopter problem, practice was obtained in the selection of weights in $\mathcal{H}_\infty$ mixed-sensitivity design, and it was seen how information about disturbances could easily be considered in the design problem.

In the aero-engine study, we applied a variety of tools to the problem of output selection and then designed a two degrees-of-freedom $\mathcal{H}_\infty$ loop-shaping controller.

The final case study was a collection of examples and exercises on the distillation process considered throughout the book. This served to illustrate the difficulties of controlling ill-conditioned plants and the adverse effects of model uncertainty. The structured singular value played an important role in the robustness analysis.

You should now be in a position to move straight to Appendix B, to complete a major project on your own and to sit the sample exam.

Good luck!

APPENDIX A

MATRIX THEORY AND NORMS

The topics in this appendix are included as background material for the book, and should ideally be studied before reading Chapter 3.

After studying the appendix the reader should feel comfortable with a range of mathematical tools including eigenvalues, eigenvectors and the singular value decomposition; the reader should appreciate the difference between various norms of vectors, matrices, signals and systems, and know how these norms can be used to measure performance.

The main references are: Strang (1976) and Horn and Johnson (1985) on matrices, and Zhou et al. (1996) on norms.

A.1 Basics

Let us start with a complex scalar

$$c = \alpha + j\beta, \quad \text{where } \alpha = \text{Re } c, \ \beta = \text{Im } c$$

To compute the magnitude $|c|$, we multiply c by its conjugate $\bar{c} \triangleq \alpha - j\beta$ and take the square root, i.e.

$$|c| = \sqrt{\bar{c}c} = \sqrt{\alpha^2 - j^2\beta^2} = \sqrt{\alpha^2 + \beta^2}$$

A complex *column vector* a with m components (elements) is written

$$a = \begin{bmatrix} a_1 \\ a_2 \\ \vdots \\ a_m \end{bmatrix}$$

where a_i is a complex scalar. a^T (the transpose) is used to denote a row vector.

Now consider a complex $l \times m$ matrix A with elements $a_{ij} = \text{Re } a_{ij} + j \text{ Im } a_{ij}$. l is the number of rows (the number of "outputs" when viewed as an *operator*) and m is the number of columns ("inputs"). Mathematically, we write $A \in \mathbb{C}^{l \times m}$ if A is a complex matrix, or $A \in \mathbb{R}^{l \times m}$ if A is a real matrix. Note that a column vector a with m elements may be viewed as an $m \times 1$ matrix.

The transpose of a matrix A is A^T (with elements a_{ji}), the conjugate is $\bar{A}$ (with elements $\text{Re } a_{ij} - j \text{ Im } a_{ij}$), the conjugate transpose (or Hermitian adjoint) matrix is $A^H \triangleq \bar{A}^T$ (with elements $\text{Re } a_{ji} - j \text{ Im } a_{ji}$), the trace is $\text{tr } A$ (sum of diagonal elements), and the determinant is $\det A$. By definition, the inverse of a non-singular matrix A, denoted A^{-1},

Multivariable Feedback Control: Analysis and Design. Second Edition
S. Skogestad and I. Postlethwaite   © 2005, 2006 John Wiley & Sons, Ltd

satisfies $A^{-1}A = AA^{-1} = I$, and is given by

$$A^{-1} = \frac{\text{adj } A}{\det A} \tag{A.1}$$

where adj A is the adjugate (or "classical adjoint") of A which is the transposed matrix of cofactors c_{ij} of A,

$$c_{ij} = [\text{adj } A]_{ji} \triangleq (-1)^{i+j} \det A^{ij} \tag{A.2}$$

Here A^{ij} is a submatrix formed by deleting row i and column j of A. As an example, for a 2×2 matrix we have

$$A = \begin{bmatrix} a_{11} & a_{12} \\ a_{21} & a_{22} \end{bmatrix}; \quad \det A = a_{11}a_{22} - a_{12}a_{21}$$

$$A^{-1} = \frac{1}{\det A} \begin{bmatrix} a_{22} & -a_{12} \\ -a_{21} & a_{11} \end{bmatrix} \tag{A.3}$$

We also have

$$(AB)^T = B^T A^T, \quad (AB)^H = B^H A^H \tag{A.4}$$

and, assuming the inverses exist,

$$(AB)^{-1} = B^{-1}A^{-1} \tag{A.5}$$

A square matrix A is *symmetric* if $A^T = A$, and *Hermitian* if $A^H = A$.

A matrix A is said to be *positive definite* if all the eigenvalues of its symmetric part $(A + A^H)$ are positive or $x^H(A + A^H)x > 0$ for any non-zero vector x. When A is a Hermitian matrix, the condition for positive definiteness simplifies as $x^H A x > 0$; this is simply denoted $A > 0$. Similarly, a Hermitian matrix A is positive semi-definite $(A \geq 0)$ if $x^H A x \geq 0$. For a positive semi-definite matrix A, the matrix square root $(A^{1/2})$ satisfies $A^{1/2}A^{1/2} = A$.

A.1.1 Some useful matrix identities

Lemma A.1 The matrix inversion lemma. *Let A_1, A_2, A_3 and A_4 be matrices with compatible dimensions such that the matrices $A_2A_3A_4$ and $(A_1 + A_2A_3A_4)$ are defined. Also assume that the inverses given below exist. Then*

$$(A_1 + A_2A_3A_4)^{-1} = A_1^{-1} - A_1^{-1}A_2(A_4A_1^{-1}A_2 + A_3^{-1})^{-1}A_4A_1^{-1} \tag{A.6}$$

Proof: Postmultiply (or premultiply) the right hand side in (A.6) by $A_1 + A_2A_3A_4$. This gives the identity matrix.   □

Lemma A.2 Inverse of a partitioned matrix. *If A_{11}^{-1} and X^{-1} exist then*

$$\begin{bmatrix} A_{11} & A_{12} \\ A_{21} & A_{22} \end{bmatrix}^{-1} = \begin{bmatrix} A_{11}^{-1} + A_{11}^{-1}A_{12}X^{-1}A_{21}A_{11}^{-1} & -A_{11}^{-1}A_{12}X^{-1} \\ -X^{-1}A_{21}A_{11}^{-1} & X^{-1} \end{bmatrix} \tag{A.7}$$

where $X \triangleq A_{22} - A_{21}A_{11}^{-1}A_{12}$ is the Schur complement of A_{11} in A; also see (A.15). Similarly if A_{22}^{-1} and Y^{-1} exist then

$$\begin{bmatrix} A_{11} & A_{12} \\ A_{21} & A_{22} \end{bmatrix}^{-1} = \begin{bmatrix} Y^{-1} & -Y^{-1}A_{12}A_{22}^{-1} \\ -A_{22}^{-1}A_{21}Y^{-1} & A_{22}^{-1} + A_{22}^{-1}A_{21}Y^{-1}A_{12}A_{22}^{-1} \end{bmatrix} \tag{A.8}$$

where $Y \triangleq A_{11} - A_{12}A_{22}^{-1}A_{21}$ is the Schur complement of A_{22} in A; also see (A.16).

A.1.2 Some determinant identities

The determinant is defined only for square matrices, so let A be an $n \times n$ matrix. The matrix is non-singular if $\det A$ is non-zero. The determinant may be defined inductively as $\det A = \sum_{i=1}^{n} a_{ij} c_{ij}$ (expansion along column j) or $\det A = \sum_{j=1}^{n} a_{ij} c_{ij}$ (expansion along row i), where c_{ij} is the ij'th cofactor given in (A.2). This inductive definition begins by defining the determinant of an 1×1 matrix (a scalar) to be the value of the scalar, i.e. $\det a = a$. We then get for a 2×2 matrix $\det A = a_{11} a_{22} - a_{12} a_{21}$ and so on. From the definition we directly get that $\det A = \det A^T$. Some other determinant identities are given below:

1. Let A_1 and A_2 be square matrices of the same dimension. Then

$$\det(A_1 A_2) = \det(A_2 A_1) = \det A_1 \cdot \det A_2 \tag{A.9}$$

2. Let c be a complex scalar and A an $n \times n$ matrix. Then

$$\det(cA) = c^n \det(A) \tag{A.10}$$

3. Let A be a non-singular matrix. Then

$$\det A^{-1} = 1/\det A \tag{A.11}$$

4. Let A_1 and A_2 be matrices of compatible dimensions such that both matrices $A_1 A_2$ and $A_2 A_1$ are square (but A_1 and A_2 need not themselves be square). Then

$$\det(I + A_1 A_2) = \det(I + A_2 A_1) \tag{A.12}$$

This is actually a special case of Schur's formula given in (A.14). Equation (A.12) is useful in the field of control because it yields $\det(I + GK) = \det(I + KG)$.

5. The determinant of a triangular or block-triangular matrix is the product of the determinants of the diagonal blocks:

$$\det \begin{bmatrix} A_{11} & A_{12} \\ 0 & A_{22} \end{bmatrix} = \det \begin{bmatrix} A_{11} & 0 \\ A_{21} & A_{22} \end{bmatrix} = \det(A_{11}) \cdot \det(A_{22}) \tag{A.13}$$

6. **Schur's formula** for the determinant of a partitioned matrix:

$$\det \begin{bmatrix} A_{11} & A_{12} \\ A_{21} & A_{22} \end{bmatrix} = \det(A_{11}) \cdot \det(A_{22} - A_{21} A_{11}^{-1} A_{12})$$

$$= \det(A_{22}) \cdot \det(A_{11} - A_{12} A_{22}^{-1} A_{21}) \tag{A.14}$$

where it is assumed that A_{11} and/or A_{22} are non-singular.

Proof: Note that A has the following decomposition if A_{11} is non-singular:

$$\begin{bmatrix} A_{11} & A_{12} \\ A_{21} & A_{22} \end{bmatrix} = \begin{bmatrix} I & 0 \\ A_{21} A_{11}^{-1} & I \end{bmatrix} \begin{bmatrix} A_{11} & 0 \\ 0 & X \end{bmatrix} \begin{bmatrix} I & A_{11}^{-1} A_{12} \\ 0 & I \end{bmatrix} \tag{A.15}$$

where $X = A_{22} - A_{21} A_{11}^{-1} A_{12}$. The first part of (A.14) is proved by evaluating the determinant using (A.9) and (A.13). Similarly, if A_{22} is non-singular,

$$\begin{bmatrix} A_{11} & A_{12} \\ A_{21} & A_{22} \end{bmatrix} = \begin{bmatrix} I & A_{12} A_{22}^{-1} \\ 0 & I \end{bmatrix} \begin{bmatrix} Y & 0 \\ 0 & A_{22} \end{bmatrix} \begin{bmatrix} I & 0 \\ A_{22}^{-1} A_{21} & I \end{bmatrix} \tag{A.16}$$

where $Y = A_{11} - A_{12} A_{22}^{-1} A_{21}$, and the last part of (A.14) follows. $\square$

A.2   Eigenvalues and eigenvectors

Definition A.1 Eigenvalues and eigenvectors. *Let A be a square $n \times n$ matrix. The eigenvalues λ_i, $i = 1, \ldots, n$, are the n solutions to the n'th-order characteristic equation*

$$\det(A - \lambda I) = 0 \tag{A.17}$$

The (right) eigenvector t_i corresponding to the eigenvalue λ_i is the non-trivial solution $(t_i \neq 0)$ to

$$(A - \lambda_i I)t_i = 0 \quad \Leftrightarrow \quad At_i = \lambda_i t_i \tag{A.18}$$

The corresponding left eigenvectors q_i satisfy

$$q_i^H(A - \lambda_i I) = 0 \quad \Leftrightarrow \quad q_i^H A = \lambda_i q_i^H \tag{A.19}$$

When we just say eigenvector *we mean the right eigenvector.*

Remark 1 Note that if t is an eigenvector then so is αt for any constant α. Therefore, the eigenvectors are usually normalized to have unit length, i.e. $t_i^H t_i = 1$.

Remark 2 From (A.19) we get $A^H q_i = \bar{\lambda}_i q_i$. Thus, for computations, we may obtain the left eigenvectors q_i from the right eigenvectors of A^H. However, notice that we have the conjugate of the eigenvalue, $\bar{\lambda}_i$, so the order of the complex eigenvalues may be different for A and A^H.

Remark 3 The eigenvalues are sometimes called characteristic gains. The set of eigenvalues of A is called the spectrum of A. The largest of the absolute values of the eigenvalues of A is the *spectral radius* of A,

$$\boxed{\rho(A) \triangleq \max_i |\lambda_i(A)|} \tag{A.20}$$

An important result for eigenvectors is that *eigenvectors corresponding to distinct eigenvalues are always linearly independent.* For repeated eigenvalues, this may not always be the case; that is, not all $n \times n$ matrices have n linearly independent eigenvectors (these are the so-called "defective" matrices).

The eigenvectors may be collected as columns in the matrix T and the eigenvalues $\lambda_1, \lambda_2, \ldots, \lambda_n$ as diagonal elements in the matrix Λ:

$$T = \{t_1, t_2, \ldots, t_n\}; \quad \Lambda = \mathrm{diag}\{\lambda_1, \lambda_2, \ldots, \lambda_n\} \tag{A.21}$$

We may then write (A.18) as $AT = T\Lambda$. If the eigenvectors are linearly independent such that T^{-1} exists, we then have that A may be "diagonalized" as follows:

$$\Lambda = T^{-1}AT \tag{A.22}$$

This always happens if the eigenvalues are distinct, and may also happen in other cases, e.g. for $A = I$. For distinct eigenvalues, we also have that the right and left eigenvectors are mutually orthogonal, and we may scale the columns in Q such that they are also mutually orthonormal,

$$q_i^H t_j = \begin{cases} 1 & \text{if } i = j \\ 0 & \text{if } i \neq j \end{cases}$$

Then we have the following *dyadic expansion* or *spectral decomposition* of the matrix A in terms of its right and left eigenvectors:

$$A = \sum_{i=1}^{n} \lambda_i t_i q_i^H \tag{A.23}$$

Remark. The case where the eigenvalues are not distinct (i.e. repeated) is much more complicated, both theoretically and computationally. Fortunately, from a practical point of view it is sufficient to understand the case where the eigenvalues are distinct.

A.2.1 Eigenvalue properties

Let λ_i denote the eigenvalues of A in the following properties:

1. The sum of the eigenvalues of A is equal to the trace of A (sum of the diagonal elements): $\text{tr} A = \sum_i \lambda_i$.
2. The product of the eigenvalues of A is equal to the determinant of A: $\det A = \prod_i \lambda_i$.
3. The eigenvalues of an upper or lower triangular matrix are equal to the diagonal elements of the matrix.
4. For a real matrix the eigenvalues either are real, or occur in complex conjugate pairs.
5. A and A^T have the same eigenvalues (but in general different eigenvectors).
6. The inverse A^{-1} exists if and only if all eigenvalues of A are non-zero. The eigenvalues of A^{-1} are then $1/\lambda_1, \ldots, 1/\lambda_n$.
7. The matrix $A + cI$ has eigenvalues $\lambda_i + c$.
8. The matrix cA^k where k is an integer has eigenvalues $c\lambda_i^k$.
9. Consider the $l \times m$ matrix A and the $m \times l$ matrix B. Then the $l \times l$ matrix AB and the $m \times m$ matrix BA have the same non-zero eigenvalues. To be more specific, assume $l > m$. Then the matrix AB has the same m eigenvalues as BA plus $l - m$ eigenvalues which are identically equal to zero.
10. Eigenvalues are invariant under similarity transformations; that is, A and DAD^{-1} have the same eigenvalues.
11. The same eigenvector matrix diagonalizes the matrix A and the matrix $(I + A)^{-1}$. (*Proof:* $T^{-1}(I + A)^{-1}T = (T^{-1}(I + A)T)^{-1} = (I + \Lambda)^{-1}$.)
12. *Gershgorin's theorem.* The eigenvalues of the $n \times n$ matrix A lie in the union of n circles in the complex plane, each with centre a_{ii} and radius $r_i = \sum_{j \neq i} |a_{ij}|$ (sum of off-diagonal elements in row i). They also lie in the union of n circles, each with centre a_{ii} and radius $r_i' = \sum_{j \neq i} |a_{ji}|$ (sum of off-diagonal elements in column i).
13. The eigenvalues of a Hermitian matrix (and hence of a symmetric matrix) are real.
14. A Hermitian matrix is positive definite ($A > 0$) if and only if all its eigenvalues are positive.

From the above properties we have, for example, that

$$\lambda_i(S) = \lambda_i((I + L)^{-1}) = \frac{1}{\lambda_i(I + L)} = \frac{1}{1 + \lambda_i(L)} \tag{A.24}$$

In this book, we are sometimes interested in the eigenvalues of a real (state) matrix A, and in other cases in the eigenvalues of a complex transfer function matrix evaluated at a given frequency, e.g. $L(j\omega)$, as in (A.24). It is important to appreciate this difference.

A.2.2 Eigenvalues of the state matrix

Consider a system described by the linear differential equations

$$\dot{x} = Ax + Bu \tag{A.25}$$

Unless A is diagonal this is a set of coupled differential equations. For simplicity, we assume that the eigenvectors t_i of A are linearly independent and introduce the new state vector $z = T^{-1}x$, i.e. $x = Tz$. We then get

$$T\dot{z} = ATz + Bu \quad \Leftrightarrow \quad \dot{z} = \Lambda z + T^{-1}Bu \tag{A.26}$$

which is a set of uncoupled differential equations in terms of the new states z. The unforced solution (i.e. with $u = 0$) for each state z_i is $z_i = z_{0i}e^{\lambda_i t}$ where z_{0i} is the value of the state at $t = 0$. If λ_i is real, then we see that this mode is stable ($z_i \to 0$ as $t \to \infty$) if and only if $\lambda_i < 0$. If $\lambda_i = \text{Re}\lambda_i + j\text{Im}\lambda_i$ is complex, then we get $e^{\lambda_i t} = e^{\text{Re}\lambda_i t}(\cos(\text{Im}\lambda_i t) + j\sin(\text{Im}\lambda_i t))$ and the mode is stable ($z_i \to 0$ as $t \to \infty$) if and only if $\text{Re}\lambda_i < 0$. The fact that the new state z_i is complex is of no concern since the real physical states $x = Tz$ are of course real. Consequently, a linear system is stable if and only if all the eigenvalues of the state matrix A have real parts less than zero; that is, lie in the open left-half plane.

A.2.3 Eigenvalues of transfer functions

The eigenvalues of the loop transfer function matrix, $\lambda_i(L(j\omega))$, evaluated as a function of frequency, are sometimes called the characteristic loci, and to some extent they generalize $L(j\omega)$ for a scalar system. In Chapter 8, we make use of $\lambda_i(L)$ to study the stability of the $M\Delta$-structure where $L = M\Delta$. Even more important in this context is the spectral radius, $\rho(L(j\omega)) = \max_i |\lambda_i(L(j\omega))|$.

A.3   Singular value decomposition

Definition A.2 Unitary matrix. *A (complex) matrix U is unitary if*

$$U^H = U^{-1} \tag{A.27}$$

All the eigenvalues of a unitary matrix have absolute value equal to 1, and all its singular values (as we shall see from the definition below) are equal to 1.

Definition A.3 SVD. *Any complex $l \times m$ matrix A may be factorized using the SVD*

$$A = U\Sigma V^H \tag{A.28}$$

where the $l \times l$ matrix U and the $m \times m$ matrix V are unitary, and the $l \times m$ matrix Σ contains a diagonal matrix Σ_1 of real, non-negative singular values, σ_i, arranged in a descending order as in

$$\Sigma = \begin{bmatrix} \Sigma_1 \\ 0 \end{bmatrix}; \quad l \geq m \tag{A.29}$$

or

$$\Sigma = [\Sigma_1 \quad 0]; \quad l \leq m \tag{A.30}$$

where

$$\Sigma_1 = \text{diag}\{\sigma_1, \sigma_2, \ldots, \sigma_k\}; \quad k = \min(l, m) \tag{A.31}$$

and

$$\bar{\sigma} \equiv \sigma_1 \geq \sigma_2 \geq \cdots \geq \sigma_k \equiv \underline{\sigma} \tag{A.32}$$

The unitary matrices U and V form orthonormal bases for the column (output) space and the row (input) space of A. The column vectors of V, denoted v_i, are called right or input singular vectors and the column vectors of U, denoted u_i, are called left or output singular vectors. We define $\bar{u} \equiv u_1, \bar{v} \equiv v_1, \underline{u} \equiv u_k$ and $\underline{v} \equiv v_k$.

Note that the decomposition in (A.28) is not unique. For example, for a square matrix, an alternative SVD is $A = U'\Sigma V'^H$, where $U' = US, V' = VS, S = \text{diag}\{e^{j\theta_i}\}$ and θ_i is any real number. However, the singular values, σ_i, are unique.

The singular values are the positive square roots of the $k = \min(l, m)$ largest eigenvalues of both AA^H and $A^H A$. We have

$$\sigma_i(A) = \sqrt{\lambda_i(A^H A)} = \sqrt{\lambda_i(AA^H)} \tag{A.33}$$

Also, the columns of U and V are unit eigenvectors of AA^H and $A^H A$, respectively. To derive (A.33) we write

$$AA^H = (U\Sigma V^H)(U\Sigma V^H)^H = (U\Sigma V^H)(V\Sigma^H U^H) = U\Sigma\Sigma^H U^H \tag{A.34}$$

or equivalently since U is unitary and satisfies $U^H = U^{-1}$ we get

$$(AA^H)U = U\Sigma\Sigma^H \tag{A.35}$$

We then see that U is the matrix of eigenvectors of AA^H and $\{\sigma_i^2\}$ are its eigenvalues. Similarly, we have that V is the matrix of eigenvectors of $A^H A$.

A.3.1 Rank

Definition A.4 *The **rank** of a matrix is equal to the number of non-zero singular values of the matrix. Let* $\text{rank}(A) = r$, *then the matrix A is called rank deficient if* $r < k = \min(l, m)$, *and we have singular values* $\sigma_i = 0$ *for* $i = r + 1, \ldots, k$. *A rank-deficient square matrix is a singular matrix (non-square matrices are always singular).*

The rank of a matrix is unchanged after left or right multiplication by a non-singular matrix. Furthermore, for an $l \times m$ matrix A and an $m \times p$ matrix B, the rank of their product AB is bounded as follows (Sylvester's inequality):

$$\text{rank}(A) + \text{rank}(B) - m \leq \text{rank}(AB) \leq \min(\text{rank}(A), \text{rank}(B)) \tag{A.36}$$

A.3.2 Singular values of a 2×2 matrix

In general, the singular values must be computed numerically. For 2×2 matrices, however, an analytic expression is easily derived. From (A.33), $\sigma_i(A) = \sqrt{\lambda_i(A^H A)}$. We introduce

$$b \triangleq \text{tr}(A^H A) = \sum_{i,j} |a_{ij}|^2, \quad c \triangleq \det(A^H A)$$

Now the sum of the eigenvalues of a matrix is equal to its trace and the product is equal to its determinant, so

$$\lambda_1 + \lambda_2 = b, \quad \lambda_1 \cdot \lambda_2 = c$$

Upon solving for λ_1 and λ_2, and using $\sigma_i(A) = \sqrt{\lambda_i(A^H A)}$, we get

$$\bar{\sigma}(A) = \sqrt{\frac{b + \sqrt{b^2 - 4c}}{2}}; \quad \underline{\sigma}(A) = \sqrt{\frac{b - \sqrt{b^2 - 4c}}{2}} \tag{A.37}$$

For example, for $A = \begin{bmatrix} 1 & 2 \\ 3 & 4 \end{bmatrix}$ we have $b = \sum |a_{ij}|^2 = 1 + 4 + 9 + 16 = 30$, $c = (\det A)^2 = (-2)^2 = 4$, and we find $\bar{\sigma}(A) = 5.465$ and $\underline{\sigma}(A) = 0.366$. Note that for singular 2×2 matrices (with $\det A = 0$ and $\underline{\sigma}(A) = 0$) we get $\bar{\sigma}(A) = \sqrt{\sum |a_{ij}|^2} \triangleq \|A\|_F$ (the Frobenius norm), which is actually a special case of (A.127).

A.3.3 SVD of a matrix inverse

Since $A = U\Sigma V^H$ we get, provided the $m \times m$ matrix A is non-singular, that

$$A^{-1} = V\Sigma^{-1}U^H \tag{A.38}$$

This is the SVD of A^{-1} but with the order of the singular values reversed. Let $j = m - i + 1$. Then it follows from (A.38) that

$$\sigma_i(A^{-1}) = 1/\sigma_j(A), \quad u_i(A^{-1}) = v_j(A), \quad v_i(A^{-1}) = u_j(A) \tag{A.39}$$

and in particular

$$\bar{\sigma}(A^{-1}) = 1/\underline{\sigma}(A) \tag{A.40}$$

A.3.4 Singular value inequalities

The singular values bound the magnitude of the eigenvalues (also see (A.117)):

$$\underline{\sigma}(A) \leq |\lambda_i(A)| \leq \bar{\sigma}(A) \tag{A.41}$$

The following is obvious from the SVD definition:

$$\bar{\sigma}(A^H) = \bar{\sigma}(A) \quad \text{and} \quad \bar{\sigma}(A^T) = \bar{\sigma}(A) \tag{A.42}$$

The next important property is proved below (see (A.98)):

$$\bar{\sigma}(AB) \leq \bar{\sigma}(A)\bar{\sigma}(B) \tag{A.43}$$

For a non-singular A (or B) we also have a lower bound on $\bar{\sigma}(AB)$

$$\underline{\sigma}(A)\bar{\sigma}(B) \leq \bar{\sigma}(AB) \quad (\text{or} \quad \bar{\sigma}(A)\underline{\sigma}(B) \leq \bar{\sigma}(AB)) \tag{A.44}$$

For non-singular A and B, we also have a lower bound on the minimum singular value

$$\underline{\sigma}(A)\underline{\sigma}(B) \leq \underline{\sigma}(AB) \tag{A.45}$$

For a partitioned matrix, $M = \begin{bmatrix} A \\ B \end{bmatrix}$ or $M = [\, A \quad B \,]$, the following inequalities are useful:

$$\max\{\bar{\sigma}(A), \bar{\sigma}(B)\} \leq \bar{\sigma}(M) \leq \sqrt{2}\max\{\bar{\sigma}(A), \bar{\sigma}(B)\} \tag{A.46}$$

$$\bar{\sigma}(M) \leq \bar{\sigma}(A) + \bar{\sigma}(B) \tag{A.47}$$

$$\underline{\sigma}(M) \leq \min\{\underline{\sigma}(A), \underline{\sigma}(B)\} \tag{A.48}$$

The following equality for a block-diagonal matrix is used extensively in the book:

$$\bar{\sigma} \begin{bmatrix} A & 0 \\ 0 & B \end{bmatrix} = \max\{\bar{\sigma}(A), \bar{\sigma}(B)\} \tag{A.49}$$

Another very useful result is Fan's theorem (Horn and Johnson, 1991, p. 140 and p. 178):

$$\sigma_i(A) - \bar{\sigma}(B) \leq \sigma_i(A + B) \leq \sigma_i(A) + \bar{\sigma}(B) \tag{A.50}$$

Two special cases of (A.50) are

$$|\bar{\sigma}(A) - \bar{\sigma}(B)| \leq \bar{\sigma}(A + B) \leq \bar{\sigma}(A) + \bar{\sigma}(B) \tag{A.51}$$

$$\underline{\sigma}(A) - \bar{\sigma}(B) \leq \underline{\sigma}(A + B) \leq \underline{\sigma}(A) + \bar{\sigma}(B) \tag{A.52}$$

Relation (A.52) yields

$$\underline{\sigma}(A) - 1 \leq \underline{\sigma}(I + A) \leq \underline{\sigma}(A) + 1 \tag{A.53}$$

On combining (A.40) and (A.53) we get a relationship that is useful when evaluating the amplification of closed-loop systems:

$$\underline{\sigma}(A) - 1 \leq \frac{1}{\bar{\sigma}(I + A)^{-1}} \leq \underline{\sigma}(A) + 1 \tag{A.54}$$

A.3.5   SVD as a sum of rank 1 matrices

Let r denote the rank of the $l \times m$ matrix A. We may then consider the SVD as a decomposition of A into r $l \times m$ matrices, each of rank 1. We have

$$A = U\Sigma V^H = \sum_{i=1}^{r} \sigma_i u_i v_i^H \tag{A.55}$$

The remaining terms from $r + 1$ to $k = \min(l, m)$ have singular values equal to zero and give no contribution to the sum. The first and most important submatrix is given by $A_1 = \sigma_1 u_1 v_1^H$. If we now consider the residual matrix

$$A^1 = A - A_1 = A - \sigma_1 u_1 v_1^H \tag{A.56}$$

then

$$\sigma_1(A^1) = \sigma_2(A) \tag{A.57}$$

That is, the largest singular value of A^1 is equal to the second singular value of the original matrix. This shows that the direction corresponding to $\sigma_2(A)$ is the second most important direction, and so on.

A.3.6   Singularity of matrix $A + E$

From the left inequality in (A.52) we find that

$$\bar{\sigma}(E) < \underline{\sigma}(A) \quad \Rightarrow \quad \underline{\sigma}(A + E) > 0 \tag{A.58}$$

and $A + E$ is non-singular. On the other hand, there always exists an E with $\bar{\sigma}(E) = \underline{\sigma}(A)$ which makes $A + E$ singular, e.g. choose $E = -\underline{u}\,\underline{\sigma}\,\underline{v}^H$; see (A.55). Thus the smallest singular value $\underline{\sigma}(A)$ measures how near the matrix A is to being singular or rank deficient. This test is often used in numerical analysis, and it is also an important inequality in the formulation of robustness tests.

A.3.7   Economy-size SVD

Since there are only $r = \text{rank}(A) \leq \min(l, m)$ non-zero singular values, and since only the non-zero singular values contribute to the overall result, the SVD of A is sometimes written as an economy-size SVD, as follows:

$$A^{l \times m} = U_r^{l \times r} \, \Sigma_r^{r \times r} \, (V_r^{m \times r})^H \tag{A.59}$$

where the matrices U_r and V_r contain only the first r columns of the matrices U and V introduced above. Here we have used the notation $A^{l \times m}$ to indicate that A is an $l \times m$ matrix. The economy-size SVD is used for computing the pseudo-inverse, see (A.62).

Remark. The "economy-size SVD" presently used in Matlab is not quite as economic as the one given in (A.59) as it uses m instead of r for Σ.

A.3.8   Pseudo-inverse (generalized inverse)

Consider the set of linear equations

$$y = Ax \tag{A.60}$$

with a given $l \times 1$ vector y and a given $l \times m$ matrix A. A *least squares* solution to (A.60) is an $m \times 1$ vector x such that $\|x\|_2 = \sqrt{x_1^2 + x_2^2 + \cdots + x_m^2}$ is minimized among all vectors for which $\|y - Ax\|_2$ is minimized. The solution is given in terms of the pseudo-inverse (Moore–Penrose generalized inverse) of A:

$$x = A^\dagger y \tag{A.61}$$

The pseudo-inverse may be obtained from an SVD of $A = U\Sigma V^H$ by

$$A^\dagger = V_r \Sigma_r^{-1} U_r^H = \sum_{i=1}^{r} \frac{1}{\sigma_i(A)} v_i u_i^H \tag{A.62}$$

where r is the number of non-zero singular values of A. We have that

$$\underline{\sigma}(A) = 1/\bar{\sigma}(A^\dagger) \tag{A.63}$$

Note that $A^\dagger$ exists for any matrix A, even for a singular square matrix and a non-square matrix. The pseudo-inverse also satisfies

$$AA^\dagger A = A \quad \text{and} \quad A^\dagger AA^\dagger = A^\dagger$$

Note the following cases (where r is the rank of A):

1. $r = l = m$, i.e. A is non-singular. In this case $A^\dagger = A^{-1}$ is the inverse of the matrix.
2. $r = m \le l$, i.e. A has full column rank. This is the "conventional least squares problem" where we want to minimize $\|y - Ax\|_2$, and the solution is

$$A^\dagger = (A^H A)^{-1} A^H \tag{A.64}$$

In this case $A^\dagger A = I$, so $A^\dagger$ is a *left inverse* of A.
3. $r = l \le m$, i.e. A has full row rank. In this case we have an infinite number of solutions to (A.60) and we seek the one that minimizes $\|x\|_2$. We get

$$A^\dagger = A^H (A A^H)^{-1} \tag{A.65}$$

In this case $A A^\dagger = I$, so $A^\dagger$ is a *right inverse* of A.
4. $r < k = \min(l, m)$ (general case). In this case both matrices $A^H A$ and $A A^H$ are rank deficient, and we have to use (A.62) to obtain the pseudo-inverse. In this case A has neither a left nor a right inverse.

Principal component regression (PCR)

We note that the pseudo-inverse in (A.62) may be very sensitive to noise and "blow up" if the smallest non-zero singular value, σ_r, is small. In the PCR method one avoids this problem by using only the $q \le r$ first singular values which can be distinguished from the noise. The PCR pseudo-inverse then becomes

$$A^\dagger_{PCR} = \sum_{i=1}^{q} \frac{1}{\sigma_i} v_i u_i^H \tag{A.66}$$

Remark. This is similar in spirit to the use of Hankel singular values for model reduction.

A.3.9   Condition number

The *condition number* of a matrix is defined in this book as the ratio

$$\gamma(A) = \sigma_1(A)/\sigma_k(A) = \bar{\sigma}(A)/\underline{\sigma}(A) \tag{A.67}$$

where $k = \min(l, m)$. A matrix with a large condition number is said to be ill-conditioned. This definition yields an infinite condition number for rank-deficient matrices. For a non-singular matrix we get from (A.40)

$$\gamma(A) = \bar{\sigma}(A) \cdot \bar{\sigma}(A^{-1}) \tag{A.68}$$

Other definitions for the condition number of a non-singular matrix are also in use, e.g.

$$\gamma_p(A) = \|A\|_p \cdot \|A^{-1}\|_p \tag{A.69}$$

where $\|A\|_p$ denotes any matrix norm. If we use the induced 2-norm (maximum singular value) then this yields (A.68). From (A.68) and (A.43), we get for non-singular matrices

$$\gamma(AB) \le \gamma(A)\gamma(B) \tag{A.70}$$

The *minimized condition number* is obtained by minimizing the condition number over all possible scalings. We have

$$\gamma^*(A) \triangleq \min_{D_I, D_O} \gamma(D_O A D_I) \tag{A.71}$$

where D_I and D_O are (complex) diagonal scaling matrices. If we allow scaling only on one side then we get the input and output minimized condition numbers:

$$\gamma_I^*(A) \triangleq \min_{D_I} \gamma(A D_I), \quad \gamma_O^*(A) \triangleq \min_{D_O} \gamma(D_O A) \tag{A.72}$$

As shown in (A.79) and (A.80), the minimized condition number is closely related to the norm of the RGA-matrix.

Remark. To compute these minimized condition numbers we define

$$H = \begin{bmatrix} 0 & A^{-1} \\ A & 0 \end{bmatrix} \tag{A.73}$$

Then we have, as proven by Braatz and Morari (1994),

$$\sqrt{\gamma^*(A)} = \min_{D_I, D_O} \bar{\sigma}(DHD^{-1}), \quad D = \text{diag}\{D_I^{-1}, D_O\} \tag{A.74}$$

$$\sqrt{\gamma_I^*(A)} = \min_{D_I} \bar{\sigma}(DHD^{-1}), \quad D = \text{diag}\{D_I^{-1}, I\} \tag{A.75}$$

$$\sqrt{\gamma_O^*(A)} = \min_{D_O} \bar{\sigma}(DHD^{-1}), \quad D = \text{diag}\{I, D_O\} \tag{A.76}$$

These convex optimization problems may be solved using available software for the upper bound on the structured singular value $\mu_{\text{up}}(H)$; see (8.87) and Example 12.4. In calculating $\mu_{\text{up}}(H)$, we use for $\gamma^*(A)$ the structure $\Delta = \text{diag}\{\Delta_{\text{diag}}, \Delta_{\text{diag}}\}$, for $\gamma_I^*(A)$ the structure $\Delta = \text{diag}\{\Delta_{\text{diag}}, \Delta_{\text{full}}\}$, and for $\gamma_O^*(A)$ the structure $\Delta = \text{diag}\{\Delta_{\text{full}}, \Delta_{\text{diag}}\}$.

A.4   Relative gain array

The relative gain array (RGA), see section 3.4 (page 82), was introduced by Bristol (1966). Many of its properties were stated by Bristol, but they were not proven rigorously until the work by Grosdidier et al. (1985). The RGA of a complex non-singular $m \times m$ matrix A, denoted RGA(A) or $\Lambda(A)$, is a complex $m \times m$ matrix defined by

$$\text{RGA}(A) \equiv \Lambda(A) \triangleq A \times (A^{-1})^T \tag{A.77}$$

where the operation $\times$ denotes element-by-element multiplication (Hadamard or Schur product). If A is real then $\Lambda(A)$ is also real. For calculating RGA using Matlab, use `rga = a.*pinv(a).'`; see also Table 3.1 on page 87.
Example:

$$A_1 = \begin{bmatrix} 1 & -2 \\ 3 & 4 \end{bmatrix}, \ A_1^{-1} = \begin{bmatrix} 0.4 & 0.2 \\ -0.3 & 0.1 \end{bmatrix}, \ \Lambda(A_1) = \begin{bmatrix} 0.4 & 0.6 \\ 0.6 & 0.4 \end{bmatrix}$$

A.4.1 Algebraic properties of the RGA

Most of the properties below follow directly if we write the RGA elements in the form

$$\lambda_{ij} = a_{ij} \cdot \tilde{a}_{ji} = a_{ij} \frac{c_{ij}}{\det A} = (-1)^{i+j} \frac{a_{ij} \det A^{ij}}{\det A} \tag{A.78}$$

where $\tilde{a}_{ji}$ denotes the ji'th element of the matrix $\tilde{A} \triangleq A^{-1}$, A^{ij} denotes the matrix A with row i and column j deleted, and $c_{ij} = (-1)^{i+j} \det A^{ij}$ is the ij'th cofactor of the matrix A.

For any non-singular $m \times m$ matrix A, the following properties hold:

1. $\Lambda(A^{-1}) = \Lambda(A^T) = \Lambda(A)^T$.
2. Any permutation of the rows and columns of A results in the same permutation in the RGA. That is, $\Lambda(P_1 A P_2) = P_1 \Lambda(A) P_2$ where P_1 and P_2 are permutation matrices. (A permutation matrix has a single 1 in every row and column and all other elements equal to 0.) $\Lambda(P) = P$ for any permutation matrix.
3. The sum of the elements in each row (and each column) of the RGA is 1. That is, $\sum_{i=1}^{m} \lambda_{ij} = 1$ and $\sum_{j=1}^{m} \lambda_{ij} = 1$.
4. $\Lambda(A) = I$ if A is a lower or upper triangular matrix; and in particular the RGA of a diagonal matrix is the identity matrix.
5. The RGA is scaling invariant. Therefore, $\Lambda(D_1 A D_2) = \Lambda(A)$ where D_1 and D_2 are diagonal matrices.
6. The RGA is a measure of sensitivity to relative element-by-element uncertainty in the matrix. More precisely, the matrix A becomes singular if a single element in A is perturbed from a_{ij} to $a'_{ij} = a_{ij}(1 - \frac{1}{\lambda_{ij}})$; see Theorem 6.6 on page 251.
7. The norm of the RGA is closely related to the minimized condition number γ^* defined in (A.71). For a 2×2 matrix (Grosdidier et al., 1985) and a real 3×3 matrix (Liang, 1992):

$$\gamma^* + 1/\gamma^* = \|\Lambda\|_m \tag{A.79}$$

In general, for a (complex) matrix of any size (Nett and Manousiouthakis, 1987):

$$\gamma^* + 1/\gamma^* \geq \|\Lambda\|_m \tag{A.80}$$

Here $\|\Lambda\|_m \triangleq 2\max\{\|\Lambda\|_{i1}, \|\Lambda\|_{i\infty}\}$ is two times the maximum row or column sum of the RGA (the matrix norms are defined in Section A.5.2). (A.80) shows that a matrix with large RGA elements always has a large minimized condition number. The reverse has also been conjectured (Nett and Manousiouthakis, 1987), but it does not hold for 4×4 matrices or larger as shown by the following counterexample motivated by Liang (1992):

$$A = \begin{bmatrix} 1 & -1 & -1 & 1 \\ 1 & 1 & 1 & 1 \\ 1 & 1 & -1 & -1 \\ 1 & -1 & 1 & -1 \end{bmatrix} \begin{bmatrix} k & 0 & 0 & 0 \\ 0 & k & 0 & 0 \\ 0 & 0 & 1 & 0 \\ 0 & 0 & 0 & 1 \end{bmatrix} \begin{bmatrix} 1 & 1 & 1 & -1 \\ -1 & 1 & 1 & 1 \\ 1 & -1 & 1 & 1 \\ 1 & 1 & -1 & 1 \end{bmatrix} = 2 \begin{bmatrix} k & 1 & -1 & -k \\ 1 & k & k & 1 \\ -1 & k & k & -1 \\ k & -1 & 1 & -k \end{bmatrix}$$

has $\gamma^*(A) = \gamma(A) = k$ (which can be arbitrary large), but for any k all RGA elements are 0.25 so $\|\Lambda(A)\|_m = 2$.

8. The diagonal elements of the matrix ADA^{-1} are given in terms of the corresponding row elements of the RGA (Skogestad and Morari, 1987c; Nett and Manousiouthakis, 1987). For any diagonal matrix $D = \text{diag}\{d_i\}$ we have

$$[ADA^{-1}]_{ii} = \sum_{j=1}^{m} \lambda_{ij}(A)d_j \tag{A.81}$$

$$[A^{-1}DA]_{ii} = \sum_{i=1}^{m} \lambda_{ij}(A)d_i \tag{A.82}$$

9. It follows from Property 3 that Λ always has at least one eigenvalue and one singular value equal to 1.

Proofs of some of the properties: Property 3: Since $AA^{-1} = I$ it follows that $\sum_{j=1}^{m} a_{ij}\hat{a}_{ji} = 1$. From the definition of the RGA we then have that $\sum_{j=1}^{m} \lambda_{ij} = 1$. *Property 4*: If the matrix is upper triangular then $a_{ij} = 0$ for $i > j$. It then follows that $c_{ij} = 0$ for $j > i$ and all the off-diagonal RGA elements are zero. *Property 5*: Let $A' = D_1 A D_2$. Then $a'_{ij} = d_{1i}d_{2j}a_{ij}$ and $\hat{a}'_{ij} = \frac{1}{d_{2j}}\frac{1}{d_{1i}}\hat{a}_{ij}$ and the result follows. *Property 6*: The determinant can be evaluated by expanding it in terms of any row or column, e.g. by row i, $\det A = \sum_i (-1)^{i+j} a_{ij} \det A^{ij}$. Let A' denote A with a'_{ij} substituted for a_{ij}. By expanding the determinant of A' by row i and then using (A.78) we get

$$\det A' = \underbrace{\det A - (-1)^{i+j}\frac{a_{ij}}{\lambda_{ij}} \det A^{ij}}_{\det A} = 0$$

Property 8: The ii'th element of the matrix $B = ADA^{-1}$ is $b_{ii} = \sum_j d_j a_{ij}\hat{a}_{ji} = \sum_j d_j \lambda_{ij}$. $\qquad\square$

Example A.1

$$A_2 = \begin{bmatrix} 56 & 66 & 75 & 97 \\ 75 & 54 & 82 & 28 \\ 18 & 66 & 25 & 38 \\ 9 & 51 & 8 & 11 \end{bmatrix} ; \quad \Lambda(A_2) = \begin{bmatrix} 6.16 & -0.69 & -7.94 & 3.48 \\ -1.77 & 0.10 & 3.16 & -0.49 \\ -6.60 & 1.73 & 8.55 & -2.69 \\ 3.21 & -0.14 & -2.77 & 0.70 \end{bmatrix} \tag{A.83}$$

In this case, $\gamma(A_2) = \bar{\sigma}(A_2)/\underline{\sigma}(A_2) = 207.68/1.367 = 151.9$, $\gamma^*(A_2) = 51.73$ (obtained numerically using (A.74)), $\gamma_I^*(A_2) = 118.70$ and $\gamma_O^*(A_2) = 92.57$. Furthermore, $\|\Lambda\|_m = 2\max\{22.42, 19.58\} = 44.84$, so (A.80) is satisfied. The matrix A_2 is non-singular and the $1,3$ element of the RGA is $\lambda_{13}(A_2) = -7.94$. Thus from Property 6 the matrix A_2 becomes singular if the $1,3$ element is perturbed from 75 to $75(1 - \frac{1}{-7.94}) = 84.45$.

Additional examples on the properties of RGA are given in Section 3.4.

A.4.2   RGA of a non-square matrix

The RGA may be generalized to a non-square $l \times m$ matrix A by use of the pseudo-inverse $A^\dagger$ defined in (A.62). We have

$$\Lambda(A) = A \times (A^\dagger)^T \tag{A.84}$$

Properties 1 (transpose and inverse) and 2 (permutations) of the RGA also hold for non-square matrices, but the remaining properties do not apply in the general case. However, they partly apply if A is either of full row rank or full column rank.

1. *A has full row rank*, $r = \text{rank}(A) = l$ (i.e. A has at least as many inputs as outputs, and the outputs are linearly independent). In this case $AA^\dagger = I$, and the following properties hold:

   (a) The RGA is independent of output scaling, i.e. $\Lambda(DA) = \Lambda(A)$.
   (b) The elements in each row of the RGA sum to 1, $\sum_j^m \lambda_{ij} = 1$.

(c) The elements of column j of the RGA sum to the square of the 2-norm of the j'th row in V_r,

$$\sum_{i=1}^{l} \lambda_{ij} = \|e_j^T V_r\|_2^2 \leq 1 \qquad (A.85)$$

Here V_r contains the first r input singular vectors for G, and e_j is an $m \times 1$ basis vector for input u_j; $e_j = [0 \cdots 0 \ 1 \ 0 \cdots 0]^T$ where 1 appears in position j.

(d) The diagonal elements of $B = ADA^\dagger$ are $b_{ii} = \sum_{j=1}^{m} d_j a_{ij} \widehat{a}_{ji} = \sum_{j=1}^{m} d_j \lambda_{ij}$, where $\widehat{a}_{ji}$ denotes the ji'th element of $A^\dagger$ and D is any diagonal matrix.

2. *A has full column rank*, $r = \text{rank}(A) = m$ (i.e. A has no more inputs than outputs, and the inputs are linearly independent). In this case $A^\dagger A = I$, and the following properties hold:

(a) The RGA is independent of input scaling, i.e. $\Lambda(AD) = \Lambda(A)$.

(b) The elements in each column of the RGA sum to 1, $\sum_i^l \lambda_{ij} = 1$.

(c) The elements of row i of the RGA sum to the square of the 2-norm of the i'th row in U_r,

$$\sum_{i=1}^{m} \lambda_{ij} = \|e_i^T U_r\|_2^2 \leq 1 \qquad (A.86)$$

Here U_r contains the first r output singular vectors for G, and e_i is an $l \times 1$ basis vector for output y_i; $e_i = [0 \cdots 0 \ 1 \ 0 \cdots 0]^T$ where 1 appears in position i.

(d) The diagonal elements of $B = A^\dagger DA$ are equal to $b_{jj} = \sum_{i=1}^{l} \widehat{a}_{ji} d_i a_{ij} = \sum_{i=1}^{l} d_i \lambda_{ij}$, where $\widehat{a}_{ji}$ denotes the ji'th element of $A^\dagger$ and D is any diagonal matrix.

3. *General case.* For a general square or non-square matrix which has neither full row nor full column rank, identities (A.85) and (A.86) still apply.

From this it also follows that the rank of any matrix is equal to the sum of its RGA elements. Let the $l \times m$ matrix G have rank r, then

$$\sum_{i,j} \lambda_{ij}(G) = \text{rank}(G) = r \qquad (A.87)$$

Proofs of (A.85) and (A.86): We will prove these identities for the general case. Write the SVD of G as $G = U_r \Sigma_r V_r^H$ (this is the economy-size SVD from (A.59)) where Σ_r is invertible. We have that $g_{ij} = e_i^H U_r \Sigma_r V_r^H e_j$, $[G^\dagger]_{ji} = e_j^H V_r \Sigma_r^{-1} U_r^H e_i$, $U_r^H U_r = I_r$ and $V_r^H V_r = I_r$, where I_r denotes the identity matrix of dim $r \times r$. For the row sum (A.86) we then get

$$\sum_{j=1}^{m} \lambda_{ij} = \sum_{j=1}^{m} e_i^H U_r \Sigma_r V_r^H e_j e_j^H V_r \Sigma_r^{-1} U_r^H e_i$$

$$= e_i^H U_r \Sigma_r V_r^H \underbrace{\sum_{j=1}^{m} e_j e_j^H}_{I_m} V_r \Sigma_r^{-1} U_r^H e_i = e_i^H U_r U_r^H e_i = \|e_i^H U_r\|_2^2$$

The result for the column sum (A.85) is proved in a similar fashion. □

Remark. The extension of the RGA to non-square matrices was suggested by Chang and Yu (1990) who also stated most of its properties, although in a somewhat incomplete form. More general and precise statements are found, for example, in Cao (1995).

A.5   Norms

It is useful to have a single number which gives an overall measure of the size of a vector, a matrix, a signal, or a system. For this purpose we use functions which are called norms. The most commonly used norm is the Euclidean vector norm, $\|e\|_2 = \sqrt{|e_1|^2 + |e_2|^2 + \cdots + |e_m|^2}$. This is simply the distance between two points y and x, where $e_i = y_i - x_i$ is the difference in their i'th coordinates.

Definition A.5 *A norm of e (which may be a vector, matrix, signal or system) is a real number, denoted $\|e\|$, that satisfies the following properties:*

1. *Non-negative:* $\|e\| \geq 0$.
2. *Positive:* $\|e\| = 0 \Leftrightarrow e = 0$ *(for semi-norms we have $\|e\| = 0 \Leftarrow e = 0$).*
3. *Homogeneous:* $\|\alpha \cdot e\| = |\alpha| \cdot \|e\|$ *for all complex scalars α.*
4. *Triangle inequality:*

$$\|e_1 + e_2\| \leq \|e_1\| + \|e_2\| \tag{A.88}$$

More precisely, e is an element in a vector space V over the field $\mathbb{C}$ of complex numbers, and the properties above must be satisfied $\forall e, e_1, e_2 \in V$ and $\forall \alpha \in \mathbb{C}$.

In this book, we consider the norms of four different objects (norms on four different vector spaces):

1. e is a constant vector.
2. e is a constant matrix.
3. e is a time-dependent signal, $e(t)$, which at each fixed t is a constant scalar or vector.
4. e is a "system", a transfer function $G(s)$ or impulse response $g(t)$, which at each fixed s or t is a constant scalar or matrix.

Cases 1 and 2 involve *spatial* norms and the question that arises is: how do we average or sum up the channels? Cases 3 and 4 involve function norms or *temporal* norms where we want to "average" or "sum up" as a function of time or frequency. Note that the first two are finite-dimensional norms, while the latter two are infinite-dimensional.

Remark. Notation for norms. The reader should be aware that the notation on norms in the literature is not consistent, and one must be careful to avoid confusion. First, in spite of the fundamental difference between spatial and temporal norms, the same notation, $\|\cdot\|$, is generally used for both of them, and we adopt this here. Second, the same notation is often used to denote entirely different norms. For example, consider the infinity norm, $\|e\|_\infty$. If e is a constant vector, then $\|e\|_\infty$ is the largest element in the vector (we often use $\|e\|_{\max}$ for this). If $e(t)$ is a scalar time signal, then $\|e(t)\|_\infty$ is the peak value of $|e(t)|$ as a function of time. If E is a constant matrix then $\|E\|_\infty$ may denote the largest matrix element (we use $\|A\|_{\max}$ for this), while other authors use $\|E\|_\infty$ to denote the largest matrix row sum (we use $\|E\|_{i\infty}$ for this). Finally, if $E(s)$ is a stable proper system (transfer function), then $\|E\|_\infty$ is the $\mathcal{H}_\infty$ norm which is the peak value of the maximum singular value of E, $\|E(s)\|_\infty = \max_w \bar{\sigma}(E(j\omega))$ (which is how we mostly use the ∞-norm in this book).

A.5.1 Vector norms

We will consider a vector a with m elements; that is, the vector space is $V = \mathbb{C}^m$. To illustrate the different norms we will calculate each of them for the vector

$$b = \begin{bmatrix} b_1 \\ b_2 \\ b_3 \end{bmatrix} = \begin{bmatrix} 1 \\ 3 \\ -5 \end{bmatrix} \tag{A.89}$$

We will consider three norms which are special cases of the vector p-norm

$$\|a\|_p = \left(\sum_i |a_i|^p \right)^{1/p} \tag{A.90}$$

where we must have $p \geq 1$ to satisfy the triangle inequality (property 4 of a norm). Here a is a column vector with elements a_i and $|a_i|$ is the absolute value of the complex scalar a_i.

Vector 1-norm (or sum norm). This is sometimes referred to as the "taxi-cab norm", as in two dimensions it corresponds to the distance between two places when following the "streets" (New York style). We have

$$\|a\|_1 \triangleq \sum_i |a_i| \qquad (\|b\|_1 = 1 + 3 + 5 = 9) \tag{A.91}$$

Vector 2-norm (Euclidean norm). This is the most common vector norm, and corresponds to the shortest distance between two points

$$\|a\|_2 \triangleq \sqrt{\sum_i |a_i|^2} \qquad (\|b\|_2 = \sqrt{1 + 9 + 25} = 5.916) \tag{A.92}$$

The Euclidean vector norm satisfies the property

$$a^H a = \|a\|_2^2 \tag{A.93}$$

where a^H denotes the complex conjugate transpose of the vector a.

Vector ∞-norm (or max norm). This is the largest-element magnitude in the vector. We use the notation $\|a\|_{\max}$ so that

$$\|a\|_{\max} \equiv \|a\|_\infty \triangleq \max_i |a_i| \qquad (\|b\|_{\max} = |-5| = 5) \tag{A.94}$$

Since the various vector norms only differ by constant factors, they are often said to be *equivalent*. For example, for a vector with m elements

$$\|a\|_{\max} \leq \|a\|_2 \leq \sqrt{m} \, \|a\|_{\max} \tag{A.95}$$

$$\|a\|_2 \leq \|a\|_1 \leq \sqrt{m} \, \|a\|_2 \tag{A.96}$$

In Figure A.1 the differences between the vector norms are illustrated by plotting the contours for $\|a\|_p = 1$ for the case with $m = 2$.

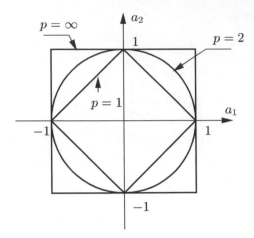

Figure A.1: Contours for the vector p-norm, $\|a\|_p = 1$ for $p = 1, 2, \infty$. (Mathematically: The unit ball on $\mathbb{R}^2$ with three different norms.)

A.5.2 Matrix norms

We will consider a constant $l \times m$ matrix A. The matrix A may represent, for example, the frequency response, $G(j\omega)$, of a system $G(s)$ with m inputs and l outputs. For numerical illustrations we will use the following 2×2 matrix example:

$$A_0 = \begin{bmatrix} 1 & 2 \\ -3 & 4 \end{bmatrix} \qquad (A.97)$$

Definition A.6 *A norm on a matrix $\|A\|$ is a* **matrix norm** *if, in addition to the four norm properties in Definition A.5, it also satisfies the multiplicative property (also called the consistency condition):*

$$\|AB\| \le \|A\| \cdot \|B\| \qquad (A.98)$$

Property (A.98) is very important when combining systems, and forms the basis for the small-gain theorem. Note that there exist *norms on matrices* (thus satisfying the four properties of a norm), which are not *matrix norms* (thus *not* satisfying (A.98)). Such norms are sometimes called *generalized matrix norms*. The only generalized matrix norm considered in this book is the largest-element norm, $\|A\|_{\max}$.

Let us first examine three norms which are direct extensions of the definitions of the vector p-norms.

Sum matrix norm. This is the sum of the element magnitudes

$$\|A\|_{\mathrm{sum}} = \sum_{i,j} |a_{ij}| \qquad (\|A_0\|_{\mathrm{sum}} = 1 + 2 + 3 + 4 = 10) \qquad (A.99)$$

Frobenius matrix norm (or Euclidean norm). This is the square root of the sum of the squared element magnitudes

$$\|A\|_F = \sqrt{\sum_{i,j} |a_{ij}|^2} = \sqrt{\mathrm{tr}(A^H A)} \qquad (\|A_0\|_F = \sqrt{30} = 5.477) \qquad (A.100)$$

The trace tr is the sum of the diagonal elements, and A^H is the complex conjugate transpose of A. The Frobenius norm is important in control because it is used for summing up the channels, e.g. when using LQG optimal control.

Max element norm. This is the largest-element magnitude,

$$\|A\|_{\max} = \max_{i,j} |a_{ij}| \qquad (\|A_0\|_{\max} = 4) \tag{A.101}$$

This norm is *not* a matrix norm as it does not satisfy (A.98). However, note that $\sqrt{lm}\,\|A\|_{\max}$ *is* a matrix norm.

The above three norms are sometimes called the 1-, 2- and ∞-norm, respectively, but this notation is *not* used in this book to avoid confusion with the more important induced p-norms introduced next.

Induced matrix norms

Figure A.2: Representation of (A.102)

Induced matrix norms are important because of their close relationship to signal amplification in systems. Consider the following equation which is illustrated in Figure A.2:

$$z = Aw \tag{A.102}$$

We may think of w as the input vector and z as the output vector and consider the "amplification" or "gain" of the matrix A as defined by the ratio $\|z\|/\|w\|$. The maximum gain for all possible input directions is of particular interest. This is given by the *induced norm* which is defined as

$$\|A\|_{ip} \triangleq \max_{w \neq 0} \frac{\|Aw\|_p}{\|w\|_p} \tag{A.103}$$

where $\|w\|_p = (\sum_i |w_i|^p)^{1/p}$ denotes the vector p-norm. In other words, we are looking for a direction of the vector w such that the ratio $\|z\|_p/\|w\|_p$ is maximized. Thus, the induced norm gives the largest possible "amplifying power" of the matrix. The following equivalent definition is also used:

$$\|A\|_{ip} = \max_{\|w\|_p \leq 1} \|Aw\|_p = \max_{\|w\|_p = 1} \|Aw\|_p \tag{A.104}$$

For the induced 1-, 2- and ∞-norms the following identities hold:

$$\|A\|_{i1} = \max_j (\sum_i |a_{ij}|) \quad \text{"maximum column sum"} \tag{A.105}$$

$$\|A\|_{i\infty} = \max_i (\sum_j |a_{ij}|) \quad \text{"maximum row sum"} \tag{A.106}$$

$$\|A\|_{i2} = \bar{\sigma}(A) = \sqrt{\rho(A^H A)} \quad \text{"singular value or spectral norm"} \tag{A.107}$$

where the spectral radius $\rho(A) = \max_i |\lambda_i(A)|$ is the largest eigenvalue of the matrix A. Note that the induced 2-norm of a matrix is equal to the (largest) singular value, and is often called the spectral norm. For the example matrix in (A.97) we get

$$\|A_0\|_{i1} = 6; \quad \|A_0\|_{i\infty} = 7; \quad \|A_0\|_{i2} = \bar{\sigma}(A_0) = 5.117 \tag{A.108}$$

Theorem A.3 *All induced norms $\|A\|_{ip}$ are matrix norms and thus satisfy the multiplicative property*

$$\|AB\|_{ip} \leq \|A\|_{ip} \cdot \|B\|_{ip} \tag{A.109}$$

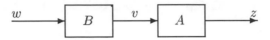

Figure A.3: Representation of (A.110)

Proof: Consider the following set of equations which is illustrated graphically in Figure A.3:

$$z = Av, \quad v = Bw \quad \Rightarrow z = ABw \tag{A.110}$$

From the definition of the induced norm we get by first introducing $v = Bw$, then multiplying the numerator and denominator by $\|v\|_p \neq 0$, and finally maximizing each term involving w and v independently, that

$$\|AB\|_{ip} \triangleq \max_{w \neq 0} \frac{\|A \overbrace{Bw}^{v}\|_p}{\|w\|_p} = \max_{w \neq 0} \frac{\|Av\|_p}{\|v\|_p} \cdot \frac{\|Bw\|_p}{\|w\|_p} \leq \max_{v \neq 0} \frac{\|Av\|_p}{\|v\|_p} \cdot \max_{w \neq 0} \frac{\|Bw\|_p}{\|w\|_p}$$

and (A.109) follows from the definition of an induced norm. ☐

Implications of the multiplicative property

For matrix norms the multiplicative property $\|AB\| \leq \|A\| \cdot \|B\|$ holds for matrices A and B of any dimension as long as the product AB exists. In particular, it holds if we choose A and B as vectors. From this observation we get:

1. Choose B to be a vector, i.e. $B = w$. Then for any matrix norm we have from (A.98) that

$$\|Aw\| \leq \|A\| \cdot \|w\| \tag{A.111}$$

We say that the "matrix norm $\|A\|$ is compatible with its corresponding vector norm $\|w\|$". Clearly, from (A.103) any induced matrix p-norm is compatible with its corresponding vector p-norm. Similarly, the Frobenius norm is compatible with the vector 2-norm (since when w is a vector $\|w\|_F = \|w\|_2$).
2. From (A.111) we also get for any matrix norm that

$$\|A\| \geq \max_{w \neq 0} \frac{\|Aw\|}{\|w\|} \tag{A.112}$$

Note that the induced norms are defined such that we have equality in (A.112). The property $\|A\|_F \geq \bar{\sigma}(A)$ then follows since $\|w\|_F = \|w\|_2$.

3. Choose both $A = z^H$ and $B = w$ as vectors. Then using the Frobenius norm or induced 2-norm (singular value) in (A.98) we derive the Cauchy–Schwarz inequality

$$|z^H w| \leq ||z||_2 \cdot ||w||_2 \qquad (A.113)$$

where z and w are column vectors of the same dimension and $z^H w$ is the Euclidean inner product between the vectors z and w.

4. The inner product can also be used to define the angle ϕ between two vectors z and w

$$\phi = \cos^{-1}\left(\frac{|z^H w|}{||z||_2 \cdot ||w||_2} \right) \qquad (A.114)$$

Note that with this definition, ϕ is between $0°$ and $90°$.

A.5.3   The spectral radius

The spectral radius $\rho(A)$ is the magnitude of the largest eigenvalue of the matrix A,

$$\rho(A) = \max_i |\lambda_i(A)| \qquad (A.115)$$

It is *not* a norm, as it does not satisfy norm properties 2 and 4 in Definition A.5. For example, for

$$A_1 = \begin{bmatrix} 1 & 0 \\ 10 & 1 \end{bmatrix}, \quad A_2 = \begin{bmatrix} 1 & 10 \\ 0 & 1 \end{bmatrix} \qquad (A.116)$$

we have $\rho(A_1) = 1$ and $\rho(A_2) = 1$. However, $\rho(A_1 + A_2) = 12$ and $\rho(A_1 A_2) = 101.99$, which satisfy neither the triangle inequality (property 4 of a norm) nor the multiplicative property in (A.98).

Although the spectral radius is not a norm, it provides a lower bound on any matrix norm, which can be very useful.

Theorem A.4 *For any matrix norm (and in particular for any induced norm)*

$$\rho(A) \leq ||A|| \qquad (A.117)$$

Proof: Since $\lambda_i(A)$ is an eigenvalue of A, we have that $At_i = \lambda_i t_i$ where t_i denotes the eigenvector. We get

$$|\lambda_i| \cdot ||t_i|| = ||\lambda_i t_i|| = ||At_i|| \leq ||A|| \cdot ||t_i|| \qquad (A.118)$$

(the last inequality follows from (A.111)). Thus for any matrix norm $|\lambda_i(A)| \leq ||A||$ and since this holds for all eigenvalues the result follows.   ☐

For our example matrix in (A.97) we get $\rho(A_0) = \sqrt{10} \approx 3.162$ which is less than all the induced norms ($||A_0||_{i1} = 6$, $||A_0||_{i\infty} = 7$, $\bar{\sigma}(A_0) = 5.117$) and also less than the Frobenius norm ($||A||_F = 5.477$) and the sum norm ($||A||_{sum} = 10$).

A simple physical interpretation of (A.117) is that the eigenvalue measures the gain of the matrix only in certain directions (given by the eigenvectors), and must therefore be less than that for a matrix norm which allows any direction and yields the maximum gain, recall (A.112).

A.5.4   Some matrix norm relationships

The various norms of the matrix A are closely related as can be seen from the following inequalities from Golub and van Loan (1989, p. 15) and Horn and Johnson (1985, p. 314). Let A be an $l \times m$ matrix, then

$$\bar{\sigma}(A) \leq \|A\|_F \leq \sqrt{\min(l, m)}\, \bar{\sigma}(A) \tag{A.119}$$

$$\|A\|_{\max} \leq \bar{\sigma}(A) \leq \sqrt{lm}\, \|A\|_{\max} \tag{A.120}$$

$$\bar{\sigma}(A) \leq \sqrt{\|A\|_{i1}\|A\|_{i\infty}} \tag{A.121}$$

$$\frac{1}{\sqrt{m}}\|A\|_{i\infty} \leq \bar{\sigma}(A) \leq \sqrt{l}\, \|A\|_{i\infty} \tag{A.122}$$

$$\frac{1}{\sqrt{l}}\|A\|_{i1} \leq \bar{\sigma}(A) \leq \sqrt{m}\, \|A\|_{i1} \tag{A.123}$$

$$\max\{\bar{\sigma}(A), \|A\|_F, \|A\|_{i1}, \|A\|_{i\infty}\} \leq \|A\|_{\text{sum}} \tag{A.124}$$

All these norms, except $\|A\|_{\max}$, are matrix norms and satisfy (A.98). The inequalities are tight; that is, there exist matrices of any size for which the equality holds. Note from (A.120) that the maximum singular value is closely related to the largest element of the matrix. Therefore, $\|A\|_{\max}$ can be used as a simple and readily available estimate of $\bar{\sigma}(A)$.

An important property of the Frobenius norm and the maximum singular value (induced 2-norm) is that they are invariant with respect to unitary transformations, i.e. for unitary matrices U_i, satisfying $U_i U_i^H = I$, we have

$$\|U_1 A U_2\|_F = \|A\|_F \tag{A.125}$$

$$\bar{\sigma}(U_1 A U_2) = \bar{\sigma}(A) \tag{A.126}$$

From an SVD of the matrix $A = U\Sigma V^H$ and (A.125), we then obtain an important relationship between the Frobenius norm and the singular values, $\sigma_i(A)$, namely

$$\|A\|_F = \sqrt{\sum_i \sigma_i^2(A)} \tag{A.127}$$

The Perron–Frobenius theorem, which applies to a square matrix A, states that

$$\min_D \|DAD^{-1}\|_{i1} = \min_D \|DAD^{-1}\|_{i\infty} = \rho(|A|) \tag{A.128}$$

where D is a diagonal "scaling" matrix, $|A|$ denotes the matrix A with all its elements replaced by their magnitudes, and $\rho(|A|) = \max_i |\lambda_i(|A|)|$ is the Perron root (Perron–Frobenius eigenvalue). The Perron root is greater than or equal to the spectral radius, $\rho(A) \leq \rho(|A|)$.

A.5.5 Matrix and vector norms in Matlab

The following Matlab commands are used for matrices:

$$\bar{\sigma}(A) = \|A\|_{i2} \quad \texttt{norm(A,2)} \text{ or } \texttt{max(svd(A))}$$
$$\|A\|_{i1} \quad \texttt{norm(A,1)}$$
$$\|A\|_{i\infty} \quad \texttt{norm(A,'inf')}$$
$$\|A\|_F \quad \texttt{norm(A,'fro')}$$
$$\|A\|_{\text{sum}} \quad \texttt{sum (sum(abs(A)))}$$
$$\|A\|_{\text{max}} \quad \texttt{max(max(abs(A)))} \text{ (which is not a matrix norm)}$$
$$\rho(A) \quad \texttt{max(abs(eig(A)))}$$
$$\rho(|A|) \quad \texttt{max(eig(abs(A)))}$$
$$\gamma(A) = \bar{\sigma}(A)/\underline{\sigma}(A) \quad \texttt{cond(A)}$$

For vectors:

$$\|a\|_1 \quad \texttt{norm(a,1)}$$
$$\|a\|_2 \quad \texttt{norm(a,2)}$$
$$\|a\|_{\text{max}} \quad \texttt{norm(a,'inf')}$$

A.5.6 Signal norms

We will consider the temporal norm of a time-varying (or frequency-varying) signal, $e(t)$. In contrast with spatial norms (vector and matrix norms), we find that the choice of temporal norm makes a big difference. As an example, consider Figure A.4 which shows two signals, $e_1(t)$ and $e_2(t)$. For $e_1(t)$ the ∞-norm (peak) is 1, $\|e_1(t)\|_\infty = 1$, whereas since the signal does not "die out" the 2-norm is infinite, $\|e_1(t)\|_2 = \infty$. For $e_2(t)$ the opposite is true.

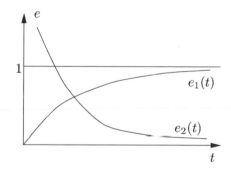

Figure A.4: Signals with entirely different 2-norms and ∞-norms

For signals we may compute the norm in two steps:

1. "Sum up" the channels at a given time or frequency using a vector norm (for a scalar signal we simply take the absolute value).
2. "Sum up" in time or frequency using a temporal norm.

Recall from above that the vector norms are "equivalent" in the sense that their values differ only by a constant factor. Therefore, it does not really make too much difference which norm we use in step 1. We normally use the same p-norm for both the vector and the signal, and

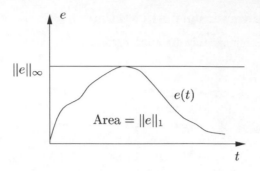

Figure A.5: Signal 1-norm and ∞-norm

thus define the temporal p-norm, $\|e(t)\|_p$, of a time-varying vector as

$$L_p \text{ norm:} \quad \|e(t)\|_p = \left(\int_{-\infty}^{\infty} \sum_i |e_i(\tau)|^p d\tau \right)^{1/p} \tag{A.129}$$

The following temporal norms of signals are commonly used:
1-norm in time (integral absolute error (IAE), see Figure A.5):

$$\|e(t)\|_1 = \int_{-\infty}^{\infty} \sum_i |e_i(\tau)| d\tau \tag{A.130}$$

2-norm in time (quadratic norm, integral square error (ISE), "energy" of signal):

$$\|e(t)\|_2 = \sqrt{\int_{-\infty}^{\infty} \sum_i |e_i(\tau)|^2 d\tau} \tag{A.131}$$

∞**-norm in time** (peak value in time, see Figure A.5):

$$\|e(t)\|_\infty = \max_\tau \left(\max_i |e_i(\tau)| \right) \tag{A.132}$$

In addition, we will consider the power norm or rms norm (which is actually only a semi-norm since it does not satisfy norm property 2)

$$\|e(t)\|_{\text{pow}} = \lim_{T \to \infty} \sqrt{\frac{1}{2T} \int_{-T}^{T} \sum_i |e_i(\tau)|^2 d\tau} \tag{A.133}$$

Remark 1 In most cases we assume $e(t) = 0$ for $t < 0$ so the lower value for the integration may be changed to $\tau = 0$.

Remark 2 To be mathematically correct we should have used $\sup_\tau$ (least upper bound) rather than $\max_\tau$ in (A.132), since the maximum value may not actually be achieved (e.g. if it occurs for $t = \infty$).

A.5.7 Signal interpretation of various system norms

Two system norms are considered in Section 4.10. These are the $\mathcal{H}_2$ norm, $\|G(s)\|_2 = \|g(t)\|_2$, and the $\mathcal{H}_\infty$ norm, $\|G(s)\|_\infty$. The main reason for including this subsection is to show that there are many ways of evaluating performance in terms of signals, and to show that the $\mathcal{H}_2$ and $\mathcal{H}_\infty$ norms are useful measures in this context. This in turn will be useful in helping us to understand how to select performance weights in controller design problems. The proofs of the results in this subsection require a good background in functional analysis and can be found in Doyle et al. (1992), Dahleh and Diaz-Bobillo (1995) and Zhou et al. (1996).

Consider a system G with input d and output e, such that

$$e = Gd \tag{A.134}$$

For performance we may want the output signal e to be "small" for any allowed input signals d. We therefore need to specify:

1. What d's are allowed. (Which set does d belong to?)
2. What we mean by "small". (Which norm should we use for e?)

Some possible input signal sets are:

1. $d(t)$ consists of impulses, $\delta(t)$. These generate step changes in the states, which is the usual way of introducing the LQ objective and gives rise to the $\mathcal{H}_2$ norm.
2. $d(t)$ is a white noise process with zero mean.
3. $d(t) = \sin(\omega t)$ with fixed frequency, applied from $t = -\infty$ (which corresponds to the steady-state sinusoidal response).
4. $d(t)$ is a set of sinusoids with all frequencies allowed.
5. $d(t)$ is bounded in energy, $\|d(t)\|_2 \leq 1$.
6. $d(t)$ is bounded in power, $\|d(t)\|_{\text{pow}} \leq 1$.
7. $d(t)$ is bounded in magnitude, $\|d(t)\|_\infty \leq 1$.

The first three sets of inputs are specific signals, whereas the latter three are classes of inputs with bounded norm. The physical problem at hand determines which of these input classes is the most reasonable.

To measure the output signal one may consider the following norms:

1. 1-norm, $\|e(t)\|_1$
2. 2-norm (energy), $\|e(t)\|_2$
3. ∞-norm (peak magnitude), $\|e(t)\|_\infty$
4. power, $\|e(t)\|_{\text{pow}}$

Other norms are possible, but, again, it is engineering issues that determine which norm is the most appropriate. We will now consider which system norms result from the definitions of input classes, and output norms, respectively. That is, we want to find the appropriate system gain to test for performance. The results for SISO systems in which $d(t)$ and $e(t)$ are scalar signals are summarized in Tables A.1 and A.2. In these tables $G(s)$ is the transfer function and $g(t)$ is its corresponding impulse response. Note in particular that

$$\mathcal{H}_\infty \text{ norm:} \quad \|G(s)\|_\infty \triangleq \max_\omega \bar{\sigma}(G(j\omega)) = \max_{d(t)} \frac{\|e(t)\|_2}{\|d(t)\|_2} \tag{A.135}$$

and

$$\mathcal{L}_1 \text{ norm:} \quad \|g(t)\|_1 \triangleq \int_{-\infty}^{\infty} g(t)dt = \max_{d(t)} \frac{\|e(t)\|_\infty}{\|d(t)\|_\infty} \qquad (A.136)$$

(where the two right equalities are not by definition; these are important results from functional analysis). We see from Tables A.1 and A.2 that the $\mathcal{H}_2$ and $\mathcal{H}_\infty$ norms appear in several positions. This gives some basis for their popularity in control. In addition, the $\mathcal{H}_\infty$ norm results if we consider $d(t)$ to be the set of sinusoids with all frequencies allowed, and measure the output using the 2-norm (not shown in Tables A.1 and A.2, but discussed in Section 3.3.5). Also, the $\mathcal{H}_2$ norm results if the input is white noise and we measure the output using the 2-norm.

Table A.1: System norms for two specific input signals and three different output norms

	$d(t) = \delta(t)$	$d(t) = \sin(\omega t)$
$\|e\|_2$	$\|G(s)\|_2$	∞ (usually)
$\|e\|_\infty$	$\|g(t)\|_\infty$	$\bar{\sigma}(G(j\omega))$
$\|e\|_{\text{pow}}$	0	$\frac{1}{\sqrt{2}}\bar{\sigma}(G(j\omega))$

Table A.2: System norms for three sets of norm-bounded input signals and three different output norms. The entries along the diagonal are induced norms.

	$\|d\|_2$	$\|d\|_\infty$	$\|d\|_{\text{pow}}$
$\|e\|_2$	$\|G(s)\|_\infty$	∞	∞ (usually)
$\|e\|_\infty$	$\|G(s)\|_2$	$\|g(t)\|_1$	∞ (usually)
$\|e\|_{\text{pow}}$	0	$\leq \|G(s)\|_\infty$	$\|G(s)\|_\infty$

The results in Tables A.1 and A.2 may be generalized to MIMO systems by use of the appropriate matrix and vector norms. In particular, the induced norms along the diagonal in Table A.2 generalize if we use for the $\mathcal{H}_\infty$ norm $\|G(s)\|_\infty = \max_\omega \bar{\sigma}(G(j\omega))$, and for the L_1 norm we use $\|g(t)\|_1 = \max_i \|g_i(t)\|_1$, where $g_i(t)$ denotes row i of the impulse response matrix. The fact that the $\mathcal{H}_\infty$ norm and L_1 norm are induced norms makes them well suited for robustness analysis; for example, using the small-gain theorem. The two norms are also closely related as can be seen from the following bounds for a proper scalar system:

$$\|G(s)\|_\infty \leq \|g(t)\|_1 \leq (2n+1) \cdot \|G(s)\|_\infty \qquad (A.137)$$

where n is the number of states in a minimal realization. We have the following generalization for a multivariable $l \times m$ system (Dahleh and Diaz-Bobillo, 1995, p. 342):

$$\|G(s)\|_\infty \leq \sqrt{l} \cdot \|g(t)\|_1 \leq \sqrt{lm} \cdot (2n+1) \cdot \|G(s)\|_\infty \qquad (A.138)$$

A.6   All-pass factorization of transfer function matrices

Consider a plant model G with N_z RHP-zeros at z and associated input and output zero directions u_z and y_z, respectively. Then G can be factored as follows:

$$G = G^1\mathcal{B}_1 \qquad \mathcal{B}_1 = I - \frac{2\mathrm{Re}(z_1)}{s + \bar{z}_1}\widehat{u}_{z_1}\widehat{u}_{z_1}^H \qquad (A.139)$$

where $\widehat{u}_{z_1}$ is the input zero direction of z_1. With this factorization, z_1 is not a zero of G^1. By repeated application of (A.139) on G^i, $i = 1 \cdots N_z - 1$, G can be factored into a minimum-phase part and an all-pass filter as

$$G = G_{mi}\mathcal{B}_{zi} \qquad \mathcal{B}_{zi} = \prod_{i=1}^{N_z}\left(I - \frac{2\mathrm{Re}(z_i)}{s + \bar{z}_i}\widehat{u}_{z_i}\widehat{u}_{z_i}^H\right) \qquad (A.140)$$

In (A.140), G_{mi} is minimum phase with the RHP-zeros of G mirrored across the imaginary axis and $\mathcal{B}_{zi}$ is an all-pass filter. Note that except for the direction associated with the zero factored first, $\widehat{u}_{z_i}$ differs from u_{z_i}, as it is calculated based on $G^{(i-1)}$ and not G. The RHP-zeros can be alternatively factored at system's output similarly:

$$G = \mathcal{B}_{zo}G_{mo} \qquad \mathcal{B}_{zo} = \prod_{i=N_z}^{1}\left(I - \frac{2\mathrm{Re}(z_i)}{s + \bar{z}_i}\widehat{y}_{z_i}\widehat{y}_{z_i}^H\right) \qquad (A.141)$$

When G has N_p RHP-poles at p, these poles can also be factored into a stable part and an all-pass filter on the input and output side as follows:

$$G = G_{si}\mathcal{B}_{pi}^{-1} \qquad \mathcal{B}_{pi}^{-1} = \prod_{i=N_p}^{1}\left(I - \frac{2\mathrm{Re}(p_i)}{s - p_i}\widehat{u}_{p_i}\widehat{u}_{p_i}^H\right) \qquad (A.142)$$

$$G = \mathcal{B}_{po}^{-1}G_{so} \qquad \mathcal{B}_{po}^{-1} = \prod_{i=1}^{N_p}\left(I - \frac{2\mathrm{Re}(p_i)}{s - p_i}\widehat{y}_{p_i}\widehat{y}_{p_i}^H\right) \qquad (A.143)$$

For SISO systems, (A.140)–(A.143) simplify as

$$\mathcal{B}_{zi} = \mathcal{B}_{zo} = \prod_{i=1}^{N_z}\frac{s - z_i}{s + \bar{z}_i} \qquad (A.144)$$

$$\mathcal{B}_{pi}^{-1} = \mathcal{B}_{po}^{-1} = \prod_{i=1}^{N_p}\frac{s + \bar{p}_i}{s - p_j} \qquad (A.145)$$

In Chapters 5 and 6, we use the all-pass factorizations of RHP-zeros and RHP-poles to derive bounds on peaks of important closed-loop transfer functions. $\mathcal{B}_{zi}, \mathcal{B}_{zo}, \mathcal{B}_{pi}$ and $\mathcal{B}_{po}$ are collectively called *Blaschke products*. A collection of many useful properties of Blaschke products can be found in Havre (1998).

Remark. In the first edition of this book (Skogestad and Postlethwaite, 1996), the Blaschke products are defined as the inverse of the more conventional definitions used here. However, note that the alternative definitions used in the first edition have no effect on any ensuing analysis.

A.7   Factorization of the sensitivity function

Consider two plant models, G a nominal model and G' an alternative model, and assume that the same controller is applied to both plants. Then the corresponding sensitivity functions are

$$S = (I + GK)^{-1}, \quad S' = (I + G'K)^{-1} \tag{A.146}$$

A.7.1   Output perturbations

Assume that G' is related to G by either an output multiplicative perturbation E_O, or an inverse output multiplicative perturbation E_{iO}. Then S' can be factorized in terms of S as follows:

$$S' = S(I + E_O T)^{-1}; \quad G' = (I + E_O)G \tag{A.147}$$

$$S' = S(I - E_{iO}S)^{-1}(I - E_{iO}); \quad G' = (I - E_{iO})^{-1}G \tag{A.148}$$

For a square plant, E_O and E_{iO} can be obtained from a given G and G' by

$$E_O = (G' - G)G^{-1}; \quad E_{iO} = (G' - G)G'^{-1} \tag{A.149}$$

Proof of (A.147):

$$I + G'K = I + (I + E_O)GK = (I + E_O \underbrace{GK(I + GK)^{-1}}_{T})(I + GK) \tag{A.150}$$

□

Proof of (A.148):

$$
\begin{aligned}
I + G'K &= I + (I - E_{iO})^{-1}GK = (I - E_{iO})^{-1}((I - E_{iO}) + GK) \\
&= (I - E_{iO})^{-1}(I - E_{iO} \underbrace{(I + GK)^{-1}}_{S})(I + GK)
\end{aligned} \tag{A.151}
$$

□

Similar factorizations may be written in terms of the complementary sensitivity function (Horowitz and Shaked, 1975; Zames, 1981). For example, by writing (A.147) in the form $S = S'(I + E_O T)$ and using the fact that $S - S' = T' - T$, we get

$$T' - T = S'E_O T \tag{A.152}$$

A.7.2   Input perturbations

For a square plant, the following factorization in terms of input multiplicative uncertainty E_I is useful:

$$S' = S(I + GE_I G^{-1}T)^{-1} = SG(I + E_I T_I)^{-1}G^{-1}; \quad G' = G(I + E_I) \tag{A.153}$$

where $T_I = KG(I + KG)^{-1}$ is the input complementary sensitivity function.

Proof: Substitute $E_O = GE_I G^{-1}$ into (A.147) and use $G^{-1}T = T_I G^{-1}$.   □

Alternatively, we may factor out the controller to get

$$S' = (I + TK^{-1}E_I K)^{-1}S = K^{-1}(I + T_I E_I)^{-1}KS \tag{A.154}$$

Proof: Start from $I + G'K = I + G(I + E_I)K$ and factor out $(I + GK)$ to the left.   □

A.7.3 Stability conditions

The next lemma follows directly from the generalized Nyquist theorem and the factorization (A.147):

Lemma A.5 *Assume that the negative feedback closed-loop system with loop transfer function $G(s)K(s)$ is stable. Suppose $G' = (I + E_O)G$, and let the number of open-loop unstable poles of $G(s)K(s)$ and $G'(s)K(s)$ be P and P', respectively. Then the negative feedback closed-loop system with loop transfer function $G'(s)K(s)$ is stable if and only if*

$$\mathcal{N}(\det(I + E_O T)) = P - P' \tag{A.155}$$

where $\mathcal{N}$ denotes the number of clockwise encirclements of the origin as s traverses the Nyquist D-contour in a clockwise direction.

Proof: Let $\mathcal{N}(f)$ denote the number of clockwise encirclements of the origin by $f(s)$ as s traverses the Nyquist D-contour in a clockwise direction. For the encirclements of the product of two functions we have $\mathcal{N}(f_1 f_2) = \mathcal{N}(f_1) + \mathcal{N}(f_2)$. This together with (A.150) and the fact $\det(AB) = \det A \cdot \det B$ yields

$$\mathcal{N}(\det(I + G'K)) = \mathcal{N}(\det(I + E_O T)) + \mathcal{N}(\det(I + GK)) \tag{A.156}$$

For stability we need from Theorem 4.9 that $\mathcal{N}(\det(I + G'K)) = -P'$, but we know that $\mathcal{N}(\det(I + GK)) = -P$ and hence Lemma A.5 follows. The lemma is from Hovd and Skogestad (1994); similar results, at least for stable plants, have been presented by, for example, Grosdidier and Morari (1986) and Nwokah and Perez (1991). □

In other words, (A.155) tells us that for stability $\det(I + E_O T)$ must provide the required additional number of clockwise encirclements. If (A.155) is not satisfied then the negative feedback system with $G'K$ must be unstable. We show in Theorem 6.7 how the information about what happens at $s = 0$ can be used to determine stability.

A.8 Linear fractional transformations

Linear fractional transformations (LFTs), as they are currently used in the control literature for analysis and design, were introduced by Doyle (1984). Consider a matrix P of dimension $(n_1 + n_2) \times (m_1 + m_2)$ and partition it as follows:

$$P = \begin{bmatrix} P_{11} & P_{12} \\ P_{21} & P_{22} \end{bmatrix} \tag{A.157}$$

Let the matrices Δ and K have dimensions $m_1 \times n_1$ and $m_2 \times n_2$, respectively (compatible with the upper and lower partitions of P, respectively). We adopt the following notation for the lower and upper linear fractional transformations:

$$F_l(P, K) \triangleq P_{11} + P_{12}K(I - P_{22}K)^{-1}P_{21} \tag{A.158}$$

$$F_u(P, \Delta) \triangleq P_{22} + P_{21}\Delta(I - P_{11}\Delta)^{-1}P_{12} \tag{A.159}$$

where subscript l denotes lower and subscript u upper. In the following, let R denote a matrix function resulting from an LFT.

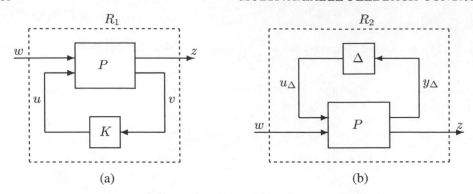

Figure A.6: (a) R_1 as lower LFT in terms of K. (b) R_2 as upper LFT in terms of Δ.

The lower fractional transformation $F_l(P, K)$ is the transfer function R_1 resulting from wrapping (positive) feedback K around the lower part of P as illustrated in Figure A.6(a). To see this, note that the block diagram in Figure A.6(a) may be written as

$$z = P_{11}w + P_{12}u, \quad v = P_{21}w + P_{22}u, \quad u = Kv \tag{A.160}$$

Upon eliminating v and u from these equations we get

$$z = R_1 w = F_l(P, K)w = [P_{11} + P_{12}K(I - P_{22}K)^{-1}P_{21}]w \tag{A.161}$$

In words, R_1 is written as a lower LFT of P in terms of the parameter K. Similarly, in Figure A.6(b) we illustrate the upper LFT, $R_2 = F_u(P, \Delta)$, obtained by wrapping (positive) feedback Δ around the upper part of P.

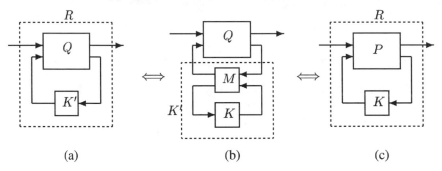

Figure A.7: An interconnection of LFTs yields an LFT

A.8.1   Interconnection of LFTs

An important property of LFTs is that any interconnection of LFTs is again an LFT. Consider Figure A.7 where R is written in terms of a lower LFT of K', which again is a lower LFT of K, and we want to express R directly as an LFT of K. We have

$$R = F_l(Q, K') \quad \text{where} \quad K' = F_l(M, K) \tag{A.162}$$

and we want to obtain the P (in terms of Q and M) such that

$$R = F_l(P, K) \tag{A.163}$$

We find

$$P = \begin{bmatrix} P_{11} & P_{12} \\ P_{21} & P_{22} \end{bmatrix}$$

$$= \begin{bmatrix} Q_{11} + Q_{12}M_{11}(I - Q_{22}M_{11})^{-1}Q_{21} & Q_{12}(I - M_{11}Q_{22})^{-1}M_{12} \\ M_{21}(I - Q_{22}M_{11})^{-1}Q_{21} & M_{22} + M_{21}Q_{22}(I - M_{11}Q_{22})^{-1}M_{12} \end{bmatrix}$$

$$\tag{A.164}$$

Similar expressions apply when we use *upper* LFTs. For

$$R = F_u(M, \Delta') \quad \text{where} \quad \Delta' = F_u(Q, \Delta) \tag{A.165}$$

we get $R = F_u(P, \Delta)$ where P is given in terms of Q and M by (A.164).

A.8.2 Relationship between F_l and F_u

F_l and F_u are obviously closely related. If we know $R = F_l(M, K)$, then we may directly obtain R in terms of an upper transformation of K by reordering M. We have

$$F_u(\widetilde{M}, K) = F_l(M, K) \tag{A.166}$$

where

$$\widetilde{M} = \begin{bmatrix} 0 & I \\ I & 0 \end{bmatrix} M \begin{bmatrix} 0 & I \\ I & 0 \end{bmatrix} \tag{A.167}$$

A.8.3 Inverse of LFTs

On the assumption that all the relevant inverses exist we have

$$(F_l(M, K))^{-1} = F_l(\widetilde{M}, K) \tag{A.168}$$

where $\widetilde{M}$ is given by

$$\widetilde{M} = \begin{bmatrix} M_{11}^{-1} & -M_{11}^{-1}M_{12} \\ M_{21}M_{11}^{-1} & M_{22} - M_{21}M_{11}^{-1}M_{12} \end{bmatrix} \tag{A.169}$$

This expression follows easily from the matrix inversion lemma in (A.6).

A.8.4 LFT in terms of the inverse parameter

Given an LFT in terms of K, it is possible to derive an equivalent LFT in terms of K^{-1}. If we assume that all the relevant inverses exist we have

$$F_l(M, K) = F_l(\widehat{M}, K^{-1}) \tag{A.170}$$

where $\widehat{M}$ is given by

$$\widehat{M} = \begin{bmatrix} M_{11} - M_{12}M_{22}^{-1}M_{21} & -M_{12}M_{22}^{-1} \\ M_{22}^{-1}M_{21} & M_{22}^{-1} \end{bmatrix} \tag{A.171}$$

This expression follows from the fact that $(I+L)^{-1} = I - L(I+L)^{-1}$ for any square matrix L.

APPENDIX B

PROJECT WORK AND SAMPLE EXAM

B.1 Project work

Students are encouraged to formulate their own project based on an application they are working on. Otherwise, the project is given by the instructor. In either case, a preliminary statement of the problem must be approved before starting the project; see the first item below.

A useful collection of benchmark problems for control system design is provided in Davison (1990). The helicopter, aero-engine and distillation case studies in Chapter 13, and the chemical reactor in Example 6.17, also provide the basis for several projects. These models are available over the Internet.

1. *Introduction: Preliminary problem definition.*

    (i) Give a simple description of the engineering problem with the aid of one or two diagrams.

    (ii) Discuss briefly the control objectives.

    (iii) Specify the exogenous inputs (disturbances, noise, setpoints), the manipulated inputs, the measurements and the controlled outputs (exogenous outputs).

    (iv) Describe the most important sources of model uncertainty.

    (v) What specific control problems do you expect, e.g. due to interactions, RHP-zeros, saturation, etc.?

    The preliminary statement of no more than three pages must be handed in and approved before starting the project.
2. *Plant model.* Specify all parameters, operating conditions, etc., and obtain a linear model of the plant. Comment: You may need to consider more than one operating point.
3. *Analysis of the plant.* For example, compute the steady-state gain matrix, plot the gain elements as a function of frequency, obtain the poles and zeros (both of the individual elements and the overall system), compute the SVD and comment on directions and the condition number, perform an RGA analysis, a disturbance analysis, etc. Does the analysis indicate that the plant is difficult to control?
4. *Initial controller design.* Design at least two controllers using, for example,

    (i) Decentralized control (PID).

Multivariable Feedback Control: Analysis and Design. Second Edition
S. Skogestad and I. Postlethwaite © 2005, 2006 John Wiley & Sons, Ltd

(ii) Centralized control (LQG, LTR, $\mathcal{H}_2$ (in principle same as LQG, but with a different way of choosing weights), $\mathcal{H}_\infty$ loop shaping, $\mathcal{H}_\infty$ mixed sensitivity, etc.).

(iii) A decoupler combined with PI control.

5. *Simulations.* Perform simulations in the time domain for the closed-loop system.
6. *Robustness analysis using μ.*

(a) Choose suitable performance and uncertainty weights. Plot the weights as functions of frequency.

(b) State clearly how RP is defined for your problem (using block diagrams).

(c) Compute μ for NP, RS and RP.

(d) Perform a sensitivity analysis. For example, change the weights (e.g. to make one output channel faster and another slower), move uncertainties around (e.g. from input to output), change the Δ's from a diagonal to full matrix, etc.

Comment: You may need to move back to step (a) and redefine your weights if you find out from step (c) that your original weights are unreasonable.

7. *Optional: $\mathcal{H}_\infty$ or μ-optimal controller design.* Design an $\mathcal{H}_\infty$ or μ-optimal controller and see if you can improve the response and satisfy RP. Compare simulations with previous designs.
8. *Discussion.* Discuss the main results. You should also comment on the usefulness of the project as an aid to learning and give suggestions on how the project activity might be improved.
9. *Conclusion.*

B.2 Sample exam

A Norwegian-style five-hour exam.

Problem 1 (35%). Controllability analysis.
Perform a controllability analysis (compute poles, zeros, RGA ($\lambda_{11}(s)$), check for constraints, discuss the use of decentralized control (pairings, etc.) for the following four plants. You can assume that the plants have been scaled properly.

1. 2×2 plant:

$$G(s) = \frac{1}{(s+2)(s-1.1)} \begin{bmatrix} s-1 & 1 \\ 90 & 10(s-1) \end{bmatrix} \qquad (B.1)$$

2. SISO plant with disturbance:

$$g(s) = 200\frac{-0.1s+1}{(s+10)(0.2s+1)}; \quad g_d(s) = \frac{40}{s+1} \qquad (B.2)$$

3. Plant with two inputs and one output:

$$y(s) = \frac{s}{0.2s+1}u_1 + \frac{4}{0.2s+1}u_2 + \frac{3}{0.02s+1}d \qquad (B.3)$$

4. Consider the following 2×2 plant with one disturbance given in state-space form:

$$\dot{x}_1 = -0.1x_1 + 0.01u_1$$

$$\dot{x}_2 = -0.5x_2 + 10u_2$$

$$\dot{x}_3 = 0.25x_1 + 0.25x_2 - 0.25x_3 + 1.25d$$

$$y_1 = 0.8x_3; \quad y_2 = 0.1x_3$$

(a) Construct a block diagram representation of the system with each block in the form $k/(1 + \tau s)$.

(b) Perform a controllability analysis.

Problem 2 (25%). General control problem formulation.

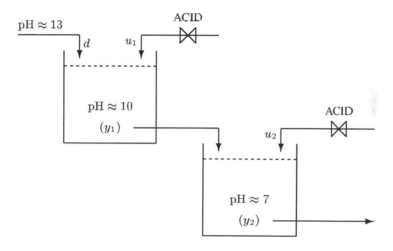

Figure B.1: Neutralization process

Consider the neutralization process in Figure B.1 where acid is added in two stages. Most of the neutralization takes place in tank 1 (left) where a large amount of acid is used (input u_1) to obtain a pH of about 10 (measurement y_1). In tank 2 the pH is fine-tuned to about 7 (output y_2) by using a small amount of acid (input u_2). This description is just to give you some idea of a real process; all the information you need to solve the problem is given below.

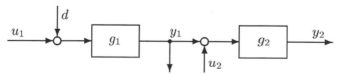

Figure B.2: Block diagram of neutralization process

A block diagram of the process is shown in Figure B.2. It includes one disturbance, two inputs and two measurements (y_1 and y_2). The main control objective is to keep $y_2 \approx r_2$. In addition, we would like to reset input 2 to its nominal value; that is, we want $u_2 \approx r_{u_2}$ at low frequencies. Note that there is no particular control objective for y_1.

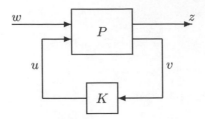

Figure B.3: General control configuration

(a) Define the general control problem: that is, find z, w, u, v and P (see Figure B.3).

(b) Define an $\mathcal{H}_\infty$ control problem based on P. Discuss briefly what you want the unweighted transfer functions from d to z to look like, and use this to say a little about how the performance weights should be selected.

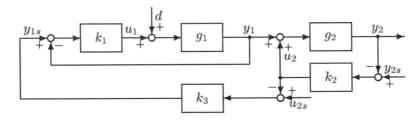

Figure B.4: Proposed control structure for neutralization process

(c) A simple practical solution based on single loops is shown in Figure B.4. Explain briefly the idea behind this control structure, and find the interconnection matrix P and the generalized controller $K = \text{diag}\{\, k_1, \quad k_2, \quad k_3 \,\}$. Note that u and y are different in this case, while w and z are the same as in (a).

Problem 3 (40%). Various.

Give brief answers to each of the following questions:

(a) Consider the plant

$$\dot{x}(t) = a(1 + 1.5\delta_a)x(t) + b(1 + 0.2\delta_b)u(t); \quad y = x$$

where $|\delta_a| \leq 1$ and $|\delta_b| \leq 1$. For a feedback controller $K(s)$ derive the interconnection matrix M for robust stability.

(b) For the above case consider using the condition $\min_D \bar{\sigma}(DMD^{-1}) < 1$ to check for robust stability (RS). What is D (give as few parameters as possible)? Is the RS condition tight in this case?

(c) When is the condition $\rho(M\Delta) < 1$ necessary and sufficient for robust stability? Based on $\rho(M\Delta) < 1$, derive the RS condition $\mu(M) < 1$. When is this last condition necessary and sufficient?

(d) Let

$$G_p(s) = \begin{bmatrix} g_{11} + w_1 \Delta_1 & g_{12} + w_2 \Delta_2 \\ g_{21} + w_3 \Delta_1 & g_{22} \end{bmatrix}, \quad |\Delta_1| \le 1, |\Delta_2| \le 1$$

Represent this uncertainty as $G_p = G + W_1 \Delta W_2$ where Δ is diagonal. Determine the corresponding $M\Delta$-structure and derive the RS condition.

(e) Let

$$G_p(s) = \frac{1 - \theta s}{1 + \theta s}; \quad \theta = \theta_0(1 + w\Delta), \quad |\Delta| < 1$$

and consider the controller $K(s) = c/s$. Put this into the $M\Delta$-structure and find the RS condition.

(f) Show by a counterexample that in general $\bar{\sigma}(AB)$ is not equal to $\bar{\sigma}(BA)$. Under what conditions is $\mu(AB) = \mu(BA)$?

(g) The PRGA matrix is defined as $\Gamma = G_{\text{diag}} G^{-1}$. What is its relationship to the RGA?

BIBLIOGRAPHY

Alstad, V. (2005). *Studies on Selection of Controlled Variables*, PhD thesis, Norwegian University of Science and Technology, Trondheim.

Alstad, V. and Skogestad, S. (2004). Combinations of measurements as controlled variables: Application to a Peltyuk distillation column, *Proceedings of the 7th International Symposium on ADCHEM*, Hong Kong, P. R. China, pp. 249 254.

Anderson, B. D. O. (1986). Weighted Hankel-norm approximation: Calculation of bounds, *Systems & Control Letters* **7**(4): 247–255.

Anderson, B. D. O. and Liu, Y. (1989). Controller reduction: Concepts and approaches, *IEEE Transactions on Automatic Control* **AC-34**(8): 802–812.

Anderson, B. D. O. and Moore, J. B. (1989). *Optimal Control: Linear Quadratic Methods*, Prentice Hall, Upper Saddle River, NJ.

Ariyur, K. B. and Krstic, M. (2003). *Real-Time Optimization by Extremum-Seeking Control*, John Wiley & Sons, Hoboken, NJ.

Balas, G., Chiang, R., Packard, A. and Safonov, M. (2005). *Robust Control Toolbox User's Guide*, 3.0.1 edn, MathWorks, South Natick, MA.

Balas, G. J. (2003). Flight control law design: An industry perspective, *European Journal of Control* **9**(2–3): 207–226.

Balas, G. J., Doyle, J. C., Glover, K., Packard, A. and Smith, R. (1993). *μ-Analysis and Synthesis Toolbox User's Guide*, MathWorks, South Natick, MA.

Balchen, J. G. and Mumme, K. (1988). *Process Control. Structures and Applications*, Van Nostrand Reinhold, New York.

Bates, D. and Postlethwaite, I. (2002). *Robust Multivariable Control of Aerospace Systems*, Delft University Press, The Netherlands.

Bode, H. W. (1945). *Network Analysis and Feedback Amplifier Design*, Van Nostrand, New York.

Boyd, S. and Barratt, C. (1991). *Linear Controller Design — Limits of Performance*, Prentice Hall, Upper Saddle River, NJ.

Boyd, S. and Desoer, C. A. (1985). Subharmonic functions and performance bounds in linear time-invariant feedback systems, *IMA Journal of Mathematical Control and Information* **2**: 153–170.

Boyd, S., Ghaoui, L. E., Feron, E. and Balakrishnan, V. (1994). *Linear Matrix Inequalities in System and Control Theory*, Society for Industrial and Applied Mathematics (SIAM), Philadelphia, PA.

Braatz, R. D. (1993). *Robust Loopshaping for Process Control*, PhD thesis, California Institute of Technology, Pasadena, CA.

Braatz, R. D. and Morari, M. (1994). Minimizing the Euclidean condition number, *SIAM Journal on Control and Optimization* **32**(6): 1763–1768.

Braatz, R. D., Morari, M. and Skogestad, S. (1996). Loopshaping for robust performance, *International Journal of Robust and Nonlinear Control* **6**(8): 805–823.

Braatz, R. D., Young, P. M., Doyle, J. C. and Morari, M. (1994). Computational complexity of μ calculation, *IEEE Transactions of Automatic Control* **AC-39**(5): 1000–1002.

Bristol, E. H. (1966). On a new measure of interactions for multivariable process control, *IEEE Transactions on Automatic Control* **AC-11**(1): 133–134.

Campo, P. J. and Morari, M. (1990). Robust control of processes subject to saturation nonlinearities, *Computers and Chemical Engineering* **14**(4–5): 343–358.

Campo, P. J. and Morari, M. (1994). Achievable closed-loop properties of systems under decentralized control: Conditions involving the steady-state gain, *IEEE Transactions on Automatic Control* **AC-39**(5): 932–942.

Cao, Y. (1995). *Control Structure Selection for Chemical Processes Using Input–output Controllability Analysis*, PhD thesis, University of Exeter.

Chang, J. W. and Yu, C. C. (1990). The relative gain for non-square multivariable systems, *Chemical Engineering Science* **45**(5): 1309–1323.

Chen, C. T. (1984). *Linear System Theory and Design*, Holt, Rinehart and Winston, New York.

Chen, J. (1995). Sensitivity integral relations and design trade-offs in linear multivariable feedback systems, *IEEE Transactions on Automatic Control* **AC-40**(10): 1700–1716.

Chen, J. (2000). Logarithmic integrals, interpolation bounds, and performance limitations in MIMO feedback systems, *IEEE Transactions on Automatic Control* **AC-45**(6): 1098–1115.

Chen, J. and Middleton, R. H. (2003). New developments and applications in performance limitation of feedback control, *IEEE Transactions on Automatic Control* **AC-48**(8): 1297.

Chiang, R. Y. and Safonov, M. G. (1992). *Robust Control Toolbox User's Guide*, MathWorks, South Natick, MA.

Chilali, M. and Gahinet, P. (1996). $\mathcal{H}_\infty$ design with pole placement constraints: An LMI approach, *IEEE Transactions on Automatic Control* **AC-41**(3): 358–367.

Churchill, R. V., Brown, J. W. and Verhey, R. F. (1974). *Complex Variables and Applications*, McGraw-Hill, New York.

Cui, H. and Jacobsen, E. W. (2002). Performance limitations in decentralized control, *Journal of Process Control* **12**(4): 485–494.

Dahl, H. J. and Faulkner, A. J. (1979). Helicopter simulation in atmospheric turbulence, *Vertica* pp. 65–78.

Dahleh, M. and Diaz-Bobillo, I. (1995). *Control of Uncertain Systems. A Linear Programming Approach*, Prentice Hall, Englewood Cliffs, NJ.

Daoutidis, P. and Kravaris, C. (1992). Structural evaluation of control configurations for multivariable nonlinear processes, *Chemical Engineering Science* **47**(6): 1091–1107.

Davison, E. J. (ed.) (1990). *Benchmark Problems for Control System Design*, Report of the IFAC Theory Committee, International Federation of Automatic Control, Laxenberg, Austria.

Desoer, C. A. and Vidyasagar, M. (1975). *Feedback Systems: Input–Output Properties*, Academic Press, New York.

Downs, J. J. and Vogel, E. F. (1993). A plant-wide industrial process control problem, *Computers Chem. Engng.* **17**: 245–255.

Doyle, J. C. (1978). Guaranteed margins for LQG regulators, *IEEE Transactions on Automatic Control* **AC-23**(4): 756–757.

Doyle, J. C. (1982). Analysis of feedback systems with structured uncertainties, *IEE Proceedings, Part D Control Theory and Applications* **129**(6): 242–250.

Doyle, J. C. (1983). Synthesis of robust controllers and filters, *Proceedings of the IEEE Conference on Decision and Control*, San Antonio, TA, USA, pp. 109–114.

Doyle, J. C. (1984). *Lecture Notes on Advances in Multivariable Control*, ONR/Honeywell Workshop, Minneapolis, USA.

Doyle, J. C. (1986). Redondo Beach lecture notes, Internal Report, Caltech, Pasadena, CA.

Doyle, J. C., Francis, B. and Tannenbaum, A. (1992). *Feedback Control Theory*, Macmillan, New York.

Doyle, J. C., Glover, K., Khargonekar, P. P. and Francis, B. A. (1989). State-space solutions to standard $\mathcal{H}_2$ and $\mathcal{H}_\infty$ control problems, *IEEE Transactions on Automatic Control* **AC-34**(8): 831–847.

Doyle, J. C. and Stein, G. (1979). Robustness with observers, *IEEE Transactions on Automatic Control* **AC-24**(4): 607–611.

Doyle, J. C. and Stein, G. (1981). Multivariable feedback design: Concepts for a classical/modern synthesis, *IEEE Transactions on Automatic Control* **AC-26**(1): 4–16.

Eaton, J. W. and Rawlings, J. B. (1992). Model-predictive control of chemical processes, *Chemical Engineering Science* **47**(4): 705–720.

Engell, S. (1988). *Optimale Lineare Regelung*, Vol. 18 of *Fachberichte Messen, Steuern, Regeln*, Springer-Verlag, Berlin.

Enns, D. (1984). Model reduction with balanced realizations: An error bound and a frequency weighted generalization, *Proceedings of the 23rd IEEE Conference on Decision and Control*, Las Vegas, NV, USA, pp. 127–132.

Fernando, K. V. and Nicholson, H. (1982). Singular perturbational model reduction of balanced systems, *IEEE Transactions on Automatic Control* **AC-27**(2): 466–468.

Finsler, P. (1937). Uber das Vorkommen definiter und semi-definiter Formen in Scharen quadratischer Formen, *Comentarii Mathematica Helvetici* **9**: 192–199.

Fisher, W. R., Doherty, M. F. and Douglas, J. M. (1985). Steady-state control as a prelude to dynamic control, *Chemical Engineering Research & Design* **63**: 353–357.

Foss, A. S. (1973). Critique of chemical process control theory, *AIChE Journal* **19**(2): 209–214.

Foster, N. P., Spurgeon, S. K. and Postlethwaite, I. (1993). Robust model-reference tracking control with a sliding mode applied to an ACT rotorcraft, *19th European Rotorcraft Forum*, Italy.

Francis, B. (1987). *A course in $\mathcal{H}_\infty$ control theory*, Vol. 88 of Lecture Notes in Control and Information Sciences, Springer-Verlag, Berlin.

Francis, B. A. and Zames, G. (1984). On $\mathcal{H}_\infty$ optimal sensitivity theory for SISO feedback systems, *IEEE Transactions on Automatic Control* **AC-29**(1): 9–16.

Frank, P. M. (1968a). Vollständige Vorhersage im stetigen Regelkreis mit Totzeit, Teil I, *Regelungstechnik* **16**(3): 111–116.

Frank, P. M. (1968b). Vollständige Vorhersage im stetigen Regelkreis mit Totzeit, Teil II, *Regelungstechnik* **16**(5): 214–218.

Freudenberg, J. S. and Looze, D. P. (1985). Right half planes poles and zeros and design tradeoffs in feedback systems, *IEEE Transactions on Automatic Control* **AC-30**(6): 555–565.

Freudenberg, J. S. and Looze, D. P. (1988). *Frequency Domain Properties of Scalar and Multivariable Feedback Systems*, Vol. 104 of Lecture Notes in Control and Information Sciences, Springer-Verlag, Berlin.

Gahinet, P. and Apkarian, P. (1994). A linear matrix inequality approach to $\mathcal{H}_\infty$ control, *International Journal of Robust and Nonlinear Control* **4**: 421–448.

Gahinet, P., Nemirovski, A., Laub, A. and Chilali, M. (1995). *LMI Control Toolbox*, MathWorks, South Natick, MA.

Georgiou, T. T. and Smith, M. C. (1990). Optimal robustness in the gap metric, *IEEE Transactions on Automatic Control* **AC-35**(6): 673–686.

Gjøsæter, O. B. (1995). *Structures for Multivariable Robust Process Control*, PhD thesis, Norwegian University of Science and Technology, Trondheim.

Glover, K. (1984). All optimal Hankel-norm approximations of linear multivariable systems and their L^∞-error bounds, *International Journal of Control* **39**(6): 1115–1193.

Glover, K. (1986). Robust stabilization of linear multivariable systems: Relations to approximations, *International Journal of Control* **43**(3): 741–766.

Glover, K. and Doyle, J. C. (1988). State-space formulae for all stabilizing controller that satisfy an $\mathcal{H}_\infty$ norm bound and relations to risk sensitivity, *Systems and Control Letters* **11**(3): 167–172.

Glover, K. and McFarlane, D. (1989). Robust stabilization of normalized coprime factor plant descriptions with $\mathcal{H}_\infty$ bounded uncertainty, *IEEE Transactions on Automatic Control* **AC-34**(8): 821–830.

Glover, K., Vinnicombe, G. and Papageorgiou, G. (2000). Guaranteed multi-loop stability margins and the gap metric, *Proceedings of the 39th IEEE Conference on Decision and Control*, Sydney, Australia, pp. 4084–4085.

Goddard, P. (1995). *Performance Preserving Controller Approximation*, PhD thesis, Trinity College, Cambridge.

Golub, G. H. and van Loan, C. F. (1989). *Matrix Computations*, Johns Hopkins University Press, Baltimore, MD.

Goodwin, G. C., Salgado, M. E. and Silva, E. I. (2005). Time-domain performance limitations arising from decentralized architectures and their relationship to the rga, *Int. J. Control* **78**: 1045–1062.

Goodwin, G. C., Salgado, M. E. and Yuz, J. I. (2003). Performance limitations for linear feedback systems in the presence of plant uncertainty, *IEEE Transactions on Automatic Control* **AC-48**(8): 1312–1319.

Govatsmark, M. S. (2003). *Integrated Optimization and Control*, PhD thesis, Norwegian University of Science and Technology, Trondheim.

Grace, A., Laub, A. J., Little, J. N. and Thompson, C. M. (1992). *Control System Toolbox*, MathWorks, South Natick, MA.

Green, M. and Limebeer, D. J. N. (1995). *Linear Robust Control*, Prentice Hall, Upper Saddle River, NJ.

Grosdidier, P. and Morari, M. (1986). Interaction measures for systems under decentralized control, *Automatica* **22**(3): 309–319.

Grosdidier, P., Morari, M. and Holt, B. R. (1985). Closed-loop properties from steady-state gain information, *Industrial and Engineering Chemistry Fundamentals* **24**(2): 221–235.

Haggblom, K. E. and Waller, K. V. (1988). Transformations and consistency relations of distillation control structures, *AIChE Journal* **34**(10): 1634–1648.

Halvorsen, I. J., Skogestad, S., Morud, J. C. and Alstad, V. (2003). Optimal selection of controlled variables, *Industrial & Engineering Chemistry Research* **42**(14): 3273–3284.

Hanus, R., Kinnaert, M. and Henrotte, J. (1987). Conditioning technique, a general anti-windup and bumpless transfer method, *Automatica* **23**(6): 729–739.

Havre, K. (1998). *Studies on Controllability Analysis and Control Structure Design*, PhD thesis, Norwegian University of Science and Technology, Trondheim.

Havre, K. and Skogestad, S. (1998). Effect of RHP zeros and poles on the sensitivity functions in multivariable systems, *Journal of Process Control* **8**(3): 155–164.

Havre, K. and Skogestad, S. (2001). Achievable performance of multivariable systems with unstable zeros and poles, *International Journal of Control* **74**(11): 1131–1139.

Havre, K. and Skogestad, S. (2003). Selection of variables for stabilizing control using pole vectors, *IEEE Transactions on Automatic Control* **AC-48**(8): 1393–1398.

Helton, J. W. (1976). Operator theory and broadband matching, *Proceedings of the 14th Annual Allerton Conference on Communications, Control and Computing*, Monticello, USA.

Herrmann, G., Turner, M. C. and Postlethwaite, I. (2003). Discrete time anti-windup - part 2: Extension to sampled data case, *Proceedings of the European Control Conference, Cambridge, UK*.

Herrmann, G., Turner, M. C., Postlethwaite, I. and Guo, G. (2003). Application of a novel anti-windup scheme to a HDD-dual-stage actuator, *Proceedings of the American Control Conference, Denver, CO, USA*.

Holt, B. R. and Morari, M. (1985a). Design of resilient processing plants V — The effect of deadtime on dynamic resilience, *Chemical Engineering Science* **40**(7): 1229–1237.

Holt, B. R. and Morari, M. (1985b). Design of resilient processing plants VI — The effect of right plane zeros on dynamic resilience, *Chemical Engineering Science* **40**(1): 59–74.

Hori, E. S., Skogestad, S. and Kwong, W. H. (2005). Use of perfect indirect control to minimize the state deviations, *in* L. Puigjaner and A. Espuna (eds), *European Symposium on computer-aided process engineering (ESCAPE) 15, Barcelona, Spain*, Elsevier.

Horn, R. A. and Johnson, C. R. (1985). *Matrix Analysis*, Cambridge University Press, Cambridge.

Horn, R. A. and Johnson, C. R. (1991). *Topics in Matrix Analysis*, Cambridge University Press, Cambridge.

Horowitz, I. M. (1963). *Synthesis of Feedback Systems*, Academic Press, London.

Horowitz, I. M. (1991). Survey of quantitative feedback theory (QFT), *International Journal of Control* **53**(2): 255–291.

Horowitz, I. M. and Shaked, U. (1975). Superiority of transfer function over state-variable methods in linear time-invariant feedback system design, *IEEE Transactions on Automatic Control* **AC-20**(1): 84–97.

Hovd, M. (1992). *Studies on Control Structure Selection and Design of Robust Decentralized and SVD Controllers*, PhD thesis, Norwegian University of Science and Technology, Trondheim.

Hovd, M., Braatz, R. D. and Skogestad, S. (1997). SVD controllers for $\mathcal{H}_2$-, $\mathcal{H}_\infty$-, and μ-optimal control, *Automatica* **33**(3): 433–439.

Hovd, M. and Skogestad, S. (1992). Simple frequency-dependent tools for control system analysis, structure selection and design, *Automatica* **28**(5): 989–996.

Hovd, M. and Skogestad, S. (1993). Procedure for regulatory control structure selection with application to the FCC process, *AIChE Journal* **39**(12): 1938–1953.

Hovd, M. and Skogestad, S. (1994). Pairing criteria for decentralised control of unstable plants, *Industrial & Engineering Chemistry Research* **33**(9): 2134–2139.

Hoyle, D., Hyde, R. A. and Limebeer, D. J. N. (1991). An $\mathcal{H}_\infty$ approach to two degree of freedom design, *Proceedings of the 30th IEEE Conference on Decision and Control*, Brighton, UK, pp. 1581–1585.

Hung, Y. S. and MacFarlane, A. G. J. (1982). *Multivariable Feedback: A Quasi-Classical Approach*, Vol. 40 of Lecture Notes in Control and Information Sciences, Springer-Verlag, Berlin.

Hyde, R. A. (1991). *The Application of Robust Control to VSTOL Aircraft*, PhD thesis, University of

Cambridge.

Hyde, R. A. and Glover, K. (1993). The application of scheduled $\mathcal{H}_\infty$ controllers to a VSTOL aircraft, *IEEE Transactions on Automatic Control* **AC-38**(7): 1021–1039.

Iwasaki, T. and Skelton, R. E. (1994). All controllers for the general $\mathcal{H}_\infty$ control problem: LMI existence conditions and state space formulae, *Automatica* **30**(8): 1307–1317.

Jaimoukha, I. M., Kasenally, E. M. and Limebeer, D. J. N. (1992). Numerical solution of large scale Lyapunov equations using Krylov subspace methods, *Proceedings of the 31st IEEE Conference on Decision and Control, Tucson, AZ, USA.*

Jensen, J. B. and Skogestad, S. (2005). Optimal operation of an Ammonia refrigration cycle, *Technical report*, Norwegian University of Science and Technology, Trondheim, Norway. (Available on book home page).

Johnson, C. R. and Shapiro, H. M. (1986). Mathematical aspects of the relative gain array ($A \circ A^{-T}$), *SIAM Journal on Algebraic and Discrete Methods* **7**(4): 627–644.

Kailath, T. (1980). *Linear Systems*, PrenticeHall, Engelwood Cliffs, NJ.

Kalman, R. E. (1964). When is a linear control system optimal?, *Journal of Basic Engineering — Transaction on ASME — Series D* **86**: 51–60.

Kariwala, V. (2004). *Multi Loop Controller Synthesis and Performance Analysis*, PhD thesis, University of Alberta, Edmonton.

Kariwala, V., Forbes, J. F. and Meadows, E. S. (2003). Block relative gain: Properties and pairing rules, *Industrial & Engineering Chemistry Research* **42**(20): 4564–4574.

Khalil, H. (1996). *Nonlinear Systems*, Prentice Hall, Englewood Cliffs, NJ.

Khargonekar, P. and Rotea, M. A. (1991). Mixed $\mathcal{H}_2/\mathcal{H}_\infty$ control: A convex optimization approach, *IEEE Transactions on Automatic Control* **AC-36**(7): 824–837.

Kline, R. (1993). Harold Black and the negative-feedback amplifier, *IEEE Control Systems Magazine* **13**(4): 82–85.

Kothare, M. V., Balakrishnan, V. and Morari, M. (1996). Robust constrained model predictive control using linear matrix inequalities, *Automatica* **32**(10): 1361–1379.

Kouvaritakis, B. (1974). *Characteristic Locus Methods for Multivariable Feedback Systems Design*, PhD thesis, University of Manchester Institute of Science and Technology, Manchester.

Kwakernaak, H. (1969). Optimal low-sensitivity linear feedback systems, *Automatica* **5**(3): 279–285.

Kwakernaak, H. (1985). Minimax frequency domain performance and robustness optimization of linear feedback systems, *IEEE Transactions on Automatic Control* **AC-30**(10): 994–1004.

Kwakernaak, H. (1993). Robust control and $\mathcal{H}_\infty$-optimization — Tutorial paper, *Automatica* **29**(2): 255–273.

Kwakernaak, H. and Sivan, R. (1972). *Linear Optimal Control Systems*, Wiley Interscience, New York.

Larsson, T. (2000). *Studies on Plantwide Control*, PhD thesis, Norwegian University of Science and Technology, Trondheim.

Larsson, T. and Skogestad, S. (2000). Plantwide control: A review and a new design procedure, *Modeling, Identification and Control* **21**: 209–240.

Laub, A. J., Heath, M. T., Page, C. C. and Ward, R. C. (1987). Computation of system balancing transformations and other applications of simultaneous diagonalization algorithms, *IEEE Transactions on Automatic Control* **AC-32**(2): 115–122.

Laughlin, D. L., Jordan, K. G. and Morari, M. (1986). Internal model control and process uncertainty – mapping uncertainty regions for SISO controller-design, *International Journal of Control* **44**(6): 1675–1698.

Laughlin, D. L., Rivera, D. E. and Morari, M. (1987). Smith predictor design for robust performance, *International Journal of Control* **46**(2): 477–504.

Leon de la Barra S., B. A. (1994). On undershoot in SISO systems, *IEEE Transactions on Automatic Control* **AC-39**(3): 578–581.

Levine, W. (ed.) (1996). *The Control Handbook*, CRC Press, Boca Raton, FL.

Liang, Q. (1992). Is the relative gain array a sensitivity measure?, *IFAC Workshop on Interactions between Process Design and Process Control*, London, UK, pp. 133–138.

Limebeer, D. J. N. (1991). The specification and purpose of a controller design case study, *Proceedings of the IEEE Conference on Decision and Control*, Brighton, UK, pp. 1579–1580.

Limebeer, D. J. N., Kasenally, E. M. and Perkins, J. D. (1993). On the design of robust two degree of freedom controllers, *Automatica* **29**(1): 157–168.

Liu, Y. and Anderson, B. D. O. (1989). Singular perturbation approximation of balanced systems, *International Journal of Control* **50**(4): 1379–1405.

Lundström, P. (1994). *Studies on Robust Multivariable Distillation Control*, PhD thesis, Norwegian University of Science and Technology, Trondheim.

Lundström, P., Skogestad, S. and Doyle, J. C. (1999). Two degrees of freedom controller design for an ill-conditioned plant using μ-synthesis, *IEEE Transactions on Control System Technology* **7**(1): 12–21.

Lunze, J. (1992). *Feedback Control of Large-Scale Systems*, Prentice-Hall, New York, NY.

MacFarlane, A. G. J. and Karcanias, N. (1976). Poles and zeros of linear multivariable systems: A survey of algebraic, geometric and complex variable theory, *International Journal of Control* **24**: 33–74.

MacFarlane, A. G. J. and Kouvaritakis, B. (1977). A design technique for linear multivariable feedback systems, *International Journal of Control* **25**: 837–874.

Maciejowski, J. M. (1989). *Multivariable Feedback Design*, Addison-Wesley, Wokingham.

Manness, M. A. and Murray-Smith, D. J. (1992). Aspects of multivariable flight control law design for helicopters using eigenstructure assignment, *Journal of American Helicopter Society* **37**(3): 18–32.

Manousiouthakis, V., Savage, R. and Arkun, Y. (1986). Synthesis of decentralized process control structures using the concept of block relative gain, *AIChE Journal* **32**: 991–1003.

Marlin, T. (1995). *Process Control*, McGraw Hill, New York.

McFarlane, D. and Glover, K. (1990). *Robust Controller Design Using Normalized Coprime Factor Plant Descriptions*, Vol. 138 of Lecture Notes in Control and Information Sciences, Springer-Verlag, Berlin.

McMillan, G. K. (1984). *pH Control*, Instrument Society of America, Research Triangle Park, NC.

Meinsma, G. (1995). Unstable and nonproper weights in $\mathcal{H}_\infty$ control, *Automatica* **31**(11): 1655–1658.

Meyer, D. G. (1987). *Model Reduction via Factorial Representation*, PhD thesis, Stanford University, Stanford, CA.

Middleton, R. H. (1991). Trade-offs in linear control system design, *Automatica* **27**(2): 281–292.

Middleton, R. H. and Braslavsky, J. H. (2002). Towards quantitative time domain design tradeoffs in nonlinear control, *Proceedings of the American Control Conference*, Anchorage, AK, USA, pp. 4896–4901.

Middleton, R. H., Chen, J. and Freudenberg, J. S. (2004). Tracking sensitivity and achievable $\mathcal{H}_\infty$ performance in preview control, *Automatica* **40**(8): 1297–1306.

Moore, B. C. (1981). Principal component analysis in linear systems: Controllability, observability and model reduction, *IEEE Transactions on Automatic Control* **AC-26**(1): 17–32.

Morari, M. (1983). Design of resilient processing plants III – A general framework for the assessment of dynamic resilience, *Chemical Engineering Science* **38**(11): 1881–1891.

Morari, M. and Zafiriou, E. (1989). *Robust Process Control*, Prentice Hall, Englewood Cliffs, NJ.

Nett, C. N. (1986). Algebraic aspects of linear control system stability, *IEEE Transactions on Automatic Control* **AC-31**(10): 941–949.

Nett, C. N. (1989). A quantitative approach to the selection and partitioning of measurements and manipulations for the control of complex systems, *Presentation at Caltech Control Workshop*, Pasadena, USA, January.

Nett, C. N. and Manousiouthakis, V. (1987). Euclidean condition and block relative gain: Connections, conjectures, and clarifications, *IEEE Transactions on Automatic Control* **AC-32**(5): 405–407.

Nett, C. N. and Minto, K. D. (1989). A quantitative approach to the selection and partitioning of measurements and manipulations for the control of complex systems, Copy of transparencies from talk at *American Control Conference*, Pittsburgh, PA, USA, June.

Niemann, H. and Stoustrup, J. (1995). Special Issue on Loop Transfer Recovery, *International Journal of Robust and Nonlinear Control* **7**(7): November.

Nwokah, O. D. I. and Perez, R. (1991). On multivariable stability in the gain space, *Automatica* **27**(6): 975–983.

Owen, J. G. and Zames, G. (1992). Robust $\mathcal{H}_\infty$ disturbance minimization by duality, *Systems & Control Letters* **19**(4): 255–263.

Packard, A. (1988). *What's New with μ*, PhD thesis, University of California, Berkeley, CA.

Packard, A. and Doyle, J. C. (1993). The complex structured singular value, *Automatica* **29**(1): 71–109.

Packard, A., Doyle, J. C. and Balas, G. (1993). Linear, multivariable robust-control with a μ-perspective, *Journal of Dynamic Systems Measurement and Control — Transactions of the ASME*

115(2B): 426–438.

Padfield, G. D. (1981). Theoretical model of helicopter flight mechanics for application to piloted simulation, *Technical Report 81048*, Defence Research Agency (now QinetiQ), UK.

Perkins, J. D. (ed.) (1992). *IFAC Workshop on Interactions Between Process Design and Process Control*, (London, September), Pergamon Press, Oxford.

Pernebo, L. and Silverman, L. M. (1982). Model reduction by balanced state space representations, *IEEE Transactions on Automatic Control* **AC-27**(2): 382–387.

Poolla, K. and Tikku, A. (1995). Robust performance against time-varying structured perturbations, *IEEE Transactions on Automatic Control* **AC-40**(9): 1589–1602.

Postlethwaite, I., Foster, N. P. and Walker, D. J. (1994). Rotorcraft control law design for rejection of atmospheric turbulence, *Proceedings of IEE Conference, Control 94*, Warwick, UK, pp. 1284–1289.

Postlethwaite, I. and MacFarlane, A. G. J. (1979). *A Complex Variable Approach to the Analysis of Linear Multivariable Feedback Systems*, Vol. 12 of Lecture Notes in Control and Information Sciences, Springer-Verlag, Berlin.

Postlethwaite, I., Prempain, E., Turkoglu, E., Turner, M. C., Ellis, K. and Gubbles, A. W. (2005). Design and flight testing of various $\mathcal{H}_\infty$ controllers for the Bell 205 helicopter, *Control Engineering Practice* **13**: 383–398.

Postlethwaite, I., Samar, R., Choi, B.-W. and Gu, D.-W. (1995). A digital multi-mode $\mathcal{H}_\infty$ controller for the Spey turbofan engine, *3rd European Control Conference*, Rome, Italy, pp. 3881–3886.

Postlethwaite, I., Smerlas, A., Walker, D. J., Gubbels, A. W., Baillie, S. W., Strange, M. E. and Howitt, J. (1999). $\mathcal{H}_\infty$ control of the NRC Bell 205 fly-by-wire helicopter, *Journal of American Helicopter Society* **44**(4): 276–284.

Postlethwaite, I. and Walker, D. J. (1992). Advanced control of high performance rotorcraft, *Institute of Mathematics and Its Applications Conference on Aerospace Vehicle Dynamics and Control*, Cranfield Institute of Technology, UK, pp. 615–619.

Prempain, E. and Postlethwaite, I. (2004). Static $\mathcal{H}_\infty$ loop shaping of a fly-by-wire helicopter, *Proceedings of the 43rd IEEE Conference on Decision and Control*, Bahamas.

Qiu, L. and Davison, E. J. (1993). Performance limitations of non-minimum phase systems in the servomechanism problem, *Automatica* **29**(2): 337–349.

Rosenbrock, H. H. (1966). On the design of linear multivariable systems, *Third IFAC World Congress*, London, UK. Paper 1a.

Rosenbrock, H. H. (1970). *State-space and Multivariable Theory*, Nelson, London.

Rosenbrock, H. H. (1974). *Computer-Aided Control System Design*, Academic Press, New York.

Safonov, M. G. (1982). Stability margins of diagonally perturbed multivariable feedback systems, *IEE Proceedings, Part D* **129**(6): 251–256.

Safonov, M. G. and Athans, M. (1977). Gain and phase margin for multiloop LQG regulators, *IEEE Transactions on Automatic Control* **AC-22**(2): 173–179.

Safonov, M. G. and Chiang, R. Y. (1989). A Schur method for balanced-truncation model reduction, *IEEE Transactions on Automatic Control* **AC-34**(7): 729–733.

Safonov, M. G., Limebeer, D. J. N. and Chiang, R. Y. (1989). Simplifying the $\mathcal{H}_\infty$ theory via loop-shifting, matrix-pencil and descriptor concepts, *International Journal of Control* **50**(6): 2467–2488.

Samar, R. (1995). *Robust Multi-Mode Control of High Performance Aero-Engines*, PhD thesis, University of Leicester.

Samar, R. and Postlethwaite, I. (1994). Multivariable controller design for a high performance aero engine, *Proceedings of IEE Conference, Control 94*, Warwick, UK, pp. 1312–1317.

Samar, R., Postlethwaite, I. and Gu, D.-W. (1995). Model reduction with balanced realizations, *International Journal of Control* **62**(1): 33–64.

Samblancatt, C., Apkarian, P. and Patton, R. J. (1990). Improvement of helicopter robustness and performance control law using eigenstructure techniques and $\mathcal{H}_\infty$ synthesis, *16th European Rotorcraft Forum*, Scotland. Paper No. 2.3.1.

Sato, T. and Liu, K.-Z. (1999). LMI solution to general $\mathcal{H}_2$ suboptimal control problems, *Systems and Control Letters* **36**(4): 295–305.

Seborg, D. E., Edgar, T. F. and Mellichamp, D. A. (1989). *Process Dynamics and Control*, John Wiley & Sons, New York.

Sefton, J. and Glover, K. (1990). Pole–zero cancellations in the general $\mathcal{H}_\infty$ problem with reference to a two block design, *Systems & Control Letters* **14**(4): 295–306.

Seron, M., Braslavsky, J. and Goodwin, G. (1997). *Fundamental limitations in filtering and control*, Springer-Verlag, Berlin.

Shamma, J. S. (1994). Robust stability with time-varying structured uncertainty, *IEEE Transactions on Automatic Control* **AC-39**(4): 714–724.

Shinskey, F. G. (1967). *Process Control Systems*, 1st edn, McGraw-Hill, New York.

Shinskey, F. G. (1984). *Distillation Control*, 2nd edn, McGraw-Hill, New York.

Shinskey, F. G. (1996). *Process Control Systems: Application, Design and Tuning*, 4th edn, McGraw-Hill, New York.

Skogestad, S. (1996). A procedure for SISO controllability analysis — with application to design of pH neutralization process, *Computers & Chemical Engineering* **20**(4): 373–386.

Skogestad, S. (1997). Dynamics and control of distillation columns - a tutorial introduction, *Transactions of IChemE (UK)* **75**(A). Plenary paper from symposium Distillation and Absorption 97, Maastricht, Netherlands, September.

Skogestad, S. (2000). Plantwide control: The search for the self-optimizing control structure, *Journal of Process Control* **10**: 487–507.

Skogestad, S. (2003). Simple analytic rules for model reduction and PID controller tuning, *Journal of Process Control* **13**: 291–309. Also see *corrections* in **14**, 465 (2004).

Skogestad, S. (2004a). Control structure design for complete chemical plants, *Computers & Chemical Engineering* **28**: 219–234.

Skogestad, S. (2004b). Near-optimal operation by self-optimizing control: From process control to marathon running and business systems, *Computers & Chemical Engineering* **29**(1): 127–137.

Skogestad, S. and Havre, K. (1996). The use of the RGA and condition number as robustness measures, *Computers & Chemical Engineering* **20**(S): S1005–S1010.

Skogestad, S., Lundström, P. and Jacobsen, E. (1990). Selecting the best distillation control configuration, *AIChE Journal* **36**(5): 753–764.

Skogestad, S. and Morari, M. (1987a). Control configuration selection for distillation columns, *AIChE Journal* **33**(10): 1620–1635.

Skogestad, S. and Morari, M. (1987b). Effect of disturbance directions on closed-loop performance, *Industrial & Engineering Chemistry Research* **26**(10): 2029–2035.

Skogestad, S. and Morari, M. (1987c). Implications of large RGA elements on control performance, *Industrial & Engineering Chemistry Research* **26**(11): 2323–2330.

Skogestad, S. and Morari, M. (1988a). Some new properties of the structured singular value, *IEEE Transactions on Automatic Control* **AC-33**(12): 1151–1154.

Skogestad, S. and Morari, M. (1988b). Variable selection for decentralized control, *AIChE Annual Meeting*, Washington, DC. Paper 126f. Reprinted in *Modeling, Identification and Control*, 1992, Vol. **13**, No. 2, 113–125.

Skogestad, S. and Morari, M. (1989). Robust performance of decentralized control systems by independent designs, *Automatica* **25**(1): 119–125.

Skogestad, S., Morari, M. and Doyle, J. C. (1988). Robust control of ill-conditioned plants: High-purity distillation, *IEEE Transactions on Automatic Control* **AC-33**(12): 1092–1105.

Skogestad, S. and Postlethwaite, I. (1996). *Multivariable Feedback Control: Analysis and Design*, 1st edn, Wiley, Chichester.

Skogestad, S. and Wolff, E. A. (1991). TANKSPILL - A process control game, *CACHE News* **32**: 1–4. Published by CACHE Corporation, Austin, TX, USA.

Smerlas, A., Walker, D. J., Postlethwaite, I., Strange, M. E., Howitt, J. and Gubbels, A. W. (2001). Evaluating $\mathcal{H}_\infty$ controllers on the NRC Bell 205 fly-by-wire helicopter, *Control Engineering Practice* **9**(1): 1–10.

Sourlas, D. D. and Manousiouthakis, V. (1995). Best achievable decentralized performance, *IEEE Transactions on Automatic Control* **AC-40**(11): 1858–1871.

Stanley, G., Marino-Galarraga, M. and McAvoy, T. J. (1985). Short cut operability analysis. 1. The relative disturbance gain, *Industrial and Engineering Chemistry Process Design and Development* **24**(4): 1181–1188.

Stein, G. (2003). Respect the unstable, *IEEE Control Systems Magazine* **23**(4): 12–25.

Stein, G. and Athans, M. (1987). The LQG/LTR procedure for multivariable feedback control design,

IEEE Transactions on Automatic Control **AC-32**(2): 105–114.

Stein, G. and Doyle, J. C. (1991). Beyond singular values and loopshapes, *AIAA Journal of Guidance and Control* **14**: 5–16.

Stephanopoulos, G. (1984). *Chemical Process Control*, Prentice Hall, Englewood Cliffs, NJ.

Strang, G. (1976). *Linear Algebra and Its Applications*, Academic Press, New York.

Takahashi, M. D. (1993). Synthesis and evaluation of an $\mathcal{H}_2$ control law for a hovering helicopter, *Journal of Guidance, Control and Dynamics* **16**: 579–584.

Tøffner-Clausen, S., Andersen, P., Stoustrup, J. and Niemann, H. H. (1995). A new approach to μ-synthesis for mixed perturbation sets, *Proceedings of 3rd European Control Conference*, Rome, Italy, pp. 147–152.

Toker, O. and Ozbay, H. (1998). On the complexity of purely complex μ computation and related problems in multidimensional systems, *IEEE Transactions on Automatic Control* **AC-43**(3): 409–414.

Tombs, M. S. and Postlethwaite, I. (1987). Truncated balanced realization of a stable non-minimal state-space system, *International Journal of Control* **46**: 1319–1330.

Tsai, M. C., Geddes, E. J. M. and Postlethwaite, I. (1992). Pole–zero cancellations and closed-loop properties of an $\mathcal{H}_\infty$ mixed sensitivity design problem, *Automatica* **28**(3): 519–530.

Turner, M. C., Herrmann, G. and Postlethwaite, I. (2003). Discrete time anti-windup - part 1: stability and performance, *Proceedings of the European Control Conference, Cambridge, UK* .

Turner, M. C., Herrmann, G. and Postlethwaite, I. (2004). An introduction to linear matrix inequalities in control, *University of Leicester Department of Engineering Technical Report no 02-04* .

Turner, M. C. and Postlethwaite, I. (2004). A new perspective on static and low order anti-windup compensator synthesis, *International Journal of Control* **77**(1): 27–44.

Turner, M. C. and Walker, D. J. (2000). Linear quadratic bumpless transfer, *Automatica* **36**(8): 1089–1101.

Van de Wal, M. (1994). Control structure design for dynamic systems: A review, *Technical Report WFW-94-084*, Eindhoven University of Technology, Eindhoven, The Netherlands.

Van de Wal, M. and de Jager, B. (2001). A review of methods for input/output selection, *Automatica* **37**(4): 487 510.

van Diggelen, F. and Glover, K. (1994a). A Hadamard weighted loop shaping design procedure for robust decoupling, *Automatica* **30**(5): 831–845.

van Diggelen, F. and Glover, K. (1994b). State-space solutions to Hadamard weighted $\mathcal{H}_\infty$ and $\mathcal{H}_2$ control-problems, *International Journal of Control* **59**(2): 357–394.

Vidyasagar, M. (1985). *Control System Synthesis: A Factorization Approach*, MIT Press, Cambridge, MA.

Vidyasagar, M. (1988). Normalized coprime factorizations for non-strictly proper systems, *IEEE Transactions on Automatic Control* **AC-33**(3): 300–301.

Vinnicombe, G. (1993). Frequency domain uncertainty and the graph topology, *IEEE Transactions on Automatic Control* **AC-38**(9): 1371–1383.

Vinnicombe, G. (2001). *Uncertainty and feedback: $\mathcal{H}_\infty$ loop-shaping and the ν-gap metric*, Imperial College Press, London.

Walker, D. J. (1996). On the structure of a two degrees-of-freedom $\mathcal{H}_\infty$ loop-shaping controller, *International Journal of Control* **63**(6): 1105–1127.

Walker, D. J. and Postlethwaite, I. (1996). Advanced helicopter flight control using two degrees-of-freedom $\mathcal{H}_\infty$ optimization, *Journal of Guidance, Control and Dynamics* **19**(2): March–April.

Walker, D. J., Postlethwaite, I., Howitt, J. and Foster, N. P. (1993). Rotorcraft flying qualities improvement using advanced control, *American Helicopter Society/NASA Conference*, San Francisco, USA.

Wang, Z. Q., Lundström, P. and Skogestad, S. (1994). Representation of uncertain time delays in the $\mathcal{H}_\infty$ framework, *International Journal of Control* **59**(3): 627–638.

Weston, P. and Postlethwaite, I. (2000). Linear conditioning for systems containing saturating actuators, *Automatica* **36**(9): 1347–1354.

Whidborne, J. F., Postlethwaite, I. and Gu, D. W. (1994). Robust controller design using $\mathcal{H}_\infty$ loop shaping and the method of inequalities, *IEEE Transactions on Control Systems Technology* **2**(4): 455–461.

Willems, J. (1970). *Stability Theory of Dynamical Systems*, Nelson, London.

Wolff, E. A. (1994). *Studies on Control of Integrated Plants*, PhD thesis, Norwegian University of Science and Technology, Trondheim.

Wonham, M. (1974). *Linear Multivariable Systems*, Springer-Verlag, Berlin.

Youla, D. C., Bongiorno, J. J. and Lu, C. N. (1974). Single-loop feedback stabilization of linear multivariable dynamical plants, *Automatica* **10**(2): 159–173.

Youla, D. C., Jabr, H. A. and Bongiorno, J. J. (1976). Modern Wiener-Hopf design of optimal controllers, part II: The multivariable case, *IEEE Transactions on Automatic Control* **AC-21**(3): 319–338.

Young, P. M. (1993). *Robustness with Parametric and Dynamic Uncertainties*, PhD thesis, California Institute of Technology, Pasadena, CA.

Young, P. M. (1994). Controller design with mixed uncertainties, *Proceedings of the American Control Conference*, Baltimore, MD, USA, pp. 2333–2337.

Young, P. M. and Doyle, J. C. (1997). A lower bound for the mixed μ problem, *IEEE Transactions on Automatic Control* **42**(1): 123–128.

Young, P. M., Newlin, M. and Doyle, J. C. (1992). Practical computation of the mixed μ problem, *Proceedings of the American Control Conference*, Chicago, USA, pp. 2190–2194.

Yu, C. C. and Fan, M. K. H. (1990). Decentralized integral controllability and D-stability, *Chemical Engineering Science* **45**(11): 3299–3309.

Yu, C. C. and Luyben, W. L. (1987). Robustness with respect to integral controllability, *Industrial & Engineering Chemistry Research* **26**(5): 1043–1045.

Yue, A. and Postlethwaite, I. (1990). Improvement of helicopter handling qualities using $\mathcal{H}_\infty$ optimization, *IEE Proceedings - D Control Theory and Applications* **137**: 115–129.

Zafiriou, E. (ed.) (1994). *IFAC Workshop on Integration of Process Design and Control*, (Baltimore, MD, June), Pergamon Press, Oxford. See also special issue of *Computers & Chemical Engineering*, Vol. **20**, No. 4, 1996.

Zames, G. (1981). Feedback and optimal sensitivity: Model reference transformations, multiplicative seminorms, and approximate inverse, *IEEE Transactions on Automatic Control* **AC-26**(2): 301–320.

Zames, G. and Bensoussan, D. (1983). Multivariable feedback, sensitivity and decentralized control, *IEEE Transactions on Automatic Control* **AC-28**(11): 1030–1035.

Zhou, K., Doyle, J. and Glover, K. (1996). *Robust and Optimal Control*, Prentice Hall, Upper Saddle River, NJ.

Ziegler, J. G. and Nichols, N. B. (1942). Optimum settings for automatic controllers, *Transactions of the A.S.M.E.* **64**: 759–768.

Ziegler, J. G. and Nichols, N. B. (1943). Process lags in automatic-control circuits, *Transactions of the A.S.M.E.* **65**: 433–444.

INDEX

---

[1] Page numbers in *italic* refer to definitions.

The End Has No End

The Strokes
From the album "Room on Fire"
October 2003

Printed and bound by CPI Group (UK) Ltd, Croydon, CR0 4YY

26/06/2023

03230169-0001